ROUTLEDGE HANDBOOK OF PRIVATE LAW AND SUSTAINABILITY

The *Routledge Handbook of Private Law and Sustainability* reflects on how the law can help tackle the current environmental challenges and make our societies more resilient to future crises.

Sustainability has been high on the political agenda since the approval of the Sustainable Development Goals in 2015 and the EU Green Deal in 2019. The Green Agenda aims at making Europe the first climate-neutral continent by 2050, but humanity persists in an ecological overshoot that puts at risk the survival of species, including that of our own. Drawing together a selection of leading thinkers in the field, this Handbook provides a curated overview of the most recent and relevant discussions for private lawyers related to environmental and sustainability concerns. The authors delve into case study examples from 20 countries in Europe and beyond and discuss a wide range of issues, including new property law and consumer law paradigms, the use of legal tech for promoting sustainable property management, strategies for fighting planned obsolescence, eco-design, the servitisation economy, advances on corporate climate litigation and mandated green private sludges. Overall, the volume is designed to empower new generations of legal scholars to take an active role in the transition to a more sustainable future. It will also assist policymakers in producing better policy, through pinpointing the main legal issues that need to be addressed and offering a comparative overview of legal solutions and best practices.

Divided into six key parts and overseen by a team of internationally recognised expert editors, this Handbook will be an essential resource for students, scholars, private lawyers and policymakers who wish to have a comprehensive, fundamental overview of how environmental sustainability concerns reflect on private law.

Marta Santos Silva is a Postdoctoral Researcher at the Research Center for Justice and Governance at the University of Minho, and an Affiliated Researcher at the Maastricht European Private Law Institute and the Centre for European Law and Politics at the University of Bremen. Prior to her current role, she worked at Maastricht University and as an Affiliated Senior Researcher at KU Leuven researching at the intersection between private law and sustainability studies. She gained experience working in Brussels for a consultancy working with the European institutions. She was appointed a member of the European Commission's Expert Group on Liability and New Technologies (Product Liability Formation) and is an elected European Law Institute Council member.

Andrea Nicolussi is a Full Professor of Civil Law at the Law Faculty of the Catholic University of Milan, where he teaches comparative private law and family law. His published works focus on unjustified enrichment, contract law and tort law, as well as biolaw and family law. He has taken part in international collaborations such as the Study Group on a European Civil Code and the Eusoco project and was a member of the National Bioethics Committee.

Christiane Wendehorst is a Professor of Law at the University of Vienna. She is a member of the Academia Europaea, the Austrian Academy of Sciences, the International Academy of Comparative Law, the American Law Institute and numerous international research groups and advisory bodies. Before going to Vienna, she held full professorships in Germany and served as the Managing Director of the Sino-German Institute for Legal Studies. Professor Wendehorst was the President of the ELI, was one of its Founding Members and is today the ELI's Scientific Director.

Pablo Salvador Coderch studied law and economics at the University of Barcelona. He is a Cofounder of and Emeritus Professor at the Pompeu Fabra University, Barcelona. Since 1995, he is of counsel at Cuatrecasas Abogados. He is a member of the ALI, the ELI and the International Academy of Comparative Law and founder of InDret. He writes and teaches civil law, civil constitutional law, legislative drafting and comparative law of torts and contracts.

Marc Clément is, since September 2018, the presiding judge at the Administrative Tribunal of Lyon (France). In addition, he is, since 2014, a member of the French Environmental Authority. He has experience working as a lawyer at the DG Environment of the European Commission and as a legal adviser of the European Environment Agency. He published in 2010 the book *Environment European Law* and contributed extensively to books in the field of environmental law.

Fryderyk Zoll is a Doctor Honoris Causa by the West-Ukrainian National University in Ternopil, Professor at the Chair of Civil Law at Jagiellonian University in Kraków and the European Legal Studies Institute, University of Osnabrück. He is the leading researcher of numerous Polish, German and EU grants, a member of numerous academic associations and a vice-chairperson of the Committee for the Defence of Democracy (KOD) regional board in Lesser Poland. He is a former member of the European Law Institute's Executive Committee, the Common Frame of Reference and the Committee of the Acquis Group drafting teams, as well as the Codification Commission at the Polish Ministry of Justice. He has authored more than 200 scientific publications in European private law, ecologisation of private law, consumer law, comparative law, legal education and judicial independence.

ROUTLEDGE HANDBOOK OF PRIVATE LAW AND SUSTAINABILITY

Edited by Marta Santos Silva, Andrea Nicolussi, Christiane Wendehorst, Pablo Salvador Coderch, Marc Clément and Fryderyk Zoll

Routledge
Taylor & Francis Group
LONDON AND NEW YORK

earthscan
from Routledge

Designed cover image: © Baac3nes / Getty Images

First published 2024
by Routledge
4 Park Square, Milton Park, Abingdon, Oxon OX14 4RN

and by Routledge
605 Third Avenue, New York, NY 10158

Routledge is an imprint of the Taylor & Francis Group, an informa business

British Library Cataloguing-in-Publication Data
A catalogue record for this book is available from the British Library

ISBN: 978-1-032-66200-8 (hbk)
ISBN: 978-1-032-66202-2 (pbk)
ISBN: 978-1-032-66204-6 (ebk)

DOI: 10.4324/9781032662046

Typeset in Times New Roman
by codeMantra

CONTENTS

CONTRIBUTORS

Lucila de Almeida holds the Abreu Chair in ESG Impact and is the Director of the NOVA Green Lab at NOVA School of Law, CEDIS – R&D Centre on Law and Society, NOVA University of Lisbon. She is also a Part-time Assistant Professor at the European University Institute (EUI) and a member of the Coordination Committee at the Florence School of Regulation. Her research fields are energy, environmental and climate law. She holds a PhD in European Law (EUI), an LLM in Comparative, European and International Law (EUI), and a Master in Law and Development from FGV, São Paulo, Brazil. Before joining the faculty, she was an Edmond J. Safra Research Fellow at the Center for Ethics at Tel Aviv University (2019–2020), a Postdoctoral Researcher at the Faculty of Law, University of Helsinki (2017–2019), and a Research Associate at the Law Department of the EUI (2013–2016).

Guido Alpa is a Lawyer and Full Professor of Civil Law at the Università di Roma 'La Sapienza' and currently serves as the Director of the Master di Diritto privato europeo at the same university, a position that he has been holding since 2004. He is a Visiting Professor at the School of Law of the Universities of Oregon, California (Berkeley), London, Mannheim, Trento, Coimbra, Barcelona, Granada, Oxford, Malta, and King's College London, as well as at the Institute of Advance Legal Studies (London). He has been awarded an honorary degree by the Universidad Complutense de Madrid, the Universidad Nacional Mayor de San Marcos de Lima y la Universidad de Buenos Aires. His research interests include civil law (contract and liability law), commercial law, banking law and comparative law.

Wojciech Bańczyk is a PhD in law with distinction at the Jagiellonian University in Kraków, UJ, Assistant at the Chair of Civil Law UJ, Director of the American Law Program (The Catholic University of America in Washington DC and UJ) and Attorney at Law in Kraków. He is the winner of the first prize in the contest of the General Counsel to the Republic of Poland for the best master thesis in court law and the Z. Radwański competition for the best PhD thesis on Polish private law. He is a member of the European Society of Comparative Legal History and the ELI. He was awarded scholarships by the DAAD, the Foundation of Polish Science and the Ministry of Science and Higher Education. He is a Co-investigator and principal investigator of grants of the

National Science Centre and a Co-investigator of the Polish-German Academic Foundation and the German Federal Bank.

Maciej Bujalski is a PhD candidate at the Doctoral School in the Social Sciences, Jagiellonian University in Kraków, collaborating with the Civil Law Chair at the Jagiellonian University. He completed with distinction a Master's degree in Law at the Jagiellonian University in Poland and a Master 2 in Private law (Master 2 Droit privé) at the University of Orleans in France. He is a member of the editorial team at the 'Transformacje Prawa Prywatnego' (*Transformation of Private Law*) Journal and publishes in the fields of contract law, ecologisation of private law, consumer and banking law, as well as comparative law.

Roberta Calvano is a Full Professor of Constitutional Law at Unitelma Sapienza University. Roberta Calvano has a PhD in Constitutional Law and Public Law. A Lecturer since 2002 in Constitutional Law at the Faculty of Law, University of Rome 'La Sapienza', she is a member of the editorial board of several Italian scientific journals, among which are Giurisprudenza Costituzionale and Democrazia e diritto, where she also published. She is the author of four books concerning constitutional problems of European integration and constitutional issues implied in University and School legislation. Professor Calvano is a member of the teaching board of the PhD program in Constitutional Law and Public Law at the Faculty of Law, University of Rome 'La Sapienza' and she has been teaching undergraduate and post-graduate courses of Constitutional and Public Law and EU Constitutional Law at several Italian universities. She was audited twice as an expert by the Italian Parliament in 2023, and she was a member of a ministerial committee on Regionalism in 2019/2020.

Jorge Morais Carvalho is an Associate Professor and Vice Dean of the NOVA School of Law, where he also serves as a Coordinator of the Undergraduate Programme. He is a Researcher at CEDIS – Centre for Research on Law and Society and Director of the NOVA Consumer Lab. He has authored or co-authored approximately 140 books and scientific articles spanning a wide array of legal disciplines, including private law, commercial law, consumer law, civil procedural law, alternative dispute resolution, and comparative law. He has been invited to share his expertise at numerous events, having participated in over 160 seminars, conferences, webinars, congresses, scientific colloquia, and training sessions, spanning more than a dozen countries. He is also one of the editors of the *EuCML – Journal of European Consumer and Market Law*.

Marc Clément is, since September 2018, the presiding judge at the Administrative Tribunal of Lyon (France). In addition, he is, since 2014, a member of the French Environmental Authority. He is the Vice-chair of the Compliance Committee of the Aarhus Convention. He has experience working as a lawyer at the DG Environment of the European Commission and as a legal adviser of the European Environment Agency. He published in 2010 the book *Environment European Law* (fourth edition in 2021) and contributed extensively to books in the field of environmental law.

Carl Dalhammar is an Associate Professor at IIIEE, Lund University in Sweden. In his research, he has analysed numerous environmental laws and policies, including the Ecodesign Directive, Circular Economy policies, right-to-repair policies, planned obsolescence, EU energy efficiency laws, government procurement programs, and product destruction. He has led several research projects related to circular economy developments in the industry. Carl Dalhammar teaches environmental law and politics in several European and Swedish master's and undergraduate programs. He is a member of the Research Council at the Swedish Competition Authority (*Konkurrensverket*), and a board member of the Swedish national innovation program 'Re:Source'.

Fabrizio Esposito is a Bocconi and EUI alumnus (2018) and Assistant Professor of Private at the NOVA School of Law, Lisbon. In his scholarship, doctrinal analysis, especially of EU consumer and regulatory law, is enriched by insights from economics, psychology, and other social sciences. Fabrizio has more than 40 publications, including three co-edited volumes (*Research Methods in Consumer Law*, *Economics in Legal Reasoning*, and *The Cambridge Handbook on Algorithmic Price Personalization and the Law*). His book *The Consumer Welfare Hypothesis in Law and Economics* opens a new path in the economic approach to law, where consumer welfare and sovereignty are the cornerstone of market law.

Irene-marié Esser is a Professor of Corporate Law and Governance in the School of Law at the University of Glasgow and Dean of the Graduate School, College of Social Sciences. Since 2020, she is an Extraordinary Professor at Stellenbosch University, South Africa, and also a Visiting Professor at the National Law University Jodhpur, India. She was admitted as an Attorney of the High Court of South Africa in 2005. She currently teaches Corporate Governance, Corporate Social Responsibility and Company Law at the University of Glasgow and supervises postgraduate research students in the United Kingdom and South Africa. Her research spans doctrinal and empirical approaches, covering the UK, EU and South Africa in the field of corporate governance, company law and corporate social responsibility. Prof. Esser is part of the Corporate and Financial Law research group.

Michael Faure has studied law (1982) and criminology (1983) and has a master of laws from the University of Chicago Law School (1985) and a doctorate from the Albert Ludwigs Universität in Freiburg im Breisgau (Germany, 1989). Since 1991, he has been the director of the Maastricht European Institute for Transnational Legal Research at the Faculty of Law of Maastricht University. He still holds that position today, but in February 2008, he joined the Erasmus School of Law in Rotterdam halftime. There, he is the director of the Rotterdam Institute of Law and Economics and of the European Doctorate in Law and Economics (EDLE). He specializes in environmental law and policy, accident law and the economic analysis of law. Within the field of energy law, he worked on liability and insurance issues related *inter alia* to nuclear power, marine oil pollution, offshore oil drilling and carbon capture and gas.

Tomàs Gabriel García-Micó holds a PhD in Law, currently a SECTI postdoctoral researcher at the Universitat de Barcelona since December 2022. From September 2023 to January 2024, he was a visiting postdoctoral researcher at the Centro de Investigação em Justiça e Governação (JusGov) at the Escola de Direito, Universidade do Minho. He is a member of the Consultive Committee of the ELI project on Guiding Principles and Model Rules on Algorithmic Contracts. In July 2022, he was awarded a PhD in Law with the distinction of Excellent Cum Laude. His dissertation titled 'Surgical Robots and Tort Law' explored the intersections between law and new surgical technologies, specifically surgical robots and medical artificial intelligence, from a comparative law perspective. During his PhD, he was a visiting researcher at the China-EU School of Law (Beijing, September 2019) and the Dickson Poon School of Law at King's College London (London, January–April 2020).

Rosa M. Garcia-Teruel is an Assistant Professor of Civil Law at the University of Barcelona in Spain. She holds a master's degree in Business and Contract Law and, since 2018, an international PhD from Rovira i Virgili University in Tarragona, Spain. She has been a researcher in several competitive research projects and has carried out research stays at the University of Bremen, the University of Leiden and the University of Groningen. Her fields of research are real estate law (from the point of view of obligations, contracts and property law), tort law, inheritance law and the law of new technologies.

Mireia Artigot Golobardes is a Ramon y Cajal Researcher at Pompeu Fabra Law School and a teaching fellow of the Europaeum Program (Oxford). She focuses on contracts, with particular interest in consumer contracts and the disruption resulting from the digital economy, AI and ADM in consumer markets and justice systems. Her research of the law draws from the inputs provided by microeconomics and macroeconomics. She has been a visiting professor of Brooklyn Law School, the Institut für Wirtschaftsrecht at Kassel Universität, the Montenegro University Economics Department and Law School, the University of Trento Law School and the Center for South African Studies of the University of Western Cape. She is the coordinator of the European project Justice Fundamental rights and Artificial Intelligence (JuLIA)(https://www.julia-project.eu/) and the principal investigator of AlgorithmLaw and iConsumers funded by the Spanish Research Agency.

Maria Raquel Guimarães is a tenured Associate Professor of Law at the Faculty of Law, University of Porto, Portugal, teaching Civil Law (General Part) and Contract Law on the grade course, and Contracts & Consumer's Law on the master course. She obtained her PhD in Civil Law at the University of Porto and a Master in Banking and Corporate Law at University of Coimbra, Portugal. She is a researcher at CIJ—Centre for Legal Research, and General Editor of RED—Revista Electrónica de Direito. She has conducted research in the areas of civil and contract law, and her main research fields of interest are consumer contracts, payment services, personality rights and digital law. She has authored and co-edited several books on electronic banking, contract and civil law plus contributed chapters in Portuguese and European law books and has published widely on payments services issues in leading Portuguese and European law journals.

Janja Hojnik is a Full Law Professor of the University of Maribor (Slovenia), Vice-Dean for Research (2015–2018, 2023–2027) and Head of the European Legal Studies Programme (since 2019). She is a former Vice-Rector of University of Maribor (2018–2022) and a Global Law Professor at KU Leuven Faculty of Law and Criminology (holding a course on Regulatory Aspects of Servitization in the EU) in 2023. She is a member of an international and interdisciplinary experts group dealing with new technologies (Brown Bag Lunch Group). She is an Advisory Board member of the Common Market Law Review. She teaches and researches in the field of EU Law (internal market law, focusing on goods and services, price regulation, servitization, international trade, digital economy). She is also interested in topics of legal research assessment, integrity in research and conflicts of interests in academia. Up to date, she is a holder of three Jean Monnet projects. Between 2011 and 2023, she has been voted seven times among ten most influential lawyers in Slovenia.

Björn Hoops is Full Professor and holds the Chair of Private Law and Sustainability at the University of Groningen. He read German and Dutch law as well as economics at the universities of Bremen, Groningen, Hagen, and Oldenburg. He obtained a PhD with distinction from the

University of Groningen in 2017. His research focuses on land law and governance in the energy transition. From April 2022 to March 2024, Björn was an EU-funded Marie Curie Fellow at the University of Turin, conducting the research project 'Private Law and the Energy Commons' on the private-law aspects of citizen initiatives in the energy sector. Björn has been a visiting scholar at the Max-Planck-Institute of Comparative and International Private Law, Columbia Law School, and the universities of Cape Town, Cologne, Leuven, Monterrey, Stellenbosch and York. Björn has been serving as a legal advisor to the Notaries' Office 'Olenz Notarissen' in Veenendaal, the Netherlands.

Dario Hug earned his Bachelor's degree at the University of Neuchâtel (2010) and his Master's degree at the University of Lausanne (2012). He wrote his PhD thesis on 'The formation of the Consumer Contract under Swiss Law' (2019). He was a PostDoc at KU Leuven (2021–2022) and the University of Zurich (2023). His favourite research topics are Obligation and Contract Law, Consumer Law and the Circular Economy. He is the author and co-author of various scientific papers in these research fields. Additionally, Dario was *Dozent* in Obligation Law at the University of Zurich (2023) and is a lecturer in Consumer and Contract Law at the University of Neuchâtel. He is currently also working at the Institute of Civil Law of the University of Bern. He also has practical working experience as an independent attorney-at-law in Switzerland and as a law clerk at the Swiss Administrative Federal Court.

Tatjana Josipović is a tenured Professor of Civil Law at the Faculty of Law, University of Zagreb. She has written many scholarly and professional papers and several books on civil law, property law, land registration law and European private law. She was a visiting professor at many European universities. Professor Josipović is a fellow of the Academy of Legal Sciences of Croatia, an associate member of the International Academy of Comparative Law, a member of the ELI, the Academia Europea, the Croatian Society for Civil Law Science and Practice and SECOLA. She has also held several offices within the higher education system in Croatia, in the Academic community and staff services of different state bodies. She is a recipient of the Croatian Academy of Arts and Sciences award (1990) and the State Award for Social Sciences (2018). She is a researcher on several EU e-justice projects.

Miha Juhart is a Professor of Civil and Commercial Law at the Faculty of Law at the University of Ljubljana in Slovenia. His research interests include dealing primarily with the right to property, registration of real estate, and transfer of claims and contracts. He was a Vice-Rector for Research at the University of Ljubljana and Dean of the Faculty of Law. His scientific *oeuvre* comprises more than 100 scientific articles, articles in monographs and international research conferences. He is the president of the Association of Slovenian Lawyers' Associations. He is one of the most cited authors in court case law in Slovenia.

Candida Leone is an Associate Professor of Private Law at the University of Amsterdam, The Netherlands. Her PhD dissertation, written at the University of Amsterdam (UvA) in 2020, studied judicial and statutory responses to unfair terms in consumer and employment contracts against the discussion on the so-called 'instrumentalisation' of European private law. Her most recent work discusses sustainable legal education and private law theory, flexibilization terms and commodification, as well as the protection of consumer debtors (in particular against aggressive debt collection). She teaches EU and Dutch consumer law and an introduction to legal research for LLM students at the University of Amsterdam.

Carlos Gómez Ligüerre is a Professor of private law at the Universitat Pompeu Fabra in Barcelona, Spain. He holds a degree in law (1996) and a degree in business administration (2006) and a doctorate in law (2005) from Universitat Pompeu Fabra University. He is a specialist in contract law and civil liability. He has developed his research work at the Max-Planck-Institut für ausländisches und internationales Privatrecht (Germany) and at the University of Pennsylvania School of Law (United States). He has been a visiting professor at Universidad Nacional del Sur (Argentina), at the IDC-School of Law (Israel) and at the University of Illinois (United States). He is a regular speaker at conferences and courses organized by universities, professional associations and research centres. He has been a fellow of the Alexander von Humboldt Stiftung and his research has focused on the interaction between contract law and tort liability rules.

Iain MacNeil is the Alexander Stone Professor of Commercial Law at the University of Glasgow. He began his academic career after a decade working in investment banking in the City of London. He has undertaken research and collaborated with colleagues in Australia, Canada, China, Hong Kong and the United States. He is a Trustee of the British Institute of International and Comparative Law and Chair of the International Securities Regulation Committee of the International Law Association. He was the Deputy Chair of the Hong Kong RAE Law Panel in 2020 and of the UK REF Law Panel in 2021. He has acted as a Senior Adviser on several projects examining national compliance with EU financial sector directives.

Eléonore Maitre-Ekern dr. juris is a postdoctoral researcher in Sustainability Law and the Circular Economy. Eléonore's PhD thesis, which she wrote at the Department of Public and International Law, University of Oslo, is entitled 'Towards a Circular Economy for Products: A legal analysis of Europe's policy and regulatory framework from an ecological perspective'. She successfully defended her PhD thesis in 2019. At the time of publication, she is finishing her postdoctoral fellowship with the research group Futuring Sustainable Nordic Business Models at the Law Faculty, University of Oslo. Eléonore has published several articles on various aspects of the circular transition in the EU and Norway, notably on premature obsolescence, sustainable consumption, durability and the right to repair. As of 2024, she will be working as a Researcher at NIVA, the Norwegian Research Institute for Water and the Environment.

Victor Mehnert studied law at the Universities of Mainz and Cologne. He then worked for a law firm and the University of Cologne before taking up a position as a research assistant at the Chair of Civil Law, Law of Data Economy, Data Protection, Digitalisation and Artificial Intelligence (Prof. Dr. Louisa Specht-Riemenschneider). There, he is involved in the research project 'From Smart Laws to Smart Technologies to Smart Laws: Comparative Perspectives from Germany and the United Kingdom', funded by the DFG and the AHRC. Among other publications, he is a co-author of the policy brief Right to Repair published in 2022 by the German Advisory Council for Consumer Affairs. He is writing his dissertation on the right to repair.

Lela Mélon, an expert in law and economics, specializes in EU law, corporate conduct, and sustainability. A former Marie Curie Research Fellow at Pompeu Fabra University, she received the 2021 Emerging Sustainability Leader Award. She served as the Executive Director of the Planetary Wellbeing Institutional Framework at Pompeu Fabra University. Currently, she directs the postgraduate program Sustainability Transition Management and Climate Emergency at ESCI-UPF. Her publications include 'Shareholder Primacy and Global Business' (Routledge 2018) and 'Sustainability in Public Procurement, Corporate Law and Higher Education' (Routledge 2023).

Presently, her research centers on policy coherence for sustainability at the EU level, emphasizing corporate law policies. Dr. Mélon's influence extends to reshaping academic curricula with sustainable corporate law courses and collaborating with industry as a senior sustainability consultant at msg global solutions and CER Slovenia, demonstrating a commitment to positive change in law, economics, and sustainability.

Ralf Michaels is the Director at the Max Planck Institute for Comparative and International Private Law in Hamburg, Germany, Chair of Global Law at Queen Mary University in London (UK), and Professor of Law at Hamburg University, in Germany. Until 2019, he was the Arthur Larson Professor at Duke University School of Law (Durham, North Carolina, USA). Ralf Michaels is a widely published scholar of private international law, comparative law, and legal theory. His current research focuses on decolonial comparative law, theoretical foundations of private international law and global legal plurality, sustainability in private and private international law, and the law of combatting antisemitism.

Hans-W. Micklitz was a Professor at the Department of Law at the European University Institute (EUI) in Florence until 2019, currently a part-time professor at the Robert Schuman Centre for Advanced Studies. He was also the Head of the Advisory Board of the Max Planck Institute for Innovation and Competition, Jean Monnet Chair of European Economic Law at the University of Bamberg/Erlangen for Civil and Commercial Law, Head of the Institute for European and Consumer Law in Bamberg and Professor at the Berlin School of Economics, in addition to multiple visiting professorships and advisory positions at institutions across Europe and the USA, he was a member of the German Advisory Council for Consumer Affairs until 2022. He was elected a member of the Academia Europaea in 2016. His research focuses on European and transnational economic and private law and legal theory.

Maria Miguel Oliveira da Silva is currently a PhD candidate in Law and a teaching assistant at NOVA School of Law in Lisbon, Portugal. She is the legal advisor of the NOVA Consumer Lab, a research associate at NOVA Business, Human Rights and the Environment and a researcher at CEDIS – Centre for Research on Law and Society. She is also the author of a dozen published works on consumer law, criminal law, sustainability, and human rights. Her doctoral research is funded by the FCT – Fundação para a Ciência e a Tecnologia. In the past, she was a conflict mediator at the National Center for Information and Arbitration of Consumer Conflicts in Portugal.

Sandra Passinhas is a PhD from the European University Institute (2010) and the University of Coimbra (2016). She is a Professor of Civil and Consumer Law at the University of Coimbra and Executive Director of the Legal Research Center on Consumer Law of the Faculty of Law of the University of Coimbra. She is a Portuguese Member of the DSA Committee and the DSA Expert Group and a member of the General Council of the Portuguese Judiciary School. She is a researcher at the University of Coimbra Institute for Legal Research and coordinator of the UC Team on JuLIA Justice, Fundamental Rights and Artificial Intelligence. She is a member of the Portuguese Medico-Legal Council and a former researcher on the judicial training project FRICoRE (Project on Fundamental Rights in Courts and Regulation), the training and research project GoInEU (Governing Inheritance Statutes after the Entry into Force of EU Succession Regulation) and GoInEUplus (Integration, migration, transnational relationships, Governing Inheritance Statutes after the entry into force of EU Succession Regulations).

Denis Philippe is a Professor at the University of Louvain and a Visiting Professor at the University of Paris Ouest and the Vrije Universiteit Brussel. He is a lawyer at the Brussels and Luxembourg bar and managing partner of Philippe & Partners. He is one of the ELI's founding members. His research and professional interests encompass Banking and Financial Law, Commercial Law, Contract Law, Corporate and M&A Law, European Law, Insurance Law, Real Estate, Torts and Liability and Energy. He is the Editor-in-Chief of Revue de Droit des Affaires, member of the Editorial Committee Revue de Jurisprudence Liège, série Handelscontract of Groupe Larcier, Revue Générale de Droit civil (Contrats et Obligations) and member of the Scientific Committee of Businesslawtele.net.

Katarzyna Południak-Gierz is an Assistant Professor at the Jagiellonian University's Civil Law Chair. She defended her PhD thesis 'The usefulness of the regulation of defects of consent to contracts concluded on the consumer online market' with distinction in 2019. Dr Południak-Gierz's academic interests focus on the interplay between private law and technology. She researches private law reactions toward personalization techniques in consumer contracts and the possibility of using granularity to increase the ecological efficiency of sales law. She is a member of the editorial team at the 'Transformacje Prawa Prywatnego' (*Transformation of Private Law*) Journal.

Barbara Pozzo is a Full Professor of Private Comparative Law at the Law School of the University of Insubria (Como) and President of the Italian Society for Research in Comparative Law. She holds a JD from the University of Milan and a PhD from the University of Florence. She completed her education in France (Faculté Internationale de Droit Comparé, Strasbourg), Germany (Max-Planck-Institut, Hamburg), United States (UC Berkeley School of Law). She is a Titular Member of the International Academy of Comparative Law, a Member of the Board of the European Law Faculties Association and of the European Law Institute. Her main fields include the private aspects of environmental law, the evolution of legal systems and the construction of European private law, with a focus on legal translation problems in the context of European multilingualism. In 2019, she was awarded the UNESCO Chair on Gender Equality and Women's Rights in the Multicultural Society.

Rebecca Ravalli has a PhD in law from the European University Institute (EUI), which followed a Master of Laws at the International University College of Turin, a study visit at the Centre for Transnational Legal Studies (Georgetown University), and a Master of Laws in EU Law at the EUI. Her research focuses on the contractual structure of global value chains and the related implications for consumers and sustainability objectives. Since 2022, she has been working in the private sector as a law and sustainability specialist with a particular focus on the textile sector and the financial sector. She also works *pro bono* at Rén Collective and Torino Stratosferica.

Maria Regina Redinha undertook graduate and post-graduate studies at the University of Coimbra and obtained her PhD in Labour Law at Porto University, Portugal. She is currently a tenured professor at the University of Porto, for Labour and Civil Law as well as a senior researcher for CIJ (Centre for Legal Research, University of Porto). She has conducted research in the areas of civil, employment and labour law, and her main research fields of interest are atypical employment, employment contract qualification, the digital workplace, workers' fundamental and personal rights. Professor Redinha serves as the Portuguese labour law review editor ('Questões Laborais'), as an editorial board member of RED—Revista Electrónica de Direito and as a vice-president of the Portuguese Labour Law Society. She has authored and edited several books on Portuguese labour law, co-authored a book on Angolan labour law, plus contributed chapters in Portuguese and European law books and law reviews.

Ewa Rott-Pietrzyk, prof. dr hab., is employed at the Institute of Legal Sciences, Faculty of Law and Administration, University of Silesia in Katowice (Poland). She is a lecturer in private law and an expert in contract law, as well as an arbitrator. She is the coordinator and member of many national and international projects, namely coordinator of projects financed by the DPWS (Deutsch-Polnische Wissenschaftsstiftung). She has authored several studies on private law and private international law (mainly in general private law institutions, contract law, including commercial intermediary agreements), Chinese law, and comparative law, publishing both in Polish and in foreign outlets.

Hugo-Maria Schally holds a doctorate of law from the University of Graz/Austria and is a graduate of the Vienna Diplomatic Academy. He worked for the United Nations Development Programme, the Austrian Ministry for Foreign Affairs and the Organisation for Security and Cooperation in Europe before joining the European Commission in 1998, where he has held positions in the Directorates General for External Relations, Development and Environment. Since 1990, he has been involved in policy making and legislative work, in particular on environment policy and related bilateral and multilateral international negotiations. More recently, he was responsible, in the Directorate General for the Environment, for the development and implementation of the EU Circular Economy Action Plans and related policies and legislation. Currently, he serves as the EU lead negotiator for the Convention on Biological Diversity and for a new Global Agreement on Plastic Pollution.

Kristina Siig is a Professor of Transport Law and Law of the Sea at the University of Southern Denmark and Professor II at the Scandinavian Institute of Maritime Law at the University of Oslo in Norway. She obtained the degree of Doctor Juris from the Scandinavian Institute of Maritime Law in Oslo in 2003 with a thesis on international commercial arbitration.

She specializes in the interaction between private and public international laws, focusing particularly on how the private actor and private law-making, including contractual practices and standards, interact with international public law rules. She chairs the interdisciplinary maritime and marine network BlueSDU at the University of Southern Denmark, as well as the Parliamentary Investigation Task Force on the Scandinavian Star ferry disaster.

Marta Santos Silva is a Postdoctoral Researcher at the Research Center for Justice and Governance at the University of Minho, and an Affiliated Researcher at the Maastricht European Private Law Institute and the Centre for European Law and Politics at the University of Bremen. Prior to her current role, she worked at Maastricht University and as an affiliated senior researcher at KU Leuven researching at the intersection between private law and sustainability studies. She provided legal advice for a Brussels-based consultancy working with the European institutions. She was appointed a member of the European Commission's Expert Group on Liability and New Technologies (Product Liability Formation) and is an elected European Law Institute Council member.

Verica Trstenjak is a former Advocate General of the European Court of Justice in Luxembourg, a former judge at the General Court of the EU and a professor of EU law. She teaches in Austria (University of Vienna, Sigmund Freud Private University Vienna) and Slovenia (University of Ljubljana). Since 2019, she has been a member of the Permanent Court of Arbitration in The Hague and an arbiter at the ICSID-International Centre for Settlement of Investment Disputes.

She is an external scientific member of the Max Planck Institute Luxembourg for International and European Procedural Law. Since 2020, she has been a substitute member of the Venice Commission (Council of Europe). She has published several books and more than 300 articles and has been a lecturer at universities worldwide and international conferences. She is a member of the editorial boards of several important legal journals, e.g. the European Law Review.

Willemijn van Doorn-Hoekveld currently works as a Researcher/Consultant at Deltares. She is engaged in the fields of water safety, water quality, soil and climate adaptation. Previously, she worked as an Assistant Professor at Utrecht University. She taught courses in the fields of environmental law, state liability law and public and constitutional law. Willemijn completed a doctoral dissertation on distributional justice and flood risk management in the Netherlands, Flanders and France. She is a guest researcher at Utrecht University (Utrecht University Centre for Water, Oceans and Sustainability Law). Her research focus is on distributive issues, particularly in the context of climate adaptation and transition management.

Marleen van Rijswick is a Professor of European and Dutch Water Law and is the Director of the Utrecht University Centre for Water, Oceans and Sustainability Law (see http://ucwosl.rebo.uu.nl/). She researches the question of how law can contribute to sustainable and equitable use of water based on shared responsibilities. She was a visiting professor of European environmental law at the Université Panthéon-Assas (Paris II) in Paris and in the Master Energy, Environmental and Climate Law at the University of Leuven and Malta. In 2015, she was a visiting professor of international and European environmental law at the Research Institute of Environmental Law of Wuhan University, China and in 2018 at the Sun Yat-Sen University, Guangzhou, China. She published in many international and national journals and books. She is a substitute judge at the district court of Oost-Brabant.

Tim van Zuijlen is a Lecturer and a PhD candidate at the Faculty of Law of the University of Groningen in The Netherlands. He is currently preparing a dissertation on a comparative legal analysis of consumer law in cases of planned obsolescence. He is a member of the board of IT Law (portfolio education), of the Alumni committee of the Faculty of Law of the University of Groningen in The Netherlands and also a member of the editorial board of the IT law paper series. His research interests include IT law, technology in a legal context (particularly legal knowledge-based systems, artificial intelligence and legal tech), and private law (particularly sustainability and consumer law).

Radosveta Vassileva is a Visiting Research Fellow at Middlesex University in the UK. She holds a PhD in Law from University College London in the UK, a Master's degree in Economic Law from Sciences Po Paris in France, a Master's degree in Global Business Law from Université de Paris 1 Panthéon-Sorbonne in France) and a BA in International Relations from Tufts University in the USA. She publishes extensively in the areas of comparative private law, international commercial law, and the rule of law, and she has recently authored *Bulgarian Private Law at Crossroads*, which was published by Intersentia in 2022. In addition to being a legal scholar, Dr Vassileva is also a social advocate. She runs a blog on the rule of law and human rights and on occasion publishes articles in various media, such as the *Verfassungsblog*, *New Eastern Europe*, the *EUobserver*, *Euronews*, and *The Brussels Times*.

Jasper Verstappen is an Assistant Professor in Technology Law at the University of Groningen in The Netherlands. He holds a PhD from the Maastricht University in The Netherlands. Before joining the University of Groningen, he worked as a senior financial services consultant for a large international advisory firm. His research interests include the interaction between emerging technologies and private laws, shifting power balances through new technology and protection of consumers in European law, and digital assets in European systems of property law. Jasper's expertise includes European systems of property law, European systems of contract law, consumer law, and comparative law.

Petra Weingerl is an Associate Professor at the Faculty of Law, University of Maribor. Petra completed her doctorate and Magister Juris at the University of Oxford. Before commencing her DPhil, she worked at the Court of Justice of the EU in Luxembourg and the Appellate Court in Ljubljana. She was a visiting researcher at the European University Institute in Florence and the Max Planck Institute in Hamburg. As an adjunct professor, she taught EU Law and Policy courses at the University of Trieste. She is a PI in two projects: the module holder of Jean Monnet Module on Fundamental Rights in EU Business Law and Policy (FURBUS) and the PI of ARIS (Slovene Research Agency) postdoctoral research project entitled 'Deterrence as a concept in private enforcement of EU law'.

Yayun Shen holds a PhD in Law and Economics from Erasmus University Rotterdam, the Netherlands. She defended her thesis in 2019, and it was entitled 'Green Building in the US and China: A Law and Economics Perspective'. Yayun Shen has been working on several research projects concerning smart instrument mixes for environmental governance, with a particular focus on the impact of nudges on pro-environmental/green behaviours. She is a recognised expert in environmental standardisation and private standards, as well as smart mixes to promote green buildings in China. She has published extensively and her academic works have been cited in China, Europe and beyond.

Francesco Zecchin graduated in Law from Università Cattolica del Sacro Cuore (Milano), then he earned a PhD in Civil Law at the International Doctoral School of Law and Economics 'Tullio Ascarelli'-Università Roma TRE. In 2011, he was a Visiting Scholar for one term at the Law School of the University of East Anglia (Norwich – UK). He was Research Fellow in Private Law at the Faculty of Economics and Law of the Università Cattolica (Piacenza), and he is currently Assistant Professor in the same scientific-disciplinary sector. He is a member of the teaching staff of the PhD programme in Persona e Ordinamenti giuridici at the Università Cattolica del Sacro Cuore (Milano). He serves on the editorial board of the journals *JusOnline* and *Europa e diritto privato*. He is a member of the Unione dei Privatisti, of the ELI (European Law Institute), and of the SISDiC (Società Italiana Studiosi del Diritto Civile).

Samuel Zeh studied law from 2011 to 2017 at the University of Heidelberg and King's College London. During his period of study, he was a student assistant at the Institute for Comparative Law, Conflicts of Law and International Business Law in Heidelberg. In 2017, he sat for the first state examination in law. From 2017 to 2019, he completed his legal traineeship at the Higher Regional Court of Hamburg. He was a research associate at the Max Planck Institute for Comparative and International Private Law from 2018 to 2023 and is currently a doctoral student at the University of Hamburg.

Fryderyk Zoll is a doctor honoris causa by the West-Ukrainian National University in Ternopil, Professor at the Chair of Civil Law at Jagiellonian University in Kraków and the European Legal Studies Institute, University of Osnabrück. He is the leading researcher of numerous Polish, German, and EU grants, a member of numerous academic associations. He is a former vice-chairperson of the Committee for the Defence of Democracy (KOD) regional board in Lesser Poland and a former member of the European Law Institute's Executive Committee, the Common Frame of Reference and the Committee of the Acquis Group drafting teams, as well as the Codification Commission at the Polish Ministry of Justice. He has authored more than 200 scientific publications in European private law, ecologisation of private law, consumer law, comparative law, legal education and judicial independence.

Jakub Błażej Zwierzchowski, Master of Laws, Jakub is currently LLM Comparative Law student at China University of Political Science and Law and a PhD candidate at the University of Silesia in Katowice, where he prepares his dissertation on environmental solutions within the Chinese Civil Code. Between 2023 and 2024, he was a visiting scholar at the Max Planck Institute for Comparative and International Private Law, supported by the DAAD (Deutscher Akademischer Austauschdienst) PhD scholarship grant. Jakub was also NAWA (Polish National Agency for Academic Exchange) scholarship holder, for a one-year study at Beijing Foreign Studies University. His research centers on exploring the possibilities of implementation of environmental sustainability into private law, with a focus on the law of obligations.

PREFACE: FORGING A PATH TO ENVIRONMENTAL SUSTAINABILITY IN EUROPEAN PRIVATE LAW

It is with great pleasure and gratitude that I introduce this compelling and comprehensive volume, *Routledge Handbook on Private Law and Sustainability*.

As we delve into its pages, we embark on a journey through the challenges, complexities, and potential solutions at the intricate intersection of private law and environmental sustainability—a journey made possible by the collective efforts of a remarkable group of editors, authors, assistants, and reviewers.

The chapters within this volume, contributed by a distinguished cadre of academics, judges and policymakers, are a testament to the dedication that each author has poured into their work.

Behind the scenes, our journey has been shaped by a dedicated team, without whose invaluable contributions this book would not have been possible.

I would like to start by thanking our Publisher Routledge, for their exceptional support and collaboration throughout the creation of this handbook, as well as our sponsors, the Manuel António da Mota Foundation, whose generosity and commitment have not only facilitated the publication of this volume but also underscored the theoretical and practical importance of the diverse perspectives presented within.

A special word of thanks goes to my co-editors, whose wisdom and collaborative spirit have guided this project from its inception: Marc Clément, Andrea Nicolussi, Pablo Salvador Coderch, Christiane Wendehorst and Fryderyk Zoll. Their shared passion for this subject and the camaraderie that we have forged have made the editing process both enjoyable and enlightening.

I extend my heartfelt gratitude to the esteemed members of our Advisory Board, Ewoud Hondius (Chair), Verica Trstenjak, Christoph Busch, Christian Twigg-Flesner and Ignacio Herrera Anchustegui, whose guidance has played a pivotal role in ensuring the success and scholarly quality of this collaborative effort.

The editorial assistants who have supported this project have also been instrumental in ensuring that it ran smoothly and efficiently. Giuseppe Portonera, Hennu Malmi, Jakub (Kuba) Błażej Zwierzchowski, Katarzyna (Kasia) Południak-Gierz, Laura Zsarnai, Luisa Cortat, Matilda Titeca, Mehmet Cagatay Mavice and Tea Mustać have generously offered their editorial assistance while juggling demanding professional affairs. Their dedication to the project was deeply appreciated. A particular word of thanks goes to Tomàs Gabriel García-Micó and Nataly Machado, whose unwavering daily commitment and tireless efforts over many months have been the cornerstone of this project's success. Their diligence, meticulous attention to detail and exceptional teamwork have been invaluable in bringing this endeavour to fruition.

Equally deserving of our gratitude are the external peer-reviewers who dedicated their time and expertise to thoroughly evaluate and comment on our manuscript during the double-blind peer review process: Alberto De Franceschi, Alberto Quintavalla, Aleš Ferčič, Anaïs Michel, Anna Beckers, Antonio Davola, Antonis Karampatzos, Bert Keirsbilck, Berte-Elen R. Konow, Bonnie Holligan, Bram Akkermans, Brigitta Lurger, Carlos Nóbrega, Caroline Cauffman, Christian Twigg-Flesner, Christiana Galvão Ferreira de Freitas, Cristina Poncibò, Daniela Caruso, Daniela Farto Baptista, David Markworth, Dimitar Stoimenov, Emilia Mišćenić, Eric Tjong Tjin Tai, Esther Arroyo Amayuelas, Eva Micheler, Sónia Moreira, Eve Truilhé, Eveline Ramaekers, Evelyne Terryn, Fabrizio Cafaggi, Fátima Castro Moreira, Felix Ekardt, Felix Pflücke, Filippo Valguarnera, Geraint Howells, Harry Slachmuylders, Henrique Sousa Antunes, Ivano Alogna, Joana Campos Carvalho, Karin Sein, Katarzyna Kryla-Cudna, Kurt Xerri, Laura Valle, Leonie Reins, Lorena Bachmaier, Luisa Antoniolli, Lydia Velliscig, Maria Irene da Silva Ferreira Gomes, Marie Jull Sørensen, Martin Schmidt-Kessel, Mateja Durovic, Mateusz Grochowski, Matija Damjan, Michele Graziadei, Monica Navarro-Michel, Niko Soininen, Olaf Meyer, Paola Iamiceli, Patrícia Jerónimo, Paul Craig, Pedro Coutinho, Per Norberg, Piotr Machnikowski, Prashant Sabharwal, Przemysław Pałka, Riccardo Omodei Salè, Rosa Milà Rafel, Rui Miguel Prista Patrício Cascão, Sergio Nasarre-Aznar, Sjef van Erp, Taina Pihlajarinne, Tamas Szabados, Thalia Kruger, Teresa Rodriguez de las Heras Ballell, Tom Bouwman, Ulrich G. Schroeter, Vanessa Mak, Vibe Ulfbeck and Virginie Rouas. Their insightful feedback and constructive criticism have improved the quality and rigour of our work.

Last but not least, my deepest appreciation goes to the authors themselves. It was their research and commitment to advancing private law and its role in sustainability that have brought this book to life. Their willingness to share their knowledge and expertise has enriched the discourse on this vital topic, and I am honoured to have had the privilege of collaborating with each of them.

This volume stands as a testament to the power of collective endeavour, intellectual exploration and shared dedication to addressing one of the most pressing challenges of our time.

In closing, I would like to express my wish that this book will inspire further research, policy development and informed decision-making aimed at achieving environmental sustainability in Europe and beyond.

Marta Santos Silva,
Lead Editor *Routledge Handbook on Private Law and Sustainability*

Funded by FMAM

PART 1

Greening Private Law and in Particular the Law of Obligations

1

PRIVATE LAW AND ENVIRONMENTAL SUSTAINABILITY

Barbara Pozzo

1.1 Introduction

At a time when sustainability seems to have become the watchword in any post-pandemic policy, we might wonder what contribution private law can make in setting up an efficient system of environmental protection.[1]

The comparatist's gaze goes beyond the national horizon and analyses the evolution of the law in different contexts, considering the challenges that environmental protection and climate change pose compared to classical canons of intervention, where dialogue with the social sciences is increasingly present and the overlapping of international, European and national sources punctuates the work of the interpreter.[2]

The idea of sustainable development is deeply connected with the debate around the concept of well-being that our society acknowledges. If the idea of social well-being was strongly linked with the ideology of 'development' as economic growth in the first part of the 20th century, since the end of World War II that ideology began to fade away and leave space to a more complex vision of the world, where respect of the environment gained a central place.

While the debate around the possibility of reconciling the needs of development with the protection of the environment is not new, however, it is only in recent times that the concept of sustainable development started to be elaborated in its actual form, assuming an increasingly important role in the development of various policies at the EU and international levels.

Sustainable development also relies on a commitment to equity towards future generations. The idea to leave future generations at least the same level of opportunities as we ourselves have, although simple in theory, seems much more difficult to reach in practice: what legislative measures should be adopted to achieve this magical point of equilibrium between the needs of today and those of tomorrow? This question leads us to investigate which legal tools allow us to implement sustainable development goals in an effective way.

From a private law perspective, the revisiting of traditional remedies to cope with the new demands arising from increased sensitivity to environmental issues has been the focus of an experience common to both *civil law* and *common law* systems.

To overcome the bottlenecks of tradition, interpreters have sought to free the structure of traditional institutions to adapt it to new requirements. Experience shows the presence of a common

 DOI: 10.4324/9781032662046-2

effort to extend as far as possible the scope of traditional norms, generally placed to protect traditional goods, to provide at least indirect protection for the environment.

1.2 The Adaptability of the Private Law System to New Environmental Issues

As is well known, torts are for private parties among the richest instruments in terms of social planning. It should come as no surprise that precisely in this area, scholars and courts have engaged in elaborations taking in consideration new cases of environmental risk within the framework of traditional torts, developing innovative theories about standing, causation, proof of damage and damage assessment.

In common law systems, courts resorted to revisiting traditional doctrines that were based on the torts[3] system. The question about the adaptability of the traditional common law system to the new requirements arising from the so-called *environmental risk* has not only concerned the American courts, causing no small amount of ink to flow in academia as well.

Between the 1970s and early 1990s, a whole series of articles appearing in major American law reviews identified the advantages and disadvantages of the 'traditional tort system', resulting in a 'private law v. public law' debate on the regulation of new environmental issues.[4]

Proponents of the private law approach have argued that private, rather than public, action would be best suited to deal with the problems of risks arising from the new scientific activities of modern society. Courts, rather than other institutions, would have the advantage of being able to apply particularly ductile methods for determining the defendant's liability.

In contrast, those who argued for the substantial inapplicability of the *tort system* invoked in particular its weaknesses with respect to the specific peculiarities of environmental damage: the fact that environmental damage can come to light long after the occurrence of the harmful action, challenging the requirement of the *remoteness of damage* and the traditional criteria for causation; the super-individual nature of environmental interests, which undermine the common rules on standing and finally, the fact that pollution phenomena can occur as *mass torts*, in which neither all the injurers nor all the injured parties can be easily identified.[5]

In civil law systems, a similar debate can be found about the possibility of extending the rules governing civil liability to new cases of environmental damage.

In France, the development of the theory of *troubles de voisinage* considers environmental issues within a discipline initially anchored in the legal status of land ownership. In the course of many years of development since the 1970s, case law and scholars[6] have affirmed that in *troubles de voisinage*, a relationship of liability existed not only between owners of different plots of land, but – also – between the victim of a tort and a polluting activity.[7]

Also in Germany, §823 BGB on tort liability and §906 BGB on nuisances were used as a basis for developing environmental protection.

As early as the 1980s, German scholars debated how to extend liability rules to new tort cases arising from environmental damage.[8] §823 BGB establishes an obligation to pay damages in cases where one of the goods specifically mentioned, namely life, health, body, liberty, property or another right ('*ein sonstiges Recht*'), is harmed with intention or negligence and in an unlawful way.[9]

In this perspective, part of German scholars argued that the first paragraph of §823 BGB could apply to pollution phenomena, considering the right to a clean environment as a personality right – as a *Persönlichkeitsrecht*[10] – interpreting the phrase *another right* ('*sonstiges Recht*') extensively and including in this rather broad term the concept of environment.[11]

Other German scholars focused on the possible reinterpretation of §906 BGB related to nuisances[12] to deal with environmental pollution. In this case, a provision that was originally conceived at the end of the 19th century, when the German BGB was promulgated, to foster the development of industry, was reinterpreted to meet new environmental protection requirements.

Specifically, §906 BGB originally stated:

> The owner of a plot of land may not prohibit the introduction of gases, steam, smells, smoke, soot, warmth, noise, vibrations and similar influences emanating from another plot of land to the extent that the influence does not interfere with the use of his plot of land or interferes with it only to an insignificant extent.

The 1900 version of the rule further provided that the owner would still have to bear the nuisance even if 'significant', if it resulted from the customary use of the other property, having regard to the situation of the area ('*ortsübliche Benutzung des anderen Grundstücks*').

Over the years, the provision was reworked by both case law and legislation. The fact that a landowner had to put up with all nuisances that did not, or did in a nonessential way, impair the use of the land, and in any case all nuisances that, even if essential, were deemed customary in the situation of those places, led to grave injustices, especially where *significant* but *customary* nuisances prejudiced the owners of neighbouring land in their economic activity. This was especially the case when entire agricultural areas bordered on industrial areas, where it was effectively impossible to decide which of the two activities was more usual for the respective area.

To solve such conflicts, German courts, in the late 1950s, developed the principle of *nachbarliches Gemeinschaftsverhältnis*, or *neighbourhood community relationship*,[13] according to which if in a given area both industry and agriculture could be considered *customary* activities for those places, unavoidable damages would have to be compensated according to equity.

This case law elaboration was later incorporated by legislation, which in 1960 introduced a reform of the text of §906 BGB,[14] so that today the text of the rule reads

> The same applies to the extent that a material interference is caused by use of the other plot of land that is customary in the location and cannot be prevented by measures that are financially reasonable for users of this kind. Where the owner is obliged to tolerate an influence under these provisions, he may require from the user of the other plot of land reasonable compensation in money if the influence impairs the use of the owner's plot of land that is customary in the location or its income beyond the degree that the owner can be expected to tolerate.

In conclusion, one can observe a reversal in the interpretation in favour of a greater focus on the protection of the environment of certain terms and concepts that the German legislator of 1900 had certainly intended in favour of industrial development.

Similarly, in Italy, Article 844 of the Civil Code,[15] which regulates nuisances, and Article 2043 on tortious liability[16] have been reinterpreted by both doctrine and case law to achieve better environmental protection.

Italian Courts, when applying Article 844 of the Civil Code, began to consider Article 32 of the 1948 Constitution,[17] which protects the right to health, establishing that nuisances harmful to health should be considered as in any case intolerable, excluding the possibility of continuing the harmful activity in exchange for compensation and granting the pollution victim an injunction.[18]

Since the 1970s, the Italian Court of Cassation had ruled that physical health is a subjective right

> which the system specifically protects against dangerous and harassing nuisances (Article 844 of the Civil Code) and where appropriate through the rule of Article 2043 of the Civil Code against injuries that are caused to it through intentional or negligent conduct of others.[19]

More generally, we can see that also in the Italian context the debate concerning the task of private law in the protection of the environment[20] led to a reinterpretation of some Civil Code provisions, especially of Article 844,[21] where case law extended the availability of an injunction to parties other than the owner.[22]

Further, the issue of the possible extension of the code provisions to situations other than those originally envisaged by legislation came to the fore when Article 844 was submitted to the Constitutional Court by a first instance Court of Bologna in the now distant 1972.[23] The reasons that the Bologna Court put forward for submitting Article 844 of the Civil Code to the Constitutional Court were, first, that the provision violated the principle of equality of citizens affected by nuisances, as it protected only owners, and second, that the provision did not sufficiently protect the fundamental right to health. The Constitutional Court rejected the plea of unconstitutionality in 1974, for, while admitting that through Article 844 of the Civil Code it was not possible to arrive at a broader protection of those parties affected than those who were to be considered neighbouring owners, it brought the applicability of Article 844 of the Civil Code back into the realm of proprietary discipline.[24] According to what the Court said, the protection of health and the environment would be entrusted to the relevant criminal and administrative legislation, 'without prejudice in any case to the application of Article 2043'.[25] This ruling practically marked a turning point in the history of the interpretation of Article 844 of the Civil Code, which was definitively placed in a proprietary dimension, thus difficult to extend to other cases not directly covered by the rule itself.

Finally, in all the legal systems mentioned, after the attempt to adapt private rules to new environmental problems, with the gradual development of specific legislation on the subject, new problems arose concerning the relationship between public and private laws.

1.3 The Emergence of Environmental Legislation and the Focus on Administrative Law Instruments

Since the 1970s, administrative law has become the core of environmental policy and natural resource management.[26] Administrative law instruments are considered effective in the prevention of environmental impairments and therefore particularly appropriate in the environmental sector, where there is great concern for preventing damage, rather than curing it.

This occurred in different contexts at different times.

In the United States, awareness of environmental problems has been developing since the 1960s, beginning with the publication of Rachel Carson's *Silent Spring* in 1962, and leading to the celebration of the first Earth Day in 1970 with the participation of more than 20 million American citizens.[27] A highly committed environmental movement and a Democratic-dominated Congress led President Nixon and his successor Gerald Ford to enact a series of environmental statutes between 1969 and 1976 that were vital to the subsequent development of U.S. environmental law: the *National Environmental Policy Act* in 1969, the *Clean Air Act Amendments* in 1970, the Federal *Environmental Pesticide Control Act* and the *Marine Mammal Protection Act* in 1972, the

Endangered Species Act in 1972, the *Toxic Substance Control Act* in 1976 and the *Resource Conservation and Recovery Acts* in 1976.[28]

To complement this important legislative system, the *Environmental Protection Agency (EPA)*[29] was established in 1970, which had a lasting influence on the development of environmental law.

Subsequently, Democratic President Jimmy Carter, supported by a Democratic majority in Congress, succeeded in enacting major amendments to the *Clean Air Act* and the *Clean Water Act in 1977*, and enacted the *Comprehensive Environmental Response Compensation and Liability Act (CERCLA)* in 1980, which became the backbone of the entire environmental liability system.

With Reagan's election in 1980, the new President initially sought to mitigate environmental burdens on American industries. However, public interest in environmental issues remained stable throughout the 1980s, so, although regulation of the sector did not expand at the rates of the past, Congress' further environmental protection was promoted by the *Solid Waste Amendments* of 1984, a reform of the *Resource Conservation and Recovery Act*, strengthening CERCLA in 1986 with the *Superfund Amendments and Reauthorization Act (SARA)*, and finally enacting the *Global Climate Protection Act* in 1987 and the *Ocean Dumping Ban* in 1988.

With attention to environmental issues remaining high until the late 1980s, Republican candidate George Bush ran his campaign in 1988 based on a hard-hitting environmental programme and – once elected – collaborated with the Democratic Congress to enact amendments to the *Clean Air Act* of 1990, which addressed two important issues that were also hotly debated internationally at the time: the ban on Chlorofluorocarbons (CFCs), which had long been used as refrigerants in refrigerators and in air conditioning systems, believed to be responsible for the ozone hole, which had been made the subject of the Montreal Protocol that the United States had already ratified in 1987,[30] the solution to acid rain problems that had been the source of a long conflict with Canada.[31]

Between the 1960s and late 1980s, the United States were an enthusiastic promoter of international environmental agreements,[32] playing a leading role in the development of international environmental law.[33]

However, from 1992 onwards, the U.S. role in the environmental field began a downward parabola. In the absence of effective support from environmentalists and in the face of numerous criticisms from the industrial world, President Bush took a much more detached attitude towards environmental issues and was the only major leader not to participate in this initiative the 1992 *Earth Summit* in Rio.[34]

The different positions of environmental *lobbyists* became even clearer after the election of President Clinton. Clinton proposed a package of environmental reforms and signed the Convention on Biological Diversity adopted in Rio but failed to secure Senate ratification or the adoption of any specific environmental legislation.[35] In 1997, when the Kyoto Protocol was opened for signature, Clinton signed the international commitment, but it was never ratified by the Senate.

Later, President George Bush Jr. distanced himself from the international climate change negotiations, preferring not to impose new environmental burdens on American industries.

In Europe, with the provision of a specific environmental competence in the Single European Act in 1986, there is a whole series of measures aimed at protecting the environment in all its forms.

In less than 20 years, more than 200 measures in the environmental field have been enacted, based on command-and-control.

However, command-and-control standards, while an effective tool in preventing risks, do not always provide efficient incentives for potential polluters, who tend to comply with the standard, with no valid incentive to introduce cleaner technologies.[36]

Moreover, the effectiveness of command-control regulations depends largely on the quality of control, which in the environmental sector is delegated to bodies that are not always equipped with the necessary expertise for this purpose.[37]

As a result, controls do not take place at all or only with considerable delay; sometimes, this is due to the lack of adequate technical tools, not to mention cases in which even the financial resources to improve the bureaucratic apparatus itself are lacking.

1.4 The New Instruments of Environmental Policy

In a context characterized by an awareness of these limitations, and in the face of the recognition of major new challenges, such as those posed by climate change, there has been a renewed discussion on the most appropriate means of tackling environmental problems.[38]

In recent decades, environmental policy objectives have led to the development of new instruments to cope with new ecological demands.[39] The discussion of tools to be adopted in this area has distant roots. The debate arose in the United States as early as the 1980s, highlighting how *market-based* instruments could offer a viable alternative to the centralized regulatory system.[40]

In the following years, the focus was on formulating a 'new generation' of market-based instruments, which would have to cope with a sector now characterized by new phenomena,[41] including the evolution of a global environmental law and the rapid circulation of legal models.[42]

If the debate on new instruments in the environmental field in the United States has been guided by scholars and courts,[43] in Europe it was the EU Commission that marked the stages of a new approach to the subject.

This development emerged clearly from the Fifth Environmental Action Program adopted by the Commission in 1992.[44]

To drive a change in the approach to environmental issues, the Commission was advocating a broader range of instruments that would make the business community part of a process of raising awareness and consciousness of environmental issues.[45]

Alongside the traditional environmental policy instruments, market instruments were introduced, including taxes, tariffs, specific environmental incentives, allowances trading systems, eco-labelling and environmental budgeting systems, the regulation environmental of liability and, finally, environmental agreements.[46]

Subsequently, with the Sixth Environmental Action Program,[47] the Commission indicated that one of the objectives of environmental policy should be to induce the market to work for the environment through better collaboration with the business community, by introducing reward programmes for companies with the best environmental performance, by promoting a switch to greener products and processes, and by encouraging the adoption of eco-labels that allow consumers to compare similar products on the basis of their environmental performance.[48]

In the latest Seventh Environmental Action Program,[49] the Commission combines the reasons of the environment with those of the market, within the framework of a conception of well-being, in which the economic prosperity and well-being of the Union are closely dependent on its natural capital.[50]

This reflects the Union's commitment to transforming into an inclusive green economy that fosters growth and development, protects people's health and well-being, creates decent jobs, reduces inequalities and invests in biodiversity.[51]

To achieve these ambitious goals, the Commission emphasized that an '*appropriate mix of policy instruments*' is needed,[52] aimed at enabling economic actors and consumers to gain a better understanding of the environmental impact of their activities and to manage it. This approach encompasses public law and private law instruments, including economic incentives, market-based instruments, information obligations and voluntary measures and instruments that engage stakeholders at various levels, complementing those imposed by legislation.[53]

1.5 Reviewing Private Law Instruments for Environmental Protection

The current trend in countries with advanced economies is to promote consumption and production models that internalize environmental impacts through a combination of various instruments, such as civil liability for environmental damage, proprietary mechanisms used in the regulation of tradable emission allowances and reform of contractual regulation.

As in other fields, environmental law is experiencing the reallocation of regulatory powers to other actors, giving rise to a phenomenon of co-regulation or self-regulation by stakeholders,[54] which contributes to the development of the phenomenon known as *transnational private regulation* (TPR), i.e. that set of rules created by private actors in areas where regulation by national or international public actors is lacking, or supplementing, or even replacing, the latter.[55]

The appearance of a private regulator alongside a public regulator implies a mixture of different logics and categories that makes the instruments used in this context characterized by a certain degree of 'hybridisation' between private and public elements.[56]

1.5.1 Liability for Environmental Harm

Since the 1960s, a debate has developed on the introduction of liability to cope with the consequences of industrialization.[57] Environmental liability has been presented as a useful tool to guide corporate behaviour as a deterrent[58] against environmentally harmful activities, as well as an effective means to achieve the internalization of environmental diseconomies.

Comparative law analysis revealed that liability can indeed be a valuable tool in this area[59] and that the rapid circulation of models has spread this idea almost everywhere.[60]

In 1980, the United States Congress passed the *Comprehensive Environmental Response Compensation and Liability Act* (CERCLA),[61] introducing federal legislation that provided a specific definition of *environmental damage*, of the liable parties, of the standing issues and the strict liability regime, while introducing detailed criteria for repairing environmental damage that should facilitate the insurability of environmental risk.[62]

During the 1980s and 1990s, the idea that civil liability could be a useful tool to prevent environmental damage spread throughout Europe: in 1986, an Italian law (no. 349/86)[63] was enacted, in 1987 a Portuguese law,[64] in 1990 a German law,[65] to providing specific regulation in this field.

This fragmentation prompted the European Commission to develop several initiatives aimed at harmonizing the different regimes in force at the national level in this area, first publishing a Green Paper in 1993[66] and then a White Paper,[67] which regularly took American legislation as a model for comparison. Finally, in 2004, Directive 2004/35 on environmental liability[68] was enacted, providing a concept of environmental damage that covers only selected environmental resources,[69] excluding the compensability of private parties.[70]

The spread of environmental liability does not stop at the *Western Legal Tradition*, as the circulation of the model appears successful in other contexts as well.

China has introduced specific rules on liability for environmental damage since 1986, when the *General Principles of Civil Law* were promulgated,[71] which provided in Article 124 a general principle that 'Whoever pollutes the environment and causes damage to third parties in violation of state regulations designed to protect the environment and prevent pollution, shall be held liable in accordance with the law'.[72] An interesting form of model circulation can be noted here, as the 1986 *General Principles* were drafted following the German model contained in the General Part (*Allgemeiner Teil*) of the German Civil Code (*Bürgerliches Gesetzbuch*, BGB).[73] However, the provision in Article 124, which introduces a specific provision on pollution liability, is not found in the German archetype and must be considered a novelty introduced by the Chinese legislature.[74]

On July 1, 2010, China's new law on *civil liability* came into force, which devoted a specific *Chapter VIII* to the problem of environmental liability,[75] containing four norms (Articles 65–68) of paramount importance for the problem of civil liability in this context, where again it is possible to note a clear focus on the model introduced by the German legislature in 1990.[76]

In India, the system inherited from the English common *law of torts* is developed in an original way by introducing *public interest litigation* on the one hand and, on the other, by introducing the *National Green Tribunal* as a specialized court in this area.[77]

In Japan, the notion of environmental damage appears to focus on the human being as a victim, where therefore – unlike in other contexts – the environment is not seen as an entity but acquires value only insofar as there are human subjects capable of interacting with it.[78]

The civil liability model is also spreading in Central and South America,[79] where some jurisdictions consider the model of the 2004 European directive to innovate national legislation, as in the case of the Mexican federal law of 2013,[80] while others take up the example of the French legislation and introduce specific provisions in the Civil Code in this regard, as in the case of Peru.[81]

1.5.2 Property Rights and Environmental Protection

Property is at the centre of a similar debate regarding its functions in the environmental sphere.

On the one hand, it has been pointed out that in certain cases, property becomes a source of obligations if it is found to be polluted,[82] as in the case of the U.S. *Comprehensive Environmental Response Compensation and Liability Act*, and later partly taken up in Europe.[83]

On the other hand, the idea of property has been used to develop special instruments to deal with pollution, as in the case of *tradable pollution rights*.[84]

Tradable pollution allowances seek to curb pollution by setting an industry limit and then allowing industry players to determine how much they are willing to pay to pollute. What is traded in the market consists of portions of the pollution allowed in each area, requiring a clear allocation of the portion itself, which is often done in terms of 'property of the portion'. This policy gives flexibility to the industry to regulate its own pollution, by making them compete for the right to pollute.[85]

Although they have also been used in areas such as water pollution, the most widespread field of application of transferable pollution rights is air pollution, for the control of which the location of sources is often negligible.

At the international level, tradable pollution rights have received great attention since the Kyoto Protocol[86] provided – among others – the introduction of an *international emissions trading scheme*.[87]

In 2000, the Commission adopted a *Green Paper on Greenhouse Gas Emissions Trading in the EU*,[88] which was followed by the adoption of Directive 2003/87/EC, establishing a scheme for greenhouse gas emission allowances trading within the Community, the first European implementation of one of the mechanisms under the Kyoto Protocol.[89]

Directive 2003/87/EC thus introduces a new concept of property, with the aim of regulating emissions trading to implement emission reductions throughout the community where it costs less.

1.5.3 Contracts

To increasingly involve businesses in the dynamics of ecological policies, the so-called *voluntary agreements* have been developed in recent times, which are concluded between businesses and the public administration aimed at achieving specific objectives in the field of environmental protection which have developed in recent times.[90] These are instruments that innovate from the classic *top-down* approach to environmental regulation and aim to find common ground between business and the regulator, to improve environmental protection through a shared approach.

In the United States, the *Environmental Protection Agency (EPA)* launched in 1991 a specific programme aimed at involving businesses in reducing pollution.[91] At that time, the concept of a voluntary agreement was relatively new to industry, although when the programme ended in 1996, more than 1,300 companies that had joined recognized the success of the EPA's initiative for establishing government-industry partnerships to reduce pollution and promote the development of innovative pollution reduction techniques.[92]

At the European level, as early as 1996, the Commission adopted a Communication to the Council and the European Parliament on environmental agreements,[93] which were recognized as a new policy tool to complement regulatory measures.

The Communication recognized that environmental agreements could have several potential advantages over mandatory regulation, including encouraging a more participatory attitude on the part of industry, the possibility of arriving at tailor-made and cost-effective solutions, and finally a more rapid achievement of environmental objectives.[94]

Among the agreements that have been developed in environmental matters are those of the European, Japanese and Korean automobile manufacturers' associations on the reduction of CO_2 emissions of passenger cars, which were recognized in some of the Commission's recommendations[95] and were later supplemented by the European Parliament and Council Decision establishing a system for monitoring the average specific CO_2 emissions of new passenger cars.[96]

An extensive survey of the use of this instrument was carried out by the Organization for Economic Cooperation and Development (OECD) and published in 1999,[97] where it was concluded that environmental agreements are most effective when used as part of a *policy mix*, in combination with economic and legislative instruments.

In 2002, the Commission published a new Communication on environmental agreements,[98] in which 'voluntary environmental agreement' was presented as instruments that could best respond to new environmental policy needs as outlined by the Fifth Environmental Action Program onwards.

The logic of consensus underlying these agreements appears successful in various contexts because it allows overcoming the conflicting view of the relationship between the environment and economic development, and thus also facilitates a change in the way industry is conceived, from the cause of the problem to the protagonist of its solution. This approach has been incorporated in Italy[99] and other national legal systems.[100] In China, too, the role of environmental agreements has recently come to the fore.[101]

Contracts have further developed in the environmental context in other respects. In fact, environment and sustainability have long penetrated the realm of contracts between private parties and particularly contracts between businesses.

In recent decades, *statutes of associations* of economic operators have spread that establish as requirements for acquiring and maintaining membership the observance of environmentally sustainable behaviour or, again, that provide for the awarding of 'prizes' for members who achieve certain levels of environmental sustainability.[102]

Further, the adoption of *codes of conduct*, as a part of the broader phenomenon of corporate social responsibility that began in the United States in the 1960s–1970s and then moved to Europe from the 1980s to 1990s, is spreading now all over the world, mainly in response to campaigns denouncing the violation of workers' rights, relocation policies and environmental impacts of multinational companies operating in a wide range of sectors, from chemicals to clothing.[103] These codes are voluntary instruments of self-regulation by which companies, in addition to setting their own reference values, establish a whole series of behavioural rules that often have the purpose of managing the impact that the activity of companies may have on the environment.[104] These rules of conduct can refer both to internal company relations and to relations with external parties – *business-to-business* and *business-to-consumer* – as well as with categories of stakeholders, particularly environmental and social.

Codes of conduct are expressed at different levels of regulation, as they can be corporate codes, category codes and even codes developed by public and private bodies or institutions of international standing to which companies can voluntarily adhere. In fact, since one of the major limitations of corporate codes of conduct is the lack of an independent monitoring system to ensure that they are not mere image management tools, but actual attempts to improve corporate sustainability, NGOs and trade associations have long since started to develop 'model codes' that involve the simultaneous participation of various stakeholders and include appropriate independent monitoring and verification procedures.[105]

However, in multinational companies, which are often characterized by a very small direct operational structure and carry out their activities through subsidiaries or outsourcing, codes are drawn up by the parent company and are not only disseminated within the legal sphere of the parent company and its subsidiaries, but also have an impact in the context of relations with third parties, producer-suppliers, who do not formally belong to the company, but whom the company asks to respect the rules contained in the codes, thus affecting the entire value chain.[106]

Although it may happen that codes of conduct partly overlap with what is already provided for in legislation at the national or supranational level, most of the standards they contain are voluntary commitments that go beyond pre-existing obligations, enabling companies to operate in different national contexts. Codes of conduct thus become a source of private production of environmental and social standards, taking on a role of supplementing and replacing public regulation, especially where the latter is lacking for various reasons, so much that through them multinational companies have become part of the process of defining global issues, such as the protection of human rights and the environment.

The phenomenon is interesting from the point of view of legal transplants, in that, given the supranational nature of companies and their global dimension, the production of codes of conduct and their dissemination within the network of business relations necessarily produces a circulation of principles and values from the 'central' company to the entire network, albeit with varying intensity. In this case, however, the circulation takes on different contours from the traditional ones, leading to the creation of a kind of private sub-regime within the different orders in which the enterprise value chain operates.[107]

Precisely for this reason, interest in this area has increased. With respect to international business contracts, the United Nations Global Compact (UNGC) expressly requests companies to exert their influence in the area of environmental protection throughout their supply chains.[108] The UNGC practical guide on supply chain sustainability advises companies to clearly formulate their expectations towards their suppliers in a code of conduct and subsequently implement the principles of Code through their 'integration (…) into supplier contracts'. In 2011, the UN Special Representative of the Secretary General for Business and Human Rights presented the '*Principles for Responsible Contracts*': a guide for negotiators and others on how to ensure the management of human rights and environmental risks is integrated into State-investor contract negotiations.

The increasing use of contractual instruments in the environmental field entails, on the one hand, an evolution of environmental law, which becomes more flexible, negotiated and the result of the consensus of the addressees of the rules themselves, and, on the other hand, a transformation also of contract law, which takes into account not only the interest of the contracting parties, but also the general interest in environmental protection, acquiring a new 'solidaristic' dimension.

Moreover, the involvement of a plurality of actors and an increasing expansion of the scope of contractual instruments in the environmental field are evident. In this regard, we can observe on the one hand a tendency towards the contractualization of environmental law by public authorities to implement a more flexible and at the same time more effective environmental protection policy, resorting to the contract either as a direct substitute for public regulation or as an accessory function of the latter to supplement it or enforce it. On the other hand, private actors also play an increasingly important role in environmental protection, either as parties involved in contracting by public authorities or as producers of their own contractual mechanisms with a direct or indirect environmental protection function.

In 2002, the OECD adopted a *Recommendation on Green Public Procurement* (GPP) that recognizes that well-designed public procurement systems can contribute to achieving policy goals such as environmental protection,[109] followed by a collection of good practices in 2015.[110] Within the EU, the potential of GPP was first highlighted in the 2003 *Commission Communication on Integrated Product Policy*, where Member States were recommended to adopt national action plans for GPP by the end of 2006, followed in 2008 by a new *Communication on Green Public Procurement*,[111] where the Commission identifies 'greener' goods on the ground of a life-cycle approach that affects the whole supply chain, ranging from the use of raw materials and production methods to the types of packaging used and the respect of certain take-back conditions, suggesting that these criteria can equally inform private procurement practices.

With regard to the broadening of the scope of application of contractual instruments in the environmental field, it should be emphasized that the contractualization of environmental law is deeply influenced by the goal of sustainable development, which ends up affecting a multitude of economic activities that may have an impact on the environment, such as public procurement, supply and subcontracting contracts, consumer contracts and even contracts in the real estate and construction sectors.[112]

For example, Directive 2012/27/EU, on energy efficiency, provides at Art. 7 (c) that

> Member States shall encourage public bodies, including at regional and local level, and social housing bodies governed by public law, with due regard for their respective competences and administrative set-up to use, where appropriate, energy service companies, and energy performance contracting to finance renovations and implement plans to maintain or improve energy efficiency in the long term.[113]

However, *consumer contract law* is and will be increasingly also influenced by sustainability issues. Consumer contract law has always had the objective of protecting consumers by allowing them a right of withdrawal if they have purchased goods at a distance. This rule, combined with the widespread free-return policy, whereby not even the only costs that may possibly be charged to the consumer – those of (re)shipping the goods back to the sender – are borne by the latter, determines a result that, by facilitating the transport, production and consumption of goods, is 'regrettable' from the point of view of sustainability. More generally, it could be argued that the discipline aimed at satisfying the interest of those who purchase goods or services is conflicting with that aimed at a use of resources commensurate with the needs of the environment or of future generations.[114]

Nonetheless, it appears that these divergencies must be reconciled, as sustainability is 'one of the founding principles of consumer law'.[115] In the new Circular Economy Action Plan of 2020,[116] the Commission proposes a revision of EU consumer law to ensure that consumers receive trustworthy and relevant information on products at the point of sale, including on their lifespan and on the availability of repair services, spare parts and repair manuals. The Commission is also considering further strengthening consumer protection against greenwashing and premature obsolescence, setting minimum requirements for sustainability labels/logos and for information tools.[117] The EU Parliament Resolution of November 25, 2020 *Towards a more sustainable single market for business and consumers* goes in the same direction,[118] focusing on consumer rights and the reduction of planned obsolescence, facilitating repair and reuse.

The directives that come into play to realize a new sustainable framework for consumer contract law are Directive 2005/29/EC concerning unfair business-to-consumer commercial practices in the internal market,[119] Directive 2011/83/EU on consumer rights[120] and Directive 2019/771 on certain aspects concerning contracts for the sale of goods.[121]

As scholars[122] have put forward, the potential of the Sale of Goods Directive to serve as a vehicle for achieving sustainability is evident. Nonetheless, the current version of the Sale of Goods Directive falls short regarding the potential fostering of sustainability, although extensive studies on the possibility of reform have currently been published,[123] including that by the European Law Institute.[124]

A *Proposal for a directive of the European Parliament and of the Council amending Directives 2005/29/EC and 2011/83/EU as regards empowering consumers for the green transition through better protection against unfair practices and better information*[125] was presented in March 2022. On September 23, 2023, the Council and the Parliament have reached a provisional political agreement on this proposal. The text of the document foresees achieving an improved participation of consumers in the circular economy by providing better information on the durability and reparability of certain products to consumers before concluding the contract and stepping up the protection of consumers against unfair commercial practices that prevent sustainable purchases.[126]

1.6 The Greening of Civil Codes

Environmental protection and sustainability issues have not only characterized the interpretation of traditional private law institutions but have also entered civil codes, introducing new paradigms and rules: a tangible sign of the extent to which private law is taking on new characteristics in order to meet the new demands that society now considers fundamental to its survival.[127]

In France, the Law of August 8, 2016[128] has introduced into the *Civil Code, le code civil par excellence,* in Articles 1246–1252,[129] provisions aimed at recognizing the notion of *préjudice écologique* and regulating its reparation.[130]

Another notable example can be seen in the provision of Article 603 of the Romanian Civil Code (2009), which specifies the landowner's obligation to operate in an appropriate manner regarding '*environmental protection and to ensure good neighbourly relations*'.[131]

Examples multiply throughout the world, although the presence of verbatim provisions in civil codes does not guarantee their concrete application in practice, as comparative lawyers are aware of the difference between the *law in the books* and the *law in action.*

In this perspective, the example provided by the Chinese Civil Code, which has been identified by some scholars as the 'green civil code', appears interesting, even though China's rapid economic development has put a strain on the environment and natural resources.[132] An emblematic example is the obligation to protect the environment that was incorporated into Chinese civil law by Article 9 of the General Part of the Civil Code of the PRC: 'when engaged in civil activities, all persons shall be aware of the need to save resources and protect the environment'.

On the other side of the world, the new 2014 Argentinian Civil and Commercial Code marks a turning point in sustainable consumption. In particular, Article 1094 of the Argentinian Civil Code states: 'the norms that govern consumer relations must be applied and interpreted according to the principle of consumer protection and that of access to sustainable consumption'.[133]

The Brazilian Civil Code (2002) in regulating property introduces new values, such as that of 'environmental beauty', as well as that of 'ecological balance'. According to Art. 1228:

> The owner has the right to use, enjoy and dispose of the property and the right to recover it from whoever unjustly possesses or detains it. §1 The right of ownership must be exercised in accordance with its economic and social purposes and in such a way that the flora, fauna, natural beauty, ecological balance and historic and artistic heritage are preserved, in accordance with that established by special law, and that air and water pollution is avoided.[134]

In the same line is the Draft of the new Colombian Civil Code (2020),[135] where new values are introduced, such as 'ecological coexistence' (Article 276),[136] or the conservation of the environment and 'social well-being' (Article 268),[137] that will have to be taken into consideration in order to be weighed against 'economic activities and in relationships and development of markets' (Article 38).[138]

This brief review is particularly interesting from the perspective of comparative law, as it shows that in the field of environmental protection, we are witnessing a rapid development in the direction of introducing new environmental values into civil codes.

1.7 Conclusions

The current historical moment seems to highlight some important variables in the evolution of private rules and institutions that have undergone a metamorphosis to meet the new requirements of environmental protection.

Everywhere there is an evident recourse to the instruments offered by private law to solve at least some of the problems that our time imposes on our attention.

In civil law systems, not only does the environment enter the civil codes with specific rules, but it is the civil codes themselves that are reinterpreted from an environmental perspective.[139]

In those belonging to the common law, the law of torts appears to be the area that before others has offered important cues for finding innovative solutions to then leave room for new designs in both the proprietary and contractual fields.

It should be however noted that the concern to give a supranational discipline to problems that cannot be delimited within national boundaries, and which has necessarily led to the evolution of a 'global' discipline of certain aspects of the protection of natural resources, has re-proposed two specific phenomena in this area as well: on the one hand, the consolidation of rules at the supranational level which are strongly affected by the solutions developed in the different national contexts, but which at the same time innovate with respect to national models by creating new taxonomies; on the other hand, the tension between global and local dimensions, since despite the growing number of international standards in the different areas of environmental protection, it must be remembered that their concrete application is strongly and necessarily affected by the different *legal processes* existing at the local level.

In this context, characterized by a strong push for legal transplants, as well as the imposition of 'global' models of environmental protection, the revision of private law instruments, aimed at involving all stakeholders in the development of environmental protection policies, leads to the blurring of the classic distinction between public and private laws.

In the perspective of overcoming this dichotomy, the evolution of environmental law highlights that hybridization of the instruments employed in environmental policies is currently underway: property, tort and contract share this fate.

The adaptation of classic private law instruments to deal with new ecological issues goes hand in hand with the abandonment of the classic rhetoric about the pre-eminence of public law instruments over private law instruments in environmental policies.

What appears to be at the centre of scholars' attention is the *effectiveness* of environmental policies, often invoking the concept of the *performance* of regulations in this area. The problem is not whether the norms are of public law or private law, but whether in practice they comply with and succeed in changing the behaviour of the consociates in a way that is consistent with environmental goals.

In a comparative law analysis, it is important to complete the study of systemic variables affecting the practical application of the instruments in question, since – as is well known – one and the same formula could give rise to entirely different practical solutions if transplanted in a legal process that does not reflect all the variables of the system of origin.

This analysis appears even more important in a context characterized by rapid legal transplants. This analysis appears even more important in a context characterized by rapid legal transplants. Therefore, it underscores the importance of the systemic variables that in some cases have decreed the success of certain innovative solutions, such as the existence of special courts in India.[140]

Private law in the environmental sector has not yet finished its evolutionary journey, which – necessarily – will also accompany us in the years to come.

Notes

1 M. Mekki, É. Naim-Gesbert (eds), *Droit public et droit privé de l'environnement: Unité dans la diversité?* (Paris, LGDJ 2016).

2 M. Mekki (ed), *Les notions fondamentales de droit privé à l'épreuve des questions environnementales* (Brussels, Bruylant 2016).

3 F.E. Maloney, 'Judicial Protection of the Environment: A New Role for Common Law Remedies' (1972) 25 (1) Vanderbilt L.R., 145 <https://scholarship.law.vanderbilt.edu/vlr/vol25/iss1/14> accessed 25 June 2023; G. Alpa, M. Bessone, A. Gambaro, 'Aspetti privatistici della tutela dell'ambiente: l'esperienza americana e francese' in A. Gambaro (ed), *La tutela degli interessi diffusi nel diritto comparato* (Milano, Giuffrè 1976) 295.

4 T. A. Brennan, 'Environmental Torts' (1993) 46 (1) Vanderbilt L.R., <https://scholarship.law.vanderbilt.edu/vlr/vol46/iss1/1> accessed 14 July 2023; W.B. Alcorn, 'Liability Theories for Toxic Torts' (1988) 3 (2) Natural Resources & Environment, 3; R.L. Rabin, 'Environmental Liability and the Tort System' (1987) 24 Houston L.R., 27; P.J. Strand, 'The Inapplicability of Traditional Tort Analysis to Environmental Risk: The Example of Toxic Waste Pollution Victim Compensation' (1983) 35 Stanford Law Review, 575.

5 J.G. Fleming, *The American Tort Process* (Oxford, A Clarendon Press Publication 1990) 252; John G. Fleming, 'Mass Torts' (1994) 42 (3) The American Journal of Comparative Law, 507.

6 P. Girod, *La réparation du dommage écologique* (Paris, Librairie générale de droit et de jurisprudence 1974) 90.

7 J.B. Blaise, 'Responsabilité et obligations coutumières dans les rapports de voisinage' (1965) t. LXIII Revue Trimestrielle de Droit Civil, 275.

8 R. Ganten, M. Lemke, 'Haftungsprobleme im Umweltbereich' (1989) UPR, 1.

9 §823 BGB, paragraph 1, reads: 'Liability in damages»: (1) A person who, intentionally or negligently, unlawfully injures the life, body, health, freedom, property or another right of another person is liable to make compensation to the other party for the damage arising from this'.

10 H. Forkel, *Immissionschutz und Persönlichkeitsrecht* (Köln, Heymanns 1968) 47; G.H. Roth, Materiellrechtliche und prozessuale Aspekte eines privatrechtlichen Umweltschutzes (Neue Juristische Wochenschrift 1972) 921.

11 J. Köndgen, 'Überlegungen zur Fortbildung des Umwelthaftpflichtrechts' (1983) UPR, (Jehle) 345; criticized by D. Medicus, *Zivilrecht und Umweltschutz, JuristenZeitung* (1986) (Mohr Siebeck) 17, 778 and by U. Diederichsen, *Die Haftung für Umweltschäden, Betriebs-Berater* (1973) (Verlagsgesellschaft Recht und Wirtschaft MBH Heidelberg) 485.

12 F. Baur, *Lehrbuch des Sachenrechts* (IV, § 25, 15ª ed., München, 1989) 228; F. Baur, 'Zur Entstehung des Umweltschutzes aus dem Sachenrecht des BGB' (1987) 42 (7) JuristenZeitung, 317–322; H. Westermann, *Die Funktion des Nachbarrechts. Zugleich eine Untersuchung der Bedeutung eines Immissionsschutzgesetzes für das Privatrecht, Festschrift für Karl Larenz zum 70. Geburstag* (München, Beck 1973) 1003; O. Jauernig, 'Zivilrechtlicher Schutz des Grundeigentums in der neueren Rechtsentwicklung' (1986) 41 (13) JuristenZeitung, 605; U. Diederichsen, 'Schadenersatz und Entschädigung im Umweltsschutz' in: *Schriftenreihe der Deutschen Richterakademie, Bd. 5: Natur- und Umweltschutz* (Heidelberg, 1989) 19, 23; M. Ronellenfitsch, R. Wolf, 'Ausbau des Individualschutzes gegen Umwelt belastungen als Aufgabe des bürgerlichen und des öffentlichen Rechts?' (1986) 1960 Neue Juristische Wochenschrift; See further A. Procida Mirabelli di Lauro, *Immissioni e "rapporto proprietario* (Napoli, Edizioni Scientifiche Italiane, 1984) 74.

13 B. Kleindienst, *Der privatrechtliche Immissionenschutz nach §906 BGB* (Tübingen, J.C.B Mohr (Paul Siebeck), 1964); see further BGH NJW 1959, 97; BGHZ 30, 273.

14 By law of December 22, 1959 (BGBl. I 781) which came into effect on June 1, 1960.

15 Art. 844 Civil Code of Italy: 'The owner of land cannot prevent the emission of smoke, heat, fumes, noises, vibrations or similar propagation from the land of a neighbor unless they exceed normal tolerability, with regard to the condition of the sites. In applying this rule, the court shall reconcile the requirements of production with rights of ownership. It can also take account of the priority of a given use'.

16 Art. 2043 Civil Code of Italy: 'Whatever malicious or negligent misconduct, or negligence, which causes un unfair damage to another person, obliges who committed the fact to pay the compensation'.

17 Art. 32 of the Italian Constitution: 'The Republic safeguards health as a fundamental right of the individual and as a collective interest and guarantees free medical care to the indigent'.

18 Cfr. G. Alpa, M. Bessone, A. Gambaro, 'Aspetti privatistici della tutela dell'ambiente: l'esperienza americana e francese' in A. Gambaro (ed), *La tutela degli interessi diffusi nel diritto comparato* (Milano, Giuffrè 1976) 391. In case law see Cass. July 30, 1984 No. 4523, *Giur.it.*, I, 1 (1985) 1585, where the principle was affirmed that 'if the emissions cause injury to the right to health and tranquility and the right to normal family life within a dwelling house, the solution consisting in the payment of compensation is not applicable'.

19 Cass. sez. un. 9 April 1973, n. 999, *Foro it*., (1974) I, c. 843.

20 On the private protection of the environment see S. Patti, 'Environmental Protection in Italy: The Emerging Concept of a Right to a Healthful Environment' (1984) 24 Natural Resources Journal, 535; G. Morbidelli, 'Strumenti privatistici contro l'inquinamento delle acque interne' (1971) III Foro amm., 369; P. Paradiso, 'Inquinamento delle acque interne e strumenti privatistici di tutela' (1977) Riv. trim. dir. proc. civ., 1391; C. Salvi, *Le immissioni industriali: rapporti di vicinato e tutela dell'ambiente* (Milano, Giuffrè, 1979); F. Di Giovanni, *Strumenti privatistici e tutela dell'ambiente* (Padova, CEDAM,1982); A. Procida Mirabelli di Lauro, *Immissioni e 'rapporto proprietario'* (Napoli, ESI, 1984) 398; F. Nappi, *Le regole proprietarie e la teoria delle immissioni* (Napoli, ESI 1986) 170.

21 F. Forte, *Per una lettura alternativa dell'art. 844 del codice civile* (Dir. e Giur. 1976) 650; F. Nappi, *Le regole proprietarie e la teoria delle immissioni* (Napoli, ESI 1986) 7.

22 F. Di Giovanni, *Strumenti privatistici e tutela dell'ambiente* (Padova, CEDAM 1982) 92; G. Acerbis, 'Le immissioni, con particolare riferimento alla tutela della salute e dell'ambiente' (1989) Parte II, *Quadrimestre rivista di diritto privato*, Giuffrè Editore, 151.

23 Pret. Bologna 18 May 1972 (1973) I (2) Giur. it., 798. See C. Salvi, 'Immissioni, ecologia, norme costituzionali' (1973) I (2) Giur. it., 798; L. Porcinari, 'In tema di legittimità costituzionale del divieto di immissioni nocive previste dall'art.844 c.c.' (1973) Giur. agr. it, 310; M. Longo, 'Assurdi dubbi della costituzionalità dell'art.844 c.c.' (1974) Riv. dir. agr, 608.

24 A. Iannelli, 'Sulla tutela delle immissioni industriali e sulla non operatività dell'art. 844' (1980) Rass. dir. civ., 371; G. Visintini, 'Il divieto di immissioni ed il diritto alla salute nella giurisprudenza odierna e nei rapporti con le recenti leggi ecologiche' (1980) II Riv. dir. civ., 249.

25 Corte Cost. 23 luglio 1974, *Giur. Cost.* (1974) 2371.

26 C. Rose, 'Rethinking Environmental Controls: Management Strategies for Common Resources' (1991) 1 Duke Law Journal, 1.

27 K. Sale, *The Green Revolution. The American Environmental Movement 1962–1992* (New York, Hill and Wang 1993) 24.

28 R.D. Kelemen, D. Vogel, 'Trading Places: The US and EU in International Environmental Politics' (2010) 43 (4) Journal Comparative Political Studies, 427–456.

29 With the Executive Order *Reorganization Plan No. 3* (35 FR 15623, 84 Stat. 2086) President Nixon presented at the Congress of 9th July 1970 the proposal to establish the *Environmental Protection Agency* (EPA) and at the same time the *National Oceanic and Atmospheric Administration* (NOAA).

30 G. Epstein et al., 'Governing the Invisible Commons: Ozone Regulation and the Montreal Protocol' (2014) 8 (2) International Journal of the Commons, 337–360 <http://doi.org/10.18352/ijc.407> accessed 25 June 2023.

31 R.D. Kelemen, D. Vogel, 'Trading Places: The US and EU in International Environmental Politics' (2010) 43 (4) Journal Comparative Political Studies, 14.

32 P. Sands, 'The "Greening" of International Law: Emerging Principles and Rules' (1994) 1 (2) Indiana Journal of Global Studies, 293.

33 A.D. Tarlock, 'The Influence of International Environmental Law on U.S. Pollution Control Law' (1997) 21 Vermont L. Rev., 759.

34 S. Hopgood, *American Foreign Environmental Policy and the Power of the State* (Oxford, Oxford University Press 1998) 140.

35 R.D. Kelemen, D. Vogel, 'Trading Places: The US and EU in International Environmental Politics' (2010) 43 (4) Journal Comparative Political Studies, 19.

36 M. Bresso, *Per un'economia ecologica* (Roma, Carocci 1993) 210.

37 As the United Nations has been highlighting since 1997, with the *Global State of the Environment Report*, Chapter 3: Policy Responses and Directions <http://www.unep.org/geo/geo1/ch/toc.htm> accessed 25 June 2023.

38 C. Rose, 'Rethinking Environmental Controls: Management Strategies for Common Resources' (1991) 1 Duke Law Journal, 1–38.

39 M. Falque, *Introduction générale, in Droits de propriété et environnement* (Paris, Dalloz, 1997) 17.

40 B.R. Ackerman, R.B. Stewart, 'Reforming Environmental Law, Stanford Law Review' (1985) 37 Stanford Law Review, 1333; D.J. Dudek, J. Palmisano, 'Emissions Trading: Why Is This Thoroughbred Hobbled' (1987) 13 Columbia Journal of Environmental Law, 217.

41 J.B. Hockenstein, R.N. Stavins, B.W. Whitehead, 'Crafting the Next Generation of Market-Based Environmental Tools' (1997) 39 (4) Environment: Science and Policy for Sustainable Development, 12–33.

42 B. Pozzo, 'Modelli notevoli e circolazione dei modelli giuridici in campo ambientale: tra imitazione e innovazione' in U. Mattei, A. Candian, B. Pozzo, A. Monti, C. Marchetti (eds), *Studi in onore di Antonio Gambaro: un giurista di successo* (Milano, Giuffrè, 2017) 351–377.

43 G.G. Neustadter, 'The Role of the Judiciary in the Confrontation with the Problems of Environmental Quality' (1970) 17 UCLA, 1070; F.E. Maloney, 'Judicial Protection of the Environment: A New Role for Common Law Remedies' (1972) 25 Vanderbilt L.R., 145; B.R. Furrow, 'Governing Science: Public Risks and Private Remedies' (1983) 131 University of Pennyslvania Law Review, 1403; D.R. Hodas, 'Private Actions for Public Nuisance: Common Law Citizen Suits for Relief From Environmental Harm' (1989) 16 Ecology Law Quarterly, 883; D. Dewees, M.J. Trebilcock, 'The Efficacy of the Tort System and Its Alternatives: A Review of Empirical Evidence' (1992) 30 (1) Osgood Hall Law Journal; T.A. Brennan, 'Environmental Torts' (1993) 46 (1) Vanderbilt Law Review <https://scholarship.law.vanderbilt.edu/vlr/vol46/iss1/1> accessed 25 June 2023.

44 Commission, 'Fifth Action Program' OJEC C 138/5, May 17, 1993.

45 New environmental policy instruments and the concept of "shared responsibility" are examined in item 7 of the Fifth Action Program, OJEC C 138/70, May 17, 1993.
46 Commission, 'Europe's Environment: What Directions for the Future? The Global Assessment of the European Community Programme of Policy and Action in Relation to the Environment and Sustainable Development, "Towards Sustainability"', COM (1999) 543 def.
47 See the Commission, 'Sixth Environmental Action Programme' [Communication from the Commission to the Council, the European Parliament, the Economic and Social Committee and the Committee of the Regions] COM (2001) 31 def.
48 See the sixth recital of Decision 1600/2002 establishing the Sixth Environmental Action Programme, Council Decision No 1600/2002/EC of the European Parliament and of the Council of 22 July 2002 laying down the Sixth Community Environment Action Programme [2002] OJ L 242/1.
49 Council Decision No. 1386/2013/EU of 20 November 2013 on a general program of Union action in the field of environment up to 2020 [2013] OJ L 354/171.
50 Action program item 17.
51 Action program item 10.
52 Recital No. 33 of Decision No. 1386/2013/UE.
53 Recital No. 33 of Decision No.1386/2013/UE.
54 F. Cafaggi, 'Rethinking Private Regulation in European Regulatory Space' in F. Cafaggi (ed), *Reframing Self-Regulation in European Private Law* (London, Kluwer 2006) 3.
55 For an overview of the topic, see inter alia F. Cafaggi, A. Renda, R. Schmidt, 'Transnational Private Regulation' in *International Regulatory Co-operation: Case Studies, vol. 3 Transnational Private Regulation and Water Management* (Paris, OECD 2013).
56 S. Casabona, *L'accordo in materia ambientale* (Monograph, Padova: CEDAM 2008).
57 Cfr. P. Trimarchi, *Rischio e responsabilità oggettiva* (Milano, Giuffré, 1961); G. Calabresi, *The Costs of Accidents: A Legal and Economic Analysis* (Yale, Yale University Press, 1970).
58 On the deterrent function of civil liability cfr. G. Schwartz, 'Reality in the Economic Analysis of Tort Law: Does Tort Law Really Deter?' (1994) 42 (377) UCLA Law Review; W.M. Landes, R.A. Posner, 'Tort Law as a Regulatory Regime for Catastrophic Personal Injuries' (1984) 13 Journal of Legal Studies, 417. For an analysis of the advantages of regulation compared with the liability model, see D. Wittman, 'Prior Regulation versus Post Liability: The Choice between Input and Output Monitoring' (1977) 6 Journal of Legal Studies, 193.
59 B. Pozzo, *Danno ambientale ed imputazione della responsabilità – Esperienze giuridiche a confronto* (Milano, Giuffrè 1996).
60 B. Pozzo, Modelli notevoli e circolazione dei modelli giuridici in campo ambientale: tra imitazione e innovazione in U. Mattei, A. Candian, B. Pozzo, A. Monti, C. Marchetti (eds), *Studi in onore di Antonio Gambaro: un giurista di successo* (Milano, Giuffrè 2017) 351–377.
61 J.P.S. Maloney, 'A Legislative History of Liability under CERCLA' (1992) 16 Seton Hall Legislative Journal, 517.
62 B. Pozzo, *Danno ambientale ed imputazione della responsabilità – Esperienze giuridiche a confronto* (Milano, Giuffrè, 1996) 151 s.
63 Law of July 8, 1986, n. 349 – Istituzione del Ministero dell'ambiente e norme in materia di danno ambientale.
64 Lei n. 11/87, de 07 de Abril, *Lei de bases do ambiente*.
65 Gesetz über die Umwelthaftung vom 10 Dezember 1990.
66 Commission, Green Paper on Compensation for Environmental Damage, presented as Communication from the Commission to the Council and the Parliament and the Economic and Social Committee of the EU, submitted by the Commission of the European Communities, COM (93) 47, Brussels, May 14, 1993, OJ No. C/149, May 29, 1993.
67 Commission, White Paper on Environmental Liability, COM (2000) 66 final. See B. Pozzo, 'The New Civil Liability for Environmental Damage. Italian Issues in the Light of European Union Initiatives' (2002) 1 (2) The Global Jurist, article 2.
68 Council Directive 2004/35/EC of April 21, 2004 on environmental liability with regard to the prevention and remedying of environmental damage, in OJEC April 30 [2004] OJ L 143/56. See B. Pozzo (ed), *La responsabilità ambientale. La nuova direttiva sulla responsabilità ambientale in materia di prevenzione e riparazione del danno ambientale* (Milan, Giuffrè 2005).

69 According to Article 2.2 of Directive 2004/35 the notion of environmental damage includes: 1. damage to protected species and natural habitats, as regulated by Directives 92/43/EEC and 79/409/EEC; 2. damage to water, i.e. any damage that significantly adversely affects the status of water, as defined by Directive 2000/60/EC; 3. Land damage, i.e. any land contamination that creates a significant risk of adverse effects on human health as a result of the direct or indirect introduction of substances, preparations, organisms or microorganisms into the soil, soil or subsoil.

70 Article 3.3. of the directive states that: Without prejudice to relevant national legislation, this Directive shall not give private parties a right to compensation following environmental damage to an imminent threat of such damage.

71 *General Principles of the Civil Law*, promulgated on April 12, 1986, and came into effect on January 1, 1987.

72 Art. 124 in the English version provides: Any person who pollutes the environment and causes damage to others in violation of state provisions for environmental protection and the prevention of pollution shall bear civil liability in accordance with the law.

73 S. Xianzhong, 'Rezeption der westlichen Zivilrechtswissenschaft im modernen China: Analysen aus der rechtsvergleichenden Perspektive' (2010) 18 (5) European Review of Private Law 901–913.

74 B. Pozzo, 'Tutela ambientale e modelli giuridici: il caso cinese' (2010) Rivista Giuridica dell'Ambiente, 877, 844.

75 B. Pozzo, L. Wang, 'Liability for Environmental Pollution within the Framework of the New Chinese Tort Law' (2011) 19 (1) European Review of Private Law, 87–99.

76 B. Pozzo, L. Wang, 'Liability for Environmental Pollution within the Framework of the New Chinese Tort Law' (2011) 19 (1) European Review of Private Law, 96. See further in this Chapter under 7: The greening of civil codes.

77 Gitanjali Nain Gill, *Judicial craftsmanship: evolving environmental tortious dimensions in India* (Annuario di Diritto Comparato, 2021) 59.

78 Cfr. A. Ortolani, 'Environmental Damage Remediation in Japan: A Comparative Assessment' in Yumiko Nakanishi (ed), *Contemporary Issues in Environmental Law – The EU and Japan* (Springer Japan 2016) 185.

79 L. De la Puente Brunke, 'Responsabilidad por el daño ambiental puro y el código civil peruano' (2011) 60 Themis Revista De Derecho, 295–307; J.J. González Márquez, I. Montelongo Buenavista, A.C. Cedeño León, 'Los progresos recientes en el desarrollo del derecho ambiental mexicano: del comando y control a la reparación del daño ambiental' (2016) Alegatos-Revista Jurídica de la Universidad Autónoma Metropolitana, 94; M.A.L. Alfonsín, A. N. Martínez, 'Una mirada constitucional a la responsabilidad por daño ambiental en el Nuevo Código Civil argentino' (2015) 13 (16) Lex-revista de la facultad de derecho y ciencias políticas, 55.

80 *Ley federal de responsabilidad ambiental,* Diario Oficial de la Federación el 7 de junio de 2013. On the influence of the European directive on Mexican law, see J.I. Aguilar Torres, *La responsabilidad civil objetiva por daños al medio ambiente y su regulación en México*, Secretaría General Secretaría de Servicios Parlamentarios Dirección General del Centro de Documentación, Información y Análisis.

81 L. De la Puente Brunke, 'Responsabilidad por el daño ambiental puro y el código civil peruano' (2011) 60 Themis Revista De Derecho, 295–307.

82 M. Cenini, *La proprietà della terra inquinata. Responsabilità, circolazione e garanzie* (Milano, Giuffrè 2017).

83 B. Pozzo, B. Vanheusden, L. Bergkamp, E. Brans, 'The Remediation of Contaminated Sites and the Problem of Assessing the Liability of the Innocent Landowner: A Comparative Law Perspective' (2015) 23 (6) European Review of Private Law, 1071–1120.

84 V. Jacometti, *Lo scambio di quote di emissione: analisi di un nuovo strumento di tutela ambientale in prospettiva comparatistica* (Milan, Giuffrè 2010).

85 The central idea behind the use of property rights or tradeable pollution permits as a device to solve problems created by externalities is Coase's Theorem, which roughly says that if tradeable property rights are established and enforced, then negotiation between the parties affected by the externality will lead to a socially optimal outcome no matter who initially owns the property rights. See D. Easley, J. Kleinberg, *Networks, Crowds, and Markets: Reasoning about a Highly Connected World* (Cambridge University Press, 2010) 779.

86 See <http://unfccc.int /resource/index.html> accessed 25 June 2023.

87 D.M. Driesen, 'Sustainable Development and Market Liberalism's Shotgun Wedding: Emissions Trading under the Kyoto Protocol' (2008) 83 (1) Indiana Law Journal, 21; J.-T. Boom, 'International

Emissions Trading under the Kyoto Protocol: Credit Trading' (2001) 29 (8) Energy Policy, Elsevier, 605–613.

88 Commission 'Green Paper on greenhouse gas emissions trading within the European Union' COM (2000) 87 final.

89 Council Directive 2003/87/EC of 13 October 2003 establishing a scheme for greenhouse gas emission allowance trading within the Community and amending Council Directive 96/61/EC [2003] OJ L 275/32.

90 On this point, see S. Casabona, 2012/27/EU, *L'accordo in materia ambientale* (Monograph, Padova: CEDAM 2008); M. Hautereau-Boutonnet (dir), *Le contrat et l'environnement. Etude de droit interne, international et européen* (Aix-en-Provence, Presses Universitaires d'Aix-Marseille – P.U.A.M., 2014).

91 M. Zatz, S. Harbour, 'The United States Environmental Protection Agency's 33/50 Program: The Anatomy of a Successful Voluntary Pollution Reduction Program' (1999) 7 (1) Journal of Cleaner Production, 17–26.

92 K. Segerson, T.J. Miceli, 'Voluntary Environmental Agreements: Good or Bad News for Environmental Protection?' (1998) 36 (2) Journal of Environmental Economics and Management, 109–130.

93 Brussels, 27.11.1996 COM(96) 561 final.

94 COM (96) 561 def., p. 3.

95 Commission Recommendation 1999/125/CE of 5 February 1999 on the reduction of CO_2 emissions from passenger cars (notified under document number C(1999) 107) OJ L 40/49 , Commission Recommendation 2000/303/CE of 13 April 2000 on the reduction of CO_2 emissions from passenger cars (KAMA) (notified under document number C(2000) 801) OJ L 100/55 and Commission Recommendation of 2000/304/CE 13 April 2000 on the reduction of CO_2 emissions from passenger cars (JAMA) (notified under document number C(2000) 803) OJ L 100/ 57.

96 Decision n. 1753/2000/CE del 22 June 2000, OJ L 202 of 10 August 2000, 1.

97 Voluntary Approaches for Environmental Policy – an Assessment. OCSE 1999.

98 Commission, 'Environmental Agreements at Community Level Within the Framework of the Action Plan on the Simplification and Improvement of the Regulatory Environment' [Communication from the Commission to the European Parliament, the Council, the Economic and Social Committee and the Committee of the Regions] COM(2002) 412 def.

99 B. Pozzo, 'Les conventions environnementales en Italie' in M. Boutonnet (ed), *Le contrat et l'environnement*, Étude de droit comparé (Brussels, Bruylant 2015) 199–219.

100 M. Hautereau-Boutonnet (ed), *Le contrat et l'environnement, Etude de droit comparé* (Brussels, Bruylant 2015).

101 H.T.A. Bressers, Y. Xue, 'The Feasibility of Environmental Negotiated Agreements in China' (2007) 6 (3) International Journal of Environment and Sustainable Development, 221; U. Eichhorst, D. Bongardt, 'Towards Cooperative Policy Approaches in China-Drivers for Voluntary Agreements on Industrial Energy Efficiency in Nanjing' (2009) 37 (5) Energy Policy, Elsevier, 1855–1865. M. Bu, Q. Zhenzi, B. Liu, 'Voluntary Environmental Regulation and Firm Innovation in China' (2020) 89 (C) Economic Modelling, Elsevier, 10–18.

102 S. Landini, 'Clausole di sostenibilità nei contratti tra privati. Problemi e riflessioni' in M. Pennasilico (a cura di) *Contratto e ambiente. L'analisi 'ecologica' del diritto contrattuale* (Napoli, 2016), 343.

103 It should be mentioned, however, that there are studies that point out that forms of ethical commitments by businesses in response to the reforms of the late 19th century are observed as early as the beginning of the 20th century. On the subject, see A. Kolk, R. van Tulder, 'Setting New Global Rules? TNCs and Codes of Conduct' (2005) 14 (3) Transnational Corporations, 1; G.C. Benson, 'Codes of Ethics' (1989) 8 (5) Journal of Business Ethics, 305.

104 On codes of conduct, see R. Jenkins, *Corporate Codes of Conduct. Self-Regulation in a Global Economy*. Technology, Business and Society Programme Paper Number 2 (Geneva, UNRISD, 2001); S. Rossi, 'Luci e ombre dei codici etici d'impresa' (2008) Rivista di Diritto Societario, 23; G. Teubner, 'Self-Constitutionalizing TNCs? On the Linkage of "Private" and "Public" Corporate Codes of Conduct' (2011) 18 (2) Indiana Journal of Global Legal Studies, 617; A. Beckers, *Enforcing Corporate Social Responsibility Codes. On Global Self-Regulation and National Private Law* (Oxford, 2015).

105 See, for example: the UN Global Compact, the OECD Guidelines for Multinational Enterprises, the Code of Conduct of the Business Social Compliance Initiative (BSCI) and the Workplace Code of Conduct of US Fair Labor Organization (FLA).

106 Cfr. L.-W. Lin, 'Legal Transplants through Private Contracting: Codes of Vendor Conduct in Global Supply Chains as an Example' (2009) 57 (3) American Journal of Comparative Law, 711.

107 Cfr. T. Ferrando, 'Codes of Conduct as Private Legal Transplant: The Case of European Extractive MNEs' (2013) 19 (6) European Law Journal, 799.

108 C. Poncibò, 'The Contractualisation of Environmental Sustainability' (2016) 12 (4) European Review of Contract Law, 335–355.

109 The *OECD Recommendation of the Council on green public procurement* may be found at: <https://www.oecd.org/gov/public-procurement/OECD-Recommendation-on-Public-Procurement.pdf> accessed 25 June 2023.

110 *Going green: best practices for sustainable procurement,* may be found at <https://www.oecd.org/gov/public-procurement/Going_Green_Best_Practices_for_Sustainable_Procurement.pdf> accessed 25 June 2023.

111 Commission, 'Public procurement for a better environment', Brussels [Communication from the Commission to the European Parliament, the Council, the European Economic and Social Committee and the Committee of the Regions] COM (2008) 400 final.

112 S. Lavallée, P. Halley, M. Hautereau-Boutonnet, 'Avant-propos' (2020/3) 25 Revue juridique de l'environnement, 433–436.

113 Council Directive 2012/27/EU of 25 October 2012 on energy efficiency, amending Directives 2009/125/EC and 2010/30/EU and repealing Directives 2004/8/EC and 2006/32/EC [2017] OJ L 315/1, Art. 7 c).

114 T. Wilhelmsson, 'Consumer Law and the Environment: From Consumer to Citizen' (1998) 21(1) Journal of Consumer Policy, 45.

115 H.-W. Micklitz, 'Squaring the Circle? Reconciling Consumer Law and the Circular Economy' in B. Keirsbilck, E. Terryn (eds), *Consumer Protection in a Circular Economy*, p. 323 (Intersentia, Cambridge – Antwerp – Chicago) 345.

116 Commission, 'A new Circular Economy Action Plan for a Cleaner and More Competitive Europe' [Communication from the Commission to the European Parliament, the Council, the European and Social Committee and the Committee of the Regions] COM/2020/98 final.

117 Commission, 'A new Circular Economy Action Plan for a Cleaner and More Competitive Europe' [Communication from the Commission to the European Parliament, the Council, the European and Social Committee and the Committee of the Regions] COM/2020/98 final, 5.

118 European Parliament resolution of 25 November 2020 *Towards a more sustainable single market for business and consumers* (2020/2021(INI) (2021/C 425/03).

119 Council Directive 2005/29/EC of 11 May 2005 concerning unfair business-to-consumer commercial practices in the internal market and amending Council Directive 84/450/EEC, Directives 97/7/EC, 98/27/EC and 2002/65/EC of the European Parliament and of the Council and Regulation (EC) No 2006/2004 of the European Parliament and of the Council ('Unfair Commercial Practices Directive') [2005] OJ L 149/ 22.

120 Council Directive 2011/83/EU of 25 October 2011 on consumer rights, amending Council Directive 93/13/EEC and Directive 1999/44/EC of the European Parliament and of the Council and repealing Council Directive 85/577/EEC and Directive 97/7/EC of the European Parliament and of the Council [2011] OJ L 304/64.

121 Council Directive (EU) 2019/771 of 20 May 2019 on certain aspects concerning contracts for the sale of goods, amending Regulation (EU) 2017/2394 and Directive 2009/22/EC, and repealing Directive 1999/44/EC [2019]OJ L 136/28.

122 E. Van Gool, A. Michel, 'The New Consumer Sales Directive 2019/771 and Sustainable Consumption: A Critical Analysis' (2021) 10 (4) Journal of European Consumer and Market Law, 136–147; A. Mathios, H.-W. Micklitz, L.A. Reisch, J. Thøgersen, C. Twigg-Flesner, 'Journal of Consumer Policy's 40th Anniversary Conference: A Forward Looking Consumer Policy Research Agenda' (2020) 43 Journal of Consumer Policy, 1–9; V. Mak, E. Terryn, 'Circular Economy and Consumer Protection: The Consumer as a Citizen and the Limits of Empowerment Through Consumer Law' (2020) 43 Journal of Consumer Policy, 227–248.

123 To provide the Members of the European Parliament with evidence and recommendations for the above-mentioned report entitled, 'Towards a more sustainable single market for businesses and consumers', the IMCO Committee has requested Policy Department A to prepare an In-Depth Analysis on 'Sustainable Consumption and Consumer Protection Legislation: How can sustainable consumption and longer lifetime of products be promoted through consumer protection legislation?'

124 Augenhofer, Susanne, *European Commission's Public Consultation on Sustainable Consumption of Goods – Promoting Repair and Reuse* (Response of the European Law Institute, 2022) <https://www.

europeanlawinstitute.eu/fileadmin/user_upload/p_eli/Publications/ELI_Response_Sustainable_Consumption_of_Goods.pdf> accessed 14 July 2023.

125 Brussels, 30.3.2022 COM (2022) 143 final.

126 COM (2022) 143 final.

127 S. Lanni, 'New Civil Codes and the Environment: The Contributions of China and Argentina' in M. Graziadei, L. Zhang Editors (eds) *The Making of the Civil Codes – A Twenty-First Century Perspective* (Springer, Ius Gentium: Comparative Perspectives on Law and Justice 2022) 385; S. Lanni, *Greening the Civil Codes: Comparative Private Law and Environmental Protection* (Routledge, 2023).

128 Loi n° 2016-1088 du 8 août 2016 relative au travail, à la modernisation du dialogue social et à la sécurisation des parcours professionnels.

129 Art. 1246: 'Toute personne responsable d'un préjudice écologique est tenue de le réparer'; Art. 1247: 'Est réparable, dans les conditions prévues au présent titre, le préjudice écologique consistant en une atteinte non négligeable aux éléments ou aux fonctions des écosystèmes ou aux bénéfices collectifs tirés par l'homme de l'environnement'; Art. 1248: 'L'action en réparation du préjudice écologique est ouverte à toute personne ayant qualité et intérêt à agir, telle que l'Etat, l'Office français de la biodiversité, les collectivités territoriales et leurs groupements dont le territoire est concerné, ainsi que les établissements publics et les associations agréées ou créées depuis au moins cinq ans à la date d'introduction de l'instance qui ont pour objet la protection de la nature et la défense de l'environnement'; Art. 1249: 'La réparation du préjudice écologique s'effectue par priorité en nature. En cas d'impossibilité de droit ou de fait ou d'insuffisance des mesures de réparation, le juge condamne le responsable à verser des dommages et intérêts, affectés à la réparation de l'environnement, au demandeur ou, si celui-ci ne peut prendre les mesures utiles à cette fin, à l'Etat. L'évaluation du préjudice tient compte, le cas échéant, des mesures de réparation déjà intervenues, en particulier dans le cadre de la mise en œuvre du titre VI du livre Ier du code de l'environnement'; Art. 1250: 'En cas d'astreinte, celle-ci est liquidée par le juge au profit du demandeur, qui l'affecte à la réparation de l'environnement ou, si le demandeur ne peut prendre les mesures utiles à cette fin, au profit de l'Etat, qui l'affecte à cette même fin. Le juge se réserve le pouvoir de la liquider'; Art. 1251: 'Les dépenses exposées pour prévenir la réalisation imminente d'un dommage, pour éviter son aggravation ou pour en réduire les conséquences constituent un préjudice réparable'; Art. 1252: 'Indépendamment de la réparation du préjudice écologique, le juge, saisi d'une demande en ce sens par une personne mentionnée à l'article 1248, peut prescrire les mesures raisonnables propres à prévenir ou faire cesser le dommage'.

130 M. Hautereau-Boutonnet, *Le Code civil, un code pour l'environnement* (Paris, Dalloz 2021) 139.

131 I.A. Duşcă, 'The Civil Code and the Environment' (2013) 17 (1) AGORA International Journal of Juridical Sciences, 24–30 <www.juridicaljournal.univagora.ro> ISSN 1843-570X, E-ISSN 2067-7677.

132 T. Zhai, Y.-C. Chang, 'The Contribution of China's Civil Law to Sustainable Development: Progress and Prospects' (2019) 11 (1) Sustainability 294 <https://doi.org/10.3390/su11101029> accessed 25 June 2023.

133 Art. 1094: 'Las normas que regulan las relaciones de consumo deben ser aplicadas e interpretadas conforme con el principio de protección del consumidor y el de acceso al consumo sustentable'.

134 Art. 1.228. 'O proprietário tem a faculdade de usar, gozar e dispor da coisa, e o direito de reavê-la do poder de quem quer que injustamente a possua ou detenha. § 1° O direito de propriedade deve ser exercido em consonância com as suas finalidades econômicas e sociais e de modo que sejam preservados, de conformidade com o estabelecido em lei especial, a flora, a fauna, as belezas naturais, o equilíbrio ecológico e o patrimônio histórico e artístico, bem como evitada a poluição do ar e das águas'.

135 The draft of the new Civil Code of Colombia is available online: <http://derecho.bogota.unal.edu.co/fileadmin/Codigo_Civil/Proyecto_Codigo_Civil_de_Colombia_Primera_Version_Digital.pdf> accessed 25 June 2023.

136 Artículo 276. La ley establecerá las exigencias en las viviendas, locales, oficinas, centros comerciales, clubes, construcciones y espacios que sean necesarios para el desarrollo urbano y rural pertinente y que garanticen los elementos del ambiente sano, la vecindad y la convivencia ecológica.

137 Artículo 268: 'El propietario debe explotar su derecho conforme a su destinación económica, ecológica, cultural y social e indemnizará los perjuicios concretos o difusos causados en razón de una explotación sin interés para él o para el bienestar social. También debe respetar el derecho de los demás a un ambiente sano, el manejo y aprovechamiento legal de los recursos naturales, la integridad y el uso común del espacio público y el uso debido de los suelos, urbanos, conforme a la ley de ordenamiento territorial, a la regulación urbanística y de servicios públicos domiciliarios. La ley regulará el dominio de los objetos de interés cultural histórico, religioso, y de las minorías étnicas y lingüísticas de la población colombiana'.

138 Art. 38: 'Cada cual tiene libertad e iniciativa para adquirir y gobernar sus derechos en la forma que le plazca dentro de los límites de la ley, el orden público y las buenas costumbres. En consecuencia, en las actividades económicas y en las relaciones y desarrollo de los mercados, todas las personas gozarán, entre otras, de las libertades, derechos y protecciones consiguientes en materia de emprendimiento, empresa, producción, comercialización, competencia, consumo, contratación, importación y exportación de bienes y servicios, dentro de los límites del orden público, las buenas costumbres, el medio ambiente y los compromisos internacionales sobre la materia. Proyecto de Código Civil de Colombia. De igual manera, las personas también tendrán la responsabilidad social pertinente y aquellas que surjan por las actividades que desarrollen y por los daños que ocasionen'.

139 M. Hautereau-Boutonnet, *Le Code civil, un code pour l'environnement* (Paris, 2021) 139.

140 Gitanjali Nain Gill, *Judicial Craftsmanship: Evolving Environmental Tortious Dimensions* (India, Annuario di Diritto Comparato 2021) 59.

Bibliography

Acerbis G, 'Le immissioni, con particolare riferimento alla tutela della salute e dell'ambiente', (1989) Parte II, *Quadrimestre rivista di diritto privato*, Giuffrè Editore, 151

Ackerman B, Stewart R, 'Reforming Environmental Law, Stanford Law Review' (1985) 37 Stanford Law Review, 1333

Alcorn W, 'Liability Theories for Toxic Torts' (1988) 3 (2) Natural Resources & Environment, 3

Alfonsín M, Martínez A, 'Una mirada constitucional a la responsabilidad por daño ambiental en el Nuevo Código Civil argentino' (2015) 13 (16) Lex-revista de la facultad de derecho y ciencias políticas, 55

Alpa G, Bessone M, Gambaro A, 'Aspetti privatistici della tutela dell'ambiente: l'esperienza americana e francese' in A. Gambaro (ed), *La tutela degli interessi diffusi nel diritto comparato* (Milano, Giuffrè 1976) 391

Augenhofer S, *European Commission's Public Consultation on Sustainable Consumption of Goods – Promoting Repair and Reuse* (Response of the European Law Institute 2022) <https://www.europeanlawinstitute.eu/fileadmin/user_upload/p_eli/Publications/ELI_Response_Sustainable_Consumption_of_Goods.pdf> accessed 14 July 2023

Baur F, 'Zur Entstehung des Umweltschutzes aus dem Sachenrecht des BGB' (1987) 42 (7) JuristenZeitung, 317–322

Baur F, *Lehrbuch des Sachenrechts* (IV, § 25, 15ª ed., München, Beck C.H., 1989) 228

Beckers A, *Enforcing Corporate Social Responsibility Codes. On Global Self-Regulation and National Private Law* (Oxford, Hart Publishing, 2015)

Benson G, 'Codes of Ethics' (1989) 8 (5) Journal of Business Ethics, 305

Blaise J, 'Responsabilité et obligations coutumières dans les rapports de voisinage' (1965) t. LXIII Revue Trimestrielle de Droit Civil, 275

Boom J-T, 'International Emissions Trading under the Kyoto Protocol: Credit Trading' (2001) 29(8) Energy Policy, Elsevier, 605–613

Brennan T, 'Environmental Torts' (1993) 46 (1) Vanderbilt Law Review <https://scholarship.law.vanderbilt.edu/vlr/vol46/iss1/1> accessed 25 June 2023

Bressers H, Xue Y, 'The Feasibility of Environmental Negotiated Agreements in China' (2007) 6 (3) International Journal of Environment and Sustainable Development, 221

Bresso M, *Per un'economia ecologica* (Roma, Carocci 1993) 210

Bu M, Zhenzi Q, Liu B, 'Voluntary Environmental Regulation and Firm Innovation in China' (2020) 89 (C) Economic Modelling, Elsevier, 10–18

Cafaggi F, 'Rethinking Private Regulation in European Regulatory Space' in F. Cafaggi (ed), *Reframing Self-Regulation in European Private Law* (London, Kluwer 2006) 3

Cafaggi F, Renda A, Schmidt R, 'Transnational Private Regulation' in *International Regulatory Co-Operation: Case Studies*, vol.3 *Transnational Private Regulation and Water Management* (Paris, OECD 2013)

Calabresi G, *The Costs of Accidents: A Legal and Economic Analysis* (Yale, Yale University Press 1970)

Casabona S, *L'accordo in materia ambientale* (Monograph, Padova: CEDAM 2008)

Cenini M, *La proprietà della terra inquinata. Responsabilità, circolazione e garanzie* (Milano, Giuffrè 2017)

De la Puente Brunke L, 'Responsabilidad por el daño ambiental puro y el código civil peruano' (2011) 60 Themis Revista De Derecho, 295–307.

Dewees D, Trebilcock M, 'The Efficacy of the Tort System and Its Alternatives: A Review of Empirical Evidence' (1992) 30 (1) Osgood Hall Law Journal, 57–138.

Di Giovanni F, *Strumenti privatistici e tutela dell'ambiente* (Padova, CEDAM 1982)

Diederichsen U, *Die Haftung für Umweltschäden* (Betriebs-Berater 1973) 485
Diederichsen U, 'Schadenersatz und Entschädigung im Umweltschutz' in *Schriftenreihe der Deutschen Richterakademie, Bd. 5: Natur- und Umweltschutz* (Heidelberg, Juristischer Verlag, 1989) 19
Driesen D, 'Sustainable Development and Market Liberalism's Shotgun Wedding: Emissions Trading under the Kyoto Protocol' (2008) 83 (1) Indiana Law Journal, 21
Dudek DJ, Palmisano J, 'Emissions Trading: Why Is This Thoroughbred Hobbled' (1987) 13 Columbia Journal of Environmental Law, 217
Duşcă I, 'The Civil Code and the Environment' (2013) 17 (1) AGORA International Journal of Juridical Sciences, 24–30 <www.juridicaljournal.univagora.ro> ISSN 1843-570X, E-ISSN 2067-7677
Easley D, Kleinberg J, *Networks, Crowds, and Markets: Reasoning about a Highly Connected World* (Cambridge University Press, 2010) 779
Eichhorst U, Bongardt D, 'Towards Cooperative Policy Approaches in China-Drivers for Voluntary Agreements on Industrial Energy Efficiency in Nanjing' (2009) 37 (5) Energy Policy, Elsevier, 1855–1865
Epstein G, et al., 'Governing the Invisible Commons: Ozone Regulation and the Montreal Protocol' (2014) 8 (2) International Journal of the Commons, 337–360 <http://doi.org/10.18352/ijc.407> accessed 25 June 2023
Falque M, *Introduction générale, Droits de propriété et environnement* (Paris, Dalloz, 1997) 17
Ferrando T, 'Codes of Conduct as Private Legal Transplant: The Case of European Extractive MNEs' (2013) 19 (6) European Law Journal, 799
Fleming J, *The American Tort Process* (Oxford, A Clarendon Press Publication 1990) 252
Fleming J, 'Mass Torts' (1994) 42 (3) The American Journal of Comparative Law, 507
Forkel H, *Immissionschutz und Persönlichkeitsrecht* (Köln, Heymanns 1968) 47
Forte F, *Per una lettura alternativa dell'art. 844 del codice civile* (Dir. e Giur. 1976) 650
Furrow B, 'Governing Science: Public Risks and Private Remedies' (1983) 131 University of Pennyslvania Law Review, 1403
Gill G, *Judicial Craftsmanship: Evolving Environmental Tortious Dimensions* (Annuario di Diritto Comparato 2021) 59
Girod P, *La réparation du dommage écologique* (Paris, Librairie générale de droit et de jurisprudence 1974) 90
González Márquez J, Buenavista I, León A, 'Los progresos recientes en el desarrollo del derecho ambiental mexicano: del comando y control a la reparación del daño ambiental' (2016) Alegatos-Revista Jurídica de la Universidad Autónoma Metropolitana, vol. 94, 571–602.
Hautereau-Boutonnet M (dir), *Le contrat et l'environnement. Etude de droit interne, international et européen* (Aix-en-Provence, Presses Universitaires d'Aix-Marseille – P.U.A.M. 2014)
Hautereau-Boutonnet M (ed), *Le contrat et l'environnement, Etude de droit comparé* (Brussels, Bruylant 2015)
Hautereau-Boutonnet M, *Le Code civil, un code pour l'environnement* (Paris, Dalloz 2021) 139
Hockenstein J, Stavins R, Whitehead B, 'Crafting the Next Generation of Market-Based Environmental Tools' (1997) 39 (4) Environment: Science and Policy for Sustainable Development, 12–33
Hodas D, 'Private Actions for Public Nuisance: Common Law Citizen Suits for Relief from Environmental Harm' (1989) 16 Ecology Law Quarterly, 883
Hopgood S, *American Foreign Environmental Policy and the Power of the State* (Oxford, Oxford University Press 1998) 140
Iannelli A, 'Sulla tutela delle immissioni industriali e sulla non operatività dell'art. 844' (1980) Rass. dir. civ., 371
Jacometti V, *Lo scambio di quote di emissione: analisi di un nuovo strumento di tutela ambientale in prospettiva comparatistica* (Milan, Giuffrè 2010)
Jenkins R, *Corporate Codes of Conduct. Self-Regulation in a Global Economy*. Technology, Business and Society Programme Paper Number 2 (UNRISD, Geneva, 2001)
Kelemen R, Vogel D, 'Trading Places: The US and EU in International Environmental Politics' (2010) 43 (4) Journal Comparative Political Studies, 19
Kolk A, van Tulder R, 'Setting New Global Rules? TNCs and Codes of Conduct' (2005) 14 (3) Transnational Corporations, 1
Landes W, Posner R, 'Tort Law as a Regulatory Regime for Catastrophic Personal Injuries' (1984) 13 Journal of Legal Studies, 417
Landini S, 'Clausole di sostenibilità nei contratti tra privati. Problemi e riflessioni' in M. Pennasilico (a cura di) *Contratto e ambiente. L'analisi 'ecologica' del diritto contrattuale* (Napoli, 2016), 343

Lanni S, 'New Civil Codes and the Environment: The Contributions of China and Argentina' in M. Graziadei, L. Zhang Editors (eds) *The Making of the Civil Codes – A Twenty-First Century Perspective* (Springer, Ius Gentium: Comparative Perspectives on Law and Justice 2022) 385
Lanni S, *Greening the Civil Codes: Comparative Private Law and Environmental Protection* (Routledge, 2023 forthcoming)
Lavallée S, Paule H, Mathilde H-B, '*Avant-propos*' (2020/3) 25 Revue juridique de l'environnement, 433–436
Lin L, 'Legal Transplants through Private Contracting: Codes of Vendor Conduct in Global Supply Chains as an Example' (2009) 57 (3) American Journal of Comparative Law, 711
Longo M, 'Assurdi dubbi della costituzionalità dell'art.844 c.c.' (1974) Riv. dir. agr, 608
Mak V, Terryn E, 'Circular Economy and Consumer Protection: The Consumer as a Citizen and the Limits of Empowerment Through Consumer Law' (2020) 43 Journal of Consumer Policy, 227–248
Maloney F, 'Judicial Protection of the Environment: A New Role for Common Law Remedies' (1972) 25 (1) Vanderbilt L.R., 145 <https://scholarship.law.vanderbilt.edu/vlr/vol25/iss1/14> accessed 25 June 2023
Maloney J, 'A Legislative History of Liability under CERCLA' (1992) 16 (517) Seton Hall Legislative Journal, 517
Mathios A, H.- Micklitz W, L. Reisch A, Thøgersen J, Twigg-Flesner C, 'Journal of Consumer Policy's 40th Anniversary Conference: A Forward-Looking Consumer Policy Research Agenda' (2020) 43 Journal of Consumer Policy, 1–9
Medicus D, 'Zivilrecht und Umweltschutz' (1986) 17 *JuristenZeitung*, 778
Mekki M (ed), *Les notions fondamentales de droit privé à l'épreuve des questions environnementales* (Brussels, Bruylant 2016)
Mekki M, Naim-gesbert É (eds), *Droit public et droit privé de l'environnement: Unité dans la diversité?* (Paris, LGDJ 2016)
Micklitz H-W, 'Squaring the Circle? Reconciling Consumer Law and the Circular Economy' in B. Keirsbilck, E. Terryn (eds) *Consumer Protection in a Circular Economy*, p. 323 (Intersentia, Cambridge – Antwerp – Chicago) 345
Morbidelli G, 'Strumenti privatistici contro l'inquinamento delle acque interne' (1971) III Foro amm., 369
Nappi F, *Le regole proprietarie e la teoria delle immissioni* (Napoli, ESI 1986) 7
Neustadter G, 'The Role of the Judiciary in the Confrontation with the Problems of Environmental Quality' (1970) 17 UCLA, 1070
Ortolani A, 'Environmental Damage Remediation in Japan: A Comparative Assessment' in Yumiko Nakanishi (ed), *Contemporary Issues in Environmental Law – The EU and Japan* (Springer Japan 2016) 185
Paradiso P, 'Inquinamento delle acque interne e strumenti privatistici di tutela' (1977) Riv. trim. dir. proc. civ., 1391
Patti S, 'Environmental Protection in Italy: The Emerging Concept of a Right to a Healthful Environment' (1984) 24 Natural Resources Journal, 535
Poncibò C, 'The Contractualisation of Environmental Sustainability, European Review of Contract Law' (2016) 12 (4) European Review of Contract Law, 335–355
Porcinari L, 'In tema di legittimità costituzionale del divieto di immissioni nocive previste dall'art.844 c.c.' (1973) Giur. agr. it, 310
Pozzo B, *Danno ambientale ed imputazione della responsabilità – Esperienze giuridiche a confronto* (Milano, Giuffrè 1996)
Pozzo B (ed), *La responsabilità ambientale. La nuova direttiva sulla responsabilità ambientale in materia di prevenzione e riparazione del danno ambientale* (Milan, Giuffrè, 2005)
Pozzo B, 'Tutela ambientale e modelli giuridici: il caso cinese' (2010) Rivista Giuridica dell'Ambiente, 877, 844
Pozzo B, 'Les conventions environnementales en Italie' in M. Boutonnet (ed.) *Le contrat et l'environnement*, Étude de droit comparé (Brussels, Bruylant 2015) 199–219
Pozzo B, 'Modelli notevoli e circolazione dei modelli giuridici in campo ambientale: tra imitazione e innovazione' in U. Mattei, A. Candian, B. Pozzo, A. Monti, C. Marchetti (eds), *Studi in onore di Antonio Gambaro: un giurista di successo* (Milano, Giuffrè 2017) 351–377
Pozzo B, Vanheusden B, Bergkamp L, Brans E, 'The Remediation of Contaminated Sites and the Problem of Assessing the Liability of the Innocent Landowner: A Comparative Law Perspective' (2015) 23 (6) European Review of Private Law, 1071–1120
Pozzo B, Wang L, 'Liability for Environmental Pollution within the Framework of the New Chinese Tort Law' (2011) 19 (1) European Review of Private Law, 96

Procida Mirabelli di Lauro A, *Immissioni e "rapporto proprietario"* (Napoli, ESI1984) 398
Rabin R, 'Environmental Liability and the Tort System' (1987) 24 Houston L.R., 27
Ronellenfitsch M, Wolf R, 'Ausbau des Individualschutzes gegen Umwelt belastungen als Aufgabe des bürgerlichen und des öffentlichen Rechts?' (Neue Juristische Wochenschrift 1986, 1960–1965.
Rose C, 'Rethinking Environmental Controls: Management Strategies for Common Resources' (1991) 1 Duke Law Journal, 1–38
Rossi S, 'Luci e ombre dei codici etici d'impresa' (2008) Rivista di Diritto Societario, 23
Roth G, *Materiellrechtliche und prozessuale Aspekte eines privatrechtlichen Umweltschutzes* (Neue Juristische Wochenschrift 1972) 921
Sale K, *The Green Revolution. The American Environmental Movement 1962–1992* (New York, Hill and Wang 1993) 24
Salvi C, 'Immissioni, ecologia, norme costituzionali' (1973) I (2) Giur. it., 798
Salvi C, *Le immissioni industriali: rapporti di vicinato e tutela dell'ambiente* (Milano, Giuffrè 1979)
Sands P, 'The "Greening" of International Law: Emerging Principles and Rules' (1994) 1 (2) Indiana Journal of Global Studies, 293
Schwartz G, 'Reality in the Economic Analysis of Tort Law: Does Tort Law Really Deter?' (1994) 42 UCLA L. REV. 377, 381–387
Segerson K, Miceli T, 'Voluntary Environmental Agreements: Good or Bad News for Environmental Protection?' (1998) 36 (2) Journal of Environmental Economics and Management, 109–130
Strand P, 'The Inapplicability of Traditional Tort Analysis to Environmental Risk: The Example of Toxic Waste Pollution Victim Compensation' (1983) 35 Stanford Law Review, 575
Tarlock A, 'The Influence of International Environmental Law on U.S. Pollution Control Law' (1997) 21 Vermont L. Rev., 759
Teubner G, 'Self-Constitutionalizing TNCs? On the Linkage of "Private" and "Public" Corporate Codes of Conduct' (2011) 18 (2) Indiana Journal of Global Legal Studies, 617
Torres J, *La responsabilidad civil objetiva por daños al medio ambiente y su regulación en México*, Secretaría General Secretaría de Servicios Parlamentarios Dirección General del Centro de Documentación, Información y Análisis
Trimarchi P, *Rischio e responsabilità oggettiva* (Milano, Giuffrè, 1961)
Van Gool E, Michel A, 'The New Consumer Sales Directive 2019/771 and Sustainable Consumption: A Critical Analysis' (2021) 10 (4) Journal of European Consumer and Market Law, 136–147
Visintini G, 'Il divieto di immissioni ed il diritto alla salute nella giurisprudenza odierna e nei rapporti con le recenti leggi ecologiche' (1980) II Riv. dir. civ., 249
Westermann H, *Die Funktion des Nachbarrechts. Zugleich eine Untersuchung der Bedeutung eines Immissionschutzgesetzes für das Privatrecht, Festschrift für Karl Larenz zum 70. Geburstag* (München, Beck 1973) 1003
Wilhelmsson T, 'Consumer Law and the Environment: From Consumer to Citizen' (1998) 21 (1) Journal of Consumer Policy, 45
Wittman D, 'Prior Regulation versus Post Liability: The Choice between Input and Output Monitoring' (1977) 6 Journal of Legal Studies, 193
Xianzhong S, 'Rezeption der westlichen Zivilrechtswissenschaft im modernen China: Analysen aus der rechtsvergleichenden Perspektive' (2010) 18 (5) European Review of Private Law, 901–913
Zatz M, Harbour S, 'The United States Environmental Protection Agency's 33/50 Program: The Anatomy o[f] a Successful Voluntary Pollution Reduction Program' (1999) 7 (1) Journal of Cleaner Production, 17–2[illegible]
Zhai T, Chang Y-C, 'The Contribution of China's Civil Law to Sustainable Development: Progress and Pro[s]pects' (2019) 11 (1) Sustainability 294 <https://doi.org/10.3390/su11101029> accessed 25 June 2023

2

TOWARDS A SUSTAINABLE INTERPRETATION OF STANDARD CONTRACT TERMS?

Dario Hug[1]

2.1 Introduction

The relation between environmental sustainability[2] (hereafter: ES) and standard contract terms (hereafter: SCT) is far from obvious. They may be at odds with ES.[3] At the same time, however, they can also promote the integration of environmental objectives into the contract.[4] Contractual interpretation[5] and the control of content represent a potential anchor point for the integration of ES as an element 'external' to contractual parties' direct interests.[6] However, a strictly instrumental pproach to the interpretative process – as well as of private law as a whole[7] – entails risks for edom of contract and personal autonomy. It is thus essential to identify such heteronomous rvention.

ur descriptive, comparative,[8] and evaluative contribution builds on these preliminary find- hrough the lens of ES. First, we delve into the fairness and transparency requirements of the Contract Terms Directive[9] (hereafter: UCTD) (2.2). We then illustrate the control of the f SCT in a progressive way from the perspectives of Swiss and Belgian laws (2.3), before g (2.4).

2.2 Unfair Contract Terms Directive

a significant (regulatory) legal instrument of European contract and consumer law CT. Since its adoption in 1993, it has attracted a great deal of attention from both ECJ case law.[10]

l 6 of the UCTD makes it clear that its main purpose is to facilitate the comple- market and to safeguard the 'citizen in his role as consumer'[11] when acquiring Moreover, recitals 7 and 9 emphasize the sellers' and suppliers' 'task of selling espectively, the 'protection of the economic interests of the consumer'.[12]

absent to these considerations, which are limited to a strict *inter partes* ss as a stronger party and the consumer as a weaker one that needs to be of unfair terms to achieve the necessary trust for the cross-border internal tial. This should not come as a surprise. Environmental protection and argely seen as non-integrated topics in the 1990s (and before) on an EU level.[13] Although such legal parochialism is still largely present today,[14] recent legislative proposals from the European Commission nevertheless seek to mitigate it, especially by the development of ecodesign rules,[15] as well as in the context of consumer sales and unfair commercial practices.[16]

Considering the current climate crisis, the question is then whether and, if so, to what extent the UCTD provides an interface for integrating ES considerations into the interpretation of SCT in B2C relations.[17] Bearing in mind that the national courts, considering the circumstances of each case, must determine if a contractual term meets the following requirements,[18] we will focus on two central aspects of the UCTD: fairness (Section 2.2.1) and transparency (Section 2.2.2).

2.2.1 Fairness

Article 3 UCTD is the general legal basis for assessing the unfairness of a contractual term which is opposed to the consumer. The concept of good faith is central. Additionally, Article 4 UCTD sets out other criteria to be taken into consideration when examining the unfairness of a clause, such as the nature of the goods or services for which the contract was concluded.[19]

As an objective concept, the notion of good faith in Article 3(1) UCTD relates 'to the question of whether, in light of its content, the contract term in question is compatible with fair and equitable market practices that take the consumer's legitimate interests sufficiently into account' and is 'closely linked to the (im)balance in the rights and obligations of the parties'.[20] In other words, the traditional focus is on the relationship between two contractual partners (and the protection of the weaker), without integrating external aspects such as ES.

But could an innovative (or '*eco-literate*'[21]) interpretation lead to the following alternative basic understanding: a market can only be fair and equitable if it integrates and supports ES? And, does the current climate crisis not jeopardize the 'legitimate interests of the consumer' by compromising the *conditio sine qua non* for the blooming of her very economic interests, i.e., a healthy, safe, and sustainable environment?[22] While such a progressive understanding of fairness is conceivable – although not undisputed by ways of judicial interpretation, at least on the national level[23] – its concretization requires, however, most certainly a paradigmatic change in the current understanding of the principles of the fairness review mechanism for SCT.

That said, an eco-literate understanding might be closer to reach than one might think. By referring to the 'solidarity among users' and 'an overall evaluation of the different interests involved',[24] recital 16 of the UCTD does not peremptorily rule out the relevance of collective interest in the process of assessing the fairness of a given term, moreover in close relation with the very requirement of good faith. And, although the judicial significance of Article 11 TFEU[25] (see also Article 37 Charter of Fundamental Rights[26]) as an integration clause remains in practice rather limited, the ECJ has in the past referred to this transversal clause (*Querschnittsklausel*) to consider ecological problems.[27] When confronted with the interpretation of indeterminate legal notions – good faith being a perfect illustration for moral values to enter the law of contract[28] – the integration clause serves as a standard of interpretation.[29] Therefore, as argued by Sjåfjell:

> (…) the Court [i.e. the ECJ] should to a greater extent and also ex officio focus on what the objective of sustainable development and the reference thereto in Article 11 TFEU entails with regard to a duty to integrate environmental protection in all EU laws – also in areas where environmental concerns traditionally have not been even considered.[30]

The reference in Article 11 TFEU to promoting sustainable development[31] is in this respect also not innocuous, as this very notion 'relies on the premise that environmental protection, economic and social development are mutually compatible, rather than conflicting, objectives' (see also Article 3(3) TEU[32]).[33] The potential of integrating environmental considerations into the fairness assessment is thus already there; its concretization 'only' requires a change of perspective.

Moreover, based on the preparatory work related to the adoption of the current Article 11 TFEU, the Member States[34] have also expressed their intention that the provision is effectively applied across the board – consequently also to secondary EU Law as the UCTD – and not only to certain specific areas of competence such as agriculture, transport, and trans-European networks. In other words, the EU has an obligation to integrate environmental protection in other fields of competences such as its consumer policy (Article 12 TFEU) and the interpretation of the related legislation,[35] especially also the UCTD. However, although environmental protection had been declared by the ECJ as 'one of the community's essential objectives' already in the mid-1980s,[36] it is not yet the case for consumer law.[37] This is somewhat surprising: the wording of Article 11 TFEU expressly reflects its binding nature ('must be'), whereas other integration clauses such as in particular Article 12 TFEU are drafted in a less stringent manner ('shall be').[38] A literal interpretation of fundamental EU provisions thus already retains the need to integrate environmental protection when interpreting EU consumer (contract) law.[39]

Furthermore, Article 3(3) UCTD specifies the fairness control by referring to the UCTD's Annex containing a non-exhaustive list of terms which may be regarded as unfair. This indicative list substantiates the concept of good faith.[40] A term contained in that list is not *per se* unfair and must still be assessed against the criteria of Articles 3(1) and 4 of the UCTD.[41] Annex 1, lit. h (automatic extension of a contract of fixed duration), reveals to be interesting when approached through the lens of ES. An extension of the contract's duration is here not necessarily bad.[42] Think of a default extension towards a longer contractual period automatically increasing the typical two-year holding period of a smartphone to a four-year period, thus reducing the frequency of the consumer's entitlement to a new one and, consequently, the impact on natural resources.[43]

Hence, in view of the above general developments, it could at least be expected that the indeterminate concept of good faith provided by Article 3 UCTD, as well as the concept of fairness itself, are complemented by an 'ecological touch', especially when applied and interpreted by the ECJ.[44] For example, SCT which exclude the circular replacement of a good by a repaired, refurbished, or remanufactured one, even if such replacement meets all applicable conformity requirements and corresponds to the consumer's will, might be considered unfair; such an interpretation would be supported by Article 11 TFEU, with respect to both the Sale of Goods Directive[45] and the UCTD.

2.2.2 *Transparency*

Articles 4(2) and 5 UCTD provide the general legal basis for the transparency requirement. According to these provisions, SCT must be drafted in plain, intelligible language.[46] The ECJ has interpreted this requirement broadly: the consumer must be put in a position to evaluate the economic consequences of contractual terms ('awareness of the risks and dangers arising from the term').[47] But what about their ecological consequences?

At first sight, transparency on ES in B2C relations plays primarily a role from the perspective of the Unfair Commercial Practices Directive,[48] especially in the context of the fight against greenwashing under its Articles 6, 7, and 12, as well as 5(2) of this Directive.[49] On 30 March 2022, the EU Commission proposed to strengthen the information model by empowering the consumer to take environment-friendly choices when buying products by imposing a more explicit ban on greenwashing and planned obsolescence.[50] However, because they relate to commercial communication, marketing, and advertising, these considerations remain on a broader conceptual and practical level than transparency in the UCTD. Moreover, these market fairness aspects do not relate – at least directly – to the economic interests of the consumer[51] and, especially, not to the economic consequences of contractual terms.

So, how to eventually integrate ES when assessing the fairness of a B2C contract term through the transparency requirement of the UCTD?

The first instrument that comes to mind is the rule set out in Article 5 UCTD. In case of doubt about the meaning of a term in an action involving an individual consumer (as opposed to a collective action, see Article 7(2) UCTD), the interpretation most favourable to the consumer shall prevail.[52]

But such a *contra proferentem* rule may prove contrary to ES in cases where the most favourable interpretation for the consumer is the least favourable for the environment. With reference to Dutch law (Article 6.238, paragraph 2, BW[53]), Pavillon illustrates this hypothesis with a commercial guarantee that confers the consumer a right to a replacement but that is ambiguous as to determine if the whole good or only the broken part is covered by this right. For the consumer, getting a completely new item is traditionally seen as the most favourable interpretation[54] although, from an ES perspective, it is (often) the least desirable one.[55]

For Pavillon, it could, however, be argued that the option that maximizes the lifespan of the purchased good and protects the consumer's right to a clean environment is to her advantage. Hence, the interpretation that favours the replacement of the broken part should prevail over obtaining a new good.[56] In a similar way, one might also consider that the contractual parties' mutual interest in ES excludes any doubt about the eco-literate meaning of the clause. The *contra proferentem* rule would thus be irrelevant here because such a 'sustainable' interpretation of the ambiguous clause is simply not unfavourable to the business (at least from a long-term perspective).[57]

Through the lens of ES, we will now present some aspects related to the control of the content of SCT (2.3).

2.3 Control of Content

In this third part, we will focus on the control of content mechanisms for SCT (*Inhaltskontrolle*) from the perspective of national contract law; there are at least three reasons why it is generally (also) relevant in this respect:

First, the UCTD[58] covers important but nevertheless limited aspects of the interpretation process in B2C contracts. All other relations, as well as the formation and the validity of the contract as such, are not specifically regulated by the UCTD but by national contract law.[59]

Second, national law may provide for other control mechanisms such as the admissibility of contractual terms with public policy, morality, or rights of personal privacy. The control of content can thus constitute an(other) interface for ES and the ecological analysis into private law.[60]

Third, even if there is a tendency on an EU level to strive towards full harmonization in areas of consumer (contract) law, the harmonized field remains limited and the role of national contract law is important.[61] In other words, comparative law – which especially also includes the law of non-EU Member States with similar or even the same legal and cultural backgrounds as EU Member States – has (also) a crucial role to play for ES on this level.

We will start by exposing the general relevance of collective interests for the control of content (2.3.1) and then expose some applications under Swiss and Belgian laws (2.3.2).

2.3.1 Collective Interests

One general aspect of interest is the treatment reserved to the inclusion of collective interests in the control of the content of SCT.[62]

What constitutes such a collective interest in the context of SCT? According to Baetge, collective (or general) interests are those which are to be understood as an expression of the common good of the state or society.[63] They must be distinguished from third-party interests (e.g., those of the creditors of a customer) as individual interests which, consequently, evolve on a lower conceptual basis.[64] Even if Baetge's developments in 2002 focused on specific anthropocentric situations where collective interests are deemed relevant,[65] it seems, however, rather undisputable that the achievement of ES to safeguard the very conditions of life is (also) a typical collective interest, if not the archetype thereof.

From a methodological perspective, the challenge is to identify how ES may eventually find its way through the glass ceiling of the otherwise traditionally relative contractual relationship. Or, in other words, to what extent may such collective interests be relevant to the control mechanisms of SCT? One way is to consider that safeguarding the collective values protected by fundamental rights is an issue concerning the third-party effect of those rights.[66] However, that is not necessarily a direct integration of collective interests in the context of the balancing of interests required by the specific control of the content mechanism of SCT.[67] It might in any case be rather controversial to allow the integration of external aspects here, as the balancing of interests is traditionally only – but nevertheless – seen as an instrument of social justice on the individual level of a contractual party.[68] At least where civil law expressly provides for a balancing of interests when assessing the validity of SCT (for Germany, e.g., §307 BGB[69]), the integration of ES as a collective interest is then still far from obvious.

In turn, this poses the question of alternatives that might frame such integration. We will attempt to shed some light by highlighting potential applications from the perspectives of Swiss and Belgian laws (2.3.2).

2.3.2 Applications

2.3.2.1 The Perspective of Swiss Law: The Role of General Control Standards Provided by Contract Law

Although being a non-EU Member State, Switzerland's law on unfair contract terms is nevertheless strongly influenced by the UCTD. Its approach to the control of the content[70] of SCT is, however, often described as being 'dualistic'.[71]

On the one hand, Article 8 UCA[72] – the provision influenced by the UCTD[73] – stipulates the unfairness of terms that are contrary to good faith and that create a significant imbalance between the rights and the duties stemming from the contract to the detriment of the consumer. As the UCTD, this provision only applies to B2C relations. What has been said above on the need for a more eco-literate understanding of the very concept of market fairness under the scope of the UCTD can thus be transposed to this context.

On the other hand, general rules of national contract law apply to the control of content tests on a non-specific level (i.e., not limited to B2C contracts). As such, however, explicit and specific legal control of content clauses like §307 BGB is unknown in Swiss contract law (see Articles 19 and 20 CO[74]).[75] Except for the situation of Article 8 UCA for B2C contracts, which indeed calls – in a specific B2C context – for some balancing of interests, reference must thus (also) be made to general control standards provided by contract law.

To be valid, SCT must especially respect public order ('public policy', Article 19(2) CO[76]). The notion '*encompasses all independent values and principles immanent to the legal order*'.[77] As for the principles of sustainable development and conservation of resources, they are expressly stated in Articles 2 and 73 of the Swiss Federal Constitution.[78] To my knowledge, however, no link has been made yet between these aspects related to ES and the control of content for SCT.

This does not come as a surprise. When confronting external interests to internal – private – interests in the process of interpreting SCT too,[79] an important difficulty arises: the guarantee of a private legal order is just as much in the public interest as the public order, of which it is an integral part.[80] This circularity makes it difficult to consider whether ecological aspects should prevail as public order. To be functional and internalized on the (national) level of the control of content in a context of global challenges induced by the climate crisis, public order may thus have to be extended more specifically to an emerging[81] ecological public order. Contrary to a so-called 'economical public order', which is jeopardized by the one-sided shifting of business risks to the consumer as aimed at by SCT,[82] such an ecological public order could, however, be a tool to implement current and future[83] external interests.

The main issue here is to determine if a violation of this ecological public order should at all lead to the nullity of the contractual content and, if yes, to what extent (and obviously if such a solution is proportionate with respect to the parties' autonomy as a cardinal principle of private law). Based on past applications of public order to the control of the content of SCT in Switzerland – insofar as this is even deemed acceptable – it seems currently rather difficult – if not utopian – to envisage such an outcome.[84]

In any case, the emergence and development of an ecological public order cannot rest on private law alone, because instrumentalization is not its only function.[85] For certain private relations previously identified as having significant potential for improvement in terms of ES, it may thus be more proportionate and efficient to intervene on a sector-specific basis[86] rather than by general – sometimes indeterminate – control standards. But, similarly, as for the principle of good faith, an ecological public order could nevertheless serve as a standard of interpretation and insist on the necessity of minimal ecological conduct also on the level of private law.

2.3.2.2 The Perspective of Belgian Law: Causa as a Complementary Tool?

An interesting national example of how to potentially deal in a progressive way with an ecological public order on the level of private law might be provided by Belgian law.[87]

In the context of the recent revision of Book 5 on the Law of Obligations (entry into force on 1 January 2023), Article 5.53 Belgian Civil Code[88] provides for the retention of the notion of *causa* as the motives that determined each party to conclude the contract ('indicia of seriousness'), if they are known or ought to have been known to the other party.[89] The *causa* is considered unlawful if contrary to public order (see Article 5.56 Belgian Civil Code). Of interest here is Article 1.3 Belgian Civil Code:

> a rule of law which affects the essential interests of the State or the community, or which establishes, in private law, the legal foundations on which society is based, such as the economic, moral, social or *environmental order, is of public order.*

For an eco-literate understanding of private agreements, such a conception of the *causa* – itself a controversial issue[90] – may have an interesting function.

From an ES perspective, could a reference to the *causa* permit a social control between contracts 'that are socially worth legal enforcement and those that should not receive legal protection' and thus 'reconnect contracts with the "whole"', especially with the 'true costs of the whole production process'?[91] And, should the parties more easily benefit from legal protection (i.e., enforceability or 'actionability'),[92] in a sense similar to a presumption of validity of their contract, if their agreement concurs 'with the well-being of the whole'.[93] Inversely, should an unsustainable contract be associated with a presumption of invalidity?

As Kryla-Cudna pointed out in relation to the principle of good faith,[94] it can indeed be argued 'that the principles of proper conduct accepted in the society nowadays require an account to be taken of environmental and sustainability concerns'[95]: could the same not be said of the individual motives of contractual partners from the perspective of the *causa*?

On the one hand, it would be an obvious exaggeration to declare a contract void for the purchase of an airline ticket on this basis,[96] but, on the other hand, it is undeniable that private law – including the interpretation of SCT – does not operate in a vacuum absolutely detached from environmental considerations. It is all about balance.

Obviously, one major difficulty is to determine, in a convincing way for practice and with clear objective assessment criteria,[97] which environmental externalities related to the contractual object can have an influence on the control of the contractual content. If so, it must also (still) be determined how and with what consequences for the contractual parties,[98] to avoid 'arbitrary and irrational results'.[99]

Given the urgency of the current climate crisis, it seems, however, at least worth (also) exploring progressive understandings of the *causa*. This may give courts a complementary approach to the ordinary rules on the validity of the parties' consent and the control of the legality of the contractual object,[100] especially when confronted with the implementation of external considerations related to ES in a specific case. Admittedly, such interpretations of the *causa* – as far as admissible – still need to be circumscribed in (much) more detail to eventually ensure their acceptance and predictability. And they are, of course, not politically neutral.[101] But neither is the rejection of the *causa* to promote the development of an 'efficient' and market-friendly legal framework whose effects and externalities are recognized by some as being also partially responsible for the current climate crisis.[102]

2.4 Conclusion

So, what to remember from our above general developments?

The inclusion of ES values into contract law indeed poses 'extraordinary problems to contract law scholars'.[103] The same is obviously true for the interpretation of SCT and especially the control of their content.

However, although Nature is unfortunately not always fair, the transition towards a greener understanding of the process of contractual interpretation might sometimes be closer to reach than one might think. For the interpreter of current law, setting apart the complex questions of whether this is or not (strictly) a political issue, it often 'only' requires a change of perspective and understanding of already existing rules and concepts such as fairness and the *contra proferentem rule* to foster ES on the level of contractual interpretation. EU law, also with respect to the UCTD, has the potential to be a driving force for change through an effective application of the integration clause that is Article 11 TFEU to B2C relations: this could also reconcile consumer protection with environmental protection, as areas that are still often seen as unrelated today.

That rather optimistic part being said, much conceptualizing work remains to be done too on the level of national law (i) to eventually integrate collective interests such as ES in a methodologically convincing way into the control of the content of SCT and (ii) to circumscribe the control of the content of SCT through instruments such as their adequacy to an ecological public order or even by exploring the 'old-fashioned taboo'[104] of the *causa*.

When assessing the interpretation of SCT through the lens of ES – without questioning the potential and importance of EU law – one can also discern the need for national – local – solutions for what is in fact a global problem. This seems to indicate that the challenge of integrating aspects related to ES into the interpretation process of SCT is at least partially to be supported by national legislators too. This could also reveal a significant advantage rarely taken into consideration: the competition between national legal orders to identify the best possible solution with respect to eventually achieving ES in private law.

Ultimately, the general concept of ecological public order should nevertheless be similar or even identical in scope across different legislations (including European non-EU Member States). It is then rather a constructive coordination than a strict competition that will have to evolve to truly achieve an 'eco-literate' interpretation of SCT also on a transnational level.

Notes

1 Dr. Iur., Scientific collaborator at the University of Bern and Lecturer in Consumer and Contract Law at the University of Neuchâtel (Switzerland). I would like to thank Prof. Evelyne Terryn and Christof Koolen from CCM, KU Leuven, as well as the anonymous reviewers for their valuable feedback and suggestions on the first draft of this chapter.

2 Conceptualized as one of the three pillars of 'sustainable development', the two others being social and economic sustainability. On the so-called '*Drei-Säulen-Konzept*', A. Hellgardt and V. Jouannaud, 'Nachhaltigkeitsziele und Privatrecht' (2022) 222 *Archiv für die civilistische Praxis* 164, 167–170. See also A. Halfmeier, 'Nachhaltiges Privatrecht' (2016) 216 *Archiv für die civilistische Praxis* 717, 721–725 and R. Mahaim, 'Art. 73 Cst.', in V. Martenet and J. Dubey (eds), *Commentaire romand de la Constitution fédérale* (Helbing Lichtenhahn, Basel 2021) N 15.

3 For example, SCT can support environmentally detrimental practices (e.g., a commercial guarantee in which the right of replacement takes precedence over the possibility of repair – or even excludes it altogether – or clauses limiting a legal right to a free repair), C. Pavillon, *Herijking van consumentencontractenrecht: duurzaamheid als nieuw ijkpunt?*, in C. Pavillon and W.H. van Boom, *Privaatrechtelijke bescherming herijkt. Preadviezen 2021 Uitgebracht voor de Vereniging voor Burgerlijk Recht* (Uitgeverij Paris, Zutphen 2021) N 75–77.

4 For example, a non-individually negotiated term can enshrine duties of care, extra-quality guarantees, and other long-term obligations (e.g., service and return guarantees, obligations to update or repair within a reasonable period of time, and obligation to recycle or sustainable scrapping), C. Pavillon, 'Herijking van consumentencontractenrecht: duurzaamheid als nieuw ijkpunt?', in C. Pavillon and W.H. van Boom, *Privaatrechtelijke bescherming herijkt. Preadviezen 2021 Uitgebracht voor de Vereniging voor Burgerlijk Recht* (Uitgeverij Paris, Zutphen 2021) N 75. See also I. Bach and E.-M. Kieninger, 'Ökologische Analyse des Zivilrechts' (2021) 76 (22) *JuristenZeitung* 1069, 1096–1097 on '*Laufzeitklauseln*'.

5 R. Posner, 'The Law and Economics of Contract Interpretation' (2004) 83 *Texas Law Review* 1581, 1581–182: 'Contract interpretation is the undertaking by a judge or a jury (or an arbitrator [...]) to figure out what the terms of a contract are or should be understood to be'.

6 On the influence for the interpretation of contracts of policy considerations related to the social, political, and economic organization of a society, as well as the protection of the weaker party, L. Kaehler, 'Policy Interventions via Contract Interpretation' (2014) 22 (5) *European Review of Private Law*, 641, <https://doi.org/10.54648/erpl2014054>. On 'Principles of Interpretation' with respect to 'a contract law accommodating sustainability clauses', V. Ulfbeck and O. Hansen, 'Sustainability Clauses in an Unsustainable Contract Law?' (2020) 16 (1) *European Review of Contract Law* 186, 192–194.

7 Our paper is based on the – not undisputed – preliminary assumption that private law can (also) be adequate and suitable for the implementation of ES aspects. For further details, I. Bach and E.-M. Kieninger, 'Ökologische Analyse des Zivilrechts' (2021) 76 (22) *JuristenZeitung* 1069, A. Halfmeier, 'Nachhaltiges Privatrecht' (2016) 216 (5) *Archiv für die civilistische Praxis*, 717 and A. Hellgardt and V. Jouannaud, 'Nachhaltigkeitsziele und Privatrecht' (2022) 222 *Archiv für die civilistische Praxis* 164.

8 The paper is not a truly comparative work in substance. However, apart from a particular personal proximity to these two legal orders, I find it interesting to specifically integrate elements of Swiss and Belgian laws for the following main reasons: (i) one country is a Member State of the EU, the other is not; (ii) the two countries have different and important linguistic communities; and (iii) while Swiss law is at the

crossroads of the German and French legal traditions, Belgian law is still strongly influenced by French law. Moreover, Belgium has recently adopted a new civil code, which contains provisions on the Law of Obligations. In contrast, Swiss law still applies its 1911 Code of Obligations.

9 Council Directive 93/13/EEC of 5 April 1993 on unfair terms in consumer contracts, OJ L 95/29-34.

10 This is not the place to review all these developments, whose origins at the European level date back as far as the 1970s. For a historical and comparative report, E. Hondius, *Unfair Terms in Consumer Contracts* (Utrecht 1987). See also, with further references, R. Schulze and F. Zoll, *European Contract Law* (3rd ed, Baden-Baden 2021) N 173–203 and N. Jansen, 'Introduction before Art. 6:201', in N. Jansen and R. Zimmermann (eds), *Commentaries on European Contract Laws* (Oxford University Press, Oxford 2018) N 1–4.

11 For the reverse paradigm of the 'consumer as a citizen' in the context of the circular economy, V. Mak and E. Terryn, 'Circular Economy and Consumer Protection: The Consumer as a Citizen and the Limits of Empowerment Through Consumer Law' (2020) 43 (1) *Journal of Consumer Policy* 227. See also T. Wilhelmsson, 'Consumer Law and the Environment: From Consumer to Citizen' (1998) 21 (1) *Journal of Consumer Policy* 45.

12 See also recital 16 with reference to the 'legitimate interests' of the consumer.

13 L. Krämer, 'On the Interrelation between Consumer and Environmental Policies in the European Community' (1993) 16 (3–4) *Journal of Consumer Policy* 455; C. Kye, 'Environmental Law and the Consumer in the European Union' (1995) 7 (1) *Journal of Environmental Law* 31; M. Hedemann-Robinson, 'EC Law, the Environment and Consumers: Addressing the Challenge of Incorporating an Environmental Dimension to Consumer Protection at Community Level' (1997) 20 *Journal of Consumer Policy* 1. See also F. Jacobs, 'The Role of the European Court of Justice in the Protection of the Environment' (2006) 18 (2) *Journal of Environmental Law* 185.

14 B. Sjåfjell, 'The Environmental Integration Principle: A Necessary Step towards Policy Coherence for Sustainability', in F. Ippolito, M.E. Bartoloni and M. Condinanzi (eds), *The EU and the Proliferation of Integration Principles under the Lisbon Treaty* (Routledge 2018) 109: '(…) EU policy-making is still influenced by the compartmentalisation that perpetuates the illusion that environmental protection can be left to environmental law, while other areas of law do not have to concern themselves with environmental protection'; E. Terryn, 'The New Consumer Agenda: A Further Step Toward Sustainable Consumption?' (2021) 10 (1) *Journal of European Consumer and Market Law* 1; A. Halfmeier, 'Nachhaltiges Privatrecht' (2016) 216 *Archiv für die civilistische Praxis* 717, 749–752.

15 For an in-depth analysis of these rules, A. Michel, *Premature Obsolescence: In Search of an Improved Legal Framework* (Leuven 2022) N 87–169.

16 On 30 March 2022, the EU Commission published new proposals to make sustainable products the norm <https://ec.europa.eu/commission/presscorner/api/files/document/print/en/ip_22_2013/IP_22_2013_EN.pdf> accessed 24 May 2023 and new consumer rights for information on the durability and reparability of products, as well as a ban on greenwashing <https://ec.europa.eu/commission/presscorner/detail/en/ip_22_2098> accessed 24 May 2023. On 22 March 2023, the EU Commission adopted a new proposal on common rules promoting the repair of goods <https://ec.europa.eu/commission/presscorner/detail/en/ip_23_1794> accessed 24 May 2023 and proposed common criteria against greenwashing and misleading environmental claims <https://ec.europa.eu/commission/presscorner/detail/en/IP_23_1692> accessed 24 May 2023.

17 The answer to this question is obviously very complex and we can only touch on certain general aspects in the limited scope of this contribution. We will not address the issue through the control mechanisms provided by important harmonization and unification initiatives such as the Acquis Principles, the DCFR, or the CESL. For further details on unfair contract terms and these instruments (also with a comparative approach but without environmental considerations), R. Schulze and F. Zoll, *European Contract Law* (3rd ed, Baden-Baden 2021) N 177–203.

18 Case C-237/02, *Freiburger Kommunalbauten GmbH Baugesellschaft & Co. KG v. Ludger Hofstetter et Ulrike Hofstetter*, ECLI:EU:C:2004:209; R. Schulze and F. Zoll, *European Contract Law* (3rd ed, Baden-Baden 2021) N 174. See also Commission Notice, Guidance on the interpretation and application of Council Directive 93/13/EEC on unfair terms in consumer contracts (Text with EEA relevance), OJ C 323/4, 22.

19 R. Schulze and F. Zoll, *European Contract Law* (3rd ed, Baden-Baden 2021) N 175.

20 Commission Notice, Guidance on the interpretation and application of Council Directive 93/13/EEC on unfair terms in consumer contracts (Text with EEA relevance), OJ C 323/ 4, 30.

21 To refer to the expression coined by U. Mattei and A. Quarta, *The Turning Point in Private Law – Ecology, Technology and the Commons* (Cheltenham and Northampton 2019) N 120: 'There is a vital need for a new ecologically literate legal culture to counteract, through counter-hegemonic interpretation (…)'.

22 For a similar argument with respect to private autonomy, A. Halfmeier, 'Nachhaltiges Privatrecht' (2016) 216 *Archiv für die civilistische Praxis* 717, 743: 'Derselbe Zusammenhang zwischen Freiheit und für sie notwendigen Bedingungen besteht auch im Hinblick auf Privatautonomie und das Prinzip der Nachhaltigkeit: *Die mit dem Nachhaltigkeitsprinzip angesprochene Sicherung der natürlichen Lebensgrundlagen ermöglicht gerade erst die dauerhafte Ausübung von Freiheit und Selbstbestimmung*. Eine Freiheit im Sinne eines «Rechts auf Lebensgrundlagenzerstörung» ist nicht sinnvoll zu denken'.

23 For German law, G. Wagner, 'Klimaschutz durch Gerichte' (2021) 31 *Neue Juristische Wochenschrift* 2256: 'Die für die Lösung des Klimaproblems erforderlichen diffizilen Allokations- und Abwägungsentscheidungen sind auf der politischen Ebene zu treffen und zu verantworten'.

24 The French and German versions of said recital state 'un moyen d'évaluation globale des *différents intérêts impliqués*' and, seemingly more restrictive for the same section, 'die Möglichkeit einer globalen Bewertung der *Interessenlagen der Parteien*'.

25 Consolidated Version of the Treaty on the Functioning of the European Union, OJ C 326/47.

26 Charter of Fundamental Rights of the European Union, OJ C 326/391.

27 C. Calliess, 'Art. 11 AEUV', in C. Calliess and M. Ruffert (eds), *EUV – AEUV mit Europäischer Grundrechtecharta* (6th ed, Munich 2022) N 26 referring to Case C-17/90, *Pinaud Wieger Spedition GmbH v. Bundesanstalt für den Güterfernverkehr*, ECLI:EU:C:1991:416 and C-195/90, *Commission of the European Communities v. Federal Republic of Germany*, ECLI:EU:C:1992:219. See also, A. Hellgardt and V. Jouannaud, 'Nachhaltigkeitsziele und Privatrecht' (2022) 222 *Archiv für die civilistische Praxis* 164, 179. See also already M. Wasmeier, 'The Integration of Environmental Protection as a General Rule for Interpreting Community Law' (2001) 38 *Common Market Law Review* 159.

28 K. Kryla-Cudna, 'Sales Contracts and the Circular Economy' (2020) 28 (6) *European Review of Private Law* 1207, 1207–1214 with further references.

29 C. Calliess, 'Art. 11 AEUV', in C. Calliess and M. Ruffert (eds), EUV – AEUV mit Europäischer Grundrechtecharta (6th ed, Munich 2022) N 25 referring to Case C-513/99, Concordia Bus Finland Oy Ab, formerly Stagecoach Finland Oy A v. Helsingin kaupunki and HKL-Bussiliikenne, ECLI:EU:C:2002:495. See also A. Hellgardt and V. Jouannaud, 'Nachhaltigkeitsziele und Privatrecht' (2022) 222 *Archiv für die civilistische Praxis* 164, 172: 'Begreift man Nachhaltigkeit in diesem Sinne als konkretisierungsbedürftiges Rechtsprinzip, das auf möglichst hohe rechtliche Realisierung drängt, kann es im jeweiligen rechtlichen Kontext vom Gesetzgeber durch Erlass spezifischer Regelungen sowie im Rahmen der Auslegung und der Interpretation von unbestimmten Rechtsbegriffen und Generalklauseln durch die Verwaltung und – für das Privatrecht von größerem Interesse – die Rechtsprechung Beachtung bzw. Anwendung finden'.

30 B. Sjåfjell, 'The Environmental Integration Principle: A Necessary Step towards Policy Coherence for Sustainability', in F. Ippolito, M.E. Bartoloni and M. Condinanzi (eds), *The EU and the Proliferation of Integration Principles under the Lisbon Treaty* (Routledge 2018) N 114.

31 B. Sjåfjell, 'The Environmental Integration Principle: A Necessary Step towards Policy Coherence for Sustainability', in F. Ippolito, M.E. Bartoloni and M. Condinanzi (eds), *The EU and the Proliferation of Integration Principles under the Lisbon Treaty* (Routledge 2018) N 106 rightly advocates that the sustainable development goal 'must be defined in light of natural science' to 'mitigate the problem of sustainability being an extremely overused and broad concept with multiple definitions'.

32 Consolidated Version of the Treaty on European Union, OJ C 326, 26.10.2012, pp. 13–45.

33 J. Nowag, 'The Sky is the Limit: On the Drafting of Article 11 TFEU's Integration Obligation and Its Intended Reach', in B. Sjåfjell and A. Wiesbrock (eds), *The Greening of European Business under EU Law: Taking Article 11 TFEU Seriously* (Routledge 2015) N 1, 9.

34 For an analysis on the relation of art. 11 TFEU and the Member States of the EU, B. Sjåfjell, 'The Environmental Integration Principle: A Necessary Step towards Policy Coherence for Sustainability', in F. Ippolito, M.E. Bartoloni and M. Condinanzi (eds), *The EU and the Proliferation of Integration Principles under the Lisbon Treaty* (Routledge 2018) N 106, 116–121.

35 J. Nowag, 'The Sky Is the Limit: On the Drafting of Article 11 TFEU's Integration Obligation and Its Intended Reach', in B. Sjåfjell and A. Wiesbrock (eds), *The Greening of European Business under EU Law: Taking Article 11 TFEU Seriously* (Routledge 2015) N 2, 8 11.

36 J. Nowag, 'The Sky Is the Limit: On the Drafting of Article 11 TFEU's Integration Obligation and Its Intended Reach', in B. Sjåfjell and A. Wiesbrock (eds), *The Greening of European Business under EU*

Law: Taking Article 11 TFEU Seriously (Routledge 2015) N 4 citing case C-240/83, Procureur de la République v. Association de défense des brûleurs d'huiles usagées, ECLI:EU:C:1985:59, para 13. Critical with the ECJ's approach in relation to the sustainability potential of art. 11 TFEU, B. Sjåfjell, 'The Environmental Integration Principle: A Necessary Step towards Policy Coherence for Sustainability', in F. Ippolito, M.E. Bartoloni and M. Condinanzi (eds), *The EU and the Proliferation of Integration Principles under the Lisbon Treaty* (Routledge 2018) N 109: '(…) the Court has not taken upon itself clearly enough the role art. 11 TFEU bestows upon it: to be a guardian of the planetary boundaries. There are even some contradictory statements in the case-law'.

37 Case C-681/17, *slewo — schlafen leben wohnen GmbH v. Sascha Ledowski*, ECLI:EU:C:2019:255, para 32, 34 and 39: 'striking the right balance between a high level of consumer protection and the competitiveness of enterprises', interpreting Directive 2011/83/EU on consumer rights. For a critical review of this judgement from the perspective of environmental protection, A. Michel, 'Matelas et vêtements dans le même sac?' (2020) 126 *Droit de la consommation – Consumentenrecht*, 34.

38 J. Nowag, 'The Sky Is the Limit: On the Drafting of Article 11 TFEU's Integration Obligation and Its Intended Reach', in B. Sjåfjell and A. Wiesbrock (eds), *The Greening of European Business under EU Law: Taking Article 11 TFEU Seriously* (Routledge 2015) N 10.

39 As an example of such integration in the context of Council Directive (EU) 2019/771 of the European Parliament and of the Council of 20 May 2019 on certain aspects concerning contracts for the sale of goods the Sale of goods, amending Regulation (EU) 2017/2394 and Directive 2009/22/EC, and repealing Directive 1999/44/EC, OJ L 136/ 28 (hereafter: Sale of Goods Directive), E. Van Gool and A. Michel, 'The New Consumer Sales Directive 2019/771 and Sustainable Consumption: A Critical Analysis' (2021) 10 (4) *Journal of European Consumer and Market Law* 136, 145 suggest interpreting the 'disproportionality-test' in art. 13, paragraphs 2 and 3 of this Directive 'in a broader way that also takes the sustainability impact of a specific replacement and repair into account'. In their right opinion, this would here 'correctly reflet article 11 TFEU'.

40 R. Schulze and F. Zoll, *European Contract Law* (3rd ed, Baden-Baden 2021) 175.

41 Case C-472/10, *Nemzeti Fogyasztóvédelmi Hatóság v. Invitel Távközlési Zrt*, ECLI:EU:C:2012:242. See also Commission Notice, Guidance on the interpretation and application of Council Directive 93/13/EEC on unfair terms in consumer contracts (Text with EEA relevance), OJ C 323/4, 35.

42 Even though such a clause might – rightly so – be considered unfair, especially when this extension comes as a surprise for the consumer.

43 See I. Bach and E.-M. Kieninger, 'Ökologische Analyse des Zivilrechts' (2021) 76 (22) *JuristenZeitung* 1069, 1097: 'Bemerkenswert ist in diesem Zusammenhang allerdings, dass nicht nur kürzere, sondern auch längere Vertragslaufzeiten die Haltedauer von Handys erhöhen könnten: Bei einer Laufzeit von vier Jahren, würde der «Anspruch» auf ein neues Handy nur noch halb so oft entstehen wie bei der derzeit üblichen Zweijahresbindung. Dies spricht aus rein ökologischen Gesichtspunkten dafür, (auch) solche langen Laufzeiten AGB-rechtlich zuzulassen'. Such an interpretation obviously targets the 'smartphone as a status symbol' where 'many consumers buy a new device every two years and leave the old one lying around', see feedback from GISAD on the EU Commission's Initiative for Sustainable consumption of goods – promoting repair and reuse, <https://ec.europa.eu/info/law/better-regulation/have-your-say/initiatives/13150-Sustainable-consumption-of-goods-promoting-repair-and-reuse/F2986034_en> accessed 24 May 2023.

44 J. Nowag, 'The Sky Is the Limit: On the Drafting of Article 11 TFEU's Integration Obligation and Its Intended Reach', in B. Sjåfjell and A. Wiesbrock (eds), *The Greening of European Business under EU Law: Taking Article 11 TFEU Seriously* (Routledge 2015) 11 citing H.-P. Zils, *Die Wertigkeit des Umweltschutzes in Beziehung zu anderen Aufgaben der Europäischen Gemeinschaft: Untersuchungen zur Anwendung der Querschnittsklausel Art. 130r Abs. 2 Satz 2 EWGV im Gemeinschaftsrecht* (Decker, Heidelberg 1994) 27.

45 For the admissibility of circular replacement with respect to this Directive, see E. Van Gool, A. Michel, B. Keirsbilck and E. Terryn, CCM KU Leuven Feedback to EU Commission Consultation on Sustainable Consumption of Goods - Promoting Repair and Reuse Legislative Initiative, 4 April 2022, <https://papers.ssrn.com/sol3/papers.cfm?abstract_id=4211301> accessed 24 May 2023.

46 Commission Notice, Guidance on the interpretation and application of Council Directive 93/13/EEC on unfair terms in consumer contracts (Text with EEA relevance), OJ C 323/ 4, 25. See also recital 20: '(…) the consumer should actually be given an opportunity to examine all the terms and, if in doubt, the interpretation most favourable to the consumer should prevail'.

47 Cases C-96/14, *Jean-Claude Van Hove v. CNP Assurances SA*, ECLI:EU:C:2015:262 and C-26/13, *Árpád Kásler and Hajnalka Káslerné Rábai v. OTP Jelzálogbank Zrt*, ECLI:EU:C:2014:282. This can be seen as '*an indirect effect of the role of reasonable expectations*', R. Schulze and F. Zoll, *European Contract Law* (3rd ed, Baden-Baden 2021) N 187–188.

48 Council Directive 2005/29/CE of the European Parliament and of the Council of 11 May 2005 concerning unfair business-to-consumer commercial practices in the internal market and amending Council Directive 84/450/EEC, Directives 97/7/EC, 98/27/EC and 2002/65/EC of the European Parliament and of the Council and Regulation (EC) No 2006/2004 of the European Parliament and of the Council (Text with EEA relevance), OJ L 149/ 22.

49 Greenwashing can be defined as the use of an untrue or unverifiable environmental claim (i.e., the 'practice of suggesting or otherwise creating the impression (…) that a good or a service has a positive or no impact on the environment or is less damaging to the environment than competing goods or services'). See Commission Notice, Guidance on the interpretation and application of Council Directive 2005/29/EC of the European Parliament and of the Council concerning unfair business-to-consumer commercial practices in the internal market (Text with EEA relevance), OJ C 526/ 1, 72–84.

50 Circular Economy: Commission proposes new consumer rights and a ban on greenwashing, <https://ec.europa.eu/commission/presscorner/detail/en/ip_22_2098> accessed 24 May 2023.

51 On the one hand, admittedly, the case of premature obsolescence is the most debatable in this respect, as it has also a direct impact on the economic interests of the consumer pushed to renew its acquisitions at a shorter deadline. On the other hand, by reducing the quality of the product, premature obsolescence can lower the price of the concerned products and hence make a given product more accessible (at least in the short run). It should be recalled here, however, that according to its art. 4(2), the UCTD 'does not serve to ensure that the consumer concludes a cost-effective contract', R. Schulze and F. Zoll, *European Contract Law* (3rd ed, Baden-Baden 2021) N 188.

52 Both the duty of transparency linked to the European fairness test and the *contra proferentem* rule as a traditional aspect of contractual interpretation are 'expressions of the general European principle of transparency', N. Jansen, 'Art. 6:206', in N. Jansen and R. Zimmermann (eds), *Commentaries on European Contract Laws* (Oxford University Press, Oxford 2018) N 1.

53 Burgerlijk Wetboek of 1992.

54 For example, V. Mak and E. Terryn, 'Circular Economy and Consumer Protection: The Consumer as a Citizen and the Limits of Empowerment Through Consumer Law' (2020) 43 *Journal of Consumer Policy* 227, 236 with reference to District Court Amsterdam 8 July 2016, ECLI:NL:RBAMS:2016:4197 and District Court Amsterdam 18 April 2017, ECLI:NL:RBAMS:2017b:2519. See also Glostrup District Court, Case number BS 10E-3689/2014, 9th December 2016. The Danish Court even stated that 'the fact that environmental concerns can speak for the recycling of reusable modules and components production in general, is not relevant in determining whether a replacement product complies with the agreement'. See <https://forums.macrumors.com/threads/danish-court-rules-apple-must-replace-mans-iphone-with-new-rather-than-refurbished-model.2020632/page-11> accessed 24 May 2023.

55 C. Pavillon, 'Herijking van consumentencontractenrecht: duurzaamheid als nieuw ijkpunt?', in C. Pavillon and W.H. van Boom (eds), *Privaatrechtelijke bescherming herijkt, Preadviezen 2021 Uitgebracht voor de Vereniging voor Burgerlijk Recht* (Uitgeverij Paris, Zutphen 2021) N 77.

56 C. Pavillon, 'Herijking van consumentencontractenrecht: duurzaamheid als nieuw ijkpunt?', in C. Pavillon and W.H. van Boom (eds), *Privaatrechtelijke bescherming herijkt, Preadviezen 2021 Uitgebracht voor de Vereniging voor Burgerlijk Recht* (Uitgeverij Paris, Zutphen 2021) N 77. At the same time, however, the author expresses doubts that such an interpretation is compatible with current mandatory interpretation rules.

57 C. Pavillon, 'Herijking van consumentencontractenrecht: duurzaamheid als nieuw ijkpunt?', in C. Pavillon and W.H. van Boom (eds), *Privaatrechtelijke bescherming herijkt, Preadviezen 2021 Uitgebracht voor de Vereniging voor Burgerlijk Recht* (Uitgeverij Paris, Zutphen 2021) N 77. Moreover, the *contra proferentem* rule does not apply when each meaning of the clause has approximately equal advantages and disadvantages for each party, P. Jäggi and P. Gauch, 'Kommentar zu Art. 18 OR', in W. Schönenberger and P. Gauch (eds), *Zürcher Kommentar zum Schweizerischen Zivilgesetzbuch, V. Band, Obligationenrecht, Teilband V 1b* (Schulthess, Zürich 1980) N 456. Thus, could one also argue that the interpretation leading to the obtaining of a completely new good – as it *in fine* involves equivalent 'external' drawbacks for each contractual party (i.e., the need to extract more resources to produce a new replacement good) – implies that there is no doubt about the meaning of the clause?

58 Which is of minimal harmonization (art. 8 Directive).

59 Commission Notice, Guidance on the interpretation and application of Council Directive 93/13/EEC on unfair terms in consumer contracts (Text with EEA relevance), OJ C 323/4, 19: 'Other national rules that may apply to cases involving unfair contract terms include general provisions of contract law, in particular on the formation and validity of contracts, as well as the rules of procedure for proceedings before national courts. Such questions are not specifically regulated in the UCTD but may have a significant impact on its application'.

60 On the importance of the control through the general clause of §307 BGB in German law, I. Bach and E-M. Kieninger, 'Ökologische Analyse des Zivilrechts' (2021) 76 (22) *JuristenZeitung*, 1069, 1096–1097.

61 H.-W. Micklitz, 'The Full Harmonization Dream' (2022) 11 (4) *Journal of European Consumer and Market Law* 117, 117, 118, 121: 'The messy field of private/consumer law is the present and will be the future. This is exactly the private law beyond the nation state. "United in diversity" is the basis on which European integration is built'.

62 For the assessment of collective interests under the control of content provided by §307 of the German BGB (but without ES considerations), D. Baetge, 'Allgemeininteressen in der Inhaltskontrolle – Der Einfluss öffentlicher Interessen auf die Wirksamkeit Allgemeiner Geschäftsbedingungen' (2002) 202 *Archiv für die civilistische Praxis* 972, 973; A. Fuchs, '§307 BGB', in Ulmer/Brandner/Hensen, AGB-Recht (13th ed, Köln 2022) N 133–139; M. Wendland, '§307 BGB', Staudinger BGB, §§ 305–310; UKlaG – AGB-Recht 1 und Unterlassungsklagengesetz, (18th ed, Sellier/de Gruyter 2019) N 151. See also I. Bach and E.-M. Kieninger, 'Ökologische Analyse des Zivilrechts' (2021) 76 (22) *JuristenZeitung* 1069, 1097 (with ES considerations).

63 D. Baetge, 'Allgemeininteressen in der Inhaltskontrolle – Der Einfluss öffentlicher Interessen auf die Wirksamkeit Allgemeiner Geschäftsbedingungen' (2002) 202 *Archiv für die civilistische Praxis* 972, 973.

64 D. Baetge, 'Allgemeininteressen in der Inhaltskontrolle – Der Einfluss öffentlicher Interessen auf die Wirksamkeit Allgemeiner Geschäftsbedingungen' (2002) 202 *Archiv für die civilistische Praxis* 972, 973.

65 Such as contractual penalty clauses for a specific German trust institution (*Vertragsstrafeklauseln der Treuhandanstalt*), defence clauses against bid rigging in public contracts (*Abwehrklauseln gegen Submissionsabsprachen in öffentlichen Vergabeverträgen*), section clauses concerning the use of cadaver parts for scientific research and education (*Sektionsklauseln*) and form-based consent to the transfer of personal data (*Formularmäßige Einwilligung in die Weitergabe persönlicher Daten*).

66 That is, the civil judge's obligation to observe fundamental rights when specifying general clauses.

67 E. Kramer, 'Art. 19-20 OR', in A. Meier-Hayoz (ed), *Berner Kommentar, Bd VI, 1. Abteilung, Allgemeine Bestimmungen, 2. Teilband, Unterteilband 1a, Inhalt des Vertrages, Kommentar zu Art. 19-22 OR* (Stämpfli, Bern 1991) N 163: 'Die Generalklausel der öffentlichen Ordnung kann darüber hinaus als Aufhänger dienen, um verfassungsrechtliche Wertungen in die Kontrolle des Vertragsinhalts einfließen zu lassen; dies hauptsächlich i. S. Einer «mittelbaren» Drittwirkung von Grundrechten' (…); M. Wendland, '§307 BGB', Staudinger BGB', §§ 305–310; UKlaG – AGB-Recht 1 und Unterlassungsklagengesetz (18th ed, Sellier/de Gruyter 2019) N 151: 'nicht aber um ein «Interesse» der Gemeinschaft im Rahmen der Interessenabwägung'. See also, A. Hellgardt and V. Jouannaud, 'Nachhaltigkeitsziele und Privatrecht' (2022) 222 *Archiv für die civilistische Praxis* 164, 198 on the indirect effect of art. 20a of the German Grundgesetz.

68 M. Wendland, '§307 BGB', Staudinger BGB, §§ 305–310; UKlaG – AGB-Recht 1 und Unterlassungsklagengesetz (18th ed, Sellier/de Gruyter 2019) N 151. See also A. Fuchs, '§307 BGB', in P. Ulmer, H. E. Brandner and H.-D. Hensen (eds), AGB-Recht Kommentar, §§305–310 BGB – Besondere Vertragstypen – UKlaG (13th ed, Köln 2022) N 136. Stating the need for a stricter control of content in B2C contracts under the aspect of public order, E. Kramer, 'Art. 19-20 OR', in A. Meier-Hayoz (ed), *Berner Kommentar, Bd VI, 1. Abteilung, Allgemeine Bestimmungen, 2. Teilband, Unterteilband 1a, Inhalt des Vertrages, Kommentar zu Art. 19-22 OR* (Stämpfli, Bern 1991) N 165.

69 Bürgerliches Gesetzbuch of 1896.

70 For a general analysis of the control of content under Swiss Law, D. Hug, La formation du contrat de consommation: entre régime général et approche sectorielle – analyse et perspectives en droit suisse (Basel/Neuchâtel, 2020) N 1754–1849.

71 P. Pichonnaz, 'Art. 8 LCD', in V. Martenet and P. Pichonnaz (eds), *Commentaire romand, Loi contre la concurrence déloyale* (Helbing Lichtenhahn, Basel 2017) N 30.

72 Federal Act of 19 December 1986 against Unfair Competition (CC 241; 'Bundesgesetz gegen den unlauteren Wettbewerb', 'Loi fédérale contre la concurrence déloyale').
73 P. Pichonnaz, 'Art. 8 LCD', in V. Martenet and P. Pichonnaz (eds), *Commentaire romand, Loi contre la concurrence déloyale* (Helbing Lichtenhahn, Basel 2017) N 117.
74 Federal Act of 30 March 1911 on the Amendment of the Swiss Civil Code, Part Five: The Code of Obligations (CC 220: 'Obligationenrecht', 'Codedesobligations').
75 E. Kramer, 'Art. 19-20 OR', in A. Meier-Hayoz (ed), *Berner Kommentar, Bd VI, 1. Abteilung, Allgemeine Bestimmungen, 2. Teilband, Unterteilband 1a, Inhalt des Vertrages, Kommentar zu Art. 19-22 OR* (Stämpfli, Bern 1991) N 123: 'Der praktisch wichtigste und in Art. 19, 20 einzig geregelte Aspekt einer Kontrolle der Vertragsfreiheit ist die Inhaltskontrolle. Die in Art. 19 I verheissene Freiheit, den Inhalt des Vertrags «beliebig» festzustellen, wird (gleich wie in Art. 641 I ZGB die Eigentümerfreiheit) bereits im selben Satz («innerhalb der Schranken des Gesetzes»!) dementiert. Dieses Dementi wird in Art. 19 II und Art. 20 I durch Nennung der massgeblichen Kontrollkriterien konkretisiert, Kriterien, die (abgesehen vom speziell gelagerten Fall der anfänglichen Unmöglichkeit) jeweils die Funktion haben, die (in Art. 20 angeordnete) «Elimination» solcher Verträge zu begründen, in bezug auf die es «der Staat… nicht zulassen» kann, «dass aus» ihnen «Rechte geltend gemacht werden» (…)'.
76 The provision states that 'The law excludes the parties' agreements only when it lays down a strict rule of law, or when a departure from its text would be contrary to morality, *public policy* or rights relating to the personality'.
77 O. Guillod and G. Steffen, 'Art. 19/20 CO', in L. Thévenoz and F. Werro (eds), *Commentaire romand, Code des obligations I, Art. 1-529 CO* (Helbing Lichtenhahn, Basel 2021) N 66. This is, however, a protean definition, which is not always easy to define, E. Kramer, 'Art. 19-20 OR', in A. Meier-Hayoz (ed), *Berner Kommentar, Bd VI, 1. Abteilung, Allgemeine Bestimmungen, 2. Teilband, Unterteilband 1a, Inhalt des Vertrages, Kommentar zu Art. 19-22 OR* (Stämpfli, Bern 1991) N 151 ff.
78 Federal Constitution of the Swiss Confederation of 18 April 1999 (CC 101: 'Bundesverfassung', 'Constitution fédérale').
79 As such, SCT remain a matter of contract law (*Vertragstheorie*). For Swiss law, E. Kramer, T. Probst and R. Perrig, *Schweizerisches Recht der Allgemeinen Geschäftsbedingungen* (Bern 2016) N 241; D. Hug, *La formation du contrat de consommation : entre régime général et approche sectorielle – analyse et perspectives en droit suisse* (Basel/Neuchâtel, 2020) N 1427.
80 H. Becker, 'Art. 19 OR', in H. Becker (ed), *Berner Kommentar, Allgemeine Bestimmungen, Art. 1-183 OR* (Stämpfli, Bern 1941) N 24.
81 For an analysis of this emergence under French Law, A. Stevignon, *Le climat et le droit des obligations* (Paris 2022) N 288–293.
82 E. Kramer, 'Art. 19-20 OR', in A. Meier-Hayoz (ed), *Berner Kommentar, Bd VI, 1. Abteilung, Allgemeine Bestimmungen, 2. Teilband, Unterteilband 1a, Inhalt des Vertrages, Kommentar zu Art. 19-22 OR* (Stämpfli, Bern 1991) N 158.
83 In relation with an intergenerational contract theory, C. Poncibò, 'A Contract Law for Future Generations' (2020) 2 (2) *Revija Kopaonicke skole prirodnog prava*, 35.
84 See E. Kramer, T. Probst and R. Perrig, *Schweizerisches Recht der Allgemeinen Geschäftsbedingungen* (Bern 2016) N 277.
85 See A. Hellgardt and V. Jouannaud, 'Nachhaltigkeitsziele und Privatrecht' (2022) 222 *Archiv für die civilistische Praxis* 164, 182: 'Regulierung tritt neben andere Funktionen des Privatrechts, wie insbesondere den Interessenausgleich, der wesentlich der Verwirklichung der Privatautonomie dient und nach wie vor unbestritten die zentrale Aufgabe des Privatrechts darstellt', also with further references to divergent opinions.
86 For example, in the context of sales law, by adopting a legal right to repair for the buyer, which is currently still inexistent in Switzerland even in a B2C context.
87 Strictly speaking, the chosen example is not related to the 'control of content' of SCT (but rather to the 'contract formation', see G. Christandl, 'Art. 2:101 (1)', in N. Jansen and R. Zimmermann (eds), *Commentaries on European Contract Laws* (Oxford University Press, Oxford 2018) N 1. Nevertheless, the *causa* might highlight a complementary tool for an eco-literate analysis of the contract, as will precisely be discussed in the following paragraphs.
88 Code civil Belge du 4 avril 2019.
89 <https://www.lachambre.be/FLWB/PDF/55/1806/55K1806001.pdf> and <https://www.lachambre.be/FLWB/PDF/55/1805/55K1805001.pdf> both accessed 24 May 2023. See also G. Christandl, 'Art. 2:101

(1)', in N. Jansen and R. Zimmermann (eds), *Commentaries on European Contract Laws* (Oxford University Press, Oxford 2018) N 17.

90 Against the requirement of a *causa,* E.G. Lorenzen, 'Causa and Consideration in the Law of Contracts' (1918–1919) 28 *Yale Law Journal* 621, 644. It is interesting to note that French law on the contrary recently abandoned cause 'as a further requirement for contract formation', G. Christandl, 'Art. 2:101 (1)', in N. Jansen and R. Zimmermann (eds), *Commentaries on European Contract Laws* (Oxford University Press, Oxford 2018) N 17.

91 R. Feldkamp, A. François and L. Cornelis, 'Le droit privé au service d'une société durable : rêve ou réalité démocratique ?', in A. Van Hoe and G. Croisant (coord.), *Droit et durabilité/Recht en duurzaamheid, Actes publiés à l'occasion du colloque du 6 mai 2022 organisé par la Revue de droit commercial belge* (Bruxelles 2022) N 105, 142; U. Mattei and A. Quarta, *The Turning Point in Private Law – Ecology, Technology and the Commons* (Cheltenham and Northhampton 2019) N 93, 110, 119.

92 E.G. Lorenzen, 'Causa and Consideration in the Law of Contracts' (1918–1919) 28 *Yale Law Journal* 621, 625.

93 U. Mattei and A. Quarta, *The Turning Point in Private Law – Ecology, Technology and the Commons* (Cheltenham and Northhampton 2019) 93: 'In an ecological perspective, the parts are legally protected only as long as they concur with the well-being of the whole, so that a strong interpretation of internal limits to unrestrained selfishness such as causa is needed, not its abolition in a disruptive vision of a winner-takes-all society'.

94 The principle of good faith, however, is wider than the doctrine of *causa.* See K. Kryla-Cudna, 'Sales Contracts and the Circular Economy' (2020) 28 (6) *European Review of Private Law* 1207, 1215 points out: 'Since the principle of good faith is considered to apply at all stages of the life of a contract, including the stage of the exercise of the remedies for a breach (…)'.

95 K. Kryla-Cudna, 'Sales Contracts and the Circular Economy' (2020) 28/6 *European Review of Private Law* 1207, 1214–1215. See also E. Van Gool and A. Michel, 'The New Consumer Sales Directive 2019/771 and Sustainable Consumption: A Critical Analysis' (2021) 10 (4) *Journal of European Consumer and Market Law* 136, 145 with reference to giving a 'sustainability-driven interpretation to the principle of good-faith'.

96 See <https://www.lachambre.be/doc/FLWB/pdf/55/1805/55K1805004.pdf> accessed 24 May 2023: '(…)Il est toutefois impossible d'interdire les voyages en avion par le biais d'un article du Code civil. Le droit privé régit en effet les relations micro-économiques, tandis que le droit public est de nature macro-économique et confère au législateur le pouvoir d'action dont il a besoin. Le Code civil joue seulement un rôle de facilitateur'.

97 For example, L. Zampori and R. Pant, 'Suggestions for updating the Product Environmental Footprint (PEF) Method', *European Commission, JRC Technical Reports*, 2019, <https://eplca.jrc.ec.europa.eu/permalink/PEF_method.pdf> accessed 24 May 2023.

98 For example, should an 'unsustainable' term lead to the nullity of the whole contract, or to a partial nullity with the retention of its 'sustainable' terms, respectively, to the reduction of the 'unsustainable' clause? The same difficulties arise with respect to the UCTD. Based on the interpretation of the UCTD, the principle is setting aside unfair contract terms and a prohibition of revising them, with further ECJ case law references, Commission Notice, Guidance on the interpretation and application of Council Directive 93/13/EEC on unfair terms in consumer contracts (Text with EEA relevance), OJ C 323/ 4. But could and should an unfair contractual term be substituted by national law to eventually integrate ES considerations stemming from national law into the parties' right and obligations when the continuity of the contract is not possible without the unfair term? See Case C-26/13, *Arpád Kásler and Hajnalka Káslerné Rábai v. OTP Jelzálogbank Zrt*, ECLI:EU:C:2014:282 (an unfair term can under circumstances be substituted by a supplementary provision of national law); C-260/18, *Kamil Dziubak and Justyna Dziubak v. Raiffeisen Bank International AG*, ECLI:EU:C:2019:819 (no substitution of an unfair term with national rules of general nature) and C-80/21 to C-82/21 *E.K., S.K. v. D.B.P and B.S., W.S. v. M. and B.S., L.S. v. M.*, ECLI:EU:C:2022:646 (no interpretation of the parties' wishes when substituting an unfair term).

99 E.G. Lorenzen, 'Causa and Consideration in the Law of Contracts' (1918–1919) 28 *Yale Law Journal* 621, 644: 'The codes of Brazil, Germany, Japan and Switzerland, in omitting the requirement of a causa or consideration as an essential element of a contract, have in the estimation of the writer placed the subject of contracts upon a proper scientific basis and have freed it from doctrines which lead either to great confusion or to arbitrary and irrational results'.

100 See E.G. Lorenzen, 'Causa and Consideration in the Law of Contracts' (1918–1919) 28 *Yale Law Journal* 621, 633. See also <https://www.lachambre.be/FLWB/PDF/55/1805/55K1805001.pdf> accessed 24 May 2023.

101 The requirement of the *causa* also fulfils a policy function, G. Christandl, 'Art. 2:101 (1)', in N. Jansen and R. Zimmermann (eds), *Commentaries on European Contract Laws* (Oxford University Press, Oxford 2018) N 17. This remains the case if one discusses the need to abolish it or to (re)adopt it for environmental purposes.

102 U. Mattei and A. Quarta, *The Turning Point in Private Law – Ecology, Technology and the Commons* (Cheltenham and Northhampton 2019) N 119: 'This (i.e., the removal of the *causa*) was deemed an obstacle for the development of a "market-friendly" legal framework where corporations can benefit from environmental and social subsidies in their depredation of the global commons'. See also N. Graham, 'Teaching Private Law in A Climate Crisis' (2021) 40 (3) *University of Queensland Law Journal* 403 on the 'significative role of private law in facilitating the conditions of climate change'.

103 C. Poncibo, 'The Contractualisation of Environmental Sustainability' (2016) 12 *European Review of Contract Law* 335, 338 and further reference to K.P. Mitkidis, 'Sustainability Clauses in International Business Contracts', Den Haag (2015). See also E. Fisher, E. Scotford and E. Barritt, 'The Legally Disruptive Nature of Climate Change' (2017) 80 *The Modern Law Review* 173.

104 U. Mattei and A. Quarta, *The Turning Point in Private Law – Ecology, Technology and the Commons* (Cheltenham and Northhampton 2019) N 93.

Bibliography

Bach I and Kieninger E-M, 'Ökologische Analyse des Zivilrechts' (2021) 76 (22) *JuristenZeitung* 1069

Baetge D, 'Allgemeininteressen in der Inhaltskontrolle – Der Einfluss öffentlicher Interessen auf die Wirksamkeit Allgemeiner Geschäftsbedingungen' (2002) 202 *Archiv für die civilistische Praxis* 972

Becker H, 'Art. 19 OR', in H. Becker (ed), *Berner Kommentar, Allgemeine Bestimmungen, Art. 1-183 OR* (Stämpfli, Bern 1941)

Calliess C, 'Art. 11 AEUV', in C. Calliess and M. Ruffert (eds), *EUV – AEUV mit Europäischer Grundrechtecharta* (6th ed, Munich 2022)

Christandl G, 'Art. 2:101 (1)', in N. Jansen and R. Zimmermann (eds), *Commentaries on European Contract Laws* (Oxford University Press, Oxford 2018)

Feldkamp R, François A and Cornelis L, 'Le droit privé au service d'une société durable : rêve ou réalité démocratique ?', in A. Van Hoe and G. Croisant (coord.), *Droit et durabilité/Recht en duurzaamheid, Actes publiés à l'occasion du colloque du 6 mai 2022 organisé par la Revue de droit commercial belge* (Bruxelles 2022)

Fisher E, Scotford E and Barritt E, 'The Legally Disruptive Nature of Climate Change' (2017) 80 *The Modern Law Review* 173

Fuchs A, '§307 BGB', in P. Ulmer, H. E. Brandner and H.-D. Hensen (eds), *AGB-Recht Kommentar, §§305-310 BGB – Besondere Vertragstypen – UKlaG* (13th ed, Köln 2022)

Graham N, 'Teaching Private Law in A Climate Crisis' (2021) 40 (3) *University of Queensland Law Journal* 403

Guillod O and Steffen G, 'Art. 19/20 CO', in L. Thévenoz and F. Werro (eds), *Commentaire romand, Code des obligations I*, Art. 1-529 CO (Helbing Lichtenhahn, Basel 2021)

Halfmeier A, 'Nachhaltiges Privatrecht' (2016) 216 (5) *Archiv für die civilistische Praxis* 717

Hedemann-Robinson M, 'EC Law, the Environment and Consumers: Addressing the Challenge of Incorporating an Environmental Dimension to Consumer Protection at Community Level' (1997) 20 *Journal of Consumer Policy* 1

Hellgardt A and Jouannaud V, 'Nachhaltigkeitsziele und Privatrecht' (2022) 222 *Archiv für die civilistische Praxis* 164

Hondius E, *Unfair Terms in Consumer Contracts* (Utrecht 1987)

Hug D, *La formation du contrat de consommation : entre régime général et approche sectorielle – analyse et perspectives en droit suisse* (Basel/Neuchâtel, 2020)

Jacobs F, 'The Role of the European Court of Justice in the Protection of the Environment' (2006) 18 (2) *Journal of Environmental Law* 185

Jäggi P and Gauch P, 'Kommentar zu Art. 18 OR', in W. Schönenberger and P. Gauch (eds), *Zürcher Kommentar zum Schweizerischen Zivilgesetzbuch, V. Band, Obligationenrecht, Teilband V 1b* (Schulthess, Zürich 1980)
Jansen N, 'Art. 6:206', in N. Jansen and R. Zimmermann (eds), *Commentaries on European Contract Laws* (Oxford University Press, Oxford 2018)
Jansen N, 'Introduction before Art. 6:201', in N. Jansen and R. Zimmermann (eds), *Commentaries on European Contract Laws* (Oxford University Press, Oxford 2018)
Kaehler L, 'Policy Interventions via Contract Interpretation' (2014) 22 (5) *European Review of Private Law*, 641
Kramer E, 'Art. 19-20 OR', in A. Meier-Hayoz (ed), *Berner Kommentar, Bd VI, 1. Abteilung, Allgemeine Bestimmungen, 2. Teilband, Unterteilband 1a, Inhalt des Vertrages, Kommentar zu Art. 19-22 OR* (Stämpfli, Bern 1991)
Kramer E, Probst T, Perrig R, *Schweizerisches Recht der Allgemeinen Geschäftsbedingungen* (Bern 2016)
Krämer L, 'On the Interrelation between Consumer and Environmental Policies in the European Community' (1993) 16 (3–4) *Journal of Consumer Policy* 455
Kryla-Cudna K, 'Sales Contracts and the Circular Economy' (2020) 28 (6) *European Review of Private Law* 1207
Kye C, 'Environmental Law and the Consumer in the European Union' (1995) 7 (1) *Journal of Environmental Law* 31
Lorenzen E, 'Causa and Consideration in the Law of Contracts' (1918–1919) 28 *Yale Law Journal* 621, 644
Mahaim R, 'Art. 73 Cst.', in V. Martenet and J. Dubey (eds), *Commentaire romand de la Constitution fédérale* (Helbing Lichtenhahn, Basel 2021)
Mak V and Terryn E, 'Circular Economy and Consumer Protection: The Consumer as a Citizen and the Limits of Empowerment Through Consumer Law' (2020) 43 (1) *Journal of Consumer Policy* 227
Mattei U and Quarta A, *The Turning Point in Private Law – Ecology, Technology and the Commons* (Cheltenham and Northhampton 2019)
Michel A, 'Matelas et vêtements dans le même sac?' (2020) 126 *Droit de la consommation – Consumentenrecht*, 34
Michel A, *Premature Obsolescence: In Search of an Improved Legal Framework* (Intersentia, Leuven 2022)
Micklitz H-W, 'The Full Harmonization Dream' (2022) 11 (4) *Journal of European Consumer and Market Law* 117
Mitkidis K, *Sustainability Clauses in International Business Contracts* (Eleven International Publishing, Den Haag 2015)
Nowag J, 'The Sky Is the Limit: On the Drafting of Article 11 TFEU's Integration Obligation and Its Intended Reach', in B. Sjåfjell and A. Wiesbrock (eds), *The Greening of European Business under EU Law: Taking Article 11 TFEU Seriously* (Routledge 2015)
Pavillon C, 'Herijking van consumentencontractenrecht: duurzaamheid als nieuw ijkpunt?', in C. Pavillon and W.H. van Boom (eds), *Privaatrechtelijke bescherming herijkt. Preadviezen 2021 Uitgebracht voor de Vereniging voor Burgerlijk Recht* (Uitgeverij Paris, Zutphen 2021)
Pichonnaz P, 'Art. 8 LCD', in V. Martenet and P. Pichonnaz (eds), *Commentaire romand, Loi contre la concurrence déloyale* (Helbing Lichtenhahn, Basel 2017)
Poncibò C, 'The Contractualisation of Environmental Sustainability' (2016) 12 *European Review of Contract Law* 335
Poncibò C, 'A Contract Law for Future Generations' (2020) 2 (2) *Revija Kopaonicke skole prirodnog prava*, 35
Posner R, 'The Law and Economics of Contract Interpretation' (2004) 83 *Texas Law Review* 1581, 1582
Schulze R and Zoll F, *European Contract Law* (3rd ed, Baden-Baden 2021)
Sjåfjell B, 'The Environmental Integration Principle: A Necessary Step towards Policy Coherence for Sustainability', in F. Ippolito, M.E. Bartoloni and M. Condinanzi (eds), *The EU and the Proliferation of Integration Principles under the Lisbon Treaty* (Routledge 2018)
Stevignon A, *Le climat et le droit des obligations* (Paris 2022)
Terryn E, 'The New Consumer Agenda: A Further Step Toward Sustainable Consumption?' (2021), 10 (1) *Journal of European Consumer and Market Law* 1
Ulfbeck V and Hansen O, 'Sustainability Clauses in an Unsustainable Contract Law?' (2020) 16 (1) *European Review of Contract Law* 186

Van Gool E and Michel A, 'The New Consumer Sales Directive 2019/771 and Sustainable Consumption: A Critical Analysis' (2021) 10 (4) *Journal of European Consumer and Market Law* 136

Van Gool E, Michel A, Keirsbilck B and Terryn E, CCM KU Leuven Feedback to EU Commission Consultation on Sustainable Consumption of Goods – Promoting Repair and Reuse Legislative Initiative, 4 April 2022, <https://papers.ssrn.com/sol3/papers.cfm?abstract_id=4211301> accessed 24 May 2023

Wagner G, 'Klimaschutz durch Gerichte' (2021) 31 *Neue Juristische Wochenschrift* 2256

Wasmeier M, 'The Integration of Environmental Protection as a General Rule for Interpreting Community Law' (2001) 38 *Common Market Law Review* 159

Wendland M, '§307 BGB', Staudinger BGB, §§ 305–310

Wilhelmsson T, 'Consumer Law and the Environment: From Consumer to Citizen' (1998) 21 (1) *Journal of Consumer Policy* 45

Zampori L and Pant R, 'Suggestions for Updating the Product Environmental Footprint (PEF) Method', *European Commission, JRC Technical Reports*, 2019, <https://eplca.jrc.ec.europa.eu/permalink/PEF_method.pdf> accessed 24 May 2023

Zils H-P, *Die Wertigkeit des Umweltschutzes in Beziehung zu anderen Aufgaben der Europäischen Gemeinschaft: Untersuchungen zur Anwendung der Querschnittsklausel Art. 130r Abs. 2 Satz 2 EWGV im Gemeinschaftsrecht* (Decker, Heidelberg 1994) 27

3

ENVIRONMENTAL DELICTS AND POST-COMMUNIST PRIVATE LAW IN BULGARIA

Time for Reforms in View of UN's Sustainable Development Goals and the European Green Deal?

Radosveta Vassileva

3.1 Introduction

The right of citizens to a 'healthy and favourable environment' was enshrined in Article 55 of Bulgaria's democratic Constitution which was enacted in 1991, shortly after the fall of the country's communist regime in 1989. Nevertheless, frightening data shows that this constitutional obligation remains an empty slogan. Bulgaria traditionally faces high levels of air pollution – among the Member States of the European Union (EU); it has the highest annual mean concentration of fine particles in urban areas.[1] Local civil society is often shaken by reoccurring controversies about ecological disasters in the Black Sea.[2] Social advocates have raised awareness of how thermal power stations in Bulgaria pollute water with impunity.[3] A rare study on the quality of drinking water in Bulgarian cities by a reputed consumer association revealed that between one and three pollutants in quantities significantly above the legally acceptable limit were present in the drinking water of 14 out of the 29 examined cities.[4]

Regrettably, Bulgaria's EU membership has not led to a palpable change in the attitude of Bulgarian institutions towards the importance of environmental protection.[5] Infringement procedures by the European Commission before the Court of Justice of the European Union (CJEU), which concern the country's repeated violations of the Ambient Air Quality Directive, provide ample illustration.[6] In 2017, the CJEU held that Bulgaria had failed its obligations under this directive by exceeding the daily and annual limit values for PM^{10} 'systemically and continuously' in various parts of the country.[7] In 2020, the European Commission took Bulgaria to court again because it deemed that Bulgaria had not fulfilled its obligations to 'adopt and implement a series of measures' to comply with the 2017 CJEU judgement.[8] Meanwhile, in separate proceedings, in 2022, the CJEU established that Bulgaria had breached the directive by continuously exceeding the values of sulphur dioxide and failing to adopt appropriate measures to ensure compliance with EU requirements.[9]

More disturbingly, according to Eurostat, Bulgaria's performance is substantially below the EU average on most United Nations Sustainable Development Goals (SDGs).[10] The 17 SDGs were embraced by the 2030 Agenda for Sustainable Development which was adopted by all United

DOI: 10.4324/9781032662046-4

Nations (UN) Member States in 2015.[11] They purport to 'balance the three dimensions of sustainable development' – namely, 'the economic, social and environmental' one.[12] In turn, the SDGs have informed the European Green Deal,[13] which 'resets the [European] Commission's commitment to tackling climate and environmental-related challenges'.[14]

It has been argued that when it comes to environmental protection, 'public law is *far better placed* than private law to engage with the collective public interest [emphasis added]'.[15] However, the drama surrounding the violations of the Ambient Air Quality Directive in Bulgaria alone shows the limits of public law. Moreover, recent literature has tried to bust the myth that the relationship between environmental law and private law is 'necessarily antagonistic'.[16] It has been contended, for instance, that private law may indeed have a role to play in deterring or repairing environmental damage.[17] Furthermore, in countries which face severe rule of law backsliding[18] and rampant corruption,[19] such as Bulgaria, state institutions may be captured and may intentionally disregard the public interest.[20] Such institutions may purposefully ignore their responsibilities and turn a blind eye to violations of environmental legislation – such an example is examined in Section 3.3. In this context, private law has the potential to empower concerned citizens and civil action groups to litigate and try to push national judges – assuming at least some of them have remained independent– to take a stand, too.[21]

This chapter considers if Bulgarian private law may contribute to seeking solutions to the country's serious environmental problems and aligning Bulgaria better with SDGs, such as Clean Water and Sanitation (SDG6), Sustainable Cities and Communities (SDG11), and Life Below Water (SDG14). Unfortunately, the hands of citizens have been tied for years because the notion of environmental delict is not well understood in Bulgaria. The deficiencies of the Law on Obligations and Contracts of 1950, which governs tortious liability, as well as its messy relationship with laws governing environmental protection, may shed light on Bulgaria's lack of a steady track record of successful environmental litigation.[22] In view of promoting compliance with the SDGs and, respectively, the European Green Deal, the shortcomings of Bulgarian legislation are surely problematic.

First, this chapter highlights the importance of the procedural right of access to justice in environmental matters as it emerges from international and EU laws to better delineate the potential role which tort law can play in supporting environmental protection (Section 3.2). Then, it surveys the untidy notion of environmental delict in contemporary Bulgarian legislation, scholarly writing, and case law (Section 3.3). Finally, it suggests what reforms are necessary to help Bulgarian private law become a useful tool in promoting environmental protection and contributing to Bulgaria's better alignment with the UN's SDGs (Section 3.4).

3.2 Setting the Scene: Access to Justice in Environmental Matters and Tortious Liability

Before delving into the specific Bulgarian case, it seems relevant to explain the vital role assigned to access to justice in environmental matters by international and EU laws. One of the main goals of this procedural right of citizens is to ensure that the consequences of acts and omissions jeopardizing environmental protection can be reversed. Nevertheless, this procedural right also paves the way for citizens' holding wrongdoers liable in tort.

Considering that access to justice in environmental matters is a key value of both international and EU laws, it is not surprising that improving access to justice in environmental matters is also one of the priorities of the European Green Deal. Sadly, however, the EU's policies underestimate tort law's potential to contribute to environmental protection.

3.2.1 The Aarhus Convention

The EU and its 27 Member States are parties to the United Nations Economic Commission for Europe Convention on Access to Information, Public Participation in Decision-making, and Access to Justice in Environmental Matters (Aarhus Convention) of 1998. It has been asserted that this international instrument is 'ground-breaking in linking environmental rights and human rights'.[23] The main goal of the Aarhus Convention is to

> ...contribute to the protection of the *right* of every person of present and future generations to live in an environment adequate to his or her health and well-being...
>
> [emphasis added][24]

Article 9 defines the various rights to access to justice which should be granted to citizens. Article 9(3) of this convention, which is the most relevant to our study, requires that:

> members of the public have access to administrative or judicial procedures to challenge acts and omissions by *private persons and public authorities* which contravene provisions of its national law relating to the environment.
>
> [emphasis added]

From a private law perspective, it is surely notable that a convention promoting environmental protection purports to ensure that citizens have the right to litigate against other citizens rather than merely against public authorities. Moreover, one of the examples given in *The Aarhus Convention: An Implementation Guide* in relation to Article 9(3) is the following:

> ...individuals and environmental organizations that meet the national criteria may challenge a violation by a facility of wastewater discharge limitations in its permit. One means of challenging such a violation would be to take the owner or operator of the facility to court, claiming a violation of the law, and receive a remedy such as a court order to stop the illegal wastewater discharges.[25]

While there is no overt reference to tortious liability in the passage, the scenario described may, under certain conditions, fall in the realm of tort law in many jurisdictions – namely, if there is damage to property, a person's health, etc. Moreover, it should be remembered that injunctions are one of the two main remedies in tort along with damages in the common law. The debate on the importance of injunctive relief has paved its way to leading continental jurisdictions, too.[26] While it may sound unusual to public lawyers, the Aarhus Convention may be interpreted to indicate that tort law has some role to play in its enforcement. In fact, Section 3.3 shows an example of how Bulgarian civil society obtained injunctions against the Municipality of Sofia from a Bulgarian *civil* court, forcing the municipality to comply with environmental legislation and to take action against private actors violating the law.

It should also be stressed that the EU has adopted a number of directives and regulations which are in line with its obligations under the Aarhus Convention – the Access to Environmental Information Directive,[27] the Public Participation Directive,[28] and the so-called 'Aarhus Regulation' which implements the Aarhus Convention to EU institutions[29] and which was recently amended by another regulation.[30] Article 10 of the Aarhus Regulation allows non-governmental organizations meeting certain criteria to make requests for internal review to relevant EU institutions and bodies that have 'adopted an administrative act under environmental law' or, alternatively, have omitted to adopt such an act. The 2021 regulation, which amended the Aarhus Regulation, extended the prerogative under Article 10 to 'members of the public' meeting certain criteria.[31]

Meanwhile, Article 12 of the Aarhus Regulation states that non-governmental organizations, which have made requests for internal review, can institute proceedings before the CJEU in accordance with the relevant provisions of the EU Treaties. The most relevant provisions seem to be Articles 263 and 265 of the Treaty on the Functioning of the European Union (TFEU).[32] Pursuant to Article 263 TFEU, applicants may seek the annulment of a measure contrary to EU law. Pursuant to Article 265 TFEU, natural persons or artificial legal persons may bring an action against EU institutions or bodies for failure to act. In both scenarios, according to Article 266 TFEU, the EU institution or body found in violation is supposed to 'take the necessary measures to comply' with the CJEU judgement.

Finally, acts or omissions jeopardizing environmental protection may also pave the way to the application of the second paragraph of Article 340 TFEU in case damage has been caused:

> In the case of non-contractual liability, the Union shall, in accordance with the general principles common to the laws of the Member States, make good any damage caused by its institutions or by its servants in the performance of their duties.

The liability of the EU envisaged in this paragraph is tortious in nature. This specific type of tort is known under different names – in a study by the European Parliamentary Research Service, it was referred to as 'public tort law'.[33] In some East European jurisdictions, including Bulgaria, it is known as an administrative delict.

3.2.2 The European Commission's Recent Efforts to Foster Access to Justice in Environmental Matters on a National Level

In view of the efforts by EU institutions to foster accountability for administrative acts or omissions jeopardizing the environment on an EU level, it does not seem surprising that the European Green Deal attempts to promote access to justice and, as a consequence, compliance with the Aarhus Convention on a national level, too. After all, a chain is as strong as its weakest link.

Already in the communication on the European Green Deal, the European Commission promised to 'take action to improve [citizens'] access to justice before national courts in all Member States'.[34]

In a subsequent communication dedicated to enhancing access to justice in environmental matters, the European Commission stressed the importance of civil society in identifying potential breaches of EU law by submitting administrative complaints or taking cases to court.[35] To this end, the European Commission monitors citizens' access to justice in environmental matters via the Environmental Implementation Review (EIR) – a new reporting tool purporting to improve the implementation of EU environmental laws and policies,[36] which was embraced as playing 'a critical role in mapping the situation in each Member State' in the Communication on the European Green Deal.[37]

A quick look at the latest reports under the EIR, nevertheless, demonstrates that when monitoring access to justice in environmental matters, the European Commission has a very narrow understanding of this notion – it is primarily focused on the opportunities to challenge administrative acts and regulatory decisions in the realm of administrative law.[38] Regrettably, the European Commission does not seem to monitor the opportunities for civil society to challenge acts and omissions by 'private persons', as envisaged in Article 9(3) of the Aarhus Convention cited above.

Moreover, when it comes to tortious liability, under the EIR, the European Commission solely follows the implementation of the Environmental Liability Directive.[39] This directive enshrines a specific type of liability for damage caused to natural resources rather than to private property or a person's physical integrity and health. Here, it should be clarified that literature distinguishes between two types of environmental damage – damage to a private interest, which is the traditional subject of tort law, and '(pure) ecological damage ... where the harm is not to a private interest but to the environment itself'.[40] The directive is focused on the latter.

In other words, the EU's monitoring of access to justice in environmental matters seems to disregard the potential which traditional tort law has to foster environmental protection. This is unfortunate for in the latest communication on the EIR, the European Commission itself stresses that:

> [The EIR] is not simply one more report among many others. It raises public awareness, it builds the case for accountability, and it enables people across the EU to be part of joint effort by giving them a better understanding of what can be and/or still needs to be done at individual, local, national or EU level. The EIR seeks to empower all of these stakeholders to act coherently, but from different angles, in pursuit of the same common-sense objective: to apply the rules we have.[41]

In this light, relying on private law to help promote compliance with environmental legislation has some advantages:

- First, tort law allows injured parties to circumvent state institutions, which have abdicated from their duties to enforce environmental legislation, and directly engage the responsibility of private actors.
- Second, the threat of having to answer in damages in tort on its own may serve as an incentive for private actors to comply with environmental legislation.
- Third, as explained below, there are mechanisms at the blurry boundary between private law and administrative law, which can be employed to incentivize both public authorities and private actors to comply with their obligations via injunctions by national courts and at the threat of answering in damages.

3.3 The Untidy Notion of Environmental Delict(s) in Bulgarian Law

Now that we have examined the potential of tort law to promote compliance with environmental legislation, we focus on the amorphous notion of environmental delict, which emerges from Bulgarian legislation, scholarly writing, and case law. In the various pieces of legislation that can be relied upon to impose tortious liability for environmental damage surveyed below, delicts have different elements. This untidiness, coupled with other contextual factors, limits the opportunities of civil society to hold those responsible for environmental damage liable in tort.

3.3.1 Preliminary Remarks: The Messy Post-Communist Legal Landscape

To understand why the notion of environmental delict is untidy, however, one needs to appreciate the specific way in which Bulgarian private law developed. It has been asserted that:

> in the realm of [Bulgarian] private law, one observes a complex patchwork – creative compilations from jurisdictions that many jurists would find incompatible, which were sewn together in different political contexts.[42]

The development of Bulgarian private law can roughly be divided into three periods – re-establishment and development of the Bulgarian State after the Liberation from the Ottoman Empire (1878–1944), Communism (1944–1989), and Democracy (1989–present).[43] Each of them has left a mark on the complex legal landscape in the country today.

Moreover, it is helpful to know that Bulgarian legislators are often oblivious of the necessity for coherence, which usually leads to gaps and contradictions. One should also be aware that the line between public and private law in Bulgaria may be blurry – the main reason for this is that communism denied the Roman distinction between public and private law altogether.[44] The public/private law dichotomy was restored only partially post-1989.

3.3.2 Delicts in the Law of Obligations and Contracts

One of the vibrant patchworks, which is relevant to this study, is the current Law on Obligations and Contracts (LOC). It was enacted in 1950, during communism, for the needs of a planned economy in which there was no right to private initiative. The LOC bears a secret which has been discovered only recently due to years of communist censorship – while it harbours principles borrowed from Polish and Soviet laws, it is heavily based on the relevant sections on the law of obligations in the Italian Civil Code of 1942.[45] When reforming the law of obligations after the fall of communism in 1989 to adapt it to the needs of a market economy, legislators looked to the Bulgarian LOC of 1892, which was in force prior to communism, as well as to the contemporary laws of both East and West European jurisdictions, further enhancing the mélange.[46]

Articles 45–54 comprise the section on tortious liability in the LOC. While the rules on tort were not conceived with environmental damage in mind, three of them appear relevant to such cases. Article 45, which is regarded as a general and as a fall-back provision, states: ‘Everyone is obliged to repair the damage they have caused to others through their own fault. In all cases of tort, fault is presumed until proven otherwise’.

According to Article 49, anyone who assigns work to another is liable for damage caused by the other in performing the assignment. Article 50 enshrines the liability of owners and custodians of a thing or an animal for damage caused by the thing or the animal.

The most interesting provision from a comparative perspective is Article 49 of the LOC. I suspect that it was inspired by Article 2049 of the Italian Civil Code, which governs the liability of the ‘master’ for damage caused by his servant. One should remember, however, that the LOC’s authors had to comply with the values of communist ideology which denies such relationships of subjugation. That is why Article 49 is so neutrally phrased and refers to ‘assigned work’ which has a broad scope.

In an ironic twist, this provision started living a new life post-1989 which surely was not envisaged by the LOC’s drafters. It has been used to engage the responsibility of media which have published libellous content.[47] It has been relied upon to impose liability on hospitals for medical negligence.[48] In this light, it is interesting that contemporary Bulgarian authors consider that Article 49 of the LOC is best suited to hold wrongdoers liable for environmental damage.[49]

In essence, Article 49 of the LOC imposes liability on third parties who are not themselves at fault – it only requires (1) the existence of damage; (2) a causal link between the assigned work and the damage. Hypothetically, the defendant can either argue that the damage was not caused in relation to the assignment or that the performer of the assignment was not at fault, as required by Article 45 of the LOC. However, when it comes to large industrial facilities, which often face allegations of polluting, establishing who exactly has caused the damage may be difficult, if not impossible. Bulgarian (communist) courts have resolved such issues in favour of the claimant – pursuant to a decree on interpretation of 1959, the fact that the person who actually caused the damage cannot be identified does not preclude liability under Article 49.[50]

Because the liability of the one who delegates the work is not based on fault, Article 49 can be weaponized against artificial legal persons that do not comply with environmental legislation and pollute the environment causing damage to property or citizens' health. Article 49 may also hold the key to holding accountable powerful private actors who act with impunity because they are shielded by state institutions vested with responsibilities to enforce environmental legislation.

Unfortunately, to the best of my knowledge, Bulgaria has not seen a successful application of Article 49 of the LOC in environmental matters. At first glance, this appears suspicious. Nevertheless, it has been suggested that the main reasons for this include distrust in the judiciary and the financial expenses associated with such litigation.[51] While Bulgaria is the country with the *lowest median earnings* in the EU and more than 22% of Bulgarian citizens live below the national poverty line,[52] court fees are notoriously high – when bringing an action for damages, the claimant has to pay 4% of the amount it seeks as part of the court fees.[53] Crowdfunding is still relatively unpopular, too. Because of such contextual factors, below we will see that Bulgarian civil society has resorted to unexpected ways of enforcing environmental legislation – namely, via a class action suit against a municipality.

3.3.3 Delicts in Environmental Legislation

A survey of contemporary Bulgarian environmental legislation reveals a rich catalogue of environmental delicts, which were introduced in different contexts.[54] This section focuses on the Law on Environmental Protection (LEP) because it enshrines a general provision on liability for environmental damage.

3.3.3.1 Background

The first LEP was enacted in 1991. It was superseded by a law of the same name in 2002. The enactment of the first LEP indicated an important shift in legislative values – it was aligned with Article 55 of Bulgaria's democratic Constitution of 1991, which, as explained in the introduction to this chapter, protects the right to a 'healthy and favourable' environment.

Before the change of political regime in 1989, the attitude of the Bulgarian communist legislator to environmental damage was purely pragmatic. For instance, the Law on the Protection of Agricultural Property (the LPAP) of 1974 contains a detailed provision defining a specific type of delict to agricultural land. Its Article 29 stipulates:

> Anyone who pollutes air, water, and soil with his activity, making them unsuitable or dangerous for agricultural animals and birds or for the development of agricultural crops, perennials, and flower crops, is responsible for all the immediate consequences of the pollution.

However, one should remember that during communism, agricultural land was owned either by the state or by communist cooperatives because of the process of collectivization.[55] The verbatim report of the sitting at which the adoption of the LPAP was discussed shows that the communist legislator was mostly concerned about the fact that state-owned industrial facilities were damaging state-owned agricultural land and, as a consequence, were decreasing the production of communist cooperatives.[56] Examples given during the parliamentary debate include a mine whose activities led to arsenic levels 30 times above the norm, which, in turn, led to deaths of cattle, and 'dustiness' of air caused by a cement factory, which, in turn, was worsening the quality of the harvested plums, grapes, and apples.[57]

3.3.3.2 Liability for Environmental Damage in the LEP

The LEP, in contrast to other laws pertaining to environmental damage enacted during communism, echoes greater sensitivity towards the relationship between environmental damage and citizens' health. Article 29 of the original 1991 LEP stated: 'Whoever causes damages to another by fault because of pollution or harm to the environment shall be obliged to compensate them'.

The current 2002 LEP contains an identical provision in its Article 170(1). Moreover, the current LEP has preserved the definition of 'pollution to the environment' of the original LEP:

> the deterioration of the qualities and as a result of the occurrence and introduction of physical, chemical or biological factors from a natural or anthropogenic source in the country or outside it, *regardless of whether the norms in force in the country are exceeded.*
>
> [emphasis added][58]

Its definition of 'harm to the environment' is also the same as the definition in the original LEP:

> such a change to one or more of its constituent components, which leads *to a deterioration of the quality of life of people*, to the impoverishment of biological diversity or to the difficult restoration of natural ecosystems.
>
> [emphasis added][59]

In principle, the definitions of 'pollution to the environment' and 'harm to the environment' make Article 170(1) relevant to a wide range of factual circumstances, enabling both citizens and public authorities to seek its application. It encompasses the traditional notion of environmental damage as damage to a private interest and the more modern conception of environmental liability for pure ecological damage which made its way to the Environmental Liability Directive mentioned in Section 3.2.2.

Sadly, however, comparing the provisions of the 2002 LEP and the 1991 LEP shows that the 2002 reform sought to curtail civil society's involvement in environmental matters. Namely, Article 30 of the 1991 LEP allowed not only those who had suffered damage, but also municipal authorities, non-profit associations of citizens, or 'any citizen' to submit a claim demanding the cessation of the violation of environmental legislation and the elimination of all consequences of the pollution. Article 171 of the new LEP, in stark contrast, allows such a claim *only* by those who have suffered damage or, in case public property was damaged, by the minister of environment, the regional governor or the mayor depending on the case.

Such limitations on who can seek an injunction for cessation do not seem justified. First, it is more likely that experts or civil action groups have better knowledge of environmental legislation, which is often very technical, than ordinary citizens, so they are better placed to submit a well-argued claim regarding what injunction is necessary. Second, there are citizens who may not be directly injured but who may have a vested interest in seeking the cessation of pollution – personal convictions, such as environmental consciousness, may play a role. Third, in case public property is damaged, the minister, governor, or mayor may be negligent, so concerned citizens may be more motivated to seek injunctions.

3.3.3.3 Interpretation Difficulties

In principle, the section titled 'Civil Liability' in the current LEP, which comprises Articles 170–172, provides ample examples of the sloppiness of the Bulgarian legislator – it is short and poorly drafted. For instance, while Article 170 mentions 'fault', it does not specify if it is presumed or not. Moreover, it does not stipulate the type/degree of fault – whether there should be intentional actions/omissions, gross negligence, or simply negligence. There is no overt indication regarding the instances when damage is excused, either – force majeure is an obvious omission. In other words, courts are left with the only option of construing this section by analogy to other rules – for instance, the general rules on tort in the LOC and the case law pertaining to them. Yet, as this is not explicitly specified in the LEP, the relationship between the LEP and the LOC remains messy, which may give rise to interpretation difficulties.

An example is provided by the scarce case law on Article 170 of the LEP. In 2013, the Supreme Court of Cassation quashed a decision by the Varna Appellate Court which had granted damages to a claimant under Article 170(1) of the LEP in conjunction with Article 49 of the LOC.[60] The claimant was suing the Municipality of Varna for pollution on their land caused by construction waste discarded by a contractor of the municipality. In the eyes of the Supreme Court of Cassation, neither the LOC was applicable, nor civil courts had jurisdiction to examine the case.[61] It referred the case for re-examination to the administrative courts, arguing the facts had to be evaluated in light of Article 170 of the LEP in conjunction with Article 1(1) of the Law on the Liability of the State and Municipalities for Damage (LLSMD).

As explained in more detail in Section 3.3.5, administrative justice is a recent development in Bulgaria, so disputes regarding the jurisdiction of courts are common. It is already revealing that both the first instance and the second instance entertained a different view from the Supreme Court of Cassation regarding their jurisdiction. Moreover, the fact that Article 170 of the LEP will be interpreted by different courts (civil and administrative) and in conjunction with different general rules on tortious liability entails the danger of creating a multiplicity of standards.

Finally, the LEP has a hidden strength, which has not been tested before Bulgarian courts yet. No matter whether the liability under Article 170 is construed against the LOC or the LLSMD, Article 171 of the LEP allows both civil and administrative courts to issue injunctions for cessation of the violation. The Bulgarian legislator has extended the available remedies for environmental damage beyond the traditional damages in tort, albeit clumsily. The potential of this interpretation of Article 171 should not be underestimated. One can envisage a scenario where a local factory pollutes waters. The owner of even a small orchard, which has been damaged, may submit a strategic small claim for damages on the grounds of Article 170 of the LEP and the rules on tortious liability in the LOC, along with a demand for injunctions to cease pollution under Article 171 of the LEP – in this way, while formally defending his/her private interest, he/she will be defending the public interest in limiting pollution.

3.3.4 Class Action and Tortious Liability

The messiness of Bulgarian traditional tort law may explain why Bulgarian civil society has resorted to creative legal means to engage the responsibility of wrongdoers. The provisions of the Code of Civil Procedure (CCP) of 2007 pertaining to class action suits have proven useful in obtaining mandatory injunctions by the Sofia City Court against the Municipality of Sofia, the capital of Bulgaria, to force it to comply with and enforce environmental legislation – what became known as the *Group for Clean Air* case discussed below.

3.3.4.1 The CCP

The current CCP was drafted by a working group composed of experts from Austria and Bulgaria.[62] Yet, neither Bulgarian law prior to this reform, nor Austrian law disposed of provisions similar in spirit to Articles 379 and subsequent of the CCP governing class action suits. In principle, the CCP has been criticized for being ill-written.[63] In the field of consumer law, the application of the provisions on class action has resulted in disappointing outcomes, depriving injured parties of justice.[64] In this light, it is surprising that Articles 379 and subsequent could become a helpful tool promoting environmental protection. That is why, before explaining the details of the case, it is useful to explain why the rules on class action in the CCP must have appealed to those who initiated the proceedings.

Article 379(1) of the CCP, which defines the scope of class action, is vaguely worded: 'A class action may be brought on behalf of persons damaged by a violation, when, according to the nature of the violation, their circle cannot be determined precisely, but is determinable'.

It only requires that a determinable circle of persons be injured by a violation. Since it does not specify the nature of the violation, it is applicable to both contractual and extra-contractual contexts. In addition, Article 379(2) allows both injured parties and organizations representing them to initiate such proceedings, so the standing threshold appears low.

Furthermore, Article 379(3) of the CCP specifies:

> Persons who claim that their collective interest is damaged or threatened by a violation under paragraph 1, or an organisation, which protects the injured persons, the injured collective interest or against such violations, may bring a collective action on behalf of all injured persons against the violator for the cessation of the violation, the remedying of the consequences of the violation of the injured collective interest, or compensation for the damage caused to that interest.

There are three key aspects which seem crucial here. First, one may either prove that damage has occurred *or* that there is a mere threat of damage taking place. Second, the Article refers to a 'violator' without putting any restrictions on who they might be – that is why, while it may sound unusual, it was possible to take the Municipality of Sofia to a civil court. Third, one chooses what remedies to ask for. Relatedly, Article 385(1) of the CCP pertaining to injunctions in class action suits states: 'The court may sentence the defendant to perform a certain act, not to perform a certain act, or to pay a certain amount'.

In other words, the claimants may choose not to ask for damages but aim at prohibitive or mandatory injunctive relief instead. This is important because, as mentioned above, Bulgarian court fees are notoriously high – when bringing an action for damages, the claimant must pay 4% of the amount it seeks as part of the court fees.

3.3.4.2 The Group for Clean Air Case

The Group for Clean Air is an informal initiative, which unites civil organizations and concerned citizens who want to improve the air quality in Bulgaria. It is important to mention that Sofia, the capital of Bulgaria, has severe problems with air pollution and that the Municipality of Sofia is often accused of inaction (ignoring its obligations under the law) and alleged corrupt practices by prominent members of Bulgarian civil society.

3.3.4.2.1 BRAVE CLAIMANTS, BRAVE JUDGE

In 2017, members of the Group for Clean Air initiative submitted a claim against the Municipality of Sofia pursuant to Articles 379 and subsequent of the CCP before the Sofia City Court, arguing that the municipality had repeatedly violated the Ambient Air Quality Directive through inaction and lack of due care.[65] The claimants identified the determinable circle of injured parties as 'citizens living, working or studying' on the territory of the Sofia municipality in the period 2015–2017 as well as the heirs of those who have died because of dirty air.[66] They asked the court to establish unlawful inaction by the municipality, sentence it to put an end to the inaction, and sentence it to undertake actions to remedy or mitigate the consequences of its inaction.

Beyond the unusual choice of legal instrument used to force the Municipality of Sofia to comply with its obligations under the law, there are a few notable features of this litigation, which may be interesting in the EU context. One of the pieces of evidence used by the claimants to prove the repeated violations by the Municipality of Sofia was CJEU's 2017 judgement in Case C-488/15 establishing that Bulgaria had breached the Ambient Air Quality Directive, including in the Sofia area, cited in the introduction of this chapter.[67] As visible from the decision handed down in the *Group for Clean Air* case in 2021, the Sofia City Court relied on this CJEU judgement to conclude that it was 'objectively established' that Bulgaria had breached the directive.[68]

The Sofia City Court referred to the relevant sections of the Law on the Cleanliness of Atmospheric Air, which transposes the Ambient Air Quality Directive, which indicate that municipal bodies are competent to control and manage the activities related to ensuring the cleanliness of air on their territory.[69] Then, based on the relevant legislation and a court-appointed expert assessment, the Sofia City Court established the nature and scope of the omissions by the Municipality of Sofia which had endangered the health of the citizens, upholding most of the claim by the Group for Clean Air.[70]

While the Sofia City Court recognized that the type of measures necessary to remedy the consequences of omissions is, in principle, in the competences of the municipality, the circumstances demanded mandatory injunctions vis-à-vis such measures. The Sofia City Court not only continued the preliminary mandatory injunctions it had issued when starting to examine the case on the merits in 2019 but also issued new ones. In general, the injunctions by the Sofia City Court concern very specific details – for instance, what section on its website the Municipality of Sofia must maintain, what information it has to provide to the public, what research and planning it should undertake, how often it has to clean the streets when it has to enforce parking regulations, etc. In its decision, the court also gave the municipality a deadline for compliance.[71]

3.3.4.2.2 THE IMPLICATIONS

The fact that the Sofia City Court chose to impose concrete rather than abstract mandatory injunctions merely stating that the municipality had to remedy the consequences of its inaction is notable. In principle, the CJEU leaves the choice of measures to comply with its judgements on the violating Member State – for instance, the burden to comply with the judgement in Case C-488/15 was on the Bulgarian state authorities. However, vigilant members of the Sofia civil society anticipated that Bulgaria would be unwilling to comply with this judgement – as explained in the introduction to this chapter, the European Commission is now suing Bulgaria for not complying with the same judgement – and entered the courtroom to attempt to force at least the Municipality of Sofia to fulfil its obligations under the law. On its own, the *Group for Clean Air* case shows not only how important civil society may be in the enforcement of EU law, including CJEU judgements, but also how helpful national legislation pertinent to tortious liability is to this end.

Yet, the case also illustrates the risks of civil society involvement in a country facing rule of law backsliding. The case took so long to examine at the first instance (2017–2021) because three judges prior to the judge who ended up examining it recused themselves – this is indirect evidence of political pressure behind the scenes.

Moreover, the Municipality of Sofia tried to put tremendous financial pressure on the claimants. Because of the outcome of the case – part of the claim was upheld while part of it was not – the defendant had to compensate the claimant for the expenses incurred to sustain the successful part of the claim while the claimants had to compensate the defendant for the expenses made in defending itself against the unsuccessful part of the claim.[72] The Municipality of Sofia sought reimbursement of 108,000 Bulgarian leva (54,000 EUR) for legal advice. This is exorbitant for Bulgarian standards considering the country has the lowest median earnings in the EU – the *average monthly salary* in the country in 2021 was barely 800 EUR. Luckily for the claimants, the judge ordered that they reimburse only 2,649.57 Bulgarian leva (1,324.785 EUR).[73]

The exorbitant claim for the reimbursement of expenses by the Municipality of Sofia can easily be interpreted as an attempt to deter future claims by civil society under the CCP. The claimants were indeed fortunate that after three self-recusals, their case was distributed to a judge who was not easily influenced behind the scenes.

Finally, the rules governing class action suits in the CCP can clearly be relied upon in diverse circumstances to seek injunctions against private actors or public authorities. However, there are at least two caveats. Sofia is the capital of Bulgaria and the most affluent city, so it has an active civil society. In less developed areas, civil society is still in the process of emerging, so it is unlikely that they would take such a brave step – taking a municipality or a powerful private actor to court could be daunting. Second, if the Municipality of Sofia may abuse its right to compensation of expenses, one can reasonably suspect that other defendants may do so in bad faith too – this prospect may have a petrifying effect on those considering litigating.

3.3.5 Administrative Delicts

In Bulgaria, state institutions and municipalities are often complicit with those polluting the environment. Hence, it is worth examining if the rules on liability for administrative delict can be helpful in holding such institutions liable for environmental damage. As explained in Section 3.3.4, it is highly likely that the claimants in the *Group for Clean Air* case did not ask for damages because the court fee would have been exorbitant. This is where seeking liability under the LLSMD has a strategic advantage – claims under this law are exempt from the 4% of money claim court fee. One simply pays a small flat fee.

The LLSMD has a peculiar communist heritage. It was enacted in 1988. However, the Western reader should not be mistaken to believe that the totalitarian regime hoped to hold itself accountable – this was a mere move to pretend compliance with the International Covenant on Civil and Political Rights, which Bulgaria had ratified in 1970.[74] The LLSMD became a powerful tool for promoting accountability only after amendments in 2006. These amendments coincided with an overhaul of the Bulgarian court system, introducing two separate sets of courts – civil and administrative.

Article 1(1) of the LLSMD states:

> The state and the municipalities are responsible for the damage caused to citizens and artificial legal persons by unlawful acts, actions or inactions of their bodies and officials during or on the occasion of the performance of administrative activities, as well as for the damage caused by the operation of acts that were declared unlawful or regulations that were declared null and void.

The liability envisaged in this provision is not based on fault similarly to the liability in Article 49 of the LOC – one needs to prove damage and a causal link between the damage and the actions/omission of the institution in question. It is also noteworthy that determining the competent court (civil or administrative) does not depend on the personality of the defendant (whether they are a state institution, etc.), but primarily on whether the contested behaviour (acts, actions, inactions) falls under the scope of 'administrative activities' mentioned in Article 1(1) of the LLSMD. If it does not, the claimant must sue pursuant to the rules on tort in the LOC.

In this light, if the Municipality of Sofia does not comply with the injunctions in the *Group for Clean Air* case, it is interesting to consider if affected citizens will be able to ask for damages under the LLSMD. The answer is not straightforward because many of the injunctions in the case require actions that do not necessarily fall under the scope of administrative activity – for instance, research.

3.4 Conclusions and Recommendations for Law Reform

As a country facing serious environmental problems, Bulgaria illustrates the limits of public law in ensuring environmental protection. Even CJEU judgements have not sufficiently motivated Bulgarian institutions to enforce environmental legislation. That is why, it is worth considering if tort law may contribute to the prevention and repair of environmental damage and, as a consequence, better alignment of Bulgaria with UN's SDGs and the European Green Deal.

An examination of both the international and the EU frameworks pertaining to access to justice in environmental matters indicates that tort law may promote environmental protection by means of injunctions. The threat of having to answer in damages may further incentivize private actors to comply with environmental legislation and motivate public authorities to fulfil their obligations to enforce environmental legislation. Regrettably, while the European Commission recognizes the significant role of civil society in supporting environmental efforts, it seems to ignore the potential of tort law to discipline those harming the environment through their actions or omissions.

The legislation of Bulgaria contains a rich catalogue of delicts, including explicit provisions on environmental delicts. However, Bulgarian tort law developed in a piecemeal fashion and in different political contexts – it is messy and challenging to apply. As a post-communist country, Bulgaria needs reforms aimed at modernizing legislation and improving its coherence.

In its latest communication on the EIR, the European Commission emphasized:

> While we each have a part to play, the *political will* is the crucial ingredient for governments and decision-makers to drive the timely, correct and efficient implementation of EU environmental policies and regulations, achieve their objectives and reap their benefits.[75]

A similar conclusion can be drawn regarding the feasibility of law reform which can empower citizens to seek justice in environmental matters – such reform depends on the will of the

Bulgarian National Assembly and the Council of Ministers. While realizing that reforms aimed at developing tort law are not a high priority on Bulgaria's agenda, I argue that the following amendments to legislation and bylaws should be considered in view of improving access to justice in environmental matters and unleashing the potential of tort law to promote environmental protection.

3.4.1 Broadening the Scope of Injunctive Relief

Injunctions may play a vital role in the enforcement of environmental legislation. Nevertheless, as seen above, the Bulgarian approach to injunctive relief is piecemeal. The provision of Article 385(1) of the CCP pertaining to injunctions in class action suits, which allows the courts to sentence the defendant to perform a certain act, not to perform a certain act, or to pay a certain amount, may inspire amendments to the general rules on tort.

As explained above, the LOC envisages only damages as a remedy for tort. In many cases, however, even going beyond environmental matters, obtaining an injunction may be more valuable to a claimant than receiving damages. Introducing a provision similar in spirit to Article 385(1) of the CCP to the LOC will provide courts with more flexibility to address more adequately a wider range of factual circumstances. Relatedly, injunctive relief could be explicitly provided for administrative delicts via an amendment to the LLSMD.

3.4.2 Expanding the Circle of Those with Standing before the Court

As explained above, the 2002 LEP clipped the wings of civil society by limiting the circle of who can demand an injunction for cessation of pollution/harm to the environment. In this light, the provision of the 1991 LEP, which bestowed this right upon non-profit associations and 'any citizen' in addition to harmed citizens, is to be preferred – its reintroduction to legislation requires a minimal amendment to the 2002 LEP.

3.4.3 Reconsidering Court Fees

As argued above, the 4% of money claim court fee deters citizens from defending their rights in court. This 4% court fee, for instance, almost certainly guarantees that damages in a class action suit will not be sought. Meanwhile, as explained above, claims against a municipality under the LLSMD are exempt from such a fee, while claims against a municipality under the LOC are not. This distinction is arbitrary. Moreover, the concept of 'administrative activity' is fuzzy.

In this light, the Tariff for State Fees Collected by the Courts under the Civil Procedure Code adopted by the Council of Ministers should be revisited.[76] The Council of Ministers may consider a more nuanced approach – the proportional fee in class action suits could be replaced with a reasonable flat fee. As a sign of goodwill, the Council of Ministers may exempt all claims against state institutions and municipalities from the 4% of money claim fee and merely impose a reasonable flat fee, irrespective of the legal grounds of the claim (be it the LOC, the LLSMD, etc.).

3.4.4 Curtailing Abuses of the Rules on Reimbursement of Costs

The exorbitant claim for reimbursement of fees for legal advice by the Municipality of Sofia in the *Group for Clean Air* case calls for a reflection on how to curtail such bad faith behaviour in the future. Article 78(5) of the CCP contains a vaguely phrased provision allowing the reduction of exorbitant claims for reimbursement of expenses. However, such reduction depends on the discretion of the judge. Meanwhile, Article 248(1) of the CCP allows any of the parties to a dispute to request that a court supplement or amend the part of a judgement which concerns the reimbursement of costs. Nevertheless, the court examines such requests in camera, without the participation of the parties.

The vagueness and the non-transparency of the mechanism for challenging excessive costs seem problematic because Bulgaria faces long-standing challenges to judicial independence. Moreover, Bulgaria has an established track record of violating Article 6(1) of the European Convention on Human Rights on the right to a fair trial via exorbitant fees in legal proceedings.[77]

Thus, the objectivity and transparency of the mechanism for the appeal of orders on costs need to be enhanced. Bearing in mind the historic record of controversies, such appeals should be heard in open court. Furthermore, legislators may consider if the expertise of Bulgaria's Supreme Bar Council may not be of help in such proceedings. For instance, the CCP may be amended to require that the Supreme Bar Council and/or the local bar association submit a non-binding opinion regarding the reasonableness of the fees for legal advice to the court. While this approach may be criticized for increasing the length of the proceedings in a country already known for slow justice, it may foster public trust in the courts by partially alleviating the concerns that decisions are taken arbitrarily behind closed doors.

Notes

1 *Eurostat*, 'How Polluted Is the Air in Urban Areas?' (3 June 2021) <https://ec.europa.eu/eurostat/web/products-eurostat-news/-/edn-20210603-1> accessed 1 December 2022.

2 See, for instance, Krassen Nikolov, 'Black Sea Faces Ecological Catastrophe due to Bulgaria's Inaction' (*Euractiv*, 30 September 2021) <https://www.euractiv.com/section/politics/short_news/black-sea-faces-ecological-catastrophe-due-to-bulgarias-inactio/> accessed 1 December 2022.

3 Desislava Mikova, 'Polluting Water with Impunity' (*Offnews*, 15 June 2021) <https://offnews.bg/obshtestvo/beznakazanoto-zamarsiavane-na-vodite-koeto-niama-koj-da-spre-753595.html> accessed 1 December 2022.

4 *Active Consumers Association*, 'Water Quality in 29 Bulgarian Cities' (2015) <https://aktivnipotrebiteli.bg/%d1%82%d0%b5%d1%81%d1%82/149/%d0%a2%d0%95%d0%a1%d0%a2-%d0%bd%d0%b0-%d0%bf%d0%b8%d1%82%d0%b5%d0%b9%d0%bd%d0%b0-%d0%b2%d0%be%d0%b4%d0%b0-(2015)> accessed 1 December 2022.

5 Bulgaria joined the EU on 1 January 2007. As an EU Member State, it has transposed the environmental acquis.

6 Directive 2008/50/EC of the European Parliament and of the Council of 21 May 2008 on ambient air quality and cleaner air for Europe [2008] OJ L152/1.

7 PM stands for particulate matter. PM[10] refers to small particles with an aerodynamic diameter smaller than 10 μm; Case C-488/15 *European Commission v. Bulgaria* [2017] ECLI:EU:C:2017:267.

8 *European Commission*, 'Air Quality: Commission Decides to Refer Bulgaria to the Court of Justice over Its Failure to Comply with Previous Judgement. Press release IP/20/2150' (3 December 12 2020) <https://ec.europa.eu/commission/presscorner/detail/en/ip_20_2150> accessed 3 December 2022. At the time of writing of this chapter, the case is pending. See Case C-174/21 *European Commission v. Bulgaria.*

9 Case C-730/19 *European Commission v. Republic of Bulgaria* [2022] ECLI:EU:C:2022:382.

10 *Eurostat*, 'SDG Country Overview' <https://ec.europa.eu/eurostat/cache/infographs/sdg-country-overview/> accessed 1 December 2022.

11 United Nations, 'A/RES/70/1. Transforming Our World: the 2030 Agenda for Sustainable Development' (25 September 2015) 14 and ff.

12 United Nations, 'A/RES/70/1. Transforming Our World: the 2030 Agenda for Sustainable Development' (25 September 2015) 1.

13 In the eyes of the European Commission, 'the Green Deal is an integral part of [its] strategy to implement the United Nation's 2030 Agenda'. See Commission, 'Communication from the Commission to the

European Parliament, the European Council, the Council, the European Economic and Social Committee and the Committee of the Regions on the European Green Deal' COM (2019) 640 final, 11.12.2019, p. 3.

14 'Communication from the Commission to the European Parliament, the European Council, the Council, the European Economic and Social Committee and the Committee of the Regions on the European Green Deal' COM (2019) 640 final, 1.

15 Fiona Donson and Robert Lee, 'Environmental Protection: Public or Private Law?' [1996] 1(1) *Judicial Review* 56, 58 <https://doi.org/10.1080/10854681.1996.11426861>.

16 David Howarth, 'Ch.48: Environmental Law and Private Law' in Emma Lees and Jorge E. Viñuales (eds), *The Oxford Handbook of Comparative Environmental Law* (Oxford, Oxford University Press 2019) 1091, 1117 <https://doi.org/10.1093/law/9780198790952.003.0048>.

17 David Howarth, 'Ch.48: Environmental Law and Private Law' in Emma Lees and Jorge E. Viñuales (eds), *The Oxford Handbook of Comparative Environmental Law* (Oxford, Oxford University Press 2019) 1091, 1096 <https://doi.org/10.1093/law/9780198790952.003.0048>.

18 In the latest Rule of Law Index by the World Justice Project, Bulgaria has received the second worst score in the EU after Hungary; World Justice Project, *Rule of Law Index 2022* (Washington DC, World Justice Project 2022) <https://worldjusticeproject.org/rule-of-law-index/downloads/Index-2022.pdf> accessed 1 December 2022.

19 The latest 2021 Corruption Perceptions Index by Transparency International ranked Bulgaria 78th in the world – Bulgaria has received a similar score to Belarus; Transparency International, *Corruption Perceptions Index 2021* <https://www.transparency.org/en/cpi/2021> accessed 1 December 2022.

20 Captured states are those in which public power is exercised for private gain. For captured states in the EU, see Abbey Innes, 'The Political Economy of Captured States' [2014] 52(1) *Journal of Common Market Studies* 88–104 <https://doi.org/10.1111/jcms.12079>.

21 The lack of independence of the judiciary is a traditional source of concern in Bulgaria. See European Parliament resolution of 8 October 2020 on the rule of law and fundamental rights in Bulgaria, P9_TA(2020)026 <https://www.europarl.europa.eu/doceo/document/TA-9-2020-0264_EN.html> accessed 4 December 2022.

22 For the sake of terminological clarity, it should be noted that in this chapter, 'tort' and 'delict' are used synonymously. Bulgarian legal terminology in this field is imprecise too – in the 19th century, Bulgarian legislation referred to delicts as 'semi-crimes' (*poluprestupleniya*). In contemporary law, delicts are known as *nepozvoleno uvrejdane*, which can be translated as 'unallowed harm'. The latter appears paradoxical for harm should not be allowed in principle.

23 Áine Ryall, 'The Aarhus Convention: Standards for Access to Justice in Environmental Matters' in Stephen J. Turner et al (eds), *Environmental Rights: The Development of Standards* (Cambridge, Cambridge University Press 2019) 116 <https://doi.org/10.1017/9781108612500.006>.

24 Art. 1 of the Aarhus Convention.

25 United Nations, *The Aarhus Convention: An Implementation Guide* (2nd edn, Geneva, United Nations 2014) 199.

26 Paula Giliker, 'La cessation de l'illicite' [2019] 4 *Revue des contrats* 284.

27 Directive 2003/4/EC of the European Parliament and of the Council of 28 January 2003 on public access to environmental information and repealing Council Directive 90/313/EEC [2003] OJ L41/26.

28 Directive 2003/35/EC of the European Parliament and of the Council of 26 May 2003 providing for public participation in respect of the drawing up of certain plans and programmes relating to the environment and amending with regard to public participation and access to justice Council Directives 85/337/EEC and 96/61/EC [2003] OJ L156/17.

29 Regulation (EC) No 1367/2006 of the European Parliament and of the Council of 6 September 2006 on the application of the provisions of the Aarhus Convention on Access to Information, Public Participation in Decision-making and Access to Justice in Environmental Matters to Community institutions and bodies [2006] OJ L264/13.

30 Regulation (EU) 2021/1767 of the European Parliament and of the Council of 6 October 2021 amending Regulation (EC) No 1367/2006 on the application of the provisions of the Aarhus Convention on Access to Information, Public Participation in Decision-making and Access to Justice in Environmental Matters to Community institutions and bodies [2021] OJ L356/1.

31 Regulation (EU) 2021/1767 of the European Parliament and of the Council of 6 October 2021 amending Regulation (EC) No 1367/2006 on the application of the provisions of the Aarhus Convention on Access to Information, Public Participation in Decision-making and Access to Justice in Environmental Matters to Community institutions and bodies [2021] OJ L356/1.

32 See Consolidated Version of the Treaty on the Functioning of the European Union [2012] OJ C326/47.

33 Rafal Mańko, 'Action for Damages against the EU' (2018) *European Parliamentary Research Service* <https://www.europarl.europa.eu/RegData/etudes/BRIE/2018/630333/EPRS_BRI(2018)630333_EN.pdf> accessed 5 December 2022.

34 Commission, 'Communication from the Commission to the European Parliament, the European Council, the Council, the European Economic and Social Committee and the Committee of the Regions on the European Green Deal' COM (2019) 640 final, 23.

35 Commission, 'Communication from the Commission to the European Parliament, the Council, the European Economic and Social Committee and the Committee of the Regions on Improving Access to Justice in Environmental Matters in the EU and its Member States' COM (2020) 643 final, 3.

36 The Commission has published monitoring reports under this mechanism in 2017, 2019, and 2022.

37 Commission, 'Communication from the Commission to the European Parliament, the European Council, the Council, the European Economic and Social Committee and the Committee of the Regions on the European Green Deal' COM (2019) 640 final, 23.

38 See, for instance, Commission, 'Environmental Implementation Review 2022. Country Report – Bulgaria' SWD (2022) 262 final, 42–43.

39 Commission, 'Environmental Implementation Review 2022. Country Report – Bulgaria' SWD (2022) 262 final, 44–45; see also Directive 2004/35/CE of the European Parliament and of the Council of 21 April 2004 on environmental liability with regard to the prevention and remedying of environmental damage [2004] OJ L143/56.

40 Gerhard Wagner, 'Environmental Liability' in Jürgen Basedow et al (eds), *Max Planck Encyclopedia of European Private Law* (Germany, Max Planck Institute 2012) <https://max-eup2012.mpipriv.de/index.php/Environmental_Liability> accessed 5 December 2022.

41 Commission, 'Communication from the Commission to the European Parliament, the Council, the European Economic and Social Committee and the Committee of the Regions on Environmental Implementation Review 2022: Turning the tide through environmental compliance' COM (2022) 438 final, 22.

42 Radosveta Vassileva, *Bulgarian Private Law at Crossroads* (Cambridge, Intersentia 2022) 43.

43 Radosveta Vassileva, *Bulgarian Private Law at Crossroads* (Cambridge, Intersentia 2022) 43.

44 See Vitali Tadjer, *Civil Law of People's Republic of Bulgaria: General Part.* Section 1 (Sofia, Nauka i izkustvo 1972) 30.

45 Radosveta Vassileva, *Bulgarian Private Law at Crossroads* (Cambridge, Intersentia 2022) 53–55.

46 Radosveta Vassileva, *Bulgarian Private Law at Crossroads* (Cambridge, Intersentia 2022) 57–58.

47 Supreme Court of Cassation (27.11.2019), 3rd Civil Chamber, 1140/2019.

48 Supreme Court of Cassation (16.09.2022), 1st Civil Chamber, 4132/2021.

49 Ivaylo Vassilev, 'Liability of Merchants for Damage due to Environmental Pollution under Article 49 of the Law on Obligations and Contracts' (*Gramada*, 29 May 2015) <http://gramada.org/%D0%BE%D1%82%D0%B3%D0%BE%D0%B2%D0%BE%D1%80%D0%BD%D0%BE%D1%81%D1%82-%D0%BD%D0%B0-%D1%82%D1%8A%D1%80%D0%B3%D0%BE%D0%B2%D1%86%D0%B8%D1%82%D0%B5-%D0%B7%D0%B0-%D0%B2%D1%80%D0%B5%D0%B4%D0%B8-%D0%BF%D1%80/> accessed 8 December 2022.

50 Decree 7 of the Plenum of the Supreme Court of 1959; Decrees on interpretation are a distinctive feature of Bulgarian law evidencing the law-making powers of the judiciary. They are binding and concern questions of principle.

51 Ivaylo Vassilev, 'Liability of Merchants for Damage due to Environmental Pollution under Article 49 of the Law on Obligations and Contracts' (*Gramada*, 29 May 2015) <http://gramada.org/%D0%BE%D1%82%D0%B3%D0%BE%D0%B2%D0%BE%D1%80%D0%BD%D0%BE%D1%81%D1%82-%D0%BD%D0%B0-%D1%82%D1%8A%D1%80%D0%B3%D0%BE%D0%B2%D1%86%D0%B8%D1%82%D0%B5-%D0%B7%D0%B0-%D0%B2%D1%80%D0%B5%D0%B4%D0%B8-%D0%BF%D1%80/> accessed 8 December 2022.

52 *World Bank*, 'Poverty Headcount Ratio at National Poverty Lines (% of Population) – Bulgaria', <https://data.worldbank.org/indicator/SI.POV.NAHC?locations=BG> accessed 15 December 2022.

53 See Tariff for State Fees Collected by the Courts under the Civil Procedure Code.

54 For an overview, see S Tasev, *Delictual Liability* (Sofia, Ciela 2009) 93–126.

55 M Gruev, 'Collectivisation and Social Change in the Bulgarian Village (1940s–1960s)' in Ivalyo Znepolski (ed), *History of the People's Republic of Bulgaria: The Regime and Society* (Sofia, Ciela 2009) 179–188;

Communism initially significantly curtailed the right to private property. Subsequently, Bulgaria's Constitution of 1971 completely denied the right to private property.

56 Verbatim Report of the Sitting of the Bulgarian Parliament of 2.07.1974.
57 Report of the Sitting of the Bulgarian Parliament of 2.07.1974.
58 Additional Provisions, §1, point 5.
59 Additional Provisions, §1, point 6.
60 Supreme Court of Cassation (27.6.2012), 4th Civil Chamber, 1063/2011.
61 Supreme Court of Cassation (27.6.2012), 4th Civil Chamber, 1063/2011.
62 Silvy Vasilev Chernev, 'Civil Procedure Reform in the Republic of Bulgaria' in *International Conference 2010* (Burgas, Burgas Free University 2010) 298.
63 Silvy Vasilev Chernev, 'Civil Procedure Reform in the Republic of Bulgaria' in *International Conference 2010* (Burgas, Burgas Free University 2010) 298. See also Mario Bobatinov, 'Collective Claims Proceedings' in *National Conference Proceedings* (Sofia, New Bulgarian University 2015) 98 <https://ebox.nbu.bg/lawofright/images/Text-sbornik.pdf> accessed 6 December 2022.
64 Radosveta Vassileva, 'Monetary Appreciation and Foreign Currency Mortgages: Lessons from the 2015 Swiss Franc Surge' [2020] 281 *European Review of Private Law* 173, 183–184 <https://doi.org/10.54648/erpl2020008>.
65 Directive 2008/50/EC of the European Parliament and of the Council of 21 May 2008 on ambient air quality and cleaner air for Europe [2008] OJ L152/1.
66 The text of the claim is available on the website of the 'Group for Clean Air' initiative <https://chistvazduh.spasisofia.org/kolektivniyat-ni-isk/> accessed 5 December 2022.
67 Case C-488/15 *European Commission v. Bulgaria* [2017] ECLI:EU:C:2017:267.
68 Sofia City Court (8.11.2021), 6614/2017; Note the decision is 122-page long. Regrettably, Bulgarian court decisions neither have page numbers nor numbered paragraphs.
69 Sofia City Court (8.11.2021), 6614/2017.
70 Sofia City Court (8.11.2021), 6614/2017.
71 Sofia City Court (8.11.2021), 6614/2017.
72 Article 78 of the CCP.
73 Decision 266455 of 8.11.2021 on civ.c. 6614/2017.
74 Radosveta Vassileva, *Bulgarian Private Law at Crossroads* (Cambridge, Intersentia 2022) 127.
75 Commission, 'Communication from the Commission to the European Parliament, the Council, the European Economic and Social Committee and the Committee of the Regions on Environmental Implementation Review 2022: Turning the Tide through Environmental Compliance' COM (2022) 438 final, 23.
76 Decree 38 of the Council of Ministers of 7.02.2008.
77 See, for instance, *Chorbadziyski and Krasteva v. Bulgaria*, Application no. 54991/10, 2.04.2020.

Bibliography

Bobatinov M, 'Collective Claims Proceedings' in *National Conference Proceedings* (Sofia, New Bulgarian University 2015) 98 <https://ebox.nbu.bg/lawofright/images/Text-sbornik.pdf>

Chernev SV, 'Civil Procedure Reform in the Republic of Bulgaria' in Petko Chobanov et al (eds), *International Conference 2010* (Burgas, Burgas Free University 2010) 294–303

Donson F and Lee R, 'Environmental Protection: Public or Private Law?' [1996] 1(1) *Judicial Review* 56 <https://doi.org/10.1080/10854681.1996.11426861>

Giliker P, 'La cessation de l'illicite' [2019] 4 *Revue des contrats* 284

Gruev M, 'Collectivisation and Social Change in the Bulgarian Village (1940s–1960s)' in Znepolski I (ed), *History of the People's Republic of Bulgaria: The Regime and Society* (Sofia, Ciela 2009) 338–367

Howarth D, 'Ch.48: Environmental Law and Private Law' in Lees E and Viñuales JE (eds), *The Oxford Handbook of Comparative Environmental Law* (Oxford, Oxford University Press 2019) 1091 <https://doi.org/10.1093/law/9780198790952.003.0048>

Innes A, 'The Political Economy of Captured States' [2014] 52(1) *Journal of Common Market Studies* 88 <https://doi.org/10.1111/jcms.12079>

Ryall Á, 'The Aarhus Convention: Standards for Access to Justice in Environmental Matters' in Turner SJ et al (eds), *Environmental Rights: The Development of Standards* (Cambridge, Cambridge University Press 2019) 116 <https://doi.org/10.1017/9781108612500.006>

Tadjer V, *Civil Law of People's Republic of Bulgaria: General Part.* Section 1 (Sofia, Nauka i izkustvo 1972)
Tasev S, *Delictual Liability* (Sofia, Ciela 2009)
Transparency International, *Corruption Perceptions Index 2021* <https://www.transparency.org/en/cpi/2021>
United Nations, *The Aarhus Convention: An Implementation Guide* (2nd edn, Geneva, United Nations 2014)
Vassileva R, 'Monetary Appreciation and Foreign Currency Mortgages: Lessons from the 2015 Swiss Franc Surge' [2020] 281 *European Review of Private Law* 173 <https://doi.org/10.54648/erpl2020008>
———, *Bulgarian Private Law at Crossroads* (Cambridge, Intersentia 2022)
World Justice Project, *Rule of Law Index 2022* (Washington DC, World Justice Project 2022) <https://worldjusticeproject.org/rule-of-law-index/downloads/Index-2022.pdf>

4

PRIVATE LAW AS A TOOL TO OVERCOME SPLIT INCENTIVES ON ENERGY EFFICIENCY IN THE RESIDENTIAL RENTAL MARKET

Rosa M. Garcia-Teruel[1]

4.1 Introduction

4.1.1 Sustainability and Energy Efficiency in Residential Buildings: Policies from the European Union

The United Nations Sustainable Development Goals encourage countries to double the rate of improvement in energy efficiency by 2030. This includes residential buildings, which account for 27% of the European Union's (EU) energy consumption. The EU climate change policies[2] also focus on residential buildings, due to the potential for significant reductions in energy consumption, as well as greenhouse gas emissions.

The EU, based on its competence in energy outlined in Article 194 of the Treaty on the Functioning of the European Union,[3] enacted the energy efficiency Directives 2010/31/EU and 2012/27/EU, involving the harmonization of legislation among Member States in this field. These directives, as well as public funds intended to enhance energy efficiency works,[4] have helped to promote the use of energy-efficient practices in the residential sector, contributing to the overall energy efficiency target of 32.5% by 2030. Also, under the European Green Deal,[5] the EU Commission has proposed a set of policies aimed at reducing net greenhouse gas emissions by at least 55% by 2030, compared to 1990 levels.

The implementation of energy efficiency policies may have a direct impact on the environment, poverty, and health.

1) In terms of environmental impact, energy efficiency policies can reduce greenhouse gas emissions. Building 'green affordable housing', which is both affordable for low- and moderate-income households and reduces the environmental impact,[6] is therefore relevant for protecting the environment without leaving less-affluent households behind.

2) Poverty may result when vulnerable households cannot afford to pay energy bills because their houses are energy-deficient, leading to energy poverty. Reports[7] show that in some EU countries, a significant proportion of inhabitants are unable to keep their homes adequately warm (in Greece, up to 32.9% of the population, in Italy, 18%),[8] especially in certain rural areas.[9]

3) Finally, energy retrofits can reduce exposure to cold and air pollutants, as well as improve social and economic conditions conducive to better health.[10]

 DOI: 10.4324/9781032662046-5

Nevertheless, the state of repair of existing building stock is still a big challenge for the energy efficiency performance of buildings. According to the European Commission,[11] around 35% of buildings are over 50 years old and nearly 75% are not energy-efficient, with F and G labels being the most common in some countries, such as Spain.[12] Particularly, rented housing has, in general, a deficient energy performance: as stated in the European Green Deal (sec. 2.1.4), one of the priorities of the EU Commission is lifting regulatory barriers that prevent energy efficiency investments in rented housing.

This chapter analyses, from a private law perspective in a comparative context, the legal barriers when performing energy efficiency works in private rented dwellings. It studies the so-called 'split incentive' or 'tenant-owner dilemma', which discourages energy retrofits in rented housing, as well as the main incentives to counteract this phenomenon implemented in Germany, France, and Spain through private law measures. More specifically, this chapter focuses on assessing to what degree, in the private rental market, the landlord may increase the rent after energy efficiency works, being encouraged to retrofit; the right of the tenant to oppose these works or to claim compensation during the time the works are performed; and, in general, how the legislation promotes the agreement between landlords and tenants to implement energy efficiency measures. The legal regime of public rented housing, where tenants pay rent below the market price, is excluded from this analysis. This is justified by a lack of a harmonized concept of 'social housing' and the important differences between Member States in this field.[13]

Moreover, this chapter also aims to reinforce the idea that private law, in particular civil law, may contribute to sustainable development by increasing social and economic cohesion and the level of energy efficiency performance without investing further public funds. Sustainable development, foreseen in Article 11 TFEU and Article 3 Treaty of the European Union,[14] is considered a general legal principle by some authors,[15] yet not fully recognized in all the legal systems.[16] Civil law has been appointed to contribute to sustainability[17]: in Germany, the report of the working group 'Sustainability in civil law', where justice ministers of several German states participated, foresaw that the role of civil law in promoting sustainability is a topic of debate. In view of the general idea that contract law should only focus on balancing the interests of the contracting parties, the opinion of justice ministers suggests that contract law can be also used to encourage more sustainable consumer goods. From this perspective, the narrow understanding of civil law as solely concerned with balancing private interests is no longer tenable. This report also considers that the regime of rent increase due to modernization works in Germany (see Section 4.3) is a key example of how civil law may promote sustainability.

4.1.2 Effects of the Tenant-Owner Dilemma (Split Incentives) on the Energy Performance of Rented Housing

Regardless of the efforts by both the EU and Member States to improve the energy efficiency of existing residential buildings, there are some barriers that prevent parties in a housing rental contract to improve energetic performance. In particular, the European Parliament Resolution of 17 September 2020[18] specifically mentions the tenant-owner dilemma as a barrier to energy efficiency investments. Directive 2012/27/EU also referred to split incentives in its Article 19, but this concept was not defined until the recent Recast Energy Efficiency Directive 2023 (Article 2, s. 54).[19] According to Article 22.9 Recast Energy Efficiency Directive 2023, Member States shall take the necessary measures to remove split incentives between owners and tenants.

In its turn, the report of the Joint Research Centre 'Overcoming the split incentive barrier in the building sector'[20] confirms that split incentives can be a significant barrier to the adoption of energy efficiency measures in the building sector, as they can create a lack of alignment between the parties involved and can lead to a lack of investment in this field. To overcome this barrier, it may be necessary for policymakers and stakeholders to develop incentives or financing mechanisms that align the interests of building owners and tenants.

The tenant-owner dilemma or split incentives refers to the situation in which tenants and owners of rented housing have conflicting interests in terms of energy efficiency upgrades. The tenant-owner dilemma has been extensively studied by economists within the framework of 'game theory'.[21] These studies aim at predicting and understanding the behaviour of owners and tenants in different situations since there are several circumstances that may encourage or discourage landlords and tenants from deciding to retrofit, which shall be analysed on a case-by-case basis.[22]

According to this dilemma, landlords may not want to invest in energy efficiency because they are not directly benefiting from energy savings, as the tenant is the one in charge of paying utility bills. Moreover, depending on the jurisdiction, landlords cannot increase the rent after energy efficiency works (see below). However, tenants do not have the incentive to invest in energy efficiency measures when they will not be able to recover the investment. This will depend on the actual energy savings plus the total duration of the contract, whereas they may not have the authority or resources to make those upgrades. The tenant-owner dilemma specially applies to tenancy contracts for housing purposes, as legislation usually assigns to the tenant the obligation to pay consumptions (e.g., in Spain, Article 20 *Ley de Arrendamientos Urbanos*, LAU[23]; in Italy, Article 9 *Legge locazioni de immobili urbani* 1978[24]; in France, Articles 7 and 23 *Loi* num. 89-462*, tendant à améliorer les rapports locatifs*).[25] The tenant-owner dilemma may be of less relevance in other types of rental contracts, where the landlord pays utility bills, such as housing for students or tourist apartments, whose tenants pay rent, including other services and consumptions.

The tenant-owner dilemma may also have an impact on the energy performance and energy consumption of rented housing. Tenants are less likely to live in insulated dwellings compared to owner-occupiers, undertaking less energy efficiency investments than homeowners.[26] Charlier confirms that tenants are double penalized: they have high energy costs and therefore low available income.[27] The tenant-owner dilemma would also influence the energy efficiency performance of multi-unit buildings. According to the report 'Energy efficiency upgrades in multi-owner residential buildings',[28] a lack of flexibility in tenancy contracts would also be one of the challenges for improving the energy efficiency of multi-unit buildings. Moreover, Celinni found that 'imperfectly informed renters will not be willing to pay more for energy efficiency dwellings', as asymmetric information is a key factor contributing to the tenant-owner dilemma.[29]

Therefore, the question that arises is how private law, in particular tenancy law, may re-address this problem so that both landlords and tenants are encouraged to undertake energy efficiency works.

4.2 A Literature Review of the Tenancy Law Measures Contributing to Overcoming the Tenant-Owner Dilemma

In light of the EU's energy efficiency priority, several measures were published to harmonize the legal framework in this regard. However, the tenant-owner dilemma is caused, among other factors, by provisions of residential tenancy law. Tenancy law is subject to the legal framework of each Member State of the EU, as the TFEU does not include legal competence in 'civil law'. Nonetheless, the relationship between housing conditions (including energy efficiency performance) and residential tenancy law is complex and far-reaching. By way of example, in the United Kingdom, the introduction of assured short-hold tenancies posed a challenge: these short-term contracts (usually for six months, renewable at the landlord's discretion) make it difficult for tenants to challenge the conditions of the accommodation.[30] Furthermore, poor maintenance and energy performance can lead to a lack of emotional attachment to the home[31] and a greater likelihood that the tenant may vacate sooner than expected. Conversely, proper maintenance can help to engender a sense of security in the tenancy, and potentially increase the length of occupancy.[32]

The report 'Tenancy Law and Energy Renovation in European Comparison' (BBSR, 2016)[33] discusses the features of tenancy law that affect energy efficiency. In particular, it suggests some elements relevant to the tenant-owner dilemma:

- The stability of tenancy relationships, including provisions on the duration of the contract and grounds for termination. Long-term rental contracts, normally open-ended leases, positively correlate with energy efficiency performance: the longer a rental contract lasts, the more possibilities the landlord or the tenant invests in energy efficiency. In theory, countries with fixed-term contracts seem to provide a potential opportunity for the implementation of energy renovation measures at the end of each term. However, data indicates that this potential is not being realized, particularly in countries such as the United Kingdom. This may be explained by the fact that landlords are likely to be deterred by the insufficient rates of return associated with investing in older buildings.[34]
- Rent controls and rent increases: rent controls have been considered to worsen housing conditions. According to Kholodilin,[35] landlords subject to rent controls have fewer incentives to invest in refurbishment. This is in line with the findings of Moon and Stotsky,[36] who suggested that rent control regulations, which prevent landlords from increasing rent after renovation works, could exacerbate the tenant-owner dilemma by worsening housing conditions.
- Distribution of costs: being tenants in charge of paying utility bills, landlords may not have incentives to improve energy efficiency conditions. In general terms, tenancy law foresees that the tenant is the one in charge of electricity, gas, and water supplies (ex. Article 20 LAU, § 556 (II) BGB, or Article 1078.2 Portuguese Civil Code). However, assigning energy costs to landlords leads to more energy consumption, as tenants would not behave optimally.[37]
- The obligation of the tenant to tolerate energy refurbishment works: energy efficiency works are considered improvement works, as they go beyond the mere maintenance of the dwelling,[38] and they also have an impact on the obligation of the tenant to tolerate these works: while maintenance works do not affect the peaceful use of the dwelling, improvement works shall not be tolerated by tenants in certain jurisdictions: for example, in Spain, Article 22 LAU requires the consent of the tenant in case landlords want to undertake improvement works, except if those works cannot be deferred until the termination of the tenancy contract (e.g., works imposed by a public authority or by the condominium's board).[39] However, other legal regimes, such as Germany, do allow landlords to undertake energy efficiency works without the tenant's consent (§ 555 BGB). If the tenant does not have to tolerate energy efficiency works, the transactional costs for landlords may reinforce the tenant-owner dilemma, since they need an agreement with tenants to be able to improve energetic conditions.

In addition, according to Lange,[40] there are some measures that may contribute to alleviating the tenant-owner dilemma. Some of these measures are not related to tenancy law, such as the need to implement financial mechanisms, while others are closely linked, such as the possibility to increase the rent after renovation works, which is one of the main motivations for landlords from an economic point of view. At the same time, rent increases due to energy efficiency upgrades may not necessarily discourage tenants from having more efficient homes: Phillips[41] confirmed, in a local study in New Zealand, that 'tenants express a willingness to pay higher rent in exchange for improved energy efficiency, at a rate that would appear to make such upgrades economic for the landlord'. However, citing some economic literature, Lange et al. agree with the fact that a more than 20% rent increase may involve an exodus of tenants, so legislation should not allow, in any case, a rent increase of such an amount. This circumstance is also recognized by the European Parliament resolution of 17 September 2020, calling Member States to create a framework for overcoming split incentives, but also to avoid renovictions[42] through limits on rent increases, provided that it does not restrict the capacity to undertake energy efficiency renovations.

Lang et al.[43] undertook a review of the literature, identifying which are the key elements influencing the landlords' decision to perform energy efficiency works. Among others, they found 16 factors that increase landlords' willingness to retrofit, including the potential to raise property value, the existence of subsidies, the easiness to install retrofits, the tenant demand for energy efficiency, landlords' concern for tenant well-being, the relationship with tenants, among others. A greater influence on landlords' willingness was the possibility of increasing property value. They also found 23 elements reducing landlords' willingness to retrofit, including split incentives, future maintenance costs, difficulties in accessing government schemes or time constraints, and financial limitations being the most significant factors.

Of the factors detected by Lang et al., some of them may be addressed by tenancy law. As incentives for landlords they name: the potential to increase property value if the rent is increased after these works; assigning energy bills to landlords; tenants request for specific retrofits and the existence of energy efficiency regulations. And regarding factors that reduce the landlords' willingness, the authors name: future maintenance costs, the conflation of general upgrades with energy efficiency retrofits, rent controls, and time constraints (e.g., not only the duration of the work but also of the contract).

Apart from that, giving parties enough autonomy could also encourage retrofits in rented dwellings. This is the case with the so-called 'green leases'. Green leases are rental contracts, traditionally linked to commercial purposes, that include terms related to the management and improvement of a property's environmental performance. According to Janda et al.,[44] traditional leasing practices do not adequately address environmental concerns and may even hinder efforts to improve a property's energy efficiency. Green leases, however, aim to promote the inclusion of specific clauses in rental contracts that help to better organize and finance energy efficiency upgrades. These terms may include provisions for the allocation of costs for renovations, compensation for tenants in the event of early lease termination, and rent increases based on improved energy performance. While green leases have the potential to address the conflict between tenants and property owners, they are not yet a common practice in the real estate industry: as Collins[45] confirms, academic research on green leases is more common than real-world application. This could potentially be changed through increased awareness and education about the benefits of green leases in the conveyancing process.

Some jurisdictions, being aware of this dilemma, introduced provisions in order to enhance landlords' willingness to retrofit rented dwellings through positive or negative incentives while increasing tenants' information on the actual benefits of having an energy-efficient home. The French Act on Energy and Climate 2019[46] introduced a measure which requires tenancy agreements (and their corresponding advertisements) to provide information on actual energy consumption and estimated energy expenditures. The Energy Efficiency (England and Wales) Regulations 2015 stipulate that any housing which has a performance indicator lower than E is classified as 'sub-standard' and is therefore prohibited from being rented out (Article 23). Consequently, landlords of homes with poor energy efficiency will be obliged to carry out renovations in order to continue renting their properties. Among these provisions, in the following section, the rent increase after energy efficiency works in Spain, France, and Germany is studied, as being considered one of the key factors contributing to the landlord's willingness to retrofit. Only some EU countries specifically regulate this right for landlords.[47]

4.3 Rent Increases Due to Energy Efficiency Renovations: The Case of Spain, France, and Germany

4.3.1 Legislation on Residential Tenancies in Spain, France, and Germany

4.3.1.1 A general overview on the duration of rental contracts

Spain, France, and Germany foresee different types of tenancy law regimes. While Spain and France opted for regulating residential tenancies in special laws (in Spain, *Ley de Arrendamientos Urbanos* 1994, LAU, and in France, *Loi Rapports Locatifs* 1989), Germany introduced regulation on residential tenancies within the Civil Code (*Bürgerliches Gesetzbuch,* BGB). By its turn, the rates of the total population living in rented dwellings are considerably different[48]: in Germany, 50.5% of the citizens live in rented dwellings, the EU country with a higher rate, while in Spain, 24.2% of the population and in France, 35.3%. However, when assessing the number of people in rented dwellings at a market price, the rate reduces to 49.9% in Germany, 15.7% in Spain, and 15.2% in France. This shows the public support for rented dwellings in France, as there are an important number of people living there at a reduced price or for free (up to 20% of the population).

Also, one of the key differences among jurisdictions is the duration of rental contracts. Some countries allow for open-ended contracts, where there is no predetermined end date for the contract. This means that the contract continues as long as the parties fulfil their obligations. Germany is one of the countries recognizing open-ended contracts (§ 542 BGB), which is the preferred form to conclude residential tenancies[49]: in fact, failing to use the written form for contracts lasting more than one year, the contract is deemed for an indefinite period of time (§ 550 BGB). Nevertheless, parties may terminate the contract by giving notice under certain conditions (§ 543 BGB).

Other countries only allow time-limited contracts. In some of these countries, minimum lease terms are set to provide stability for tenants.[50] For example, Spain has a minimum duration of five years for contracts with natural persons as landlords, or seven years for contracts with legal persons as landlords (Article 9 LAU). France has a minimum duration of three to six years (Article 10 *Loi Rapports Locatifs*). These minimum durations provide some protection for tenants, ensuring that they are not forced to move frequently.

As indicated above, the duration of the contract may have an impact on the willingness of tenants to invest in energy efficiency works. If the duration of the contract is too short compared to the actual investment, the energy savings will not cover what the tenant invested. On the contrary, open-ended, or long-term contracts would allow tenants to benefit from energy efficiency investments. The length of the contract does not affect landlords' willingness to undertake energy upgrades as long as they can increase the rent during the current contract or at the end of it, although the mechanisms to increase the rent during the contractual term may not be totally beneficial for them. According to the report BBSR,[51] a fixed-term tenancy allows for the opportunity to make energy-saving improvements during tenant turnover, but data shows that these types of renovations are not commonly carried out in countries where short-term contracts are more prevalent.

4.3.1.2 Existence of Rent Controls

In general terms, contracts are governed by the principle of freedom of agreement. In the context of residential leases, it is common for legislators to impose limitations on the rent due: for example, linking annual rent updates to official indexes.

Germany, France, and Spain also introduced in recent years caps to the initial rent. Since the 1980s, Germany has implemented an optional rent control through the *Mietspiegel* system, allowing parties to update the rent based on local comparative rents (§ 558c BGB). Moreover, starting from 2015, the *Mietpreisbremse* was introduced (§ 556d BGB), imposing a cap on the initial rent in certain regions with high residential demand. The agreed rent must not exceed 10% of the comparative rent in the respective municipality. Berlin also implemented the *Mietendeckel*,[52] which involved a rent freeze and setting the initial rent based on local comparative rents, without surpassing the rent agreed with a previous tenant. However, the German Constitutional Court invalidated the *Mietendeckel* based on a lack of legislative power of the Berlin federated state.

In Spain, rent controls have primarily consisted of regulating annual updates according to official indexes (*índice de precios al consumo* and *índice de garantía de competitividad*), while allowing parties to freely determine the initial rent (Articles 17 and 18 LAU). Nevertheless, recent years have witnessed an increase in legislative efforts to limit the rent from the outset. For instance, the autonomous community of Catalonia passed Act 11/2020,[53] which bore similarities to Berlin's *Mietendeckel* system. This law restricted rents to surpass average rents fixed in the so-called *'índex de referència dels preus del lloguer'* and prohibited landlords to impose a rent higher than the one agreed with the previous tenant. Similar to the German *Mietendeckel*, Act 11/2020 was also invalidated by the Spanish Constitutional Court based on a lack of legislative power of the autonomous community of Catalonia. Furthermore, the recently enacted Spanish Housing Law (Act 12/2023[54]) modified Article 17 LAU, resulting in several changes for those municipalities qualified as 'stressed areas' (*zonas tensionadas*). First, in these municipalities, rental contracts must not exceed the previous tenant's rent (art. 17.6 LAU), with some exceptions. Additionally, if a rental contract is concluded for ten years or more, the price can be increased by up to 10%.

Second, the initial rent may also be limited, regardless of the rent agreed with previous tenants, when the contract is concluded with a 'large landlord' according to Article 3.k of the Housing Act.[55] In stressed housing market areas, a starting rent higher than the reference rent cannot be agreed upon. This limitation on initial rent applies to all landlords (even those not classified as large landlords) if approved by the Ministry of Transport, Mobility, and the Urban Agenda.

In France, rent control, in general, has been implemented by restricting rent updates solely based on the *indice de reference des loyers* (Article 17.1 *Loi rapports locatifs*).[56] Rent caps when setting up the initial rent have been regulated by the ALUR Act[57] (2014–2018) and by the ELAN Act,[58] reinstated for specific cities of more than 50,000 inhabitants.[59] The agreed starting rent shall fall within a range of a maximum of 20% above the average rent or a minimum of 30% of the average rent. Additionally, for dwellings with an energy efficiency rate of F or G, the initial rent shall not exceed the rent charged to the previous tenant (Article 17 II *Loi rapports locatifs*).

4.3.2 Increasing the Rent after Energy Efficiency Works in Spain

Generally speaking, in Spain, landlords may not undertake improvement works without the tenant's consent during the contractual term. This is established in Article 22 LAU, by which landlords may only perform works that cannot be delayed (e.g., when a public authority compels landlords to do so).[60] That is why Article 19, regulating the possibility to increase the rent by improvement works (including energy efficiency ones),[61] requires the landlord to increase the rent only after five or seven years of the contract have elapsed, depending on the landlord being a natural (five years) or a legal person (seven years).

This regime makes it difficult for landlords to undertake energy efficiency works: while they can only increase the rent after five/seven years, they need the consent of the tenant to undertake energy efficiency works at any time, except they are deemed 'urgent' (imposed by a public authority or by the condominium's board).[62] This limit was intended to give tenants enough security on the amount of rent that they shall pay during the contractual term and not to disturb the peaceful use of the dwelling.[63] However, it can be arguable that energy efficiency works are 'urgent' when the landlord needs a public subsidy to perform them. In most cases, public subsidies to help homeowners improve the level of energy efficiency are available only for a certain period of time (e.g., *ad hoc* calls). It could be understood that energy efficiency works are 'urgent'[64]; otherwise, the landlord would not be able to use the subsidy. In any case, it would be important for the lawmaker to clarify this aspect.

Trying to offer more flexibility to parties, Article 19 LAU para. 4 permits landlords and tenants to agree on the performance of improvement works as well as on the subsequent increase in rent before concluding the term of five or seven years: if parties agree, time limitations regarding the rent increase could be bypassed. Nevertheless, this agreement cannot be included within the initial rental contract: this is a measure to protect tenants from contracts harming their interests (essentially, preventing hidden renovictions), but it also involves the impossibility to sign 'green leases' *ab initio*.[65] Therefore, under Spanish tenancy law, parties may not conclude green leases at the beginning of the contract, when works are afforded by landlords.

While works are performed, tenants may ask for a rent reduction following Article 22.3 LAU. And, once the work is completed, the landlord may raise the rent. The method to calculate the rent increase, except otherwise agreed,[66] is the following: the annual rent would be increased by applying the legal interest rate plus 3 points to the costs of the works. Any kind of subsidies given to landlords in order to perform these works (e.g., NextGenerationEU schemes to help homeowners to improve energy efficiency) shall be reduced from the costs of the works. Furthermore, landlords may not increase the rent by more than 20%, even when the total costs justify it.[67]

In this context, there is a law project to regulate residential tenancies in the Autonomous Community of Catalonia, expected to be passed by 2024–2025.[68] According to draft Article 623-60 of the Catalan Civil Code, landlords may undertake works of energy efficiency without the tenants' consent. If these works reduce the agreed use of the home, causing disturbances, the tenant is entitled to a rent reduction in proportion to the loss of utility. Moreover, although in general terms tenants cannot undertake improvement works, they are entitled to do so without the landlord's consent when these works aim to improve the energy conditions (draft Article 623-67 Catalan Civil Code). Finally, draft Article 623-62 Catalan Civil Code regulates the possibility to increase the rent in similar terms to Article 20 LAU: the landlord may only increase the rent once a minimum term of three years has elapsed.

Comparatively to the LAU regime, this law project focuses specifically on energy efficiency investments. Parties do not need the consent of one another to undertake these works, and an increase in rent after three years is foreseen, compared to five/seven years of LAU.

4.3.3 Rent Increase and Allocation of Energy Efficiency Costs in France

France has regulated several provisions related to the energy efficiency of residential rented housing. Apart from general positive incentives aimed at reducing energy consumption, such as programmes to finance the replacement of oil-fired boilers with more efficient heating or domestic hot water supply equipment (program MaPrimeRénov), eco-loans (no-interest loans) to finance energy efficiency renovations, or reduced VAT (5.5%) when performing these works, France also published some coercive measures penalizing less-efficient homes. For example, according to the regime introduced in *Loi rapports locatifs* by Act 2021-1104, of 22 August 2021,[69] the landlord may not impose a higher rent than the one paid by the former tenant, when the housing has an F or G energy performance label (Article 17, II *Loi rapports locatifs*).

The *Loi rapports locatifs* also foresees specific measures for energy efficiency upgrades in terms of rent increases and the allocation to the tenant of the costs of the works.

First, Article 17-1 foresees two types of rent increases: (1) a rent increase to be undertaken each year, when the contract specifically includes the revision of the rent, which cannot exceed the variation of the rent reference index published by the *Institut national de la statistique et des études économiques*; and (2) a rent increase after the landlord undertakes improvement works. According to Article 17-1 II *Loi rapports locatifs*, when the parties have agreed on the performance of works to be undertaken by the landlord, the rental contract or an addendum to this contract may set the rent increase after finishing these works. Energy efficiency improvements could be included within this category,[70] so that landlords and tenants could agree on these works as well as the rent increase (this article, therefore, could be a way to articulate 'green leases'). Nevertheless, landlords may not increase the rent when the rented housing has an energetic label of class F or G, according to what is established in article L. 173-1-1 of the *Code construction et l'habitation*. This last provision was introduced by French Act 2021-1104 to reiterate the commitment to reduce greenhouse emissions, in line with the Paris Agreement and the European Green Deal (Article 1). This provision aligns with the French trend to sanction, from a private law perspective, those homeowners with less efficient homes: as stated in Article 6 *Loi rapports locatifs*, rented housing that does not reach a minimum efficiency rate will not be considered 'decent housing' and therefore cannot be rented. French decree 2021-19[71] establishes that decent housing shall consume less than 450-kilowatt hours of final energy per square metre of living space per year.

Second, regarding the allocation of energy efficiency costs to the tenant, landlords may oblige tenants to tolerate energy efficiency works, as considered 'improvement works' (Article 7.e *Loi rapports locatifs*), but tenants may ask for a rent reduction if the works last more than 21 days.[72] Also, Article 23 permits landlords to assign some costs to the tenant, such as services, taxes, and minor maintenance measures. This includes costs related to energy efficiency upgrades, provided that the landlord consulted with the tenant about the required works, discussed the expected energy savings, etc. In order to do so, the landlord may ask for this surcharge once the works are finished and provided that they are justified and benefit the tenant (e.g., through a reduction of consumption). Nevertheless, landlords of rented housing with an energy efficiency class of F or G may not ask for this surcharge. Tenants shall only bear the surcharge for energy efficiency works for a maximum period of 15 years and the total amount of this surcharge, which is fixed and not reviewable, cannot exceed half the amount of the estimated energy savings. Decree num. 2009-1439, of 23 November,[73] regulates the procedure to charge these costs depending on the year of construction: for buildings finished before 1948, landlords may impose a surcharge of 10–20 euros each month, depending on the number of rooms. For buildings finished between 1948 and 1990, the Decree offers two methods to calculate the surcharge: a fixed amount between 10 and 20 euros,[74] as commented above, or a method which depends on the energy savings (up to half of the energy savings). It remains doubtful whether this amount is enough to encourage landlords to undertake energy efficiency works.

4.3.4 Rent Increase after Energy Efficiency Works in Germany

Germany published several measures in 2013 intended to foster energy efficiency investments through the so-called *Mietrechtänderungsgesetz* (MietRändG).[75] Among other measures, the following related to energy efficiency may be highlighted.

4.3.4.1 Rent Increase Due to Energy Efficiency Works

Section 555b BGB foresees the regime of 'modernisation measures' (*Modernisierungsmaßnahmen*), which specifically refer to energy efficiency works. As stated in this article, modernization measures comprise structural changes[76] that include the following: (1) sustainable energy savings to the rented property; (2) the non-renewable primary energy is saved or protects the climate in the long term; (3) water consumption is significantly reduced in a sustainable way; (4) the value of the rented property increases in a sustainable way; (5) the rented property is connected for the first time to the public fibre-optic network; (6) the general living conditions are permanently improved; (7) the works are imposed on the landlord, if they do not qualify as conservation measures; and (8) additional living space is created.

These modernization measures may be unilaterally undertaken by landlords, according to § 555d BGB, upon notification three months prior to the commencement of the works (§ 555c BGB). However, tenants do not have to tolerate modernization works if the measure would mean a hardship for them or their families.

After the works are completed, landlords can raise the rent by 8% of the costs incurred (before 2019, 11%). In these costs, maintenance of modernization measures shall not be included. The monthly rent may not increase by more than 3 euros per square metre within six years, excluding increases pursuant to §§ 558 and 560 BGB. If the monthly rent before the rent increase was less than 7 euros per square metre, it may not increase more than 2 euros.

4.3.4.2 Exemption of Energy Efficiency Works from the Rent Reduction Regime

The German law of residential tenancies foresees in § 536 BGB the *Mietminderung*, i.e., the right of the tenant to reduce the rent due to disturbances. Depending on the type of disturbance, the tenant will be entitled to reduce a certain amount of rent. After the entry into force of MietRändG, if the disturbance is caused by a modernization measure (including works intended to improve energy saving, living conditions, water consumption, etc.), tenants will not be able to reduce the rent for the first three months.

In the preliminary impact assessment of proposal MietRändG 2013,[77] this exemption is considered to increase the costs for tenants: case law has usually conceded a rent reduction between 10% and 20% for these works, and this cost will be borne by tenants for three months. However, the Law also introduces a deduction in order to compensate for these costs.[78]

4.3.4.3 Energy Efficiency Works Are Excluded from the Mietpreisbremse

As of 2015, § 556d BGB regulates the so-called *Mietpreisbremse*, a rent control regime by which the initial rent cannot exceed 10% of the reference rent in the municipality where the property is located.[79] *Mietpreisbremse* applies to those municipalities located in *Länder* that have activated this measure. However, when landlords perform 'modernisation measures' before starting the new rental contract, they may impose a higher rent, calculated according to § 559 BGB.

Excluding renovated homes from the *Mietpreisbremse* has been considered to unprotected tenants. Kossman and von Wangenheim[80] argue that the political goal of protecting tenants through rent controls is either ineffective or counterproductive because exemptions are allowed in cases of modernization measures. The analysis shows that such restrictions may actually result in higher rent increases for tenants in properties undergoing modernization and make it more likely that rent increases will be so large as to effectively terminate the rental contract. Energy efficiency and affordable housing are at odds in this case. As discussed earlier, rent controls worsen the tenant-owner dilemma, so the lawmaker shall decide whether to promote environmental sustainability or to use coercive measures such as rent controls to moderate rent increases for the regulated contracts, but this would also have other negative effects.[81]

4.3.5 Comparative Analysis of the Three Regimes

As seen above, the three jurisdictions foresee rent increases after energy efficiency works in residential tenancies. However, these provisions have a different scope.

Regarding the type of works that may involve a rent increase, Germany specifically foresees in § 555 BGB the concept of 'modernisation measures', which may give landlords and tenants examples of the types of works. According to Article 23 *Loi rapports locatifs*, energy efficiency works shall be profitable for tenants, and Article 4 Decree 2009-1439 compiles a list of possible energy efficiency works, including thermal insulation, replacement of heating equipment, placing renewable energy sources, etc. On the contrary, Spain only foresees a rent increase due to 'improvement works', i.e., including works that do not aim to improve the energetic conditions. Since energy efficiency legislation can lead to a reduction of the tenant-owner dilemma,[82] the Spanish lawmaker could consider regulating specifically rent increases for energy efficiency works so as to inform stakeholders (e.g., real estate agents, lawyers, associations of owners and tenants) about this possibility.

As for the amount, landlords in Germany may increase the rent by 8% of the costs incurred, with a general limit of 3 euros per square metre; in France, the surcharge supposes an increase of 10–20 euros each month for buildings finished before 1948, or a calculation depending on energy savings, involving that the surcharge may not exceed half of the expected energy savings for the tenant. In Spain, the rent increase depends on the costs of the works and the interest rate, with a maximum amount of 20% of the rent. As a consequence, if landlords invest, for example, 30,000 euros in energy efficiency works, they may increase up to 2,400 euros per year in Germany (provided that it does not exceed 3 euros per square metre), 1,800 euros in Spain, following the current interest rate and, in France, a maximum of 240 euros per year during 15 years (20 euros per month; for buildings built before 1948, except otherwise agreed based on Article 17-1 *Loi rapports locatifs*). This demonstrates the vast difference between systems and poses the question of whether or not the total amount a landlord may charge in France contributes to reducing split incentives among landlords.

In terms of requirements, the German and French systems allow landlords to increase the rent or assign the costs to tenants by a unilateral decision, regardless of the contractual term. On the contrary, the Spanish system only allows the landlord to undertake the works and to raise the rent once five/seven years have elapsed, which may increase transactional costs. It is therefore quite difficult for a landlord in Spain to undertake these works and to increase the rent. Other complementary measures, such as exemptions from rent reductions and rent control (regulated in Germany), may also contribute to reducing the tenant-owner dilemma.

Therefore, the German regime of rent increases due to energy efficiency works is one of the most suitable to reduce split incentives, because the amount that can be updated and the rights for landlords intend to encourage them to undertake these works. At the same time, regulating open-ended leases does not discourage tenants to invest in energy efficiency measures, since they will be able to recover the investment in the form of energy savings. On the contrary, the Spanish regime relies mainly on freedom of will. This is because the rent cannot be increased during the first five/seven years, and the landlord cannot unilaterally decide to undertake energy efficiency works. Finally, the French surcharge may not be high enough to reduce the tenant-owner dilemma: this country relies on coercive measures in order to promote an efficient rental housing stock through banning less-efficient homes from the rental market.

4.4 Conclusions

The EU, in accordance with its energy competence as outlined in Article 194 TFEU, has implemented energy efficiency Directives 2010/31/EU and 2012/27/EU, which involved the harmonization of legislation among Member States in this field.

Also, following the European Green Deal, the European Commission has proposed a series of policies aimed at reducing net greenhouse gas emissions by at least 55% by 2030, compared to 1990 levels. However, the state of repair of existing building stock remains a significant challenge for the energy efficiency performance of buildings. In its resolution of 17 September 2020, the European Parliament specifically mentions the tenant-owner dilemma as a barrier to energy efficiency investments. The tenant-owner dilemma, or split incentives, refers to the situation in which the interests of tenants and landlords' conflict in terms of energy efficiency upgrades.

According to existing literature, there are several elements of the regulation of private rental markets that contribute to the tenant-owner dilemma, such as the recent trend to regulate rent controls, the existence of energy efficiency legislation, the stability of tenants, and the possibility to increase rents after energy efficiency improvements. Finding a balance between these mechanisms is a challenge for the lawmaker: even though long-term leases are granted, the possibility to increase the rent may cause hidden renovictions. It may seem that, at the end of the day, tenants are assigned the costs of achieving the energy efficiency public goal.

This chapter explored rent increases after energy efficiency improvements in Germany, Spain, and France. While France and Germany have extensive regulations on how homeowners can unilaterally increase rent and for which types of improvements, Spain relies on a voluntary regime agreed upon with the tenant. However, the surcharge that can be imposed on tenants in France is relatively low compared to the German and Spanish ones. This raises the question of whether the total amount that a landlord may charge in France effectively reduces split incentives. While the French regime protects tenants from renovictions and mainly relies on coercive measures, the German framework reinforces the EU's energetic commitments with more priority than the affordability of rental contracts. Also, the Spanish regime, relying on the agreement between landlords and tenants to undertake energy efficiency works, may not be adequate to encourage individuals to contribute to energy efficiency goals.

In any case, this chapter demonstrates how private law, specifically civil law, can contribute to sustainable development by increasing social and economic cohesion and the level of energy efficiency performance. Several elements contributing to the tenant-owner dilemma are actually regulated by private law rules (e.g., tenancy law), influencing the behaviour of landlords and tenants in regard to energy efficiency upgrades.

Notes

1 Assistant professor of Civil Law. University of Barcelona.

2 Commission, 'A framework strategy for a resilient energy union with a forward-looking climate change policy' COM (2015) 80 final.

3 Consolidated Version of the Treaty on Functioning of the European Union [2012] OJ C326/47.

4 For example, under NextGeneration Funds, Member States received funds to encourage energy efficiency investments, with the condition to meet the obligations included in grant agreements between each state and the EU.

5 Commission, 'Communication from the Commission to the European Parliament, the European Council, the Council, the European Economic and Social Committee and the Committee of the Regions. The European Green Deal' COM/2019/640 final.

6 William Bradshaw et al, 'The costs and benefits of green affordable housing' [2005] *Cambridge (MA): New Ecology* <https://subjectguides.york.ac.uk/ld.php?content_id=33345299>.

7 Eurostat. Inability to keep home adequately warm. 2014.

8 The report Sergio Nasarre-Aznar et al, 'Concrete actions for social and affordable housing in the EU' [2021] *The Foundation for European Progressive Studies (FEPS)* <https://feps-europe.eu/wp-content/uploads/2021/01/Concrete-actions-for-social-and-affordable-housing-in-the-EU.pdf> states that when affordability is a priority, other housing-related issues remain unresolved. This is especially true for vulnerable households who prioritize paying rent or mortgages over energy bills.

9 María Teresa Costa-Campi, Elisenda Jové-Llopis and Elisa Trujillo-Baute, 'La pobreza energética en España. Aproximación desde una perspectiva de ingresos' [2019] *Fundación Naturgy* <https://www.fundacionnaturgy.org/publicacion/la-pobreza-energetica-en-espana-aproximacion-desde-una-perspectiva-de-ingresos/>.

10 Wouter Poortinga et al, 'The health impacts of energy performance investments in low-income areas: A mixed-methods approach' [2018] 6(5) *Public Health Res* (https://doi.org/10.3310/phr06050). Ian Hamilton et al, 'Health effects of home energy efficiency interventions in England: A modelling study' [2015] 5(4) *BMJ Open* (https://doi.org/10.1136/bmjopen-2014-007298).

11 Commission, 'Energy efficiency buildings' <https://ec.europa.eu/energy/en/topics/energy-efficiency/buildings> accessed 27.12.2022.

12 In Spain, eight out of ten buildings are labelled E, F, or G, according to IDAE, Estado de la certificación energética de los edificios (9th report) [2020], *Ministerio para la Transición Ecológica y el Reto Demográfico* <https://energia.gob.es/desarrollo/EficienciaEnergetica/CertificacionEnergetica/Documentos/Documentos%20informativos/2021_Informe_seguimiento_9_CERTIFICACION_ENERGETICA.pdf>. In France, almost 17% of the housing stock is labelled F or G, according to the stats of Ministère de la Transition écologique et de la Cohésion des territoires, 'Chiffres clés de l'énergie. 2021' <https://www.statistiques.developpement-durable.gouv.fr/edition-numerique/chiffres-cles-energie-2021/8-efficacite-energetique> accessed 27.12.2022.

13 Núria Lambea Llop, *La gestión de la vivienda social en clave europea* (Tirant lo Blanch, Valencia 2022), 35 and ff.

14 Consolidated Version of the Treaty on European Union (2008) OJ C115/13.

15 According to Caterini, sustainable development ensures that economic growth is consistent with human dignity and puts people at the centre of the European legal system. Enrico Caterini, 'Sustainability and civil law' [2018] 4(2) *ItaLJ* (https://doi.org/10.23815/2421-2156.ITALJ).

16 Klaus Bosselmann, *The Principle of Sustainability: Transforming Law and Governance* (2nd edn, London, Routledge 2016).

17 Prue Taylor and David Grilinton, 'Chapter 1. Property rights and sustainability: Toward a new vision of property' in David Grinlinton and Prue Taylor (eds), *Property Rights and Sustainability* (Brill, Nijhoff 2011). In addition, Ulfbeck and Hansen studied the role of sustainability clauses in commercial contracts. They consider that traditional contract law was not designed to address the sustainability aspects of commercial transactions, which raises the question of whether the default rules of contract law need to be revised. Vibe Ulfbeck and Ole Hansen, 'Sustainability clauses in an unsustainable contract law?' [2020] 16(1) *European Review of Contract Law* (https://doi.org/10.1515/ercl-2020-0010).

18 European Parliament, 'Resolution of 17 September 2020 on maximising the energy efficiency potential of the EU building stock' 2020/2070(INI).

19 Directive (EU) 2023/1791 of the European Parliament and of the Council of 13 September 2023 on energy efficiency and amending Regulation (EU) 2023/955 (recast) Directive (EU) 2023/1791 of the European

Parliament and of the Council of 13 September 2023 on energy efficiency and amending Regulation (EU) 2023/955 (recast), OJ L 231.

20 Luca Castellazzi, Paolo Bertoldi and Marina Economidou, 'Overcoming the split incentive barrier in the building sector' [2017] *JRC Technical Reports, European Commission* <https://op.europa.eu/en/publication-detail/-/publication/ae5716d7-fb39-11e7-b8f5-01aa75ed71a1/language-en> accessed 24.09.2023.

21 Xin Liang, Yi Peng and Geoffrey Qiping Shen, 'A game theory-based analysis of decision making for green retrofit under different occupancy types' [2016] 137 *Journal of Cleaner Production* (https://doi.org/10.1016/j.jclepro.2016.07.200). Stefano Celinni, 'Split incentives and endogenous inattention in home retrofits uptake: A story of selection on unobservables?' [2021] 104(105656) *Energy Economics* (https://doi.org/10.1016/j.eneco.2021.105656).

22 Johannes Lange, *Die Neuordnung der energetischen Modernisierung im Recht der Wohnraummiete - Zur Umsetzung klima- und umweltpolitischer Ziele mit den Mitteln des privaten Vertragsrechts* (LIT Verlag, Münster 2018).

23 Spain. Act 29/1994, 24 November, on Urban Leases. BOE núm. 282, 25.11.1994.

24 Italy. *Legge* núm. 392, 27.7.1978.

25 France. Loi núm. 89-462, 6 julliet 1989, tendant à améliorer les rapports locatifs.

26 Hongguang Nie et al, 'Split incentive effects on the adoption of technical and behavioural energy-saving measures in the household sector in Western Europe' [2020] 140 *Energy Policy* (https://doi.org/10.1016/j.enpol.2020.111424).

27 Dorothée Charlier, 'Energy efficiency investments in the context of split incentives among French households' [2015] 87 *Energy Policy* (https://doi.org/10.1016/j.enpol.2015.09.005).

28 Joint Research Center, 'Energy efficiency upgrades in multi-owner residential buildings' [2018], *JRC Technical Reports, European Commission* <https://publications.jrc.ec.europa.eu/repository/bitstream/JRC110289/energy_efficiency_upgrades_in_multiowner_apartment_buildings_final.pdf> accessed 27.12.2022.

29 Stefano Celinni, 'Split incentives and endogenous inattention in home retrofits uptake: A story of selection on unobservables?' [2021] 104(105656) *Energy Economics* (https://doi.org/10.1016/j.eneco.2021.105656).

30 Roger Burridge and David Ormandy, 'Health and safety at home: Private and public responsibility for unsatisfactory housing conditions' [2007] 34(4) *Journal of Law and Society*.

31 Hazel Easthope, 'Making a rental property home' [2014] 29(5) *Housing Studies* (https://doi.org/10.1080/02673037.2013.873115).

32 Kath Hulse and Vivienne Milligan, 'Secure occupancy: A new framework for analysing security in rental housing' [2014] 29(5) *Housing Studies* 644 (https://doi.org/10.1080/02673037.2013.873116).

33 Federal Institute for Research on Building, Urban Affairs and Spatial Development (BBSR) within the Federal Office for Building and Regional Planning (BBR), *Tenancy law and energy renovation in European comparison*' [2016] BBSR-Online-Publikation 14/2016 <https://www.bbsr.bund.de/BBSR/EN/publications/OnlinePublications/2016/bbsr-online-14-2016-dl.pdf?__blob=publicationFile&v=1> accessed 27.12.2022.

34 Federal Institute for Research on Building, Urban Affairs and Spatial Development (BBSR) within the Federal Office for Building and Regional Planning (BBR), '*Tenancy law and energy renovation in European comparison*' [2016] BBSR-Online-Publikation 14/2016 <https://www.bbsr.bund.de/BBSR/EN/publications/OnlinePublications/2016/bbsr-online-14-2016-dl.pdf?__blob=publicationFile&v=1> accessed 27.12.2022.

35 Konstantin A. Kholodilin, 'Rent control effects through the lens of empirical research' (2022) *DIW Roundup Discussion Papers*, num. 139 <https://www.diw.de/documents/publikationen/73/diw_01.c.833177.de/diw_roundup_139_en.pdf> accessed 27.12.2022.

36 Choon-Geol Moon and Janet G. Stotsky, 'The effect of rent control on housing quality change: A longitudinal analysis' [1993] 101(6) *Journal of Political Economy*.

37 Jesse Melvin, 'The split incentives energy efficiency problem: Evidence of underinvestment by landlords' [2018] 115 *Energy Policy* (https://doi.org/10.1016/j.enpol.2017.11.069).

38 Ad Straub, 'Maintenance and repair' in Susan J. Smith (ed), *International Encyclopaedia of Housing and Home*, vol 2 (Elsevier, Amsterdam 2012).

39 Federico A. Rodríguez Morata, 'Régimen de las obras y mejoras en la Ley de Arrendamientos Urbanos de 1994' [1995] 3 *Aranzadi civil: revista quincenal* 173.

40 Johannes Lange, *Die Neuordnung der energetischen Modernisierung im Recht der Wohnraummiete - Zur Umsetzung klima- und umweltpolitischer Ziele mit den Mitteln des privaten Vertragsrechts* (LIT Verlag, Münster 2018).

41 Yvonne Phillips, 'Landlords versus tenants: Information asymmetry and mismatched preferences for home energy efficiency' [2012] 45 *Energy Policy* (https://doi.org/10.1016/j.enpol.2012.01.067).

42 A 'renoviction' is a situation in which a landlord evicts a tenant from a rental property in order to perform renovations. Renovictions can be controversial because they can result in the displacement of long-term tenants, who may have difficulty finding affordable housing in the same neighbourhood. Increasing the rent after energy efficiency works may also lead to hidden renovictions, since the tenant, who cannot afford to pay the updated rent, will have to leave the dwelling. Dominika V. Polanska and Ase Richard, 'Resisting renovictions: Tenants organizing against housing companies' renewal practices in Sweden' [2021] 3(1) *Radical Housing Journal* (https://doi.org/10.54825/BNLM3593).

43 Michaela Lang et al, 'Systematic review: Landlords' willingness to retrofit energy efficiency improvements' [2021] 303(127041) *Journal of Cleaner Production* (https://doi.org/10.1016/j.jclepro.2021.127041).

44 Kathryn B. Janda et al, 'The evolution of green leases: Towards inter-organizational environmental governance' [2016] 44(5–6) *Building Research & Information* (https://doi.org/10.1080/09613218.2016.1142811).

45 Dave Collins, 'Green leases and green leasing in theory and in practice: A state of the art review' [2018] 37(11/12) *Facilities* (https://doi.org/10.1108/F-02-2017-0023).

46 France. Loi n° 2019-1147 du 8 novembre 2019 relative à l'énergie et au climat. JORF num. 0261, 9.11.2019.

47 According to the report BBSR, only Denmark, Germany, the Netherlands, Austria, Poland, and France include specific provisions for rent increases after energy efficiency works. Spain, which has a rental regime not analysed in the report, should also be added. Federal Institute for Research on Building, Urban Affairs and Spatial Development (BBSR) within the Federal Office for Building and Regional Planning (BBR), '*Tenancy law and energy renovation in European comparison*' [2016] BBSR-Online-Publikation 14/2016 <https://www.bbsr.bund.de/BBSR/EN/publications/OnlinePublications/2016/bbsr-online-14-2016-dl.pdf?__blob=publicationFile&v=1> accessed 27.12.2022.

48 Eurostat. Distribution of population by tenure status, type of household and income group. EU-SILC survey. Data for 2021.

49 Julius von Staudingers, Kommentar zum Bürgerlichen Gesetzbuch: Staudinger BGB - Buch 2: Recht der Schuldverhältnisse: §§ 535-556g (Sellier, de Gruyter, Berlin 2021) 355.

50 Sergio Nasarre Aznar, 'Leases as an alternative to homeownership in Europe. Some key legal aspects' [2014] 22(6) *European Review of Private Law* (https://doi.org/10.54648/erpl2014064).

51 Federal Institute for Research on Building, Urban Affairs and Spatial Development (BBSR) within the Federal Office for Building and Regional Planning (BBR), '*Tenancy law and energy renovation in European comparison*' [2016] BBSR-Online-Publikation 14/2016 <https://www.bbsr.bund.de/BBSR/EN/publications/OnlinePublications/2016/bbsr-online-14-2016-dl.pdf?__blob=publicationFile&v=1> accessed 27.12.2022.

52 Germany. *Gesetz zur Mietenbegrenzung im Wohnungswesen in Berlin* (MietenWoG Bln), of 11.2.2020. GVBl. 2020, 50.

53 DOGC (Official Gazette of Catalonia) num. 8229, 21.9.2020.

54 BOE num. 124, 25.5.2023.

55 According to this article, a large landlord is the person (natural or legal) who owns ten or more dwellings or more than 1,500 sqm for residential uses. In stressed areas, large landlords may also be those ones owning five or more dwellings.

56 Nicolas Damas et al, *Droit et practique des baux d'habitation* (9th edn, Dalloz, Paris 2019) 343.

57 Act num. 2014-366, 24.3.2014, *pour l'accès au logement et un urbanisme rénové*. JORF num. 0072, 26.3.2014.

58 Act num. 2018-1021, 23.11.2014, *portant évolution du logement, de l'aménagement et du numérique*. JORF num. 0272, 24.11.2018.

59 Philippe Pelletier, *Traité Des Baux d'habitation et Professionnels* (Paris, Editions Le Moniteur 2021) 344.

60 Federico A. Rodríguez Morata, 'Régimen de las obras y mejoras en la Ley de Arrendamientos Urbanos de 1994' [1995] 3 *Aranzadi civil: revista quincenal* 173.

61 Celia Martínez Escribano, 'Art. 22. Obras de mejora' in Fernando Crespo Allué and Vicente Guilare Gutiérrez (dirs). *Comentarios a la Ley de Arrendamientos Urbanos* (Thomson Reuters, Cizur Menor 2014).

62 Joaquín Ataz López, 'Art. 19. Elevación de renta por mejoras' in Rodrigo Bercovitz Rodríguez-Cano (coord), *Comentarios a la Ley de Arrendamientos Urbanos* (6th edn., Thomson Reuters Aranzadi, Cizur Menor 2013) 632.

63 Joaquín Ataz López, 'Art. 19. Elevación de renta por mejoras' in Rodrigo Bercovitz Rodríguez-Cano (coord), *Comentarios a la Ley de Arrendamientos Urbanos* (6th edn., Thomson Reuters Aranzadi, Cizur Menor 2013) 626.

64 This is also the opinion of Vicente Guilarte Gutiérrez, 'Art. 19. Elevación de renta por mejoras' Fernando Crespo Allué and Vicente Guilare Gutiérrez (dirs). *Comentarios a la Ley de Arrendamientos Urbanos* (Thomson Reuters, Cizur Menor 2014) 414.

65 However, art. 17.5 LAU does allow agreements, within the initial contract, by which renovations (including energy efficiency works) are performed or afforded by the tenant in lieu of rent, as a way to facilitate the access to housing of vulnerable people who cannot afford to pay a rent in money, but they have construction skills. Rosa Maria Garcia Teruel, *La sustitución de la renta por la rehabilitación o reforma de la vivienda en los arrendamientos urbanos* (Tirant lo Blanch, Valencia 2019).

66 According to art. 19.1 LAU, parties may agree that performing improvement works does not involve an increase of the rent. Moreover, they could also agree on other methods to calculate this increase, provided that it is not more harmful for the tenant (art. 6 LAU). Rosa Maria Garcia Teruel, 'Renovations in lieu of rent in Spanish tenancy law' [2018] 10(2) *Journal of Property, Planning and Environmental Law* (https://doi.org/10.1108/JPPEL-02-2018-0006).

67 As an example, a landlord invests 20,000€ plus 3,000€, thanks to a public allowance in a rented dwelling whose tenant pays 1,000€ each month (12,000€ annually). Rent will be increased by applying the annual interest rate at the time of completion of the works (e.g., a 3% increase in 2022), plus 3 points (a total of 6%), to the capital invested without considering the allowance (20,000€). A 6% of 20,000€ results in 1,200€ each year; thus, the annual rent would increase by 10% (13,200€ annually, or 1,100€ each month).

68 Comissió de Codificació de Catalunya. Avantprojecte de llei d'incorporació del contracte d'arrendament de béns al llibre sisè del Codi civil de Catalunya. 1.8.2022.

69 France. Loi n° 2021-1104 du 22 août 2021 portant lutte contre le dérèglement climatique et renforcement de la résilience face à ses effets. JORF núm. 0196, 24.8.2021.

70 Repairs and maintenance are excluded. The type of work, whether improvement or repair, shall be determined on a case-by-case basis. Philippe Pelletier, *Traité Des Baux d'habitation et Professionnels* (Paris, Editions Le Moniteur 2021) 397.

71 France. Décret n° 2021-19 du 11 janvier 2021 relatif au critère de performance énergétique dans la définition du logement décent en France métropolitaine. JORF núm. 0011, 13.1.2021.

72 Art. 7.e *Loi rapports locatifs* refers to art. 1724 of the French Civil Code, regulating the reduction of the rent.

73 France. Décret n° 2009-1439 du 23 novembre 2009 pris en application de l'article 23-1 de la loi n° 89-462 du 6 juillet 1989 tendant à améliorer les rapports locatifs relatif à la contribution du locataire au partage des économies de charges issues des travaux d'économie d'énergie réalisés par un bailleur privé. JORF núm. 0273, 25.11.2009; See also Arrêté du 23 novembre 2009 relatif à la contribution du locataire au partage des économies de charges issues des travaux d'économie d'énergie réalisés par un bailleur privé. JORF núm. 0273, 25.11.2009.

74 Provided that the landlord has a maximum of three rented properties.

75 Published on the 11.3.2013, BGBL I S. 434.

76 The concept 'structural change' shall be interpreted widely, according to Julius von Staudingers, *Kommentar zum Bürgerlichen Gesetzbuch: Staudinger BGB - Buch 2: Recht der Schuldverhältnisse: §§ 535-556g* (Sellier, de Gruyter, Berlin 2021) 736.

77 Germany. Gesetzentwurf der Bundesregierung Entwurf eines Gesetzes über die energetische Modernisierung von vermietetem Wohnraum und über die vereinfachte Durchsetzung von Räumungstiteln (Miet rechtsänderungsgesetz – MietRÄndG). 15.8.2012.

78 See also the critique of Lange to the exemption of modernization measures from the *Mietminderung* regime: this exemption, according to the author, may challenge the principle of equivalence. Johannes Lange, *Die Neuordnung der energetischen Modernisierung im Recht der Wohnraummiete - Zur Umsetzung klima- und umweltpolitischer Ziele mit den Mitteln des privaten Vertragsrechts* (LIT Verlag, Münster 2018) 204 and ff.

79 Empirical studies showed that this rent control only worked for avoiding higher rents in regulated contracts, but made non-regulated contracts (e.g., rental contracts of cities not subject to rent controls) more expensive or even reduced the rented housing stock in those areas. However, according to Kholodilin et al., rents were moderated by *Mietpreisbremse* in municipalities where new rental contracts increased more than 3.9%. Konstantin A. Kholodilin, Andreas Mense and Claus Michelsen, 'Mietpreisbremse ist

besser als ihr Ruf, aber nicht die Lösung des Wohnungsmarktproblems' [2018] 85(7) *DIW Wochenbericht* (https://doi.org/10.18723/diw_wb:2018-7-1). Konstantin A. Kholodilin, Andreas Mense and Claus Michelsen, 'Die Mietpreisbremse wirkt bisher nicht' [2016] 22 *DIW Wochenbericht*.

80 Bastian Kossmann and Georg von Wangenheim, 'Failing to protect the tenant' (2015) University of Kassel <https://www.google.es/url?sa=t&rct=j&q=&esrc=s&source=web&cd=&ved=2ahUKEwi7z974u4SBAxUQVKQEHcN-DO8QFnoECBYQAQ&url=https%3A%2F%2Fwww.uni-kassel.de%2Ffb07%2Findex.php%3FeID%3DdumpFile%26t%3Df%26f%3D1781%26token%3D73d0195e7c775d07c557c1723a4519bfe68c581f&usg=AOvVaw0RyPZVLJFcOOfD98sQH2ON&opi=89978449> accessed 24.09.2023.

81 Konstantin A. Kholodilin, Andreas Mense and Claus Michelsen, 'Mietpreisbremse ist besser als ihr Ruf, aber nicht die Lösung des Wohnungsmarktproblems' [2018] 85(7) *DIW Wochenbericht* (https://doi.org/10.18723/diw_wb:2018-7-1). Konstantin A. Kholodilin, Andreas Mense and Claus Michelsen, 'Die Mietpreisbremse wirkt bisher nicht' [2016] 22 *DIW Wochenbericht*.

82 M. Michaela Lang et al, 'Systematic review: Landlords' willingness to retrofit energy efficiency improvements' [2021] 303(127041) *Journal of Cleaner Production* (https://doi.org/10.1016/j.jclepro.2021.127041).

Bibliography

Ataz López J, 'Art. 19. Elevación de renta por mejoras' in Rodrigo Bercovitz Rodríguez-Cano (coord), *Comentarios a la Ley de Arrendamientos Urbanos* (6th edn, Thomson Reuters Aranzadi, Cizur Menor 2013) 625–656

Bosselmann K, *The Principle of Sustainability: Transforming Law and Governance* (2nd edn, London, Routledge 2016)

Bradshaw W et al, 'The costs and benefits of green affordable housing' [2005] *Cambridge (MA): New Ecology* <https://subjectguides.york.ac.uk/ld.php?content_id=33345299>

Burridge R and Ormandy D, 'Health and safety at home: Private and public responsibility for unsatisfactory housing conditions' [2007] 34(4) 544–565, *Journal of Law and Society*

Castellazzi L, Bertoldi P and Economidou M, 'Overcoming the split incentive barrier in the building sector' [2017] *JRC Technical Reports, European Commission* <https://op.europa.eu/en/publication-detail/-/publication/ae5716d7-fb39-11e7-b8f5-01aa75ed71a1/language-en>

Caterini E, 'Sustainability and civil law' [2018] 4(2) *ItaLJ* (https://doi.org/10.23815/2421-2156.ITALJ)

Celinni S, 'Split incentives and endogenous inattention in home retrofits uptake: a story of selection on unobservables?' [2021] 104(105656) *Energy Economics* (https://doi.org/10.1016/j.eneco.2021.105656)

Charlier D, 'Energy efficiency investments in the context of split incentives among French households' [2015] 87 *Energy Policy* (https://doi.org/10.1016/j.enpol.2015.09.005)

Collins D, 'Green leases and green leasing in theory and in practice: A state of the art review' [2018] 37(11/12) *Facilities* (https://doi.org/10.1108/F-02-2017-0023)

Commission, 'Energy efficiency buildings' <https://ec.europa.eu/energy/en/topics/energy-efficiency/buildings>

Costa-Campi MT, Jové-Llopis E and Trujillo-Baute E, 'La pobreza energética en España. Aproximación desde una perspectiva de ingresos' [2019] *Fundación Naturgy* <https://www.fundacionnaturgy.org/publicacion/la-pobreza-energetica-en-espana-aproximacion-desde-una-perspectiva-de-ingresos/>

Damas N et al, *Droit et practique des baux d'habitation* (9th edn, Dalloz, Paris 2019)

Easthope H, 'Making a rental property home' [2014] 29(5) *Housing Studies* (https://doi.org/10.1080/02673037.2013.873115)

Federal Institute for Research on Building, Urban Affairs and Spatial Development (BBSR) within the Federal Office for Building and Regional Planning (BBR), '*Tenancy law and energy renovation in European comparison*' [2016] BBSR-Online-Publikation 14/2016 <https://www.bbsr.bund.de/BBSR/EN/publications/OnlinePublications/2016/bbsr-online-14-2016-dl.pdf?__blob=publicationFile&v=1>

Garcia Teruel RM, 'Renovations in lieu of rent in Spanish tenancy law' [2018] 10(2) *Journal of Property, Planning and Environmental Law* (https://doi.org/10.1108/JPPEL-02-2018-0006)

Garcia Teruel RM, *La sustitución de la renta por la rehabilitación o reforma de la vivienda en los arrendamientos urbanos* (Tirant lo Blanch, Valencia 2019)

Guilarte Gutiérrez V, 'Art. 19. Elevación de renta por mejoras' in Fernando Crespo Allué and Vicente Guilare Gutiérrez (dirs). *Comentarios a la Ley de Arrendamientos Urbanos* (Thomson Reuters, Cizur Menor 2014) 414–427

Hamilton I et al, 'Health effects of home energy efficiency interventions in England: A modelling study' [2015] 5(4) *BMJ Open* (https://doi.org/10.1136/bmjopen-2014-007298)

Hulse K and Milligan V, 'Secure occupancy: A new framework for analysing security in rental housing' [2014] 29(5) *Housing Studies* 644 (https://doi.org/10.1080/02673037.2013.873116)

IDAE, Estado de la certificación energética de los edificios (9th report) [2020], *Ministerio para la Transición Ecológica y el Reto Demográfico* <https://energia.gob.es/desarrollo/EficienciaEnergetica/CertificacionEnergetica/Documentos/Documentos%20informativos/2021_Informe_seguimiento_9_CERTIFICACION_ENERGETICA.pdf>

Janda KB et al, 'The evolution of green leases: Towards inter-organizational environmental governance' [2016] 44(5–6) *Building Research & Information* (https://doi.org/10.1080/09613218.2016.1142811)

Joint Research Center, 'Energy efficiency upgrades in multi-owner residential buildings' [2018], *JRC Technical Reports, European Commission* <https://publications.jrc.ec.europa.eu/repository/bitstream/JRC110289/energy_efficiency_upgrades_in_multiowner_apartment_buildings_final.pdf>

Kholodilin KA, 'Rent control effects through the lens of empirical research' (2022) *DIW Roundup Discussion Papers*, num. 139 <https://www.diw.de/documents/publikationen/73/diw_01.c.833177.de/diw_roundup_139_en.pdf>

Kholodilin KA, Mense A and Michelsen C, 'Die Mietpreisbremse wirkt bisher nicht' [2016] 22 *DIW Wochenbericht*

Kholodilin KA, Mense A and Michelsen C, 'Mietpreisbremse ist besser als ihr Ruf, aber nicht die Lösung des Wohnungsmarktproblems' [2018] 85(7) *DIW Wochenbericht* <https://doi.org/10.18723/diw_wb:2018-7-1>

Kossmann B and von Wangenheim G, 'Failing to protect the tenant' (2015) University of Kassel <https://www.google.es/url?sa=t&rct=j&q=&esrc=s&source=web&cd=&ved=2ahUKEwi7z974u4SBAxUQVKQEHcN-DO8QFnoECBYQAQ&url=https%3A%2F%2Fwww.uni-kassel.de%2Ffb07%2Findex.php%3FeID%3DdumpFile%26t%3Df%26f%3D1781%26token%3D73d0195e7c775d07c557c1723a4519bfe68c581f&usg=AOvVaw0RyPZVLJFcOOfD98sQH2ON&opi=89978449>

Lambea Llop N, *La gestión de la vivienda social en clave europea* (Tirant lo Blanch, Valencia 2022)

Lang M et al, 'Systematic review: Landlords' willingness to retrofit energy efficiency improvements' [2021] 303(127041) *Journal of Cleaner Production* (https://doi.org/10.1016/j.jclepro.2021.127041)

Lange J, *Die Neuordnung der energetischen Modernisierung im Recht der Wohnraummiete - Zur Umsetzung klima- und umweltpolitischer Ziele mit den Mitteln des privaten Vertragsrechts* (LIT Verlag, Münster 2018)

Liang X, Peng Y and Shen JQ, 'A game theory-based analysis of decision making for green retrofit under different occupancy types' [2016] 137 *Journal of Cleaner Production* (https://doi.org/10.1016/j.jclepro.2016.07.200)

Martínez Escribano C, 'Art. 22. Obras de mejora' in Fernando Crespo Allué and Vicente Guilare Gutiérrez (dirs). *Comentarios a la Ley de Arrendamientos Urbanos* (Thomson Reuters, Cizur Menor 2014)

Melvin J, 'The split incentives energy efficiency problem: Evidence of underinvestment by landlords' [2018] 115 *Energy Policy* (https://doi.org/10.1016/j.enpol.2017.11.069)

Moon C-G and Stotsky JG, 'The effect of rent control on housing quality change: A longitudinal analysis' [1993] 101(6) *Journal of Political Economy* 1114–1148

Nasarre-Aznar S, 'Leases as an alternative to homeownership in Europe. Some key legal aspects' [2014] 22(6) *European Review of Private Law* (https://doi.org/10.54648/erpl2014064)

Nasarre-Aznar S et al, 'Concrete actions for social and affordable housing in the EU' [2021] *The Foundation for European Progressive Studies (FEPS)* <https://feps-europe.eu/wp-content/uploads/2021/01/Concrete-actions-for-social-and-affordable-housing-in-the-EU.pdf>

Nie H et al, 'Split incentive effects on the adoption of technical and behavioural energy-saving measures in the household sector in Western Europe' [2020] 140 *Energy Policy* (https://doi.org/10.1016/j.enpol.2020.111424)

Pelletier P, *Traité Des Baux d'habitation et Professionnels* (Paris, Editions Le Moniteur 2021)

Phillips Y, 'Landlords versus tenants: Information asymmetry and mismatched preferences for home energy efficiency' [2012] 45 *Energy Policy* (https://doi.org/10.1016/j.enpol.2012.01.067)

Polanska DV and Richard A, 'Resisting renovictions: Tenants organizing against housing companies' renewal practices in Sweden' [2021] 3(1) *Radical Housing Journal* (https://doi.org/10.54825/BNLM3593)

Poortinga W et al, 'The health impacts of energy performance investments in low-income areas: A mixed-methods approach' [2018] 5 *Public Health Res* (https://doi.org/10.3310/phr06050)

Rodríguez Morata FA, 'Régimen de las obras y mejoras en la Ley de Arrendamientos Urbanos de 1994' [1995] 3 173 *Aranzadi civil: revista quincenal*

Straub A, 'Maintenance and repair' in Susan J. Smith (ed), *International Encyclopaedia of Housing and Home*, vol 2 (Elsevier, Amsterdam 2012) 168–172

Taylor P and Grilinton D, 'Chapter 1. Property rights and sustainability: Toward a new vision of property' in David Grinlinton and Prue Taylor (eds), *Property Rights and Sustainability* (Brill, Nijhoff 2011) 1–20

Ulfbeck V and Hansen O, 'Sustainability clauses in an unsustainable contract law?' [2020] 16(1) *European Review of Contract Law* (https://doi.org/10.1515/ercl-2020-0010)

von Staudingers J, *Kommentar zum Bürgerlichen Gesetzbuch: Staudinger BGB - Buch 2: Recht der Schuldverhältnisse: §§ 535-556g* (Sellier, de Gruyter, Berlin 2021)

5

WHO'S AFRAID OF SUSTAINABILITY? A PRIMER FOR MAINSTREAMING SUSTAINABILITY IN PRIVATE LAW EDUCATION

Candida Leone[1]

5.1 Introduction

If you are reading this book, you may have acquired a sense that private law is part of the problem – an aggravating factor in the ongoing climate and broader ecological crisis.[2] While sustainability is increasingly on the agenda of private law research, efforts to systematically integrate the theme in private law teaching appear to raise significant challenges of practical and theoretical nature. The chapter aims to address some of these challenges. First, how to square appropriate consideration of such a large-scale problem with classical tenets of legal certainty and justice between the parties, without falling into a reductionism similar to Law and Economics analysis? Second, how to include sustainability-related issues without (excessively) sacrificing traditionally 'core' content, so often put under pressure by workload, student numbers, and other curriculum transformations? And in doing the latter, how to prevent polarization? Given the central position of private law in (unsustainable) legal practices, the challenge of doing our part in educating 'climate-conscious' legal practitioners is one we cannot avoid.[3]

This chapter tackles the challenges laid out above in two parts, roughly connecting to theoretical and educational/practical questions, respectively. The first part succinctly articulates ways in which we can think of sustainability in order to make sense of its role in private law; first, it shows that 'weak' sustainability is relatively easy to incorporate once we, on the examples of leading contemporary European private law theories, make room for pluralism; second, it presents 'stronger' versions of sustainability by means of emerging sustainability-oriented, care-based approaches often neglected in general private law theory. This should make it easier to overcome the uncertainties of those students and colleagues currently convinced that private law has and should have nothing to do with sustainability. The second part will combine insights from emerging literature on integrating sustainability in education at large and from contemporary work on legal education to provide a high-level but operational and modular blueprint for incorporating sustainability in European private law curricula. While none of the very significant discussion on climate litigation (a great source of class material!) is addressed, attention will be paid to the role the Sustainable

DOI: 10.4324/9781032662046-6

Development Goals (henceforth SDGs) can play as a vehicle to 'codify' sustainability for core private law education. Key focal points on this part, however, will be not (only) questions regarding what to incorporate into the existing curricula but, in particular, issues of methodology and forms of instruction and assessment. The aim of this chapter is not forwarding one specific approach, but rather to review available insights and provide an accessible set of tools to facilitate the incorporation of sustainability by those lecturers who are *not* directly engaging with the subject in their research or core concerns.

Not all readers of – or even contributors to – this book will embrace the more radical readings presented in this chapter. The present overview, hence, seeks to provide insight for both those who embrace stronger, system-changing notions of sustainability and for colleagues who prefer 'weaker' sustainability, mainly achieved by reforming current practices to make them less unsustainable. At the same time, the chapter assumes that legal education treating the approaches as complementary is ultimately an almost unavoidable solution if students are to be meaningfully educated on sustainability within the material and (broadly intended) disciplinary boundaries of legal education.[4]

5.2 Thinking about Private Law and Sustainability

When I asked students in a course on *Principles of Private Law* to define what private law was about, an often-recurring answer was that private law is the law concerned with private interest, or even better, the law where the public interest does not matter.[5] Balancing interests between the parties marked the boundaries of the subject in these definitions. This is, of course, a widely shared intuition among private lawyers. However, it is an intuition that we should not accept as given and that deserves an open challenge if we are to incorporate sustainability in education. Fortunately, challenges abound in contemporary debates; here, I will limit myself to some that are currently prominent in the (European) English-language debate.

Discussions about the normative underpinnings of private law have been flourishing over the past decades. A major trigger for such debates has been the process of 'Europeanisation' of formerly largely self-sufficient national private law systems.[6] In this context, it is by now common to juxtapose the traditional non-instrumental orientation of national private law systems with the 'regulatory' function of the so-called 'European private law'.[7] Of course, we know that this is a simplification. Not only do the national systems display a broader array of concerns than (formal) corrective justice, but it is also becoming clear that European private law does not entirely disregard the 'private law perspective'.[8] Europeanization, however, has, without doubt, brought the tensions to the forefront.[9] While in North America the response to instrumentalization has been marked by a host of theories essentially claiming that 'private law's only goal is to be private law',[10] European theory and practice seem to have more comfortably aligned in accepting that a degree of pluralism characterizes the foundations of contemporary private laws.

5.2.1 Pluralism and Weak Sustainability

As far as European private law theory is concerned, while a considerable stream of literature has emphasized the contrast between 'traditional', corrective justice-oriented, national private laws and 'instrumental' European private law, it seems significant that recent ambitious attempts at theorization have sought to push the distinction into the background, foregrounding that *multiple reasons* characterize core questions of private law irrespective of where the legislative answers to such questions are to be located. In this way, for instance, Hesselink has shown how, as a matter of existing positive law, it is impossible to trace back current European contract law[11] to one unified goal or set of goals. Furthermore, Grundmann, Micklitz, and Renner's 'New private law theory',[12] which acknowledges politics and constitutional arrangements as both one of the ways and as one of the *reasons* giving shape to contemporary private laws, may be seen as even more pluralist in foregrounding the possibility of local variations.

These theorizations seem to ultimately resonate with more doctrinal practices. For instance, in 2008–2009, the editors of the Common Frame of Reference ('CFR') identified 'freedom, security, justice and efficiency' as the four foundational principles that would, one can imagine, guide interpretation/understanding of the model rules.[13] At the same time, they included a long list of additionally relevant principles ranging from good faith to human rights, not all of which could be easily traced back to justice between the parties and legal certainty.[14] Neither the mainstream theorizations above nor the CFR[15] give significant space to sustainability. This fact, however, should be seen as a reminder of the contextual limitations of those efforts more than as a sign that sustainability is not part of private law (theory).[16]

If any of the projects above were started today, it seems unlikely that they would omit to engage with sustainability. Given the proliferation of initiatives to make consumption, production, and exchanges less unsustainable,[17] it seems hard to claim today that at least a 'weak' notion of sustainability would not be part of the broad palette of principles and goals[18] informing contemporary private laws in Europe. Such 'weak' notion of sustainability is the one most clearly reflected in projects that aim to use private law to facilitate sustainable consumption, accelerate the development of sustainable technologies, or design good financial assets to incentivize carbon or biodiversity offsets. Reporting obligations, aimed at 'making sustainability profitable' via market signalling, seem to go in a similar direction. While acknowledging the urgency of legal, technological, and economic mobilization towards more sustainable human societies, this notion of sustainability does not foreground absolute limits or envisage deep reconsiderations of current structures.[19]

From Hesselink's perspective, it seems especially relevant to notice how these developments have elicited tense disagreements as to the exact contours and content of the binding force of contract (right to repair),[20] public policy in contract law (due diligence obligations),[21] and the protection of weaker parties (for instance, in the debate on distributing the burdens of the energy transition).[22]

Grundmann et al.,[23] in particular, place much emphasis on the relevance of constitutional frameworks for the development of contemporary private laws and theories. From their perspective, hence, the question can be asked regarding the extent to which sustainability can be considered an emerging constitutional principle. In this respect, it seems fair to say that legal systems may differ. However, it would be hard to ignore the accelerating developments at both national[24] and international levels. In particular, the 2022 United Nations declaration of the human right to a clean, healthy, and sustainable environment[25] can be expected to increasingly reverberate, through human and fundamental rights discourses, in private law systems across the continent.[26] Within the European legal order, it has further been argued that Article 11 TFEU on promoting sustainable development across EU policies could be used as a point of reference for European private law.[27] In this context, the specific parameters and targets associated with individual SDGs can be used as an anchoring point to translate the idea of sustainable development to private law doctrines (see infra 3.1). This is consistent with the so-called 'environmental integration principle'.[28]

From the comparative doctrinal perspective incorporated in the DCFR, the doctrinal 'discovery' of a principle of sustainability can also take place bottom-up, by induction. The principle could be abstracted from the ever-growing list of legislative proposals and instruments that aim to incorporate sustainability-related concerns in private relations. Where sustainability is not explicitly adopted as a principle by private law legislation, this 'avenue to sustainability' may suit the disciplinary sensitivity of some European private lawyers better than the constitutional or any other top-down approach. At the same time, sustainability seems to be emerging as an explicitly acknowledged higher-level principle *within private law*. Granted, no system in Europe seems to go as far as the Chinese civil code, which includes a provision according to which '[w]hen conducting a civil activity, a person of the civil law shall act in a manner that facilitates conservation of resources and protection of the ecological environment'.[29] While often criticized as aspirational, this provision – which is further specified in other articles of the Code – appears quite innovative and has sparked a wealth of works aiming to unravel its systematic function.[30] Some movement in a similar direction, however, seems to be taking place on the continent as well. For instance, the recent Belgian reform of property law explicitly aims to start a sustained engagement with sustainability.[31]

5.2.2 Strong Sustainability: Rethinking Private Law's Foundations

If sustainability in a weak sense is hence relatively easy to incorporate in contemporary understandings of the foundations of private law, it is important to highlight that a host of emerging theoretical perspectives foreground stronger notions of sustainability. Such is the case of contemporary critical approaches, which emphasize how climate change (and environmental degradation) challenge private law's ability to see its 'abstract conceptual foundations – the autonomous individual, legal personality, legal right, security, boundary, alienability, exclusion, fault, agreement, limited liability, etc. – [as] theoretically unaffected by materiality'.[32] Sustainability best resonates with projects emphasizing care and obligation over liberal individualist frameworks, which have been predominant in modern private laws.[33] Obligation, in this context, is not cast as a top-down, vertical relation in the sense of state-centred or religious orders. Rather, it is presented and constructed as an acknowledgement of our human vulnerability and hence as 'obedience to the needs of ecological solidarity', which explicates its functions in horizontal interactions in a way that is even morally prior to political obligation.[34] In contrast to accounts based on autonomy and granularity (emphasizing privity of contract, strict relativity of tortious liability, and the absolute nature of ownership, etc.), thus, contemporary critique grounded in sustainability centres expanded notions of obligations and solidarity.[35] However, by embracing broad notions of mutual and systemic interdependence (often 'entanglement') that consider human and non-human lives as part of the relevant systems,[36] these accounts also go beyond traditional discourses of solidarity in private law[37] as well as beyond current attempts to draw 'relational' liberal approaches.[38] Such approaches all embrace 'strong' notions of sustainability which foreground the need for human activity to be clearly positioned within earth systems and function within planetary boundaries.[39] In these accounts, avoiding climate collapse is but one prominent example of urgent reasons to rethink humans' terms of engagement with the planet and each other.[40]

5.3 Sustainable Alignment

Let us assume that by this point we have accepted that sustainability is sufficiently relevant to private law to deserve a place in education. How to go about it? It must be first acknowledged that this debate does not happen in a vacuum. On the one hand, there is growing literature on the broad topic of integrating sustainability in education; on the other hand, debates on the goals and contents of (academic) legal education also seem on the rise across systems. In this chapter, while aware that other approaches are conceivable,[41] I tentatively place the broad contribution of legal education to integrating sustainability into higher education within the 'sustainability competencies' approach relied on, for instance, by KU Leuven.[42] I also rely on certain assumptions concerning legal education and sustainability. Both approaches will be briefly discussed below.

At the institutional level, educating students towards sustainability has two main prongs that can be identified.[43] Firstly, developing and offering *specialization* through courses and programmes specifically centred on sustainability. And secondly, incorporating sustainability into the 'core' of existing programmes, in alignment with fundamental subjects offered in that programme. If education is to meaningfully contribute to a sustainable transformation, both prongs need to be pursued at the same time. This is necessary in order to overcome the idea of the environment, and hence environmental problems, as being 'outside of human communities'. For law, this means overcoming the idea that 'environmental questions' are essentially a matter for 'environmental law' and hence irrelevant to other (i.e., most) legal domains.[44]

Private law education, given its position at the ideal and factual core of most legal curricula – in particular, at the LLB level – bears a particularly pressing responsibility when it comes to overcoming the separation between human activities and environmental concerns and hence 'mainstreaming' sustainability. While specialized courses or even programmes like 'Private law and sustainable development' are certainly conceivable, our prior efforts should be directed to fundamental courses that all students take in their first years and that contribute to shaping their most basic understandings of how the law relates to the social – and planetary! – whole. Hence, in line with the reference to 'mainstreaming' in the chapter's title, the following sections are written with mainly *core law programme* courses in mind – the ones that all students at a certain faculty are expected to pass in order to get a basic (academic) law degree.

5.3.1 Background: Sustainability Competencies and Developments in Legal Education

'Sustainability competencies', as a term and strategy, seek to capture and organize various sorts of learning objectives connected to sustainability.[45] Law as a discipline more or less inherently reflects two such competencies, namely, systems thinking and normative competencies.[46] Wiek et al. define systems thinking as 'the ability to collectively analyse complex systems across different domains (society, environment, economy, etc.) and across different scales (local to global)'. Such ability is necessary in order to understand, in essence, what factors keep a certain state of things in place and what factors make it 'move' in one or the other direction (that is, to more or less sustainable transformations).[47] This requires understanding 'structure, function, cause-effect relations, but also perceptions, motives, decisions, and regulations'.[48] Normative competence, in turn, is defined as the ability to 'collectively map, specify, apply, reconcile and negotiate sustainability values, principles, goals, and targets'.[49] Thus understood, normative competence is indeed crucial to navigating the implications of the climate crisis for societies: values and goals must be recognized, articulated, and consciously negotiated.[50] Environmental degradation is acknowledged by most as a problematic fact. However, because is a phenomenon on a global scale, it is not uncontroversial to claim that it directly justifies normative conclusions regarding ethical obligations for individuals.

At the same time, the existence of discord as to such normative conclusions in contemporary societies requires an ability to mediate between interests, goals, and values. As such, private law is a particularly fertile ground to see how these tensions play out.

As lawyers, we are familiar with both notions of systematic reasoning and normative exercises. Compared to the way these are defined in the context of sustainability, in law, we usually conceive these competencies in a more self-referential way of course – as related to the legal system as such and to a specific type of disagreement about norms (most often, their interpretation). However, in terms of the build-up of such competencies – requiring knowledge, skills, and a certain attitude-forming[51] – there is no essential difference between what we do and what we should do to begin to incorporate sustainability education throughout the curriculum. This, at least, is true to the extent that we understand legal education to entail more than knowledge of selected fields of positive law.

I have argued elsewhere[52] that sustainable legal education, applied to private law, requires openings in line with the ones advocated for by Grundmann, Micklitz, and Renner[53] for private law theory. These include

a openness to interdisciplinarity, meaning in this case selecting reading materials going beyond standard handbook/style, in order to clearly show the connection between private law and ecological questions;
b awareness of comparative insights, meaning taking cases and developments *from wherever they happen* as starting point for discussion;
c problem-oriented incorporation of theory: while discussions should focus on concrete questions in order to 'land' (see infra), students' learning should be supported by at least an understanding that practical developments happen against a wide host of debates questioning the nature and roles of law and even science at large;
d avoiding state-centredness, that is, taking relevant codes of conduct, model bylaws for sustainable enterprises, internal rules for commons management, and so forth, when available, just as seriously as code rules;
e taking critique of private law seriously, that is, at the very least reminding students that, for over a century now, the idea that private law rules are distributionally neutral has been challenged, and this is well beyond the later regulated domains of consumer and labour contracts.[54]

Such opening facilitates both systems thinking – by showing interactions and interdependences – and normative competencies – by 'naturally' presenting students with disagreements, diverging paths and sometimes stark conflicts. Classical nationally oriented, purely positivist handbooks purporting to authoritatively present (and exhaust) the state of the discussion cannot suffice, as mainstreaming sustainability at the necessary tempo requires learning from all the available knowledge sources and jurisdictions.[55]

The five tenets above mean, essentially, that lecturers should choose (but also: should feel free to choose!) materials that fit thematically without distinguishing too sharply between current positive law and 'future perspectives' or local and seemingly 'exotic' practices. Doing so will likely mean choices concerning what is 'core' content in institutional courses will have to be partially reconsidered – if public order or causation are taking on different meanings, it could be that other topics will have to take less space. While making such choices may be difficult and at times painful, it should not scare us – important issues have always had to be left out and priorities have already shifted in past times.

Crucially, the call for diversification that emerges from the picture painted in this section does not stand in a vacuum. For, besides growing appeals to incorporate sustainability in the law school curriculum in general, more explicit emphasis on 'knowledge, skills and attitude' is a broad trend in legal education.[56] In particular, persistent trends to incorporate professional skills have revealed the implicit forms of skills training characterizing different varieties of legal education and hence challenge us to articulate them more explicitly. Additionally, such engagement with 'external' actors also pushes law schools to openly ask who their interlocutors should be – and to realize that traditional legal professions are but one part of their societal counterparts. This calls for a broadly plural legal education.[57]

In the following sub-sections, the elements above are no longer individually addressed; rather, I present three 'levels' of engagement with sustainability in core private law courses, followed by a short reflection on the need for, and possible approaches to, examination.

Starting point is that every private law course will usually begin with introducing fundamental principles, e.g., for contract law, freedom of contract, reliance/*pacta sunt servanda*, relativity, or privity. On the one hand, the *critique* of these principles with an eye to sustainability is a helpful starting point for students to be put in a position to make sense of further connections.[58] Starting with pointing out the unsustainability of currently employed solutions rather than jumping to normative claims about what sustainability would require seems like a prerequisite to avoid immediate polarization in the classroom.[59] On the other hand, from a pluralist perspective, sustainability is clearly an emerging normative concern (see above) affecting the legislative and judicial development of private legal orders. Thus, evolving notions of public order[60] would be one relatively seamless way to introduce the Chinese 'green principle' example with a comparative approach. While educators may not all agree on the specific direction and extent of changes that confronting unsustainability will require, opening to critique and any approaches, such as comparison, that suggest the contingency of current arrangements, will help empower students to think that change is possible.[61]

5.3.2 One Well-Rounded Session

In order for students to take sustainability seriously while acknowledging their multiple diverging sources of pressure and motivation, an idea is emerging in the literature that it helps if the angle is not presented as external but internal to the legal framework – it should be presented as part of the contemporary legal panorama and not as a side thought or fringe perspective.[62] Ong has suggested that an 'incorporation of no more than two or three lecture contact hours within each of the "core"/compulsory subjects of the standard English university law curriculum' would already be a meaningful starting point in integrating sustainability.[63] While in most core courses it is going to be impossible to entirely redraw a module (or significant portions thereof) in line with sustainability ambitions, it is wise to make sure that the discussion does not remain at the level of general principles. Several experiences point out the need for focusing on examples that allow us to both illustrate and problematize the relationship between the broad subject discussed and sustainability.[64]

Introducing small case studies is a feasible approach that requires little time and makes the connections visible and relatable. In the common law context, Ong mentions, for instance, 'the potential for nuisance and/or negligence to remedy environmental damage',[65] property law and 'nature conservation covenants',[66] contract law, and the incorporation of environmental undertakings within contracts.[67] In light of the EU's commitment to sustainable development (see above), private law courses across the Union should be able to concretize sustainability standards through the United Nations 'SDGs' and their specific targets. In particular, SDG 12 concerning 'Sustainable consumption and production patterns' seems relevant to a broad range of private law topics. To give but a few examples:

- the connection between target 12.5 on waste and property regimes, including the so-called 'product stewardship' or extended producer responsibility, could be integrated into courses on property law[68];

- target 12.6 on sustainable practices and sustainability reporting for (larger) companies could be the starting point for discussing developments in company law;
- target 12.8 on (consumer) information and awareness concerning sustainable lifestyles could be integrated into various ways in contract law, especially in the general parts discussing, e.g., notions of contractual information;
- the intellectual property and especially licensing implications of target 12.a on knowledge transfer between richer and 'developing' countries[69] would help make the notions of 'special contracts' more relatable for young students.

In general, all these examples require a somewhat in-context approach but still have the advantage of making the subject matter salient and relatable for students. In some cases, they will also allow students to see forms of interdisciplinary literature review integrated into legal research.[70]

For education systems that rely more on application-oriented instruction, cases based on ongoing or recent litigation, domestically and (see Section 5.3.1) abroad, can be used as examples or help build case-based questions to be discussed in class or assigned as practice (see Section 5.3.5 on the crucial matter of examination). While connecting to global trends and goals such as the SGDs contributes to pointing at the systemic position of private law in sustainability transitions, students should be clearly invited to recognize the normative tensions underlying such transitions: who will bear the brunt once low-quality, non-repairable products disappear from European markets?

5.3.3 Developing a Theme

In some cases, the above could be used as starting point for longer-spanning engagement. The best way to make students engage with sustainability and fully appreciate the connection between private law and the environmental crisis would likely be a well-designed form of problem-solving-based learning. For instance, students could be tasked with drafting enforceable contract conditions for a university that seeks to make its ICT compartment more sustainable but struggles to bind its suppliers to circular business models.[71]

Consistent incorporation can also take place in more traditional, topic-based teaching. Depending on the course, it may be advisable to choose one specific theme (e.g., circular economy and movables or owner obligations and land law) to be developed over time, or to include several examples (an overview course on institutions of private law could fish from several of the suggestions above). One thing that is important, however, is to consider how connections that seem obvious to the instructors may not be as clear to students – an observation that calls for making clear choices even at the cost of sacrificing very real complexity.[72]

How the theme is to be developed depends at this stage on the sensitivities, background, and resources of individual (groups of) instructors. While a one-off session will probably only leave enough space to reflect with students on what *is* happening in private law in reaction to increased awareness of the ongoing ecological breakdown, sustained engagement with a theme could give more space for open discussions on what *should* happen: even with a large group, students could be prompted to discuss in groups (outside the classroom!) and propose in class *one* change to the current rules, inspired by comparative examples or literature presented in the course.

5.3.4 Taking the Critique into Practice

Today's students will 'conduct their entire practice in a web of climate law'[73] – and in general in a world where sustainability is no longer a niche subject. This doesn't take away that the practice of private law is not providing a continuously sustained contribution to the maintenance and flourishing of unsustainable economic activities.[74] While both the strategies indicated above can be implemented by merely introducing students to weak notions of sustainability, this may not be enough to educate sustainable legal professionals and legally trained citizens. Acknowledging this requires taking a critical look at the relationship between academic legal education and – in particular, corporate – legal practice. Questions of legal ethics may need to be discussed within substantive private law courses: which clients, whose interests are to be represented[75] are questions that private lawyers are less familiar with than, to give an obvious example, those involved in criminal law practice. Asking students to reflect on these questions would be a way to make them engage with the literature on 'strong' sustainability in a way that is both meaningful and 'graspable' from the perspective of non-academics. While we know that many graduates end up elsewhere than in the traditional legal professions, these classical figures are likely to feature strongly in the imagination of students. Is it professionally acceptable to claim, as Shell's defence once did, that climate change does not necessarily represent a human rights threat because of air conditioning?[76] What would the students have done as judges in a controversial case?

5.3.5 Assessment

As lecturers, we can learn from day-to-day interaction that many students are intrinsically interested and motivated when it comes to environmental crises and sustainability.[77] With this fact in mind, the good news is that incorporating sustainability can improve engagement and learning outcomes: it aligns the learning environment with something which is important for many students outside the classroom.[78] Generic interest, however, does not necessarily translate into academic engagement with assigned material, nor does it, in any case, apply to all students, especially in the context of institutions of mass higher education.[79] Thus, to make their incorporation meaningful, sustainability-related components should be just as examinable as the rest of the course. As observed by a student in the context of a recent experiment, 'climate change isn't optional'; hence, it is misleading and ultimately counterproductive to treat climate/sustainability teaching as non-examinable material.[80] This will sound obvious to some but it goes directly against a common intuition concerning the status of 'new' or 'diverging' material – for instance, in the course/experiment just mentioned the content had been marked as mandatory but non-examinable.

Depending on the level, cohort size, and methods of instruction, different forms of examination should be put in place. Not all are equally apt to test systems thinking and normative competencies as outlined above, but all will contribute to sustainability literacy to an extent. In some cases, it will be the usual competencies that are tested, but then with sustainability as integrated content; only certain more advanced forms of examinations, like research papers (systems thinking) or open-ended questions and take-home exams (normative competence), seem capable of integrating key sustainability competences in testing. Instructors should be aware of these limitations – which may well be encapsulated in institutional policies as to levels of learning attainments during different years of instruction[81] – when outlining their learning goals and assessment structure.

A 'case' question is going to work well in certain settings, e.g., where relevant case law has been discussed in class, including when such case law was originally situated in a different legal system. How would a case with similar facts be treated in the system the student is situated in? A question on the assigned literature can work in other settings, e.g., where instruction generally focuses more on conceptual structures than on case-solving. A personal reflection, which the student is required to connect to the readings or class discussion, can offer an 'easy' way to test when the examination structure offers no space for either of the former (e.g., in return for a bonus point!). These approaches could be combined. The so-called 'Plymouth model', for example, includes a full-fledged workbook with different types of practice assignments suitable for bachelor's instruction.[82] At Master/advanced levels, in contrast, normative essay questions would seem more appropriate with a view to the different learning goals normally associated with graduate-level studies. A case note could be explicitly required to address the judgement's sustainability dimension, provided sufficient background has been established. Finally, while 'role-playing' is often seen as something belonging to the realm of simulations or experiential learning, it can also be considered as a way to draft engaging exam questions for individual students.[83] Implicitly, what has been said so far is a call for plural examination and assessment techniques, not necessarily within one course but throughout the (basic) degree. This is, again, broadly in line with general trends in legal education, and as in other contexts it is significantly a matter of resources, as such depending on local contexts.

5.4 Conclusion

A concern with sustainability features in any plausible version of private law to be shared with students in contemporary law programmes. While weak sustainability can be readily embraced as an emerging legal principle, attention to critical accounts of private law points to stronger versions of sustainability that would call for a more far-reaching reimagining of the discipline. Sustainability-trained law students should acquire both a sufficiently articulate understanding of the multiple ways in which private law intersects with our ability to bring human societies to work within planetary boundaries and of the normative stakes involved in the necessary transitions. No individual private law course should carry all the burden for the necessary work. The chapter, however, has suggested that at least a weak understanding of sustainability can be meaningfully furthered with small interventions on conditions that they be made relatable and accessible and openly address the *stakes* and competing interests involved. While stronger notions of sustainability may still be controversial, the chapter suggests an initial step towards most systemically sustainable legal education would be to focus on undergoing shifts in professional ethics within core private law courses. Finally, while students (*qua young people*) are often intrinsically motivated when it comes to climate change, meaningful incorporation in education will require inclusion in the assessment structure. We are literally being put through an exacting test anyway.

Notes

1 I would like to thank, in particular, Marija Bartl, Yannick van den Berg, Francesca Episcopo, and the two anonymous peer reviewers for comments on various versions of the chapter. Thanks also to the editors for their work in putting together the volume, screening submissions, and keeping us on schedule despite the obvious challenges of a large collection. All remaining errors and unaddressed limitations remain my own only.

2 Arguably, this goes well beyond private law – recent critiques have focussed on the overall idea of liberal *rights* as being fundamentally inapt to ground viable legal approaches to the Anthropocene. In this sense, several contributions have sought to re-conceptualize notions of *obligation* instead – see Kathleen Birrell and Daniel Matthews, 'Re-Storying Laws for the Anthropocene: Rights, Obligations and an Ethics of Encounter' [2020] 31 *Law and Critique* 275 (https://doi.org/10.1007/s10978-020-09274-8); Scott Veitch, *Obligations: New Trajectories in Law* (London, Routledge 2021).

3 Kim Bouwer et al., '"Climate Change Isn't Optional": Climate Change in the Core Law Curriculum' [2022] 43(2) *Legal Studies* 240 (https://doi.org/10.1017/lst.2022.35).

4 See the overview and reflections in Geneviève Saumier, 'SDG 12: Sustainable Consumption and Production' in Ralf Michaels et al (eds), *The Private Side of Transforming our World* (Cambridge, Intersentia 2021).

5 Similarly, but not uncritically, Jan M Smits, *Advanced Introduction to Private Law* (Cheltenham, Edward Elgar Publishing 2016) 1.

6 See, e.g., Ralf Michaels and Nils Jansen, 'Private Law beyond the State? Europeanization, Globalization, Privatization' [2006] 54 *The American Journal of Comparative Law* 843; Martijn W Hesselink, *The New European Legal Culture* (Alphen aan den Rijn, Kluwer Law International BV 2002).

7 Christoph U Schmid, 'The Thesis of the Instrumentalisation of Private Law by the EU in a Nutshell' in Tommi Ralli and Christian Joerges (eds), *European Constitutionalism without Private Law: Private Law without Democracy, Recon Report* (2011); Ralf Michaels, 'Of Islands and the Ocean: The Two Rationalities of European Private Law' in Roger Brownsword and others (eds), *The Foundations of European Private Law* (London, Hart Publishing 2011); Olha O Cherednychenko, 'Islands and the Ocean: Three Models of the Relationship between EU Market Regulation and National Private Law' [2021] 84(6) *The Modern Law Review* 1294 (https://doi.org/10.1111/1468-2230.12664).

8 Grigorios Bacharis and Szymon Osmola, 'Rethinking the Instrumentality of European Private Law' [2022] 30(3) *European Review of Private Law* 457 (https://doi.org/10.54648/erpl2022023); Candida Leone, 'The Missing Stone in the Cathedral: Of Unfair Terms in Employment Contracts and Coexisting Rationalities in European Contract Law' (PhD Law thesis, University of Amsterdam 2020).

9 See already Daniela Caruso, 'The Missing View of the Cathedral: The Private Law Paradigm of European Legal Integration' [1997] 3 *European Law Journal* 3 (https://doi.org/10.1111/1468-0386.00017).

10 Ernest J Weinrib, *Corrective Justice* (Oxford, Oxford University Press 2012); Ernest J Weinrib, *The Idea of Private Law* (Oxford, Oxford University Press 2012).

11 Martijn W Hesselink, *Justifying Contract in Europe: Political Philosophies of European Contract Law* (Oxford, Oxford University Press 2021). While Hesselink's analysis focusses on contract law emanating from the EU, several of the doctrines and questions considered in the book go beyond the matters regulated by European instruments.

12 Stefan Grundmann, Hans Micklitz and Moritz Renner, *New Private Law Theory: A Pluralist Approach* (Cambridge, Cambridge University Press 2021).

13 Christian Von Bar, Eric Clive and Hans Schulte-Nölke, *Principles, Definitions and Model Rules of European Private Law: Draft Common Frame of Reference* (Munich, Sellier European Law Publishers 2009) 13 but see overall reflection on principles at p 9 ff.

14 Christian Von Bar, Eric Clive and Hans Schulte-Nölke, *Principles, Definitions and Model Rules of European Private Law: Draft Common Frame of Reference* (Munich, Sellier European Law Publishers 2009) 13 but see overall reflection on principles at p 9 ff.

15 One could, arguably, read sustainability as a consequence of combined reading of the mentioned principles of solidarity and respect for human rights – but this would already represent a somewhat charged move.

16 Candida Leone, 'A Private Law Theory for Sustainable Legal Education?' [2022] 23(6) *German Law Journal* 881 (https://doi.org/10.1017/glj.2022.54).

17 Think of initiatives concerning the so-called 'right to repair' <https://www.europarl.europa.eu/RegData/etudes/BRIE/2022/698869/EPRS_BRI(2022)698869_EN.pdf> accessed 26 April 2023 and the French repairability index rules <https://repair.eu/news/the-french-repair-index-challenges-and-opportunities/>, the Corporate Sustainability Reporting Directive: Directive (EU) 2022/2464 of the European Parliament and of the Council of 14 December 2022 amending Regulation (EU) No 537/2014, Directive 2004/109/EC, Directive 2006/43/EC and Directive 2013/34/EU, as regards corporate sustainability reporting [2022] OJ L322/15, and so forth.

18 More would need to be said about the possible need to distinguish between principles and goals and where sustainability would fit in such distinction. However, taking a pragmatic approach here, I will assume that principles and goals are to an extent interchangeable (a strong assumption) and that if efficiency can be named as a principle (like in the DCFR), then the same applies to sustainability.

19 See, in the context of sustainable production and consumption, Geneviève Saumier, 'SDG 12: Sustainable Consumption and Production' in Ralf Michaels and others (eds), *The Private Side of Transforming our World* (Cambridge, Intersentia 2021). As indicated above, the author claims that the approaches could be seen as complementary or that a 'third way' could be devised in line with further literature. For the sake of analytical clarity, in this chapter, I abide by the starker distinction. See further Oksana Mont, *A Research Agenda for Sustainable Consumption Governance* (Cheltenham, Edward Elgar Publishing 2019).

20 See reactions to the delay caused by, among others, a negative assessment by the Regulatory scrutiny board <https://www.repaircafe.org/en/eu-delays-right-to-repair-for-at-least-six-months/> accessed 26 April 2023. The Commission's proposal has finally been published on 22 March 2023: Commission, 'Proposal for a Directive of the European Parliament and of the Council on common rules promoting

the repair of goods and amending Regulation (EU) 2017/2394, Directives (EU) 2019/771 and (EU) 2020/1828' COM/2023/155 final.

21 See for a reconstruction of the debate Marija Bartl, 'Towards the Imaginary of Collective Prosperity in the European Union (EU): Reorienting the Corporation' [2022] 1(4) *European Law Open* 957 (https://doi.org/10.1017/elo.2023.5).

22 See most recently *BEUC*, 'Reaping the Benefits of Renewables for Consumers' <https://www.beuc.eu/sites/default/files/publications/BEUC-X-2023-035%20BEUC_position_paper_key_asks_electricity_market_design.pdf> accessed 26 April 2023.

23 Stefan Grundmann, Hans Micklitz and Moritz Renner, *New Private Law Theory: A Pluralist Approach* (Cambridge, Cambridge University Press 2021).

24 See, for instance, the recent amendment to article 9(3) of the Italian constitution, according to which the Republic 'protects the environment, biodiversity and ecosystems, also in the interest of future generations' (*'Tutela l'ambiente, la biodiversità e gli ecosistemi, anche nell'interesse delle future generazioni'*).

25 United Nations General Assembly, 'Resolution 76/300' (28 July 2022) <https://documents-dds-ny.un.org/doc/UNDOC/GEN/N22/442/77/PDF/N2244277.pdf?OpenElement> accessed 26 April 2023.

26 Chantal Mak, 'Human Rights in Private Law' (2022) Amsterdam Law School Research Paper No. 2022-51, Amsterdam Centre for Transformative private law Working Paper No. 2022-10 <https://ssrn.com/abstract=4304817 or http://dx.doi.org/10.2139/ssrn.4304817>.

27 Evelyne Terryn and Elias Van Gool, 'The Role of European Consumer Regulation in Shaping the Environmental Impact of E-Commerce' [2021] 10(3) *Journal of European Consumer and Market Law* 89; Charlotte Pavillon and Willem Van Boom, *Privaatrechtelijke Bescherming Herijkt* (Zutphen, Uitgeverij Paris 2021).

28 David Mohan Ong, 'Prospects for Integrating an Environmental Sustainability Perspective within the University Law Curriculum in England' [2016] 50(3) *The Law Teacher* 276 (https://doi.org/10.1080/03069400.2016.1262988).

29 Article 9, General Provisions on Civil Law <http://www.npc.gov.cn/englishnpc/c23934/202012/f627aa3a4651475db936899d69419d1e/files/47c16489e186437eab3244495cb47d66.pdf> accessed 26 April 2023.

30 Tiantian Zhai and Yen-Chiang Chang, 'The Contribution of China's Civil Law to Sustainable Development: Progress and Prospects' [2019] 11(1) *Sustainability* 294 (https://doi.org/10.3390/su11010294); Jie Ouyang [presentation 2022], on file with the author.

31 Durabilité en droit, 'La logique de durabilité dans le nouveau droit des biens – par Marie-Sophie de Clippele – Jubel' (*Jubel*, 10 March 2023) <https://www.jubel.be/fr/la-logique-de-durabilite-dans-le-nouveau-droit-des-biens/> accessed 26 April 2023.

32 Nicole Graham, 'Teaching Private Law in a Climate Crisis' [2021] 40(3) *The University of Queensland Law Journal* 403 (https://doi.org/10.38127/uqlj.v40i3.6047).

33 See, for instance, Scott Veitch, *Obligations: New Trajectories in Law* (London, Routledge 2021).

34 Matt Stone, 'The Structures & Subjects of Obligation: On Veitch's Obligation' (*Critical Legal Thinking*, 15 December 2021) <https://criticallegalthinking.com/2021/12/15/the-structures-and-subjects-of-obligation-on-veitchs-obligation/> accessed 16 February 2023. For a more radical reading see Margaret Davies, 'Eco-Legal Bonds: On Veitch's Obligations' (*Critical Legal Thinking*, 16 December 2021) <https://criticallegalthinking.com/2021/12/16/eco-legal-bonds-on-veitchs-obligations/> accessed 16 February 2023.

35 Kathleen Birrell and Daniel Matthews, 'Re-Storying Laws for the Anthropocene: Rights, Obligations and an Ethics of Encounter' [2020] 31 *Law and Critique* 275 (https://doi.org/10.1007/s10978-020-09274-8); Peter D Burdon, 'Obligations in the Anthropocene' [2020] 31 *Law and Critique* 309 (https://doi.org/10.1007/s10978-020-09273-9).

36 Margaret Davies, 'Can Property Be Justified in an Entangled World?' [2020] 17(7) *Globalizations* 1104 (https://doi.org/10.1080/14747731.2019.1650696); Margaret Davies, 'Material Subjects and Vital Objects—Prefiguring Property and Rights for an Entangled World' [2016] 22(2) *Australian Journal of Human Rights* 37 (https://doi.org/10.1080/1323238X.2016.11910941).

37 Such as, to only mention the French context, the discussions referred to by Denis Mazeaud, 'Loyauté, Solidarité, Fraternité: La Nouvelle Devise Contractuelle?' [1999] *Mélanges Terré* 603; Christophe Jamin, 'Plaidoyer Pour Le Solidarisme Contractuel' [2001] *Le contrat au début du XXIème siècle* 441.

38 See, e.g., Hanoch Dagan, *A Liberal Theory of Property* (Cambridge, Cambridge University Press 2021); Hanoch Dagan and Avihay Dorfman, 'Justice in Contracts' [2022] 67(1) *The American Journal of Jurisprudence* 1 (https://doi.org/10.1093/ajj/auac001).

39 Louis J Kotzé and others, 'Earth System Law: Exploring New Frontiers in Legal Science' (2022) 11 Earth System Governance 100126.

40 Particularly powerful: Nancy Fraser, *Cannibal Capitalism: How Our System Is Devouring Democracy, Care, and the Planet and What We Can Do About It* (Verso Books 2022).

41 See several references in Kim Bouwer et al., '"Climate Change Isn't Optional": Climate Change in the Core Law Curriculum' [2022] 43(2) *Legal Studies* 240 (https://doi.org/10.1017/lst.2022.35). Also, for a very comprehensive collection of materials, topics and insights <https://www.sustainabilityexchange.ac.uk/sorted_guide_to_sustainability_in_further_educa> accessed 26 April 2023.

42 See <https://duurzaamheidseducatie.paddlecms.net> for an overview and a broad set of tools offered to educators – unfortunately on condition that they read Flemish/Dutch. I wish to thank Evelyne Terryn for pointing me to the University's well-developed sustainability policy, in particular with reference to education – see also <https://www.kuleuven.be/duurzaamheid/sustainability/education/spotlight> accessed 26 April 2023.

43 See, for instance, the 2021 'White Paper on Sustainability' by the University of Amsterdam, with two education-related prongs: number 3, on integration of sustainability within broad programmes, and number 4 on specialization. Similarly for law, Michael Mehling and others, 'Teaching Climate Law: Trends, Methods and Outlook' [2020] 32(3) *Journal of Environmental Law* 417, 419 (https://doi.org/10.1093/jel/eqz036).

44 Klaus Bosselmann, 'From Reductionist Environmental Law to Sustainability Law' in Peter D Burdon (ed), *Exploring wild law: The philosophy of earth jurisprudence* (South Australia, Wakefield Press Kent Town 2011) 204.

45 Aaron Redman and Arnim Wiek, 'Competencies for Advancing Transformations towards Sustainability' [2021] 6 *Frontiers in Education* 1 (https://doi.org/10.3389/feduc.2021.785163). The authors, relying on further literature, describe competencies as 'a complex combination of knowledge, skills, understanding, values, attitudes and desire which lead to effective, embodied human action in the world, in a particular domain'. The definition originates from Ruth Deakin Crick, 'Key Competencies for Education in a European Context: Narratives of Accountability or Care' [2008] 7 *European Educational Research Journal* 311 (https://doi.org/10.2304/eerj.2008.7.3.311). The framework's suitability has been recently assessed in conversation with educators and the core tenets have been confirmed – see Katja Brundiers et al., 'Key Competencies in Sustainability in Higher Education—toward an Agreed-upon Reference Framework' [2021] 16 *Sustainability Science* 13 (https://doi.org/10.1007/s11625-020-00838-2).

46 Other competencies identified as key are anticipatory competence, strategic competence, and interpersonal competence.

47 See for an accessible introduction Donella H Meadows, *Thinking in Systems: A Primer* (London, Chelsea Green Publishing 2008).

48 Arnim Wiek, Lauren Withycombe and Charles L Redman, 'Key Competencies in Sustainability: A Reference Framework for Academic Program Development' [2011] 6 *Sustainability Science* 203, 208 (https://doi.org/10.1007/s11625-011-0132-6).

49 Arnim Wiek, Lauren Withycombe and Charles L Redman, 'Key Competencies in Sustainability: A Reference Framework for Academic Program Development' [2011] 6 *Sustainability Science* 203, 208–209 (https://doi.org/10.1007/s11625-011-0132-6).

50 See for a plea in this direction the Dutch *Wetenschappelijke Raad voor Regeringsbeleid* (Academic Council for Government Policy) in its 2023 report *Rechtvaardigheid in klimaatbeleid. Over de verdeling van klimaatkosten* (Justice in climate policy. On the distribution of climate costs), calling for a more explicit consideration of broadly speaking distributive questions associated with climate policy.

51 See Simon Chesterman, 'Doctrine, Perspectives and Skills for Global Practice' in William van Caenegem and Mary Hiscock (eds), *The Internationalisation of Legal Education* (Cheltenham, Edward Elgar Publishing 2014).

52 Candida Leone, 'A Private Law Theory for Sustainable Legal Education?' [2022] 23(6) *German Law Journal* 881 (https://doi.org/10.1017/glj.2022.54).

53 Stefan Grundmann, Hans Micklitz and Moritz Renner, *New Private Law Theory: A Pluralist Approach* (Cambridge, Cambridge University Press 2021).

54 Robert L Hale, 'Coercion and Distribution in a Supposedly Non-Coercive State' [1923] 38 *Political Science Quarterly* 470.

55 Margaret A Young, 'Climate Change and Law: A Global Challenge for Legal Education' [2021] 40(3) *The University of Queensland Law Journal* 351 (https://doi.org/10.38127/uqlj.v40i3.6045).

56 See, for example, Emma Jones and Fiona Cownie, Key Directions in Legal Education: National and International Perspectives (London, Routledge 2020); Christopher Gane and Robin Hui Huang (eds), Legal Education in the Global Context: Opportunities and Challenges (London, Routledge & CRC Press 2016).
57 For further elaboration, see Marija Bartl and Candida Leone, 'The Politics of Legal Education', *The Politics of European Legal Research* (Cheltenham, Edward Elgar Publishing 2022).
58 Here, for instance, the work mentioned above under Section 2 could be of use.
59 See Anneleen Kenis and Matthias Lievens, 'Searching for 'the Political' in Environmental Politics' [2014] 23(4) *Environmental Politics* 531 (https://doi.org/10.1080/09644016.2013.870067). In the classroom context, <https://www.lesgevenvoorenoverduurzaamheid.be/hoofdstuk-1/14-duurzaamheid-geen-eenduidig-verhaal/141-routines-en-waarden-in-vraag-gesteld> (in Flemish).
60 See for an extended inquiry adopting this perspective Pierre Lequet, *L'ordre public environnemental et le contrat de droit privé* (Paris, LGDJ 2022) <https://www.lgdj.fr/l-ordre-public-environnemental-et-le-contrat-de-droit-prive-9782275108384.html> accessed 17 January 2023.
61 See Nicole Graham, 'Teaching Private Law in a Climate Crisis' [2021] 40(3) *The University of Queensland Law Journal* 403 (https://doi.org/10.38127/uqlj.v40i3.6047).
62 Jason Lowther and Joanne Sellick, 'Embedding Sustainability Literacy in the Legal Curriculum: Reflections on the Plymouth Model' [2016] 50(3) *The Law Teacher* 307 (https://doi.org/10.1080/03069400.2016.1240919).
63 David Mohan Ong, 'Prospects for Integrating an Environmental Sustainability Perspective within the University Law Curriculum in England' [2016] 50(3) *The Law Teacher* 276 (https://doi.org/10.1080/03069400.2016.1262988).
64 David Mohan Ong, 'Prospects for Integrating an Environmental Sustainability Perspective within the University Law Curriculum in England' [2016] 50(3) *The Law Teacher* 276 (https://doi.org/10.1080/03069400.2016.1262988); Jason Lowther and Joanne Sellick, 'Embedding Sustainability Literacy in the Legal Curriculum: Reflections on the Plymouth Model' [2016] 50(3) *The Law Teacher* 307 (https://doi.org/10.1080/03069400.2016.1240919).
65 The ongoing German case of Lluya v RWE <https://rwe.climatecase.org/en/legal#timeline> is an interesting example in this respect.
66 See for a recent contribution Laura Burgers and Kinanya Pijl, 'Collective Ownership and Representation in a Sustainable City' [2022] 15(3) *Erasmus Law Review* 212 (https://doi.org/10.5553/ELR.000230) <http://www.erasmuslawreview.nl/tijdschrift/ELR/2022/3%20(incomplete)/ELR-D-22-00036> accessed 21 February 2023.
67 See 'Nederlands Tijdschrift voor Burgerlijk Recht, De doelmatigheid van contractuele toezeggingen in de verduurzaming van handelsketens | Navigator' <https://new.navigator.nl/document/id24da6fff9fe949fd8a1648bcd731ea7b?ctx=WKNL_CSL_87&tab=tekst> accessed 21 February 2023; Jennifer Ramos, 'Shifting the Mindset of Commercial Lawyers to Rewire Contracts, to Mitigate Climate Change More Effectively in Practice: The Chancery Lane Project' [2021] 23(1) *Environmental Law Review* 3 (https://doi.org/10.1177/14614529211000152).
68 <https://www.pbl.nl/en/publications/extended-producer-responsibility>.
69 See, e.g., Des Gasper, Amod Shah and Sunil Tankha, 'The Framing of Sustainable Consumption and Production in SDG 12' [2019] 10(S1) *Global Policy* 83 (https://doi.org/10.1111/1758-5899.12592).
70 Evelyne Terryn and Elias Van Gool, 'The Role of European Consumer Regulation in Shaping the Environmental Impact of E-Commerce' [2021] 10(3) *Journal of European Consumer and Market Law* 89; Charlotte Pavillon and Willem Van Boom, *Privaatrechtelijke Bescherming Herijkt* (Zutphen, Uitgeverij Paris 2021).
71 I would like to thank Yannick van den Berg for recently bringing this example to my attention.
72 See Bouwer's experience trying to incorporate a broad contestation/problematization of core notions of land law.
73 Kim Bouwer et al., '"Climate Change Isn't Optional": Climate Change in the Core Law Curriculum' [2022] 43(2) *Legal Studies* 240, 257 (https://doi.org/10.1017/lst.2022.35).
74 Catherine Rocchi and Camila Bustos, 'Lawyers Exacerbate the Climate Crisis: Here's How We Can Help' (American Bar Association, 28 August 2022) <https://www.americanbar.org/groups/environment_energy_resources/publications/trends/2022-2023/september-october-2022/lawyers-exacerbate-the-climate-crisis/> accessed 26 April 2023.
75 Ruth Green, 'Climate Crisis: Law Firms Can Have Major Impact through Client Work' (International Bar Association, 28 April 2022) <https://www.ibanet.org/Climate-crisis-Law-firms-can-have-major-impact-through-client-work> accessed 26 April 2023.

76 See para 4.4.8, Rechtbank Den Haag (26 May 2021), *ECLI:NL:RBDHA:2021:5* <https://uitspraken.rechtspraak.nl/#!/details?id=ECLI:NL:RBDHA:2021:5339>.

77 Kim Bouwer et al., '"Climate Change Isn't Optional": Climate Change in the Core Law Curriculum' [2022] 43(2) *Legal Studies* 240 (https://doi.org/10.1017/lst.2022.35).

78 Graham Ferris and Rebecca Huxley-Binns, 'What Students Care about and Why We Should Care' in Paul Maharg and Caroline Maughan (eds), *Affect and Legal Education* (Cheltenham, Routledge 2016). In this way, it may help ease the feeling expressed by many bachelor's students that private law (in particular) is a rather *arid* subject.

79 See Emma Jones and Fiona Cownie, *Key Directions in Legal Education: National and International Perspectives* (London, Routledge 2020); Christopher Gane and Robin Hui Huang (eds), *Legal Education in the Global Context: Opportunities and Challenges* (London, Routledge & CRC Press 2016). In the books Introduction (p. xviii), the authors aptly define students as being 'at the centre of a web of relations and connections throughout their legal studies' – essentially the challenge here is to make sure that, for all students, sustainability becomes one of the nodes in that web.

80 Kim Bouwer et al., '"Climate Change Isn't Optional": Climate Change in the Core Law Curriculum' [2022] 43(2) *Legal Studies* 240 (https://doi.org/10.1017/lst.2022.35).

81 My institution for one prescribes ranges for different levels in the Bloom taxonomy <https://bloomstaxonomy.net> that differ per year of instruction.

82 Jason Lowther and Joanne Sellick, 'Embedding Sustainability Literacy in the Legal Curriculum: Reflections on the Plymouth Model' (2016) 50 *The Law Teacher* 307.

83 Graham Ferris and Rebecca Huxley-Binns, 'What Students Care about and Why We Should Care' in Paul Maharg and Caroline Maughan (eds), *Affect and Legal Education* (Cheltenham, Routledge 2016) 207.

Bibliography

Bacharis G and Osmola S, 'Rethinking the Instrumentality of European Private Law' [2022] 30(3) *European Review of Private Law* 457 (https://doi.org/10.54648/erpl2022023)

Bartl M, 'Towards the Imaginary of Collective Prosperity in the European Union (EU): Reorienting the Corporation' [2022] 1(4) *European Law Open* 957 (https://doi.org/10.1017/elo.2023.5)

Bartl M and Leone C, 'The Politics of Legal Education', in *The Politics of European Legal Research* (Cheltenham, Edward Elgar Publishing 2022) 159

Birrell K and Matthews D, 'Re-Storying Laws for the Anthropocene: Rights, Obligations and an Ethics of Encounter' [2020] 31 *Law and Critique* 275 (https://doi.org/10.1007/s10978-020-09274-8)

Bosselmann K, 'From Reductionist Environmental Law to Sustainability Law' in Peter D Burdon (ed), *Exploring Wild Law: The Philosophy of Earth Jurisprudence* (South Australia, Wakefield Press Kent Town 2011) 204.

Bouwer K, John E, Luke O and Rozhan A '"Climate Change Isn't Optional": Climate Change in the Core Law Curriculum' [2022] 43(2) *Legal Studies* 240 (https://doi.org/10.1017/lst.2022.35)

Brundiers K, Barth M, Cebrián G, Cohen M, Diaz L, Doucette-Remington S, Dripps W, Habron G, Harré N, Jarchow M, Losch K, Michel J, Mochizuki Y, Rieckmann M, Parnell R, Walker P and Zint M, 'Key Competencies in Sustainability in Higher Education—Toward an Agreed-Upon Reference Framework' [2021] 16 *Sustainability Science* 13 (https://doi.org/10.1007/s11625-020-00838-2)

Burdon PD, 'Obligations in the Anthropocene' [2020] 31 *Law and Critique* 309 (https://doi.org/10.1007/s10978-020-09273-9)

Burgers L and Pijl K, 'Collective Ownership and Representation in a Sustainable City' [2022] 15(3) *Erasmus Law Review* 212 (https://doi.org/10.5553/ELR.000230)

Caruso D, 'The Missing View of the Cathedral: The Private Law Paradigm of European Legal Integration' [1997] 3 *European Law Journal* 3 (https://doi.org/10.1111/1468-0386.00017)

Cherednychenko OO, 'Islands and the Ocean: Three Models of the Relationship between EU Market Regulation and National Private Law' [2021] 84(6) *The Modern Law Review* 1294 (https://doi.org/10.1111/1468-2230.12664)

Chesterman S, 'Doctrine, Perspectives and Skills for Global Practice' in van Caenegem W and Hiscock M (eds), *The Internationalisation of Legal Education* (Cheltenham, Edward Elgar Publishing 2014)

Dagan H, *A Liberal Theory of Property* (Cambridge, Cambridge University Press 2021)

Dagan H and Dorfman A, 'Justice in Contracts' [2022] 67(1) *The American Journal of Jurisprudence* 1 (https://doi.org/10.1093/ajj/auac001)
Davies M, 'Material Subjects and Vital Objects—Prefiguring Property and Rights for an Entangled World' [2016] 22(2) Australian Journal of Human Rights 37 (https://doi.org/10.1080/1323238X.2016.11910941)
———, 'Can Property Be Justified in an Entangled World?' [2020] 17(7) *Globalizations* 1104 (https://doi.org/10.1080/14747731.2019.1650696)
———, 'Eco-Legal Bonds: On Veitch's Obligations' (*Critical Legal Thinking*, 16 December 2021) <https://criticallegalthinking.com/2021/12/16/eco-legal-bonds-on-veitchs-obligations/>
Deakin Crick R, 'Key Competencies for Education in a European Context: Narratives of Accountability or Care' [2008] 7 *European Educational Research Journal* 311 (https://doi.org/10.2304/eerj.2008.7.3.311)
Ferris G and Huxley-Binns R, 'What Students Care about and Why We Should Care' in Maharg P and Maughan C (eds), *Affect and Legal Education* (Cheltenham, Routledge 2016) 195
Gane C and Hui Huang R, *Legal Education in the Global Context: Opportunities and Challenges* (London, Routledge & CRC Press 2016)
Gasper D, Shah A and Tankha S, 'The Framing of Sustainable Consumption and Production in SDG 12' [2019] 10(S1) *Global Policy* 83 (https://doi.org/10.1111/1758-5899.12592)
Graham N, 'Teaching Private Law in a Climate Crisis' [2021] 40(3) *The University of Queensland Law Journal* 403 (https://doi.org/10.38127/uqlj.v40i3.6047)
Green R, 'Climate Crisis: Law Firms Can Have Major Impact through Client Work' (International Bar Association, 28 April 2022) <https://www.ibanet.org/Climate-crisis-Law-firms-can-have-major-impact-through-client-work>
Grundmann S, Micklitz H and Renner M, *New Private Law Theory: A Pluralist Approach* (Cambridge, Cambridge University Press 2021)
Hale RL, 'Coercion and Distribution in a Supposedly Non-Coercive State' [1923] 38 *Political Science Quarterly* 470
Hesselink MW, *The New European Legal Culture* (Alphen aan den Rijn, Kluwer Law International BV 2002)
———, *Justifying Contract in Europe: Political Philosophies of European Contract Law* (Oxford, Oxford University Press 2021)
Jamin C, 'Plaidoyer Pour Le Solidarisme Contractuel' [2001] *Le contrat au début du XXIème siècle* 441
Jones E and Cownie F, *Key Directions in Legal Education: National and International Perspectives* (London, Routledge 2020)
Kenis A and Lievens M, 'Searching for 'the Political' in Environmental Politics' [2014] 23(4) *Environmental Politics* 531 (https://doi.org/10.1080/09644016.2013.870067)
Leone C, 'The Missing Stone in the Cathedral: Of Unfair Terms in Employment Contracts and Coexisting Rationalities in European Contract Law' (PhD Law thesis, University of Amsterdam 2020)
———, 'A Private Law Theory for Sustainable Legal Education?' [2022] 23(6) *German Law Journal* 881 (https://doi.org/10.1017/glj.2022.54)
Lequet P, *L'ordre public environnemental et le contrat de droit privé* (Paris, LGDJ 2022)
Lowther J and Sellick J, 'Embedding Sustainability Literacy in the Legal Curriculum: Reflections on the Plymouth Model' [2016] 50(3) *The Law Teacher* 307 (https://doi.org/10.1080/03069400.2016.1240919)
Mak C, 'Human Rights in Private Law' (2022) Amsterdam Law School Research Paper No. 2022-51, Amsterdam Centre for Transformative private law Working Paper No. 2022-10 <https://ssrn.com/abstract=4304817 or http://dx.doi.org/10.2139/ssrn.4304817>
Mazeaud D, 'Loyauté, Solidarité, Fraternité: La Nouvelle Devise Contractuelle?' [1999] *Mélanges Terré* 603
Meadows DH, *Thinking in Systems: A Primer* (London, Chelsea Green Publishing 2008)
Mehling M, van Asselt H, Kulovesi K and Morgera E, 'Teaching Climate Law: Trends, Methods and Outlook' [2020] 32(3) *Journal of Environmental Law* 417, 419 (https://doi.org/10.1093/jel/eqz036)
Michaels R, 'Of Islands and the Ocean: The Two Rationalities of European Private Law' in Brownsword R, Micklitz HG, Niglia L and Weatherill S (eds), *The Foundations of European Private Law* (London, Hart Publishing 2011) 139
Michaels R and Jansen N, 'Private Law beyond the State? Europeanization, Globalization, Privatization' [2006] 54 *The American Journal of Comparative Law* 843
Mohan Ong D, 'Prospects for Integrating an Environmental Sustainability Perspective within the University Law Curriculum in England' [2016] 50(3) *The Law Teacher* 276 (https://doi.org/10.1080/03069400.2016.1262988)

Mont O, *A Research Agenda for Sustainable Consumption Governance* (Cheltenham, Edward Elgar Publishing 2019)
Pavillon C and Van Boom W, *Privaatrechtelijke Bescherming Herijkt* (Zutphen, Uitgeverij Paris 2021)
Ramos J, 'Shifting the Mindset of Commercial Lawyers to Rewire Contracts, to Mitigate Climate Change More Effectively in Practice: The Chancery Lane Project' [2021] 23(1) *Environmental Law Review* 3 (https://doi.org/10.1177/14614529211000152)
Redman A and Wiek A, 'Competencies for Advancing Transformations towards Sustainability' [2021] 6 *Frontiers in Education* 1 (https://doi.org/10.3389/feduc.2021.785163)
Rocchi C and Bustos C, 'Lawyers Exacerbate the Climate Crisis: Here's How We Can Help' (*American Bar Association*, 28 August 2022) <https://www.americanbar.org/groups/environment_energy_resources/publications/trends/2022-2023/september-october-2022/lawyers-exacerbate-the-climate-crisis/>
Saumier G, 'SDG 12: Sustainable Consumption and Production' in Michaels R, Ruiz Abou-Nigm V and van Loon H (eds), *The Private Side of Transforming our World* (Cambridge, Intersentia 2021) 383
Schmid CU, 'The Thesis of the Instrumentalisation of Private Law by the EU in a Nutshell' in Ralli T and Joerges C (eds), *European Constitutionalism Without Private Law: Private Law Without Democracy, Recon Report* (2011)
Smits JM, *Advanced Introduction to Private Law* (Cheltenham, Edward Elgar Publishing 2016)
Stone M, 'The Structures & Subjects of Obligation: On Veitch's Obligation' (*Critical Legal Thinking*, 15 December 2021) <https://criticallegalthinking.com/2021/12/15/the-structures-and-subjects-of-obligation-on-veitchs-obligation/>
Terryn E and Van Gool E, 'The Role of European Consumer Regulation in Shaping the Environmental Impact of E-Commerce' [2021] 10(3) *Journal of European Consumer and Market Law* 89
Veitch S, *Obligations: New Trajectories in Law* (London, Routledge 2021)
Von Bar C, Clive E and Schulte-Nölke H, *Principles, Definitions and Model Rules of European Private Law: Draft Common Frame of Reference* (Munich, Sellier European Law Publishers 2009)
Weinrib EJ, *Corrective Justice* (Oxford, Oxford University Press 2012)
———, *The Idea of Private Law* (Oxford, Oxford University Press 2012)
Wiek A, Lauren Withycombe and Charles L Redman, 'Key Competencies in Sustainability: A Reference Framework for Academic Program Development' [2011] 6 *Sustainability Science* 203 (https://doi.org/10.1007/s11625-011-0132-6)
Young MA, 'Climate Change and Law: A Global Challenge for Legal Education' [2021] 40(3) *The University of Queensland Law Journal* 351 (https://doi.org/10.38127/uqlj.v40i3.6045)
Zhai T and Chang Y, 'The Contribution of China's Civil Law to Sustainable Development: Progress and Prospects' [2019] 11(1) *Sustainability* 294 (https://doi.org/10.3390/su11010294)

PART 2

Fostering Environmental Protection through Consumer Law

6

VARIOUS APPROACHES TO 'GREENING' CONSUMER SALES LAW[1]

Fryderyk Zoll, Katarzyna Południak-Gierz, Wojciech Bańczyk and Maciej Bujalski

6.1 Introduction

In recent years, we are finally seeing European Union legislation adopting a new approach. Environmental care, and in particular green transition, is becoming one of the more standard aims of a whole range of measures, especially in areas of public law.[2] In addition, the impact of ESG[3] on matters of corporate governance and capital markets is growing.[4] The same is true about the environmental aspects of sustainable development.[5]

Consumer law is seen as an important weapon in the fight for greener European markets.[6] The environmental costs of consumer markets can (at least potentially) be mitigated by an appropriately balanced sales law, especially concerning rules on the seller's liability. Sustainable consumption and circular economy are mentioned in Recitals 32 and 48 of Directive (EU) 2019/771 of the European Parliament and of the Council of 20 May 2019 on certain aspects concerning contracts for the sale of goods.[7] Nevertheless, the current provisions of the directive fail to sufficiently correspond to this objective.[8] As a result, just shortly after the directive was adopted and already before its provisions became applicable a discussion on possible amendments to Directive 2019/771 began.[9]

The aim of this chapter is to present possible regulatory approaches to increasing the ecological efficiency of the revised consumer sales law. For this purpose, two proposals are compared, namely the Academic Draft Amendment of Directive 2019/771[10] and the European Law Institute's Response to the European Commission's Public Consultation on Sustainable Consumption of Goods – Promoting Repair and Reuse.[11] Even though both of these initiatives are built upon the same foundation and have the same objective, the proposed outcome differs substantially, as the regulatory area within which the intervention is claimed to be needed varies, as does the intensity of the discussed amendments.

The conducted analysis focuses on the main differences between the presented proposals. Those appear in the context of the level of harmonisation, the way in which the right to repair is viewed, the role of price reduction, the requirements of conformity – especially the durability of goods, the period of the seller's liability, and the possibility of introducing a direct producer's liability. The analysis also looks at the different ways in which the proposals find a balance between promoting sustainability, protecting consumers, and achieving a proper risk distribution within a sales

DOI: 10.4324/9781032662046-8

contract, as far as the above matters are concerned.[12] On the other hand, some similarities between the approaches may be observed. Consequently, the results of this analysis should facilitate finding regulatory solutions that are relatively uncontroversial and may help to focus further academic and political debate on solving issues, where adjusting legal instruments to meet the challenges of ecological transition is more problematic.

Finally, as the European Commission's position on the matter has been published shortly before submitting the final version of this article,[13] the approach taken by the EU will also be briefly commented. As expected, the aim of this initiative is 'to increase the repair and reuse of viable defective goods purchased by consumers within and beyond the legal guarantee'.[14]

To prevent differentiation of national rules on repair, full harmonisation of some of its aspects (especially those being outside of the seller's liability regime provided for by Directive 2019/771) is proposed. Environmental aims are hoped to be achieved through synergy between several initiatives – the proposed Directive on common rules promoting the repair of goods, the proposed Ecodesign for Sustainable Products Regulation,[15] the expected revision of Directive 2019/771,[16] and the Directive on empowering consumers for the green transition[17] that should, in principle, cover the entire lifecycle of a product and have a cumulative effect on the consumer market. Contrary to what has been proposed by the Academic Proposal and the majority of recommendations formed in the ELI Response, the main focus of the strategy is to promote the repair of defective items, and not to increase the ecological efficiency of consumer sales law.[18]

6.2 Level of Harmonisation

Directive 2019/771 is aimed at full harmonisation (Article 4) in order to maintain the same level of consumer protection across the European Union and to minimise the transactional costs for sellers who want to benefit from the free movement of goods, persons, services and capital on the single market.[19] However, choosing full harmonisation may increase the environmental costs of regulatory solutions.

That will be the case if we consider that maximum harmonisation may impede national legislators from introducing norms that exceed the standard set by Directive 2019/771, concerning both consumer protection and seller's guarantees. However, some Member States may still be inclined to opt for an eco-friendlier transposition of Directive 2019/771, following the principle of common but differentiated responsibilities.[20] On the one hand, as ecology is a transnational issue and the degradation of the environment is an irreversible process, initiatives aimed at granting sustainability should not be suppressed in order to achieve full harmonisation. On the other hand, the maximum harmonisation method may, in some contexts, be better positioned to greening consumer sales law than the minimum harmonisation. The hierarchy of remedies can serve as an example here. If repair (assuming that it is the most eco-friendly remedy) is prioritised over other remedies in a minimum harmonisation directive, then national legislators could nevertheless place all remedies for non-conformity at the same level. In that manner, they would grant consumers better protection than the directive itself at the expense of the environment, which would not be the case if maximum harmonisation was introduced.[21]

This problem has been highlighted in both commented approaches. The Academic Proposal strongly recommends a mixed harmonisation approach. It states that, while consumer protection should remain a subject of maximum harmonisation, the directive: 'shall not prevent the Member States from maintaining or introducing deviating provisions of national law that increase the environmental added value without lowering the protection provided for the consumer in this Directive'.[22]

In the meantime, the ELI's Response simply indicates that there are two possibilities (namely, maintaining full harmonisation or returning to the minimum harmonisation approach in order to allow Member States to increase the ecological effectiveness of consumer sales law) without opting definitively for either of them.[23] As mentioned, this problem should be regarded from a wider perspective. The duality of the level of harmonisation may be an effective measure to promote sustainable legislation. The provisions guaranteeing the same level of consumer protection should remain fully harmonised. In contrast, in the context of provisions that may strongly influence the sustainable development and circular economy, the minimum harmonisation method should be applied to allow Member States to find measures that are best suited from an environmental perspective, while taking into consideration such things as the regional particularities and variety of possibilities to contribute to better environmental protection. This kind of competitiveness between Member States is hoped to achieve a higher level of environmental protection in general.

Although both proposals recognise that it is too early for the maximum harmonisation regarding issues that have a direct impact on the environmental costs of consumer sales law, the EU, once again, opts for maximum harmonisation when addressing the right to repair. The reasons behind this choice are of economic nature: legal fragmentation dissuades consumers from using repair services, and full harmonisation is hoped to encourage cross border movement of goods (including spare parts) and the development of repair services, as it reduces transaction costs for business.[24]

6.3 Right to Repair as a Central Point of the New Approaches

The European Commission has been focusing on one of the remedies that has been present in the European consumers' sales law since Directive 1999/44,[25] namely the right to repair. This is no longer viewed solely as a remedy for non-conformity within the framework of Directive 2019/771, but has become a general, standalone right.[26] The idea is quite simple: the environmental costs are lower when the same good is circulating on the market, as opposed to when a new one is produced, with the first good delivered usually going to waste.[27] In addition, in principle, the environmental costs of repairing an already existing good are low compared to the costs of producing a new one. This is why the right to repair is regarded as an eco-friendly solution if viewed as a self-standing right[28] or if compared with other remedies available in case of non-conformity of the good during the period of the seller's liability under Directive 2019/771.[29]

The question now is how to promote this remedy.[30] The most intuitive and already broadly discussed solution is to design the hierarchy of the remedies of Directive 2019/771 accordingly. That is why the Academic Proposal indicates that right to repair should become the primary and main remedy, which would then no longer be at the same level as the right to a replacement.[31] In addition, the ELI's Response focuses on the right to repair by arguing that its introduction as a primary remedy 'would increase the longevity of a product and in most cases lead to less consumption of new products, thereby – in the longer term – to less production, and this could help in saving resources'.[32]

It is worth mentioning that the ELI's Response also poses an important condition: the primary nature of the right to repair should respect the principle of proportionality, which means that, from the seller's side, the costs (both environmental and economic) related to the repair cannot prevail over those associated with a replacement.[33]

For sure, the enforcement of the right to repair would become the main measure of greening the consumer sales law. However, it is not as simple as that. One must be aware of all consequences of such an approach, which in certain circumstances may be even more damaging from a sustainability point of view. This is the case when the repair of a good is more harmful for the environment than its replacement (e.g., because of the transportation or utilisation of replaced parts).[34] In addition, if the replacement is a new generation model of a good, replacing a defective good might be more ecologically efficient than repairing it.

Another important aspect is that if the Sales of Goods Directive is not amended, that is to incentivise consumers to choose repair over replacement. Even though consumers tend to express a mostly positive attitude towards the protection of the environment,[35] in order not to discourage them from choosing an eco-friendly remedy, the legislative measures must meet consumers' expectations related to protecting their interests. In particular, the period needed to perform repairs, according to Article 14(1)(b) of Directive 2019/771, should be reasonable. One issue is that no fixed time frame can be introduced in this regard. This is because, depending on the nature of the good, the severity of its defect and other circumstances of the case, the time reasonably needed to perform repair may vary, making it challenging to shape the legal framework in a way that performing the right to repair does not adversely affect the consumer. One suggestion is to introduce the consumer's right to obtain 'a loaner' (a substitute for the good that is being repaired) if the repair is likely to last for longer than, e.g., two weeks.[36] Here, however, the need to promote competitiveness in the market cannot be forgotten. Granting the consumer a right to a loaner would oblige the sellers to have a substantial amount of spare items at their disposal, to fulfil the consumer's demands. This, on the one hand, increases consumption (including the production of loaners), and on the other may place small enterprises in a weaker position.[37] That is why some further options ought to be considered.

Alternatively, the same result may be achieved through claims for damages that already exist in national legal systems. In practice, it may be more convenient for a buyer to demand compensation for rental costs incurred. From an ecological point of view, it represents an advantage, as this measure does not require the sellers to maintain a given amount of replacement goods. For example, in French[38] and German[39] case law it is assumed that requesting payment of the rental costs, even if no rental has been contracted, is admissible.[40] In these cases, there is a fixed rate for damages, such as rental costs. By this, incentives for obtaining more expensive (or more ecologically inefficient) rentals are minimised.

Another issue under discussion is the admissibility of self-repair under the sellers' liability regime. *Prima facie*, a repair performed by consumers themselves may seem more efficient from an ecological point of view. First and foremost, a good can then be repaired in the consumers' vicinity, which will limit environmental costs related to transportation.[41] However, it has been noted that the legislator does not have the measures to guarantee that self-repair will comply with environmental standards.[42] As a matter of fact, a 'homemade repair' may generate more detrimental effects on the environment than a well-designed process on the seller's or producer's side. That is why the amendments of Directive 2019/771 should promote a repair performed directly by the liable entity. Thus, the seller may often pass the goods demanding repair directly to the producer, but also often direct producers' liability against consumers is considered (see part 7).

The growing importance of repair increases the number of goods that are repaired or rebuilt on the market, which forces legislators to regulate their status. Firstly, as the current text of Directive 2019/771 allows its application to second-hand goods to be excluded, it is necessary to determine whether a good, once repaired, becomes a second-hand good. In principle, the response should be negative. This is why both commented papers propose to introduce provisions on 'refurbished goods'.[43] The notion will cover goods whose key elements were rebuilt or repaired[44] by the seller or producer and reintroduced on the market.[45] Refurbishing is an important aspect of an ecologically effective circular economy. Still, for this strategy to be effective, consumers cannot be afraid of buying refurbished products. The solution could be to grant them the same position as if they bought a new product.[46] In addition, consumers that buy refurbished goods often view them as new ones.[47] A desirable spill-over effect would be that, if the new and refurbished goods start to be seen as equal from the perspective of seller's liability, then sellers should be allowed to offer the buyer a refurbished good as a replacement for a new good.[48]

For the time being, one of the most significant obstacles when it comes to promoting repair is that some goods cannot be repaired in the event of their non-conformity, e.g., because of their design or the technology used for their production. A solution for this matter is introducing a criterion of repairability as a requirement for conformity.[49] The ELI's Proposal postulates that incoming legislation should motivate businesses to produce repairable goods as much as possible.[50] The Proposal of Ecodesign for Sustainable Products Regulation,[51] where *inter alias* requirements regarding product design and availability of spare parts are set, may play a significant part in this process. However, neither the Academic Proposal nor the ELI's Proposal tackle the interplay between the right to repair (either as a standalone right or as one of the rights of the consumer under the seller's liability for non-conformity regime) and intellectual property rights. This matter deserves attention, especially if the right to self-repair or the right to repair outside the seller's liability regime was introduced as envisaged by the Proposal on the Directive on repair. The scope and intensity of protection of intellectual property rights may need to be rethought, as extensive protection of IP rights may limit the availability of repair services and increase its costs due to the need of obtaining adequate licences by repairers (being either sellers or third parties).[52]

The above recommendations only marginally influenced the Proposal for a Directive on common rules promoting the repair of goods. The sole explicit change when it comes to the right to repair under the seller's liability regime was that consumer choice between repair and replacement was limited to situations where replacement is cheaper than repair (Rec. 28, Article 12 of Proposal of Directive on common rules promoting the repair of goods). Adding the derogation from Article 13(2) Directive 2019/771 is supposed to prioritise the remedy of repair without increasing strain on small and medium enterprises with limited resources to maintain repair services infrastructure. However, this approach does not necessarily translate into lower environmental costs, because the EU fails to specify whether environmental costs should also be taken into consideration when comparing costs of repair and replacement. Also, the EU overlooks the fact that sometimes it is more ecologically efficient to withdraw a certain product (group of products) from the market than to allow its further use – that is especially the case if the functioning of the good poses a threat or is disproportionately burdensome from the environmental perspective.

6.4 Price Reduction

The remedy of price reduction, despite being present in Directive 2019/771, is not a subject of the European Commission initiative and is not considered in the ELI's Response.[53] Nevertheless, it is significant when considering the ecological effectiveness of the European consumer sales law. The Academic Proposal pays greater attention to this remedy and its position in the hierarchy, which should result from the amendments of the Sales of Goods Directive.

Price reduction may play a significant role when enforcing the sustainability of private law. Firstly, it is a remedy that presents the greatest neutrality from an ecological point of view.[54] It differs from all the other remedies under Directive 2019/771 as it does not lead to any direct environmental costs being generated: packing or transporting the good is not required, the defective good is not disposed of and no replacement needs to be produced. Instead, money is transferred (most often through the digitised banking system). From this perspective, its preponderance over termination seems to be obvious, and if the analysis ended here, it could be concluded that promoting price reduction over other remedies is a highly effective manner of greening the consumer sales law. However, this general assumption as to the ecological friendliness of price reduction may be misleading.

The Academic Proposal suggests offering a price reduction of up to 5% of price as one of the primary remedies (the other being the right to repair). However, its application should be limited. The consumer may choose between the mentioned remedies on the condition that the continued use of the non-complying good does not present a risk of environmental damage.[55] By contrast, reducing the price by more than 5% may in many cases lead to similar effects as termination, raising the likelihood of the buyer purchasing alternative goods. Consequently, a price reduction exceeding 5% is placed among second-tier remedies together with termination.

However, even though price reduction as a remedy itself achieves ecological neutrality, its consequences for the environment may sometimes be detrimental.[56] This is the case where the lack of conformity of goods is related to their influence on the environment. The price reduction may indirectly trigger unnecessary environmental costs: the goods that should be repaired or removed from the market because of increased environmental costs of their use remain on the market. The facts of the 'Dieselgate' case[57] may serve to illustrate this issue. In such circumstances, a price reduction could be regarded as a remedy that is – paradoxically – the worst one from the perspective of the environment.[58] This is reflected in the Academic Proposal, which proposes amending Article 15 of Directive 2019/771 in a way that excludes price reduction if the continued use of good is harmful for the environment.[59]

If a consumer decides to exercise price reduction, it does not necessarily mean that they will then not repair the good to meet expectations. Sometimes, repair is impossible (e.g., if the defect is purely aesthetical). However, in many cases the item will still be repaired, e.g., by the self-repair mentioned earlier. In cases where repair is quite simple,[60] it should be regarded as a desirable solution as far as self-repair generates relatively low environmental costs. In this manner, the consumer can bring the good to conformity without incurring any associated costs (from an economic perspective, the costs of repair are borne by the seller, as the repair is paid for from the money returned to the consumer following the price reduction).[61]

6.5 Ecological Lack of Conformity

The Academic Proposal represents a holistic approach to making European consumer law greener through amending Directive 2019/771. It aims at promoting sustainability and circular economy in different manners and not only by reforming the hierarchy and shape of the remedies for non-conformity.[62] The need to redefine a lack of conformity is raised so that it corresponds with the new environmentally friendly approach. It can be argued that the existing framework of consumer sales law already accommodates non-economic interests, such as expectations about the environmental features of a good, both directly, by setting durability requirement, and indirectly – by the reference to existing Union and national law, technical standards or, in the absence of such technical standards, applicable sector-specific industry codes of conduct,[63] and to consumer legitimate expectations. Consumers' expectations as to eco-related features of the good may be based on circumstances of that particular contract but can also be shaped by the market or legal standards.[64] Nevertheless, under the Academic Proposal a more explicit reference to environmental standards is recommended. It is argued that given that a lack of conformity triggers the seller's liability, the conformity requirements need to be precise, and that is why some further specifications of eco-related requirements must be provided. Those issues are not yet the subject of the European Commission's initiative or the ELI's Response.[65]

Conformity requirements under Directive 2019/771 include both subjective and objective requirements. This division should be maintained after the directive is amended, but the eco-related requirements of conformity should not appear in the context of just one of those categories. Consumers' expectations as to eco-related features of the good may be based on circumstances of that particular contract, but can also be shaped by the market or legal standards.[66] Sometimes eco-related features might be expected as a result of a statement made by the seller to a buyer, though sometimes such a quality is generally expected by the public from certain goods, especially considering the raising of environmental awareness among consumers.[67] Such an assertion was accepted by the Academic Proposal, which introduces the environmental criteria in both categories.[68]

As already said, the European sales law act may contribute to achieving sustainability in consumer commerce by promoting a more circular economy. The main role in this transformation should be played by the goods themselves (i.e., by their producers). The goods must represent the quality that will be least detrimental to the environment. A certain durability of the good was already required under Directive 1999/44 (it was considered one of the goods' characteristics, even though not explicitly so).[69] Directive 2019/771 expressly recognised the durability requirement in its Article 2(13), making it one of the objective requirements of conformity in Article 7.1.d. The lifespan of a good is, therefore, a significant aspect that needs to be addressed in order to contribute to increasing the sustainability of the consumer market, especially given the necessity to combat planned obsolescence practices.[70] At the same time, however, it needs to be underlined that there are instances where increasing the durability of goods may be detrimental to the environment, in particular, if the next 'generation' of such goods is more ecological, and the environmental cost of disposal is not significant.[71]

Furthermore, other qualities may also influence environmental costs produced by consumer goods throughout their lifecycle. Those include reusability, recyclability or the already mentioned reparability.[72] These should also be taken into consideration. According to the Academic Proposal, it is also necessary to introduce notions into the Sales of Goods Directive that will describe these other eco-related qualities and will thereby shape the scope of the seller's liability for any lack of conformity. The Academic Proposal suggests incorporating into the directive the terms of 'adaptability', 'reparability', 'reusability', 'recyclability' and the general term of 'environmental added value'.[73] Under the Academic Proposal, if the good does not have these qualities, either subjective or objective requirements are not met and thus it shall be viewed as not in conformity with the contract.[74]

One of the key concepts under the Academic Proposal is environmental added value, understood as a reduction in the strain on the environment, and/or to preserve the environment in compliance with sustainable development, including but not limited to the removal, prevention, reduction and mitigation of pollutants released into the environment, the restoration of damage to the environment or the use of natural resources in a more efficient and sustainable manner. According to the authors, achieving environmental added value should be considered as a goal of public interests, regardless of consumer knowledge or expectations.[75] Hence the seller's exoneration resulting from Article 7(2) of Directive 2019/771 would not apply in cases where the lack of conformity appears because the good does not correspond with statements regarding its environmental added value. What is more, the parties to a contract would be prohibited from deviating from the intended environmental added value.[76]

Another crucial aspect that concerns the ecological lack of conformity of a good is whether it meets the technical standards for environmental protection.[77] This problem, appearing also in the context of other, ecology unrelated standards, has been widely discussed under German law since the 'Dieselgate' case, where it was admitted that the failure to achieve those standards can be regarded as non-conformity, as a consumer can reasonably expect that these standards will be met.[78] This approach seems accurate. On the one hand, it guarantees a proper level of consumer protection, while on the other, the risk of liability motivates producers to offer goods that conform to the established standards.[79] This approach also appears in the Academic Proposal for amending the Sales of Goods Directive.[80]

6.6 Periods of Liability and Burden of Proof

Another key aspect that is inextricably bound with promoting durability and a long lifespan of goods is the length of the consumer protection period. The link between these two has been noted in both the Academic Proposal and the ELI's Response. Possible amendments to the Sales of Goods Directive aimed, among other things, at combating the phenomenon of planned obsolescence.[81] A longer period of the sellers' liability should discourage businesses from artificially limiting a product's lifetime. The Academic Proposal points out that the protection of consumers resulting from Directive 2019/771 is currently insufficient and underlines the need to tie the period of seller's liability with a previously announced or even impliedly assured durability of the good subject to a contract.[82] As a result, the duration of the seller's liability will differ depending on the type of the product. A similar approach is presented in the ELI's Response, with an even further-reaching solution being proposed. Namely, it is argued that the EU legislator should directly provide for different periods of the seller's liability for various categories of goods.[83] The proposal indicates the Ecodesign Directive[84] as a place where such divisions could be made.[85]

Furthermore, the right to repair could play an important part in greening European consumer sales law, provided it becomes a primary remedy under Directive 2019/771. However, for this measure to be as effective as intended, exercising repair should not have a – direct or indirect – negative impact on the consumer's situation. If it does, individuals would be discouraged from exercising their rights through this remedy. One of the main issues that reduces the attractiveness of repair is the risk that the repair will fail or that the good brought into conformity will remain functional only for a short while. That is why both commented approaches underline that the period of the legal guarantee (seller's liability) should be renewed after a repair has been performed.[86]

Last but not least, it must be noted that measures allowing consumers to actually exercise their rights are required. One of these is a presumption that the non-conformity already existed at the time of delivery in the event that the defect manifested shortly after the contract was concluded (e.g., under the minimum harmonisation Directive 1999/44, six months; under the maximum harmonisation Directive 2019/771, one year, which may be extended to two years), resulting in a reversed burden of proof.[87] However, the importance of that mechanism is also related to sustainability. It should be pointed out that the period of the reversed burden of proof and the difficulties for the business resulting from it will motivate producers to ensure the conformity of a good for that period, which may be described as the actual durability of a product.[88] For those reasons, both the Academic Proposal and the ELI's Response find it crucial to link the extension of the periods of sellers' liability with the appropriate extension of the period in which a consumer will benefit from the reversed burden of proof,[89] regardless of the fact that the function of the burden of proof is broader than just ensuring durability.[90]

6.7 Direct Producer's Liability

As long as the right to repair remains the central point of the European Commission's approach to making consumer sales law greener, the following question arises: should the seller be the only one held liable for repairing the good? Both the Academic Proposal and the ELI's Response opt to introduce the direct liability of producers,[91] but only in the form of a duty to repair.[92] Under the Academic Proposal, the seller's and the producer's liability are at the same level – the consumer may freely choose between seeking a repair from the seller or from the producer, as long as the good fails to meet objective conformity criteria (Article 17a of the Academic Proposal). In addition, if the seller refuses or cannot repair the good, the consumer still has a claim against the producer.[93] The ELI's Response goes even further, stating that: '[the] primary addressee of an obligation to repair products should be the manufacturer'.[94]

The idea is well founded. It is evident that the producer is better prepared from the technological point of view to perform repair and has the easiest and direct access to spare parts. In addition, the producer's direct liability will frequently limit the duration of repair,[95] as in practice the sellers often send a defective good to the producer to get it repaired anyway (probably exercising the right of redress under Article 18 of the Sales of Goods Directive). Furthermore, such a solution may indirectly increase the environmental effectiveness (lower transportation costs, etc.) of the right to repair.

The risks to competitiveness in the market, touching mostly the SMEs, resulting from the primacy of right to repair were already mentioned. The producer's direct liability may balance out these consequences.[96] If consumers were able to claim repairs directly from the producer, the burden of maintaining infrastructure and spare parts would be limited on the sellers' side. Another possibility to protect competition is to oblige producers to produce spare parts and provide them to other market participants.[97]

Furthermore, the introduction of producer's liability in the Sales of Goods Directive may have other positive effects, not directly connected with the environmental costs of exercising remedies for non-conformity in consumer transactions. Since Directive 1999/44, European consumer sales law has been trying to shift the costs and burdens resulting from non-conformity onto the producer (the least cost avoider).[98] The same aim might be achieved if the producer's direct liability is introduced. This solution might be even more efficient, as, for example, in Poland, specific provisions transposing the right to redress resulting from Directive 1999/44 were not used in practice.[99] The suggested solution may also minimise the number of litigations between sellers and producers.

Finally, one may ask if the Sales of Goods Directive would be the right place to introduce such a provision since this is a directive constituting the core of European sales law, traditionally founded on a civil contract of sale that binds only its parties. It could be argued that the introduction of producer's direct liability will disturb this balance. However, this argument is not very convincing, as contract law under, e.g., Directive 1999/44 already knew instruments with similar effects to the proposed amendments (e.g., *action directe*).[100] Furthermore, introducing the producer's direct liability in the Sales of Goods Directive can be seen as an imminent consequence of prioritising the right to repair. It would result in a unified regime for this consumer-friendly remedy, so that the consumers would not have to face too many legal complexities and could truly benefit from the remedies provided. On the other hand, in cases where the producer is an entity from outside the European Union, the ELI's Response suggests that rules similar to those in the Product Liability Directive,[101] e.g., importer's liability, should be introduced.[102]

In any event, the proposal for a Directive on common rules promoting the repair of goods which imposes on the producers the duty to repair the good may be seen as proof that the need to involve the producers in the fight for a greener consumer market is also recognised at the EU level. However, the role that this group of market actors may play within the liability for a non-conformity regime still remains neglected.

6.8 Conclusions

The proposal for a Directive on common rules promoting the repair of goods does not necessarily green consumer sales law, as the EU remains reluctant to comprehensively revise[103] Directive 2019/771 so that it actually corresponds to the principles of sustainability and circular economy. Certainly, it does not yet consider the possibility of introducing new green consumer law.[104] Instead, it only promotes repair services and gives rise to producers' repair obligations beyond and outside the scope of Directive 2019/771. Even though granting the consumer the possibility to repair the good is, in principle, positive from the environmental perspective, it is just a small step towards increasing the ecological efficiency of the consumer market.

Big changes in European consumer law, aimed at greening consumer market, are yet to come. From the perspective of the private law framework, these would not only mean redrafting numerous provisions but also forcing a revolution within private law goals. Private law would no longer be just an instrument of balancing and executing the private interests of formally equal parties. Quite the opposite, public interests, such as environmental protection, would also be taken into consideration. This would then influence functions of this branch of law as well as affect the position of parties to a contract.[105] Thus, an axiological overturn may be observed.[106]

With that in mind, it might be reasonable to think also about revising the Sales of Goods Directive accordingly. For instance, as suggested in the Academic Proposal, sustainability and circular economy could be expressly mentioned in the black letter text of the directive as a goal of the European sales law.[107] Such a change would not only be an axiological manifestation but would also have direct legal effects. Currently, given that recitals have no binding power,[108] sustainability only indirectly affects the interpretation of the provisions of Directive 2019/771.[109] From the environmental perspective, the seller's liability regime under Directive 2019/771 does not differ too much from previous one, governed by Directive 1999/44. The proposed amendments would change the situation and make sustainability an important criterion in applying particular provisions of the Sales of Goods Directive.

This chapter commented on proposed possible ways of amending the Sales of Goods Directive. The Academic Proposal and the ELI's Response present important similarities that may become the core of a forthcoming revision of Directive 2019/771. A pivotal role will be played by the right to repair which should become the primary remedy. Furthermore, periods of legal guarantees and of the reversed burden of proof should be lengthened. Finally, the producer's direct liability for the objective non-conformity of a good should be introduced.

On the other hand, there are some differences between the presented approaches. For example, the ELI's proposal argues that the consumer should be granted a right to obtain a replacement good for the time of repair ('a loaner') and advocates for the direct categorisation of goods for the purposes of determining the liability period, while the Academic Proposal touches upon the issue of self-repair. Furthermore, following the Academic Proposal, a more holistic approach, that is not limited to the right to repair and its modalities, should be applied. Especially, the ecological effectiveness of price reduction should be reviewed. Also, an explicit regulation on an ecological lack of conformity is crucial for granting sustainability, as it may truly motivate businesses to produce goods that conform to established standards and present a less detrimental influence on the environment. For sure, these ideas should be examined further by the EU, along with reconsidering the level of harmonisation of the Sales of Goods Directive at least in respect to certain matters.

To conclude, the variety, number and nature of possible environmentally driven changes are bound to trigger further debate within legal scholarship and at the regulatory level. However, it needs to be remembered that the aforementioned possibilities were construed and proposed by lawyers, and the future shape of the Sales of Goods Directive should only be determined after taking into consideration the needs and risks raised by environmental specialists, producers and sellers, as well as consumers themselves. The specific nature of particular goods, examples of their lack of conformity as well as the technically feasible ways to address such deficiencies should be considered. Once all these aspects are examined and all groups of interests are considered, the law is bound to react and provide a system based on deeper granularity, at least when it comes to remedies for non-conformity and the expected durability of goods. Only then can the European consumer sales law actually become greener.

Notes

1 This chapter is part of the research carried out under the project '*Homo consumens, Homo ecologicus*–ecological efficiency test of the new directive on certain aspects concerning contracts for the sale of goods' financed by the Narodowe Centrum Nauki (NCN) [National Science Centre] in Krakow, Poland – number 2019/34/A/ HS5/00124.

2 Fryderyk Zoll, 'Ekologiczne prawo sprzedaży—bardzo wstępne uwagi' in Katarzyna Bilewska and Dorota Krekora-Zając (eds) *Księga jubileuszowa prof. dr. hab. Adama Brzozowskiego* (C.H. Beck Poland 2021) 641. Cf., e.g., Directive 2014/94/EU of the European Parliament and of the Council of 22 October 2014 on the deployment of alternative fuels infrastructure [2014] OJ L 307 1–20; Directive (EU) 2018/2001 of the European Parliament and of the Council of 11 December 2018 on the promotion of the use of energy from renewable sources (recast) [2018] OJ L 328, 82–209.

3 Environmental, social, and corporate governance, which is a way of managing companies while considering certain social and environmental issues. For more, see Anne-Marie Weber, Zofia Mazur, Aleksandra Szczęsna, 'Zrównoważony ład korporacyjny (sustainable corporate governance) – kierunek ewolucji' (2022) 6 Przegląd Prawa Handlowego 20, 22–23 and literature cited therein.

4 Cf., e.g., Regulation (EU) 2020/852 of the European Parliament and of the Council of 18 June 2020 on the establishment of a framework to facilitate sustainable investment, and amending Regulation (EU) 2019/2088 [2020] OJ L 198 13–43; European Commission's Proposal for a Directive of the European Parliament and of the Council on Corporate Sustainability Due Diligence and amending Directive (EU) 2019/1937 (COM(2022) 71 final) <https://eur-lex.europa.eu/resource.html?uri=cellar:bc4dcea4-9584-11ec-b4e4-01aa75ed71a1.0001.02/DOC_1&format=PDF> accessed 18 January 2023.

5 Fryderyk Zoll, Katarzyna Południak-Gierz, Wojciech Bańczyk, 'Sustainable Consumption and Circular Economy in the Directive 2019/771' (2020) 4 Pravovedenie 526, 528–529.

6 Evelyne Terryn, Elias Van Gool, 'The Role of European Consumer Regulation in Shaping the Environmental Impact of e-Commerce' (2021) 3 Journal of European Consumer and Market Law 89.

7 Directive (EU) 2019/771 of the European Parliament and of the Council of 20 May 2019 on certain aspects concerning contracts for the sale of goods, amending Regulation (EU) 2017/2394 and Directive 2009/22/EC, and repealing Directive 1999/44/EC [2019] OJ L 136 28–50 (hereinafter: Sales of Goods Directive, Directive 2019/771).

8 Aneta Wiewiórowska-Domagalska, Fryderyk Zoll, Katarzyna Południak-Gierz, Wojciech Bańczyk, 'Transpozycja dyrektywy Parlamentu Europejskiego i Rady UE 2019/771 z dnia 20 maja 2019 r. w sprawie niektórych aspektów umów sprzedaży towarów' (2021) 4 Kwartalnik Prawa Prywatnego 915, 931–932; Fryderyk Zoll, Katarzyna Południak-Gierz, Wojciech Bańczyk, 'Sustainable Consumption and Circular Economy in the Directive 2019/771' (2020) 4 Pravovedenie 526, 529–530; Jorge Morais Carvalho, 'Sale of Goods and Supply of Digital Content and Digital Services—Overview of Directives 2019/770 and 2019/771' (2019) 5 Journal of European Consumer and Market Law 194, 198; Elias Van Gool, Anaïs Michel, 'The New Consumer Sales Directive 2019/771 and Sustainable Consumption: A Critical Analysis' (2021) 4 Journal of European Consumer and Market Law 136.

9 Jorge Morais Carvalho, 'The Premature Obsolescence of the New Deal for Consumers' (2021) 3 Journal of European Consumer and Market Law 87.

10 Fryderyk Zoll, Jonathon Watson, Katarzyna Południak-Gierz, Wojciech Bańczyk, Gesa Richter, Judit Estifanos, 'Academic Proposal for Amending the Directive (EU) 2019/771 of the European Parliament

and of the Council of 20 May 2019 on Certain Aspects Concerning Contract for the Sale of Goods' (2022) 4 Transformacje Prawa Prywatnego 109 (hereinafter: Academic Proposal).

11 Susanne Augenhofer, 'EC's Public Consultation on Sustainable Consumption—Promoting Repair and Reuse, Response of the European Law Institute' (2022) <https://www.europeanlawinstitute.eu/fileadmin/user_upload/p_eli/Publications/ELI_Response_Sustainable_Consumption_of_Goods.pdf> accessed 17 January 2023 (hereinafter: ELI's Response).

12 On whether balancing the sustainability principle with other values encompassed by the private law, especially consumer sales law, is a feasible task, see Fryderyk Zoll, Katarzyna Południak-Gierz, Wojciech Bańczyk, 'Towards an Environment-Friendly Law of Obligations' in André Janssen, Matthias Lehmann and Reiner Schulze (eds) *The Future of European Private Law* (Nomos/Beck/Hart 2023) 195–224; Katarzyna Południak-Gierz, 'Eco-Reasonableness. Possibility of Incorporating Green Principles into General Private Law Clauses' in Maren Heidemann and Mads Andenas (eds) *Quo Vadis Commercial Contract? Reflections on Sustainability, Ethics and Technology in the Emerging Law and Practice of Global Commerce* (Springer 2023) 110–125.

13 Proposal for a Directive of the European Parliament and of the Council on common rules promoting the repair of goods and amending Regulation (EU) 2017/2394, Directives (EU) 2019/771 and (EU) 2020/1828 (COM(2023) 155 final). See the progress of the European Commission's initiative at <https://ec.europa.eu/info/law/better-regulation/have-your-say/initiatives/13150-Sustainable-consumption-of-goods-promoting-repair-and-reuse_en> accessed 24 March 2023. Some changes to the Proposal were suggested by Council of the European Union, see: Outcome of proceedings on the European Commission's Proposal for a Directive of the European Parliament and of the Council on common Rules promoting the repair of goods and amending Regulation (EU) 2017/2394, Directives (EU) 2019/771 and (EU) 2020/1828, Brussels, 14.02.2024, 2023/0083(COD), <https://data.consilium.europa.eu/doc/document/ST-6461-2024-INIT/en/pdf> (accessed: 12.04.2024).

14 Proposal for a Directive of the European Parliament and of the Council on common rules promoting the repair of goods and amending Regulation (EU) 2017/2394, Directives (EU) 2019/771 and (EU) 2020/1828 (COM(2023) 155 final), Explanatory Memorandum, 1.

15 Proposal for a Regulation establishing a framework for setting ecodesign requirements for sustainable products and repealing Directive 2009/125/EC English, COM(2022) 142 final.

16 The scope of amendments is very limited: Art. 12 of Proposal for a Directive of the European Parliament and of the Council on common rules promoting the repair of goods and amending Regulation (EU) 2017/2394, Directives (EU) 2019/771 and (EU) 2020/1828, COM(2023) 155 final.

17 Directive (EU) 2024/825 of the European Parliament and of the Council of 28 February 2024 amending Directives 2005/29/EC and 2011/83/EU as regards empowering consumers for the green transition through better protection against unfair practices and through better information [2024] OJ L 825).

18 Critically: Susanne Augenhofer, Yeşim Atamer, Katarzyna Południak-Gierz, 'Feedback of the European Law Institute on the European Commission's Proposal for a Directive on Common Rules Promoting the Repair of Goods (COM(2023) 155 final)' (2023) <https://www.europeanlawinstitute.eu/fileadmin/user_upload/p_eli/Publications/ELI_Feedback_Right_to_Repair.pdf> accessed 6 September 2023.

19 Fryderyk Zoll, Katarzyna Południak-Gierz, Wojciech Bańczyk, 'Sustainable Consumption and Circular Economy in the Directive 2019/771' (2020) 4 Pravovedenie 526, 530.

20 Zuzana Selementová, 'Common but Differentiated Responsibilities for Developed and Developing States: A South African Perspective' in Piotr Szwedo, Richard Peltz-Steele and Dai Tamada (eds), *Law and Development. Balancing Principles and Values* (Springer 2019) 86–87.

21 For example, the Polish legislator, when implementing Directive 1999/44 (a minimum harmonisation directive that introduced a hierarchy of remedies) to the Polish Civil Code in 2014, decided to place all the remedies at the same level. See articles 560 § 1 and 561 § 1 the Polish Civil Code.

22 Fryderyk Zoll, Jonathon Watson, Katarzyna Południak-Gierz, Wojciech Bańczyk, Gesa Richter, Judit Estifanos, 'Academic Proposal for Amending the Directive (EU) 2019/771 of the European Parliament and of the Council of 20 May 2019 on Certain Aspects Concerning Contract for the Sale of Goods' (2022) 4 Transformacje Prawa Prywatnego 109, 124.

23 Susanne Augenhofer, 'EC's Public Consultation on Sustainable Consumption—Promoting Repair and Reuse, Response of the European Law Institute' (2022) 15.

24 Proposal for a Directive of the European Parliament and of the Council on common rules promoting the repair of goods and amending Regulation (EU) 2017/2394, Directives (EU) 2019/771 and (EU) 2020/1828 (COM(2023) 155 final), Explanatory Memorandum 3.
25 Directive 1999/44/EC of the European Parliament and of the Council of 25 May 1999 on certain aspects of the sale of consumer goods and associated guarantees [1999] OJ L 171 12–16 (hereinafter: Directive 1999/44).
26 'Right to repair: MEPs want more durable and more easily repairable products' (7 April 2022) <https://www.europarl.europa.eu/news/en/press-room/20220401IPR26537/right-to-repair-meps-want-more-durable-and-more-easily-repairable-products> accessed 21 January 2023.
27 Limited amount of waste is the main goal of circular economy; Katarzyna Kryla-Cudna, 'Sales Contracts and the Circular Economy' (2020) 6 European Review of Private Law 1207, 1213.
28 Evelyne Terryn, 'A Right to Repair? Towards Sustainable Remedies in Consumer Law' (2019) 4 European Review of Private Law 851, 853, 858; Elias Van Gool, Anaïs Michel, 'The New Consumer Sales Directive 2019/771 and Sustainable Consumption: A Critical Analysis' (2021) 4 Journal of European Consumer and Market Law 136, 145.
29 Katarzyna Klimkiewicz, Anna Dubel, Katarzyna Południak-Gierz, 'Supporting Environmentally Conscious Consumer Sales Law by Life-Cycle Thinking' (2023) 17 Contemporary Economics 174.
30 Evelyne Terryn, 'A Right to Repair? Towards Sustainable Remedies in Consumer Law' (2019) 4 European Review of Private Law 851, 853–854.
31 Fryderyk Zoll, Jonathon Watson, Katarzyna Południak-Gierz, Wojciech Bańczyk, Gesa Richter, Judit Estifanos, 'Academic Proposal for Amending the Directive (EU) 2019/771 of the European Parliament and of the Council of 20 May 2019 on Certain Aspects Concerning Contract for the Sale of Goods' (2022) 4 Transformacje Prawa Prywatnego 109, 139–144.
32 Susanne Augenhofer, 'EC's Public Consultation on Sustainable Consumption—Promoting Repair and Reuse, Response of the European Law Institute' (2022) 9.
33 Susanne Augenhofer, 'EC's Public Consultation on Sustainable Consumption—Promoting Repair and Reuse, Response of the European Law Institute' (2022) 10.
34 Fryderyk Zoll, 'Ekologiczne prawo sprzedaży—bardzo wstępne uwagi' in Katarzyna Bilewska and Dorota Krekora Zając (eds) *Księga jubileuszowa prof. dr. hab. Adama Brzozowskiego* (C.H. Beck Poland 2021) 644.
35 Cf. the plentiful feedback on the European Commission's initiative <https://ec.europa.eu/info/law/better-regulation/have-your-say/initiatives/13150-Sustainable-consumption-of-goods-promoting-repair-and-reuse/feedback_en?p_id=27653994> accessed 21 January 2023.
36 Susanne Augenhofer, 'EC's Public Consultation on Sustainable Consumption—Promoting Repair and Reuse, Response of the European Law Institute' (2022) 9.
37 Fryderyk Zoll, Katarzyna Południak-Gierz, Wojciech Bańczyk, 'Sustainable Consumption and Circular Economy in the Directive 2019/771' (2020) 4 Pravovedenie 526, 536–537.
38 Roman Trzaskowski, Witold Borysiak, 'Dochodzenie roszczeń o naprawienie szkody w postaci kosztów najmu pojazdu zastępczego w ramach ubezpieczenia OC posiadaczy pojazdów mechanicznych (perspektywa prawnoporównawcza)' (Raport Instytutu Wymiaru Sprawiedliwości 2016) 37.
39 Eeugeniusz Kowalewski, 'Uprawnienie poszkodowanego w wypadku komunikacyjnym do pojazdu zastępczego – ujęcie prawnoporównawcze' (2011) 1 Wiadomości ubezpieczeniowe 13–14.
40 Similar postulates start to appear also, e.g., in Polish doctrine: Fryderyk Zoll, Katarzyna Południak-Gierz, Wojciech Bańczyk, 'State Liability for Damage Caused by COVID-19 Restrictions Under Polish Law', in Ewoud Hondus, Marta Santos Silva, Andrea Nicolussi, Pablo Salvador-Coderch, Christiane Wenfehorst and Fryderyk Zoll (eds) *Coronavirus and the Law in Europe* (Intersentia 2021) 359; Fryderyk Zoll, Katarzyna Południak-Gierz, Wojciech Bańczyk, 'Zakres odpowiedzialności odszkodowawczej Skarbu Państwa za szkody majątkowe spowodowane ograniczeniami wprowadzonymi w związku z COVID-19' (2021) 3 Studia Prawa Prywatnego 1, 17.
41 Fryderyk Zoll, Katarzyna Południak-Gierz, Wojciech Bańczyk, 'Sustainable Consumption and Circular Economy in the Directive 2019/771' (2020) 4 Pravovedenie 526, 540.
42 Fryderyk Zoll, Jonathon Watson, Katarzyna Południak-Gierz, Wojciech Bańczyk, Gesa Richter, Judit Estifanos, 'Academic Proposal for Amending the Directive (EU) 2019/771 of the European Parliament and of the Council of 20 May 2019 on Certain Aspects Concerning Contract for the Sale of Goods' (2022) 4 Transformacje Prawa Prywatnego 109, 142–143.

43 Susanne Augenhofer, 'EC's Public Consultation on Sustainable Consumption—Promoting Repair and Reuse, Response of the European Law Institute' (2022) 11; Fryderyk Zoll, Jonathon Watson, Katarzyna Południak-Gierz, Wojciech Bańczyk, Gesa Richter, Judit Estifanos, 'Academic Proposal for Amending the Directive (EU) 2019/771 of the European Parliament and of the Council of 20 May 2019 on Certain Aspects Concerning Contract for the Sale of Goods' (2022) 4 Transformacje Prawa Prywatnego 109, 135–137.

44 Katarzyna Kryla-Cudna, 'Sales Contracts and the Circular economy' (2020) 6 European Review of Private Law 1207, 1222.

45 Cf. the implementation of refurbishing made by Apple: <https://www.apple.com/uk/shop/refurbished/about> accessed 22 January 2023; and Amazon: <https://www.amazon.co.uk/Certified-Refurbished-on-Amazon/b?ie=UTF8&node=8362590031> accessed 22 January 2023.

46 Susanne Augenhofer, 'EC's Public Consultation on Sustainable Consumption—Promoting Repair and Reuse, Response of the European Law Institute' (2022) 11; Fryderyk Zoll, Jonathon Watson, Katarzyna Południak-Gierz, Wojciech Bańczyk, Gesa Richter, Judit Estifanos, 'Academic Proposal for Amending the Directive (EU) 2019/771 of the European Parliament and of the Council of 20 May 2019 on Certain Aspects Concerning Contract for the Sale of Goods' (2022) 4 Transformacje Prawa Prywatnego 109, 135–137; Elias Van Gool, Anaïs Michel, 'The New Consumer Sales Directive 2019/771 and Sustainable Consumption: A Critical Analysis' (2021) 4 Journal of European Consumer and Market Law 136, 138.

47 Elias Van Gool, Anaïs Michel, 'The New Consumer Sales Directive 2019/771 and Sustainable Consumption: A Critical Analysis' (2021) 4 Journal of European Consumer and Market Law 136, 138.

48 Fryderyk Zoll, Jonathon Watson, Katarzyna Południak-Gierz, Wojciech Bańczyk, Gesa Richter, Judit Estifanos, 'Academic Proposal for Amending the Directive (EU) 2019/771 of the European Parliament and of the Council of 20 May 2019 on Certain Aspects Concerning Contract for the Sale of Goods' (2022) 4 Transformacje Prawa Prywatnego 109, 145–146. However, the EU Council's stand on the matter, presented on 14.02.2024, is that the seller should be allowed to fulfill his obligation to replace the good by providing a refurbished good only upon explicit request by the consumer (Art. 14(1) cb).

49 Johanna Croon-Gestefeld, 'Die nachhaltige Beschaffenheit der Kaufsache' (2022) 8 Neue Juristische Wochenschrift 497, 501.

50 Susanne Augenhofer, 'EC's Public Consultation on Sustainable Consumption – Promoting Repair and Reuse, Response of the European Law Institute' (2022) 13.

51 COM(2022) 142 final.

52 The matter is addressed in the Art. 19(1) of Proposal for a Directive of the European Parliament and of the Council on the legal protection of designs (recast), COM(2022) 667 final, 28 November 2022 which states that protection shall not be conferred on a registered design which constitutes a component part of a complex product, upon whose appearance the design of the component part is dependent, and which is used within the meaning of article 16(1) for the sole purpose of the repair of that complex product so as to restore its original appearance. Denying design protection to 'must-match' spare parts is encouraged to grant third-party repairers' easy access to the latter. Susanne Augenhofer, Yeşim Atamer, Katarzyna Południak-Gierz, 'Feedback of the Europe-an Law Institute on the European Commission's Proposal for a Directive on Common Rules Promoting the Repair of Goods (COM(2023) 155 final)' (2023). For a general overview, see Circular economy versus copyright protection of computer programs in the EU: Bohdan Widła, 'Challenges and lessons from the CJEU's judgment in Top System' (2023) 18(5) Journal of Intellectual Property Law & Practice 353.

53 This approach, however, has evolved, and the ELI position presented under Susanne Augenhofer, Yeşim Atamer, Katarzyna Południak-Gierz, 'Feedback of the Europe-an Law Institute on the European Commission's Proposal for a Directive on Common Rules Promoting the Repair of Goods (COM(2023) 155 final)' (2023) resembles the one argued for in the Academic Proposal.

54 Fryderyk Zoll, Katarzyna Południak-Gierz, Wojciech Bańczyk, 'Sustainable Consumption and Circular Economy in the Directive 2019/771' (2020) 4 Pravovedenie 526, 542–543.

55 Fryderyk Zoll, Jonathon Watson, Katarzyna Południak-Gierz, Wojciech Bańczyk, Gesa Richter, Judit Estifanos, 'Academic Proposal for Amending the Directive (EU) 2019/771 of the European Parliament and of the Council of 20 May 2019 on Certain Aspects Concerning Contract for the Sale of Goods' (2022) 4 Transformacje Prawa Prywatnego 109, 139.

56 Fryderyk Zoll, Jonathon Watson, Katarzyna Południak-Gierz, Wojciech Bańczyk, Gesa Richter, Judit Estifanos, 'Academic Proposal for Amending the Directive (EU) 2019/771 of the European Parliament and of the Council of 20 May 2019 on Certain Aspects Concerning Contract for the Sale of Goods' (2022) 4 Transformacje Prawa Prywatnego 109, 143.

57 Case C-343/19 Verein für Konsumenteninformation v Volkswagen AG [2020] ECR, ECLI:EU:C:2020:534.
58 Cf. Fryderyk Zoll, 'Czy nowa dyrektywa o sprzedaży zapowiada ekologiczną rewolucję? Uwagi na kanwie kilku niemieckich orzeczeń odnoszących się do kwestii ochrony środowiska w związku z odpowiedzialnością z tytułu rękojmi' in Krzysztof Oplustil, Piotr Tuleja and Kamil Prokopowicz (eds) *Nauka prawa a wyzwania XXI wieku. Księga jubileuszowa z okazji 170-lecia istnienia Towarzystwa Biblioteki Słuchaczów Prawa Uniwersytetu Jagiellońskiego* (Wydawnictwo Uniwersytetu Jagiellońskiego 2023) 344–345.
59 Fryderyk Zoll, Jonathon Watson, Katarzyna Południak-Gierz, Wojciech Bańczyk, Gesa Richter, Judit Estifanos, 'Academic Proposal for Amending the Directive (EU) 2019/771 of the European Parliament and of the Council of 20 May 2019 on Certain Aspects Concerning Contract for the Sale of Goods' (2022) 4 Transformacje Prawa Prywatnego 109, 146.
60 As indicated, the process of self-repair may be more harmful for the environment than the repair proposed by the seller/producer. For those reasons, the Academic Proposal provides a limit on price reduction of 5% of the price, to encourage consumer to benefit from the repair made by the seller if the process of bringing back the conformity is more complicated; cf. Fryderyk Zoll, Jonathon Watson, Katarzyna Południak-Gierz, Wojciech Bańczyk, Gesa Richter, Judit Estifanos, 'Academic Proposal for Amending the Directive (EU) 2019/771 of the European Parliament and of the Council of 20 May 2019 on Certain Aspects Concerning Contract for the Sale of Goods' (2022) 4 Transformacje Prawa Prywatnego 109, 142–143.
61 Fryderyk Zoll, Katarzyna Południak-Gierz, Wojciech Bańczyk, 'Sustainable Consumption and Circular Economy in the Directive 2019/771' (2020) 4 Pravovedenie 526, 543.
62 Fryderyk Zoll, Jonathon Watson, Katarzyna Południak-Gierz, Wojciech Bańczyk, Gesa Richter, Judit Estifanos, 'Academic Proposal for Amending the Directive (EU) 2019/771 of the European Parliament and of the Council of 20 May 2019 on Certain Aspects Concerning Contract for the Sale of Goods' (2022) 4 Transformacje Prawa Prywatnego 109, 111–113.
63 The case law on car software manipulating the emission-testing results is an excellent example of that. See Case C-343/19 *Verein für Konsumen-teninformation v Volkswagen AG* [2020] ECR ECLI:EU:C:2020:534, Judgement of the Polish Supreme Court of 3 July 2019, II CSK 306/18, with commentary by Maciej Bujalski (2022) 4 Osteuropa Recht 538.
64 For more, see Katarzyna Południak-Gierz, 'Eco-Reasonableness. Possibility of Incorporating Green Principles into General Private Law Clauses' in Maren Heidemann and Mads Andenas (eds) *Quo Vadis Commercial Contract? Reflections on Sustainability, Ethics and Technology in the Emerging Law and Practice of Global Commerce* (Springer 2023) 115–121.
65 Even though they are incorporated in the later ELI position presented under Susanne Augenhofer, Yeşim Atamer, Katarzyna Południak-Gierz, 'Feedback of the Europe – An Law Institute on the European Commission's Proposal for a Directive on Common Rules Promoting the Repair of Goods (COM(2023) 155 final)' (2023).
66 For more, see Katarzyna Południak-Gierz, 'Eco-Reasonableness. Possibility of Incorporating Green Principles into General Private Law Clauses' in Maren Heidemann and Mads Andenas (eds) *Quo Vadis Commercial Contract? Reflections on Sustainability, Ethics and Technology in the Emerging Law and Practice of Global Commerce* (Springer 2023) 115–121.
67 Cristina Poncibò, 'The Contractualisation of Environmental Sustainability' (2016) 12.4 European Review of Consumer Law 353, 342.
68 Fryderyk Zoll, Jonathon Watson, Katarzyna Południak-Gierz, Wojciech Bańczyk, Gesa Richter, Judit Estifanos, 'Academic Proposal for Amending the Directive (EU) 2019/771 of the European Parliament and of the Council of 20 May 2019 on Certain Aspects Concerning Contract for the Sale of Goods' (2022) 4 Transformacje Prawa Prywatnego 109, 125–132.
69 Alberto De Franceschi, 'Planned Obsolescence Challenging the Effectiveness of Consumer Law and the Achievement of a Sustainable Economy. The Apple and Samsung Cases' (2018) 6 Journal of European Consumer and Market Law 217, 219; Elias Van Gool, Anaïs Michel, 'The New Consumer Sales Directive 2019/771 and Sustainable Consumption: A Critical Analysis' (2021) 4 Journal of European Consumer and Market Law 136, 138; judgement of *Oberster Gerichtshof* (Austria) of 23 April 2015, 1 Ob 71/15.
70 Alberto De Franceschi, 'Planned Obsolescence Challenging the Effectiveness of Consumer Law and the Achievement of a Sustainable Economy. The Apple and Samsung Cases' (2018) 6 Journal of European Consumer and Market Law 217, 220; Jorge Morais Carvalho, 'The Premature Obsolescence of the New Deal for Consumers' (2021) 3 Journal of European Consumer and Market Law 85, 85.

71 Katarzyna Klimkiewicz, Anna Dubel, Katarzyna Południak-Gierz, 'Supporting Environmentally Conscious Consumer Sales Law by Life-Cycle (2023) 17(2) Contemporary Economics 187.

72 Cf. Thomas Nyström, Derek Diener, 'A Companion for the Design of Future-Adaptable Products' (RISE Research Institutes of Sweden 2018) <https://www.ri.se/sites/default/files/2019-10/A%20companion%20for%20the%20design%20of%20future-adaptable%20products%20for%20use%20in%20circular%20business%20models.pdf> accessed 22 January 2023.

73 Fryderyk Zoll, Jonathon Watson, Katarzyna Południak-Gierz, Wojciech Bańczyk, Gesa Richter, Judit Estifanos, 'Academic Proposal for Amending the Directive (EU) 2019/771 of the European Parliament and of the Council of 20 May 2019 on Certain Aspects Concerning Contract for the Sale of Goods' (2022) 4 Transformacje Prawa Prywatnego 109, 115–122.

74 Fryderyk Zoll, Jonathon Watson, Katarzyna Południak-Gierz, Wojciech Bańczyk, Gesa Richter, Judit Estifanos, 'Academic Proposal for Amending the Directive (EU) 2019/771 of the European Parliament and of the Council of 20 May 2019 on Certain Aspects Concerning Contract for the Sale of Goods' (2022) 4 Transformacje Prawa Prywatnego 109, 126, 129. The EU Council on 14.02.2024 suggested including an explicit reference to reparability within objective conformity criteria (see Art. 7(1) d).

75 Fryderyk Zoll, Jonathon Watson, Katarzyna Południak-Gierz, Wojciech Bańczyk, Gesa Richter, Judit Estifanos, 'Academic Proposal for Amending the Directive (EU) 2019/771 of the European Parliament and of the Council of 20 May 2019 on Certain Aspects Concerning Contract for the Sale of Goods' (2022) 4 Transformacje Prawa Prywatnego 109, 132.

76 Fryderyk Zoll, Jonathon Watson, Katarzyna Południak-Gierz, Wojciech Bańczyk, Gesa Richter, Judit Estifanos, 'Academic Proposal for Amending the Directive (EU) 2019/771 of the European Parliament and of the Council of 20 May 2019 on Certain Aspects Concerning Contract for the Sale of Goods' (2022) 4 Transformacje Prawa Prywatnego 109, 130–132.

77 For example, emission standards resulting from Regulation (EU) 2019/631 of the European Parliament and of the Council of 17 April 2019 setting CO_2 emission performance standards for new passenger cars and for new light commercial vehicles, and repealing Regulations (EC) 443/2009 and (EU) 510/2011 [2019] OJ L 111 13–53

78 Judgement of BGH (Germany) of 25 May 2020, VI ZR 252/19, paras 54–54; Christoph Schmid, 'Germany' in Marco Frigessi di Rattalma (ed) *The Dieselgate. A Legal Perspective* (Springer, 2017) 36; Carl-Heinz Witt, 'Der Dieselskandal und seine kauf- und deliktsrechtlichen Folgen' (2017) 51 Neue Juristische Wochenschrift 3681, 3682. See also Fryderyk Zoll, 'Czy nowa dyrektywa o sprzedaży zapowiada ekologiczną rewolucję? Uwagi na kanwie kilku niemieckich orzeczeń odnoszących się do kwestii ochrony środowiska w związku z odpowiedzialnością z tytułu rękojmi' in Krzysztof Oplustil, Piotr Tuleja and Kamil Prokopowicz (eds) *Nauka prawa a wyzwania XXI wieku. Księga jubileuszowa z okazji 170-lecia istnienia Towarzystwa Biblioteki Słuchaczów Prawa Uniwersytetu Jagiellońskiego* (Wydawnictwo Uniwersytetu Jagiellońskiego 2023) 340–343. Similarly, under Polish law in the judgement of SN (Poland) of 3 July 2019, II CSK 306/18.

79 Another possibility is the invalidity of the contract due to national law and its critics see Fryderyk Zoll, Katarzyna Południak-Gierz, Wojciech Bańczyk, 'Sustainable Consumption and Circular Economy in the Directive 2019/771' (2020) 4 Pravovedenie 526, 537–539.

80 Fryderyk Zoll, Jonathon Watson, Katarzyna Południak-Gierz, Wojciech Bańczyk, Gesa Richter, Judit Estifanos, 'Academic Proposal for Amending Directive (EU) 2019/771 of the European Parliament and of the Council of 20 May 2019 on Certain Aspects Concerning Contract for the Sale of Goods' (2022) 4 Transformacje Prawa Prywatnego 109, 128.

81 For more, see Alberto De Franceschi, 'Planned Obsolescence Challenging the Effectiveness of Consumer Law and the Achievement of a Sustainable Economy. The Apple and Samsung Cases' (2018) 6 Journal of European Consumer and Market Law 217–221; Jorge Morais Carvalho, 'The Premature Obsolescence of the New Deal for Consumers' (2021) 3 Journal of European Consumer and Market Law 85–89.

82 Fryderyk Zoll, Jonathon Watson, Katarzyna Południak-Gierz, Wojciech Bańczyk, Gesa Richter, Judit Estifanos, 'Academic Proposal for Amending the Directive (EU) 2019/771 of the European Parliament and of the Council of 20 May 2019 on Certain Aspects Concerning Contract for the Sale of Goods' (2022) 4 Transformacje Prawa Prywatnego 109, 134–136.

83 Susanne Augenhofer, 'EC's Public Consultation on Sustainable Consumption—Promoting Repair and Reuse, Response of the European Law Institute' (2022) 10. Similarly, also BEUC 'Sustainable Consumption of Goods—Promoting the Right to Repair and Reuse. Accompanying paper to the BEUC

response to the public consultation' (5 April 2022) 6 <https://www.beuc.eu/sites/default/files/publications/beuc-x-2022-034_public_consultation_on_right_to_repair.pdf> accessed 25 January 2023.

84 Directive 2009/125/EC of the European Parliament and of the Council of 21 October 2009 establishing a framework for the setting of ecodesign requirements for energy-related products [2009] OJ L 285 10–35 which is currently planned to be revisioned. See European Commission's Proposal for a Regulation of the European Parliament and of the Council establishing a framework for setting ecodesign requirements for sustainable products and repealing Directive 2009/125/EC (COM(2022) 142 final) <https://eur-lex.europa.eu/legal-content/EN/TXT/?uri=CELEX:52022PC0142> accessed 25 January 2023.

85 In accordance with the postulates of legal scholars: Elias Van Gool, Anaïs Michel, 'The New Consumer Sales Directive 2019/771 and Sustainable Consumption: A Critical Analysis' (2021) 4 Journal of European Consumer and Market Law 136, 142; Evelyne Terryn, 'A Right to Repair? Towards Sustainable Remedies in Consumer Law' (2019) 4 European Review of Private Law 851, 872.

86 Fryderyk Zoll, Jonathon Watson, Katarzyna Południak-Gierz, Wojciech Bańczyk, Gesa Richter, Judit Estifanos, 'Academic Proposal for Amending the Directive (EU) 2019/771 of the European Parliament and of the Council of 20 May 2019 on Certain Aspects Concerning Contract for the Sale of Goods' (2022) 4 Transformacje Prawa Prywatnego 109, 134; Susanne Augenhofer, 'EC's Public Consultation on Sustainable Consumption—Promoting Repair and Reuse, Response of the European Law Institute' (2022) 10. The Council on 14.02.2024 stated that if the good is repaired under seller's liability regime, the liability period should be extended once by twelve months (see Art. 10(2a)).

87 Cf. Rupprecht Podszun, 'Procedural Autonomy and Effective Consumer Protection in Sale of Goods Liability: Easing the Burden for Consumers (Even If They Aren't Consumers). Comment on Case C-497/13 Froukje Faber v. Autobedrijf Hazet Ochten BV, Judgment of the Court of Justice (First Chamber) of 4 June 2015' (2015) 4 Journal of European Consumer and Market Law 149, 152.

88 Ivo Bach, Maren Wöbbeking, 'Das Haltbarkeitserfordernis der Warenkauf-RL als neuer Hebel für mehr Nachhaltigkeit?' (2020) 37 Neue Juristische Wochenschrift 2672–2675.

89 Fryderyk Zoll, Jonathon Watson, Katarzyna Południak-Gierz, Wojciech Bańczyk, Gesa Richter, Judit Estifanos, 'Academic Proposal for Amending the Directive (EU) 2019/771 of the European Parliament and of the Council of 20 May 2019 on Certain Aspects Concerning Contract for the Sale of Goods' (2022) 4 Transformacje Prawa Prywatnego 109, 137–138; Susanne Augenhofer, 'EC's Public Consultation on Sustainable Consumption—Promoting Repair and Reuse, Response of the European Law Institute' (2022) 10.

90 Fryderyk Zoll, Katarzyna Południak-Gierz, Wojciech Bańczyk, 'Sustainable Consumption and Circular Economy in the Directive 2019/771' (2020) 4 Pravovedenie 526, 535.

91 The ELI's Proposal uses the notion of 'manufacturer', but the meaning remains the same. However, the notion of producer suits better and is already defined in article 2(4) of Directive 2019/771.

92 Fryderyk Zoll, Jonathon Watson, Katarzyna Południak-Gierz, Wojciech Bańczyk, Gesa Richter, Judit Estifanos, 'Academic Proposal for Amending the Directive (EU) 2019/771 of the European Parliament and of the Council of 20 May 2019 on Certain Aspects Concerning Contract for the Sale of Goods' (2022) 4 Transformacje Prawa Prywatnego 109, 151; Susanne Augenhofer, 'EC's Public Consultation on Sustainable Consumption—Promoting Repair and Reuse, Response of the European Law Institute' (2022) 14.

93 Fryderyk Zoll, Jonathon Watson, Katarzyna Południak-Gierz, Wojciech Bańczyk, Gesa Richter, Judit Estifanos, 'Academic Proposal for Amending the Directive (EU) 2019/771 of the European Parliament and of the Council of 20 May 2019 on Certain Aspects Concerning Contract for the Sale of Goods' (2022) 4 Transformacje Prawa Prywatnego 109, 141.

94 Susanne Augenhofer, 'EC's Public Consultation on Sustainable Consumption—Promoting Repair and Reuse, Response of the European Law Institute' (2022) 14.

95 Susanne Augenhofer, 'EC's Public Consultation on Sustainable Consumption—Promoting Repair and Reuse, Response of the European Law Institute' (2022) 14.

96 Susanne Augenhofer, 'EC's Public Consultation on Sustainable Consumption—Promoting Repair and Reuse, Response of the European Law Institute' (2022) 14.

97 Susanne Augenhofer, 'EC's Public Consultation on Sustainable Consumption—Promoting Repair and Reuse, Response of the European Law Institute' (2022) 14.

98 See article 18 of Directive 2019/771, which replaced article 4 of Directive 1999/44.

99 That is why the Polish legislator decided to abandon a particular regulation on the right to redress while transposing Directive 2019/771 and has claimed that general rules on contractual liability will be

sufficient; cf. the justification of the draft act amending the act on consumer rights and certain other acts (5 January 2022) <https://legislacja.gov.pl/docs//2/12341810/12752766/12752770/dokument539139.pdf> accessed: 23 January 2023. On the right to redress under Directive 1999/44 see Fryderyk Zoll, *Rękojmia. Odpowiedzialność sprzedawcy* (C.H. Beck Poland 2018) 383–396.

100 This instrument was also indicated as a possible measure of transposing the right to redress resulting from article 4 of Directive 1999/44; Fryderyk Zoll, *Rękojmia. Odpowiedzialność sprzedawcy* (C.H. Beck Poland 2018) 383–384. The solution was adopted in French law. See Michel Cannarsa, Olivier Moreteau, 'The French "Action Directe": The Justification for Going beyond Privity' in Martin Ebers, André Janssen and Olaf Meyer (eds) *European Perspectives on Producers' Liability. Direct Producers' Liability for Non-conformity and the Sellers' Right of Redress* (Sellier 2009) 311–321.

101 Council Directive 85/374/EEC of 25 July 1985 on the approximation of the laws, regulations and administrative provisions of the Member States concerning liability for defective products [1985] OJ L 210 29–33.

102 Susanne Augenhofer, 'EC's Public Consultation on Sustainable Consumption—Promoting Repair and Reuse, Response of the European Law Institute' (2022) 14.

103 Evelyne Terryn, 'A Right to Repair? Towards Sustainable Remedies in Consumer Law' (2019) 4 European Review of Private Law 851, 861; Fryderyk Zoll, Jonathon Watson, Katarzyna Południak-Gierz, Wojciech Bańczyk, Gesa Richter, Judit Estifanos, 'Academic Proposal for Amending the Directive (EU) 2019/771 of the European Parliament and of the Council of 20 May 2019 on Certain Aspects Concerning Contract for the Sale of Goods' (2022) 4 Transformacje Prawa Prywatnego 109, 109.

104 Hans-Wolfgang Micklitz, 'Squaring the Circle? Reconciling Consumer Law and the Circular Economy' (2019) 8 Journal of European Consumer & Market Law 229, 229.

105 Fryderyk Zoll, 'Ekologiczne prawo sprzedaży—bardzo wstępne uwagi' in Katarzyna Bilewska and Dorota Krekora-Zając (eds) *Księga jubileuszowa prof. dr. hab. Adama Brzozowskiego* (C.H. Beck Poland 2021) 639–641.

106 For example, on the example of ownership law, see Wojciech Bańczyk, 'Economic and Social Development in the Republic of South Africa's New Model of Mineral Rights: Balancing Private Ownership, Community Rights, and Sovereignty' in Piotr Szwedo, Richard Peltz-Steele and Dai Tamada (eds) *Law and Development. Balancing Principles and Values* (Springer 2019) 215.

107 Fryderyk Zoll, Jonathon Watson, Katarzyna Południak-Gierz, Wojciech Bańczyk, Gesa Richter, Judit Estifanos, 'Academic Proposal for Amending the Directive (EU) 2019/771 of the European Parliament and of the Council of 20 May 2019 on Certain Aspects Concerning Contract for the Sale of Goods' (2022) 4 Transformacje Prawa Prywatnego 109, 114–115.

108 Aleksandra Kunkiel-Kryńska, *Metody harmonizacji prawa konsumenckiego w Unii Europejskiej i ich wpływ na procesy implementacyjne w państwach członkowskich* (Wolters Kluwer Poland 2013) 81.

109 For more on pro-ecological interpretation, see Fryderyk Zoll, Katarzyna Południak-Gierz, Wojciech Bańczyk, 'Sustainable Consumption and Circular Economy in the Directive 2019/771' (2020) 4 Pravovedenie 526, 530–532. Generally, on a value-oriented interpretation of EU law Jorg Neuner, 'Die Rechtsfortbildung' in Karl Riesenhuber (ed) *Europäische Methodenlehre* (De Gruyter 2010) 392.

Bibliography

Augenhofer S., 'EC's Public Consultation on Sustainable Consumption—Promoting Repair and Reuse, Response of the European Law Institute' (2022) <https://www.europeanlawinstitute.eu/fileadmin/user_upload/p_eli/Publications/ELI_Response_Sustainable_Consumption_of_Goods.pdf> accessed 17 January 2023.

Augenhofer S., Atamer Y., Południak-Gierz K., 'Feedback of the European Law Institute on the European Commission's Proposal for a Directive on Common Rules Promoting the Repair of Goods (COM(2023) 155 final)' (2023) <https://www.europeanlawinstitute.eu/fileadmin/user_upload/p_eli/Publications/ELI_Feedback_Right_to_Repair.pdf> accessed 6 September 2023.

Bach I., Wöbbeking M., 'Das Haltbarkeitserfordernis der Warenkauf-RL als neuer Hebel für mehr Nachhaltigkeit?' (2020) 37 Neue Juristische Wochenschrift 2672.

Bańczyk W., 'Economic and Social Development in the Republic of South Africa's New Model of Mineral Rights: Balancing Private Ownership, Community Rights, and Sovereignty' in P. Szwedo, R. Peltz-Steele and D. Tamada (eds) *Law and Development. Balancing Principles and Values* (Springer 2019) 215.

BEUC 'Sustainable Consumption of Goods—Promoting the Right to Repair and Reuse. Accompanying paper to the BEUC response to the public consultation' (5 April 2022) 6 <https://www.beuc.eu/sites/default/files/publications/beuc-x-2022-034_public_consultation_on_right_to_repair.pdf> accessed 25 January 2023.

Bujalski M., 'Commentary to the Judgement of the Polish Supreme Court of 3 July 2019, II CSK 306/18' (2022) 4 Osteuropa Recht 538.

Cannarsa M., Moreteau O., 'The French "Action Directe": The Justification for Going Beyond Privity' in M. Ebers, A. Janssen and O. Meyer (eds) *European Perspectives on Producers' Liability. Direct Producers' Liability for Non-Conformity and the Sellers' Right of Redress* (Sellier 2009) 311–321.

Carvalho J.M., 'Sale of Goods and Supply of Digital Content and Digital Services—Overview of Directives 2019/770 and 2019/771' (2019) 5 Journal of European Consumer and Market Law 194.

Carvalho J.M., 'The Premature Obsolescence of the New Deal for Consumers' (2021) 3 Journal of European Consumer and Market Law 87.

Croon-Gestefeld J., 'Die nachhaltige Beschaffenheit der Kaufsache' (2022) 8 Neue Juristische Wochenschrift 497.

De Franceschi A., 'Planned Obsolescence Challenging the Effectiveness of Consumer Law and the Achievement of a Sustainable Economy. The Apple and Samsung Cases' (2018) 6 Journal of European Consumer and Market Law 217.

Klimkiewicz K., Dubel A., Południak-Gierz K., 'Supporting Environmentally Conscious Consumer Sales Law by Life-cycle Thinking' (2023) 17 Contemporary Economics 174.

Kowalewski E., 'Uprawnienie poszkodowanego w wypadku komunikacyjnym do pojazdu zastępczego – ujęcie prawnoporównawcze' (2011) 1 Wiadomości ubezpieczeniowe 7.

Kryla-Cudna K., 'Sales Contracts and the Circular Economy' (2020) 6 European Review of Private Law 1207.

Kunkiel-Kryńska A., *Metody harmonizacji prawa konsumenckiego w Unii Europejskiej i ich wpływ na procesy implementacyjne w państwach członkowskich* (Wolters Kluwer Poland 2013).

Micklitz H.-W., 'Squaring the Circle? Reconciling Consumer Law and the Circular Economy' (2019) 8 Journal of European Consumer & Market Law 229.

Neuner J., 'Die Rechtsfortbildung' in K. Riesenhuber (ed) *Europäische Methodenlehre* (De Gruyter 2010) 392.

Nyström T., Diener D., 'A Companion for the Design of Future-Adaptable Products' (RISE Research Institutes of Sweden 2018) <https://www.ri.se/sites/default/files/2019-10/A%20companion%20for%20the%20design%20of%20future-adaptable%20products%20for%20use%20in%20circular%20business%20models.pdf> accessed 22 January 2023.

Podszun R., 'Procedural Autonomy and Effective Consumer Protection in Sale of Goods Liability: Easing the Burden for Consumers (Even If They Aren't Consumers). Comment on Case C-497/13 Froukje Faber v. Autobedrijf Hazet Ochten BV, Judgment of the Court of Justice (First Chamber) of 4 June 2015' (2015) 4 Journal of European Consumer and Market Law 149.

Południak-Gierz K., 'Eco-Reasonableness. Possibility of Incorporating Green Principles into General Private Law Clauses' in M. Heidemann and M. Andenas (eds) *Quo Vadis Commercial Contract? Reflections on Sustainability, Ethics and Technology in the Emerging Law and Practice of Global Commerce* (Springer 2023) 115–121.

Poncibò C., 'The Contractualisation of Environmental Sustainability' (2016) 12.4 European Review of Consumer Law 342.

Schmid C., 'Germany' in M. Frigessi di Rattalma (ed) *The Dieselgate. A Legal Perspective* (Springer, 2017) 36.

Selementová Z., 'Common but Differentiated Responsibilities for Developed and Developing States: A South African Perspective' in P. Szwedo, R. Peltz-Steele and D. Tamada (eds) *Law and Development. Balancing Principles and Values* (Springer 2019) 86–87.

Terryn E., 'A Right to Repair? Towards Sustainable Remedies in Consumer Law' (2019) 4 European Review of Private Law 851.

Terryn E., Van Gool E., 'The Role of European Consumer Regulation in Shaping the Environmental Impact of e-Commerce' (2021) 3 Journal of European Consumer and Market Law 89.

Trzaskowski R., Borysiak W., 'Dochodzenie roszczeń o naprawienie szkody w postaci kosztów najmu pojazdu zastępczego w ramach ubezpieczenia OC posiadaczy pojazdów mechanicznych (perspektywa prawnoporównawcza)' (Raport Instytutu Wymiaru Sprawiedliwości 2016).

Van Gool E., Michel A., 'The New Consumer Sales Directive 2019/771 and Sustainable Consumption: A Critical Analysis' (2021) 4 Journal of European Consumer and Market Law 136.

Weber A.-M., Mazur Z., Szczęsna A., 'Zrównoważony ład korporacyjny (sustainable corporate governance) – kierunek ewolucji' (2022) 6 Przegląd Prawa Handlowego 20.
Widła B., 'Challenges and Lessons from the CJEU's Judgment in Top System' (2023) 18(5) Journal of Intellectual Property Law & Practice 353.
Wiewiórowska-Domagalska A., Zoll F., Południak-Gierz K., Bańczyk W., 'Transpozycja dyrektywy Parlamentu Europejskiego i Rady UE 2019/771 z dnia 20 maja 2019 r. w sprawie niektórych aspektów umów sprzedaży towarów' (2021) 4 Kwartalnik Prawa Prywatnego 915.
Witt C.-H., 'Der Dieselskandal und seine kauf- und deliktsrechtlichen Folgen' (2017) 51 Neue Juristische Wochenschrift 3681.
Zoll F., *Rękojmia. Odpowiedzialność sprzedawcy* (C.H. Beck Poland 2018).
Zoll F., 'Ekologiczne prawo sprzedaży—bardzo wstępne uwagi' in K. Bilewska and D. Krekora-Zając (eds) *Księga jubileuszowa prof. dr. hab. Adama Brzozowskiego* (C.H. Beck Poland 2021) 641.
Zoll F., 'Czy nowa dyrektywa o sprzedaży zapowiada ekologiczną rewolucję? Uwagi na kanwie kilku niemieckich orzeczeń odnoszących się do kwestii ochrony środowiska w związku z odpowiedzialnością z tytułu rękojmi' in K. Oplustil, P. Tuleja and K. Prokopowicz (eds) *Nauka prawa a wyzwania XXI wieku. Księga jubileuszowa z okazji 170-lecia istnienia Towarzystwa Biblioteki Słuchaczów Prawa Uniwersytetu Jagiellońskiego* (Wydawnictwo Uniwersytetu Jagiellońskiego 2023) 340–343.
Zoll F., Południak-Gierz K., Bańczyk W., 'Sustainable Consumption and Circular Economy in the Directive 2019/771' (2020) 4 Pravovedenie 526.
Zoll F., Południak-Gierz K., Bańczyk W., 'State Liability for Damage Caused by COVID-19 Restrictions Under Polish Law', in E. Hondus, M. Santos Silva, A. Nicolussi, P. Salvador-Coderch, C. Wenfehorst and F. Zoll (eds) *Coronavirus and the Law in Europe* (Intersentia 2021) 359.
Zoll F., Południak-Gierz K., Bańczyk W., 'Zakres odpowiedzialności odszkodowawczej Skarbu Państwa za szkody majątkowe spowodowane ograniczeniami wprowadzonymi w związku z COVID-19' (2021) 3 Studia Prawa Prywatnego 1.
Zoll F., Południak-Gierz K., Bańczyk W., 'Towards an Environment-Friendly Law of Obligations' in A. Janssen, M. Lehmann and R. Schulze (eds) *The Future of European Private Law* (Nomos/Beck/Hart 2023) 195–224.
Zoll F., Watson J., Południak-Gierz K., Bańczyk W., Richter G., Estifanos J., 'Academic Proposal for Amending the Directive (EU) 2019/771 of the European Parliament and of the Council of 20 May 2019 on Certain Aspects Concerning Contract for the Sale of Goods' (2022) 4 Transformacje Prawa Prywatnego 109.

7

THE (UN)SUSTAINABILITY OF THE SALE OF GOODS IN DIRECTIVE (EU) 2019/771

Maria Miguel Oliveira da Silva and Jorge Morais Carvalho

7.1 Sustainability: A New Player in the Sale of Consumer Goods Legislation

Directive (EU) 2019/771 of the European Parliament and of the Council of 20 May 2019 on certain aspects concerning contracts for the sale of goods, amending Regulation (EU) 2017/2394 and Directive 2009/22/EC, and repealing Directive 1999/44/EC met its final text on 20 May 2019; social, political, and legislative concerns around sustainability were not yet on the current scale. Despite not being an absolutely innovative or disruptive issue at the time, it was, however, a subject that was beginning to gain tangible expression in Europe and worldwide.

At an international level, Resolution 70/1 of the United Nations General Assembly, which set out the 17 sustainable development goals to be achieved by 2030,[1] had already appeared on 25 September 2015. Goal 12 is dedicated to ensuring 'sustainable consumption and production patterns'. At the European level, the most representative piece of attention to the issue is found in the European Ecological Pact of 11 December 2019.[2] More recently, with a focus on consumers and the ecological transition, we can find the New Consumer Agenda of 13 November 2020.[3] Interestingly, Directive 2019/771, even before these two pieces of legislation were enacted, presented sustainability concerns, albeit mostly in its recitals. Both facts seem to reflect both the European impetus to leverage green legislation and its simultaneous timidity in taking significant steps in that direction.

In any case, the necessary legislative impulse aimed at reflecting in the legal regime the demands that have been made in terms of sustainability[4] has finally been launched and, above all, we are witnessing a particularly relevant legislative awareness. The intelligibility of the consumer is seen as a crucial part in the achievement of sustainability. This is probably the most significant change of paradigm that has accompanied consumer legislation until today: a model that has favoured the possibility of consumers, individually,[5] to consume as much and everything as they want.[6]

The purpose of this article is to analyse the amendments introduced by Directive 2019/771, distinguishing between their theoretical and practical suitabilities to achieve the results it seems to be pursuing.

 DOI: 10.4324/9781032662046-9

7.2 (Still) the Hierarchy of Remedies in the Sale of Consumer Goods Directives: A Double-Edged Sword

The Directive 1999/44/EC of the European Parliament and of the Council of 25 May 1999 on certain aspects of the sale of consumer goods and associated guarantees established a hierarchy of rights to be exercised by the consumer in the event of lack of conformity of the goods. The result of this provision was that consumers could, in case of lack of conformity, ask for the repair or the replacement of the good. Article 3(3) was therefore peremptory: 'in the first place, the consumer may require the seller to repair the goods or he may require the seller to replace them, in either case, free of charge, unless this is impossible or disproportionate'. In this sense, only if repair or replacement were impossible or would involve disproportionate costs could the consumer ask a price reduction or the rescission of the contract.

However, being a minimum harmonisation Directive, Member States were able to go beyond its scope, conferring more and better rights to consumers,[7] for instance, by not implementing the hierarchy, as occurred in Portugal.

In the most recent Directive on the sale of consumer goods (Directive 2019/771), the lawmakers chose to maintain the ancestral hierarchy between the remedies. Article 13 of the Directive now stipulates that national provisions, if they had not already done so, must give precedence to the right to repair and the right to a replacement, to the detriment of the right to a price reduction and the right to terminate the contract, except where this entails a disproportionate cost for the seller or where it is impossible to implement them. As this new Directive is a maximum harmonisation Directive, Member States were definitively given the task of establishing a hierarchy between rights if they had not already done so under the previous Directive.

The drafting of a new Directive on the sale of consumer goods aimed to overcome the most obsolete aspects of its predecessor, which is understandable considering the speed of global development that has been established since the end of the last century. Given the opportunity and also the sensitivity to the issue of sustainability that was expressed in Recitals (32) and (48), forward-looking legislation was expected in all its aspects.

It is therefore unclear why the hierarchy of consumer rights has been retained in the same terms as before. It is particularly puzzling that this hierarchy has remained specifically configured to give primacy to the right to repair and replacement, without, paradoxically, establishing a hierarchy between the two. It is an unknown factor, of course, especially from the perspective of sustainability; it will become more evident if the relevant criterion has been that of maintaining the contractual relationship.[8] Should the latter have been the *mens legis*, perhaps the recitals were not aligned with its own purposes.

Recital (48) of the Directive points out that the consumer should be able to choose between the repair and the replacement of the good in order to remedy the lack of conformity, as in 1999, and immediately clarifies that the option for repair (and only for repair) is related to the need to encourage a sustainable consumption, a concern to be debuted. In fact, a Directive equipped with environmental concerns in its recitals has been prepared, a touch of originality for the standard of consumer legislation. However, it kept the logic of 1999, giving the consumer the right to remedy in the first place without ranking the right to replacement and the right to repair. If, as stated in the recitals, the right to repair in a priority place in the hierarchy contributes to the achievement of sustainability, in what way can the lack of a hierarchy between repair and replacement coexist with the same intentions?

On the one hand, the drive to harmonise the hierarchy in national legislation is to be welcomed since the idea of implementing a hierarchy, in the abstract, is likely to encourage more sustainable consumption. However, the key to making this hierarchy fruitful for the sustainable cause was that the hierarchy established was not exactly that of the previous Directive. On the other hand, if the solution found was not good for sustainability, it will not have been good for the protection of consumers who, under this new Directive, are deprived of benefiting from a more beneficial regime allowed by the previous Directive.

Thus, if, at first sight, the conservation of a hierarchy between rights seemed a beneficial measure for sustainability, albeit without a more protective implementation of the consumer's position,[9] the final result of the legal regime has kept the same orientation, without concrete results for sustainability, and without a better solution for consumers.

The right to repair in case of lack of conformity has been deeply scrutinised since the implementation of the Directive by the Member States has seen the light of day. There is a consensus that repairing goods is the least harmful way to bring them into conformity, at least in terms of environmental sustainability.[10] In this sense, imposing on the consumer the prior exercise of the right to repair before exercising any of the other three rights would, with occasional exceptions, be an intelligible legal architecture from the point of view of the sustainability of the planet.

However, as we have already had the opportunity to state, requiring repair as the only of the legal remedies or even as the first of the four remedies would mean a weakening of the consumer's contractual position in relation to the previous Directive and a step backwards in terms of rights in relation to some domestic legislations.

Thus, it is important to analyse the values to be protected in the balance between sustainability and consumption. While it is true that there is an urgent need for awareness-raising measures and, perhaps going further, for the imposition of changes in current consumption patterns, alerting consumers to the environmental impact of their transactions,[11] it is also important to remember that consumer relations are private law relations. The mechanisms of reaction to a breach of contract cannot and should not be distorted. For instance, to demand the right to compensation as the sole or first remedy would be, from a purely contractual point of view, an insufficiency. And, above all, an ideological insufficiency, removing rights from consumers for a good cause, even if the cause (greatly) concerns those very same consumers.

Repair is, generally speaking, the best combination of response to the problems of both lack of conformity and the unsustainability of the planet[12] since it gives the consumer goods in conformity with the contract, allows the seller to maintain the contractual relationship, and avoids perishing and disposal of the good. If the intention was (also) to promote sustainability, it is more difficult to agree with the solution composed by the European Union when giving primacy to repair and replacement in the same hierarchical place. In fact, the right to a replacement, which may be requested by the consumer immediately in the event of lack of conformity, proves to be a very attractive remedy for the consumer,[13] not particularly preferable for the seller, but fatal for sustainability, since it will motivate the existence of two goods to be treated at their end of life, doubling the problem.

However, the right to price reduction, with an almost artificial presence in the hierarchy considering its scarce actual application, articulates, in a certain way, the interests at stake here, although it has been discarded for the second layer of remedies. The price reduction is likely to give continuity to the contractual relationship, being a positive aspect for both parties. Even so, it is not a remedy applicable to the generality of cases and can hardly rise to the category of primary remedy since maintaining a non-conforming good does not always please the consumer. It all depends, at the end of the day, on the size and typology of the lack of conformity. Moreover, the solution of reducing the price may imply, in certain situations, the purchase of a new good, aiming to meet the need that has been resurrected with the obsolescence of the good, replicating the problem of end-of-life duplication. In any case, if the motto was sustainability, it might have been interesting to equate the possibility of occupying the podium with other remedies.[14]

Finally, the right to terminate the contract, although a remedy that definitively terminates the contractual relationship, could nevertheless have been selected as a primary remedy in light of the concerns on sustainability since it proves to be a better response than the replacement itself. Naturally, for a legal regime interested in regulating private law relationships, suggesting that a remedy that strikes down that relationship is not a priority. In any case, restitution of the price paid for the good may (and probably will) entail the acquisition of a new good, achieving the same result as would be obtained with replacement. However, while it is true that the likelihood is that the consumer will acquire a new good to meet his or her needs, it is also true that his or her involvement in the situation may be liable to make him or her rethink the purchase. Furthermore, if this is one of the first remedies of the hierarchy, it would encourage the promotion of the durability of goods, especially with the producer, and therefore leverage the promotion of sustainability. Nevertheless, the right to terminate the contract, as well as the right to a replacement, are the remedies that most polarise the interests that collide here. They do not suggest an adequate response either to the maintenance of the contractual relationship or to sustainable consumption.

The hierarchy in the manner provided for in the Directive, giving priority to the right to repair and the right to a replacement, without ranking both, besides being ineffective for sustainability, also fails for some naivety, since the consumer, faced with the two options, will certainly not hesitate much in making a decision, particularly if little committed to the environmental cause.[15] As already mentioned, the current configuration of the provision does not encourage the consumer to decide in favour of repair. Besides taking more time than the replacement, repair may still be refused for economic reasons.[16]

Strictly speaking, only the mandatory exercise of the right to repair in the first place,[17] in order to establish a hierarchy, could truly assert itself as a solution guided by concerns about sustainability. The solution adopted in 2019 turned out to be the least interesting, considering the interests involved: the consumer remains, as in 1999, with only two rights in the first place; and the sustainable imperatives remain, as in 1999, *de facto* disregarded.[18]

The best configuration of a legal regime that protects the position of the consumer is still the one that does not establish a hierarchy between remedies (which has not been the orientation of any of the Directives). It allows him or her to obtain the best remedy for the concrete case. The best configuration of a legal regime promoting sustainability is the one that sets out repair as the first mandatory solution. The current regime, establishing a hierarchy between the four rights but not establishing a hierarchy between the first two, has managed to reconcile the worst of the two worlds, limiting the consumer in his or her range of options and allowing him or her, if he or she wants, to always opt for replacement. In this sense, the most recent Proposal on common rules promoting the repair of goods illustrates the need of approaching this problem in a slightly different angle in relation to the Directive.

7.3 The Maintenance of the Liability Period of the Trader, on the One Hand, and the Extension of the Period of Presumption of Anteriority of Lack of Conformity, on the Other

A relevant amendment in Directive 2019/771 is related to the extension of the period of presumption of the existence of the lack of conformity at the time of delivery, i.e., the period during which the consumer does not need to prove that the lack of conformity already existed when the good was delivered. Due to this legal orientation, the consumer may therefore benefit from the presumption of anteriority of the non-conformity, which means that the non-conformity that he claims will be presumed to have existed from the time of delivery. This circumstance clearly favours the consumer, even though it is not a new feature of this most recent Directive.

The 1999 Directive, in its Article 5(3), stated that 'any lack of conformity which becomes apparent within six months of delivery of the goods shall be presumed to have existed at the time of delivery (...)'. The new Directive, in its Article 11(1), extends this six-month period to one year.

First, it is interesting to note that the Directive does not provide for any extension of the actual period of the liability of the seller, although it leaves at the disposal of the Member States the possibility of establishing longer periods [under the terms of Article 10(3)]. This is what happened, for instance, in Spain and Portugal, with the implementation of a three-year period of liability of the seller.[19]

On the one hand, the fact that the Directive does not determine a longer period than two years may be related to the understanding that the goods do not need to be more durable, which conflicts with the very purpose of promoting that durability. On the other hand, it may contain a deeper underlying idea, related to the promotion of a certain temporal pattern of consumer transactions that stimulates the European economy in a certain sense, i.e., a motivation according to which it is expected that consumers, after two years of use of a good, will consider a new purchase. It would be understandable if the latter were one of the reasons implicit in maintaining the liability period of the seller. The European institutions are certainly not interested in changing the economic structure of an entire society; and, if they were, it would certainly not be by means of a Directive. In this sense, we believe that both producers and sellers must be encouraged to switch to sustainable consumption, even if such an incentive is, precisely, an incentive in euros. Given the profit motives that companies are currently driven by, raising the prices of (sustainable) goods will be one of the main ways of balancing the issues that come together in this matter. In this way, the producer will be more encouraged and willing to manufacture sustainable goods, the seller to sell them and the consumer to buy these goods. Regarding the latter point, the awareness to be implemented also needs to be financial, explaining the advantages of acquiring a more expensive good instead of several goods of the same type, albeit more affordable, over a lifetime.[20]

As we said, the Directive did not extend the liability period of the seller, keeping it at two years. This paralysis is not particularly interesting for the promotion of sustainability, with some honourable exceptions, since it does not encourage the production of durable goods beyond two years. Although the assessment of non-conformity is not only related to the circumstance of the durability of the good, the liability period is mainly designed for the case of a lack of conformity related to durability. In situations where functionality is not at stake, such non-conformity is usually detected much earlier. For instance, we can think of a case where the consumer wants a blue good, and an orange one is delivered.

Firstly, innovation is to be welcomed, as it enables the problem of planned obsolescence to be fought more effectively[21] (albeit timidly), and it does not burden the consumer with proving the existence of the lack of conformity at the date of delivery of the good. This is an immensely demanding task. However, although we agree, theoretically, with an extension of the period in which the presumption is applied, it should be noted that the Directive, in Article 11(2), allows Member States to extend this period beyond one year but limits them to a maximum period of two years. In our understanding, a better legislative policy would have been to give greater freedom to Member States in the transposition.

In fact, it must be admitted that the period of presumption of anteriority of the lack of conformity is at least as relevant, if not more so, as the period of liability of the seller. Although the extension of the liability period represents a factual advance in terms of consumer rights in general and sustainability in particular, a different presumption period from the liability period implies a very tangible difficulty in the exercise of consumer rights. After all, consumers hardly have the means (or, if they have them, they will hardly use them given the corresponding cost-benefit) to make efforts to prove such anteriority, which is what the law requires from them after the end of the presumption period. Unless the producer or the seller publicly admits the lack of conformity, or if the consumer is unlikely to be able to produce such proof, it will be difficult to avoid the perish of the good for reasons related to the complexity of producing proof, which is disastrous for sustainability.

7.4 The Right to Compensation after the End of the Trader's Liability Period and After-Sales Service

If the idea of allowing consumers to claim rights from the seller and the producer after the end of their period of liability might have caused surprise until now, the new text of the Directive opens hostilities.

Recital (33) makes, for the first and only time, three references to spare parts. Part of the provision now reads as follows:

> While this Directive should not impose an obligation on sellers to ensure the availability of spare parts throughout a period of time as an objective requirement for conformity, it should not affect other provisions of national law obliging the seller, the producer or other person constituting a link in the chain of transactions, to ensure that spare parts are available or to inform consumers about such availability.

Portugal already had rules in this sense but reformulated them when transposing the new Directive. The current legal solution now requires the availability of parts for a period of ten years after the last unit of the good has been placed on the market.[22] In many situations, probably the majority, the ten-year period will constitute an extension of what the consumer could obtain via the Consumer Protection Act, which contained an indeterminate concept that could be filled according to the type of good. Even so, in some situations, the ten-year period narrows the consumer's possibilities, namely, for instance, in what concerns the automobile sector, compromising in such cases the promotion of sustainability, rendering the lack of conformity of the good useless. Thus, it seems that the solution set forth in the previous Portuguese Consumer Protection Law, in this matter, is probably the best of those available to promote sustainability, especially if conjugated with the new paradigm related to the concept of durability. The solution that does not seem to be at all aligned with sustainability will certainly be the one determined by the Directive, which offers the possibility for the Member States that do not provide after-sales assistance solutions to continue to do so.

Another innovative aspect implemented by the Directive, albeit in a minor way, is found in Recital (48). By invoking the concept of durability, suggesting that the option of repair in the event of lack of conformity should encourage sustainable consumption and 'may' contribute to the durability of the goods, it seems that the legal regime is moving towards the assumption that a sustainable good is a good that, rather than being durable, must be repairable.[23] We follow this understanding in the sense that, even after the end of the liability period of the seller, the consumer must be able to demand that the good is repaired, either from the seller, the producer, or another trader (although, in all three cases, subject to the payment of a price) or by repairing the good by herself or himself.[24] In the latter case, the producer should, of course, include repair instructions.

7.5 Second-Hand Goods, Second-Hand Rules

Despite the growing concern with sustainability, not only at the European legislative level but also on the part of consumers and other stakeholders, and the subsequent increase in circular consumption intentions, the new Directive does not implement significant changes in relation to its predecessor as far as second-hand goods are concerned. But, at the time, it was 1999. Similarly to the previous Directive, Article 10(6) establishes that in the case of second-hand goods, Member States can determine that the seller and the consumer may, by agreement, reduce the period of the liability of the seller up to one year. This has been the case in the transposition of 15 of the 27 Member States.[25]

The possibility given to Member States to shorten the liability period of the seller in the case of second-hand goods is tainted with a certain prejudice against this type of goods, confirming the iconic remark by Hans-Wolfgang Micklitz about the general feeling towards these goods: 'old used products are second hand = second best products. They are waste, they are not raw material'.[26]

It is understandable that a 1999 Directive allows the reduction of the liability period for second-hand goods. However, a 2019 Directive, to take effect in 2022, cannot maintain the *status quo* in such a decisive matter.[27] Moreover, a configuration like the one it ended up offering to the Member States even seems to conflict with the very intentions of ensuring a logic of durability of goods.[28]

The first reference in the Directive to the concept of durability appears in Recital (32), stating that this requirement should be included as an objective criterion for assessing the conformity of the goods. Recital (32) is, moreover, quite exhaustive concerning the concept of durability and even provides a definition of it.[29] The inclusion of the concept of 'durability' as an objective requirement of conformity, provided for in Article 7(1)(d), even seems to be a premeditated innovation since the previous Directive already allowed the allegation of lack of conformity for reasons of durability, namely through Article 2(2). However, the new Directive seems to have wanted to ensure that the requirement of durability is even an aspect to be considered in the equation of the lack of conformity of the goods. Thus, if a good must be as durable as possible, it is not understandable that a used good might enjoy a protection period of only one year. The argument would be better understood if it concerned goods purchased before the current Directive came into force. On the contrary, a good purchased in January 2022, which is subsequently sold second-hand in June 2022, must be durable enough to reach at least 2024, but the Directive allows it not to be.

The distinction between new and second-hand goods is not a new issue. On the one hand, the most resonant argument in favour of establishing a shorter liability period relates to the different characteristics of new and used goods, especially those related to the prolonged use of a good. On the other hand, another argument in favour of such a reduction relates to the need to balance the interests of both the consumer and the trader[30] since, in theory, the latter may be discouraged from operating in the second-hand goods market and, moreover, they do not always control the characteristics of the goods they sell. The protection of the position of the trader is an aspect to be considered in the drafting of the provision since too onerous a regime would lead them to abandon the commercialisation of this type of goods which would also not be a good solution from the point of view of sustainability.

The Directive clarifies in Recital (43) that 'although a liability or limitation period of two years or more usually reconciles the interests of both the seller and the consumer, this might not be the case with regard to second-hand goods'. Once again, a surreptitious understanding of the devaluation of second-hand goods, which, by diminishing the protection of the position of consumers of second-hand goods, only reaffirms the inferiority of the latter,[31] gathering more and more antipathy to this market.

Even if there are understandable reasons for creating different legal solutions depending on whether the goods are new or second-hand, such reasoning does not seem to extend particularly to shortening the liability period for second-hand goods. It is incomprehensible that a consumer who buys a second-hand television from a trader should be less protected than a consumer who buys one first-hand in a situation where, a year and a half after delivery, it stops working. Nor should a consumer who buys a book that has already belonged to someone else be less protected than a consumer who buys a new book if he notices that page 5 is missing one year after the purchase.

Like the legal regime, the vast majority of consumers seem to consider that second-hand goods are less durable and less economically attractive.[32] However, while it is clear that certain types of goods do indeed wear out through use, it is not equally true that this is a universal rule for all consumer goods. A few examples can be listed for illustrative purposes: mirrors, pictures, books, lamps, or even musical instruments fulfil their own purpose, even if they have been used before. In fact, the fact that a good is used does not automatically devalue it but may increase its value precisely because of being second-handed. Still, even in cases in which the use effectively diminishes the value of the good, it is not the circumstance that the good is used that devalues it. It is, on the contrary, the circumstance of the passage of time, previous misuse, or the poor quality of the materials originally used by the producer.[33] These aspects, as a general rule, can be controlled by the trader selling second-hand goods: a second look is enough.

It may, in any case, be argued that the reason behind the decrease of the liability period of the trader is related to the fact that they are in a risky position when selling used goods. However, we would always say that also in the sale of new goods, the trader has no significant control over the goods they sell unless they are the producer. Their decision is whether to contract with a certain producer and whether or not to sell certain goods. This possibility also applies to the seller of second-hand goods, who can choose which goods to sell and from whom to buy. It can also be argued that in this case sustainability may be compromised, as there will be fewer professionals interested in selling highly deteriorated goods. Again, it should be considered that the seller, fulfilling a diligent duty of information, has no reason to fear selling a highly 'used up' good.

Thus, if a distinction between periods of liability for new and used goods were to be implemented, it would be better to implement a regime that differentiated between consumable and non-consumable goods or depreciable and non-depreciable goods instead of maintaining the traditional distinction between new and used goods.[34] At least, in this case, it would ensure what is really important to ensure in this matter: the degree of 'consumability' of the good. Furthermore, a distinction based on the consumable or perishable nature would also contribute to the desired protection of the position of the trader since, by providing unequivocal information to the consumer on the factual situation of the good, it would contemplate the most unfavourable position in which the consumer may find herself or himself.

Also, in Recital (45), seeking to clarify the possibility of shortening the liability period in these cases, the following is stated:

> leaving this question to a contractual agreement between the parties increases contractual freedom and ensures that the consumer has to be informed both about the nature of the good as a second-hand good, and the shortened liability or limitation period.

The new Directive is able to respond perfectly to the concerns expressed in this passage by establishing objective (and subjective) conformity requirements. In any case, the obligations to inform the consumer about the used nature of the good and the respective shortening of the liability period already arise, by law, at least from the right to information.[35]

A key point in favour of reshaping the legal framework for second-hand goods remains. This aspect relates to the inadequacy of a liability period reduced to a minimum of one year. Based on the current understanding, consumers of second-hand goods are less protected in terms of time because these goods have already been used, which makes it impossible to expect them to last at least two years, the minimum required by the Directive for new goods. Let us take the example of the purchase of a tire. The good, bought in 2022 as a new good, must be functional at least until 2024. If in 2024 it is sold as a second-hand good, the law will require it, if nothing is said to the contrary, to be usable until at least 2025. However, there is no reason to believe that the tire will remain functional for that long if it is in normal use. In this and many other cases, even a deadline of one year is not adequate to deal with the actual situation. As we have seen above, the one-year time limit is not adequate to deal with the situation of goods that last much longer than one year by their own characteristics, nor is it, as we have now seen, adequate to deal with the situation of goods that (may) last much less than one year. Even if we consider the controversial situation of clothing or electronic devices, there is no reason to believe that the liability period of one year protects the trader since if the good is tainted with some defect that cannot be controlled by it, there is a high degree of probability that this problem will manifest itself in the very first year.

7.6 Right to Reject Sustainability: Is This What We Are Talking about in Article 3(7)?

Article 3(7) of the Directive 2019/771, under an article dedicated to the scope of application, reads as follows: 'This Directive shall not affect the freedom of Member States to allow consumers to choose a specific remedy, if the lack of conformity of the goods becomes apparent within a period after delivery, not exceeding 30 days. In addition, this Directive shall not affect national rules not specific to consumer contracts providing for specific remedies for certain types of defects that were not apparent at the time of conclusion of the sales contract'.

The Directive introduces the possibility of implementing the right to reject, which is a new feature in most of the Member States' domestic legislations. The right to reject finds its origin in the legal system of the United Kingdom (known as the short-term right to reject), with the Consumer Rights Act 2015 providing, in its Article 19(3)(a), that if a good does not conform to certain characteristics (quality, lack of fitness for purpose or does not correspond to the description made of it or to the sample or model made available), the consumer may reject the good within 30 days, returning it or asking the professional to collect it and obtaining a full refund of the amount paid.[36] This option is somehow surprising. At a time when the United Kingdom was already preparing to definitively leave the European Union, the European institutions decided to open the door to the implementation of a rather English rule. Perhaps in the name of sustainability? We will try to find out.

The provision foresees that Member States might allow consumers to exercise rights other than those listed in Article 13. Member States may grant the consumers the right to a price reduction, the termination of the contract, or, indeed, any other right immediately, in the event of lack of conformity. This would have been an excellent initiative to start the experience of implementing other rights different from the usual quartet, testing the receptivity of consumers to demand more sustainable rights.[37] However, specifically in the Portuguese case, what ended up happening was that the lawmakers found in this mechanism the possibility to bring back to life part of the regime previously established in what concerns the hierarchy of remedies. The consumer can ask for immediate replacement or termination of the contract if the lack of conformity becomes apparent in the first 30 days after delivery. These are the two solutions with the worst consequences for sustainability.

On the one hand, the motivation behind this European innovation is to be considered: the standard seems to want to protect the position of the consumer, which is understandable. However, it seems that it was also designed to protect the consumers from the hierarchy of remedies it proposes itself. If that was the motivation, it would have been better to simply remove the hierarchy.

On the other hand, the provision, by implementing a kind of 'right of withdrawal' for non-conforming goods, besides not contributing in any measure to sustainability, distorts the very architecture of the legal regime.[38] In the first 30 days after delivery, a result is obtained that is apparently against the ratio of the Directive as a whole. As already mentioned, the termination of the contract is not a particularly sustainable remedy (even so, in the limit, more sustainable than the replacement of the good) since, even if, in the best hypothesis, the consumer does not acquire a new good with the return of the paid price, an end will always have to be assigned to the returned goods, which may not be resalable. Such a situation entails an environmental burden, from the outset, in the double transportation (for the acquisition and return of the good)[39] and, finally, in the disposal of the good, which may, in the worst-case scenario, imply the acquisition of a new product, which will also require a destination at the end of its lifetime.

7.7 The Relevance of the Producer in Promoting Sustainability

Although the focus of the consumer relationship is mostly placed on the interaction between the consumer and the seller, the consumer legislation also considers the producer, although not to a very significant extent.

The new Directive has taken a significant step towards making the producer liable, which proves to be a relevant direction for the promotion of consumer rights and, at least in theory, may mean a significant step towards sustainability. In this context, the most interesting aspect concerns the strengthening of the concept of commercial guarantee, which is interconnected with the concept of durability itself.

Article 17(1) of the Directive states, inter alia, that

> where a producer offers to the consumer a commercial guarantee of durability for certain goods for a certain period of time, the producer shall be liable directly to the consumer, during the entire period of the commercial guarantee of durability for repair or replacement of the goods in accordance with Article 14.

The provision, suggesting the self-responsibility of the producer regarding the goods it produces, is to be commended since it manifests the intention to encourage the producer to produce in a sustainable way. Traders who embrace such legal responsibility for conformity are encouraged, as a rule, to produce durable goods, so as not to bear the logistical and legal consequences that arise from the lack of conformity of the goods.

However, again in Article 17(1), the Directive provides that 'the producer may offer to the consumer more favourable conditions in the commercial guarantee of durability statement'. This openness creates the possibility that the producer may be more lenient about possibilities of remedies offered to the consumer in case of a lack of conformity than the legal regime itself, which, in theory, is not a policy that should be looked down upon. However, if the producer intends to be lenient with regard to the policy for dealing with non-conforming goods, it could make a commitment to immediately replace the good regardless of the situation if the consumer so wishes, discouraging resources to repair and placing us before a threat in terms of sustainability.[40]

7.8 Social Responsibility Is Also Sustainable Consumption

The control of supply chains, from the perspective of social responsibility, is a subject that is not usually dealt with in depth in consumer law. More recently, with the advent of greenwashing and bluewashing practices, the violation of human rights in production chains has gained unprecedented expression in the private law relations context.

The 21st century has been fruitful in demonstrating the relevance of what happens along global value chains, a problem that is intrinsically related to social sustainability. It also underlines the absolutely unavoidable role of the producers in consumer relations, quite different from the secondary role that has been attributed to them over time. The Directive 2019/771 should have shown more sensitivity to the subject, especially in view of the overcoming of its predecessor, which was always guided by the primordial regulation of the direct relationship between the seller and the consumer.

In this aspect, if the Directive already fell short in terms of environmental sustainability, it is less energetic in terms of social sustainability. In fact, it does not make any reference to the imposition of a 'dialogue' between the producer and the consumer (as it happened, for instance, in France[41]) in order to guide the consumer in the transition to sustainable consumption.[42] This imposition would have been relevant, and even the appropriate place, to potentiate the change of the current unsustainable consumption paradigm. Note, therefore, that the producer, much more so than the seller, is in an excellent position to highlight the environmentally positive and negative aspects of the good it produces and the socially benign or malign consequences of the production method and conditions it uses.[43] In fact, the producer is the main, if not the only, agent capable of promoting socially responsible consumption, as it is the only one with a monopoly on the conditions of production in its sphere of knowledge. This matter should be mandatorily embodied in the right to information, even as a conformity requirement. This is a topic deeply connected with the sale of consumer goods.

7.9 Eco-Labelling: An Absence

From the point of view of the pursuit of environmental sustainability, the implementation of compulsory ecolabelling of consumer goods would have been truly useful. Some authors believe that the mere transmission of information to consumers is not enough to change their decision to contract (and is, therefore, insufficient, by itself, in the process of transition to sustainable consumption[44]). It may be that they are right. Even if this is understood, it is not easy to understand the absence of a reference to eco-labels or other forms of labelling.

In fact, the only reference to labelling is found in Article 7(1)(d), concerning the objective conformity requirements, and only contemplating situations in which the trader, on its own initiative, states that the good has a certain characteristic. In other words, the trader ends up having the advantage of not certifying absolutely anything beyond what the law imposes, including information on sustainability.

On the one hand, this is understandable. The system on which European and global labels are currently based is not a particularly interesting system from the point of view of reliability and suitability. Since most labels attesting to certain characteristics are awarded by private organisations, naturally driven by economic and possibly lucrative interests, they do not have the neutrality needed to generate confidence among the most sceptical consumers. In addition, being the certification systems, in general, optional, the traders, who are intended to encourage the use of ecological labelling (read, those who trigger harmful consequences through their goods/means of production), will simply avoid the certification process.[45]

On the other hand, if lawmakers are suspicious of certification schemes, this should at least encourage Member States to promote the use of labelling as a mechanism for communicating with the consumer about the sustainability of the good, since this may discourage some consumers from buying unsustainable goods.[46] More than nothing, yet no revolution. Moreover, if the European institutions are suspicious of existing certification systems, perhaps they should subject goods to the scrutiny of their own environmental and social impact monitoring systems, resolving several difficulties at a stroke.

We should not overlook existing specific European legislation imposing labelling in certain cases.[47] Indeed, this obligation already results from the very wording of Article 7 and, naturally, from the specific legal regimes. We are also not indifferent to the fact that the treatment of the question may not be particularly relevant in this Directive. In any case, at least a reference to the respective regulations would have been justified. For instance, a paragraph could have been included in Article 7(1), referring precisely to the need to include specific information on labels. In Article 6(1), a paragraph could refer to the soft law mechanisms available and optional for professionals to adhere to,[48] reinforcing the central role that labelling, for the time being, for want of a better mechanism, still has.

7.10 Concluding Remarks

The Directive 2019/771 has embedded the concept of sustainability in consumer goods sales legislation for the first time, which in itself is a positive aspect. The introduction of this concern, even if only in the text of the recitals, demonstrates the impetus of the European institutions to continue the green movement by also seeking to promote sustainability through the role of the consumer.

However, given the emerging need to push sustainable consumption as an urgent imperative, the Directive does not go as far as it could have gone within its scope of action, rather revealing a trend towards the implementation of homeopathic remedies. Thus, from the perspective of sustainability, the final result of the text presents apprehensions in the recitals that are not truly reflected in the solutions provided by the provisions.

Some aspects are worth highlighting. The extension of the period of presumption of anteriority of the lack of conformity of the good, providing the consumer with greater accessibility to the exercise of his or her rights, is a positive aspect, although insufficient. Requiring repair in case of lack of conformity of the good, as one of the first remedies of the hierarchy, is another positive aspect from the point of view of sustainability. Finally, the introduction of the concept of durability is an interesting innovation towards sustainability.

However, the equal treatment of the right of repair and the right of replacement at the same hierarchical level, the system of second-hand goods, the opening to the implementation of a right of rejection, the poor use made of the role of the producer and the silence on ecological and social responsibility labelling are negative aspects to be pointed out.

The choice of prioritising the hierarchy of remedies in the event of lack of conformity in the previous Directive was not particularly interesting for protecting the position of the consumer, so maintaining it in its current form in the new Directive is not particularly useful, especially if we think that sustainability concerns are adjacent to it. The possibility of assigning different liability periods for new and used goods, favouring consumers of first-hand goods, is inconsistent with the intentions of ensuring sustainable consumption. Furthermore, opening the door to the possibility of the consumer exercising other rights in the event of lack of conformity within the first 30 days after delivery, particularly the right of termination, seems to us a bad solution. Finally, the failure to legislate in such a way as to invite the producer to discuss the matter, in particular by suggesting ways of communicating with the consumer through labelling and certification schemes, was a missed opportunity.

Notes

1 United Nations General Assembly, Resolution A/RES/70/1 (Transforming Our World: The 2030 Agenda for Sustainable Development) 25 September 2015 accessed 18 November 2022.
2 Commission, 'The European Green Deal' (Communication from the Commission to the European Parliament, the European Council, the Council, the European Economic and Social Committee and the Committee of the Regions), Brussels COM (2019) 640 final.
3 Commission, 'New Consumer Agenda' (Communication from the Commission to the European Parliament and the Council) Brussels COM (2020) 696 final.
4 Highly interesting is the perspective of Mateusz Grochowski, 'European Consumer Law after the New Deal: A Tryptich' (2020) 39 Yearbook of European Law 1, 387, 408, who states that consumers, by choosing certain goods or not buying them, not only express their position in relation to political values and objectives but also vote with their money, using their funds to directly or indirectly subsidise specific political agendas.
5 William Cornetta, 'Obsolescência: Da Origem ao Problema Social e Seus Reflexos à Sociedade' (2019) 12 Estudos de Direito do Consumidor 313, 315–319.
6 Ludwig Krämer, 'On the Interrelation between Consumer and Environmental Policies in the European Community' (1993) 16 Journal of Consumer Policy 455.
7 In the words of the Directive, 'Member States may adopt or maintain in force more stringent provisions, compatible with the Treaty in the field covered by this Directive, to ensure a higher level of consumer protection' [art. 8(2)].
8 Following Rónán R. Condon, 'Sustainability, the Circular Economy and Consumer Law in Ireland' (2020) 9 EuCML 5, 218, 220, we understand that private law remedies focus on the individual contract, and it is quite challenging to integrate environmental externalities within its scope adequately.
9 Geraint Howells *et alii*, *Rethinking EU Consumer Law* (Routledge 2018) 186.
10 Christa Liedtke *et alii*, *Right to Repair -Publications of the Advisory Council for Consumer Affairs* (Advisory Council for Consumer Affairs at the Federal Ministry for the Environment, Nature Conservation, Nuclear Safety and Consumer Protection, 2022) 47.
11 As, for instance, Anaïs Michel *et alii*, 'The New Consumer Sales Directive 2019/771 and Sustainable Consumption' (2021) 10 EuCML 4, 136, 145, suggest by instilling the possibility of reversing the logic of art. 14(4) of the Directive by enabling the seller to charge the consumer a fee for the period the consumer used the good before its replacement.
12 Repair is even more energy efficient than recycling itself. See, on this topic, Lori Duvall *et alii*, 'The Fixer Movement: A Key Piece of the Circular Economy Revolution' in *Empowering Repair* (Ellen MacArthur Foundation, October 2016) 5–6.
13 Hans-Wolfgang Micklitz, 'Squaring the Circle? Reconciling Consumer Law and the Circular Economy' (2019) 8 EuCML 6, 229, 236.
14 However, it should be noted that price reduction does not always correspond to sustainability imperatives since sometimes the lack of conformity itself is the cause of more significant environmental damage. This rationale is pointed out by Fryderyk Zoll *et alii*, 'Academic Proposal for Amending the Directive (EU) 2019/771 of the European Parliament and of the Council of 20 May 2019 on Certain Aspects Concerning Contracts for the Sale of Goods' (2022) 3 Transformacje Prawa Prywatnego 146.
15 As stated by Anaïs Michel *et alii*, 'The New Consumer Sales Directive 2019/771 and Sustainable Consumption' (2021) 10 EuCML 4, 136, 144, repair is only attractive to a minority of consumers, specifically those with either strong environmental motivations or emotional attachments to goods.
16 Christa Liedtke *et alii*, *Right to Repair – Publications of the Advisory Council for Consumer Affairs* (Advisory Council for Consumer Affairs at the Federal Ministry for the Environment, Nature Conservation, Nuclear Safety and Consumer Protection, 2022) 47.
17 On this point, we agree with Eléonore Maitre-Ekern *et alii*, 'Towards a Hierarchy of Consumption Behaviour in the Circular Economy' (2019) 26 Maastricht Journal of European and Comparative Law 3, 394, 419. https://doi.org/10.1177/1023263X19840943.

18 In the sense that the consumer's choice between repair and replacement does not encourage the consumer to choose sustainable remedies, Petra Weingerl, 'Sustainability, the Circular Economy and Consumer Law in Slovenia' (2020) 9 EuCML 3, 129, 131, https://kluwerlawonline.com/api/Product/CitationPDFURL?file=Journals\EuCML\EuCML2020054.pdf, accessed 18 November 2022; and Bert Keirsbilck, 'Sustainability, the Circular Economy and Consumer Law in Belgium' (2020) 9 EuCML 4, 165, 167, https://kluwerlawonline.com/api/Product/CitationPDFURL?file=Journals%5CEuCML%5CEuCML2020062.pdf, accessed 18 November 2022.

19 See art. 12(1) of Decree-Law No. 84/2021, of 18 October.

20 In any case, most consumers seem willing to pay more for more durable products. Cf. ICF (on behalf of the Consumer Policy Evaluation Consortium), *Study on the Costs and Benefits of Extending Certain Rights under the Consumer Sales and Guarantees Directive 1999/94/EC* (Justice and Consumers, March 2017) 30–31; and LE Europe *et alii*, *Behavioural Study on Consumers' Engagement in the Circular Economy – Final Report* (Justice and Consumers, October 2018) 165–167.

21 On the problem of planned obsolescence, see, among many others, Alberto de Franceschi, 'Planned Obsolescence challenging the Effectiveness of Consumer Law and the Achievement of a Sustainable Economy: The Apple and Samsung Cases' (2018) 7 EuCML 6, 217. See also Adèle Chasson, *et alii*, *Durable and Repairable Products: 20 Steps to a Sustainable Europe: HOP's Public Policy Guide to End Premature Obsolescence in the European Union* (HOP, November 2020).

22 After-sales assistance was already provided for in the Portuguese legal system, namely in the Consumer Protection Act (*Lei de Defesa do Consumidor*), in its art. 9(5): 'the consumer has the right to after-sales assistance, including the supply of spare parts and accessories, for the normal average duration of the products supplied'. After the transposition of the new Directive, art. 21(1) of Decree-Law No. 84/2021, of 18 October, now states that the producer must 'make available the necessary spare parts for the repair of the goods purchased by the consumer for a period of 10 years after the last unit of the respective good has been placed on the market'. In addition, art. 21(3) provides that the professional must, in the case of movable goods subject to public register, guarantee after-sales assistance under appropriate market conditions for a period of ten years after the last unit of the good has been placed on the market. Finally, paragraph 4 of the same article states that 'the trader must inform the consumer, (…) in the case of movable goods subject to public register, of the existence and duration of the duty to provide after-sales assistance'.

23 In the sense that 'durability' includes the very idea of 'reparability', cf. Anaïs Michel *et alii*, 'The New Consumer Sales Directive 2019/771 and Sustainable Consumption' (2021) 10 EuCML 4, 136, 138–139.

24 Evelyne Terryn, 'A Right to Repair? Towards Sustainable Remedies in Consumer Law' (2019) 27 European Review of Public Law 4, 851, 864–869, https://kluwerlawonline.com/api/Product/CitationPDFURL?file=Journals\ERPL\ERPL2019044.pdf, accessed 18 November 2022; Anaïs Michel *et alii*, 'The New Consumer Sales Directive 2019/771 and Sustainable Consumption' (2021) 10 EuCML 4, 136, 145.

25 Marco Loos, 'Rubbish or Recycling? Furthering Sustainability by Developing Specific Provisions for the Sale of Second-Hand Goods' (06 July 2022) *Amsterdam Law School Research Paper No. 15, Amsterdam Centre for Transformative Private Law Working Paper No. 02*, https://papers.ssrn.com/sol3/papers.cfm?abstract_id=4155445, accessed 18 November 2022.

26 Hans-Wolfgang Micklitz, 'Squaring the Circle? Reconciling Consumer Law and the Circular Economy' (2019) 8 EuCML 6, 229, 230.

27 In the same vein, Marco Loos, 'Rubbish or Recycling? Furthering Sustainability by Developing Specific Provisions for the Sale of Second-Hand Goods' (06.07.2022) *Amsterdam Law School Research Paper No. 15, Amsterdam Centre for Transformative Private Law Working Paper No. 02*https://papers.ssrn.com/sol3/papers.cfm?abstract_id=4155445, accessed 18 November 2022.

28 Esther Arroyo Amayuelas, 'The Implementation of the EU Directives 2019/770 and 2019/771 in Spain' (2022) 11 EuCML 1, 35, 38.

29 'Durability in this Directive should refer to the ability of the goods to maintain their required functions and performance through normal use', which is also the definition provided in art. 2(13) of the same Directive.

30 Case C-133/16 *Ferenschild* [2017] ECLI:EU:C:2017:284, Opinion of AG Szpunar, para 81.

31 In the same vein, Marco Loos, 'Rubbish or Recycling? Furthering Sustainability by Developing Specific Provisions for the Sale of Second-Hand Goods' (06.07.2022) *Amsterdam Law School Research Paper No. 15, Amsterdam Centre for Transformative Private Law Working Paper No. 02*, https://papers.ssrn.com/sol3/papers.cfm?abstract_id=4155445, accessed 8 November 2022.

32 With due exceptions for goods customarily acquired for collecting purposes or, for example, when originally used by prominent public figures.

33 In this sense, Anaïs Michel *et alii*, 'The New Consumer Sales Directive 2019/771 and Sustainable Consumption' (2021) 10 EuCML 4, 136, 137–138.

34 As stated in ANEC-BEUC, *Making More Sustainable Products the New Normal: Consumer Recommendations for a Meaningful EU Sustainable Product Initiative* (2021) 13, currently, all consumer goods benefit from the same liability period, so for the purpose of making the circular economy a reality, liability periods for durable goods should last considerably longer. However, due to the different characteristics of goods, the provisions of the Directive may be inadequate to define liability periods.

35 Marco Loos, 'Rubbish or Recycling? Furthering Sustainability by Developing Specific Provisions for the Sale of Second-Hand Goods' (06.07.2022) *Amsterdam Law School Research Paper No. 15, Amsterdam Centre for Transformative Private Law Working Paper No. 02*, https://papers.ssrn.com/sol3/papers.cfm?abstract_id=4155445, accessed 18 November 2022.

36 On this topic, see European Commission, Consumer Market Study on the Functioning of Legal and Commercial Guarantees for Consumers in the EU: United Kingdom (Justice and Consumers, Brussels, 2015) 7–9.

37 In the same sense, Anaïs Michel *et alii*, 'The New Consumer Sales Directive 2019/771 and Sustainable Consumption' (2021) 10 EuCML 4, 136, 143.

38 Fryderyk Zoll *et alii*, 'Academic Proposal for Amending the Directive (EU) 2019/771 of the European Parliament and of the Council of 20 May 2019 on Certain Aspects Concerning Contracts for the Sale of Goods' (2022) 3 Transformacje Prawa Prywatnego 123–124.

39 Fryderyk Zoll *et alii*, 'Academic Proposal for Amending the Directive (EU) 2019/771 of the European Parliament and of the Council of 20 May 2019 on Certain Aspects Concerning Contracts for the Sale of Goods' (2022) 3 Transformacje Prawa Prywatnego, 123–124.

40 In this sense, Evelyne Terryn, 'A Right to Repair? Towards Sustainable Remedies in Consumer Law' (2019) 27 European Review of Public Law 4, 851, 862.

41 Which made it compulsory for producers to inform consumers about the durability, reparability, and environmental and social impacts of certain goods. See arts 13, 15, 19, 20, 27, 33, and 37 of Law No. 2020-105 of the French Republic, 10 February 2020.

42 On the need to introduce obligations regarding sustainability-related reporting, see Eléonore Maitre-Ekern, 'The Choice of Regulatory Instruments for a Circular Economy' in Klaus Mathis and Bruce Huber (eds.), *Environmental Law and Economics* (Springer 2017) 305, 315–328, <https://papers.ssrn.com/sol3/papers.cfm?abstract_id=3087105> accessed 18 November 2022.

43 Interestingly, Dara O'Rourke *et alii*, 'The Impact of Sustainability Information on Consumer Decision Making' (2015) 20 Journal of Industrial Ecology 4, 882 <https://doi.org/10.1111/jiec.12310> accessed 18 November 2022 conclude that, with certain exceptions, the environmental/social scoring system does not show any change vis-à-vis consumer purchasing behaviour. It is unclear whether this is due to the associated price, the lack of credibility of the scoring system, or the consumer's indifference to the issue, which seems unlikely nowadays.

44 Dara O'Rourke *et alii*, 'The Impact of Sustainability Information on Consumer Decision Making' (2015) 20 Journal of Industrial Ecology 4, 882, <https://doi.org/10.1111/jiec.12310> accessed 18 November 2022; Anaïs Michel *et alii*, 'The New Consumer Sales Directive 2019/771 and Sustainable Consumption' (2021) 10 EuCML 4, 136, 139–141; Evelyne Terryn, 'The New Consumer Agenda: A Further Step Toward Sustainable Consumption?' (2021) 10 EuCML 1, 1, 2.

45 Regarding labelling in the textile sector, Maria Miguel Oliveira da Silva, 'A Exploração Laboral no Setor Têxtil e o Direito do Consumidor à Informação: Que Convergência?' (2021) 25 Revista Electrónica de Direito 2, 335, 347–352.

46 Anaïs Michel *et alii*, 'The New Consumer Sales Directive 2019/771 and Sustainable Consumption' (2021) 10 EuCML 4, 136, 139.

47 Examples include Council Regulation (EU) 2017/1369 of 4 July 2017, setting a framework for energy labelling [2017] OJ L 198/ 1; Council Regulation (EU) 1007/2011 of 27 September 2011 on textile fibre names and related labelling and marking of the fibre composition of textile products [2011] OJ L 272/1; Council Directive 94/11/EC of 23 March 1994 on the approximation of the laws, regulations, and administrative provisions of the Member States relating to the labelling of the materials used in the main components of footwear for sale to the consumer [1994] OJ L 100/37; and Council Directive 2012/19/EU of 4 July 2012 on waste electrical and electronic equipment (WEEE) [2012] OJ L 197/38.

48 For instance, the EU Ecolabel or Fairtrade certification systems.

Bibliography

Amayuelas E, 'The Implementation of the EU Directives 2019/770 and 2019/771 in Spain' (2022) 11 EuCML 1, 35, 38

ANEC-BEUC, *Making More Sustainable Products the New Normal: Consumer Recommendations for a Meaningful EU Sustainable Product Initiative* (ANEC/BEUC, 2021) 13

Chasson A, *et alii*, *Durable and Repairable Products: 20 Steps to a Sustainable Europe: HOP's Public Policy Guide to End Premature Obsolescence in the European Union* (HOP, November 2020)

Condon R, 'Sustainability, the Circular Economy and Consumer Law in Ireland' (2020) 9 EuCML 5, 218, 220

Cornetta W, 'Obsolescência: Da Origem ao Problema Social e Seus Reflexos à Sociedade' (2019) 12 Estudos de Direito do Consumidor 313, 315–319

Duvall L, *et alii*, 'The Fixer Movement: A Key Piece of the Circular Economy Revolution' in Herrmann S, *et alii* (ed), *Empowering Repair* (Ellen MacArthur Foundation, October 2016) 5–6

Franceschi Al, 'Planned Obsolescence Challenging the Effectiveness of Consumer Law and the Achievement of a Sustainable Economy: The Apple and Samsung Cases' (2018) 7 EuCML 6, 217

Grochowski M, 'European Consumer Law after the New Deal: A Tryptich' (2020) 39 Yearbook of European Law 1, 387, 408

Howells G, *et alii*, *Rethinking EU Consumer Law* (Routledge 2018) 186

ICF (on behalf of the Consumer Policy Evaluation Consortium), *Study on the Costs and Benefits of Extending Certain Rights under the Consumer Sales and Guarantees Directive 1999/94/EC* (Justice and Consumers, March 2017)

Keirsbilck B, 'Sustainability, the Circular Economy and Consumer Law in Belgium' (2020) 9 EuCML 4, 165, 167, <https://kluwerlawonline.com/api/Product/CitationPDFURL?file=Journals%5CEuCML%5CEuCML2020062.pdf> accessed 18 November 2022.

Krämer L, 'On the Interrelation Between Consumer and Environmental Policies in the European Community' (1993) 16 Journal of Consumer Policy, 455

LE Europe *et alii*, *Behavioural Study on Consumers' Engagement in the Circular Economy – Final Report* (Justice and Consumers, October 2018) 165–167

Liedtke C, *et alii*, *Right to Repair – Publications of the Advisory Council for Consumer Affairs* (Advisory Council for Consumer Affairs at the Federal Ministry for the Environment, Nature Conservation, Nuclear Safety and Consumer Protection, 2022), 47

Loos M, 'Rubbish or Recycling? Furthering Sustainability by Developing Specific Provisions for the Sale of Second-Hand Goods' (06 July 2022) *Amsterdam Law School Research Paper No. 15, Amsterdam Centre for Transformative Private Law Working Paper No. 02*, <https://papers.ssrn.com/sol3/papers.cfm?abstract_id=4155445> accessed 18 November 2022

Maitre-Ekern E, 'The Choice of Regulatory Instruments for a Circular Economy' in Klaus Mathis and Bruce Huber (eds.), *Environmental Law and Economics* (Springer 2017) 305, 315–328, <https://papers.ssrn.com/sol3/papers.cfm?abstract_id=3087105> accessed 18 November 2022

Maitre-Ekern E, *et alii*, 'Towards a Hierarchy of Consumption Behaviour in the Circular Economy' (2019) 26 Maastricht Journal of European and Comparative Law 3, 394, 419, <https://doi.org/10.1177/1023263X19840943>

Michel A, *et alii*, 'The New Consumer Sales Directive 2019/771 and Sustainable Consumption' (2021) 10 EuCML 4, 136, 145

Micklitz H, 'Squaring the Circle? Reconciling Consumer Law and the Circular Economy' (2019) 8 EuCML 6, 229, 236

O'Rourke D, *et alii*, 'The Impact of Sustainability Information on Consumer Decision Making' (2015) 20 Journal of Industrial Ecology 4, 882, <https://doi.org/10.1111/jiec.12310> accessed 18 November 2022

Silva M, 'A Exploração Laboral no Setor Têxtil e o Direito do Consumidor à Informação: Que Convergência?' (2021) 25 Revista Electrónica de Direito *2*, 335, 347–352

Terryn E, 'The New Consumer Agenda: A Further Step Toward Sustainable Consumption?' (2021) 10 EuCML 1, 1, 2

Terryn E, 'A Right to Repair? Towards Sustainable Remedies in Consumer Law' (2019) 27 European Review of Public Law 4, 851, 864–869, <https://kluwerlawonline.com/api/Product/CitationPDFURL?file=Journals\EuCML\EuCML2020054.pdf> accessed 18 November 2022

Weingerl P, 'Sustainability, the Circular Economy and Consumer Law in Slovenia' (2020) 9 EuCML 3, 129, 131, <https://kluwerlawonline.com/api/Product/CitationPDFURL?file=Journals\EuCML\EuCML2020054.pdf> accessed 18 November 2022

Zoll F, *et alii*, 'Academic Proposal for Amending the Directive (EU) 2019/771 of the European Parliament and of the Council of 20 May 2019 on Certain Aspects Concerning Contracts for the Sale of Goods' (2022) 3 Transformacje Prawa Prywatnego 146

8

THE BLINDING EFFECT OF EU CONSUMER POLICY OVERSHADOWS THE ROLE OF CONSUMER LAW IN DELIVERING THE GREEN TRANSITION

Lucila de Almeida and Fabrizio Esposito

8.1 Introduction

Implementing lasting social change is a challenging task, especially when it comes to transforming our production and consumption patterns. It is essential to pursue societal justice while also ensuring a habitable planet for future generations. In response to this need, the European Union has introduced an ambitious policy plan called the 'Green Deal', which has been further developed through subsequent initiatives. The announcement of the Green Deal was accompanied by a powerful narrative, with European Commission President Ursula von der Leyen describing it during a presentation to EU MPs as Europe's 'man on the moon moment'.[1]

Indeed, using emphatic language and creating a strong narrative can be an effective way to guide and coordinate individual actions, ultimately contributing to lasting social change. From a technical perspective, public narratives have the ability to create focal points, which are a powerful tool for coordinating individuals towards a common goal. By providing a clear vision and inspiring people to take action, a strong narrative can create a focal point, thereby generating a sense of collective purpose and motivating individuals to work together towards achieving a shared objective.[2] In the environmental context, Greta Thunberg is an excellent example of a focal point.[3] *Fridays for Future* have eventually mobilised millions and millions of people around the globe and contributed to creating the political space for ambitious environmental reform.

The European Union's recent history offers a compelling case study of how a persuasive policy narrative can wield significant influence. Back in 2012, the Euro Zone was gripped by a severe financial crisis that seemed insurmountable. Mario Draghi took over as President of the European Central Bank on November 1, 2011, and the ECB quickly implemented assertive debt relief measures. Despite these efforts, the situation remained worrisome by July 2012. It was on July 23rd of that year when Draghi delivered his now famous 'whatever it takes' speech, which contained

DOI: 10.4324/9781032662046-10

the following pivotal statement: 'Within our mandate, the ECB is ready to do whatever it takes to preserve the euro. And believe me, it will be enough'.

People believed him. Economic research shows that by itself, Draghi's 'whatever it takes' speech 'had large beneficial effects on the stock prices and credit spreads of banks and sovereigns. By making banks safer, it also made them less prone to take risks'. To put it differently, the act of declaring that the European Central Bank was willing to take any necessary measures to rescue the Euro led to a significant societal impact on its own. Mario Draghi's memorable statement has become ingrained in our collective consciousness. As evidence, the phrase 'whatever it takes' has been included in the Italian language's Treccani dictionary as an Italian neologism for the English expression.[4] In essence, effective policy communication is a critical aspect of a policymaker's ethical approach.[5]

Building on the findings of prior work of these authors, the chapter highlights a worrying discrepancy between the European Commission's policy communication and the proposed changes to EU law that will affect consumption patterns and lifestyles of consumers in the European Union. While policy communications of the European Green Deal – the new Consumer Agenda, the new Circular Economy Action Plan, and the Fit for 55 – emphasise empowering consumers to make environmentally friendly choices, the proposed legislations restrict consumers′ freedom of choice and impact their lifestyles (Section 8.2). These include the Single-use Plastics Directive (SUPD), the proposed Regulation on batteries and waste batteries, the proposed common charger for mobile phones and portable devices, and the reviews of the Energy Efficiency Directive and the Energy Performance of Buildings Directive. The findings suggest that the European Commission's policy communication is missing an opportunity to create a unified narrative to promote sustainable consumption. This lack of clarity is concerning for the reasons summarised in the next paragraph and the one presented in Section 8.3.

Elsewhere, we have investigated at length and criticised this mismatch from two perspectives.[6] First, behavioural studies, particularly behavioural ethics, suggest that the just-mentioned mismatch is likely to undermine the effectiveness of the Green Deal. Second, but relatedly, this mismatch also raises legitimacy issues regarding the Commission's action. On these grounds, we have recommended that the European Union's communication strategy should also emphasise consumers' responsibility towards the environment. Hence, consumers need to be made aware that the EU is committed to achieving the Green Transition, which may require them to make sacrifices.

In this chapter, we show another concerning aspect of the mismatch between the Commission's policy implementation and the parallel policy communication. The mismatch is having a blinding effect on a significant part of private law scholarship reflecting on the role of consumer law in making the Green Transition a success.

The importance of this chapter to the debate about the relationship between private law and sustainability is twofold. First, analytically, the research analyses the measure of EU law regulating goods and services that are ultimately objects of contractual relationships between producers, sellers, or service providers on the one side and consumers on the other, a relationship that is of private law nature. Although the debate over the public-private law divide falls outside the scope of this chapter,[7] the authors acknowledge their endorsement of the private law concept as aligned with the concept of regulatory private law and, more precisely, European regulatory private law.[8] The expansion of EU law towards regulated markets governing subjects inherited to contractual relations, property, and torts equally calls for a definition of private law beyond codified civil law. Second, epistemically, the chapter advances a critical assessment of the legal discourse led by private law scholars, and more specifically, consumer law scholars.

To support its claims about the just-mentioned mismatch between the Green Transition's policy communication and its implementation as well as its detrimental impact on private law scholarship, this chapter is organised as follows. Section 8.2 demonstrates the disparity between the European Commission's policy communication and the proposed legislative measures. Section 8.3 highlights the blinding effect of this mismatch on leading scholars' key stakeholders, namely that some of the themes tend to focus on empowering consumers instead of making them accountable. Finally, Section 8.4 concludes by discussing the limitations of the study and potential areas for future research.

8.2 Between Saying and Doing: Promising Consumer Empowerment while Restricting Consumers' Choice Set

For some time now,[9] the Commission has been envisioning EU consumer law and policy as a means to promote consumers' interests in various interconnected ways. This broader view can be categorised as an 'empowerment paradigm' in which the primary goal is to empower consumers. The concept behind this paradigm is that consumers 'can actively participate in the market and make it work for them by exercising their power of choice and by having their rights properly enforced'.[10] Consumers shall be in the 'driving seats' of the economy.[11] In this context, consumer empowerment refers directly to those situations where the law focuses on 'helping consumers to help themselves'.[12] Providing information to help consumers make better choices or the ability to withdraw from a contract are examples of empowering methods.

In contrast, consumer protection (in the narrow sense) focuses more particularly on cases where consumers cannot reasonably be expected to safeguard their own interests.[13] To support consumers, EU consumer law employs a combination of empowering and protective measures. Empowering measures aim to give consumers more control over their decisions, such as by providing information and the right to withdraw from a contract. Protective measures focus on cases where consumers are unable to protect themselves, such as in instances of unfair contract terms or commercial practices. By prohibiting these behaviours, the law helps to safeguard consumers' interests and supports empowering measures. Overall, the aim of EU consumer law is to give powers to consumers through empowering techniques and protect them when necessary, ensuring that they have the tools and resources they need to make informed decisions in the market.

In order to address the issue of climate change, it is insufficient to rely solely on consumers' individual choices to go green. Indeed, as we show in Section 8.2.2, legislative measures limit consumers' choices while also promoting empowerment. However, Section 8.2.1 shows the EU policy communication emphasises the importance of consumer empowerment in promoting the green transition. The absence of a narrative that justifies this limitation of freedom is a gap in current policy, and this chapter terms the emerging limitation as 'responsibility'.[14] The responsibility that is being referred to here pertains to ensuring a sustainable planet for the generations to come. It is important to note that this responsibility is not merely about choosing to make more environmentally friendly choices. This would boil down to asking consumers to make more green choices. This is not consumer responsibility; it is just consumer empowerment.

The focal point of this discussion is the legal imposition of obligations on consumers that they are expected to fulfil for the benefit of the planet and future generations. The key element of this legal framework is not based on the idea of power, but rather on the concept of duty. Given that individuals, in their roles as consumers, bear the responsibility of ensuring the sustainability of the planet, it is reasonable to enforce positive and negative duties that are oriented towards sustainability. These duties may be directly imposed on consumers themselves, or they may be indirectly imposed by limiting the freedom of economic activity of professionals. For instance, the ban on the sale of single-use plastics restricts consumers' freedom indirectly since they are no longer able to purchase such products. As we shall see in Section 8.2.2, EU legislative interventions are systematically limiting consumers' freedom of choice in order to achieve sustainability-related objectives.

In summary, consumers are accountable for contributing to the green transition, and this accountability justifies the imposition of duties and restrictions on their freedom of choice.

8.2.1 European Commission's Official Policy Communications within the European Green Deal

After providing an overview of the conceptual framework of the analysis, the next step involved examining the European Commission's official policy communications to understand its narrative regarding the role of consumers and consumer law in the Green Transition. The policy communications taken into consideration are the European Green Deal, the New Consumer Agenda, the Circular Economy Action Plan, and the Renovation Wave. These documents are analysed with the aim to determine whether the Commission proposed legal reforms to empower, protect, or make consumers responsible by limiting their freedom of choice.

The European Green Deal,[15] announced by Ursula von der Leyen in December 2019, is the EU's most ambitious economic sustainable growth strategy to date. It aims to transform the economic model by promoting efficient resource use and moving towards a clean and circular economy, cutting pollution, and reversing biodiversity loss, with reforms across various sectors such as transport, energy, agriculture, building, finance, and industries. The policy considers both production and consumption, with the Commission's Communication on the Green Deal, referring to consumers and consumption seven times throughout its 24 pages. The Green Deal presents policies aligned with empowering consumers who want to make green choices like demanding response measures, as well as consumer protection, such as promises to secure affordable clean energy for consumers and addressing the risk of energy poverty. However, the Green Deal also refers to policies aligned with ensuring effective price signals to consumers, such as carbon pricing signal effectiveness and tax reforms to incentivise sustainable behaviour. These policies are framed in terms of producers internalising their own costs rather than consumer responsibility.

The new Consumer Agenda[16] is a policy framework for EU consumer law and policy from 2020 to 2025, focusing on five key priority areas, including green transition.

On the one hand, actions four, five, and six of the Consumer Agenda aim to introduce new legislation to empower consumers for the green transition, such as better information to consumers on products' sustainability, and protecting consumers against practices such as greenwashing and early obsolescence. The spotlight is on giving all consumers the opportunity to make the green choice if they so wish, and on being protected from unfair commercial practices. On the other hand, the new Agenda also refers to other Commission's Communications, such as the new Circular Economy Strategic Plan, Renovation Wave, Chemicals Strategy for Sustainability, Farm to Fork Strategy, and EU Biodiversity Strategy, among others, to ensure that goods and services offered to EU consumers fit the declared objectives. This chapter will look closer at two of those policy documents, namely Circular Economy Strategic Plan and Renovation Wave, being the latter part of the Fit for 55.

The European Commission released a new Circular Economy Action Plan[17] in March 2020 as part of the European Green Deal. The plan aims to prioritise sustainability criteria in the production of all goods, empower consumers and public buyers, reduce waste, and take specific actions in sectors with high resource use. The plan refers to consumer law and policies twelve times, with more than half focusing on empowering consumers and public buyers through trustworthy and relevant information on products, strengthening protection against greenwashing and premature obsolescence, setting minimum requirements for sustainability labels and logos, and facilitating consumer involvement in waste directives. Sector-specific actions focus on empowering consumers through guidance for battery recycling, preventing misleading biodegradable or compostable plastics labelling, encouraging the use of eco-design textile materials, and facilitating consumer involvement in waste management through common bin colours, harmonised waste symbols, product labels, information campaigns, and economic instruments. The plan only subtly suggests potential consumer responsibility in environmental taxation, with a reference to the use of VAT to promote circular economy activities targeting final consumers, particularly repair services.

The Fit for 55[18] is a package of proposals by the European Commission aimed at reducing net emissions by at least 55% by 2030. While it refers less to consumers than previous policies, it acknowledges their responsibility towards climate change and highlights the need to empower them with relevant information. The package includes emission trading schemes and energy efficiency measures that could affect consumers' consumption patterns and prices. However, the Commission argues that the weight of emission trading schemes will be imposed on upstream fuel suppliers rather than individual households or road transport users. The Fit for 55 emphasises the benefits of the green transition, such as clean air and lower energy bills, rather than focusing on consumer responsibility. It also acknowledges the need for a socially fair transition that protects vulnerable consumers in the energy sector and highlights the correlation between the reduction of energy consumption and emission and energy costs, which will benefit both consumers and industries.[19]

Having in mind the four above-mentioned European Commission policy Communications on the green transition, we argue that the European Commission endorses three narratives regarding consumers: empowerment, protection from greenwashing, and avoidance of irrefutable statements that the transition creates costs or mandates choices for consumers. The Commission provides accurate information enabling voluntary sustainable behaviour but avoids discussing explicit costs, restrictions, or consumer responsibility. As a result, the policy narrative is detached from the legislative framework, as shown in the subsequent section.

8.2.2 Actual Legislation and Legislative Proposals

The EU's most ambitious sustainable economic growth plan since its inception was unveiled in the form of the European Green Deal. This triggered a series of detailed action plans, such as those listed in the Circular Economy Action Plan Annex or the new Consumer Agenda, outlining the Commission's intended tasks to achieve its policy objectives of the green transition. These tasks involve proposing new legislation, reviewing existing ones, and ensuring that the proper implementation of recently approved laws aligns with the Green Deal's goals. Therefore, implementing the European Green Deal involves reviewing and reinforcing the existing legal framework.

This section analyses whether existing and proposed EU legislation is consistent with the European Commission's narrative of consumer empowerment in the green transition. While the Commission claims to empower consumers to make better choices, many of the EU's mandatory measures restrict consumer choice. To reach this conclusion, we considered the Directives and Regulations referred to in the New Consumer Agenda, Circular Economy Action Plan, and Fit for 55 that have already been proposed or listed to be reviewed. The list includes the Directive on Single-use Plastics, the Commission proposal for a Regulation on batteries and waste batteries, the Commission proposal for a common charger for mobile devices, and reviews of the Energy Efficiency Directive and Energy Performance of Buildings Directive.

The EU's Circular Economy Action Plan includes proposals for reviewing existing Directives and introducing new legislative acts between 2020 and 2022. The plan emphasises the importance of implementing the SUPD[20] in a timely manner. The SUPD aims to reduce the long-term environmental impact of plastic products that are used once and for a short period. The Directive classifies plastic products into five categories and imposes different measures for each category. Only the plastics listed in Part D, which are subject to market requirements, are subject to measures aligned with consumer empowerment. The other four categories have measures that either restrict their placement on the market or completely ban their commercialisation, thus limiting consumers' freedom to choose. Part A requires Member States to reduce consumption by 2026, while Part B bans certain products. Part C has restricted design requirements, and Part E extends producer responsibility for the costs of waste collection and litter clean-up, which are likely to be transferred to consumers.

Also, within the new EU's Circular Economy Action Plan, the proposed Regulation on batteries and waste batteries aims to widen the scope for battery regulation,[21] covering the entire battery life cycle. The proposed measures are divided into four categories: product manufacturing requirements, labelling and information, extension of economic operators' responsibility, and end-of-life management. The proposed regulation sets sustainability and safety standards, such as restricting hazardous substances and carbon footprint standards. The labelling and information requirements empower consumers to make greener choices. The extension of economic operators' responsibility and lifetime management imposes obligations on producers, importers, distributors, service providers, and end-users. If approved, the new Regulation would impose obligations on consumers to discard waste batteries only in designated separate collection points, in line with the specific arrangement concluded with producers or a producer responsibility association and prohibit discarding batteries in mixed municipal waste.

Like the new Circular Economic Action Plan, many of the action plans listed in the latest Consumer Agenda concerning the green transition have not been implemented yet. However, the Commission did release two proposals: one for amending Directive 2014/53/EU on the harmonisation of the laws of the Member States relating to the making available on the market of radio equipment[22]; the other Proposal for amending Directives 2005/29/EC and 2011/83/EU as regards empowering consumers for the green transition through better protection against unfair practices and better information.[23] The latter has an obvious correlation with the empowerment narrative of the Commission, but not the former, the reason why we look in-depth into the proposal for a common charger for mobile phones and portable devices.

The proposal for a common charger for mobile phones and portable devices is brief and straightforward. It aims to mitigate the 'fragmentation of the market for charging interfaces and charging communication protocols for mobile phones and other similar radio equipment', which 'results in a lack of consumer convenience and an increase in e-waste'.[24] If approved, the use of a single charger would, therefore, protect consumers from the producers' commercial practices that, to some extent, force them to use variable chargers or restrict their choices according to the type of charger. Considering the protection of consumers is a means for their empowerment, the measures proposed by the amendment of the Directive 2014/53/EU are aligned with the Commission's narrative.

The Fit for 55 package includes two proposals for revision and amendment of EU legislation aimed at reducing greenhouse gas emissions: the Energy Efficiency Directive recast and the Energy Performance Buildings Directive review. These proposals target energy consumption from both industrial and individual customers to improve energy efficiency, which is essential for reducing emissions. The Energy Efficiency Directive of 2018 aims to reduce energy consumption by at least 32.5% by 2030, and the Commission proposes to raise this to 39% for primary and 36% for final consumption. The directive relies on three types of measures: mandatory obligations applied to public authorities while purchasing goods and services, obligatory energy efficiency schemes, and alternative policy measures. The latter includes individual metering systems and accurate bill information to empower consumers to reduce energy consumption. The Directive also leaves it to the Member States to decide which final customers should be subject to the energy efficiency target, but the choice must be non-discriminatory and should promote efficient energy use by small energy customers, including domestic customers. Therefore, consumers share the responsibility for reducing energy consumption. The Energy Performance Buildings Directive aims to ensure energy efficiency improvements in buildings by setting performance standards and ensuring regular inspections of heating and air conditioning systems. The proposal for the revised Directive also includes measures to increase the use of renewable energy sources and promote the renovation of existing buildings. The Fit for 55 package shows the EU's commitment to reducing greenhouse gas emissions and the central role of energy consumption in achieving this goal.

The analysis shows that the current and proposed EU legislation not only empowers consumers to make green choices but also forces them to change their consumption patterns. Some products, such as certain single-use plastics, are prohibited from marketing, and consumers are obliged to consume less or renovate their residences to meet energy efficiency standards. Consumer responsibility is a key aspect of the legislation implementing the Commission's policy communication, which focuses solely on empowerment. Hence, there is a mismatch between the Commission's narrative and the existing and emerging legislative framework.

Section 8.3 explains why this mismatch is a problem for private law scholarship.

8.3 Blinding Spotlight: Scholars Falling for the Empowerment Narrative

Having shown that a mismatch exists between the policy communication and the legislative measures regarding the consumers' role in implementing the Green Transition, Sections 8.3 and 8.4 explain why this mismatch is a problem for the effectiveness of this policy. More precisely, this section discusses the 'blinding effect' of this narrative on experts, and in particular consumer law scholars.

At the outset, it is relevant to emphasise that EU law's architecture welcomes a narrative of consumer responsibility towards the environment. In fact, 'Environmental protection requirements must be integrated into the definition and implementation of the Union's policies and activities, in particular with a view to promoting sustainable development' (Article 11 TFEU). It follows that when consumption policy meets environmental policy, it is entirely possible for consumer law to be shaped in order to make the environment prevail over the interests of consumers.[25] Indeed, this is what happens in reality, as seen in Section 8.2.2.

Yet, recent EU consumer law scholarship has failed to look outside the empowerment paradigm. Leading scholars have neither considered what already exists outside the boundaries of a narrow set of legal measures nor offered useful guidance for future developments beyond empowerment. For example, outside this more traditional area of research, one could consider restricting traditional consumer rights (e.g., right to withdraw and to repair) or introducing more intrusive regulatory private law interventions along the lines we surveyed in Section 8.2.2. Consider the following hypothetical case. A new drug is discovered that helps people lose weight easily and without significant health countereffects; this drug is likely to contribute to increased food and beverage consumption in the developed world with negative environmental effects. While it is hard to see any consumer-centred reason to restrict access to the drug, the negative environmental effects might be considered concerning enough to introduce access restrictions to this drug.

Part of the literature focuses on specific topics. For example, how to address premature obsolescence[26] and counter greenwashing.[27] This investigation is fully compatible with the aim of empowering consumers who want to make green choices and protecting them against exploitation. Another topic that has fascinated academics is consumer sales law and in particular the notion of lack of conformity and the connected remedies. This is an area where one could easily cogitate limits to the consumer interest in favour of the environment. Recently, Van Gool and Michel have analysed the new consumer sales directive[28] exactly from this perspective. They propose several interventions where the interests of consumers and the environment are aligned; however, they also note that certain rights and remedies are detrimental from an environmental perspective and explicitly call for their restriction and even elimination.[29]

However, other scholars investigating the same issues have not ventured outside the empowerment paradigm. For example, Beckers, in relation to remedies, focuses on the claim that we must 'accept the ethical (in this case environmentally concerned) consumer as part of the threshold for consumer protection in EU law and his or her right to remedy'.[30] Without denying the beneficial impact for both green consumers and the environment of such a proposal, it is useful to point out that the argument supporting this claim rests on a careful doctrinal investigation showing how 'Consumer sales law may then become an area that is not detached from the environmental objectives, but that is considered as an integral part of the implementation of environmental directives'.[31] It is clear that, on these doctrinal bases, it would be possible to transcend the empowerment paradigm.

A similar pathway can be found in a paper by Terryn focusing on the right to repair. The author moves from the normative premise that 'consumer law must not only take into account consumer protection aims but it should balance them with sustainability goals as this is in the long-term interest of the consumers it aims to protect'.[32] Yet, when looking at the right to repair, the investigation focuses on empowering consumers to make the green choice if they want and does not consider the opportunity to, for example, give it mandatory priority over substitution. This goal is achieved by ensuring that repairing a product is feasible and easy. Comparing the article by Van Gool and Michel with the one by Terryn illustrates how the empowerment narrative can contribute to limiting the scope of the investigation even of environmentally committed consumer lawyers.[33]

Moving beyond consumer sales law, it is particularly telling to consider recent consumer law scholarship discussing the big picture. Even this scholarship shows that it is easy to lose sight of the many ways in which the law can and shall restrict consumer freedom for environmental reasons. Two essays, one by Micklitz and the other by Vanessa Mak and Terryn are particularly telling given the background of these leading EU consumer law scholars. These scholars have been vocally critical of the empowerment focus of EU law in the past.[34] Micklitz, moreover, has constantly warned EU law scholars about the risks of 'silos' thinking.[35] Yet, when it came to sustainable consumption, even he has had difficulty in avoiding silos thinking.

Micklitz begins his narrative by observing the need for consumer lawyers to exit their 'comfort zone' and have a 'debate on what the role of consumer protection and consumer law could be and how a consumer law that meets the needs of a circular economy should be designed'.[36] To start this conversation, Micklitz reviews an impressive series of dimensions of this complex problem: technology, economics, ethics, law beyond the nation-state, and responsibility. It is in relation to responsibility that one can see the blinding effect of the empowerment paradigm. Micklitz observes that, 'The responsibilisation of business is at the forefront of the debate. Rightly so'.[37] With regard to consumers, Micklitz complains about the empowerment paradigm because environmental goals depend upon acts of 'individual autonomy'.[38] Up to this point, his analysis is fully in line with ours.

Then, the two perspectives diverge. In Micklitz's view, three directions are in sight to improve the situation: labelling; price internalisation; consumer responsibility. In particular consumer responsibility, would lead to nothing less than a 'tectonic shift'.[39] What would this tectonic shift amount to? The first shockwave would be suing a company when it does not live up to its 'claims about ethical and sustainable conduct in its supply chain'.[40] This is not really making the consumer responsible; on the contrary, this proposal presupposes that consumers have an actionable right to ethical and sustainable conduct in the supply chain. Ultimately, this means that consumers choose to transact with green traders and are entitled to sue them if the traders do not deliver on their promises. Consumers are not made responsible in the same sense we have been using so far, or in the same sense Micklitz uses when referring to the 'responsibilisation' of business. In our terminology, consumers are empowered to enforce the chosen level of sustainability.

The blinding effect of the empowerment paradigm is even more apparent in the other option considered by Micklitz, namely, regulation to 'steer the consumer'.[41] The author dismisses this option as unlikely to work because neither traders nor consumers would cooperate:

> Is it thinkable that industry is ready to produce such products without a stick behind the door and without consumers who are willing to accept that they will have to keep the very same car perhaps for the rest of their lives?.[42]

We can see the blinding effect at play because the analysis under consideration ignores all the regulatory interventions that – as seen in Section 8.2.2 – are in place and are planned to steer the consumer in the direction of sustainable lifestyles.

Vanessa Mak and Terryn's analysis follows a similar path. Their narrative is expressed in terms of citizenship.[43] One would expect that citizenship implies duties that one owes to the rest of the community. However, the focus of their analysis is 'ensuring the engagement of consumers as citizens'.[44] The authors note that the interest of consumers and the environment sometimes will conflict and that there is a need to rethink EU consumer law. However, they point to future directions for the development of green consumer law: 'Information rights to facilitate 'green' choices, repair as a high-quality and indeed also a green option, and servitisation as a means to increase the shared use of goods and thereby reduce the impact of consumption on the environment'.[45] Again, the realm of future interventions remains confined to the empowerment paradigm. The blinding effect of the empowerment paradigm is particularly evident because the narrative is formulated in terms of citizenship and as noted above, citizenship invokes ideas of responsibility and duty. However, in the analysis under consideration, these ideas remain buried within the empowerment silos.

8.4 Conclusions

This chapter highlights a mismatch between the European Commission's policy communication and implemented legislative measures regarding sustainable consumption, with the former emphasising consumer empowerment and the latter including measures that limit consumer freedom. This mismatch blinds leading consumer law scholars and key stakeholders, who tend to focus on updating existing consumer law to enable green choices while neglecting measures that limit consumer freedom for the sake of contributing to the Green Transition.

While evidence of blinding effects among experts is limited, it serves as a warning about the potential dangers of being blinded by an empowering narrative. The Green Transition requires a focus on both empowering consumers and making sure they take responsibility. The existence of these two approaches creates trade-offs that require policymakers to weigh the costs and benefits of each. It is important to consider how the blinding effect of policy communication may contribute to the neglect of responsibility-aimed measures during the legislative process. Therefore, experts should pay more attention to responsibility-aimed techniques. An evidence-based regulatory system must consider all aspects to achieve the desired results.

Being transparent and honest about the situation is an important aspect of a sound policy, and transparency is a central tenet of EU consumer policy. If the Green Transition is considered our generation's '(hu)man on the Moon moment', then informing consumers that the EU is willing to make any necessary sacrifices to achieve it, including asking consumers to take responsibility, is a small step for them but a significant leap for humankind.

Notes

1 U. von der Leyen, EU Commission President, 'Press Remarks by President von der Leyen on the Occasion of the Adoption of the European Green Deal Communication' (Speech at the European Commission, Brussels, 11 December 2019) <https://ec.europa.eu/commission/presscorner/detail/en/SPEECH_19_6749> accessed on 29 October 2021.

2 R. McAdams, *The Expressive Powers of Law* (Cambridge, Harvard University Press, 2015); K. Basu, *The Republic of Beliefs: A New Approach to Law and Economics* (Princeton, Princeton University Press 2018); A. Sardo and F. Esposito, 'Homo Ludicus: Expected Strategies and Jurisprudence' (2019) 2 (19) TCRS, 95–121, DOI: 10.7413/19705476006.

3 Empirical research has documented the existence of a 'Greta Thunberg effect', showing that familiarity with Greta Thunberg positively correlates with willingness to act to reduce global warming. See A. Sabherwal, M.T. Ballew, S. van der Linden, A. Gustafson, M.H. Goldberg, E.W. Maibach, J.E. Kotcher, K. Swim, S.A. Rosenthal, and A. Leiserowitz, 'The Greta Thunberg Effect: Familiarity with Greta Thunberg Predicts Intentions to Engage in Climate Activism in the United States' (2021) Journal of Applied Social Psychology, DOI: 10.1111/jasp.12737.

4 *Vocabolario Treccani* (online edn, 2020) <https://www.treccani.it/vocabolario/whatever-it-takes_%28Neologismi%29/> accessed on 29 October 2022.

5 S. Papathanassopoulos and R.M. Negrine, *Communications Policy: Theories and Issues* (Red Globe Press, 2010).

6 L. de Almeida and F. Esposito, 'Consumers and the Green Transition between Saying and Doing: Promising Consumer Empowerment While Restricting Consumers' Choices Is Dangerous' (2023, forthcoming) Oxford Yearbook of European Law.

7 Legal scholars have long debated the public and private law divides. With no intention of providing an exhaustive list, see, for instance, C. Harlow, '"Public" and "Private" Law: Definition without Distinction' (1980) 43 (3) The Modern Law Review, 241–265. N. Reich, 'The Public/Private Divide in European Law' in Hans-W. Micklitz and Fabrizio Cafaggi (eds), *European Private Law after the Common Frame of Reference* (Edward Elgar Publishing, 2010). J.W.F Allison, 'Variation of View on English Legal Distinctions between Public and Private' (2007) 66 (3) The Cambridge Law Journal, 698–711. A. Jakab, 'Public Law–Private Law Divide?' in *European Constitutional Language* (Cambridge, Cambridge University Press, 2016) 387–400. C. Michelon, 'The Public, the Private, and the Law' in Cormac Mac Amhlaigh, Claudio Michelon, and Neil Walker (eds), *After Public Law* (Oxford University Press, 2013).

8 See H-W. Micklitz, 'European Regulatory and Private Law – Between Neoclassical Elegance and Postmodern Pastiche' in M. Kuhli and M. Schmidt (eds), *Vielfalt im Recht: Diversity in Law* (Berlin: Duncker & Humblot, 2022). H-W. Micklitz, 'European Private Law' in D. Patterson AND A. Södersten (eds), A *Companion to European Union Law and International Law* (John Wiley & Sons, Ltd, 2016) 262–284. L. de Almeida, 'Rise and Fall of the Classic Concept of Private Law: Lessons from the Legal Consciousness of European Private Law' (13 December 2018) 1 (1) DESC – Direito, Economia e Sociedade Contemporânea, 95–128. L. de Almeida and F. Esposito, 'In Search of a Grand Theory of European Private Law: Social Justice, Access Justice, Societal Justice and Energy Markets' in L. de Almeida, M. Cantero Gamito, M. Durovic, and K-P. Purnhagen (eds), *The Transformation of Economic Law: Essays in Honour of Hans-W. Micklitz* (Hart Publishing, 2021). G. Comparato, 'Public Policy through Private Law: Introduction to a Debate on European Regulatory Private Law' (2016) 22 (5) European Law Journal, 621–626. O. Cherednychenko, 'Rediscovering the Public/Private Divide in EU Private Law' (2020) 26 (1–2) European Law Journal, 27–47.

9 More at length on this topic, see F. Esposito and S.M. Grundmann, 'Investor-Consumer or Overall Welfare: Searching for the Paradigm of Recent Reforms in Financial Services Contracts' (2017) EUI Law Department Research Paper Series 2017/5; F. Esposito and A-L. Sibony, 'In Search of the Theory of Harm in EU Consumer Law: Lessons from the Consumer Fitness Check' in K. Mathis and A. Tor (eds),

Consumer Law and Economics in Europe (Dordrecht, Springer 2020) 251–281, 266–267 and the sources therein.

10 Commission, 'A European Consumer Agenda – Boosting confidence and growth' COM (2012) 225 final, 5.

11 Commission, 'EU Consumer Policy Strategy 2007–2013. Empowering Consumers, Enhancing their Welfare, Effectively Protecting them' COM (2007) 99 final, 5.

12 Commission, 'Consumer Policy Action Plan 1999–2001' COM (1999) 696 final, 10.

13 Commission, 'Healthier, safer, more confident citizens: a Health and Consumer Protection Strategy' COM (2005) 115 final, 3; Commission, 'EU Consumer Policy Strategy 2007–2013. Empowering Consumers, Enhancing their Welfare, Effectively Protecting them' COM (2007) 99 final, 5.

14 The term 'responsibility' is used in a similar fashion by the European Commission (see Commission, '"Fit for 55": Delivering the EU's 2030 Climate Target on the way to climate neutrality' COM (2021) 550 final, 1 and by H-W. Micklitz, 'Squaring the Circle? Reconciling Consumer Law and the Circular Economy' (2019) 8 (6) Journal of European Consumer and Market Law, 229–238, 233 and 235 and ff.

15 Commission, 'The European Grean Deal' COM (2019) 640 final.

16 Commission, 'New Consumer Agenda: Strengthening consumer resilience for sustainable recovery' COM (2020) 696 final.

17 Commission, 'A New Circular Economy Action Plan for a Cleaner and more Competitive Europe' COM (2020) 98 final.

18 Commission, '"Fit for 55": Delivering the EU's 2030 Climate Target on the way to climate neutrality' COM (2021) 550 final, 1.

19 I.M. Hoffman, G. Rybka, G. Leventis, C.A. Goldman, L.C. Schwartz, M.A. Billingsley and S.R. Schiller, 'The Total Cost of Saving Electricity Through Utility Customer-Funded Energy Efficiency Programs: Estimates at the National, State, Sector and Program Level' (2015) 2 Berkely Lab, 1–22.

20 Council Directive (EU) 2019/904 of the European Parliament and of the Council of 5 June 2019 on the reduction of the impact of certain plastic products on the environment [2019] OJ L 155/1.

21 Commission, 'Proposal for a Regulation of the European Parliament and of the Council concerning batteries and waste batteries, repealing Directive 2006/66/EC and amending Regulation (EU) No 2019/1020' COM (2020) 798 final.

22 Commission, 'Proposal for a Regulation of the European Parliament and of the Council concerning batteries and waste batteries, repealing Directive 2006/66/EC and amending Regulation (EU) No 2019/1020' COM (2020) 798 final.

23 Commission, 'Proposal for a Regulation of the European Parliament and of the Council amending Directives 2005/29/EC and 2011/83/EU as regards empowering consumers for the green transition through better protection against unfair practices and better information' COM (2022) 143 final.

24 Commission, 'Proposal for a Directive of the European Parliament and of the Council amending Directive 2014/53/EU on the harmonisation of the laws of the Member States relating to the making available on the market of radio equipment', COM (2021) 547 final, Recital 5, 12.

25 See, for example, C. Kye, 'Environmental Law and the Consumer in the European Union' (1995) 7 (1) Journal of Environmental Law, 31–54; K. Tonner, 'Consumer Protection and Environmental Protection: Contradictions and Suggested Steps towards Integration' (2000) 23 (1) Journal of Consumer Policy, 63–78.

26 E. Maitre-Ekern and C. Dalhammar, 'Regulating Planned Obsolescence: A Review of Legal Approaches to Increase Product Durability and Reparability in Europe' (2016) 25 (3) RECIEL, 378–394; A-L. Sibony and S. Becker, 'Confronting Product Obsolescence' (2021) 27 (2) Columbia Journal of European Law, 97–147; A. De Franceschi, 'Planned Obsolescence challenging the Effectiveness of Consumer Law and the Achievement of a Sustainable Economy: The Apple and Samsung Cases' (2018) 7 (6) Journal of European Consumer and Market Law, 217–221.

27 S. Stoerk Ekstrand and K. Lilholt Nilsson, 'Greenwashing?' (2011) 6 (3) European Food & Feed Law Review, 167–173; B. Naderer, D. Schmuck, and J. Matthes, 'Greenwashing: Disinformation through Green Advertising' in G. Siegert, M. B. von Rimscha and S. Grubenmann (eds), *Commercial Communication in the Digital Age* (De Gruyter Saur, 2017), 105–120.

28 Council Directive (EU) 2019/771 of the European Parliament and of the Council of 20 May 2019 on certain aspects concerning contracts for the sale of goods, amending Regulation (EU) 2017/2394 and Directive 2009/22/EC, and repealing Directive 1999/44/EC [2019] OJ L 136/28.

29 E. Van Gool and A. Michel, 'The New Consumer Sales Directive 2019/771 and Sustainable Consumption: A Critical Analysis' (2021) 10 (4) Journal of European Consumer and Market Law, 136–148.
30 A. Beckers, 'Environmental Protection Meets Consumer Sales' (2018) 14 (2) European Review of Contract Law, 157–189, 184.
31 A. Beckers, 'Environmental Protection Meets Consumer Sales' (2018) 14 (2) European Review of Contract Law, 157–189, 187.
32 E. Terryn, 'A Right to Repair? Towards Sustainable Remedies in Consumer Law' (2019) 27 (4) European Review of Private Law, 851–874, 872. It is noteworthy that Terryn tries to stretch the notion of the interest of (today's) consumers as to include also environmental concerns.
33 The good news is that the 2020 report co-authored by both Terryn and Michael (among others) for the European Parliament on the matter succeeds in transcending the empowerment paradigm. See, B. Keirsbilck, E. Terryn, A. Michel and I. Alogna, 'Sustainable Consumption and Consumer Protection Legislation' (2020) In-Depth Analysis Requested by the IMCO committee, 20–22 <https://www.europarl.europa.eu/RegData/etudes/IDAN/2020/648769/IPOL_IDA(2020)648769_EN.pdf>.
34 H-W. Micklitz, 'The Expulsion of the Concept of Protection from the Consumer Law and the Return of Social Elements in the Civil Law – A Bittersweet Polemic' (2012) 35 (3) Journal of Consumer Policy, 283–296; V. Mak, 'The Myth of the "Empowered Consumer": Lessons from Financial Literacy Studies' (2012) 4 (1) Journal of European Consumer and Market Law, 254–263.
35 H-W. Micklitz, 'The Visible Hand of European Regulatory Private Law—The Transformation of European Private Law from Autonomy to Functionalism in Competition and Regulation' (2009) 28 (1) Yearbook of European Law, 3–59.
36 H-W. Micklitz, 'Squaring the Circle? Reconciling Consumer Law and the Circular Economy' (2019) 8 (6) Journal of European Consumer and Market Law, 229–238, 230.
37 H-W. Micklitz, 'Squaring the Circle? Reconciling Consumer Law and the Circular Economy' (2019) 8 (6) Journal of European Consumer and Market Law, 229–238, 235.
38 -W. Micklitz, 'Squaring the Circle? Reconciling Consumer Law and the Circular Economy' (2019) 8 (6) Journal of European Consumer and Market Law, 229–238, 235.
39 -W. Micklitz, 'Squaring the Circle? Reconciling Consumer Law and the Circular Economy' (2019) 8 (6) Journal of European Consumer and Market Law, 229–238, 235.
40 H-W. Micklitz, 'Squaring the Circle? Reconciling Consumer Law and the Circular Economy' (2019) 8 (6) Journal of European Consumer and Market Law, 229–238, 236.
41 H-W. Micklitz, 'Squaring the Circle? Reconciling Consumer Law and the Circular Economy' (2019) 8 (6) Journal of European Consumer and Market Law, 229–238, 236.
42 H-W. Micklitz, 'Squaring the Circle? Reconciling Consumer Law and the Circular Economy' (2019) 8 (6) Journal of European Consumer and Market Law, 229–238, 236.
43 Similarly, see T. Wilhelmsson, 'Consumer Law and the Environment: From Consumer to Citizen' (1998) 21 (1) Journal of Consumer Policy, 45–70.
44 V. Mak and E. Terryn, 'Circular Economy and Consumer Protection: The Consumer as a Citizen and the Limits of Empowerment through Consumer Law' (2020) 43 (1) Journal of Consumer Policy, 227–248, 243.
45 V. Mak and E. Terryn, 'Circular Economy and Consumer Protection: The Consumer as a Citizen and the Limits of Empowerment through Consumer Law' (2020) 43 (1) Journal of Consumer Policy, 227–248, 244.

Bibliography

Allison J, 'Variation of View on English Legal Distinctions between Public and Private' (2007) 66 (3) The Cambridge Law Journal, 698–711

Almeida L, 'Rise and Fall of the Classic Concept of Private Law: Lessons from the Legal Consciousness of European private Law' (13 December 2018) 1 (1) DESC – Direito, Economia e Sociedade Contemporânea, 95–128

Almeida L and Esposito F, 'In Search of a Grand Theory of European Private Law: Social Justice, Access Justice, Societal Justice and Energy Markets' in L. de Almeida, M. Cantero Gamito, M. Durovic, and K-P. Purnhagen (eds), *The Transformation of Economic Law: Essays in Honour of Hans-W. Micklitz* (Hart Publishing, 2021), 153–172

Almeida L and Esposito F, 'Consumers and the Green Transition Between Saying and Doing: Promising Consumer Empowerment While Restricting Consumers' Choices Is Dangerous' (2023) Oxford Yearbook of European Law, yead011

Basu K, *The Republic of Beliefs: A New Approach to Law and Economics* (Princeton, Princeton University Press, 2018)

Beckers A, 'Environmental Protection meets Consumer Sales' (2018) 14 (2) European Review of Contract Law, 157–189

Cherednychenko O, 'Rediscovering the Public/Private Divide in EU Private Law' (2020) 26 (1–2) European Law Journal, 27–47

Comparato G, 'Public Policy through Private Law: Introduction to a Debate on European Regulatory Private Law' (2016) 22 (5) European Law Journal, 621–626

De Franceschi A, 'Planned Obsolescence challenging the Effectiveness of Consumer Law and the Achievement of a Sustainable Economy: The Apple and Samsung Cases' (2018) 7 (6) Journal of European Consumer and Market Law, 217–221

Ekstrand S and Nilsson K, 'Greenwashing?' (2011) 6 (3) European Food & Feed Law Review, 167–673

Esposito F and Grundmann S, 'Investor-Consumer or Overall Welfare: Searching for the Paradigm of Recent Reforms in Financial Services Contracts' (2017) EUI Law Department Research Paper Series 2017/5

Esposito F and Sibony A-L, 'In Search of the Theory of Harm in EU Consumer Law: Lessons from the Consumer Fitness Check' in K. Mathis and A. Tor (eds), *Consumer Law and Economics in Europe* (Dordrecht, Springer 2020) 251–281

Harlow C, '"Public" and "Private" Law: Definition without Distinction' (1980) 43 (3) The Modern Law Review, 241–265

Hoffman I, Rybka G, Leventis G, Goldman C, Schwartz L, Billingsley M and Schiller S, 'The Total Cost of Saving Electricity Through Utility Customer-Funded Energy Efficiency Programs: Estimates at the National, State, Sector and Program Level' (2015) 2 Berkely Lab, 1–22

Jakab A, *European Constitutional Language* (Cambridge, Cambridge University Press, 2016)

Keirsbilck B, Terryn E, Michel A and Alogna I, 'Sustainable Consumption and Consumer Protection Legislation' (2020) In-Depth Analysis Requested by the IMCO Committee, 20–22 <https://www.europarl.europa.eu/RegData/etudes/IDAN/2020/648769/IPOL_IDA(2020)648769_EN.pdf>

Kye C, 'Environmental Law and the Consumer in the European Union' (1995) 7 (1) Journal of Environmental Law, 31–54

Maitre-Ekern E and Dalhammar C, 'Regulating Planned Obsolescence: A Review of Legal Approaches to Increase Product Durability and Reparability in Europe' (2016) 25 (3) RECIEL, 378–394

Mak V, 'The Myth of the "Empowered Consumer": Lessons from Financial Literacy Studies' (2012) 4 (1) Journal of European Consumer and Market Law, 254–263

Mak V and Terryn E, 'Circular Economy and Consumer Protection: The Consumer as a Citizen and the Limits of Empowerment through Consumer Law' (2020) 43 (1) Journal of Consumer Policy, 227–248, 243

McAdams R, *The Expressive Powers of Law* (Cambridge, Harvard University Press, 2015)

Michelon C, 'The Public, the Private, and the Law' in Cormac Mac Amhlaigh, Claudio Michelon, and Neil Walker (eds), *After Public Law* (Oxford University Press, 2013)

Micklitz H-W, 'The Visible Hand of European Regulatory Private Law—The Transformation of European Private Law from Autonomy to Functionalism in Competition and Regulation' (2009) 28 (1) Yearbook of European Law, 3–59

Micklitz H-W, 'The Expulsion of the Concept of Protection from the Consumer Law and the Return of Social Elements in the Civil Law – A Bittersweet Polemic' (2012) 35 (3) Journal of Consumer Policy, 283–296

Micklitz H-W, 'European Private Law' in D. Patterson and A. Södersten (eds), A *Companion to European Union Law and International Law* (John Wiley & Sons, Ltd, 2016) 262–284

Micklitz H-W, 'Squaring the Circle? Reconciling Consumer Law and the Circular Economy' (2019) 8 (6) Journal of European Consumer and Market Law, 229–238

Micklitz H-W, 'European Regulatory and Private Law – Between Neoclassical Elegance and Postmodern Pastiche' in M. Kuhli and M. Schmidt (eds), *Vielfalt im Recht: Diversity in Law* (Berlin: Duncker & Humblot, 2022), 75–99

Naderer B, Schmuck D and Matthes J, 'Greenwashing: Disinformation through Green Advertising' in G. Siegert, M.B. von Rimscha and S. Grubenmann (eds), *Commercial Communication in the Digital Age* (De Gruyter Saur, 2017), 105–120

Papathanassopoulos S and Negrine R, *Communications Policy: Theories and Issues* (Red Globe Press, 2010)

Reich N, 'The Public/Private Divide in European Law' in Hans-W. Micklitz and Fabrizio Cafaggi (eds), *European Private Law after the Common Frame of Reference* (Edward Elgar Publishing, 2010), 56–89

Sabherwal A, Ballew M, van der Linden S, Gustafson A, Goldberg M, Maibach E, Kotcher J, Swim K, Rosenthal, S and Leiserowitz A, 'The Greta Thunberg Effect: Familiarity with Greta Thunberg Predicts Intentions to Engage in Climate Activism in the United States' (2021) Journal of Applied Social Psychology, DOI: 10.1111/jasp.12737

Sardo A and Esposito F, 'Homo Ludicus: Expected Strategies and Jurisprudence' (2019) 2 (19) TCRS, 95–121, DOI: 10.7413/ 19705476006

Sibony A-L and Becker S, 'Confronting Product Obsolescence' (2021) 27 (2) Columbia Journal of European Law, 97–147

Terryn E, 'A Right to Repair? Towards Sustainable Remedies in Consumer Law' (2019) 27 (4) European Review of Private Law, 851–874, 872

Tonner K, 'Consumer Protection and Environmental Protection: Contradictions and Suggested Steps towards Integration' (2000) 23 (1) Journal of Consumer Policy, 63–78

Van Gool E and Michel A, 'The New Consumer Sales Directive 2019/771 and Sustainable Consumption: A Critical Analysis' (2021) 10 (4) Journal of European Consumer and Market Law, 136–148

Wilhelmsson T, 'Consumer Law and the Environment: From Consumer to Citizen' (1998) 21 (1) Journal of Consumer Policy, 45–70

9

REVISITING EUROPEAN CONSUMER PROTECTION THROUGH THE LENS OF SUSTAINABLE MARKETS

Mireia Artigot Golobardes[1]

9.1 Consumer Protection and Sustainable Markets: Aligning Private Choices with Sustainability Goals

The environmental crisis requires a reassessment of a wide range of production and consumption choices at different levels, from basic resources like energy, water, and food, all the way up to complex consumer products, transportation planning, the services sector, and the ways in which consumer preferences are shaped. In many ways, the climate crisis demands a fundamental reassessment of how we live.

Sustainable markets, an essential element of sustainable economies, are an important concern of European Member States, European institutions, and international organisations. The United Nations General Assembly and various UN bodies have adopted instruments laying down commitments towards achieving sustainable economies. These include the 2030 Agenda for Sustainable Development,[2] with its Sustainable Development Goal 12 ("Ensure Sustainable Consumption and Production Patterns"), which sets sustainability targets and actions to be taken by governments, businesses, and consumers.[3] They also include the UN Guidelines for Consumer Protection,[4] and the Global Guidance on Making Effective Environmental, Social and Economic Claims, To Empower and Enable Consumer Choice.[5] These instruments, while not binding, set targets and actions to be adopted to achieve sustainable consumption and sustainable markets, and they provide a legal framework and guidelines for the enhancement of consumer protection legislation and for reducing waste.

European Member States have assumed the targets and guidelines established by the United Nations, but the European institutions have also aimed at playing a key role in fighting climate change through sustainable markets. The proposals presented by European Union institutions in order to achieve sustainable markets have adopted general perspectives but have also specifically focused on the demand side of the market, that is, on consumers. Consumer protection is one of the fundamental values of the European Union[6] and the European Union institutions have engaged in reassessing and redesigning consumer law through the lens of sustainable markets.

From a general policy perspective, the first European Union initiatives were adopted in 2015, when the European Commission published an EU action plan for the circular economy,[7] which

DOI: 10.4324/9781032662046-11

aimed at promoting practices that could contribute to a more circular economy in order to enhance sustainable consumption. The European Parliament also issued two resolutions regarding sustainable markets: the Resolution on a longer lifetime for products: benefits for consumer and companies,[8] and the Resolution on the implementation of the Ecodesign directive.[9] The most ambitious initiative, though, arrived in 2019 with the publication of the European Green Deal,[10] which was then reformulated in 2020 by the New Circular Economy Action Plan.[11]

The European Green Deal announced a programme on various aspects of sustainable markets, including legislative proposals for sustainable products (2021), for sustainable product policy (2021), for empowering consumers in the green transition (2020), on substantiating green claims (2020), and on legislative and non-legislative measures establishing the right to repair (2021). These legislative proposals, in a way, aim at impacting the supply side as well as the demand side of the market. In other words, they aim to provide new legal rules on contractual remedies and new instruments so that consumers can pursue legal claims arising from breaches of the obligations arising from sustainability goals.[12]

Microeconomic approaches like these private law instruments, however, are insufficient to address sustainability issues that have impacts well beyond those felt by the contracting parties themselves. Following the life of a product in the European market, this chapter presents the pre-market, pre-contractual, contractual, and post-contractual phases and discusses how the negative externalities generated by production processes cannot be addressed only with private law instruments. Revisiting consumer law through the lens of sustainability goals[13] illuminates the weaknesses of microeconomic instruments like private law for solving issues with broad impacts. This chapter proposes how microeconomic individual instruments used in consumer law, particularly in European consumer law, should be reassessed and redefined so that they can better address sustainability challenges.

This chapter argues that sustainable markets require a reinterpretation of current applicable consumer law as well as a reassessment of its normative basis and a reconsideration of its substantive design. The complex anatomy of sustainable markets requires a reassessment of normative structures and strategies. Private law, specifically consumer law, can contribute to achieving sustainable markets but cannot be, by itself, the main driving force for sustainability.

9.2 Real Costs, Real Prices, and the Necessary – Albeit Insufficient – Role of Private Law

The complex environmental challenges we face require a global and multidimensional approach that includes reassessing some of the fundamental values on which our societies have been based. Sustainable markets are one of the crucial elements for addressing the environmental crisis but introducing sustainability as a fundamental value in consumer markets is not an easy task. Today, our consumption decisions and the price we pay for our products and services do not reflect the real cost of production, distribution, and delivery involved.

Prices are an instrument that convey information about the costs of production, packaging, transportation, and delivery of a product or service, about the guarantees the consumer is entitled to, and the profit the seller may be able to obtain. However, most of the costs embodied in prices are private ones, and do not reflect the environmental costs imposed on society. Environmental costs are negative externalities generated by market transactions that are not reflected by the prices of products and services exchanged in the market.[14] From this perspective, the price of products and services do not reflect and hence, do not communicate, the total costs involved in their production. That is, our current production processes, distribution chains, transportation means, and in some cases, door-to-door delivery to the ultimate consumer impose negative costs on natural resources and on the environment. Generally understood this can have profound consequences on society's well-being in terms of pollution, use of energy, and of natural resources with broad effects on global warming and unsustainability of the life as we know it in our planet. Consequently, the *real* price of the goods and services exchanged in consumer markets is higher than the prices consumers pay for them.

On the other side of the market, consumers choose the goods and services that best fit their preferences and needs. The consumer's preferences, together with the information they receive on the different goods and services available in the market, shape consumer's choices. Today, sustainable products and services are growing in importance in consumer's preferences.[15] Consumers are increasingly aware and concerned about the environmental challenges the world faces and are increasingly aware that consumer habits as we have known them in the second half of the 20th century are not sustainable and will require fundamental changes to which consumers will have to adapt.

In this context, private law, specifically consumer law, appears as a valuable instrument to address the fundamental changes consumer markets will need to undertake. Consumer law can contribute to the introduction of sustainability goals in consumer markets so that market participants internalise the externalities generated by them. These sustainability requirements could involve the whole product or service life: from the moment when the product enters the European market, before it is the object of a transaction, the contract regulating the transaction and the remedies to which the consumer would be entitled to for breaching contract terms – including sustainability commitments.

Sustainable markets involve a complex task requiring a global perspective over the different decisions adopted at each step of the supply chain – from the production stage to the availability of the product for the consumer – as well as over the demand side – from the consumer's choice to the consumer's remedy for lack of conformity.[16]

Private law and, more specifically, consumer law, can be a valuable instrument to ensure that consumers are informed about the environmental impact of the products or services they consider, that sellers comply with the obligations committed to under the contract regulating the transaction, and that remedies for breach are awarded considering the product's environmental impact.

Private law, and specifically contract law, will be a necessary, albeit insufficient, instrument to require producers and sellers to produce and market sustainable products, to steer consumer choices to sustainable products and to turn sustainability into enforceable contractual obligations.[17] Consumer law will definitely play a role in making markets more sustainable but will require a policy mix in order to achieve this goal. Consumers can actively contribute to sustainable markets through their choices. However, sustainable products will be chosen by consumers if sustainable production processes are encouraged through comprehensive regulation that results in internalising the environmental costs of production of the product to its sale and are ultimately enforced.[18] The unavoidable need for sustainable production processes on the supply side and sustainable consumption on the demand side will require different normative instruments: while sustainable production requires public law instruments,[19] sustainable consumption may be articulated through private law norms.

Additionally, the impact of (un)sustainable consumption not only reaches beyond the parties involved in the consumer contract and impacts society but it also has a dynamic dimension. Respect for the environment and the implementation of sustainability requirements today impacts not only our current living conditions but also the quality of life of future generations. Consequently, it will be insufficient to address only the externalities imposed today given that they represent an underestimation of the stream of costs generated by our current practices into the future. The dynamic and hence intergenerational effects of unsustainable practices are, however, beyond the scope of this chapter.

The following sections are structured as follows: Section 9.3 addresses challenges to achieving sustainable markets – sustainable production and sustainable consumer choices in markets – in light of the current consumer protection regime in the European Union. Section 9.4 presents the recent proposal directive on the right to repair, and Section 9.5 closes with final thoughts.

9.3 Sustainable Markets: The Need for Sustainable Production and Sustainable Consumer Choices

Turning traditional ordinary consumer markets as we have known them up to today into sustainable consumer markets is a remarkably complex endeavour. Even though structurally similar to the challenges presented by the internalisation of any negative externality, the design and implementation of sustainability as one of the bases of consumer markets is complicated. The first layer of complexity comes from the fact that the goal of achieving sustainable markets depends on a plurality of parties the decisions of which directly impact the ultimate sustainable outcome. At the same time, the decisions of this plurality of agents are correlated with each other and their effects could and, in some cases surely are, cumulative. Finally, there are market forces, particularly in digital markets, that create environments that favour non sustainable consumption models.

The first layer of complexity comes from the structure of supply chains, particularly global supply chains where products result from the assembly of parts that are produced in different locations subject to different regulations. The assembly of product components to turn them into the final product might be done in a different location that in order to be viable has required transporting the component parts. At the same time, global supply chains distribute final products across different markets that also require imports, and in some cases, distribution and additional transportation to the providers that ultimately will make these final products and services available to consumer stores that ultimately will result in the consumer's transaction.[20]

Each agent at each phase of the supply chain has decisions to adopt that affect the sustainability of consumer markets. Sustainable markets imply sustainable supply chains ranging from sustainable production processes, sustainable distribution, and sustainable transportation until the product or service reaches the final consumer.[21]

Additionally, the impact of the decisions of each of the agents participating in global supply chains for sustainable markets has a correlated effect that often is compounded. The decisions of producers in making their production processes more sustainable affect the use of resources involved in, for example, the use of raw materials and component parts and their transportation so that the final product may be produced. The correlation between the decisions of the different agents involved in global supply chains and their ultimate impact on sustainability makes its regulation remarkably complex given that placing the regulatory focus on one agent or on a specific phase of the production line might not be enough. It will be necessary not only to assess the sustainability impact of the different decisions of the agents involved in global supply chains but also how these decisions condition the impact on the sustainability of the subsequent agents and phases involved in the global supply chain.[22]

Finally, consumer markets, particularly digital consumer markets, function in a way where consumers are encouraged to consume cheaper and more but not better in, for example, sustainability terms.[23] Market environments that consumers find themselves in are, in the vast majority of cases, environments where consumers are encouraged to buy as cheap as possible, as much as possible and to neglect variables that could be of the consumers' interest – such as the environmental impact of the product they are considering to purchase.[24]

In sum, introducing sustainability goals in consumer markets involves a complex task that requires a comprehensive perspective of markets, including global supply chains and consumer dynamics, so that sustainability requirements can be imposed where their impact can be of the most importance.

9.3.1 Using Public Law to Regulate the Multilayer Structure of Sustainable Supply Chains

As noted above, one of the major complexities of achieving sustainable global supply chains lays is their multivariate and compound nature. Additionally, market dynamics and market forces steer consumers to adopt decisions based on variables such as price and quantity that often do not involve sustainability concerns. This mismatch is challenging to address.[25]

Quantifying the environmental impact of the decisions of the different agents involved in the various phases of global supply chains often located in different jurisdictions is a remarkably complex task. This is due to the highly specialised and highly complex production systems that make regulators and enforcement agencies not to be best fit for designing legal requirements, processes, thresholds, and potential sanctions for lack of compliance. Regulators do not have the best information to detect and quantify the environmental impact of the decisions involved in the different phases of the agents participating in global supply chains. In some cases, this information may not be available or even not exist, while in others, regulators may not be in a position to obtain it.

Technical knowledge about the specific functioning of global supply chains and the potential room for improvement is of fundamental importance for reducing environmental costs in terms of, for example, reducing carbon footprint of production process as well as improving the efficiency of the resources used so that it is possible to reduce waste. At the same time, requiring sustainable production processes and sustainable global supply chains could – and most likely would – have effects on business profits and on market structures.

So, achieving more sustainable supply chains it is not only an issue of identifying and quantifying environmental impact.[26] Even assuming that public agencies or regulators could identify and quantify this environmental impact generated by the different agents in the different phases of global supply chains, regulators are not in the best position to suggest alternatives that would reduce such impact or that would result in global supply chains being ideally environmentally neutral. Global supply chains are too sophisticated, too complex, and involve so many heterogeneous agents located in different places adopting highly technical decisions that leave regulators – in this case European – with a very limited capacity to impose structures, designs and production flows that eventually could become more sustainable if not fully sustainable.

These issues have a direct impact on the regulatory strategy that should be followed in achieving environmentally sustainable global supply chains. Further, highly complex global supply chains have a direct impact on the limited effects concrete legal measures might have and on the speed at which sustainable markets might be achieved. In light of these challenges contracts seem a valuable, albeit limited, instrument to achieve sustainability.[27] The highly sophisticated technical structure of global supply chains conditions the regulatory strategies to implement that most likely cannot be top-down but will have to be somehow bottom up – or a combination of both – so that the agents subject to regulations aiming at sustainable global supply chains will be the ones providing the information and reacting to their self-interest so that their decisions and processes comply with sustainability goals.[28]

As of today, the European Union counts with two norms applicable to the supply side of the market: first, the Ecodesign directive[29] and second, the Energy labelling regulation.[30] Both instruments aim at improving the energy efficiency and sustainability of products marketed in the European internal market.

The Ecodesign directive provides rules for minimising the environmental impact of certain products sold within the European Union by requiring them to have a minimum level of energy efficiency. On the one hand, the directive applies to some household appliances, information and communication technologies, and engineering products and is expected to contain product requirements on durability, repairability, and recyclability.[31] The Energy labelling regulation, on the other hand, complements the Ecodesign directive by introducing mandatory labelling requirements regarding the environmental impact of the energy used by certain products.

Additionally, on February 23rd, 2022, the European Commission adopted a proposal for their action on corporate sustainability due diligence that aims at fostering sustainable and responsible corporate behaviour throughout global supply chains.[32] Under this proposed directive, companies[33] will be required to identify and address the impact of their activities on human rights as well as on the environment. The proposal is well intended. However, it presents important loopholes in terms of concretion. For example, regarding the scope of the due diligence and regarding the allocation of liability resulting from a lack of compliance with the provisions of the obligation of due diligence. Would liability be on corporate managers? On the board of directors? On administrators?[34] These issues, within the scope of the decision of the Member States, still remain to be determined.[35]

Complementing the ecodesign rules, standards are being developed under Regulation (EU) 1025/2012. This EU Regulation on European standardisation includes sustainability requirements through imposing obligations on aspects of material efficiency. These rules are further complemented by product-specific regulation, harmonised standards, and market surveillance.

This normative structure combining general standards with product-specific requirements has a long tradition in European product regulation. Specifically, technical requirements[36] and harmonised standards[37] have been part of the European product safety regulation for the last 20 years. This normative body is expanded today to include sustainable design requirements under the Ecodesign directive and sustainable consumer information regarding the energy consumption of energy-related products.

The regulation of the supply side of the market so that products are sustainable presents a difficult trade-off: regulations cannot be product-specific, given that technology evolves very rapidly, and regulators do not have the best information on optimal production processes while broad general horizontal regulation becomes a regulatory benchmark not generating the optimal incentives across industrial sectors. This difficult trade-off unavoidably results in suboptimal regulatory strategies and regulatory measures. This does not mean, however, that pre-market or pre-transaction requirements should be discarded. Public law should be used to provide the environmental requirements that products should comply with when entering the European Union's internal market. However, these norms should be complemented by effective enforcement mechanisms so that producers can be held accountable if they do not comply with the environmental requirements imposed on the products they market.[38] Today, we are still missing an effective system of public law information and administrative measures that could complement reputational, and non-contractual sanctions in order to ensure that products placed in the European Union market comply with the environmental requirements imposed on them so that a European Union sustainable internal market is achieved.

9.3.2 Greening Private Law – Consumer Law

As explained above, sustainable markets will require not only that sustainable products are present and available in markets but also that consumers know which products are produced in a sustainable way and that ultimately, they choose to consume these sustainable products. Furthermore, is it necessary that the consumer has access to remedies to seek redress for breaches of non-compliance with pre-contractual information or contractual commitments regarding the sustainability of the product or service object of the contract. Additionally, a change of consumption culture will be required. The consumption of a lower number of products, with a longer life and more sustainable is unavoidable. However, the unavoidable change of the consumption culture for environmental reasons is beyond the scope of this chapter.

Consumers are at the core of European regulation on consumer protection that aims at affording a high degree of protection to consumers.[39] The main goal of European consumer protection regulation is ensuring consumer's sovereignty in market choices and market transactions through which consumers should be able to maximise their welfare. It is structured around three different phases of consumer's market participation: (1) ensuring consumer's sovereignty when taking choices, (2) ensuring consumer contracts reflect the distribution of risks and surplus the consumer expected from the transaction and (3) ensuring the consumer has effective legal remedies to seek redress for breach of the consumer contract as well as access to justice either individually or collectively. The focus of the European consumer protection regulation has been the micro approach, where the consumer, as an individual, is able to decide and enter into transactions when market choices match or accommodate his individual preferences. Hence, macro, global values beyond the consumer's individual focus have not been part of the consumer regulatory agenda in Europe and thus, until now, have not been considered.[40]

The role of consumers in European law, though, has been experiencing a shift of focus from a purely individual based perspective to a more collective and political role of consumers as drivers of social changes. The reform on consumer law that took place in 2018/2019 raised the focus of the role of the consumer in the market to go beyond the position of the consumer in the market and include a collective dimension of the consumer with a political agenda and social effects of preferences and decisions.[41] Consumers, as a group in the market, is the group in which the European legislator in some way trusts to successfully implement community policies on sustainability and the environment – the so-called New Deal.[42] Today, it is still too early to anticipate the impact and possible success or failure of incorporating the collective dimension of the consumer into European regulations.

The current regulatory landscape in European consumer law requires, on the one hand, revisiting consumer law norms so that they can be interpreted to contribute to achieving sustainable markets but, on the other hand, they also require the adoption of additional legislation mandating sustainability to be included in product-design and product disposability, in information requirements, in contractual design, in remedies for breach of contract as well as in the enforcement mechanisms available for breaching sustainability obligations and commitments.

The shift in consumer behaviour towards sustainable consumption requires information, awareness, and making sustainable choices easy to find and attractive to buy.[43] At the same time, obligations, sanctions, and effective enforcement mechanisms for providing misleading pre-contractual information regarding the sustainability of the product or service will be necessary.[44] European consumer protection is structured around three major steps that combine public and private law instruments: first, pre-contractual information duties, second, contractual control and third, contractual remedies for breach.[45] As of today, sustainability is not one of the core binding elements present in the applicable consumer law. However, a reinterpretation of these norms to include sustainability and in some cases, some legislative amendments could become a valuable – albeit insufficient – instrument to contribute achieving sustainable markets.

At the pre-contractual phase, once a product is placed on the market sellers are subject to pre-contractual information duties. Pre-contractual obligations are structured around mandated pre-contractual information disclosures to the consumer that, as of now, do not include binding obligations on sustainability. That is, producers are not required to inform of the environmental impact of the products they manufacture and sustainability is only included through the prohibition of general green claims.[46] However, the lack of an explicit obligation for producers and sellers does not mean that manufacturers and sellers do not ever inform consumers of the environmental impact of the products they produce or sell. Market forces, specifically reputation and a shift in consumer's preferences,[47] function as an incentive for producers to inform of their production processes when they are environmentally friendly. So today, we can't see information on sustainable products and sustainable processes such as "climate-friendly" or "eco" unless the product can prove its excellent environmental performance.[48]

Information and pre-contractual information duties currently only refer to the information that may be relevant to the consumer regarding the main characteristics of the product or service[49] which is the object of the transaction and that could have an effect on its economic decision of purchasing the product.[50] As of today, on the one hand, sellers must advertise[51] and provide consumers with accurate information about the product or service they consider purchasing in a way that reflects its characteristics, features, and functionalities.[52] There is no explicit obligation to provide information of any element with an impact beyond the parties strictly involved in the transaction. On the other hand, at least on a theoretical level, the product's environmental impact could be considered within the main characteristics of the product.[53] There is ample room to redefine what is considered the main characteristics of the product or service beyond their strict functionalities and involve the product's environmental impact or the externalities it generates.

Pre-contractual information duties also include omissions that could affect the consumer's decision. Again, omissions on products' environmental impact could also be considered misleading.[54] If that was the case, failing to inform, for example, on the durability or the repairability of the product could eventually be qualified as a misleading omission.[55]

In sum, the main basis of pre-contractual norms regarding misleading advertisement and misleading commercial practices relies on ensuring the consumer's economic behaviour is not affected by misleading advertisement or by untruthful information that could distort the consumer's economic behaviour by making them take economic decisions they would not have taken otherwise.[56] However, would it be possible to revisit the scope and definition of misleading advertisements so that it could be considered misleading not to explain to consumers the environmental impact of the product they are about to buy? Could it be considered an unfair omission under the unfair commercial practices directive not disclosing the environmental impact of the production process of the product, its transportation, or its distribution when that matters for consumer choice? Today, the EU legislator is working on two important normative instruments that aim at re-shaping pre-contractual information obligations to include information on environmental risks and sustainability. First, the Greenwashing Directive,[57] endorsed by the European Parliament last January 2024, that aims at ensuring that traders provide clear, relevant and reliable information regarding the environmental impact of the products they supply and second, the Green claims directive,[58] proposed on March 2023, that will require suppliers to substantiate their voluntary green claims by securing the truthfulness and compliance with their claims from an independent verifier.

Our perspective on sustainability has evolved and today consumers increasingly care about the environmental impact of the products and services they consider buying.[59] Sustainability today is a variable part of the consumer's preferences. Consequently, requiring informing consumers about elements they care for and that could influence their markets decisions would be welfare enhancing.[60]

At the contractual level, consumer protection is articulated through product warranties[61] and the control of unfair standard contract terms in consumer contracts.[62] The ultimate goal of the contractual control is that informed consumers are able to obtain the contract surplus they expected to get from the contracts they entered into.[63]

The ex post contractual control laid down by Directive 93/13 is applicable to standard contract terms included in business to consumer contracts.[64] Hence, business to business contracts are left out from its scope of application. In order to be subject to the control laid down by the Directive, contract terms have to be non-negotiated.[65] If individually negotiated by the parties, the terms are not considered standard terms subject to the Directive's provisions. Directive 93/13 looks at non-negotiated contract terms both as to their transparency and fairness. For the purposes of the tests to be applied on both dimensions, the Directive considers as a relevant factor whether the standard term refers to the definition of the main subject matter of the contract or not.[66] The combination of these two dimensions included in the Directive and developed by the Court of Justice of the European Union (CJEU) involves different legal outcomes and implications. Directive 93/13 establishes two layers of contractual control. A first layer, applicable to all standard contract terms, regardless of their role within the contract refers to formal transparency. The underlying idea of the formal transparency control is to ensure that consumer contracts are drafted in a plain and intelligible language so that the individual consumer is aware of the meaning of the content of the standard terms.[67] Given the case law of the CJEU,[68] it can be argued that the additional controls may touch two different dimensions. First, the level or degree of awareness by the consumer of the impact and influence of the term on the overall effects of the contract for the consumer. Second, it also refers squarely to the "content" of the term and the relative rights and obligations that the term generates for the parties. The type, focus, and legal consequences of such controls depends, in the EU overall legal design, on the role of the standard term within the contract. If the contract term would refer to the definition of the essential elements of the contract, the additional control will seem to be exclusively content-based, and focus on whether the substantive content of the term creates an imbalance[69] between the contracting parties to the detriment of the consumer and against the requirements of good faith.[70] If, instead, the contract term would not define an essential element of the contract the Directive 93/13 provides for an assessment of unfairness.[71] Article 3 of Directive 93/13 (in what constitutes the general unfairness test or quality control over content of the EU regime) provides that a non-negotiated contract term will be considered unfair if, contrary to good faith, it causes a significant imbalance in the parties' rights and obligations arising under the contract, to the detriment of the consumer.[72]

The ex post contractual controls, of general application in business to consumer contracts could also be an instrument for sustainability. If sustainability could be considered an essential element of the contract it could then be subject to a content-based control and be considered unfair if non-compliance with contractual obligations on sustainability would create an imbalance between the contracting parties to the detriment of the consumer – considering the higher environmental impact of the product compared to the contractual provisions. If contractual obligations on sustainability would not constitute an essential element of the contract unfairness controls could be applicable and contract terms could be unfair if they would create a significant imbalance in the contracting parties to the detriment of the consumer.

As explained above, contractual obligations on sustainability have a limited role in consumer contracts and when present, their enforcement, as of today, is relatively weak. There is room to improve the controls and enforcement of contractual obligations regarding sustainability, but the design of Directive 93/13 does not seem the best instrument to make them most effective.

In 2018/2019 the European legislature undertook a major reform of consumer protection[73] regulation to modernise it with respect to two different dimensions. The first was the need to standardise consumer contractual protection regardless of the nature of the object of the contract – with or without digital elements – and the face-to-face or virtual environment of the contract. The second was to modernise the rights of consumers in sales contracts, essentially represented by Directive 2019/771 on certain aspects of contracts for the sale of goods[74] and by Directive 2019/2161 on the sale of goods.[75] Directive 2019/771 did not overrule the regulation of the sale of goods established by Directive 1999/44 but regulated the aspects related to the conformity of goods sold,[76] the rights of consumers due to lack of conformity and the commercial warrantees on the objects of contracts. Directive 2019/771 fully harmonised[77] some essential aspects of consumer sales while keeping the general regime in the hands of the member states.[78] The main focus of Directive 2019/771 was strengthening consumer protection and the internal market, particularly considering the growth of digital markets. This reform/modernisation has focused very little, if any, on sustainable consumption.[79]

The modernisation of consumer protection regulation, the second objective of the 2018/2019 reform, has been articulated through Directive 2019/2161, which has modified different directives on consumer protection[80] and pursues two major goals. First, the transformation and harmonisation of the sanctioning regime applicable to infringements of consumer protection regulations and, second, the strengthening of consumers' right of information, which translates into greater transparency duties for sellers or service providers before the transaction takes place. It also incorporates mechanisms to provide greater legal certainty, coherence, and clarity in the consumer protection system.

With respect to the right of transparency and information, Directive 2019/2161 updates and expands the obligations provided for in the Unfair Commercial Practices Directive, which considers deceptive practices that substantially limit the information relevant to make an informed decision about a transaction for the average consumer.[81] Directive 2019/2161 broadens the scope of the concept of substantial information in relation to relevant elements of the transaction and requires that this information be provided by the seller and/or the digital platform, so an omission is considered a misleading practice.[82]

There is ample room for consumer law to become a central instrument for sustainability. As of today, sustainability is neither a fundamental part of the pre-contractual obligations' sellers face with respect to consumers nor it is clear whether a violation of contractual obligations on sustainability would constitute a variable for assessing unfairness of standard contract terms in consumer contracts.[83]

European consumer protection legislation continues with the same approach it has had in the last 40 years: placing consumers at the centre and ensuring their consumption decisions are sovereign and reflect their preferences while at the same time their expectations about the transaction are fulfilled. As of today, sustainability is not one of the fundamental values or driving forces of European consumer law.[84]

If we want European consumer law to become one of the bases for sustainable markets the European legislator should adopt a broader and more global perspective that combines different normative instruments with different legal natures – public and private law – and look beyond the current focus on pre-contractual information[85] and on contractual warranties and ex post contractual controls to embrace sustainability goals.

The next section illustrates the European contract-based approach under which sustainability is articulated through limiting the consumer's choice of contractual remedies to the ones considered to impose lower environmental costs.

9.4 The Proposal for Sustainable Consumption through Contract Law. The Right to Repair: A Missed Opportunity?

A nice illustration of the partial and limited contractual approach adopted in the European Union in order to achieve sustainable markets is reflected in the recent proposal for a directive on common rules promoting the repair of goods that was presented last March 2023.[86]

This proposal is part of the legal framework for achieving sustainable markets within the objectives of the European Green Deal[87] that provides, as one of its goals, the reduction of waste. The proposal on the right to repair is part of the New Consumer Agenda[88] and the Circular Economy Action Plan[89] and it is based on the idea that repair is a remedy that can contribute to achieving more sustainable markets because it reduces the amount of waste generated. The underlying idea is that even though it is a remedy available to consumers it has not been the preferred one because of its inconvenience, the lack of transparency regarding its availability as well as the difficulty in having repair services readily available. One of the driving forces of the proposal is the assumption that if consumers are aware of the availability of products that can be easily repaired, they will choose those products thus contributing to the sustainable market through their consumption decisions.

Currently, under the consumer sales Directive,[90] consumers had three available remedies when confronted with non-conforming goods: repair, replace, and ultimately, refund. Under the consumer sales Directive these remedies had a hierarchy under which the consumer does not have a free choice between repair, replace, and refund of the non-conforming product but first needs to accept the seller's choice of repairing or replacing the product.[91] Consumers can choose whether they want the product to be repaired or whether they would rather, if possible, have it replaced and/or this does not impose a disproportionate burden on sellers.[92] In cases where repair or replacement was not available or the seller did not complete it within a reasonable time, the consumer would have the right to demand a refund.[93] Once again, sustainability is not factored in this norm when structuring remedies for non-conformity that only consider the expected private cost of these remedies for the seller and neglect the social costs imposed by these remedies in terms of, for example, sustainability.

Replacement has often been preferred over repair given that in many cases that has been more attractive to consumers that could have a new product for free[94] while it has not imposed excessive burdens on sellers given that considering products with global mass production chains such as, for example, electronics or house cleaning products like vacuum cleaners, the costs of repair have often been similar, if not higher, than replacing them.

Given that the negative externalities generated by the production and sale of these products were not internalised by their production distribution and sales processes, the private costs of replacement were often equivalent if not lower than the cost of repair. From a purely private perspective, the consumers' choice of replacement over repair was welfare enhancing. Consumers often preferred the remedy that provided a brand-new product while sellers might have been indifferent, if not better off, when replacing the product instead of repairing it. The result of replacement was often a Pareto optimal outcome. However, this result does not consider the social costs imposed by this model. The lack of internalisation of the negative externality in terms of sustainability allowed for a privately welfare enhancing solution at the expense of the costs of sustainability born by society as a whole. We are at a point in history where sustainability costs undoubtedly need to be internalised and this choice purely based on private considerations should not be available.

Sustainable consumption is one of the areas the European Green Deal[95] places an emphasis on.[96] However, as explained above, in order to have sustainable markets it is necessary that products are produced in a sustainable way as well as consumers need to choose sustainable products. A proposal to promote product repairs as the primary remedy for non-conformity has been presented in May 2023 with the purpose of using European consumer private law as an instrument to contribute to sustainable markets.[97] The core of the proposed directive is the reduction of waste through the enlargement of the life of the product by promoting product repair. The scheme of the proposed directive aims at changing the relative prices of replacing and repair so that repair becomes relatively more attractive to consumers while still being economically effective to sellers.[98]

In order to be entitled to the repair of the product, the proposed directive differentiates between the period within the legal guarantee and the period beyond the legal guarantee. In the first case, when the product is non-conforming within the first two years after a purchase, repair will be the preferred remedy sellers will have to offer as long as it is not more expensive than replacement. After the period of the legal warranty, that is, more than two years after purchase, the proposed directive includes a set of mechanisms to make repair a more attractive and easy remedy but not the preferred one. First, the proposed directive gives consumers the right to claim repair – as long as the product can be repaired. The threshold of producers for rejecting product repairs is higher under the proposed directive then under the Directive 2019/771 on consumer sales. Under the proposed directive producers are required to repair the product upon the consumer's request unless repair is impossible. Consequently, the cost of the product repair, compared to the cost of the alternative remedies for non-conformity, is not the relevant variable to determine the producer's obligation for repairing the product. The relevant variable to determine the producer's obligation is whether repair is possible. Only when repair is impossible the producer will be exempted from the obligation. It should be noted that in some cases repair will not fall on the producer but will fall on the consumer. Sometimes it will be for the consumer to repair the product.[99] In these cases, producers have an obligation to inform consumers that they will have to repair the product themselves. In order to ensure more transparency and more information, and make repair an easier remedy for consumers, the proposed directive provides for the creation of an online matchmaking repair platform that will connect consumers with professionals that eventually could repair their products. Additionally, the conditions, price, and terms of the repair will have to be provided to the consumer. The proposed directive provides for the creation of a European repair information form that professionals will have to provide to consumers so that they can compare the terms of the repair they have been offered.

In order to make the right to repair an effective remedy, the proposed directive gives the right to consumers to request repair regardless of the location of the producer.[100] The proposed directive seems to intend to have extraterritorial effect so that the European Union contributes to sustainable markets beyond the European boundaries: repair, as a remedy available to consumers sold in the European Union market will affect products produced outside of the European Union borders. From this perspective European authorities hope that the proposed directive will have extraterritorial effects in encouraging the production of repairable products anywhere in the world.

The proposed directive, despite its good intentions, presents some questions regarding its effectiveness that hopefully will be amended through the legislative process.

9.4.1 Limiting the Right to Repair through Broad Exceptions

Article 5 of the proposed directive is well intended but contains important loopholes. First, the obligation to repair is only available for products the repairability of which is listed in Annex II of the proposal. At the same time, the right to repair may not necessarily be free for the consumer and ultimately the producer may reject to repair the product as long as it can claim repair is impossible. The limited scope of application of the right to repair and the potentially broad scope of the exceptions suggests a limited effectiveness of the right to repair as drafted in the proposal.

9.4.2 No New Warranty for Repaired Products

As explained earlier consumers have the right to repair their product within warranty – unless repair is more expensive than replacement – and can request their product to be repaired beyond the warranty period. This structure, providing a different set of rights within and beyond the warranty, is similar to the normative scheme established in the different directives regulating the sale of goods.[101]

The proposed directive, however, is silent regarding a potential warranty period after the product has been repaired.[102] The proposed directive is cautious in that respect.

However, the lack of commitment to a minimum period of product life after repair might remarkably dilute the incentives consumers could have when requesting a product repair. Consumers may face the risks of a short-lived product after repair as well as of the costs of repairing the product without any certainty on the lifespan of this investment. These two elements have a direct impact on consumers' incentives to exercise their right to repair.

9.4.3 The Endless Faith in Information of the European Private Law Model

The proposed directive relies on providing consumers with pre-contractual information on the right to repair as well as on giving consumers a European repair information form that producers will have to issue before the consumers enter into a contract for repair services.[103]

This structure is consistent with the traditional view of European private law, specifically European consumer law. The European model of consumer law relies on providing information to the consumer assuming that this information will be (1) read, (2) understood, and (3) ultimately used by the consumer when adopting their decisions. However, today ample empirical evidence shows that consumers most often do not read the information provided, and when they do, they do not understand, and ultimately, they do not care about the amount of information they are provided with at a pre-contractual and at a contractual stage of the transaction.[104] Consumers, for the most part, do not understand or read the contracts they are presented with and ultimately enter into. It is widely accepted that private autonomy and pure freedom of contract are not possible in the context of consumer contracts.

This information-based model is further illustrated by the European repair information form producers have to provide before the consumer enters into a repair service contract. Does this mean that producers, regardless of their size and location, will have to contact consumers to provide a form in the consumer's language informing of these specificities regarding the repair of the product? The requirement to issue this form involves a remarkable additional cost to the right to repair. Will this cost be borne by producers or by consumers? Even though the proposal is silent on the allocation of this cost it seems reasonable to expect that if the right to repair does not necessarily have to be for free, the cost of the information and of the form will ultimately be borne by consumers when requesting product repair. Despite the efforts of the proposed directive in making the right to repair easier and cheaper that cost might introduce a barrier to the exercise of this right to consumers that could limit the effectiveness of the right to repair.

9.4.4 Making Member States Co-Responsible for Implementing and Enforcing the Consumer's Right to Repair?

Member States have a limited role in developing the conceptual framework established by the proposed directive. The European Commission does not seem to rely on Member States in terms of enhancing the obligations provided under the proposed directive and in terms of providing content to some of the obligations established by the proposed directive.

The proposed directive is a maximum harmonisation directive. This approach might show that the European Commission aims at ensuring that rights provided under the proposed directive will be given full content while it might not trust the Member States to have incentives to set higher standards and additional rights and obligations than those provided by the proposed directive. However, some Member States have already adopted measures that go beyond those provided by the proposed directive[105] and could have incentives to do so in the future once other directives are in force. Achieving more sustainable markets is an incredibly complex task that ideally should actively involve Member States as well as subnational authorities so that they join forces to achieve this goal.

At the same time, the proposed directive includes a generic obligation for Member States to provide ideas on how the information about the right to repair should be provided to consumers. The obligation, however, relates only on how this information needs to be provided but not on the content of this information and more specifically, to the scope of the obligation to repair. That is, on the specificities of the reparability of the product and eventually of its limits. The role of the Member States is limited to a formal and documental requirement that forgets the substance of the obligation of the right to repair which is to increase consumer's awareness of the content of this right and the effects of repair the product's use and life. This could ultimately condition the effectiveness of the right to repair given that Member States ultimately have to be the ones encouraging and enforcing the full content and scope of this right.

9.4.5 Focusing Only on a Contractual Remedy in Consumer Contracts: A Limited and Narrow View on Sustainable Markets

The proposed directive on the right to repair is coherent and fits perfectly with the approach of the European model of consumer protection that relies on consumer contracts, particularly on consumer information as the mechanism to ensure consumer sovereignty in markets.

One of the first issues that arises with the proposed directive is whether the right to repair is an instrument to achieve sustainable markets. The right to repair reflects a partial, narrow, and limited view of sustainability with a potentially limited impact.

First, because it assumes that enlarging the life of a product contributes to market sustainability. The proposed directive implicitly assumes that enlarging the life of the product and hence reducing waste will result in more sustainable markets. The first part of the statement, might be true: enlarging the life of consumer products might reduce waste. If consumers continue to use the products regardless of how old they are and hence do not throw them away as long as they work, the amount of waste might reasonably be reduced. The implication of this measure, though, might not be having more sustainable markets necessarily. Repairing products, the production of which might have had a high negative environmental impact might not necessarily be the best solution in terms of sustainable markets even if repair reduces the amount of waste generated.

Second, the right to repair relies on a contractual instrument. However, sustainability has implications well beyond the consumer contract. The right to repair, as configured in the proposed directive, is an autonomous remedy that only looks at the object and at the contract regardless of any other instruments it could be combined with. As explained above, sustainability requires a comprehensive global perspective.[106] Product repair is a good example. Should we repair products that use high amounts of energy while they are being used? Should the right to repair be only available to products that are efficient from an energy consumption perspective? Should we adopt measures in cases where repair was available and instead the consumer decided to buy the same product again? Should we reward repairing products? The proposed directive includes the right to repair as a horizontal and general remedy that comes from the nature of the product and for the consumer contract. The right to repair is assumed to be generally beneficial regardless of any consideration beyond the consumer contract.

The literature has broadly discussed and shown that purely contractual remedies are highly ineffective in consumer contracts.[107] The proposed directive is a nice example of the limited view of the instruments used in order to achieve sustainable markets: the directive focuses on the post-contractual phase and aims at maximising the use of product repair as a cornerstone of sustainable consumption and hence of sustainable markets. The focus on this important but narrow instrument may limit its effectiveness. The proposed directive still includes ample room for exceptions to the obligation to repair and for making consumers ultimately pay for the cost of exercising their right.

The adoption of the proposed directive on the right to repair may be well intended but sustainable markets are a complex phenomenon with so many implications beyond the consumer contract that focusing on a specific remedy alone may not achieve the result it aims at.[108]

9.5 Final Thoughts

Sustainability requires a more comprehensive approach through the combination of instruments, namely public law rules with private law instruments involving pre-contractual information duties, contractual obligations as well as post-contractual actions. This perspective would reflect the complex structure of sustainable markets ranging pre-market obligations requiring sustainable production processes and sustainable supply chains to sustainable consumer choices, and consumer rights.

Achieving sustainable markets requires analysing the negative externalities generated at each phase of the life of the product while providing obligations to internalise these costs through combining the different legal instruments available for regulating markets. The European approach, placing consumers at the centre, is still based on traditional pre-contractual information requirements and ex post remedies for breach. A global phenomenon requires a comprehensive normative approach that hopefully we will see European authorities to adopt in the coming future.

Notes

1 This contribution has received the support of Protecting consumers in digital platforms PID2021-127258NA-I00/AEI/10.13039/501100011033, UE, Spanish Ministry of Science and Innovation, Spanish Research Agency and the European Regional Development Fund, RYC2018-026249-I Spanish Research Agency and PW2021-PR02, Greening consumer welfare: towards a new perspective on consumer protection (UPF).

2 *The Agenda 2030 for Sustainable Development* <https://www.un.org/sustainabledevelopment/development-agenda/> accessed 22 August 2023.
3 Resolution adopted by the General Assembly A/RES/70/1 (25 September 2015) <https://documents-dds-ny.un.org/doc/UNDOC/GEN/N15/291/89/PDF/N1529189.pdf> accessed 22 August 2023.
4 UN, 'Guidelines for Consumer Protection' <https://unctad.org/system/files/official-document/ditccplpmisc2016d1_en.pdf> accessed 8 August 2023.
5 United Nations Environment Programme, 'Guidelines for Providing Product Sustainability Information: Global Guidance on Making Effective Environmental, Social and Economic Claims, to Empower and Enable Consumer Choice' <https://wedocs.unep.org/20.500.11822/22180> accessed 8 August 2023.
6 Private law as such is not part of the exclusive competences of the EU included in Article 3, or among the supportive competences of the EU of Article 6 Consolidated version of the Treaty on the Functioning of the European Union [2012] OJ C326/12 (hereinafter TFEU). Private law falls within the regulatory powers the EU has over certain competences shared with the Member States (article 4(2) TFEU) regarding internal market, consumer protection, and the area of freedom, security and justice. At the same time, articles 114 and 115 TFEU confer the EU with powers to harmonise rules regarding the establishment and functioning of the internal market. This harmonisation powers clearly include private law rules that might impact the functioning of the market.
7 European Commission, 'Communication on a new circular economy action plan for a cleaner and more competitive Europe' *(European Commission,* 11 March 2020) <https://eur-lex.europa.eu/legal-content/EN/TXT/?qid=1583933814386&uri=COM:2020:98:FIN> accessed 8 August 2023.
8 European Parliament, 'Report on a longer lifetime for products: benefits for consumers and companies' (*European Parliament,* 9 June 2017) <https://www.europarl.europa.eu/doceo/document/A-8-2017-0214_EN.html> accessed 10 August 2023.
9 European Parliament, 'Resolution of 31 May 2018 on the implementation of the ecodesign directive' (*European Parliament*, 9 March 2020) <https://eur-lex.europa.eu/legal-content/EN/TXT/?uri=OJ:JOC_2020_076_R_0023> accessed 10 August 2023.
10 *Communication on the European Green Deal* <https://eur-lex.europa.eu/resource.html?uri=cellar:b828d165-1c22-11ea-8c1f-01aa75ed71a1.0002.02/DOC_1&format=PDF> accessed 10 August 2023.
11 A New Circular Economy Action Plan for a Cleaner and More Competitive Europe, COM(2020)98.
12 BEUC, 'Restoring consumer confidence in green claims' <https://www.beuc.eu/sites/default/files/publications/beuc-x-2020-116_getting_rid_of_green_washing.pdf> accessed 10 August 2023.
13 See also Hans-Wolfgang Micklitz, 'Squaring the circle? Reconciling consumer law and the circular economy' [2019] 8 *EuCML* 229, 230–231.
14 Hal Varian, *Intermediate Microeconomics: A modern approach* (9th edn, New York, W. W. Norton & Company 2014) 666.
15 In 2021 the European Commission conducted a special Eurobarometer that showed that 93% of respondents considered climate change to be a serious problem – 78% of them considered it to be very serious. See *Special Eurobarometer 513 Climate Change* <https://climate.ec.europa.eu/system/files/2021-07/report_summary_2021_en.pdf> accessed 12 May 2023. See Douglas Kysar, 'Preferences for processes: the process/product distinction and the regulation of consumer choice' [2004] 118 *Harvard Law Review* 525, discussing the normative relevance of the distinction between products and production processes and arguing that the consumer's choice is formed by both elements.
16 See Vanessa Mak and Evelyn Terryn, 'Circular economy and consumer protection: the consumer as a citizen and the limits of empowerment through consumer law' [2020] 43 *Journal of Consumer Policy* 227–248 (https://doi.org/10.1007/s10603-019-09435-y).
17 See Bert Keirsbilck et al, 'Sustainable consumption and consumer protection legislation: how can sustainable consumption and longer lifetime of products be promoted through consumer protection legislation?' (*Policy Department for Economic, Scientific and Quality of Life Policites. Directorate-General for Internal Policies*, 2020) <https://data.europa.eu/doi/10.2861/173816> accessed 12 May 2023.
18 Vanessa Mak and Evelyn Terryn, 'Circular economy and consumer protection: the consumer as a citizen and the limits of empowerment through consumer law' [2020] 43 *Journal of Consumer Policy* 227 (https://doi.org/10.1007/s10603-019-09435-y).
19 See Fabrizio Cafaggi, 'Regulation through contracts: supply-chain contracting and sustainability standards' [2016] 12(3) *European Review of Contract Law* 218 (https://doi.org/10.1515/ercl-2016-0013), discussing the role of contracts – business to business contracts – in achieving sustainability in the context of supply chains.

20 Tamim Bayoumi, Jelle Barkema and Diego Cerdeiro, 'The inflexible structure of global supply chains' (2019) IMF Working Papers 1 <https://www.imf.org/en/Publications/WP/Issues/2019/09/13/The-Inflexible-Structure-of-Global-Supply-Chains-48562>, showing that the structure of global supply chains is remarkably rigid in the short and in the long turn.

21 See Fabrizio Cafaggi, 'Regulation through contracts: supply-chain contracting and sustainability standards' [2016] 12(3) *European Review of Contract Law* 218 (https://doi.org/10.1515/ercl-2016-0013) noting that contracts between the different participants in supply chain can have a role for introducing sustainability as a contractual obligation and enforcing it.

22 Silvia Di Ponte et al, 'Environmental impacts of global supply chain: a review of scientific, policy and legal components for including environmental and climate challenges within the international trade patterns' [2013] *JCR Science for Policy Report. European Commission* <https://data.europa.eu/doi/10.2788/799> accessed 26 April 2023. See also Fabrizio Cafaggi and Paola Iamiceli, 'The limits of contract laws. The control of contractual power in trade practices and the preservation of freedom of contract within agri-food global supply chains' (2022) SSRN Working Paper <https://ssrn.com/abstract=4048571> accessed 11 September 2023, arguing that contracts present limits to achieve sustainable supply chains in the context of the agri-food industry. The arguments presented in the paper are applicable to more general contexts of non agri-food supply chains.

23 Erik Hermann, 'Artificial intelligence in marketing: friend or foe of sustainable consumption?' [2021] *AI & Soc* (https://doi.org/10.1007/s00146-021-01227-8), explaining that artificial intelligence can be a double-edged sword in markets given that it can lead to increased consumption and its (negative) externalities while also promoting supply and demand side sustainability efforts.

24 See Robert Evan Sanders, 'Dynamic pricing and organic waste bans: a study of grocery retailers' incentives to reduce food waste' (2020) SSRN Working Paper <https://ssrn.com/abstract=2994426 > showing that dynamic pricing has a welfare enhancing effect larger than introducing waste bans. Consequently, artificial intelligence and dynamic pricing models may also have a positive impact on sustainability.

25 See Rocco Caferra et al, 'Consumer analysis and the role of information in sustainable choices: a natural experiment' [2022] 1 *Frontiers in Environmental Economics* 1044206 (https://doi.org/10.3389/frevc.2022.1044206), noting that steering towards sustainable markets requires consumers' involvement and arguing for using visual information for raising awareness of and for increasing sustainable consumption choices.

26 For example, energy-intensive industries such as chemicals, steel, cement, aluminium, non-ferrous metals, pulp and paper, or ceramics accounted for 17% of the EU's total greenhouse gas emissions in 2019. These industries are crucial for the European economy but have a remarkable environmental impact that directly affects the quality of life of citizens (see <https://research-and-innovation.ec.europa.eu/research-area/industrial-research-and-innovation/sustainable-production-processes_en> accessed 12 May 2023).

27 See Fabrizio Cafaggi, 'Regulation through contracts: supply-chain contracting and sustainability standards' [2016] 12(3) *European Review of Contract Law* 218 (https://doi.org/10.1515/ercl-2016-0013).

28 The bottom-up approach is not new in European consumer protection regulation. European Product Safety regulation already remarkably relies on self-certifications where producers, importers or distributors are the ones certifying that the products they manufacture comply with the safety requirements – product-specific or general – established by the European Product Safety regulation. See, for example, the Regulation (EC) 765/2008 of July 2008 setting out the requirements for accreditation and market surveillance relating to the marketing of products and repealing Regulation No 339/93 [2008] OJ L218/30.

29 See Directive (EC) 2009/125 establishing a framework for the setting of ecodesign requirements for energy-related products (recast) [2009] OJ L285/10.

30 See Regulation (EU) 2017/1369 setting a framework for energy labelling and repealing Directive 2010/30/EU [2017] OJ L198/1.

31 See Ecodesign working plan 2016–2019, COM(2016) 773 final.

32 Proposal for a Directive on Corporate Sustainability Due Diligence and amending Directive (EU) 2019/1937 [2022].

33 Small and medium enterprises are not subject to the duties under the corporate sustainability due diligence directive. Considering the structure of corporations in the European Union market, where 99% of the corporations are small and medium enterprises as defined by Recommendation 2003/261 concerning

the definition of micro-, small-, and medium-sized enterprises [2003] OJ L124/36, the expected impact of the proposed corporate sustainability due diligence directive might be remarkably limited.

34 The Council eliminated the Commission's original proposal that provided for the due diligence to be within the scope of the director's duty of care and provided for a variable remuneration on directors' contribution to sustainability. See the Proposal for a Directive on Corporate Sustainability Due Diligence and amending Directive (EU) 2019/1937, COM/2022/71 final.

35 Member States will be responsible for supervising compliance with the duties included in the due diligence proposed directive and for imposing fines in case of non-compliance with them. See the Proposal for a Directive on Corporate Sustainability Due Diligence and amending Directive (EU) 2019/1937, COM/2022/71 final.

36 See Directive (EU) 2015/1535 laying down a procedure for the provision of information in the field of technical regulations and of rules on Information Society services (codification) [2015] OJ L241/1.

37 Directive (EC) 2001/95 on general product safety [2002] OJ L11/4.

38 See BEUC, Getting rid of green washing: restoring consumer confidence in green claims (2020) <https://www.beuc.eu/position-papers/getting-rid-green-washing-restoring-consumer-confidence-green-claims> accessed 3 May 2023, arguing for establishing a pre-approval scheme for all green claims by an EU authority.

39 Article 169 of the TFEU.

40 Maurie J Cohen, *The Future of Consumer Society: Prospects for Sustainability in the New Ec*onomy (Oxford, Oxford University Press 2017).

41 Mateusz Grochowski, 'European consumer law after the new deal: a tryptich' [2020] 39 *Yearbook of European Law* 387 (https://doi.org/10.1093/yel/yeaa016).

42 Commission, 'Communication from the Commission to the European Parliament, the Council and the European Economic and Social Committee: A New Deal for Consumers' COM/2018/0183 final (hereinafter: 'New Deal').

43 See Vanessa Mak and Evelyn Terryn, 'Circular economy and consumer protection: the consumer as a citizen and the limits of empowerment through consumer law' [2020] 43 *Journal of Consumer Policy* 227, 231 (https://doi.org/10.1007/s10603-019-09435-y).

44 Today, the level of enforcement of green claims if very weak because of the lack of uniform methodology for assessing them. See Vanessa Mak and Evelyn Terryn, 'Circular economy and consumer protection: the consumer as a citizen and the limits of empowerment through consumer law' [2020] 43 *Journal of Consumer Policy* 227, 231 (https://doi.org/10.1007/s10603-019-09435-y).

45 Additionally, consumers are also entitled to claim non-contractual compensation for damages suffered. See the Council Directive 85/374/EEC on the approximation of the laws, regulations and administrative provisions of the Member States concerning liability for defective products [1985] OJ L210/29.

46 See BEUC, 'Getting rid of green washing: restoring consumer confidence in green claims' (2020) <https://www.beuc.eu/position-papers/getting-rid-green-washing-restoring-consumer-confidence-green-claims> accessed 3 May 2023.

47 Douglas Kysar, 'Preferences for processes: the process/product distinction and the regulation of consumer choice' [2004] 118 *Harvard Law Review* 525

48 Directive (EC) 2005/29 concerning unfair business-to-consumer commercial practices in the internal market ('Unfair Commercial Practices Directive') [2005] OJ L 149/22.

49 Directive (EC) 2005/29 concerning unfair business-to-consumer commercial practices in the internal market ('Unfair Commercial Practices Directive') [2005] OJ L 149/22, art 7(4).

50 Directive (EC) 2005/29 concerning unfair business-to-consumer commercial practices in the internal market ('Unfair Commercial Practices Directive') [2005] OJ L 149/22, art 5.

51 Directive (EC) 2006/114 concerning misleading and comparative advertising [2006] OJ L376/21.

52 Directive (EC) 2005/29/EC concerning unfair business-to-consumer commercial practices in the internal market ('Unfair Commercial Practices Directive') [2005] OJ L149/22.

53 See Directive (EC) 2001/83 arts 5(1)(a) and 6(a)(1), the Consumer Rights Directive [2011] OJ L304/64 requiring to provide consumers with the information on all product characteristics.

54 See Directive (EC) 2005/29 concerning unfair business-to-consumer commercial practices in the internal market ('Unfair Commercial Practices Directive') [2005] OJ L 149/22, art 7.

55 See Guidance on the implementation/application of the Directive, COM(2016) 320 final

56 See Directive (EC) 2005/29 concerning unfair business-to-consumer commercial practices in the internal market ('Unfair Commercial Practices Directive') [2005] OJ L 149/22, art 2(e).

57 Proposal for a Directive amending Directives 2005/29/EC and 2011/83/EU as regards empowering consumers for the green transition through better protection against unfair practices and better information COM/2022/143 final.

58 As of March 5th 2024, the text of the proposal is not available yet. See https://www.europarl.europa.eu/thinktank/en/document/EPRS_BRI(2023)753958#:~:text=The%20proposed%20directive%20would%20require,hands%20of%20the%20co%2Dlegislators.

59 Eurobarometer 2021 showed that 77% of Europeans want to react and limit climate change <https://europa.eu/eurobarometer/surveys/detail/2672> accessed 3 May 2023

60 See Geraint Howells, 'The potential and limits of consumer empowerment by information' (2005) 32 *Journal of Law and Society* 349 (https://doi.org/10.1111/j.1467-6478.2005.00328.x), noting on the limited role of information in turning consumer's decisions into informed decisions.

61 See Directive (EU) 2019/771 on certain aspects concerning contracts for the sale of goods [2019] OJ L136/28.

62 Council Directive 93/13/EEC on unfair terms in consumer contracts [1993] OJ L95/29. See also See Nils Jansen, 'Unfair contract terms' in Nils Jansen and Reinhard Zimmermann (eds), *Commentaries on European Contract Laws* (Oxford, Oxford University Press 2018) 926–927 noting the inherent limits of harmonisation of standard contract terms in Europe.

63 See Case C-105/17 *Komisia za zashtita na potrebitelite v Evelina Kamenova* [2018], where the CJEU established that the qualification of a contractual party as a professional should be done on a case-by-case basis and in contrast to the position of consumers that have incomplete information and without capacity to negotiate the contract terms.

64 Hence, business to business contracts are left out from its scope of application.

65 For instance, Spanish Supreme Court Judgment, Civil Chamber (9 March 2017), 788/2017. See also, as to other Spanish courts, Provincial Court of Asturias (21 December 2016), 2017/14758 where the court held unfair a floor clause in a mortgage contract and hence void but completely valid the same clause in a subsequent mortgage contract that was a novation of the initial mortgage contract because such clause had been individually negotiated together with other contract clauses as part of the negotiation of the contract novation. See Karolina Lyczkowska, 'Una de cal y otra de arena: La AP de Asturias confirma la retroactividad total de la cláusula suelo nula, pero califica como válida la que fue fruto de una novación contractual' [2017] *Centro de Estudios de consumo*.

66 See Council Directive 93/13/EEC of 5 April 1993 on unfair terms in consumer contracts [1993] OJ L95/29. See also Spanish Supreme Court Judgment, Civil Chamber (9 March 2017), 788/2017.

67 See Preamble and article 5 of Council Directive 93/13/EEC of 5 April 1993 on unfair terms in consumer contracts [1993] OJ L95/29. The ECJ has interpreted that a standard term is drafted in plain and intelligible language as long as it was comprehensible without legal advice by an "average consumer". See Case C-26/13 *Árpád Kásler v. OTP Jezálogbank Zrt* "[…]*the average consumer* [should be] *in a position to evaluate, on the basis of clear, intelligible criteria, the economic consequences for him which derive from it* [the standard term]".

68 That this second layer covers two dimensions, substantive transparency, on the one hand, and the content of the terms, on the other, is not at all explicit in the text of the Directive, and it was not that clear in older case law of the ECJ: Case C-144/99 *Commission v. the Kingdom of the Netherlands*.

69 See Nils Jansen, 'Unfair contract terms' in Nils Jansen and Reinhard Zimmermann (eds), *Commentaries on European Contract Laws* (Oxford, Oxford University Press 2018) 944 suggesting that the imbalance criterion relates to how far the standard contract term deviates from the standards of fairness of the relevant contract law.

70 See Nils Jansen, 'Unfair contract terms' in Nils Jansen and Reinhard Zimmermann (eds), *Commentaries on European Contract Laws* (Oxford, Oxford University Press 2018) 944 arguing that the good faith requirement refers to the question of whether the party providing the contract term could expect the other party to agree to it. Jansen notes that the concept of good faith introduces in European law elements of hypothetical consent, implied conditions and economic considerations (contract risks would be allocated to the party who could prevent them or minimise their impact in a cheaper way).

71 See Council Directive 93/13/EEC of 5 April 1993 on unfair terms in consumer contracts [1993] OJ L95/29, article 4(2) providing: "Assessment of the unfair nature of the terms shall relate neither to the definition of the main subject matter of the contract nor to the adequacy of the price and remuneration, on the one hand, as against the services or goods supplies in exchange, on the other, in so far as these terms are in plain intelligible language".

72 Council Directive 93/13/EEC of 5 April 1993 on unfair terms in consumer contracts [1993] OJ L95/29, art 3. See also ECJ Case C-226/12 *Constructora Principado SA v José Ignacio Menéndez Álvarez* [2014] whereby the European Court of Justice notes that the concept of imbalance should not refer to the relationship between the costs to be borne by the consumer and the overall costs involved in the transaction, but should refer to a restriction of rights, an obstacle for their exercise or the imposition of additional obligations not included in the member states' legal framework. See Ricardo Pazos, *El control de las cláusulas abusivas de los contratos con consumidores* (Aranzadi 2017) 504–505; Sergio Cámara Lapuente, 'Transparencias, desequilibrios e ineficacias en el régimen de las cláusulas abusivas' [2015] 55 *Anales de la Academia Matritense del Notariado* 549, 623–624.
73 Most of the Directives adopted in the consumer protection reform of 2018–2019 entered into force in 2022.
74 Directive (EU) 2019/771 on certain aspects concerning contracts for the sale of goods [2019] OJ L136/28.
75 Directive 2019/2161 as regards the better enforcement and modernisation of Union consumer protection rules [2019] OJ L328/7.
76 Directive (EU) 2019/771 on certain aspects concerning contracts for the sale of goods [2019] OJ L136/28, articles 6 and 7.
77 Directive (EU) 2019/771 on certain aspects concerning contracts for the sale of goods [2019] OJ L136/28, article 4.
78 Directive (EU) 2019/771 on certain aspects concerning contracts for the sale of goods [2019] OJ L136/28, article 3(6).
79 See Elias Van Gool and Anaïs Michel, 'The new consumer sales directive 2019/771 and sustainable consumption' [2021] 10(4) *EuCML* 136. See also Anna Beckers, 'Environmental protection meets consumer sales: the influence of environmental market communication on consumer contracts and remedies' [2018] 14(2) *ERCL* 157, 168–170 (https://doi.org/10.1515/ercl-2018-1009) noting that the remedies for lack of conformity should be interpreted with a public policy objective such as sustainability.
80 Directive (EU) 2019/2161 modifies the following directives: Directive 93/13 on unfair contract terms in consumer contracts, Directive 98/6 on the protection of consumers in terms of indicating the prices of products offered to consumers, Directive 2005/29 concerning unfair business-to-consumer commercial practices in the internal market, and Directive 2011/83 on consumer rights.
81 Or when this information is provided in an unclear, ambiguous, intelligible or in an inadequate moment. Directive (EC) 2005/29 concerning unfair business-to-consumer commercial practices in the internal market ('Unfair Commercial Practices Directive') [2005] OJ L149/22, article 7.
82 Directive (EC) 2005/29 concerning unfair business-to-consumer commercial practices in the internal market ('Unfair Commercial Practices Directive') [2005] OJ L149/22, annex 1.
83 See Elias Van Gool and Anaïs Michel, 'The new consumer sales directive 2019/771 and sustainable consumption' [2021] 10(4) *EuCML* 136, arguing that Directive 2019/771 maintains the status quo in terms of sustainability goals which contrasts with EU sustainable goals.
84 Ludwig Krämer, 'On the interrelation between consumer and environmental policies in the European Community' [1993] 16 *Journal of Consumer Policy* 455, 457–458 and 464–466 (https://doi.org/10.1007/BF01018767).
85 See Dara O'Rourke and Abraham Ringer, 'The impact of sustainability information on consumer decision making' [2015] 20 *Journal of Industrial Ecology* 882 (https://doi.org/10.1111/jiec.12310) noting that providing more or better information has a limited impact on changing consumer's behaviour unless this information is directly related into existing decision-making processes.
86 Proposal for a Directive of the European Parliament and of the Council on common rules promoting the repair of goods and amending Regulation 2017/2394, Directives 2019/771 and 2020/1828, COM\2023\155 final.
87 Commission, 'Communication from the Commission to the European Parliament, the European Council, the Council, the European economic and social committee and the committee of the regions, The European Green Deal' COM/2019/640 final.
88 Commission, 'Communication from the Commission to the European Parliament, the European Council, New Consumer Agenda Strengthening consumer resilience for sustainable recovery' COM/2020/696 final.
89 Commission, 'Communication from the Commission to the European Parliament, the European Council, the Council, the European economic and social committee and the committee of the regions, A new Circular Economy Action Plan For a cleaner and more competitive Europe' COM/2020/98 final.

90 Directive (EU) 2019/771 on certain aspects concerning contracts for the sale of goods [2019] OJ L136/28.

91 Directive (EU) 2019/771 on certain aspects concerning contracts for the sale of goods [2019] OJ L136/28, arts 13 and 14.

92 When repair and replacement impose a disproportionate cost on the seller, he may refuse to bring the product to conformity. See Directive (EU) 2019/771 on certain aspects concerning contracts for the sale of goods [2019] OJ L136/28, article 13(3).

93 Sanne Jansen, *Consumer Sales Remedies in US and EU Comparative Perspective* (Cambridge, Intersentia 2018).

94 Eurobarometer 2021 showed that 77% of Europeans want to react and limit climate change <https://europa.eu/eurobarometer/surveys/detail/2672> accessed 3 May 2023, but chose replacement over repair generating remarkable environmental effects in terms of the use of resources, waste generation and gas emissions <https://ec.europa.eu/commission/presscorner/detail/es/qanda_23_1795> accessed 3 May 2023.

95 Commission, 'Communication from the Commission to the European Parliament, the European Council, the Council, the European economic and social committee and the committee of the regions, The European Green Deal' COM/2019/640 final.

96 Sustainable production and sustainable consumption are considered to be one of the cornerstones for the European Commission so that the European Union becomes the first climate neutral continent by 2050. See the Commission, 'Communication from the Commission to the European Parliament, the European Council, the Council, the European economic and social committee and the committee of the regions, The European Green Deal' COM/2019/640 final.

97 Proposal for a Directive of the European Parliament and of the Council on common rules promoting the repair of goods and amending Regulation 2017/2394, Directives 2019/771 and 2020/1828, COM\2023\155 final.

98 See Ece Gulserliler, Atalay Atasu and Luk Van Wassenhove, 'Business model choice under right-to -repair: economic and environmental consequences' [2022] SSRN Working Paper <https://ssrn.com/abstract=4011640> accessed 12 May 2023, noting that the right to repair may have unintendedly lead to producers changing the business model in order to avoid the risk of cloning, for example, to avoid competition from secondary markets and to retain ownership.

99 Proposal for a Directive of the European Parliament and of the Council on common rules promoting the repair of goods and amending Regulation 2017/2394, Directives 2019/771 and 2020/1828, COM\2023\155 final, article 6.

100 Proposal for a Directive of the European Parliament and of the Council on common rules promoting the repair of goods and amending Regulation 2017/2394, Directives 2019/771 and 2020/1828, COM\2023\155 final, article 5.

101 See Directive (EU) 2019/771 on certain aspects concerning contracts for the sale of goods [2019] OJ L136/28 and Directive (EU) 2019/2161 as regards the better enforcement and modernisation of Union consumer protection rules [2019] OJ L328/7.

102 Susanne Augenhofer, *European Commission's Public consultation on Sustainable Consumption of Goods- Promoting Repair and Reuse* (Vienna, European Law Institute 2022).

103 Proposal for a Directive of the European Parliament and of the Council on common rules promoting the repair of goods and amending Regulation 2017/2394, Directives 2019/771 and 2020/1828, COM\2023\155 final, article 4.

104 Yanis Bakos, Florencia Marotta-Wurgler and David R. Trossen, 'Does anyone read the fine print? Consumer attention to standard-form contracts' [2014] 43(1) *The Journal of Legal Studies*.

105 See the Austrian initiative to grant consumers a voucher covering 50% of the repair costs with a cap of 200 euros <https://austrianpress.com/2022/04/26/the-repair-bonus-starts-today-in-austria-how-to-get-your-200-euro-voucher/> accessed 3 May 2023.

106 See Chen Jin, Luyi Yang and Cungen Zhu, 'Right to repair: pricing, welfare, and environmental implications' [2022] 69(2) *Management Science* 1017 (https://doi.org/10.1287/mnsc.2022.4401), showing of that the environmental impact of products should be analysed from the perspective of their life cycle because of the unavoidable link between the repair market and the product market.

107 Yanis Bakos, Florencia Marotta-Wurgler and David R. Trossen, 'Does anyone read the fine print? Consumer attention to standard-form contracts' [2014] 43(1) *The Journal of Legal Studies*.

108 See Chen Jin, Luyi Yang and Cungen Zhu, 'Right to repair: pricing, welfare, and environmental implications' [2022] 69(2) *Management Science* 1017 (https://doi.org/10.1287/mnsc.2022.4401), arguing that

the right to repair even when made easier and more affordable may result in non-monotone higher prices for low-end products that may induce to non-monotone changes in the consumer surplus, social welfare and environmental impact and result in a lose-lose-lose outcome in terms of manufacturing profits, a reduction of the consumer surplus and environmental impact.

Bibliography

Augenhofer S, *European Commission's Public Consultation on Sustainable Consumption of Goods-Promoting Repair and Reuse* (Vienna, European Law Institute 2022)

Bakos Y, Marotta-Wurgler F and Trossen DR, 'Does anyone read the fine print? Consumer attention to standard-form contracts' [2014] 43(1) *The Journal of Legal Studies* 22

Bayoumi T, Barkema J and Cerdeiro D, 'The inflexible structure of global supply chains' [2019] IMF Working Papers 1 <https://www.imf.org/en/Publications/WP/Issues/2019/09/13/The-Inflexible-Structure-of-Global-Supply-Chains-48562> last accessed 23 August 2023

Beckers A, 'Environmental protection meets consumer sales: the influence of environmental market communication on consumer contracts and remedies' [2018] 14(2) *ERCL* 157, 168–170 (https://doi.org/10.1515/ercl-2018-1009)

Cafaggi F, 'Regulation through contracts: supply-chain contracting and sustainability standards' [2016] 12(3) *European Review of Contract Law* 218 (https://doi.org/10.1515/ercl-2016-0013)

Cafaggi F and Iamiceli P, 'The limits of contract laws. The control of contractual power in trade practices and the preservation of freedom of contract within agrifood global supply chains' (2022) SSRN Working Paper <https://ssrn.com/abstract=4048571> last accessed 23 August 2023

Caferra R et al, 'Consumer analysis and the role of information in sustainable choices: a natural experiment' [2022] 1 *Frontiers in Environmental Economics* 1044206 (https://doi.org/10.3389/frevc.2022.1044206)

Cámara Lapuente S, 'Transparencias, desequilibrios e ineficacias en el régimen de las cláusulas abusivas' [2015] 55 *Anales de la Academia Matritense del Notariado* 549

Cohen MJ, *The Future of Consumer Society: Prospects for Sustainability in the New Economy* (Oxford, Oxford University Press 2017)

Di Ponte S et al, 'Environmental impacts of global supply chain: a review of scientific, policy and legal components for including environmental and climate challenges within the international trade patterns' [2013] *JCR Science for Policy Report. European Commission* <https://data.europa.eu/doi/10.2788/799> last accessed 23 August 2023

Grochowski M, 'European consumer law after the new deal: a tryptich' [2020] 39 *Yearbook of European Law* 387 (https://doi.org/10.1093/yel/yeaa016)

Gulserliler E, Atasu A and Van Wassenhove L, 'Business model choice under right-to-repair: economic and environmental consequences' [2022] SSRN Working Paper <https://ssrn.com/abstract=4011640> last accessed 23 August 2023

Hermann E, 'Artificial intelligence in marketing: friend or foe of sustainable consumption?' [2021] *AI & Soc* (https://doi.org/10.1007/s00146-021-01227-8)

Howells G, 'The potential and limits of consumer empowerment by information' [2005] 32 *Journal of Law and Society* 349 (https://doi.org/10.1111/j.1467-6478.2005.00328.x)

Jansen N, 'Unfair contract terms' in Nils Jansen and Reinhard Zimmermann (eds), *Commentaries on European Contract Laws*, pp. 919–994 (Oxford, Oxford University Press 2018)

Jansen S, *Consumer Sales Remedies in US and EU Comparative Perspective* (Cambridge, Intersentia 2018)

Jin C, Yang L and Zhu C, 'Right to repair: pricing, welfare, and environmental implications' [2022] 69(2) *Management Science* 1017 (https://doi.org/10.1287/mnsc.2022.4401)

Keirsbilck B et al, 'Sustainable consumption and consumer protection legislation: how can sustainable consumption and longer lifetime of products be promoted through consumer protection legislation?' in *Policy Department for Economic, Scientific and Quality of Life Policites* (Directorate-General for Internal Policies, 2020) <https://data.europa.eu/doi/10.2861/173816> last accessed 23 August 2023

Krämer L, 'On the interrelation between consumer and environmental policies in the European Community' [1993] 16 *Journal of Consumer Policy* 455, 457–458 and 464–466 (https://doi.org/10.1007/BF01018767)

Kysar D, 'Preferences for processes: the process/product distinction and the regulation of consumer choice' [2004] 118 *Harvard Law Review* 525

Lyczkowska K, 'Una de cal y otra de arena: La AP de Asturias confirma la retroactividad total de la cláusula suelo nula, pero califica como válida la que fue fruto de una novación contractual' [2017] *Centro de*

Estudios de consumo, pages 1–2 (23 febreuray 2017) (https://centrodeestudiosdeconsumo.com/images/artics6mz/Una-de-cal-y-otra-de-arena.pdf)

Mak V and Terryn E, 'Circular economy and consumer protection: the consumer as a citizen and the limits of empowerment through consumer law' [2020] 43 *Journal of Consumer Policy* 227 (https://doi.org/10.1007/s10603-019-09435-y)

Micklitz H-W, 'Squaring the circle? Reconciling consumer law and the circular economy' [2019] 8 *EuCML* 229

O'Rourke D and Ringer A, 'The impact of sustainability information on consumer decision making' [2015] 20 *Journal of Industrial Ecology* 882 (https://doi.org/10.1111/jiec.12310)

Pazos R, *El control de las cláusulas abusivas de los contratos con consumidores* (Aranzadi 2017)

Sanders RE, 'Dynamic pricing and organic waste bans: a study of grocery retailers' incentives to reduce food waste' [2020] SSRN Working Paper <https://ssrn.com/abstract=2994426> last accessed 23 August 2023

Van Gool E and Michel A, 'The new consumer sales directive 2019/771 and sustainable consumption' [2021] 10(4) *EuCML* 136

Varian H, *Intermediate Microeconomics: A Modern Approach* (9th edn, W. W. Norton & Company 2014)

10

SUSTAINABILITY CHALLENGES IN SLOVENIAN PRIVATE LAW

A Focus on the Nature of Goods

Petra Weingerl[1]

10.1 Introduction

Ensuring sustainable consumption and production is key to sustaining the livelihoods of current and future generations.[2] It is listed under Goal 12 of the UN Sustainable Development Goals (SDG). This contribution examines sustainable consumption and production in Slovenia by assessing the current policy and legal framework to improve resource efficiency, reduce waste and pollution in Slovenia. It focuses on the nature of goods as the overarching theme in examining the development of law and policy on achieving sustainability with private law instruments. To this end, it recognises two key issues that private law should consider in addressing sustainability – the durability of tangible goods and the perishability of foodstuffs.

Goods can be either durable or non-durable.[3] Non-durable goods are goods that by nature deteriorate, such as perishable products (e.g., flowers) or goods which are only intended for a single use.[4] While in principle, the same set of rules applies to durable and non-durable goods, including foodstuffs, their suitability to address sustainability issues differs based on the nature of the goods in question.

The common denominator of sustainable product policy should be reducing waste significantly.[5] The current legal and policy framework to encourage sustainable production and consumption rely on the awareness-raising model in the pre-contractual stage and on the empowered consumer who will buy eco-friendly goods, produce zero food waste and resort to eco-friendly remedies in the case of non-conformity in the contractual stage, most notably the right to repair and replacement with refurbished goods. The awareness-raising model in the pre-contractual stage is relevant to all goods. The same is not true for the concept of durability and related private law measures in the performance stage, such as the right to repair. They are, by definition, designed for durable goods only. The nature of non-durable goods, for example, food, which is spoiled sooner or later, prompts different considerations to contribute to sustainability goals.

Hitherto, the most obvious measures to prevent food waste are information requirements, i.e., labelling rules and food safety rules on date marking. Food waste prevention strategies currently do not envisage private law measures. However, there are rules on unfair trading and commercial practices with important private law implications that can be applied in certain instances and will be discussed below. It will be argued that there is a potential for strengthening the sustainability

 DOI: 10.4324/9781032662046-12

potential of rules on food waste prevention through the reinforcement of their link with private law instruments applied to sanction violations of those rules. Those private law sanctions are, for example, the right to compensation, the right to withdrawal, price reduction and nullity.

In what follows, the contribution will first briefly address the relevance of the nature of goods for applying private law measures designed to redress non-conforming or defective goods. Second, it will provide a general overview of law and policy approaches to encourage sustainable production and consumption in Slovenia. Third, it will consider how sustainability goals have been pursued through the legal framework as regards durable goods. To this end, the relevant measures are found in the field of consumer law. Fourth, it will review attempts to achieve sustainable production and consumption of food with law and policies in the field of agricultural law and food law.

10.2 The Relevance of Nature of Goods for Law and Policy

In principle, EU law does not put an emphasis on different types of goods based on their nature. In the Court of Justice of the EU's (hereinafter CJEU) case law on the free movement of goods, the term 'goods' is to be understood as meaning 'products which can be valued in money, and which are capable, as such, of forming the subject of commercial transactions'.[6] Thus, this concept comprises both durable and non-durable goods. The Sale of Goods Directive (hereinafter SGD 2019), which provides consumers with a number of remedies for lack of conformity, defines 'goods' as *any tangible movable items*.[7] The same definition of 'goods' is found in Directive 2011/83 on Consumer Rights.[8] The SGD 2019 also covers 'any tangible movable items that incorporate or are inter-connected with digital content or a digital service in such a way that the absence of that digital content'.[9] Tangible goods, covered by the SGD 2019, can be either durable or non-durable too. The Directive does not make any formal distinction in that regard, and it does not mention durability or perishability in this context. However, the preamble mentions that some goods, by nature, deteriorate and that this must be considered in establishing the lack of conformity. It states in recital 45 that '(...) in some cases the presumption that the lack of conformity existed at the relevant time for establishing conformity could be incompatible with the nature of the goods or the nature of the lack of conformity. The former could be the case with goods that by nature deteriorate, such as perishable products, for example, flowers, or goods which are only intended for a single use'.[10]

Interestingly, even the Product Liability Directive from 1985 applies to all products, also to agricultural products.[11] This Directive establishes a strict liability system for damages caused by defective products. At first, the Product Liability Directive excluded primary agricultural products from the definition of 'product', although the Member States retained the option to include them under the scope of product liability rules. In 1999, Directive 1999/34 extended the scope of the Product Liability Directive to agricultural and fishery products.[12] Thus, consumers can claim compensation for damage to health caused by defective agricultural products. However, although the nature of goods is not decisive for the definition of a product and thus for the applicability of the Directive, the expected lifespan of a product is one of the factors considered when determining liability for damages caused by faulty goods. Under this Directive, a product is considered defective when it does not provide the level of safety that one can reasonably expect.[13] Thus, the product's intended use is taken into account to determine its expected durability and safety standards.

When it comes to the applicability of certain private law measures, a distinction between durable and non-durable goods is made in Directive 2011/83 on Consumer Rights. It provides that the right of withdrawal in respect of distance and off-premises contracts does not cover the supply of goods which are liable to deteriorate or expire rapidly.[14] In the past, a distinction based on the nature of goods was made in the EU legislation on price indications on consumer products.[15] However, today Directive 98/6/EC requires the selling price and the unit price of *all products* traders offer to consumers to be clearly indicated to improve consumer information and enable price comparisons. It makes no distinction between foodstuffs and non-food products.[16] However, such distinctions are made in the area of product safety. The General Product Safety Regulation, which was adopted in May 2023, applies to non-food products.[17] It provides rules on recalls of dangerous products, and governs consumer remedies in the form of the repair, replacement or an adequate refund of the value of the recalled product.[18] The General Product Safety Regulation makes a distinction between these remedies in the event of a recall, which 'serve both to ensure elimination of dangerous products from the market and an adequate remedy for the consumer', and contractual remedies.[19] To this end, the Regulation does not contain rules on contractual remedies for non-conformity or additional types of private law measures.[20] While these are examples of possible differentiation between goods based on their nature, no such differentiation exists in EU legislation governing private law measures for non-conforming or defective goods as aforementioned.

In Slovenia, the legislation needs to be in accordance with EU law. Thus, in principle, also relevant Slovenian legislation applies equally to durable and non-durable goods. However, the Consumer Protection Act (hereinafter CPA 2022)[21] imposes a special obligation on traders selling non-durable goods. Article 12 CPA 2022 governs 'care of goods' and provides that the seller ensures that all the properties of the goods intended for sale are preserved, *especially goods with a limited shelf life*. However, there is no special sanction prescribed for a breach of this abstract provision.

It can be concluded that, in principle, the relevant EU and Slovenian legislation equally applies to all products. However, in practice, the nature of goods is one of the factors considered when determining measures for the lack of conformity and liability for faulty goods. This makes private law measures underpinned by the concept of durability, discussed under section 10.4, unsuitable for non-durable goods, most notably the right to repair. Thus, different avenues for the contribution of private law measures to the sustainability of non-durable goods will be explored under section 10.5.

10.3 Sustainable Production and Consumption and the Slovene Policy and Legal Framework

In Slovenia, sustainability goals, especially Goal 12 SDG, and the circular economy have found their way into the Constitution[22] and into the key strategic national documents, including the Slovenian Development Strategy 2030[23] and the Roadmap for the Transition to the Circular Economy.[24] Sustainability goals have been enshrined in the Constitution of Slovenia through the right to a healthy living environment and the right to drinking water.[25]

On a policy level, the National Action Plan on Business and Human Rights[26] explicitly addresses the link of consumer law to sustainability also in the context of the right to a healthy living environment, as it states that '[c]onsumer protection also extends to environmental protection, which is focused on the *consumers' right to a healthy environment*, and *on sustainable consumption*'.[27] Moreover, various Resolutions on the National Consumer Protection Programme underlined that the constitutional right to a healthy living environment imposes an obligation on consumers to contribute to a healthier living environment through their behaviour and purchase choices made.[28] The goal of 'environmental protection, nature conservation, sustainable development' is among the five priorities of the National Action Plan.[29]

Also at the legislative level, Slovenia has in recent years introduced a number of measures to promote sustainable choices in the production and consumption of goods and services, often as a direct result of actions taken at the EU level. These rules affect the whole life cycle of a product, from the design and production stage (eco-design) to the consumption stage (waste management and recycling). Examples are the regulation on the use of plastics, such as lightweight carrier bags, packaging, and the problem of graveside candle waste,[30] the implementation of the EU company law obligation of non-financial reporting on the environmental and social impacts of major enterprises, an emphasis on green public procurement,[31] sustainable farming and the circular economy,[32] etc. In the remainder of the contribution, I will assess whether the regulatory framework in Slovenia is favourable for achieving sustainable development goals, especially sustainable consumption, for durable goods (in the fields of consumer law) and food (in agriculture and food law).

10.4 Sustainable Production and Consumption of Durable Goods

This part of the contribution focuses on durable goods, i.e., goods that do not deteriorate with use. The EU has adopted an extensive repertoire of measures and initiatives dealing with durable goods, most notably in consumer law, to support the objectives of the European Green Deal to reduce waste.[33] To this end, the focus is on measures encouraging sustainable choices and enhancing the durability[34] of goods. Examples in pre-contractual stage are rules on empowering consumers through information requirements (the awareness-raising model): the mandatory general information requirements, labelling rules, and the regulatory framework for green claims.[35] In the performance stage, the focus is on guaranteeing durability through rules on remedies and time limits. In 2019, the EU adopted two directives dealing with conformity of goods, digital content, or digital services with the contract,[36] and remedies in the event of a lack of such conformity – the aforementioned Twin Directives. In Slovenia, they were transposed with CPA 2022.

In March 2023, the Commission tabled several relevant proposals for measures promoting more sustainable consumption by increasing the repair and reuse of goods both within and outside the legal guarantee, most notably a Proposal for a Directive empowering consumers for the green transition through better protection against unfair practices and better information[37] and a Proposal for a Directive on common rules promoting the repair of goods.[38] The former provides better information on the durability and reparability of goods at the point of sale, and thereby enabling consumers to make sustainable purchasing decisions. The latter aims to increase the repair and reuse of viable defective goods purchased by consumers within and beyond the legal guarantee of SGD 2019.[39]

The contribution considers how sustainability goals have been pursued by the rules found in the new CPA 2022, focusing on the pre-contractual stage (information requirements) and the contractual stage (modalities of the right to repair and guarantees in consumer sales law in Slovenia). It depicts the transposition of the relevant rules of SGD 2019 into the Slovene legislation. Moreover, the contribution will present a unique concept of obligatory guarantee governed by CPA 2022 and its potential in light of sustainability goals.

10.4.1 Information Requirements in the Pre-Contractual Stage: Raising Awareness

Reliance on information requirements has long been a prominent regulatory technique in EU measures affecting the protection of consumers' economic interests.[40] It can also serve to raise awareness about environmental issues and enable consumers to buy more sustainable goods and services. Consumers play a crucial role in the circular transition as enablers of the circular economy.[41] As provided by the EU's Circular Economy Package, '[t]he choices made by millions of consumers can support or hamper the circular economy. These choices are shaped by the information to which consumers have access, the range and prices of existing products and the regulatory framework'.[42] Apart from the mandatory general information requirements in consumer law and labelling rules, the regulatory framework for (voluntary) green claims plays an important role in raising awareness of consumers.

When green claims are false or cannot be verified, this practice is called 'greenwashing'.[43] A legal basis to ensure that traders do not present green claims in unfair ways to consumers is found in the Directive 2005/29 on Unfair Commercial Practices (hereinafter UCPD),[44] which prohibits misleading advertising.[45] In Slovenia, it is implemented with the CPA 2022.

Breaches of the UCPD lead to possibilities of public and private enforcement. In the past, its enforcement relied on public authorities and fines imposed on infringing traders. In 2019, the UCPD was amended with the Omnibus Directive. The latter introduced new forms of redress available to consumers harmed by unfair commercial practices to facilitate private enforcement.[46] Today, the consumer should have access to private law remedies in the form of compensation for damage and, where relevant, a price reduction or termination of the contract.[47] Thus, consumer redress for the use of unfair commercial practices includes both contractual and non-contractual remedies. The conditions for the application of the remedies are determined by the Member States.

In Slovenia, this is governed by Articles 55–57 CPA 2022. Regarding rules on the right to compensation, CPA 2022 refers to general civil liability rules.[48] Concerning the price reduction and the right to withdraw from the contract, the trader must return the paid amount or part of the purchase price within 14 days of receiving the request.[49] Article 57 CPA provides that the court and the competent inspection authority may, taking into account the circumstances of each case and the legitimate interests of the trader and other parties in the proceedings, require the trader to submit evidence of the accuracy of the factual statements regarding its business practice within a period that may not exceed eight days. If the trader fails to submit the evidence or if it submits insufficient evidence, it is considered that the factual statements in its business practice are not accurate.

In March 2023, the Commission published a Proposal for a Directive on Green Claims.[50] This proposal does not envisage any private enforcement measures. Thus, the possibility of redress under the UCPD will remain relevant for individuals also in the future. It is envisaged that rules on the applicability of the UCPD to green claims will be clarified with the aforementioned proposed directive empowering consumers for the green transition through better protection against unfair practices and better information.

The reliance on empowerment through information disclosure should be seen as just one element of EU consumer protection law.[51] Such approaches support sustainable consumption only if consumers themselves value products that are more sustainable,[52] which also depends on their affordability and other factors that have an impact on sustainable behavioural change.[53] It is a (re-regulatory) choice enhancing the durability of goods which reveals the EU's and Member States' concern to achieve adequate levels of protection aligned with sustainability goals for all consumers, whether empowered or not.[54] In what follows, I will address this concern focusing on the transposition of the SGD 2019 into the CPA 2022.

10.4.2 Performance Stage: Remedies for Non-Conformity

10.4.2.1 The Concept of Durability in the SGD 2019: Verbatim *Transposition*

The SGD 2019 is a maximum harmonisation directive[55] which applies to sales contracts for goods, including goods with a digital component such as a smartphone, a smart TV or a smart watch.[56]

Under the SGD 2019, goods must conform with the contract.[57] It provides consumers with a number of remedies for lack of conformity, namely the repair and replacement of goods, price reduction and termination of the sales contract. There are two sets of revised conformity requirements: subjective and objective.[58]

The objective requirements for conformity relate to objective standards of reasonable expectation, which apply regardless of the terms of the contract. The sustainability-friendly novelties of the SGD 2019 are the explicit mention of durability and the reasonable expectations of the consumer as the criteria for the assessment of objective conformity.[59] Article 2(13) of the SGD 2019 defines durability as 'the ability of the goods to maintain their required functions and performance through normal use'. Thus, in order for

> goods to be in conformity, they should possess the durability which is normal for goods of the same type and that the consumer can reasonably expect given the nature of the specific goods, including the possible need for reasonable maintenance of the goods, such as the regular inspection or changing of filters in a car, and any public statement made by or on behalf of any person constituting a link in the chain of transactions.[60]

The assessment should also take into account all other relevant circumstances, such as the price of the goods and the intensity or frequency of the use that the consumer makes of the goods. Moreover, if any pre-contractual declaration forming part of the sales contract contains specific sustainability information, the consumer should be able to refer to it as part of the subjective compliance requirements.[61]

The CPA 2022 transposed the concept of durability *verbatim*. Its definition is found in Article 4 CPA 2022. The durability as a criterion for the objective assessment of conformity of goods is governed by Article 73(4) CPA 2022.

10.4.2.2 Durability and Time Limits

According to Article 10(1) of SGD 2019, the seller is liable for any non-conformity that exists at the time of delivery and becomes apparent within two years of that time. The time limit for the liability of the seller is extended for digital content or services supplied continuously to goods with digital elements during more than two years. Slovenia did not opt for longer time limit as enabled by Article 10(3) of the SGD 2019.[62] This could have been beneficial from the sustainability perspective, since longer time limits encourage the production of goods with better durability,[63] which results in less environmental externalities from the production of new and disposal of old goods.[64] Moreover, as Van Gool and Michel argue, longer time limits for the liability of the seller may increase the chance that consumers seek repairs for their products instead of buying new ones, which again prolongs the lifespan of goods.[65]

It can be argued that a longer lifespan of goods is not necessarily the most sustainable option for all goods, as one size does not fit all.[66] Some newer goods can improve energy-efficiency or may be less environmentally hazardous.[67] Moreover, extended time limits for liability may increase prices for certain goods.[68] Thus, the EU legislator left it to the Member States to decide whether they wish to introduce longer time limits from those harmonised in the SGD 2019. The Slovene legislator did not opt for this possibility.

As regards second-hand goods, the SGD 2019 gives Member States the possibility to provide that the seller and the consumer can agree to contractual terms or agreements with a shorter liability or limitation period, provided that such shorter periods shall not be less than one year (Article 10(6) SGD 2019). This option is transposed into the CPA 2022 with Article 78(5).

10.4.2.3 Commercial and Obligatory Guarantees

While the SGD 2019 governs optional commercial guarantees, the CPA 2022 governs a one-year obligatory guarantee for technical products in addition to optional commercial guarantees.[69] The obligatory guarantee must be issued by the producer.

Legal remedies under the obligatory guarantee are the same as remedies for non-conformity (repair and replacement of goods, price reduction and termination of the sales contract). The main difference is in their availability, as there is a strict three-level hierarchy of remedies under the framework of the obligatory guarantee. Article 97(1) CPA 2022 governs repair as the primary remedy in the framework of the obligatory guarantee. Only if goods are not repaired within 30 days (this time limit was shortened, under the old Consumer Protection Act [*Zakon o varstvu potrošnikov*,[70] ZVPot, hereinafter CPA 1998], it was 45 days), the producer shall replace the goods with 'identical, new and faultless' goods free of charge. Thus, goods presented for repair should not be replaced by refurbished goods. This is regrettable since refurbishment can contribute to sustainability goals to a similar extent as repair. However, as emphasised by Kryla-Cudna, the courts have questioned its interplay with the need to protect the buyer's contractual interest.[71] Only if the producer fails to repair and replace the faulty product, the consumer can resort to price reduction or rescission of the contract (Article 97(3) CPA 2022). The guarantor can provide the consumer with free use of similar goods during the repair of goods for which an obligatory guarantee has been issued (Article 97(7) CPA 2022). If the guarantor does not provide the consumer with replacement goods for temporary use, the consumer has the right to claim the damage he or she suffered because he or she was unable to use the goods from the moment he or she requested repair or replacement until their execution (Article 97(8) CPA 2022).

These remedies under the obligatory guarantee can be exercised by any contractual party (also in business-to-business (B2B) contracts), not only the consumer (Article 97(9) CPA 2022).

The CPA 1998 also governed the obligatory guarantee for used goods, with the guarantee of at least one month.[72] The CPA 2022 did not retain this obligatory guarantee for used products. This solution strikes a reasonable balance for the producer since the guarantee is valid for one year even if the product is resold in this period. Thus, it also covers defects of used goods up to one year from the first sale. It would be seen as a disproportionate burden if the producer's obligation would be extended each time the item is resold. The situation was different under the old law, where the seller was obliged to issue the guarantee.

The CPA 2022 allows consumers to rely on the right to refuse if the lack of conformity of the goods becomes apparent within a period after delivery, not exceeding 30 days, also in the case of the obligatory guarantee.[73]

The CPA 2022 shortens the period in which the guarantor must bring goods into conformity. For products placed on the market before 25 January 2023, the 45-day period will continue to apply, while for products that will be placed on the market later, the repair period is reduced to 30 days.[74] This deadline can be extended to the shortest time necessary to complete the repair or replacement, but for a maximum of 15 days. The consumer must be informed about this before the expiry of the original period and at the same time given reasons for the extension.[75]

The obligatory guarantee has been a popular option for consumers to receive goods in conformity with the contract, since the consumer can request repair anytime by the end of the guarantee period, and he or she is not required to notify the seller of a lack of conformity in a certain time period. This is different than as regards the regulation on the conformity of goods, since in the latter case the consumer can exercise his or her rights if he or she informs the seller about the non-conformity within two months from the date on which the non-conformity was established.[76]

The shorter period in which the guarantor must bring goods into conformity could enhance the use of the obligatory guarantee instead of remedies for the non-conformity of goods. This would be positive from the sustainability perspective since the framework of obligatory guarantee put repair on the pedestal and the consumer cannot choose between repair and replacement if repair is a suitable remedy. For this reason, the obligatory guarantee is the most sustainability-friendly measure of Slovenian consumer sales law.[77] This framework could serve as a model for the system of remedies for non-conforming goods where this is deemed appropriate.[78] The Commission's proposal on common rules promoting the repair of goods heads towards this direction as it prioritises repair as a remedy for the non-conformity of goods whenever repair is cheaper or as costly as replacement.

10.5 Sustainable Production and Consumption of Perishable Goods (Food)

This part of the contribution discusses the sustainable production and consumption of perishable goods, focusing on food. It will focus on the potential of legal framework in reducing food waste through private law instruments. In this context, both agricultural law and the related food law can contribute to food waste reduction. The most obvious measures to prevent food waste are information requirements, i.e., labelling rules and food safety rules on date marking. The food waste prevention strategies currently do not envisage private law measures. However, there are rules on unfair trading and commercial practices with important private law implications that can be applied in certain instances. There is a potential for strengthening the sustainability potential of rules on food waste prevention by reinforcing their link with private law instruments applied to sanctioning the violations of those rules. This link with private law measures will be discussed in the following subsections.

Preventing and reducing food waste emerges as an important priority in the pursuit of holistic approaches to address food security, nutrition, and environmental sustainability.[79] Food waste occurs along the entire food supply chain, both in developed and in developing countries, for various reasons.[80] The Food and Agriculture Organisation of the United Nations (FAO) estimates that nearly a third (1.3 billion tonnes a year) of all the food intended for human consumption is lost or wasted between the farm and the fork.[81] Food losses and waste cost the global economy billions of euros, exacerbate food insecurity and malnutrition, and ultimately consume about a quarter of all the water used for agriculture purposes.[82] It is responsible for an estimated 8% of global greenhouse gas emissions, and also contributes to loss of biodiversity.[83] Thus, the production of food which is later wasted puts unnecessary pressure on limited natural resources.[84] There are also several other negative consequences for soil, marine eutrophication, and also economic and social consequences linked to wasting food.[85]

The EU and the EU countries are committed to meeting the Sustainable Development Goal Target 12.3 (hereinafter, SDG Target 12.3) to halve per capita food waste at the retail and consumer level by 2030 and reduce food losses along the food production and supply chains.[86] In July 2023, the Commission published its proposal to amend the Waste Framework Directive (WFD proposal), which focuses on two resource intensive sectors: textiles and food.[87] As the proposal highlights,

> food waste is one of the largest sources of inefficiency in the agri-food chain. In particular, it results in negative environmental and climate impacts. Food consumption is the main contributor to the environmental impacts and biodiversity footprint of EU consumption.[88]

The specific objectives of this proposal are, firstly, to assign clear responsibility to Member States for accelerating reduction of food waste along the food supply chain and in households, and thus make a solid contribution towards achieving SDG Target 12.3; and secondly, to ensure sufficient and consistent response by all Member States to reduce food waste.[89] This aims to support consumer behavioural change.[90]

However, consumer behavioural change is only one part of the equation. Also, agricultural production and the agrifood supply chain have a vast potential to contribute to achieving SDG Target 12.3 through better management at all stages of production, including the primary production and processing sector. While the focus on retail and consumption is crucial, it is believed that a holistic approach focussing on the entire agrifood chain could bring better results. However, in the promotion of sustainable agriculture one must also include consideration for social and human needs.[91] For any agricultural production system to be sustainable, it cannot just deal with soil and water or price and income. The farmers, their families, and the rural communities that make up the cultural structure of an agrarian system must be considered in designing legal and policy frameworks in this field.[92] This aspect goes beyond the objectives of this contribution; however, it is a regulatory challenge that needs to be tackled by new initiatives.

FAO identified the following initiatives that can help to reduce food loss along the production supply chain: capacity-building for producers (and producer organisations) to reduce loss and address quality management; closer-to-home markets that may have less stringent requirements; infrastructure upgrades; and postharvest systems development. Fiscal instruments can also be employed such as taxes, fees, incentives or subsidies to promote the diversion of waste from landfills (e.g., landfill taxes, price support mechanisms or incineration taxes).[93]

In the EU, the WFD requires Member States to prepare specific food waste prevention programmes, in line with the waste hierarchy. Moreover, it requires Member States to reduce food waste at each stage of the food supply chain, monitor food waste levels and report on progress made.[94] In 2019, the Commission adopted a common food waste measurement methodology, to be utilised as a basis for EU-wide food waste monitoring.[95]

In 2021, food waste generation amounted to just over 143,000 tonnes in Slovenia.[96] Thus, an average Slovenian threw away 68 kilograms of food. About half of the food waste was generated by households. Twenty-eight per cent of food waste was thrown away in catering and food service, 10% in grocery stores, and at least 9% in food production.[97] The Statistical Office estimates that 40% of the food waste was edible, while the rest comprised of bones, peels, eggshells and the like.[98] The fact is that the amount of wasted food could be significantly less if each of us changed our attitude towards food in terms of awareness that food is not a self-evident commodity and that it is therefore necessary to treat it thoughtfully and respectfully. The first step towards this is raising the awareness of the general public about this issue with concrete advice on how to change daily habits and reduce the amount of wasted food.[99] To raise the awareness, the Ministry of Agriculture, Forestry and Food set up a website 'Our Great Food' (*Naša super hrana*).[100] Moreover, every April, a No Food Waste Day is observed.

In what follows, the potential of laws on unfair commercial and trading practices to tackle food waste and their link with private law measures will be examined.

10.5.1 Date Marking and the Food Waste Prevention: The Potential of Law on Unfair Commercial Practices (B2C)

A prospective avenue for legislation to address food waste is through date marking as a means to reduce food waste.[101] This is especially relevant for B2C relationships.[102] A Commission study published in 2018 estimated that 10% of food waste in the EU supply chain is linked to date marking (i.e., 'use by' and 'best before' dates indicated on food labelling).[103] The EU Regulation on

Food Information to Consumers (hereinafter the FIC Regulation)[104] requires pre-packed foods to display a date mark that clarifies whether such date corresponds to the product's safety ('use by') or its quality ('best before'). This is mirrored in the EU Waste Framework Directive.[105]

The incorrect date marking, and food labelling more generally, can lead to unfair commercial practices under the UCPD. The UCPD applies to unfair B2C commercial practices before, during and after a commercial transaction in relation to a product (Article 3(1) UCPD). The 'product' is defined as 'any goods or service including immovable property, rights, and obligations' (Article 2c UCPD). There is no distinction in the applicability of specific rules in the UCPD based on the nature of goods.

Incorrect food labelling could lead to misleading omissions prohibited by Article 7 UCPD. It provides that a commercial practice shall be regarded as misleading if, in its factual context, it omits *material information* that the average consumer needs to take an informed transactional decision. Thereby it causes or is likely to cause the average consumer to take a transactional decision that he would not have taken otherwise. Thus, this article establishes 'a positive obligation on traders to provide all the information which the average consumer needs to make an informed purchasing decision'.[106] This is called 'material information'.

The UCPD provides that also information requirements in relation to commercial communication established by other EU laws are 'material'.[107] Thus, the information requirements established by the FIC Regulation are 'material' information within the meaning of Article 7(5) of the UCPD. The omission of this information could be considered misleading to the extent that it is likely to affect the transactional decisions of the average consumer.[108]

As aforementioned, breaches of the UCPD lead to possibilities of public and private enforcement. The consumer should have access to private law remedies in the form of compensation for damage and, where relevant, a price reduction or termination of the contract.[109] The conditions for the application of the remedies are determined by Member States. Here, it should be emphasised that traders sometimes take an over-cautious approach to date labelling and have safety margins.[110] Since the quality of food deteriorates before safety, 'use by' dates are linked to quality. As 'use by' dates are sometimes set where quality begins to deteriorate, food can be declared unsafe despite being safe for use.[111]

Despite the information requirements and awareness-raising methods, it is not always possible to prevent food waste. In that case, the most sustainable choice is to reuse it. Farmers, restaurants and supermarkets can give their surpluses to food banks; it can be used to feed animals, to make super fertile soil or even to generate electricity.[112] At the occasion of observing the No Food Waste Day in April 2023, the Ministry of Agriculture, Forestry and Food presented an innovative concept of the processing plant Mini factory, which meets the needs of processing surpluses that arise from farmers or small growers into other dishes.[113] Another promising action to reduce food waste is an application Plate (*Krožnik*).[114] With this application, it will be possible to view a list of stores throughout Slovenia that offer their products, such as bread, bakery products, fruit and vegetables, at half the price. This is food that would otherwise be wasted because it was not sold in time.

Actions taken at the national level must be aligned with activities at the EU level. In 2016, the Council in its conclusions on food losses and food waste emphasised that the reduction of food losses and waste delivers a threefold benefit: it alleviates pressure on climate, water and land; it has a positive economic impact on farmers, companies and households; and it means that more people can be fed with the food currently produced.[115] The Commission, in its EU action plan for the Circular Economy, announced the following actions in this field: to reduce food loss and waste generation, including the establishment of a multistakeholder platform dedicated to food waste prevention; to develop a common EU methodology to measure food waste in the entire food value chain; to clarify EU legislation on waste, food and feed in order to facilitate food donation and ensure the safe use of former foodstuffs and by-products in feed production; and to explore options to improve the understanding and use of date marking by all actors, including consumers.[116] In order to support achievement of SDG Target 12.3 on food loss and waste and maximise the contribution of all actors, the Action Plan envisaged the establishment of a Platform dedicated to food waste prevention. Thus, the EU Platform on Food Losses and Food Waste was established in 2016, bringing together EU institutions, experts from the EU countries, international organisations and relevant stakeholders selected through an open call for applications.[117] The EU Platform plays a key role in mobilising action to reduce food loss and waste across the EU as part of the Farm to Fork Strategy. The Platform was re-established in 2022 for a second mandate, until 2026.

Tackling food waste requires rethinking how we produce, market, distribute and consume food.[118] In December 2019, the EU Platform adopted the Recommendations for action in food waste prevention. Moreover, one of the initiatives of the Farm to Fork Strategy is a proposal for a legislative framework for sustainable food systems (FSFS). As announced in the Farm to Fork Strategy, it will be adopted by the Commission by the end of 2023. Part of the FSFS will be the sustainability labelling framework.

Food safety is also among the specific objectives of the Single Market Programme. They include, inter alia, support of sustainable food production and consumption by reducing food waste and informing consumers in view of ensuring high-quality standards across the EU.[119]

10.5.2 Unfair Trading Practices and Food Waste Prevention (B2B)

In Slovenia, the regulation that could indirectly affect the prevention of food waste can be found in the Agriculture Act.[120] This Act sets out sustainable food production as one of the objectives of agricultural policy and regulates provisions governing planning agricultural and rural development, agricultural policy measures, the safety of food of plant origin in primary production, food quality at all stages of production, processing and transport, and consumer protection.[121] Article 15 of the Agriculture Act expressly governs sustainable agriculture. It provides that sustainable agriculture maintains the biodiversity of animal and plant species and preserves the soil and its fertility while protecting the natural conditions for life in the soil, water, and air.

The Agriculture Act does not contain any measures that would directly address the problem of sustainable production of food and its management with the aim of prevention of food waste. A potentially promising mechanism to combat food waste and to rationalise food production is the regulation on the unfair trading practices (B2B) in the agrifood supply chain. To this end, the Agriculture Act implements the Unfair Trading Practices Directive (UTP Directive).[122] This is not a private law instrument *per se*, however, breaches of the unfair trading practices (UTPs) can lead to enforcement with private law implications. Currently, such an implication envisaged by the Agriculture Act is in the form of nullity of provisions containing UTPs, and there is also potential for the right to compensation in such instances. These rules apply to agrifood supply chains, i.e., to the sale of 'agricultural and food products' between suppliers (in practice usually farmers) and buyers. Thus, they apply to B2B relationships in the agrifood chain. The Agriculture Act expressly excludes their application to agreements between suppliers and consumers (B2C).[123]

The Agriculture Act contains a general norm on UTPs and a list of prohibited practices. The general norm, governed by Article 61.f (1), covers the conduct with which the buyer with its significant market power, especially in relation to the supplier, exploits the other contractual party contrary to good business practices.

One of the biggest retailers in Slovenia attacked a previous version of this general norm before the Constitutional Court for being contrary to the principle of legality. It is relevant for this contribution that, as regards the standard of good business practices, the Constitutional Court expressly linked the Agriculture Act provision with the general law of obligations:

> Based on the general principles of the law of obligations, participants in legal transactions are bound by the legal standard of good business practices in the civil law sphere regarding the content of their contractual relationships.

Thus, within the framework of the first paragraph of Article 61.f of ZKme-1, the legislator has only explicitly placed the above-mentioned legal standard, which already defines the limit of what is (not) permitted in legal transactions at the general level, also in the area of the food supply chain.[124]

The list of black and grey prohibited practices is found in the fourth paragraph of Article 61.f. A especially relevant prohibited practice in the context of food waste is found under point 17:

> cancellation of an order for perishable agricultural and food products in such a short period of time that the supplier cannot reasonably be expected to find an alternative way to market or use these products; cancellation within a period of less than 30 days is always considered short notice.

In Slovenia, the sanction for using UTPs is a private law sanction of nullity of contractual provisions.[125] The Agriculture Act itself does not provide for the possibility to award compensation. In the case of inflicted damage due to UTPs, the general law on civil liability applies.

To sum up, activities to prevent food waste mostly include methods building on information and awareness-raising. Here, private law remedies focusing on the durability of goods play no prominent role. However, there is potential for private law measures to strengthen the food waste prevention policy through the law on unfair commercial and trading practices.

10.6 Conclusion

The current legal and policy framework to encourage sustainable consumption and production in Slovenia relies on the awareness-raising raising model and on the empowered consumer who will buy eco-friendly goods, produce zero food waste and resort to eco-friendly remedies in the case of non-conformity. Consumer law measures designed for durable goods could be further strengthened by encouraging repair and self-repair, supporting replacement with refurbished goods, and extending the guarantee periods to encourage the production of more durable goods. Regarding perishable goods, the contribution focuses on food waste. Here, the legal framework encouraging sustainable production and consumption is still somewhat limited, as it relies on the information requirements and awareness-raising model. There are several initiatives aimed at raising awareness regarding sustainable production and consumption and food waste. Still, the amount of food waste has not decreased over the past few years.[126] Thus, to succeed in reaching sustainability goals and to contribute to safeguarding the planet for future generations, something more is needed. The question is whether and how this can be reached with private law measures. While the concept of durability and the related private law measures are, by definition, designed for durable goods, food is spoiled sooner or later. It is argued that there is a potential for strengthening the sustainability potential of information requirements rules on food waste prevention through the reinforcement of their link with private law instruments applied to sanction violations of those rules. Notably, there is a capacity to strengthen further the link between unfair commercial and trading practices and private law sanctions.

Notes

1 University of Maribor, Faculty of Law. I thank the anonymous reviewers for their insightful comments and suggestions.

2 United Nations: Sustainable Development Goals, available at <https://www.un.org/sustainabledevelopment/sustainable-consumption-production/> accessed 08 May 2023.

3 The European Central Bank report divided consumer goods in three categories, depending on their durability. Durable consumer goods typically include commodities with an expected lifetime of more than three years. Semi-durable and non-durable consumer goods have an expected lifetime of between one and three years and less than one year, respectively. ECB, Monthly Bulletin, May 2014, available at <https://www.ecb.europa.eu/pub/pdf/other/mb201405_focus06.en.pdf> accessed 01 July 2023, 58.

4 See recital 45 of the preamble to Council Directive (EU) 2019/771 of the European Parliament and of the Council of 20 May 2019 on certain aspects concerning contracts for the sale of goods, amending Regulation (EU) 2017/2394 and Directive 2009/22/EC, and repealing Directive 1999/44/EC (Text with EEA relevance.) PE/27/2019/REV/1, OJ L 136/28 (SGD 2019).

5 See also Communication from the Commission to the European Parliament, the European Council, the Council, the European Economic and Social Committee and the Committee of the Regions, 'The European Green Deal' COM(2019) 640 final (European Green Deal), 8.

6 Case 7/68, *Commission v Italy*, EU:C:1968:51; C-65/05, *Commission v Greece*, EU:C:2006:673, para. 23.

7 Art. 2(5) (a) SGD 2019.

8 Art. 2(3) Council Directive 2011/83/EU of the European Parliament and of the Council of 25 October 2011 on consumer rights, amending Council Directive 93/13/EEC and Directive 1999/44/EC of the European Parliament and of the Council and repealing Council Directive 85/577/EEC and Directive 97/7/EC of the European Parliament and of the Council Text with EEA relevance, OJ L 304/ 64 (Directive 2011/83 on Consumer Rights).

9 Art. 2(5) SGD 2019.

10 Recital 45 of the preamble to SGD 2019.

11 Council Directive 85/374/EEC of 25 July 1985 on the approximation of the laws, regulations and administrative provisions of the Member States concerning liability for defective products, OJ L 210/29 (Product Liability Directive).

12 Council Directive 1999/34/EC of the European Parliament and of the Council of 10 May 1999 amending Council Directive 85/374/EEC on the approximation of the laws, regulations and administrative provisions of the Member States concerning liability for defective products, OJ L 141/20.

13 Art. 6 Product Liability Directive.

14 Art. 16(d) Directive 2011/83 on Consumer Rights.

15 To this end, rules were found in two separate directives for foodstuff prices (Directives 79/581/EEC) and for non-food product prices (Directive 88/314/EEC). However, Directive 98/6/EC repealed those two directives and required the selling price and the unit price of *all products* offered by traders to consumers to be clearly indicated in order to improve consumer information and enable price comparisons.

16 This Directive was amended by the so-called 'Omnibus Directive', aiming to modernise EU consumer law and improve its enforcement. It provided for regulatory options for EU countries regarding goods that are liable to deteriorate or expire rapidly (in particular food). Council Directive (EU) 2019/2161 of the European Parliament and of the Council of 27 November 2019 amending Council Directive 93/13/EEC and Directives 98/6/EC, 2005/29/EC and 2011/83/EU of the European Parliament and of the Council as regards the better enforcement and modernisation of Union consumer protection rules (Text with EEA relevance) PE/83/2019/REV/1, OJ L 328/7 (Omnibus Directive).

17 Regulation (EU) 2023/988 of the European Parliament and of the Council of 10 May 2023 on general product safety, amending Regulation (EU) No 1025/2012 of the European Parliament and of the Council and Directive (EU) 2020/1828 of the European Parliament and the Council, and repealing Directive 2001/95/EC of the European Parliament and of the Council and Council Directive 87/357/EEC (Text with EEA relevance) PE/79/2022/REV/1, OJ L 135, 23.5.2023, pp. 1–51 (General Product Safety Regulation). It will replace the current General Product Safety Directive and the Food Imitating Product Directive.

18 Art. 37 General Product Safety Regulation.

19 Recital 88 General Product Safety Regulation.

20 Art. 43 and recitals 43 and 88 General Product Safety Regulation.
21 Consumer Protection Act- *ZVPot-1* (Official Gazette of the Republic of Slovenia No. 130/22); CPA 2022.
22 Arts. 72 and 70a of the Constitution of the Republic of Slovenia (Official Gazette of the Republic of Slovenia No. 33/91-I ff.). This part of the contribution to some extent summarises my article; P. Weingerl, 'Sustainability, the Circular Economy and Consumer Law in Slovenia' (2020) 9 *Journal of European Consumer and Market Law* 3, 129–232.
23 Government of the RS, 'Slovenian Development Strategy 2030', 12 October 2017, available at <http://www.vlada.si/en/media_room/newsletter/slovenia_weekly/news/article/slovenian_development_strategy_2030_60466/> accessed 08 May 2023.
24 The Government published the Roadmap in May 2018, available at <http://www.vlada.si/fileadmin/dokumenti/si/projekti/2016/zeleno/Kazipot_prehoda_v_krozno_gospodarstvo.pdf> accessed 08 May 2023.
25 Arts. 72 and 70a of the Constitution of the Republic of Slovenia (Official Gazette of the Republic of Slovenia No. 33/91-I ff). This part of the contribution to some extent summarises my article; P. Weingerl, 'Sustainability, the Circular Economy and Consumer Law in Slovenia' (2020) 9 *Journal of European Consumer and Market Law* 3, 129–132.
26 National Action Plan on Business and Human Rights (National Action Plan) was adopted in 2018. English version is available at <https://www.ohchr.org/Documents/Issues/Business/NationalPlans/Slovenia_EN.pdf> accessed 08 May 2023.
27 National Action Plan, 15. Emphasis added.
28 Resolution on National Consumer Protection Programmer 2006–2010 (Official Gazette of the Republic of Slovenia No. 114/05); Resolution on National Consumer Protection Programmer 2012–2017 (Official Gazette of the RS No. 47/12).
29 National Action Plan, 6.
30 Act Regulating Emergency Measures for Handling Waste Packaging and Graveside Candle Waste (Official Gazette of the Republic of Slovenia No. 84/18).
31 For example, Decree on green public procurement (Official Gazette of the Republic of Slovenia Nos. 51/17 and 64/19).
32 Resolution: Our food, rural areas, and natural resources after 2021(Official Gazette of the Republic of Slovenia No. 8/20).
33 See a Proposal for a Directive of the European Parliament and of the Council on common rules promoting the repair of goods and amending Regulation (EU) 2017/2394, Directives (EU) 2019/771 and (EU) 2020/1828, COM/2023/155 final.
34 Recital 32 of the preamble to the SGD 2019 provides that 'durability' should refer to the ability of the goods to maintain their required functions and performance through normal use.
35 For example, Directive 2011/83 on Consumer Rights; in March 2023, the Commission published a Proposal for a Directive on Green Claims, COM/2023/166 final.
36 Council Directive (EU) 2019/770 of the European Parliament and of the Council of 20 May 2019 on certain aspects concerning contracts for the supply of digital content and digital services (Text with EEA relevance.) PE/26/2019/REV/1, OJ L 136/1.
37 Proposal for a Directive of the European Parliament and of the Council amending Directives 2005/29/EC and 2011/83/EU as regards empowering consumers for the green transition through better protection against unfair practices and better information, COM/2022/143 final.
38 Proposal for a Directive of the European Parliament and of the Council on common rules promoting the repair of goods and amending Regulation (EU) 2017/2394, Directives (EU) 2019/771 and (EU) 2020/1828, COM/2023/155 final.
39 Proposal for a Directive of the European Parliament and of the Council on common rules promoting the repair of goods and amending Regulation (EU) 2017/2394, Directives (EU) 2019/771 and (EU) 2020/1828, COM/2023/155 final, 1.
40 S. Weatherill, *EU Consumer Law and Policy* (Edward Elgar, Cheltenham 2013) 92.
41 Report from the Commission on the implementation of the Circular Economy Action Plan, COM(2019) 190 final.
42 Communication from the Commission to the European Parliament, the Council, the European Economic and Social Committee and the Committee of the Regions, 'Closing the Loop – An EU Action Plan for the Circular Economy' COM(2015) 614 final.
43 Commission Staff Working Document, A comprehensive approach to stimulating cross-border e-Commerce for Europe's citizens and businesses, SWD(2016) 163 final, 95.

44 Council Directive 2005/29/EC of the European Parliament and of the Council of 11 May 2005 concerning unfair business-to-consumer commercial practices in the internal market and amending Council Directive 84/450/EEC, Directives 97/7/EC, 98/27/EC and 2002/65/EC of the European Parliament and of the Council and Regulation (EC) No 2006/2004 of the European Parliament and of the Council ('Unfair Commercial Practices Directive') (Text with EEA relevance), OJ L 149/22 (UCPD).
45 Criteria for fair environmental claims are included in the updated guidance on the application of UCPD. Commission Notice, Guidance on the interpretation and application of Directive 2005/29/EC of the European Parliament and of the Council concerning unfair business-to-consumer commercial practices in the internal market (Text with EEA relevance) (2021/C 526/01) (UCPD Guidance).
46 Recital 16 of the preamble to the Omnibus Directive.
47 New art. 11a UCPD, introduced with art. 3 Omnibus Directive.
48 Art. 55(2) CPA 2022.
49 Art. 56(1) CPA 2022.
50 Proposal for a Directive on Green Claims, COM/2023/166 final.
51 S. Weatherill, 'Empowerment Is Not the Only Fruit' in D. Leczykiewicz and S. Weatherill (eds), *The Images of the Consumer in EU Law: Legislation, Free Movement and Competition Law* (Hart Publishing, Oxford 2016) 203–222; for discussion on empowerment and the circular economy, see V. Mak and E. Terryn, 'Circular Economy and Consumer Protection: The Consumer as a Citizen and the Limits of Empowerment Through Consumer Law' (2020) 43 *Journal of Consumer Policy*, 227–248; V. Mak and E. Lujinović, 'Towards a Circular Economy in EU Consumer Markets: Legal Possibilities and Legal Challenges and the Dutch Example' (2019) 8(1) *Journal of European Consumer and Market Law*, 4–12.
52 See also A. Mathios, H.-W. Micklitz, L.A. Reisch, J. Thøgersen and C. Twigg-Flesner, 'Journal of Consumer Policy's 40th Anniversary Conference: A Forward-Looking Consumer Policy Research Agenda' (2020) 43 *Journal of Consumer Policy*, 1–9, https://doi.org/10.1007/s10603-019-09446-9.
53 P. Weingerl, 'Sustainability, the Circular Economy and Consumer Law in Slovenia' (2020) 9(3) *Journal of European Consumer and Market Law*, 130–131, available at SSRN: <https://ssrn.com/abstract=4331861>.
54 For the discussion on the shortcomings of the empowerment agenda in EU consumer law generally, see S. Weatherill, 'Empowerment Is Not the Only Fruit' in D. Leczykiewicz and S. Weatherill (eds), *The Images of the Consumer in EU Law: Legislation, Free Movement and Competition Law* (Hart Publishing, Oxford 2016) 203–222.
55 Art. 4 SGD 2019.
56 Art. 3 SGD 2019.
57 Art. 5 SGD 2019.
58 Arts. 6 and 7 SGD 2019, respectively.
59 Art. 7(1) (d) SGD 2019.
60 Recital 32 of the preamble to SGD 2019.
61 Recital 32 of the preamble to SGD 2019.
62 See art. 78 CPA 2022.
63 F. Gomez, 'Economic Analysis of the Directive' in M. Bianca and S. Grundmann (eds), *EU Sales Directive: Commentary* (Intersentia, Antwerp 2002) 57–59.
64 E. Van Gool and A. Michel, 'The New Consumer Sales Directive 2019/771 and Sustainable Consumption: A Critical Analysis' (2021) 4 *Journal of European Consumer and Market Law*, 141, http://dx.doi.org/10.2139/ssrn.3732976.
65 E. Van Gool and A. Michel, 'The New Consumer Sales Directive 2019/771 and Sustainable Consumption: A Critical Analysis' (2021) 4 *Journal of European Consumer and Market Law*, 141, http://dx.doi.org/10.2139/ssrn.3732976, referring to R. Van den Berge, L. Magnier and R. Mugge, 'Too Good to Go? Consumers' Replacement Behaviour and Potential Strategies for Stimulating Product Retention' (2021) 39 *Curr Op Psychology*, 66, 68.
66 See e. g. C. Montalvo, D. Peck and E. Rietveld, 'A Longer Lifetime for Products: Benefits for Consumers and Companies' (2016) European Parliament Study, available at <https://www.europarl.europa.eu/RegData/etudes/STUD/2016/579000/IPOL_STU(2016)579000_EN.pdf> accessed 09 May 2023.
67 E. Van Gool and A. Michel, 'The New Consumer Sales Directive 2019/771 and Sustainable Consumption: A Critical Analysis' (2021) 4 *Journal of European Consumer and Market Law*, 141, http://dx.doi.org/10.2139/ssrn.3732976.
68 See European Commission, Directorate-General for Justice and Consumers, Study on the costs and benefits of extending certain rights under the Consumer Sales and Guarantees Directive 1999/94/EC,

Publications Office, 2017, available at <https://data.europa.eu/doi/10.2838/590766> accessed 08 August 2023, 28–35.

69 Those technical products are defined in the Regulation on goods, for which conformity guarantee shall be issue (Official Gazette of the Republic of Slovenia No. 14/12). The CPA 2022 expressly governs also the optional commercial guarantee based on art. 17 of the SGD 2019 (see, e.g., arts. 89(2) and 92 CPA 2022).

70 Consumer Protection Act– *ZVPot* (Official Gazette of the Republic of Slovenia No. 20/98); CPA 1998. The CPA 1998 remains relevant even after the entry into force of the new CPA 2022, since it applies to contracts concluded until 25 January 2023.

71 K. Kryla-Cudna, 'Sales Contracts and the Circular Economy' (2020) 28(6) *European Review of Private Law*, 1207.

72 Art. 19 CPA 1998.

73 Art. 97(5) CPA 2022.

74 Art. 97(1) CPA 2022.

75 Art. 97(2) CPA 2022.

76 Art. 84(1) CPA 2022.

77 P. Weingerl, 'Sustainability, the Circular Economy and Consumer Law in Slovenia' (2020) 9(3) *Journal of European Consumer and Market Law*, 132, available at SSRN: <https://ssrn.com/abstract=4331861>.

78 Moreover, accompanying measures to the right to repair could be the reduction of the cost of repair services (via a reduction in the taxation of the service) and the promotion of self-repair. T. Bourgoignie, 'Contribution of Consumer Protection to Sustainable Consumption', *UNCTAD Intergovernmental Group of Experts (IGE) Meeting on Consumer Protection* (Geneva 2019), available at <https://unctad.org/meetings/en/Contribution/cicplp4_Cont_Bourgoignie.pdf> accessed 19 August 2023. See also E. Terryn, 'A Right to Repair? Towards Sustainable Remedies in Consumer Law' in B. Keirsbilck and E. Terryn (eds), *Consumer Protection in a Circular Economy* (Intersentia, Antwerp 2019) 127–147.

79 FAO, *The State of Food and Agriculture 2019. Moving Forward on Food Loss and Waste Reduction* (Rome) 156, available at <http://www.fao.org/3/ca6030en/ca6030en.pdf> accessed 09 May 2023.

80 A. Gobena, *Legislative Approaches to Sustainable Agriculture and Natural Resources Governance* (Food and Agriculture Organization of the United Nations, Rome 2020) 645.

81 As measured by weight. FAO. *Global Food Losses and Food Waste – Extent, Causes and Prevention* (Rome), available at <https://www.fao.org/3/i2697e/i2697e.pdf> accessed 15 June 2023.

82 As the European Commission highlights in the Proposal for a Directive of the European Parliament and of the Council amending Directive 2008/98/EC on waste, Brussels, 5.7.2023 COM(2023) 420 final 2023/0234 (COD) (WFD proposal) , 'the amount of water consumed to produce food that is ultimately wasted can be quantified as 342 bn m3 water eq.21, corresponding to 12% of the total impact of EU food production and consumption'. WFD proposal, p. 6.

83 Council, Food losses and food waste–Council conclusions (28 June 2016), 10730/16, 2, available at <https://food.ec.europa.eu/system/files/2017-08/fw_lib_council_food-losses-food-waste_2016.pdf> accessed 21 August 2023.

84 WFD proposal, 3.

85 WFD proposal, 3–4.

86 European Commission, EU actions against food waste, available at <https://food.ec.europa.eu/safety/food-waste/eu-actions-against-food-waste_en> accessed 08 May 2023.

87 WFD proposal, 1.

88 WFD proposal, 3.

89 WFD proposal, 6.

90 WFD proposal, 6.

91 N.D. Hamilton, 'The Role of Law in Promoting Sustainable Agriculture: Reflections on Ten Years of Experience in the United States' (1998) 3 *Drake Journal of Agricultural Law*, 428.

92 N.D. Hamilton, 'The Role of Law in Promoting Sustainable Agriculture: Reflections on Ten Years of Experience in the United States' (1998) 3 *Drake Journal of Agricultural Law*, 428.

93 A. Gobena, *Legislative Approaches to Sustainable Agriculture and Natural Resources Governance* (Food and Agriculture Organization of the United Nations, Rome 2020) 645.

94 WFD proposal, 5.

95 Commission Delegated Decision (EU) 2019/1597 of 3 May 2019 supplementing Directive 2008/98/EC of the European Parliament and of the Council as regards a common methodology and minimum

quality requirements for the uniform measurement of levels of food waste (Text with EEA relevance.), C/2019/3211OJ L 248/77.
96 Statistical Office of the Republic of Slovenia, available at <https://www.stat.si/StatWeb/en/News/Index/10967> accessed 08 May 2023.
97 Available at <https://www.gov.si/novice/2023-04-24-24-april-je-slovenski-dan-brez-zavrzene-hrane/> accessed 08 May 2023.
98 Statistical Office of the Republic of Slovenia, available at < https://www.stat.si/StatWeb/News/Index/9865> accessed 08 July 2023.
99 Available at <https://www.gov.si/novice/2023-04-24-24-april-je-slovenski-dan-brez-zavrzene-hrane/> accessed 08 May 2023.
100 Available at <https://www.nasasuperhrana.si/> accessed 08 May 2023.
101 A. Gobena, *Legislative Approaches to Sustainable Agriculture and Natural Resources Governance* (Food and Agriculture Organization of the United Nations, Rome 2020) 645.
102 Council Regulation 178/2002 on general food law establishes the protection of consumers' interests as a general principle of food law. Art. 8 of Regulation (EC) No 178/2002 of the European Parliament and of the Council of 28 January 2002 laying down the general principles and requirements of food law, establishing the European Food Safety Authority and laying down procedures in matters of food safety, OJ L 31/1.
103 Commission, Frequently asked questions: Reducing food waste in the EU, available at <https://food.ec.europa.eu/safety/food-waste/eu-actions-against-food-waste/frequently-asked-questions-reducing-food-waste-eu_en> accessed 08 May 2023.
104 Council Regulation (EU) No 1169/2011 of the European Parliament and of the Council of 25 October 2011 on the provision of food information to consumers, amending Regulations (EC) No 1924/2006 and (EC) No 1925/2006 of the European Parliament and of the Council, and repealing Commission Directive 87/250/EEC, Council Directive 90/496/EEC, Commission Directive 1999/10/EC, Directive 2000/13/EC of the European Parliament and of the Council, Commission Directives 2002/67/EC and 2008/5/EC and Commission Regulation (EC) No 608/2004, OJ L 304/18.
105 Council Directive (EU) 2018/851 of the European Parliament and of the Council of 30 May 2018 amending Directive 2008/98/EC on waste (Text with EEA relevance), E/11/2018/REV/2, OJ L 150/109.
106 UCPD Guidance, 50.
107 Art. 7(5) UCPD; UCPD Guidance, 10.
108 UCPD Guidance, 48.
109 New art. 11a UCPD, introduced with art. 3 Omnibus Directive.
110 Susanna Vince, 'Date Labelling: Diluting Food Safety Indicators' (2014), 4 available at <https://edepot.wur.nl/313034> accessed 10 July 2023.
111 Susanna Vince, 'Date Labelling: Diluting Food Safety Indicators' (2014), 4 available at <https://edepot.wur.nl/313034> accessed 10 July 2023.
112 Available at <https://www.eatresponsibly.eu/en/foodwaste/6#section-pyramid> accessed 08 May 2023.
113 Available at <https://www.gov.si/novice/2023-04-24-24-april-je-slovenski-dan-brez-zavrzene-hrane/> accessed 08 May 2023.
114 Available at <https://n1info.si/magazin/slovenska-ideja-ki-resuje-problematiko-zavrzene-hrane/> accessed 08 May 2023.
115 Council, Food losses and food waste–Council conclusions (28 June 2016), 10730/16, 5.
116 Council, Food losses and food waste–Council conclusions (28 June 2016), 10730/16, 6.
117 Commission, EU Platform on Food Losses and Food Waste, available at <https://food.ec.europa.eu/safety/food-waste/eu-actions-against-food-waste/eu-platform-food-losses-and-food-waste_en> accessed 08 May 2023.
118 Commission, Frequently asked questions: Reducing food waste in the EU, available at <https://food.ec.europa.eu/system/files/2023-04/fw_lib_reduce-food-waste-eu_faqs_0.pdf> accessed 08 May 2023.
119 Commission: Single Market Programme: Food safety, available at <https://commission.europa.eu/funding-tenders/find-funding/eu-funding-programmes/single-market-programme/overview/food-safety_en> accessed 08.05.2023; Commission: Food safety in the future EU budget (2021-2027), available at <https://food.ec.europa.eu/horizontal-topics/future-food-safety-budget-and-policy/food-safety-future-eu-budget-2021-2027_en> accessed 08 May 2023.
120 Agriculture Act (the Official Gazette of the Republic of Slovenia No. 45/08 ff.).
121 Art. 3 Agriculture Act.

122 Council Directive (EU) 2019/633 of the European Parliament and of the Council of 17 April 2019 on unfair trading practices in business-to-business relationships in the agricultural and food supply chain, PE/4/2019/REV/2, OJ L 111/59. In fact, the existing Agriculture Act contained rules applicable to UTPs already prior to the implementation of the UTP Directive. Besides the Agriculture Act, UTPs have also been covered by the general law of obligations found in the Code of Obligations (e. g. good business customs (art. 12 of Code of Obligations [Official Gazette of the Republic of Slovenia No. 97/07 ff.]) and rules on standard contract terms (found in Chapter VIII of the Code of Obligations, arts 120–121) and by the Prevention of Restriction of Competition Act–ZPOmK-1 (Official Gazette of the Republic of Slovenia No. 36/08 ff). On 29 September 2022, the new competition act was adopted, "Prevention of Restriction of Competition Act–ZPOmK-2" (Official Gazette of the Republic of Slovenia No. 130/2022) Its art. 134 contains similar definition and prohibition of unfair competition. The latter contains a special chapter on unfair competition.

123 Art. 61.m (3) of the Agriculture Act.

124 Decision of the Constitutional Court U-I-491/18-28, 6 May 2021, para. 37.

125 Art. 61.f (f) of the Agriculture Act.

126 See Eurostat: Waste generated by households by year and waste category, available at <https://ec.europa.eu/eurostat/web/waste/data/main-tables>, accessed 10 June 2023.

Bibliography

Bourgoignie T, 'Contribution of Consumer Protection to Sustainable Consumption', *UNCTAD Intergovernmental Group of Experts (IGE) Meeting on Consumer Protection*, Geneva 2019, available at <https://unctad.org/meetings/en/Contribution/cicplp4_Cont_Bourgoignie.pdf> accessed 19 August 2023

Gobena A, *Legislative Approaches to Sustainable Agriculture and Natural Resources Governance* (Food and Agriculture Organization of the United Nations, Rome 2020) 645

Gomez F, 'Economic Analysis of the Directive' in M. Bianca and S. Grundmann (eds), *EU Sales Directive: Commentary* (Intersentia, Antwerp 2002) 57–59

Hamilton N, 'The Role of Law in Promoting Sustainable Agriculture: Reflections on Ten Years of Experience in the United States' (1998) 3 *Drake Journal of Agricultural Law*, 428

Kryla-Cudna K, 'Sales Contracts and the Circular Economy' (2020) 28(6) *European Review of Private Law*, 1207

Mak V and Lujinović E, 'Towards a Circular Economy in EU Consumer Markets: Legal Possibilities and Legal Challenges and the Dutch Example' (2019) 8(1) *Journal of European Consumer and Market Law*, 4–12

Mak V and Terryn E, 'Circular Economy and Consumer Protection: The Consumer as a Citizen and the Limits of Empowerment through Consumer Law' (2020) 43 *Journal of Consumer Policy*, 227–248

Mathios A, Micklitz H-W, Reisch LA, Thøgersen J and Twigg-Flesner C, 'Journal of Consumer Policy's 40th Anniversary Conference: A Forward-Looking Consumer Policy Research Agenda' (2020) 43 *Journal of Consumer Policy*, 1–9, https://doi.org/10.1007/s10603-019-09446-9

Montalvo C, Peck D and Rietveld E, 'A Longer Lifetime for Products: Benefits for Consumers and Companies' (2016) European Parliament Study, available at <https://www.europarl.europa.eu/RegData/etudes/STUD/2016/579000/IPOL_STU(2016)579000_EN.pdf> accessed 09 May 2023

Terryn E, 'A Right to Repair? Towards Sustainable Remedies in Consumer Law' in B. Keirsbilck and E. Terryn (eds), *Consumer Protection in a Circular Economy* (Intersentia, Antwerp 2019) 127–147

Van den Berge R, Magnier L and Mugge R, 'Too Good to Go? Consumers' Replacement Behaviour and Potential Strategies for Stimulating Product Retention' (2021) 39 *Curr Op Psychology*, 66, 68

Van Gool E and Michel A, 'The New Consumer Sales Directive 2019/771 and Sustainable Consumption: A Critical Analysis' (2021) 4 *Journal of European Consumer and Market Law*, 141, http://dx.doi.org/10.2139/ssrn.3732976

Vince S, 'Date Labelling: Diluting Food Safety Indicators' (2014) 4 available at <https://edepot.wur.nl/313034> accessed 10 July 2023

Weatherill S, *EU Consumer Law and Policy* (Edward Elgar, Cheltenham 2013) 92

Weatherill S, 'Empowerment Is Not the Only Fruit' in D. Leczykiewicz and S. Weatherill (eds), *The Images of the Consumer in EU Law: Legislation, Free Movement and Competition Law* (Hart Publishing, Oxford 2016) 203–222

Weingerl P, 'Sustainability, the Circular Economy and Consumer Law in Slovenia' (2020) 9(3) *Journal of European Consumer and Market Law*, 130–131 available at SSRN: <https://ssrn.com/abstract=4331861>

PART 3

Coordinating Public and Private Sustainability Law in Europe

11

SUSTAINABILITY AS A REGULATORY PRINCIPLE AND THE RULES OF ECONOMIC DEVELOPMENT

The Italian Legal System

Guido Alpa[1]

11.1 Sustainability, a Political and Economic Term

The term 'sustainability' has only recently been introduced into the Italian legal vocabulary and, like all terms with a verb suffix in '-able', expresses the possibility – or necessity – of what is predicated in the verb. From this point of view, therefore, it is not a term that defines a certain semantic boundary, but, if anything, it expresses a path *in progress* and, at the same time, a criterion of judgement, which allows to verify whether what is stated in the verb has then been fulfilled. By way of example, the term 'possible' is often used in the civil code, which allows the judge to verify, on the basis of the situation in which the subject finds himself, whether he was able to perform what was required by the normative precept, and to measure his behaviour on the basis of standards, such as diligence, declined in its various meanings, or on the basis of general clauses, such as good faith and fairness, possibly taking into account what normally happens in similar situations. Another significant example is given, in the language of the theory of private law, by the term 'enforceable', which implies a check on the performance of obligations: according to the rules of good faith and fair dealing, the debtor may not be required to make an effort that extends to the limit of the 'impossible'. What is 'sustainable' precisely expresses a yardstick that checks what has been sustained in order to ascertain whether – in the conditions in which the party was placed – a greater effort could have been required of it or whether, under the circumstances, what was performed should be considered sufficient.

The vagueness of the term is inherent in the elasticity of general clauses, which are in themselves necessarily undefined.[2]

It is precisely for these reasons that it may be useful to reconstruct its entry into the world of law, given that, from that moment, the term becomes a *normative concept,* and as such expresses a command, which, if not observed, implies a sanction, or as we used to say, the already-mentioned yardstick, on the basis of which to establish whether or not a sanction is applicable.

The first official document in which the subject was enunciated – without, however, claiming to be imperative – was in 1972, at the United Nations Conference on the Human Environment

 DOI: 10.4324/9781032662046-14

in Stockholm. Here, attention was drawn to the fact that, in order to sustainably improve the living conditions of mankind, natural resources must be safeguarded for the benefit of all and international cooperation is required to achieve this goal. In Stockholm, emphasis was therefore placed on the need to protect the environment in which human beings live and to find a solution to environmental problems, without, however, forgetting the social, economic, and developmental aspects of populations.

The premise underlying the Concluding Declaration is that the man has a fundamental right to freedom, equality and satisfactory living conditions in an environment that allows him to live in dignity and well-being and is highly responsible for the protection and improvement of the environment before future generations.

It forms the basis for any programme of action to make the implementation of this right possible. We are in the presence of an *agreement between states,* and precisely the intentions expressed concern the behaviour of states: private individuals, men considered as such, are the recipients of the effects of that behaviour, effects that also include 'future generations'.

The text states, *inter alia*, that resources shall be made available to preserve and improve the environment, taking particular account of the special needs of developing countries, the costs they will incur in introducing environmental protection into their development programmes and the need to provide them, if they so request, with international technical and financial assistance for this purpose. And for a more rational administration of resources and thus improving the environment, States shall, in planning development, take integrated and coordinated measures to ensure that such development is compatible with the need to protect and improve the human environment for the benefit of their people.

The environment is the main focus of the Declaration, precisely because, being understood as a human *habit*, it is the framework within which the current and future survival of humanity is at stake.

In 1987, also as part of the UN's work, the President of the *World Commission on Environment and Development* (WCED), Gro Harlem Bruntland presented the report 'Our *common* future'.

The Report spans more than three hundred pages, is divided into three parts (concerns, challenges, commitments), and outlines an agenda that is handed over to states in order to control, improve and save the environment. Businesses are not the recipients of the commitments, yet States, in order to fulfil the commitments, must intervene in the economy in order to achieve 'sustainable development'.

The Brundtland report '*Our Common Future*' notes that the global environmental hotspots and problems are mainly due to the great poverty of the global South and the unsustainable production and consumption patterns of the global North. The report, therefore, highlights the need to implement a strategy that integrates development and environmental needs. This strategy has been referred to, in English, as 'sustainable development', which is currently in wide use.

'Sustainable development is a development that enables the present generation to meet its own needs without compromising the ability of future generations to meet their needs' (WCED, 1987).

The protection of nature in coordination with the expansion of human rights has been the subject of successive elaborate international documents and conferences that have set out a number of principles and objectives to be achieved by governments to promote economic development while protecting the environment. These include in In Rio de Janeiro, the UN Conference on Environment and Development (Earth Summit, 1992); in South Africa, the World Summit on Environmental Sustainability (2002); in Monterrey, Mexico, the International Conference on Financing for Development (2002); in Doha, Qatar, the Second International Conference on Financing for Development (2008); the first European document, the 'Europe 2020' strategy for smart, sustainable, and inclusive growth (2010); in Rio de Janeiro 'Rio+20', the World Conference on Sustainable Development (*The Future We Want, 2012*) and, in New York, the Summit for the Adoption of the 2030 Agenda for Sustainable Development (2015).

An examination of these documents shows how the strategy that considered only states as recipients of commitments have gradually shifted to a more extensive and all-encompassing strategy, in which companies and individuals are also involved, so that the coordinated efforts of public and private institutions can achieve the desired goal.

The direct involvement of companies means that this issue does not only develop within the framework of macroeconomic and international policy analyses but extends to encompass economic-business analyses.[3]

The debate on the rational distribution of resources, on the costs of environmental protection, and on human rights programmes has intensified across the planet.

In order to emphasise the relevance of the protected interests, some constitutions of less developed countries, such as those of South America, have included protection obligations to be observed with regard to new subjects of law.

The Constitutional Court of the Republic of Colombia recognised the Colombian Amazon Rainforest as a subject of law. In its ruling No. 4360 of 5 April 2018, the Supreme Court of Justice defines nature as a '*sujeto de derechos*' that has been granted the same rights as a natural person, including the right to life. Legal subjectivity has given this vital ecosystem for the world the right to protection, conservation, maintenance, and upkeep incumbent on the state and territorial entities of the Amazon basin, and in the ruling, urgent action was ordered to stem the deforestation of the rainforest. The ruling of the Supreme Court of Justice also focuses on analysing the Colombian Constitution from the perspective of environmental protection, considering it as a '*Constitución Ecológica*' or '*Constitución Verde*', and elevating the concept of environmental protection to a fundamental human right. In fact, in the aforementioned judgement, the Supreme Court of Justice emphasises that the environment constitutes a right of constitutional rank, contained in Chapter III of the Colombian Constitutional Magna Carta, under the heading 'collective and environmental rights' (Articles 79 and 80). In any case, 'the principle of sustainability has become a benchmark for economic and social choices on a global level'.[4]

11.2 Sustainability as a Normative Concept and the Joint-Stock Company Debate

It is more difficult to trace the path on the legal terrain, because, beyond the textual relevance of the term, which will be discussed in a moment, sustainability, i.e., the achievement of objectives that make private economic initiative compatible with the protection of human rights and the environment, lies at the heart of a debate that involves corporate *social responsibility*, corporate governance, corporate social responsibility reporting and other technically complex aspects.

On these occasions, the topic 'sustainability' can be assessed from different perspectives. The first reference is to the 'social theory of business' as a European-style theoretical basis, to be examined in the context of general business discipline.

It was already being discussed before the introduction of the civil code of 1942 (and, in particular, within the framework of corporate law) and then the discourse became more accentuated during the 1950s. It started from a basic consideration, also accepted in the text of the code, namely the distinction between the interest of the company and the interest of the shareholders.

The analysis of the interests protected in the context of the legal regulation of activities considered it unquestionable that the company had to realise private ends – the realisation of profit – because, in a free market in which private initiative could be carried out without being bent to the interests of the state, the maximisation of profit was, so to speak, the natural aim of the enterprise. Hence the reactions of those who, when faced with programmes to convert private enterprise to the realisation of public interests, spoke out against the 'functionalisation' of private enterprise.[5]

Article 41 of the Italian Constitution (1948) was to be understood as a precept that allowed the legislature to limit economic initiative for reasons of *social utility*, but this did not imply that the individual enterprise should, on its own initiative, set itself objectives to protect interests other than that of the enterprise itself. 'Social utility' is a general clause, and it has a very pregnant political meaning: it allows the legislator to introduce limits to private economic initiative for the realisation of social goals; it represents in some way a compromise between the free market and state direction of the economy.[6]

If anything, the extent to which shareholders' interests should be taken into account, in particular their right to the division of profits and the dividend, was debated.

To this question the prevailing doctrine gave an answer that qualified the company as an institution autonomous from the shareholders, the bearer of its own interest, the '*social interest*', to which the shareholders' right to receive profits was conditional upon the availability of resources after the needs of the company, functional to the attainment of its purpose, had been satisfied.

This position was overtaken in the following period by many Italian scholars who had begun to discuss the meaning and content of the 'social interest' formula.[7]

At the end of the last century, it was agreed that there is no interest of the partners other than that of the company since the partners within the company concur to develop a common interest; if anything, one could argue between the interests of majority and minority partners. And, taking into account the theories based on the American-style economic analysis of law, the conception of the company has prevailed not as a separate entity but as a 'bundle of contracts', i.e., as an expression of the contractual will of the shareholders that manifests itself in the *governance* mechanisms that are being modified to allow for the emphasis to be placed on the particular rights of individuals or groups of shareholders vis-à-vis others with respect to operational management, the definition of strategic objectives, and the company's policy of remuneration of investments.

The doctrine was also implemented in Italy with the 2003 company law reform and the expansion of the protection of the rights of individual shareholders' interests in the management dynamics of the company.

However, the regulatory framework for defining the content of the corporate interest that the company must pursue has become more complicated in relation to the extension of interests external to the shareholder structure that find affirmation and consequent protection in the legal system and that in fact complement the interests of shareholders. It is increasingly necessary to combine in the management of the company the interest in profit with the protection of widespread interests of the civil community in which the company operates with the constant development of the attention due to the care of these external interests as well.

This is not yet a theory of the socialisation of the company, but it is certainly an integration of the content of the company's own interest, which therefore stands as an autonomous entity with respect to its shareholders. In this context, the role of subjects external to the corporate structure, generically referred to as 'stakeholders', is also being affirmed, the definition of which has created and continues to create misunderstandings as to the determination of the perimeter of the subjects and interests included therein, so that we have turned towards a very broad vision of the same notion, leaving it to the directors and their discretionary evaluation to define the relative contents according to the company's concrete activity and the scope of its relations with the social and natural environment.

The result was a further evolution of the doctrine, and a return of the 'institutional' conception of the company, understood as an expression of an autonomous entity common not only to the shareholders but also to other stakeholders to be protected in the management of the company.

Recently, it has also been argued that the contractualist theory of society is to be considered abandoned.

The debate is still ongoing between the supporters of prioritising the protection of the interests of *stockholders*[8] and those who, appealing to *Corporate Social Responsibility* and the protection of stakeholders, believe that the company, even if under private law, must take on the protection of interests other than that of the shareholders, who are involved in its activities, such as employees, suppliers, customers, users, and then, of course, the environment and human rights.[9]

In other words, the *shareholders*' primacy is no longer a dogma[10] and their power must be shared with the holders of 'other' interests than those of the shareholders.

In the 2021 book on '*Stakeholder* Capitalism' by Klaus Schwab, founder of the *World Economic Forum*, a model is promoted according to which private companies are not only profit-seeking entities but also act for the benefit of the wider society (not only the shareholders). Thus, there is a shift from '*shareholder capitalism*' to '*stakeholder capitalism*' *with the* need to provide a more articulated indication of the objectives of business activity.

The debate has also penetrated into Italy, where it has become intertwined with the discussion on the notion of 'company', the purpose of the company, and the categories of protected interests, with the discussion on corporate social responsibility and the contents and prospects of the social business economy.[11]

Sustainability goals are now typified on the scale of interests.[12] They are certainly remarkable goals, but it is one thing to regard them as goals that states, on the basis of international cooperation agreements, can achieve; a quite another thing is to propose them to private individuals.

Of course, some of them are already the subject of constitutional provision, others of articulated legislation, e.g., health protection, water, or land use; others are not feasible as a goal of private individuals, but '*big corporations*' can help pursuing them. Here they are in the order indicated by the 2015 UN Declaration (*Transforming our world: the 2030 Agenda for Sustainable Development*, on the *United Nations – Sustainable Development knowledge platform*).[13]

11.3 The Normative Path of the 'Sustainability' Concept

The European Union began to address this issue in the early 2000s with documents limited to specific aspects, workers' rights, minority shareholders' rights, consumer rights and environmental protection. This economic and social policy – which goes back to the Germanic theory of the social market economy – is much debated, both by those who consider it restrictive of free economic initiative and by those who see in it an image of a capitalism 'with a human face'.

The legal framework that currently informs corporate social responsibility and translates the principle of sustainability into legal terms consists mainly of:

a Directive 2014/95/EU (the so-called 'Non-Financial Reporting Directive'), which imposes disclosure requirements with regard to the management of environmental and social risks associated with a company's activities. The related discipline is based on the vision of a global role of the company as an operator that must look at the *long term* and therefore also take care of social and environmental factors in order to gather the trust of investors. It is the application of social market theory and the affirmation of interests that must take into account and be compatible with social and environmental aspects.

b Legislative Decree No. 254 of 2016 (for the implementation of which the Consob Regulation set out in Resolution No. 20267/2018 was also issued) implementing the Non-Financial Reporting Directive (2014/95/EU). The Decree and the provisions contained therein require large companies that qualify as relevant public interest entities to prepare and publish a so-called 'non-financial statement'. This declaration must contain, for each financial year, information relating to environmental, social and personnel issues, respect for human rights and the fight against active and passive corruption, aimed at ensuring an understanding of the company's activity, its performance, results, and impact. Thus, specific rules are introduced for the disclosure of information to the public, binding on companies and directors who are required not only to manage their economic and financial activities correctly in the pursuit of profit but also to take responsibility for the external impacts of these management activities and provide adequate external information. Sanctions are foreseen for missing or inaccurate information to the market.

c Directive 2017/828/EU on shareholders' rights, transposed into our law by Legislative Decree No. 49 of 2019 (the '*Shareholders Rights 2*'), which, with a view to strengthening the pursuit of non-financial objectives, intervenes on the remuneration policies of directors and management with the obligation to give evidence of the connection of such policies with the 'corporate strategy, long-term interests and with the sustainability of the company'.

The introduction of specific regulations concerning the 'non-financial' aspects of company management complements the duties of directors and justifies the use of resources to protect interests other than those of shareholders.

These are interests that, insofar as they are protected, increase the value of the shares in the long run. The regulatory provision makes it possible to exclude the possibility that their protection can be imputed to errors or shortcomings in the management of the company, and thus shields directors and managers from actions brought by shareholders, since in the context of the economic management of the company, the use of funds intended for these purposes cannot be censured.

With the attraction of such interests to the care of the directors, they are called upon to carry out a weighting activity whose evaluation, also by the shareholders and other subjects with protected interests, should be considered as referring to the application of the '*business judgement rule*'. The assessment should in fact only concern the diligence in reasonably evaluating and weighing the elements of risk and opportunity for the care of the social and environmental impact of the business activity.

In recent times, a desire has developed in the EU to direct companies more incisively towards taking care of these interests, imposing more specific rules with a focus on the subject that has led to the approval of regulations for the pursuit of 'sustainability' in certain sectors and for financial services and investments. In this context, in harmony with the goals set forth in the UN 2030 Agenda for Sustainable Development, non-financial goals, the SDGs, have been defined, on which the activities of companies should focus, and which also represent the parameters for evaluating the actions carried out by them in pursuit of these goals, which must be reported in the Non-Financial Document.

The obligation to draw up a social balance sheet, which is reserved by the 2014 directive for listed companies, banks, and insurance companies, responds to the need to *inform the market*. There is still no obligation to allocate resources for these purposes, but it is clear that if a 'balance sheet' is to be drawn up, it is necessary to document the initiatives taken by directors and managers to achieve those purposes.

With reference to sustainable finance, the divergence in regulations and application between the various EU Member States led to the approval of the EU Regulation 2020/852 on Taxonomy, ratified on 18 June 2020, which is directly applicable in the Member States, with the establishment of a unified classification system for sustainable activities in which the criteria for determining whether an economic activity can be considered environmentally sustainable are set out.

In March 2021, Assonime[14] produced a document that also contains a summary of the legislation in force in the EU countries with an in-depth discussion of their respective contents. The Assonime document focuses in particular on the directors' responsibility for the pursuit of non-financial objectives and the most appropriate *governance* rules. Two alternative hypotheses are indicated in this regard:

a incorporation of the objectives, or even just the principle of sustainable development, in the corporate purpose stated in the company's articles of association. This would directly expand the corporate interest or purpose itself and would represent a structural change in the model that binds not only the directors but also the shareholders, to the extent of their respective competences;
b provide for the care of non-financial aspects among the directors' duties. This would add an element affecting the assessment of their diligence in the performance of their duties, but would not affect the company's purpose, which would remain the typical one of pursuing the profit to be distributed among the shareholders.

Assonime favours solution (b), which has been incorporated into the new *Corporate Governance* Code, in keeping with the policy of leaving the regulation of corporate *governance* rules to voluntary compliance with recommendations (so-called 'soft law') rather than rigid and binding rules, in order to allow for the best flexibility in defining the organisation of each company and the care of interests, also taking into account the vagueness of the expression 'sustainable success' and the content of the 17 Sustainable Development Goals (SDG's) themselves.

For some of these SDGs goals, it seems difficult to relate their achievement directly to the tasks and responsibilities of individual companies and therefore of individual directors, since in some cases they may represent mere wishes for a better world, such as the desire to end all forms of poverty in the world or to end world hunger, and in other cases, they are an orientation towards concrete actions rather than indications of concrete obligations. Still others require the necessary contribution of public initiatives that are beyond the management or intervention capacities of individual societies.

Therefore, it is a matter of examining on a case-by-case basis how and with what intensity the company's action must take care of the individual aspects of environmental and social impact and the attainment of the related objectives in order to assess whether there can be a liability of the directors for the lack or insufficiency of adequate initiatives in company management.

On 21 April 2021, the European Commission presented a proposal for a directive on the publication of corporate sustainability information amending Dir. 2013/34/EU, Dir. 2004/109/EU, Dir. 2006/43/EU and Reg. (EU) No. 537/2014. The text of the proposal was in consultation until 22 June 2021, and was subsequently transmitted to the European Parliament and the EU Council for adoption under the ordinary procedure. The proposed Directive amends the provisions of the Non-Financial Information Directive and provides, *inter alia, for the* extension of the scope of the preparation of the Non-Financial Document to medium-sized companies.

11.4 Three Paradigmatic Cases of Corporate Social Responsibility for Violation of Fundamental Rights: Vedanta, Okpabi, Milieudefensie

Within a short period of time, different courts in the United States of America, the United Kingdom, and the Netherlands have made relevant decisions on corporate social responsibility, applying rules of domestic law but taking into account resolutions and documents of international organisations, primarily the United Nations and the OECD. It could be debated whether the interests affected are, in addition to health and property, those traditionally assigned to individuals – even collectives – living in a given area or whether, in addition to these interests, the measures are concerned with protecting a broader interest aimed at defending the environment and future generations. In any case, it is relevant to emphasise how, following different paths of argumentation, the judges have sought to protect populations once subject to colonial exploitation and today to exploitation by large *corporations* by resorting to the social responsibility of the company, and the responsibility of the parent company for harmful acts of the subsidiary.

Recourse to private law instruments exempts judges from appealing to public law disciplines with detailed prescriptions, the general clauses operating (especially) in the area of civil liability being sufficient to prepare the ground on which to base injunctive or compensatory measures.

The facts of the *Vedanta* case take place in Zambia (the former Rhodesia), a country belonging to the British Commonwealth, and concern villagers who, as a result of the activities of two companies of the Vedanta group, claimed to have suffered personal injury, damage to their property, and loss of profits, together with loss of enjoyment of the environment, as a result of the dumping of copper mine waste managed by a company set up ad hoc on-site by the defendant group. The slurry had been pouring onto the soil and into the groundwater for more than a decade.

The question is primarily one of jurisdiction: should villagers turn to Zambian courts or English courts to defend their rights? Does the damage stem from the activity of the subsidiaries, which is carried out locally, and are they based in the country, or can it also involve the parent company?

The subsidiaries are owned by the holding company, which is listed in India but has its registered office in London. The plaintiffs insist on the jurisdiction of the English court on the grounds that Zambian courts are unreliable, and the Zambian court is a '*forum non conveniens*'.

The matter goes through three degrees of judgement.

The High Court upheld the Zambians' action, ruling out the possibility that they had abused EU law: since the local companies do not have adequate resources to compensate for the damage, and it is legitimate to suspect that the holding company had planned the water pollution, the court held that it is possible, in the abstract, to base an action on the non-contractual liability of the holding company since there is a *duty of care* – characterised by foreseeability, proximity and reasonableness – aimed at protecting the plaintiffs' interests. The holding company's liability – *parent liability* – arises from the holding company's control and direction over the companies operating there. And in fact, the court held, the business carried on by the holding company and the subsidiaries was the same, the holding company should reasonably have known how the business carried on by the subsidiaries would be carried on, with serious damage to the inhabitants; the holding company had greater experience and expertise than the subsidiaries, it should have realised the need for greater expertise and experience on the part of the subsidiaries.

Apart from the arguments concerning the application of the principle of *forum non conveniens* linked to Article 4 of the Brussels Convention, the court is concerned with whether the companies operating there are viable, whether they cannot run the risk of bankruptcy, whether the aggrieved populations, living in extreme poverty, are able to have recourse to other courts and whether they can be supported in defending their rights with legal aid. All these reasons militate in favour of rooting the litigation in the United Kingdom.

The Court of Appeal follows the decision of the High Court. Vedanta insists on its arguments.

The matter came before the Supreme Court, which, in a wide-ranging and educated reasoning, confirmed the previous judgements.

It is interesting, to examine, rather than the debated question of the abuse of the application of the Brussels Convention to root the issue in the United Kingdom, to follow the Court's reasoning regarding the liability of the parent company for tortious acts committed by the subsidiary.[15]

The cited precedent[16] had upheld the plaintiff's claim for damages, who had been injured by a subsidiary of the defendant. The judges of the Court of Appeal had held that, as a general rule, due to the fact that the two companies are different legal persons, the parent company does not have a duty to monitor the conduct of the subsidiary; however, when it is put in a position to monitor the conduct of *managers*, to express opinions and views on the acts performed, to be informed about the use of assets owned, it is reasonable to place on the parent company the burden of exercising due diligence to prevent the damage.

Well, in the case examined, the Court considers that Vedanta exercised such control over the subsidiary's activity that it was aware of the infiltration that had caused serious damage to the plaintiff populations.

After carefully examining Vedanta's defences, the court finds that the mining company violated local environmental protection regulations. Following a detailed analysis of the precedents, the Judge Rapporteur refutes all arguments used by Vedanta, and finds that the appeal can be argued.

For our purposes, it is worth considering an aspect of substantive law and an aspect of a formal nature.

The first concerns social justice: we are in the presence of poor populations, whose income comes from livestock and agriculture, from an economic activity that provides simple survival support, and which fits into the biological cycle and adapts to the environment perfectly. It is the outsourcing of mining that alters the environment and makes farming problematic.

It should be noted that the purpose of the action is to obtain compensation for the irreversible transformation of the land caused by the mining activity. Interruption or suspension of the activity is not required.

There is no discussion of the legal nature of the injured interest, nor, as would be the case here, whether we are in the presence of 'common property', of *ad hoc* interests, of collective or diffuse interests, it being clear, however, that the mining exploitation of the land is subsequent to its previous use for cultivation and livestock farming.

From a formal point of view, it is interesting to emphasise both the configuration of the facts as damaging protected interests within the scope of civil liability and, above all, to highlight the imputation of liability also to the parent company, on the basis of the exercise of management and *control* powers.

Hence a number of inferences: that the protection of the environment and the orographic consistency of a place can be coordinated with each other, taking into account past uses and the generational future; that private law can fill the gaps in the public system, at least in terms of reinstatement of patrimonial damage; that risks can be allocated between group companies, overcoming the problems of subjectivity and patrimonial liability, by resorting to *parental liability.*

A rather similar case was recently decided by the English Supreme Court. This time, it concerned a large number of inhabitants of the Niger delta area. The inhabitants were complaining of very serious damage – to life, health, the environment – resulting from water pollution caused by oil leaking into pipelines by the extracting companies and their subsidiaries. The parent company – *Royal Dutch Shell plc* – is registered in the United Kingdom. The extraction activity and its transportation are carried out by a subsidiary based in Nigeria. Again, the action against the parent company was based on non-contractual tort and the subsidiaries' management and control activities.

The reasoning of the judgement starts from the premise of the Vedanta case, namely the existence of a *duty of care on the part of* the parent company. The premise is not entirely secure, of course, because the existence of the parent company's liability for acts imputed to the subsidiary is still disputed in the United Kingdom. And yet the Supreme Court notes that, in the present case, it was established that the parent company had greater experience than the subsidiary, as well as greater expertise and resources to be able to verify the damage caused to life, health and the environment by the extraction and transportation of oil.

In order to prove the involvement of the parent company – in other words, the *proximity of* the UK-based company to the aggrieved persons residing in Nigeria – the rapporteur carefully examines the defences of the aggrieved persons, concerning the internal structure of the parent company and the acts of the different bodies concerning the mining activity. That is the deliberations that can be ascribed to the network of controls organised by Shell. It, therefore, examines the activities of the *corporate governance* body, sustainability reports, with particular emphasis on analyses of the effects of the activity on the environment, and also communications between parent company and subsidiary.

The facts ascertained by the appellate court and examined by the Supreme Court, however, had shown that the parent company is a holding company with a small number of employees, which does not carry out operational activities and has no particular expertise in oil extraction so that all decisions on how to carry out extraction and pipelines were the responsibility of the subsidiary.

In order to give a correct answer to the question of whether the parent company should be held liable for the damage caused by the subsidiary, the relator examines – in fact – five aspects: (i) the existence of mandatory rules imposing on the extractive company particular conduct, (ii) the imposition of obligations to take proceedings, (iii) the imposition of a system of controls, (iv) the imposition of a system of expenses, and (v) the existence of a system of direction and supervision over the subsidiary. To these questions, the rapporteur responds to the effect that the *proximity* requirement, although facilitated by the evidence, does not lead to a convincing and persuasive solution in the affirmative, since most of the controls are carried out on the basis of standard procedures, which standardise the activities of all the subsidiaries in the group. So, it is for the requirement of fairness and substantive justice, which does not seem appreciable to the reporting judge.

The opinions expressed by the other two judges are varied and all thoroughly argued. And the dissenting opinion of one of the judges of the panel insists on the existence of proximity and thus the applicability of the precedent (*Vedanta*) also to Shell.

The third court concludes that the proximity requirement is not sufficiently proven, and therefore acquits Shell.

This case also shows how complex the assessment of corporate social responsibility is, and that, although conceivable in the abstract, the parent company's liability cannot be presumed, but must be established in fact, on the basis of *duty of care*.

The third case also shows similarities with the previous ones; it was decided against the parent company Shell by the District Court of The Hague on 26 May 2021; the decision applies the Dutch Civil Code. The text was also published by the Court in an English version.[17] This is one of the first cases of violation of CO_2 emission standards: Shell was not found guilty in the past but was obliged to immediately implement all necessary measures to reduce emissions caused by the group's activities in the future.

Here again, the rules invoked are of private law and generally concern the requirements of the wrongful act.

First, the issues of legal standing. The plaintiff was an environmental protection association and acted on behalf of two other associations and four foundations engaged in the same field. The defendant was the holding company of the Shell group, as seen above registered in the United Kingdom but with its head office in the Netherlands. As the reasoning makes clear, it is for the parent company to define the policies of the subsidiaries, which, as operating companies, are engaged worldwide in the extraction, processing, and distribution of oil, through a system of service companies coordinated by the parent company. The Court dwells on the side effects of the business activity, pointing out the harmful aspects of CO_2, climate change and the progressively deteriorating global situation as represented by the UN studies on the subject. Associated with these effects are the damage to health, the level of livability, and the risks to the entire world population. For pages and pages, the explanatory statement describes the situation around the globe, in Europe and the Netherlands, enumerates all the major measures and findings taken by the UN, and resolves preliminary legal issues.

Firstly, the legal standing of associations and foundations. This concerns the application of Article III, 305a of the Civil Code, which admits the standing of associations and foundations in the promotion of *class actions*. Having to verify whether the interests defended by these entities could fall within the scope of the provision, the Court asked whether the interests of future generations could be such as to ground an obligation attributable to the defendant company as a useful safeguard of the *stakeholders*' interests.

Given the wording of the rule, it excludes that such interests can be protected because populations scattered around the world can be harmed at different times by climate change and in different ways. However, with regard to Dutch residents, the Court considers that some of the associations may be considered legitimate, taking into account their statutory purposes. The Court then wonders whether the harm suffered should be that which aggregates to the CO_2 situation worldwide or only concerns the population residing in the Netherlands, and inclines towards this alternative.

Very interesting are the arguments concerning the configuration of the tort, since the holding company had contested its liability, assuming that the definition of the '*corporate policy*' was to be considered only a preparatory act that went beyond the boundaries of the tort, since corporate policy definitions are not in themselves a cause of immediate harm. In the Court's view, it is precisely the definition of company policy that must be regarded as constituting an independent wrongful act in itself. In other words, the definition of company policy constitutes a wrongful act if it contradicts an (unwritten) standard against which to measure the diligence required to administer a company.

The examination of the factual findings concerning the CO_2 emissions in the region of the residents belonging to the plaintiff association showed, in the Court's assessment, that the activity of the Shell group was potentially harmful.

But what interests have been harmed?

The plaintiff identified the infringement of human and fundamental rights such as respect for life, *privacy,* and family: these are Articles 2 and 8 of the European Convention and the Charter of Fundamental Rights of the Union, as well as Articles 6 and 17 of the International Covenant on Civil and Political Rights approved by the United Nations in 1966 and entered into force in 1976.

Corroborating the series of documents and measures useful for the configuration of the illicit act, the Court recalls the determinations of the UN Committee on Human Rights on the deterioration of the climate and the environment, the UN Guiding Principles in which the responsibility of States is distinguished from the responsibility of companies, so that the distinction does not imply an absolution of companies, but rather their 'individual responsibility' also in monitoring the measures adopted by States. This obligation also emerges from the OECD Guide for Multinational Enterprises (Principle No. 23), which reminds companies of the necessary protection of fundamental rights, 'in order to protect the environment and health in the places where they operate by contributing to the achievement of sustainable development objectives'. Hence the obligation of companies to respect fundamental rights and to take all necessary measures to achieve this, which does not imply a passive attitude, but if anything, an active cooperation. This obligation applies to all companies of all sizes, subject of course to the principle of proportionality.

This obligation must be observed by all companies in the production and distribution chain, according to the criteria developed by the Report prepared by Oxford University in 2020.

On the basis of the factual data collected in the course of the procedure concerning the CO_2 emissions of the entire Shell group, the Court considers that the parent company has a *result obligation* (paragraph 4.4.23) to prevent and reduce the risks associated with its activity, and this in respect of the entire group over which the holding company exercises its power of direction and control. Hence, a series of measures of a technical nature is indicated by the Court to the parent company to achieve that result.

But the Court does not stop there. It also takes into account an acute argument raised by the defendant's defence, namely the futility of the measures taken because in the oil market the space freed up by Shell in complying with its obligation to perform will be occupied by its competitors. The Court does not fall into the trap: it reiterates the usefulness of the measures and the irrelevance of the conduct of competing companies in relieving Shell of its obligation. After all, individual cooperation contributes to the collective welfare.

It is through the cooperation of public bodies – states *first and foremost* – and private entities that the goals set by researchers to halt the degradation of the planet can be realised.

To achieve this, the obligations imposed on companies must be *appropriate* and *proportional*. Hence the need for Shell to adapt its corporate policies, and the order to reduce emissions proportionally immediately enforceable.

However, the Court ruled out damages as it did not consider the group's emissions up to that point to be illegal.

These cases have been widely discussed because they show how sustainability can also be taken into account in the judicial review of companies' activities.[18]

11.5 Reform of the Italian Constitution for the Protection of the Environment and Animals

On Tuesday, 8 February 2022, the Italian Chamber of Deputies approved the Constitution reform law that amended Articles 9 and 41 to introduce environmental and animal protection into the text of the Charter. The process of the constitutional law was long and troubled, starting on 3 April 2018[19]; it was endorsed by some political parties, predominantly progressive, but it also registered adhesions from individual members of conservative parties. After the second reading, it became law with a majority of 468 votes. The objective pursued is briefly described in the title, which focuses on the environment, but the approved text adds a third paragraph to c.2 of Article 9 ('[the Republic] Protects the landscape and the historical and artistic heritage of the Nation'): 'Protects the environment, biodiversity, and ecosystems, also in the interest of future generations. The law of the State regulates the ways and forms of animal protection'. One of the draft constitutional laws that later merged into the unitary text had also included the expression 'promoting sustainable development', which fell during the course of the process.[20] In addition to this paragraph, the constitutional law adds to the second paragraph of Article 41 after the word 'damage' the words 'to health and the environment' and in the third paragraph at the end the adjective 'environmental'. What emerges is a picture that outlines a long-term project: the purpose, raised to constitutional status, of our country to contribute with other states to the protection of the environment, the climate, and biodiversity, and to make this a specific purpose of the Republic; the purpose of protecting future generations; the purpose of raising animals to a status that, although not equal to that of persons, considers them a value that the Republic intends to protect; the purpose of directing economic initiative towards purposes that do not conflict with the values mentioned.

It is barely necessary to point out that these amendments – which are certainly appreciable – constitute an updating of the text that is in line with many other recent constitutions and with updates introduced in constitutions more or less contemporary to ours that refer to the environment and, in certain respects, translates into explicit rule the interpretative orientations that have long been accredited in our jurisprudence, on the basis of a complex and cultured doctrinal tradition.[21]

One might think that the previous interpretation of the provisions of the constitution, in particular the correlation of Article 9 on landscape protection with Article 41 mentioned above, made the interaction of the text unnecessary, but in fact the political significance of the integration is very relevant: it extends the regulatory scope and is specifically concerned with protecting the environment, animals, and future generations. Hence also the idea that the new text does not embrace a static idea of the environment centred on its conservation, but advocates a dynamic idea of the environment, aimed at the development of protection techniques as well as the development of a sustainable economy in the decades to come.

As for the revision, the Senate and the Chamber of Deputies have carried out textual research on the protection of the environment in the constitutions of the EU Member States,[22] while they have ignored the amendments to the constitutions of Latin American countries, where it has been preferred to protect *nature*, and in some cases even to give it *legal subjectivity*.

Interestingly, the updating or direct inclusion of these values in European constitutions is done in different contexts.

For example, in Belgium the right to the protection of a healthy environment is aggregated with the right to 'conduct a life in accordance with human dignity' (Article 23).

In Bulgaria, the environment, which is intended to be healthy and favourable, is a value connected with the maintenance and diversity of wildlife and the rational use of resources (Articles 15 and 55);

In Croatia, the protection of the environment and nature is placed on the same level as the protection of human rights and the rule of law (Article 3), it is a limitation of the protection of property (Article 50), it is a multi-factor asset that enjoys special protection (Article 52) and it is associated with healthy living (Article 70)

In Estonia, natural resources are protected (Article 5), and the protection of the environment constitutes a limitation on the free movement of persons (Article 34) and is subject to compensation for damage (Article 53).

In Finland, there is an obligation for everyone to take responsibility for the natural world and its biodiversity, environment, and cultural heritage; and public authorities must ensure everyone's right to a healthy environment (Section 20).

In Germany, Paragraph 20 of the Constitution provides for the responsibility of the state for future generations with regard to living conditions and the protection of animals.

In Greece, the State is obliged to protect the natural and cultural environment and the right of every person to its protection (Article 24), including the forest ecosystem (Article 24).

The Latvian Constitution is perhaps the one that goes the least far in protecting rights, as it refers in the preamble to the obligation of every individual to care for future generations, the environment and nature. And in Article 115, it stipulates the state's obligation to provide information on the condition of the environment, promoting its preservation and improvement.

The Lithuanian Constitution also provides for the protection of the environment in two provisions (Articles 53 and 54): the state and private individuals are obliged to protect the environment from harmful effects and the state is obliged to protect all the component factors of the environment.

The Luxembourg Constitution adopted a provision very similar to those introduced in the Italian constitutional reform, promoting sustainable development, but making no reference to private economic activity (Article 11a).

The same applies to the Maltese Constitution (Article 9). The most laconic provision is in the Dutch Constitution (Article 5), which requires public authorities to protect the habitability and the environment. Similarly, the Polish Constitution includes environmental protection in the provision protecting people's freedoms and rights (Article 5), and places limits on environmental protection that are incompatible with citizens' freedoms (Article 31).

The Portuguese Constitution elevates provisions found in other experiences in ordinary laws, with regard to popular actions, compensation for damages, and health, to a constitutional norm. In particular, Article 66 establishes a right for everyone to enjoy a healthy and ecologically balanced human living environment, but also the duty to defend it (Article 66). Moreover, it outlines an articulated programme of prevention and promotion.

The Romanian Constitution provides for a right to a healthy and ecologically balanced environment and among the limits to the free market is the restoration and preservation of the environment, as well as the maintenance of ecological balance (Articles 35, 135). Protection of property not harmful to health, protection of health and the environment, the right to a 'favourable' environment are the declinations of the Slovak Constitution (Articles 20, 23, 44, 45).

The Slovenian constitution (Article 72) speaks of a healthy natural environment, in which everyone has the right to live, and of the state's duty to guarantee this environment, including the obligation to compensate those who damage it.

The Spanish Constitution guarantees the right to use an environment suitable for the development of the person, as well as the duty to preserve it. And it entrusts the public authorities with the task of watching over the rational use of resources, the protection of the quality of life, and the defence and restoration of the environment, 'supporting the indispensable collective solidarity' (Article 45).

The Swedish Good Governance Act provides for sustainable development for future generations (Article 2) and the right of access to the natural environment (Article15).

The Hungarian Constitution guarantees everyone the right to a healthy environment and provides for compensation for environmental damage and the prohibition of transporting waste for disposal (Article XXI).

An initial, superficially literal comparison of the texts, obviously extrapolated from the constitutional context and the legislative, jurisprudential, and doctrinal experience, clearly shows that the system of balancing interests, with the enunciation of the rights recognised in the individual and of the limits to private economic initiative is only present in the Italian text of the reform.

Aggregating the assessments, it also demerits that not all Member States recognise the right to the environment as a *fundamental right*, which, moreover, is not included among the fundamental rights even in the European Charter, in which Article 37, A high level of environmental protection and the improvement of the quality of the environment must be integrated into the policies of the Union and guaranteed in accordance with the principle of sustainable development. They link the right to the environment to human health Belgium, Bulgaria, Croatia, Portugal, Romania,

Slovakia, Hungary. Many texts provide for obligations on the part of the State, for protection and information, as in Estonia, Finland, Greece, Latvia, Lithuania, the Netherlands, and Spain. Future generations are mentioned in texts in Germany, Latvia, and Sweden. Some concern themselves with constitutionalising the obligation to compensate for damage, such as Estonia and Finland.

The balancing of interests, which concerns on the one hand the protection of the person, and on the other the protection of the environment, and on the other the protection of profit from free economic initiative thus rests on the amendment of Article 41 of the Constitution. No constitutional text is as precise and as useful to the interpreter as the Italian one.

The formulation of the new articles must be considered an enrichment of the merits of our Constitution. Even if, as has been pointed out, the same results had already been reached with the doctrinal elaboration of the right to the environment by Massimo Severo Giannini[23] and Alberto Predieri (the latter having recourse to a broad notion of 'landscape'[24]) and with the jurisprudential orientations of the Constitutional Court[25] and the Court of Cassation.[26]

The affair now underway in France also presents aspects similar to the Italian one. It will be necessary to verify the outcome in order to make a fruitful comparison,[27] also following the filing of the opinion of the *Conseil Constitutionnel*[28] rendered on 14 January 2021, which appears perplexing, also on the point of considering the economic effects of a generalised recognition of the right to the environment.

The discussion of the constitutional bill to amend the first article of the French Constitution (of 1958) concerning the protection of the environment has slowed down the intended legislative measures. The bill implies a revision of the Constitution and will therefore have to be submitted to a referendum. Since deputies and senators, after a double reading, did not reach an agreement to pass it, the proposal was suspended.

Presented by the French Minister of Justice, Eric Dupont-Moretti, the text was approved by the Assembly of Deputies, then transmitted to the Senate, where it was amended, and then sent back to the Assemble, which amended it again and transmitted it to the Senate.

The text is less articulate than the formula approved by the Italian Parliament. It reads:

'It [France] guarantees the conservation of the environment'. The aim is therefore twofold: on the one hand to preserve the environment, and on the other hand to provide public remedies to achieve the goal. In his presentation of the project, the Minister for the Ecological Transition referred to the importance of guaranteeing citizens the enjoyment of a protected environment in which biodiversity is preserved and remedies against global warming are provided. Hence the commitment of public institutions to achieve this result, and correspondingly, their liability for culpable inaction and the damage caused.

The French constitutional revision is the final segment of a long journey that began with Constitutional Law No. 2005-205 of 1 March 2005, which inscribed the Environmental Charter in the Preamble of the Constitution, alongside the Declaration of the Rights of Man and of the Citizen of 1789 and the Preamble of the Constitution of 1946.

The 2004 *Charte de l'environnement* proclaims certain fundamental rights: the right to live in a balanced environment that respects health (Article 1), the right to acquire information about the environment and to participate in decisions taken to manage damage caused to the environment. The rights are also counterbalanced by duties: the duty of every person to participate in the preservation and improvement of the environment (Article 2); the duty of every person to prevent damage to the environment and to limit consequential damage (Article 3), and to contribute to the compensation of damage caused (Article 4). It also provides for the competence of public authorities to observe the precautionary principle, to assess risks and to take the necessary measures in case of environmental damage.

Education and innovation are areas where citizens' rights are further promoted.

The *Conseil Constitutionnel* specified in a decision concerning the law on genetically modified organisms (19 July 2008 no 2008-564) that the Charter has constitutional value. This decision has also been reiterated recently, with regard to restrictive measures imposed on the production and circulation of pharmaceutical products (no 2019-823 of 31 January 2020) and with a decision concerning the damage resulting from the use of sugar beet (no 2020-809 of 10 December 2020).

In this context, the constitutional revision would only have a symbolic function since the framework of rights and duties and public initiatives in this area had already been defined.

However, the perplexities that had accompanied the adoption of the Charter[29] spilled over onto the constitutional reform project, because it was felt – as was the case in Italy with the similar reform – that the formulas were too general and vague, and the public and private costs and commitments too burdensome. French jurists wondered whether the expression 'guarantees' referring to the Republic implied an obligation of result, although it was objected that the formula did not limit the freedom of Parliament to protect a constitutionally provided right in the most appropriate form and at the most appropriate time.

In the course of the debate, a more nuanced formula, suggested by environmental scientists, was also examined:

> [France] préserve l'environnement ainsi que la diversité biologique et agit contre le dérèglement climatique, dans les conditions prévues par la Charte de l'environnement de 2004.

As no political agreement was reached in the Senate between Macron's party and the Right, the previously announced referendum was abandoned. The outcome of this affair is therefore highly uncertain.

It was therefore not necessary to revise legal categories and attribute an autonomous subjectivity to the environment, as was done, in a symbolic form, in some Latin American countries, with regard to 'nature' as a subject of law.

Notes

1 Emeritus of Civil Law, University of Rome La Sapienza.

2 Vito Velluzzi, *Le clausole generali. Semantica e politica del diritto* (Giuffrè, 2012).

3 See, for all, Francesco Perrini, *Sostenibilità* (EGEA, 2018).

4 Enrico Giovanni, 'The Sustainability Principle' in *Aspenia* (Aspen Institute Italia, 2019), 76.

5 Gustavo Minervini, 'Contro la funzionalizzazione dell'impresa privata' [1958] Riv. Dir. Civ., 618 ff; Alberto Asquini, 'I battelli del Reno' [1959] in Riv. Dir. Soc., 619 ff, revised by Floriano D' Alessandro, 'Il diritto delle società, dai battelli del Reno alle navi vichinghe', in Foro it. [1988] V, 48 ff.

6 Massimo Luciani, La produzione economica privata nel sistema costituzionale (CEDAM, 1983).

7 See, in particular, Pier Giusto Jaeger, *L'interesse sociale* (Giuffrè, 1964), critically reviewed by Giuseppe Ferri, 'Jaeger Pier Giusto, L'interesse sociale' (1965) 5–6 Riv. Dir. Com. 5-6, 244.

8 Henry Hansmann, 'Corporation and Contract' [2006] 8 American Law and Economics Review Working Paper, 1; Henry Hansmann and Reinier Kraakman, 'The End of History for Corporate Law' [2001] 89 Geo. L. J., 439 ff. Henry Hansmann and Reinier Kraakmann, 'The End of History for Corporate Law', Working Paper, 2000.

9 V. Freeman and D.L. Reed, 'Stockholders and Stakeholders: A New Perspective on Corporate Governance' [1983] 25 Calif. Manag. Rev., 88 ff.

10 All Answers ltd, 'Shareholder Primacy Theory Summary' (*Lawteacher.net*, September 2023) <https://www.lawteacher.net/free-law-essays/business-law/chapter-shareholder-primacy-theory-summary-law-essays.php?vref=1> accessed 19 September 2023; N.G. Smith and D. Ronnegard, 'Shareholder Primacy, Corporate Social Responsibility and the Role of Business Schools' [2014] J. Bus. Ethics, 1 ff.; G.M. Hayden and

M.T. Boodie, *Reconstructing the Corporation. From Shareholder Primacy to Shared Governance* (CUP, 2021).

11 Among the large mass of contributions, see Carlo Angelici, 'Divagazioni sulla 'responsabilità sociale' d'impresa' [2018] Riv. soc., 3 ff.; G. Marasà, 'Purpose of Profit and Purpose of Common Benefit in Benefit Corporations' [2017] 2 Orizzonti del diritto commerciale, 8 ff.; Jr Mario Stella Richter, 'Benefit and Non-Benefit Companies' [2017] 2 Orizzonti del diritto commerciale, 1 ff.; Mario Libertini, '*Economia sociale di mercato e responsabilità soccoiale dell'impresa*' [2013] Orizzonti del diritto commerciale, 1 ff.; Giuseppe Conte, *L'impresa responsabile* (Giuffrè, 2018).

12 E.B. Barbier and J.C. Burgess, 'Sustainable Development Goals and the Systems Approach to Sustainability' [2017] 11 Economics, 1 ff.; Rita Rolli, *L'impatto dei fattori ESG sull'impresa. Modelli di governance e nuove responsabilità [The Impact of ESG Factors on Business. Governance models and new responsibilities]* (il Mulino, 2021).

13 'Overcoming poverty: end poverty in all its forms, everywhere; - overcoming hunger: end hunger, ensure food security, improve nutrition and promote sustainable agriculture; - good health: ensure a healthy life and promote wellbeing for all at all ages; - quality education: ensure inclusive education for all and promote fair and quality lifelong learning opportunities; - gender equality: achieve gender equality through the empowerment of women and girls; - clean water and sanitation ensuring the availability and sustainable management of water and sanitation services for all;- renewable and affordable energy: ensuring the availability of affordable, reliable, sustainable and modern energy services for all;- good jobs and economic growth: promoting inclusive, sustained and sustainable economic growth, full and productive employment and decent work for all;- innovation and infrastructure: building sound infrastructure, promoting inclusive and sustainable industrialisation and fostering innovation;- reducing inequalities: reduce inequalities within and between countries;- sustainable cities and communities: create sustainable cities and human settlements that are inclusive, safe and sound;- responsible use of resources: ensure sustainable consumption and production patterns;- combating climate change: take urgent action to combat climate change and its consequences;- sustainable use of the sea: conserve and sustainably use the oceans, seas and marine resources for sustainable development;- sustainable use of land: protecting, restoring and promoting the sustainable use of terrestrial ecosystems, managing forests sustainably, combating desertification, halting and reversing land degradation and halting biodiversity loss; - peace and justice: promoting sustainable development; strengthening implementation instruments and revitalising the global partnership for sustainable development'. 12. See, e.g., A. Somma, 'Il diritto privato liberista. Reflections on Private AutonomyRiflessioni sull'autonomia private stimoulate da un recente contributo' [2001] Bul. mex. der. comp., 101.

14 The association of shareholders corporations. See also Assonime, *Corporate Governance sostenibilità e dialogo con azionisti*, (issue nr. 8, 2023).

15 Vedanta Resources plc v Lungowe, [2019] 20, 10 April 2019.

16 *Chandler v Cape plc* [2012] EWCA Civ 525.

17 ECLI:RBDHA:2021:5339.

18 See other references in [2023] 2 Contratto e impresa.

19 Bill: S. 83-212-938-1203-1532-1627-1632-2160. Proposals for constitutional laws on the initiative of Senators: De Detris and Nugnes; De Petris and others; Collina and others; Perilli; Gallone; L'Abbate; Bonino; Calderoli and others: 'Amendments to Arts. 9 and 41 of the Constitution on environmental protection'.

20 On the initiative of Senator Perilli, A.S.N.1203 announced to the Presidency on 2 April 2019.

21 See N. Zorzi Galgano, 'Iniziativa economica privata e pluralità degli interessi costituzionalmente tutelati: l'ambiente e gli animali' [2023] 3 Contratto e impresa.

22 V. Senate Study Service, No. 140, October 2019; Senate Study Service, *Dossier Protection of the Environment in the Constitution*. Notes on A.S.No.83 and combined -A, No. 396, June 2021.

23 M.S. Giannini, 'Ambiente: saggio sui diversi suoi aspetti giuridici' [1973] Riv. trim. dir. pubbl., 23 ff.

24 See Alberto Predieri, 'Paesaggio' in *Enc. Dir.*, XXXI (Giuffrè, 1981).

25 Lastly, on the Ilva case, the Court carried out a balancing of the interests at stake, considering that the legislator had privileged the continuity of production to the detriment of the protection of health and the environment (Corte cost. 2018 no. 58).

26 On the environmental good protected in civil law through the direct application of the constitutional rules, see Court of Cassation 3 February 1998, no. 1087. On the notion of the environment as an intangible asset – according to M.S.Giannini's proposal – see Court of Cassation 10 October 2008, no. 25010.

27 On this point see projet de loi constitutionnelle, in jusx2036137l: https://www.legifrance.gouv.fr/dossierlegislatif/jorfdole000043022845/ (last accessed 31 October 2023); Y. Jégouzo, 'Quelques réflexions sur le projet de charte de l'environnement' [January 2004] 15 Cahiers du conseil constitutionnel (dossier: Constitution et environnement); J. Morand-Deviller, 'L'environnement dans les consittutions etrangères' [April 2014], Cahiers Const.

28 Avis sur un projet de loi constitutionnelle complétant l'article 1er de la Constitution et relatif à la préservation de l'environnement NOR: JUSX2036137L.

29 Y. Jégouzo, 'Quelques réflexions sur le projet de charte de l'environnement' [January 2004] 15 Cahiers du conseil constitutionnel (dossier: Constitution et environnement).

Bibliography

All Answers ltd, 'Shareholder Primacy Theory Summary' (*Lawteacher.net*, September 2023) <https://www.lawteacher.net/free-law-essays/business-law/chapter-shareholder-primacy-theory-summary-law-essays.php?vref=1> accessed 19 September 2023.

Angelici C, 'Divagazioni sulla "responsabilità sociale" d'impresa' [2018] Riv. soc. 1, 3 ff.

Asquini A, 'I battelli del Reno' [1959] in *Rivista di Diritto Societario*, 617 ff.

Assonime, *Corporate Governance sostenibilità e dialogo con azionisti* (issue nr. 8, 2023).

Barbier EB and Burgess JC, 'Sustainable Development Goals and the Systems Approach to Sustainability' [2017] 11 Economics, 1 ff.

Conte G, *L'impresa responsabile* (Giuffrè, 2018).

D' Alessandro F, 'Il diritto delle società, dai battelli del Reno alle navi vichinghe', in *Foro italiano* [1988], V, 48.

Ferri G, 'Jaeger Pier Giusto, L'interesse sociale' (1965) 5–6 Riv. Dir. Com. 5-6, 244.

Freeman V and Reed DL, 'Stockholders and Stakeholders: A New Perspective on Corporate Governance' [1983] 25 Calif. Manag. Rev., 88 ff.

Giannini MS, 'Ambiente: saggio sui diversi suoi aspetti giuridici' [1973] Riv. trim. dir. pubbl. 1, 23 ff.

Giovanni E, 'The Sustainability Principle' in *Aspenia* (Aspen Institute Italia, 2019).

Hansmann H, 'Corporation and Contract' [2006] 8 American Law and Economics Review Working Paper, 1

Hansmann H and Kraakman R, 'The End of History for Corporate Law' [2001] 89 Geo. L. J., 439 ff.

Hansmann H and Kraakmann R, 'The End of History for Corporate Law', Working Paper, 2000.

Hayden GM and Boodie MT, *Reconstructing the Corporation. From Shareholder Primacy to Shared Governance* (CUP, 2021).

Jaeger PG, *L'interesse sociale* (Giuffrè, 1964).

Jégouzo Y, 'Quelques réflexions sur le projet de charte de l'environnement' [January 2004] 15 Cahiers du conseil constitutionnel (dossier: Constitution et environnement).

Libertini M, '*Economia sociale di mercato e responsabilità soccoiale dell'impresa*' [2013] Orizzonti del diritto commerciale, 1 ff.

Luciani M, *La produzione economica privata nel sistema costituzionale* (CEDAM, 1983).

Marasà G, 'Purpose of Profit and Purpose of Common Benefit in Benefit Corporations' [2017] 2 Orizzonti del diritto commerciale, 8.

Minervini G, 'Contro la funzionalizzazione dell'impresa privata' [1958] Riv. dir. civ. 1, 618.

Morand-Deviller J, 'L'environnement dans les consittutions etrangères' [April 2014] *Cahiers Const*, 1.

Perrini F, *Sostenibilità* (EGEA, 2018).

Predieri A, 'Paesaggio' in *Enc. Dir.*, XXXI (Giuffrè, 1981), 506 ff.

Rolli R, *L'impatto dei fattori ESG sull'impresa. Modelli di governance e nuove responsabilità [The Impact of ESG Factors on Business. Governance models and new responsibilities]* (il Mulino, 2021).

Smith NG and Ronnegard D, 'Shareholder Primacy, Corporate Social Responsibility and the Role of Business Schools' [2014] J. Bus. Ethics, 134(3), 1.

Somma A., 'Il diritto privato liberista. Reflections on Private AutonomyRiflessioni sull'autonomia private stimoulate da un recente contributo' [2001] Bul. mex. der. comp., 101.

Stella Richter Jr M, 'Benefit and Non-Benefit Companies' [2017] 2 Orizzonti del diritto commerciale, 1 ff.

Velluzzi V, *Le clausole generali. Semantica e politica del diritto* (Giuffrè, 2012).

Zorzi Galgano N, 'Iniziativa economica privata e pluralità degli interessi costituzionalmente tutelati: l'ambiente e gli animali' [2023] 3 Contratto e impresa, 442 ff.

12

SUSTAINABILITY AND THE IMPACT OF THE COURT OF JUSTICE OF THE EU

Verica Trstenjak[1]

12.1 Introduction

Today, sustainability is a reality that pervades all areas of life, it is mostly linked to the environment, but it also concerns all areas of the economy and has various social aspects. Therefore, we can talk about several types of sustainability such as environmental, economic, and social. Sustainability covers climate change and land, air, and water pollution, among other areas. It also includes modern trends, e.g., in the field of the development of a digital society. The EU White Paper on Artificial Intelligence[2] emphasises that artificial intelligence contributes to climate change mitigation and adaptation.

The economy must take sustainability into account in production and services at all levels of business. Sustainability must also be taken into account by the state administration and universities, as well as by civil society, consumers, and others, i.e., essentially in all areas of life.[3] Thus, the requirements regarding sustainability should not be overlooked even in specialised areas, such as the agricultural sector.[4]

In addition to various initiatives and extensive theory,[5] many international commitments require respect for sustainability from all sectors of society, and this is also required for economic subjects in special areas. The UN 2030 Agenda for Sustainable Development includes 17 Sustainable Development Goals (SDGs) which should apply universally to all countries. There are also sustainability principles governing specific sectors. Within the framework of the UN Environmental Programme, there are the UN Principles for Responsible Banking,[6] which require the banking sector to accelerate a positive global transition for people and the planet and the UN Net-zero Banking Alliance initiative is also important.[7]

This chapter will mainly address the concept of sustainability as tackled by EU law. The case law of the Court of Justice of the EU (CJEU) will also be presented. It will be shown that the CJEU indirectly affects sustainability and contributed for its enforcement with its decisions.

12.2 Sustainability in the EU

The EU has adopted numerous guiding documents to address the concept of sustainability, as well as numerous legal acts (e.g., directives), which contain specific requirements and thus contribute to

DOI: 10.4324/9781032662046-15

sustainability, especially in the field of the environment. The EU advocates for a holistic approach to sustainable development while implementing the UN's 2030 Agenda for Sustainable Development together with its Member States.[8] At the global level, an important document is the Paris Agreement,[9] the first-ever universal legally binding global climate change agreement, adopted at the Paris climate conference in 2015. The EU ratified it in 2016.[10]

The EU adopted numerous policies and strategies in this area. Some of them will be discussed below, as their number shows the importance that EU institutions attach to sustainability.

The European Green Deal[11] was adopted in 2019 after the realisation that climate change and environmental degradation are a danger to the EU and the whole world. One of the goals of this initiative is to improve the well-being and health of citizens and future generations.

With the European Green Deal Investment Plan,[12] the European Commission wants to enable a framework to facilitate the public and private investments needed for the transition to a climate-neutral, green, competitive, and inclusive economy. This investment plan should mobilise at least €1 trillion of sustainable investments over the next decade.

The Commission also prepared the 2030 Climate Target Plan,[13] where the goal is, among other things, to cut greenhouse gas emissions and stimulate the creation of green jobs. This Target Plan was followed by a package of proposals for new legal acts.[14]

The EU Commission has adopted a number of documents regarding sustainability in the field of finance. Among their goals is also the mobilisation of private capital for sustainable projects, especially for infrastructure. This is one of the conditions for the transition to a more sustainable economic model. The Commission emphasises the importance of sustainable finance and states: 'Sustainable finance refers to the process of taking environmental, social and governance (ESG) considerations into account when making investment decisions in the financial sector, leading to more long-term investments in sustainable economic activities and projects'.[15]

The Commission also adopted in 2018 the Sustainable Finance Action Plan, which comprises a strategy for a financial system that supports the EU's climate and sustainable development agenda.[16]

All the mentioned documents are so-called 'soft law', i.e., guidelines. The EU has also adopted several binding legal acts. There are numerous directives that affect the realisation and achievement of the sustainability goals. These directives were adopted in the fields of environment, production (e.g., cars), as well as services (e.g., banking). Many other directives or regulations also contain requirements related to the sustainability goals, for example, Directive 2009/28 on the promotion of the use of energy from renewable sources,[17] which mentions sustainability several times.[18]

The Corporate Sustainability Reporting Directive (CSRD), adopted in December 2022, sets strict requirements for reporting.[19] Already in 2019, the Commission published demanding guidelines on reporting climate-related information. Now these requirements are increasing, as some companies will have to report on how sustainability issues affect their business and about their own impact on people and the environment.[20] In the proposal for the CSRD, the Commission explained:

> This proposal aims to ensure that the reporting requirements for companies are consistent with the taxonomy. This will be achieved above all through the proposed sustainability reporting standards. These will take into account the indicators companies have to disclose about the extent to which their activities are environmentally sustainable according to the taxonomy, and the screening criteria and do-no-significant-harm thresholds of the taxonomy.[21]

An example of a regulation is Regulation 2016/646 concerning emissions from light passenger and commercial vehicles.[22] In the financial services field, there are regulations such as Regulation 2019/2088 on sustainability-related disclosures in the financial services sector,[23] which was already amended by Regulation 2020/852 (Taxonomy Regulation) on the establishment of a framework to facilitate sustainable investment.[24]

12.3 Sustainability and the Impact on the CJEU

12.3.1 Different Procedures before the CJEU

Within the framework of various procedures, the CJEU often deals with cases related to disputes affecting the sustainability goals, especially in the field of the environment. We will highlight matters regarding motor vehicles, nitrates, and energy from renewable resources. All these three matters were dealt with through different procedures before the CJEU. Different procedures are possible at the CJEU, and the consequences of a judgement depend on the type of procedure, which will be presented below.

The most important procedure is the one in which the CJEU decides on the basis of Article 267 of the Treaty of the Functioning of the European Union (TFEU) in the preliminary reference procedure. In this procedure, which is the most common at the CJEU, the Court interprets primary or secondary EU law. The CJEU can also decide on the validity of secondary EU law within the framework of this procedure. Here, it is the national courts of the Member States that raise a question regarding the interpretation or validity of EU law.

Important proceedings at the CJEU are also based on an action for failure to fulfil obligations by Member States, in the context of which the Commission can bring actions against Member States under Articles 258 and 260 TFEU. In the case of an action according to Article 258 TFEU, only a declaratory judgement is possible. Article 260 TFEU, however, allows the CJEU to impose weighty financial sanctions, which is especially common in cases concerning the environment.[25]

Actions for annulment under Article 263 TFEU are also important for Member States, as well as companies and individuals. During this procedure the CJEU decides on the possible nullity of EU acts. The deadline for bringing this action is limited to two months, namely from the publication of a directive or regulation. It is important to note that in these cases, the General Court of the EU (General Court) often has jurisdiction, depending on the type of act and the plaintiff, which means that an appeal to the Court of Justice is possible in these cases. For example, in 2015, Austria brought an action for annulment in the case T-356/15 regarding the state aid planned for the nuclear power station Hinkley Point that the Commission allowed the United Kingdom when it was still part of the EU.[26] The action before the General Court of the EU was unsuccessful. Austria started an appeal procedure in case C-594/18 P, but the CJEU dismissed the appeal.[27]

Austria started a similar procedure in 2018 also against the European Commission and its approval of investment aid, notified by Hungary, for the construction of new nuclear reactors. In this case T-101/18, *Austria v Commission*,[28] the General Court dismissed the action brought by Austria. In February 2023, Austria brought an appeal against this decision before the Court of Justice.[29]

In October 2022, Austria brought an action for annulment before the General Court of the EU in case T-625/22 against the Commission for inclusion of nuclear energy and gas as 'green' in the Taxonomy Regulation.[30] Austria claims that the Court should annul the Commission Delegated Regulation (EU) 2022/1214 of 9 March 2022 amending Delegated Regulation (EU) 2021/2139 as regards economic activities in certain energy sectors and Delegated Regulation (EU) 2021/2178 as regards specific disclosure requirements for those economic activities.[31]

We will briefly discuss the most important matters in which the EU has, at least partially, intervened in the area of sustainability.

12.3.2 Preliminary Reference Procedure before the CJEU: The Use of Energy from Renewable Sources

The CJEU dealt with the issue of the promotion of the use of energy from renewable sources in the *L.E.G.O.* case, in which the *Consilio di Stato* from Italy referred the question in the preliminary ruling procedure according to Article 267 TFEU.[32]

The CJEU interpreted Article 18 of Directive 2009/28 on the promotion of the use of energy from renewable sources,[33] read in conjunction with the Commission Implementing Decision 2011/438 on the recognition of the 'International Sustainability and Carbon Certification' (ISCC) scheme.[34]

The request was made in the context of proceedings between Legatoria Editoriale Giovanni Olivotto (L.E.G.O.) SpA and Gestoredeiservizienergetici (GSE) SpA and different Italian ministries[35] concerning the non-submission of sustainability certificates for bioliquids used to operate L.E.G.O.'s thermal energy plant, leading to the disqualification of the plant from the green certificate incentive scheme.[36] Therefore, L.E.G.O. had to return the amounts that were paid to it from aid schemes in the years 2012–2014.[37] The pertinent question was whether the requirements imposed by Italy were too strict regarding certain sustainability certificates, especially for the ISCC. The CJEU held that the Italian regulation is compatible with EU law. Specifically, it ruled on the obligation of intermediaries which do not take physical possession of the bioliquids, objects of relevant transactions, to submit the aforementioned certificates. The CJEU decided that this obligation complies with EU law.

In this case, questions were also raised in connection with the free movement of goods and Article 34 TFEU. The CJEU reiterated the case law of the CJEU that national legislation constituting a measure having equivalent effect to quantitative restrictions on imports is, in principle, incompatible with Article 34 TFEU, unless that legislation can be objectively justified.[38]

The CJEU ruled that a limitation does exist in this case, but it is justified by the objectives of protecting the environment and combating fraud. The CJEU emphasised that the national legislation at issue in the main proceedings may also contribute to the protection of the health and life of humans, animals, and plants.[39]

The CJEU specifically emphasised sustainability in the judgement and held: 'by targeting all operators in the bioliquids supply chain, the national provision in question helps to prevent fraud in connection with the sustainability of the bioliquid'.[40] Further, the CJEU examined the proportionality of the measure and confirmed that it is a measure that is appropriate for obtaining the legitimate objective pursued.

12.3.3 Actions for Annulment: Emissions from Light Passenger and Commercial Vehicles

The proceedings in the framework of actions for annulment brought before the General Court (and later appealed to the CJEU) regarding the harmonised EU framework for the approval of motor vehicles to facilitate their registration, sale, and entry into service in the EU were widely discussed. In this field, the EU adopted the Regulation 2016/646 concerning emissions from light passenger and commercial vehicles.[41]

In 2016, the City of Paris, the City of Brussels, and the Municipality of Madrid brought actions for annulment before the General Court in joined cases T-339/16, T-352/16, and T-391/16.[42] They requested the annulment of the Regulation 2016/646.

In this case, as the plaintiffs were cities, the General Court had to deal with a procedural question of whether the contested regulation is of direct concern to the applicants within the meaning Article 263(4) TFEU. The case deals with the requirements for the real driving emission (RDE) tests which are intended to measure the pollutant emissions from light passenger and commercial vehicles in the procedures that take place before the authorisation to place new vehicles in the market. The three cities sought the annulment of the Regulation 2016/646 due to the Commission's alleged failure to adopt the not-to-exceed (NTE) values for emissions of oxides of nitrogen applied, which are higher than the limits on those emissions laid down for the Euro 6 standard.

Actions for annulment can only exceptionally be brought by individuals (natural or legal persons) against regulatory acts, i.e., only if they are directly concerned by them and do not entail implementing measures. Thus, the General Court assessed the case also in terms of its admissibility. The parties agreed that the contested regulation must be regarded as a regulatory act in light of Article 263(4) TFEU and that it did not entail implementing measures within the meaning of that provision.[43] Thus, the dispute between the parties related solely to the question whether the contested regulation is of direct concern to the applicants, which the Commission contested.[44]

The General Court partially upheld the actions and held that the contested regulation was of direct concern to the applicants, but the Commission appealed the case to the CJEU. It is worth noting that the General Court even ruled on the exception of the *ex tunc* effect of judgements,[45] but later this was no longer relevant because of the subsequent CJEU judgement. Besides the Commission, Germany and Hungary appealed the judgement of the General Court. Based on these appeals, the CJEU set aside the judgement of the General Court. The CJEU decided that since the cities of Paris, Brussels, and Madrid are not directly concerned by the regulation, their actions seeking its annulment must be dismissed as inadmissible.

Unfortunately, in this case, there was no decision on the merits of the claim, which mainly dealt with environmental protection and sustainability. It is interesting, however, that the big cities of Brussels, Madrid, and Paris initiated this procedure before the CJEU.

12.3.4 Infringement Procedure against Spain: The Protection of Waters against Pollution Caused by Nitrates from Agricultural Sources

In order to protect the environment and, consequently, the requirements of sustainability, the Commission may also decide to bring an action for failure to fulfil obligations against a Member State, i.e., infringement procedures. An example is the case C-576/22, *Commission v Spain*.[46] In December 2021, the Commission decided to bring an action against Spain before the CJEU for poor implementation of the Nitrates Directive 91/676.[47] The case is the result of a dispute in Spain, as ClientEarth made environmental claims in Spain and then reported Spain to the Commission. The action before the CJEU was brought only in August 2022, so the process will continue into 2023 and the decision of the CJEU has not been made at the time of writing.

Nevertheless, we should state that the essence of the dispute concerns water pollution and the insufficient implementation of the aforementioned directive in Spain.

The aim of the Nitrates Directive 91/676 is to protect water quality across Europe. Its objective is to prevent nitrates from agricultural sources polluting ground and surface waters. The Directive is intended to promote the use of good farming practices. The content of the Nitrates Directive 91/676 complements the objectives of the Directive 2000/60 concerning the Water Framework.[48] It stipulates that the Member States should monitor their waters and identify those waters which are, or are likely to be, affected by pollution due to nitrates from agricultural sources. According to the Commission, Spain's measures have failed to achieve the objectives of the Nitrates Directive 91/676. To remedy this situation, Spain should further designate Nitrate Vulnerable Zones in seven regions and take additional measures in some regions where the current measures are insufficient to achieve the Directive's objectives.[49]

In particular, the Commission alleges that Spain has not adopted the additional measures or reinforced actions required by Article 5(5) of the Nitrates Directive 91/676. This provision requires Member States also to take preventive action against further pollution caused by nitrates from agricultural sources.

As a common theme in all three cases discussed, we see the tendency of the CJEU to support the adopted policies and practices of the EU (as well as individual countries) regarding the achievement of sustainability goals, especially in the field of the environment. Consequently, the CJEU has a significant impact on the realisation of the adopted goals regarding sustainability.

12.4 Conclusion

Sustainability is a necessity and a challenge for society that will have implications in the future. Each individual, company, state, and also the EU has a responsibility to contribute to tackling this challenge. We, as individuals, should not rely on national, European, or international institutions alone, but must contribute to sustainability and to saving the planet ourselves. As Robert Swan said, 'The greatest threat for our planet is the belief that someone else will save it'.[50] The EU has taken many measures in the form of legal acts. Member States and their subjects must respect all these measures. In the end, there is always the possibility that the CJEU interprets possible ambiguities, or even makes a decision in a procedure regarding an action brought by the Commission against a Member State. Thus, the CJEU's role is becoming ever more important in achieving sustainability goals in all areas.

Notes

1 Former Advocate General of the Court of Justice of the EU; Professor of European Union Law; Alma mater Europaea, Maribor, Slovenia.

2 European Commission, 'White Paper on Artificial Intelligence – A European Approach to Excellence and Trust' (*European Commission*, 19 February 2020) <https://ec.europa.eu/info/sites/default/files/commission-white-paper-artificial-intelligence-feb2020_en.pdf> accessed 13 July 2023.

3 European Commission, 'Sustainable Development Goals' <https://ec.europa.eu/info/strategy/international-strategies/sustainable-development-goals_en> accessed 13 July 2023, the Commission states that '[c]ivil society and other stakeholders' engagement are key towards reaching the goals of the 2030 Agenda'.

4 Giulio Malorgio and Francesco Marangon, 'Agricultural Business Economics: The Challenge of Sustainability' (2021) 9 Agricultural and Food Economics 6, Springer Open <https://agrifoodecon.springeropen.com/articles/10.1186/s40100-021-00179-3> accessed 13 July 2023

5 John Morelli, 'Environmental Sustainability: A Definition for Environmental Professionals' (2013) 1 (1) Journal of Environmental Sustainability 1 DOI: 10.14448/jes.01.0002 accessed 12 July 2023; Nancy Bocken et al, 'A Literature and Practice Review to Develop Sustainable Business Model Archetypes' (2014) 65 Journal of Cleaner Production 42 DOI: 10.1016/j.jclepro.2013.11.039 accessed 12 July 2023; Petra Weingerl, 'Sustainability, the Circular Economy and Consumer Law in Slovenia' (2020) 9 (3) Journal of European Consumer and Market Law 3, 129–132 accessed 12 July 2023; L.M. Tomazic, 'Zeleni prehod in trajnostni razvoj' (2022) 6–7 Podjetje in delo 1001–1013.

6 UN Environment Programme, 'Principles for Responsible Banking' <https://www.unepfi.org/banking/bankingprinciples/> accessed 12 July 2023.

7 Net-Zero Alliance is part of the UN Environment Programme and 'reinforces, accelerates, and supports the implementation of decarbonisation strategies, providing an internationally coherent framework and guidelines in which to operate, supported by peer-learning from pioneering banks. It recognises the vital role of banks in supporting the global transition of the real economy to net-zero emissions'. See UN Environment Programme, 'Net-Zero Banking Alliance <https://www.unepfi.org/net-zero-banking/> accessed 13 July 2023. For example, the biggest bank in Slovenia NLB, d.d. is, since 2022, officially a member of the UN-Convened Net-Zero Banking Alliance. See NLB, 'Press release' (*NLB Communications*, 6 June 2022) <https://www.nlb.si/press-release-6-06-2022> accessed 13 July 2023; in Austria UniCredit Bank Austria, see Bank Austria, 'Press Releases' (9 May 2022) <https://www.bankaustria.at/en/about-us-press-releases_60607.jsp> accessed 13 July 2023.

8 European Commission, 'Sustainable Development Goals' <https://ec.europa.eu/info/strategy/international-strategies/sustainable-development-goals_en> accessed 12 July 2023.

9 UN, 'The Paris Agreement' <https://unfccc.int/process-and-meetings/the-paris-agreement/the-paris-agreement> accessed 12 July 2023.

10 The problem of this act is obtaining ratification in a sufficient number of countries. As stated by the Commission, for the agreement to enter into force, at least 55 countries representing at least 55% of global emissions had to deposit their instruments of ratification. For more details, see European Commission, 'Paris Agreement' <https://climate.ec.europa.eu/eu-action/international-action-climate-change/climate-negotiations/paris-agreement_en> accessed 12 July 2023.

11 European Commission, 'A European Green Deal' <https://ec.europa.eu/info/strategy/priorities-2019-2024/european-green-deal_en> accessed 12 July 2023.

12 European Commission, 'A European Green Deal Investment Plan and Just Transition Mechanism' (*European Commission*, 14 January 2020) <https://finance.ec.europa.eu/publications/european-green-deal-investment-plan-and-just-transition-mechanism_en> accessed 12 July 2023.

13 European Commission, '2030 Climate Target Plan' <https://climate.ec.europa.eu/eu-action/european-green-deal/2030-climate-target-plan_en> accessed 12 July 2023.

14 In this regard, the Commission refers to the preparation of the following proposals: Increasing the ambition of the EU Emissions Trading System; Strengthening the Market Stability Reserve linked to the review of the EU Emissions Trading System; Revision of the EU Emission Trading System Directive concerning aviation; Revision of the EU Emission Trading System Directive concerning aviation (ICAO notification); Updating Member State emissions reduction targets (Effort Sharing Regulation) in line with the 2030 climate target plan; Land Use, Forestry and Agriculture; Revision of the CO_2 emission performance standards for new passenger cars and vans and Social Climate Fund. See European Commission, 'Delivering the European Green Deal' <https://climate.ec.europa.eu/eu-action/european-green-deal/delivering-european-green-deal_en> accessed 12 July 2023.

15 European Commission, 'Overview of Sustainable Finance' <https://finance.ec.europa.eu/sustainable-finance/overview-sustainable-finance_en> accessed 12 July 2023.

16 European Commission, 'Sustainable Finance: Commission's Action Plan for a Greener and Cleaner Economy' (*European Commission*, 8 March 2018) <https://ec.europa.eu/commission/presscorner/detail/en/IP_18_1404> accessed 13 July 2023.

17 Council Directive 2009/28/EC of 23 April 2009 on the promotion of the use of energy from renewable sources and amending and subsequently repealing Directives 2001/77/EC and 2003/30/EC [2009] OJ L 140/16.

18 For example, Recital 65 of the aforementioned Directive states: 'Biofuel production should be sustainable. Biofuels used for compliance with the targets laid down in this Directive, and those that benefit from national support schemes, should therefore be required to fulfil sustainability criteria'.

19 Council Directive (EU) 2022/2464 of 14 December 2022 amending Regulation (EU) No 537/2014, Directive 2004/109/EC, Directive 2006/43/EC and Directive 2013/34/EU, as regards corporate sustainability reporting [2022] OJ L 322/15.

20 European Commission, 'Questions and Answers: Corporate Sustainability Reporting Directive Proposal' (*European Commission*, 21 April 2021) <https://ec.europa.eu/commission/presscorner/detail/ro/qanda_21_1806> accessed 12 July 2023.

21 European Commission proposal for a Directive of the European Parliament and of the Council amending Directive 2013/34/EU, Directive 2004/109/EC, Directive 2006/43/EC and Regulation (EU) No 537/2014, as regards corporate sustainability reporting [2021] 2021/0104(COD).

22 Commission Regulation (EU) 2016/646 of 20 April 2016 amending Regulation (EC) No 692/2008 as regards emissions from light passenger and commercial vehicles (Euro 6) [2016] OJ L 109/1.
23 Council Regulation (EU) 2019/2088 of 27 November 2019 on sustainability-related disclosures in the financial services sector [2019] OJ L 317/1.
24 Council Regulation (EU) 2020/852 of 18 June 2020 on the establishment of a framework to facilitate sustainable investment and amending Regulation (EU) 2019/2088 [2020] OJ L 198/13.
25 For example, Case C- 387/97 *Commission v Greece* [2000] ECR I-5092, where the CJEU decided in para 100: '2. Orders the Hellenic Republic to pay to the Commission of the European Communities, into the account "EC own resources", a penalty payment of EUR 20 000 for each day of delay in implementing the measures necessary to comply with the judgment in Case C-45/91 from delivery of the present judgment until the judgment in Case C-45/91 has been complied with'.
26 Case T-356/15 *Austria v Commission* [2018].
27 Case C-594/18 P *Austria v Commission* [2020].
28 Case T-101/18 *Austria v Commission* [2022].
29 Case C-59/23 P *Austria v Commission* [2023], Austria could start the appeal procedure within two months.
30 Case T-625/22 *Austria v Commission* [2022], <https://eur-lex.europa.eu/legal-content/EN/TXT/PDF/?uri=CELEX:62022TN0625&from=EN> last accessed 13 July 2023.
31 See the form of order sought published in the Official Journal C 24/43, Case T-625/22 *Austria v Commission* [2022].
32 Case C-242/17 *L. E. G. O.* [2018].
33 Council Directive 2009/28/EC of 23 April 2009 on the promotion of the use of energy from renewable sources and amending and subsequently repealing Directives 2001/77/EC and 2003/30/EC [2009] OJ L 140/16.
34 Commission Implementing Decision 2011/438/EU of 19 July 2011 on the recognition of the 'International Sustainability and Carbon Certification' [ISCC] scheme [2011] OJ L 190/79.
35 Case C-242/17 *L. E. G. O.* [2018], para 2 lists the Ministero dell'Ambiente e della Tutela del Mare e del Territorio (Ministry of the Environment and Territorial and Sea Protection, Italy), the Ministero dello Sviluppo Economico (Ministry of Economic Development, Italy) and the Ministero delle Politiche Agricole e Forestali (Ministry of Agricultural and Forestry Policy, Italy).
36 Case C-242/17 *L. E. G. O.* [2018], para 2.
37 Case C-242/17 *L. E. G. O.* [2018], para 17.
38 See Case C-242/17 *L. E. G. O.* [2018], para 72 and, by analogy, Case C-573/12 *Ålands Vindkraft* [2014], para 75.
39 Case C-242/17 *L. E. G. O.* [2018], para 70.
40 Case C-242/17 *L. E. G. O.* [2018], para 69.
41 Commission Regulation (EU) 2016/646 of 20 April 2016 amending Regulation (EC) No 692/2008 as regards emissions from light passenger and commercial vehicles (Euro 6) [2016] OJ L 109/1.
42 Joined Cases T-339/16, T-352/16 and T-391/16 Ville de Paris, Ville de Bruxelles and Ayuntamiento de Madrid v Commission [2018].
43 Consolidated version of the Treaty on the Functioning of the European Union [2012] OJ C326/47 Art 263(4) limits the possibilities for actions of annulment brought by individuals to the following: 'Any natural or legal person may, under the conditions laid down in the first and second paragraphs, institute proceedings against an act addressed to that person or which is of direct and individual concern to them, and against a regulatory act which is of direct concern to them and does not entail implementing measures'.
44 Joined Cases T-339/16, T-352/16 and T-391/16 Ville de Paris, Ville de Bruxelles and Ayuntamiento de Madrid v Commission [2018], para 35.
45 The third paragraph of the operative part of the judgement states: 'Orders the effects of the provision annulled pursuant to paragraph 1 of the operative part of this judgment to be maintained pending the adoption, within a reasonable period, of new legislation replacing those provisions; that period may not exceed 12 months from the date on which the present judgment takes effect'.
46 Case C-576/22 *Commission v Spain* [pending].
47 Council Directive 91/676/EEC of 12 December 1991 concerning the protection of waters against pollution caused by nitrates from agricultural sources [1991] OJ L 375/1.
48 Council Directive 2000/60/EC of 23 October 2000 establishing a framework for Community action in the field of water policy [2000] OJ L 327/1.

49 According to the Commission, these regions are Castilla y León, Extremadura, Galicia, Islas Baleares, Islas Canarias, Madrid, and Comunidad Valenciana. See European Commission, 'Water: Commission decides to refer SPAIN to the Court of Justice of the European Union for poor implementation of the Nitrates Directive' (*European Commission*, 2 December 2021) <https://ec.europa.eu/commission/presscorner/api/files/document/print/en/ip_21_6265/IP_21_6265_EN.pdf> accessed 13 July 2023.

50 Robert Swan, OBE, FRGS, Closing Speech at the 2041 Social Innovation Summit 2016, Silicon Valley, CA (16 November 2016) <https://www.youtube.com/watch?v=su0-8-yA-3M> accessed 29 October 2023.

Bibliography

Bocken N, et al, 'A Literature and Practice Review to Develop Sustainable Business Model Archetypes' (2014) 65 Journal of Cleaner Production 42 DOI: 10.1016/j.jclepro.2013.11.039 accessed 12 July 2023

Malorgio G and Marangon F, 'Agricultural Business Economics: The Challenge of Sustainability' (2021) 9 Agricultural and Food Economics 6, Springer Open <https://agrifoodecon.springeropen.com/articles/10.1186/s40100-021-00179-3> accessed 13 July 2023, accessed 12 July 2023

Morelli J, 'Environmental Sustainability: A Definition for Environmental Professionals' (2013) 1 (1) Journal of Environmental Sustainability 1 DOI: 10.14448/jes.01.0002 accessed 12 July 2023

Tomazic L, 'Zeleni prehod in trajnostnirazvoj' (2022) 6–7 Podjetje in delo 1001–1013

Weingerl P, 'Sustainability, the Circular Economy and Consumer Law in Slovenia' (2020) 9 (3) Journal of European Consumer and Market Law 3, 129–132 <https://ssrn.com/abstract=4331861> accessed 12 July 2023

13

PRIVATE LAW RESPONSES TO IMPERFECT REGULATION IN INTERNATIONAL PUBLIC LAW – THE CASE OF VESSEL RECYCLING

Kristina Siig

13.1 Introduction

While often praised as the most sustainable form of transportation when looking at volume-to-environmental impact ratios,[1] maritime transport, like any major industry, is also a source of pollution. Due to the international nature of the maritime transport industry, the international community, mainly through the United Nations' International Maritime Organization (IMO), has issued treaties to prevent negative environmental impacts and to react to them in terms of responsibility and liability, when they do occur.[2] Dramatic and devastating spills from oil tankers are probably the most known of such negative environmental impacts,[3] but the shipping industry also presents notable challenges regarding the working environment and the environment in general when the vessel reaches the end of its lifespan. In later years, this has attracted understandable attention from international and municipal legislators,[4] NGOs,[5] and private actors in the market alike.[6]

Vessels intended for scrapping/breaking or, as it is more commonly called these days, recycling, are often exported from, e.g., North-Western Europe to the Middle and Far East.[7] This makes sense in market terms as the processes involved are labor-intensive and at the present stage of technology do not lend themselves to automatization. Unfortunately, apart from affordable labor,[8] the receiving destinations also tend to offer a more lenient approach to environmental protection and physically safe working conditions for its employees than one would have wished for, oftentimes resulting in working conditions that can only be described as appalling and in pollution of the environment where the scrapping is carried out. The international community has sought to handle this challenge through international treaties or regional and national regulation.[9] Still, at present, no uniform international regulation of this issue is in place, mainly as the international regime intended to handle the issue, The Hong Kong Convention for the Safe and Environmentally Sound Recycling of Ships, 2009 (The Hong Kong Convention) will not enter into force until 26 June 2025, 16 years after its inception. Therefore, at present, the regulation is fragmented and very much dependent on coastal and flag states to effectively enforce whichever regulation is in place.

DOI: 10.4324/9781032662046-16

Unfortunately, effective enforcement within the shipping sector, where the subject of enforcement – the vessel – may easily be moved outside the scope of a state's enforcement powers, both physically and legally, presents a challenge in this regard.[10]

Against this backdrop, shipowners, traders in recycled metals, and scrapping and recycling facilities have sought to optimize revenue, legal exposure, and corporate social responsibility issues through the contracts dealing with vessels at the end of their lifespan. Here, the focus will be on the standard contracts used when selling vessels for the specific purpose of scrapping/recycling, to assess *whether these contracts, upon their proper interpretation, provide a relevant instrument that may rectify (some of) the shortcomings of the fragmented legal tapestry*, currently regulating this issue.

First, a brief overview of the development in the international public law framework, surrounding the contracts, will be presented, providing an outline international public law framework surrounding contracts for scrapping/recycling of vessels (Section 13.2). After that, three core standard contracts will be analyzed to assess if they promote sustainable practices (Section 13.3), allowing for discussion of, and conclusion on, the aptness of the standard contracts as a remedy for the shortcomings in international public law to be provided in Section 13.4.

The reader should note that the issue has not been the subject of much legal academic discussion, and that case law is sparse. The below discussion will therefore, to a great extent, rely on primary materials, and on materials that focus more on the market aspects of the issue than on its strictly legal aspects. The reader should also note that the contribution, due to its restrictions as to volume, will provide a rather condensed presentation of what is a wide-ranging topic. For further discussion of the respective issues, please refer to the footnotes.

13.2 In Brief on the International Public Law Context[11]

The first international convention of relevance to the scrapping and recycling of vessels is the Basel Convention on the Control of Transboundary Movements of Hazardous Wastes and their Disposal (1989). The convention was conceived under the auspices of the UN Environment Programme. In 1995, an amendment was made to the Convention,[12] effectively banning the export of dangerous wastes to countries outside the OECD.[13] Unless vessels are stripped of all pollutants, incl., e.g., asbestos, oils, or heavy metal sediments that occur over time, vessels will tend to be 'hazardous' in the understanding of the Basel Convention. The vessel may also constitute 'waste' under Article 2 of the Convention when it is no longer intended for use. In respect of the banning of the export of dangerous waste to non-OECD countries, the issue is, if the *decision to decommission the vessel* was made whilst the vessel was *physically present* within the territory of a contracting state; in other words: Was the decision made before the vessel exited territorial waters? It is the intention to decommission the vessel and not the vessel's physical state or potential earning capacity of the vessel that determines if the vessel is 'waste'.[14] Thus, the Basel Convention, including the Ban Amendment, theoretically covers the issue of vessel scrapping/demolition but is not made for the purpose of dealing specifically with the issue. At the European level, the EU Waste Shipment Regulation (EWSR) implemented the Basel Convention, including its 1995 amendment, in the European Community.[15]

Several initiatives were taken in the late 1990s, of both a private and a public law nature. A private initiative, an 'Industry Code of Practice on Ship Recycling', was issued in 2001,[16] and 2003 saw technical guidelines for dismantling of ships issued under the Basel Convention,[17] as well as ILO guidelines on safety and health in shipbreaking.[18] Finally, in December 2003, IMO adopted its first guideline on ship recycling, IMO Resolution A.962(23) on Ship Recycling (IMO Guidelines (2004)) incorporating aspects of the industry code of practice and Basel guidelines mentioned.[19]

The guidelines indicated 'best practice' and were as such non-binding.[20] The guidelines are centered around the Ship Recycling Plan and an inventory of hazardous wastes and substances and are as such focused primarily on the direct physical and environmental safety of the ship demolishing process. Reflecting on the above, the parties to the Basel Convention, on their meeting later in 2004, urged the IMO to work toward the 'establishment of mandatory requirements to ensure the environmentally sound management of ship dismantling'.[21,22] This eventually resulted in the conclusion of the Hong Kong Convention, in 2009. The convention applies a lifecycle perspective to the issue of ship dismantling and includes rules on how to prepare ships for dismantling. More importantly, it also *covers the operation of ship recycling facilities* and purports to introduce appropriate enforcement mechanisms.[23] In keeping with other maritime conventions intended to prevent damage to the marine environment, the convention applies a flag state principle and applies to all vessels that are *flagged within a contracting state when the decision to recycle them is made*.[24] The IMO has issued several guidelines, intended to be used together with the convention when it enters into force[25]. To 'facilitate early implementation of the Hong Kong Convention' the EU implemented the convention with minor amendments in Regulation (EU) No. 1257/2013 on ship recycling.[26] The Regulation entered into force on 31 December 2018.

In conclusion, at present, the international legal landscape is highly fragmented. The Basel Convention incl. the Ban Amendment offer guidance on certain aspects, but no effective international system is in place. Instead, regional measures, such as the EU initiatives, combined with national legislation in either flag states, coastal states, or the territories in which the reclamation yards are situated, must be relied on. The question is, though, to what extent the private law contractual layer of regulation may compensate for some of these shortcomings.

13.3 Private Law Standard Contracts for the Dismantling of Vessels

13.3.1 In Brief on Relevant Market Factors

The value of a vessel's component metals, mainly steel but also the copper piping and, e.g., aluminum and zinc, far outweighs the cost of the service of breaking down and extracting those metals. The economic focus of the shipbreaking and recycling contracts is on these values and is essentially *sales contracts*, where the value of the item for purchase is based on the calculated/expected weight of the metals it includes; see further on this in Section 13.3.2. The weight is calculated as per the builder's plan taking into consideration the degradation and oxidation processes expected to have occurred in its lifespan (the 'lightweight' of the vessel[27]). The service of dismantling the vessel and safely distributing its component parts for recycling is not quantified and paid for specifically in the contract. Recycling and shipbreaking yards that have low running costs, e.g., because they apply the beaching method,[28] and do not spend much money on wages, safety for their employees, or environmental protection, may offer far better prices for the vessel.[29,30,31]

The owner of the vessel may choose to sell it directly to the dismantling and recycling facility, or it may be sold to an intermediate trader, a so-called 'cash buyer'.[32] In case the vessel is sold to a 'cash buyer', the owner must ensure that there is consensus as to the purpose of the sale. On the one hand, as the vessel would normally fetch a higher price when sold as a working vessel. The owner would not wish to sell the vessel at the lower scrapping or dismantling price, only to see the cash buyer resell it as a working vessel at a higher price. On the other hand, the owner would not want to risk that a vessel that the owner has sold for continued use ends up being dismantled under inappropriate or illegal conditions shortly after the sale. This would at best reflect negatively on the owners' corporate social responsibility status, at worst result in legal action against the owners for illegal export of waste or dismantling of vessels.[33,34]

13.3.2 The Development of Standard Contracts on Dismantling of Vessels

As mentioned above, from the perspective of private law, contracts for the demolishing or recycling of vessels are sales contracts for specified goods. Generally, standard contracts, with relevant ad hoc amendments, are used. Larger ship-owning companies may have their own template, as will some cash buyers and shipbreaking facilities/recycling yards. Here, the starting point will be the standard documents Salescrap 87, Demolishcon 2004, and Recyclecon 2012.[35] The documents are issued by the Baltic and International Maritime Council (BIMCO), a major international lobby organization for the shipping industry.[36] As one of its services, BIMCO produces and/or approves standard documentation used in the maritime trade. The standard documents are developed after input from relevant trade interests and even if not as such 'agreed documents' they present a somewhat balanced approach. Also, BIMCO's contracts are commonly used either as they are or as template for developing ad hoc contracts and they may be seen as a good indication of the *lex mercatoria* of the market.[37]

BIMCO first produced a specific contract aimed at shipbreaking in 1987. A product of its time, the contract, Salescrap 87, focused solely on the private law issues of the breaking of vessels at the end of their lifespan, and apart from the obligation not to put the vessel into service, it contains the provisions one would expect to find in a business-to-business transaction between the seller and the buyer of larger movable assets. In case the vessel is a tanker vessel, Salescrap 87, clause 12.2, provides that a certificate should be issued that the 'tanks, cofferdams and pipe rooms are gas-free, safe for men and fire, and substantially free of residues, slops and sludges…' and thus does make provisions for the immediate physical safety of the vessel. Apart from that, the contract is focused on the quid pro quo regarding the sale and pays no heed to issues relating to sustainability or the working environment.

As mentioned above, in Section 13.2, the late nineties and early noughts saw several initiatives, resulting in the IMO Guidelines (2004).[38] In that light, Salescrap 87 was outdated and BIMCO revised it to create Demolishcon. First issued in 2001, tallying with the 2001 industry code of practice, which refers to it and suggests its use,[39] it was updated again in 2004 to reflect and incorporate the IMO Guidelines (2004). Most of the contract corresponds to the private law template of Salescrap 87, but Demolishcon 2004 does, in its preamble, indicate that the parties to the contract undertake to 'comply with the … [IMO guidelines] … in accordance with Clause 17'. The clause provides that the parties should use '*best endeavours*' to comply with IMO's recommendations. Clause 17 also requires that the Sellers' representatives shall be allowed to visit the ship recycling facility to ascertain the practices applied there. In clause 7, the contract provides that sellers produce '… an inventory, in the form recommended by the IMO Guidelines as applicable…', as well as a slightly more specific reference to the status of tanks, etc., and former hazardous cargoes. However, the contract, in clause 10, presupposes specifically that the vessel is beached. Unfortunately, beaching generally involves a real risk of substandard practices.[40]

The next step in international development was the Hong Kong Convention (2009). After its inception, the IMO issued technical and procedural guidelines to assist the application of the convention when it entered into force. Even if that has not happened, the guidelines are in place and have been adopted as soft law by the IMO assembly.[41] Overall, this means that currently much of the intended effect of the Hong Kong Convention may be achieved by implementing the soft law regulation of the IMO Guidelines directly into the contractual relationship. Also, the Hong Kong Convention text may be incorporated by reference into the contract. Both approaches have been recommended by ship owners' organizations, awaiting the entry into force of the convention.[42] The convention's text is aimed at general state obligations and is not particularly well suited for incorporation into a contract between two business partners. The IMO guidelines, however, provide for specific technical steps or processes that should be carried out. Therefore, they lend themselves better to a 'translation' from public international soft law to a real contractual obligation. In keeping with this, BIMCO used the IMO guidelines as its starting point for the post-Hong Kong update of its standard contract.

The outcome of the revision process was the BIMCO Standard Contract of the Sale of Vessels for Green Recycling: Recyclecon 2012. To provide appropriate regulation of the private law aspects of the seller's and buyer's legal relationship, the contract updates and repeats many of the provisions of the two earlier standard contracts. However, the contract should not be seen as a mere revision of earlier materials, but rather as providing a new contractual starting point, which aims at incorporating international environmental and working environmental standards directly at the contractual level.

This is achieved mainly by creating a distinct narrative around the contract that firmly integrates existing or future IMO guidelines on the issue with the contract at the definition level. Specifically, clause 1 incorporates the IMO Guidelines 2011 for the development of the inventory of hazardous materials, including any subsequent amendments to it, directly into the contract by reference.[43] Clause 1 also incorporates 'the relevant guidelines to be developed by IMO' on the issue of the ship recycling facility plan, even if they were yet to be issued when the contract was published.[44]

In addition to the IMO guidelines, the contract repeats certain provisions of the Hong Kong Convention *verbatim* in its text. Thus, Recyclecon 2012, clause 1, point 9, copies the Hong Kong Convention Article 2(10) and defines 'recycling' as '...the activity of complete or partial dismantling of ships at the Ship Recycling Facility in order to recover components and materials for reprocessing and re-use, whilst taking care of hazardous and other materials...'. Furthermore, Recyclecon 2012 gives up the (damning) word 'beaching' altogether, and instead presupposes that the vessel is recycled at an authorized facility that is '... designed, constructed, and operated in a safe and environmentally sound manner'.[45] However, it should be noted that the contract does not rule out beaching as such.[46] This is coupled with a warranty from buyers in clause 17 that the vessel will be recycled at the pinpointed ship recycling facility in keeping with the specific procedures and work plans agreed to. These plans and procedures are defined by reference to the IMO guidelines on the issue, making the guidelines the standard against which buyer's obligations as to sustainability and a safe working environment should be measured. Making the conditions under which the ship is recycled, a specific warranty under the contract means that contractual remedies from the relevant background law may be applied in case the buyer fails to live up to the IMO standards.

13.4 Comparisons and Reflections. Lip Service or Contractual Reality?

Both Demolishcon and Recyclecon have incorporated the relevant international (soft) law regulation on vessel demolishing by reference into the parties' contractual relationship, thereby aligning the contractual landscape with international public law rules aiming at protecting the environment and employees. When considered this way, the question posed in this contribution – *whether these contracts, upon their proper interpretation, provide a relevant instrument that may rectify (some of) the shortcomings of the fragmented legal tapestry* – may be answered in the affirmative:

Considering that the Hong Kong Convention is not in force and that the only international public law regulation specifically aimed at the demolishing of vessels is in the form of soft law, private law *may* enhance the de facto application of the international public law regime.

Whether private law also provides some degree of *enforcement* to otherwise unenforceable international regulation is, however, a more difficult question. To answer it, one must consider both the type of obligation undertaken by the parties and the remedies connected to its potential breach.

Demolishcon obligates the parties to carry out their 'best efforts' to adhere to the IMO guidelines. Best effort-obligations may be hard to define due to their somewhat subjective nature. Also, Demolishcon stipulates no consequences in case the 'best endeavors' applied are found to be insufficient. Compared to this, the focus on objective criteria in Recyclecon 2012, where the conditions under which the ship should be dismantled are simply stated as contractual obligations, is preferable. This particularly, as Recyclecon 2012, couples these objective criteria with a warranty from the buyers that the vessel will be recycled under the agreed conditions.

Still, it presents the question how one, under the laws of obligation, quantifies the breach of a warranty to recycle in keeping with, e.g., a specific Ship Recycling Plan, under what is in essence a sales contract. The obligations as to sustainability and (working) environment are auxiliary to the central quid pro quo of exchanging a vessel at the end of its lifespan for payment. Realistically, therefore, such a warranty is likely to be unactionable unless it is coupled with a penalty clause quantifying the breach in economic terms. It remains to be seen if future revisions of Recyclecon 2012 will include such specified remedies or in other ways make sure that sustainability and safe working conditions are reflected in the economic reality of the contract.

On a final note, the question should be posed if a contract focused on the value of goods is the best private law instrument to ensure that vessels are demolished in a safe and environmentally acceptable manner. Recyclecon 2012 already has features of a mixed contract where the sale of goods and the procurement of services are intertwined. The idea of a sale, however, still dominates the economic and legal contexts of the contract, as well as its more specified remedies for breach. This is noteworthy, as it could very well be argued that the party who delivers the characteristic performance under the contract is the reclamation yard – and not the seller of the vessel. If instead the contracts were centered around the services provided, making the very specific task of sustainably and safely demolishing or recycling a vessel the core of the contract, maybe strategic, legal, and economic focuses would shift?

Under the current contractual approach, where the reclaimed goods end up as the property of the recycling yards, the push for the 'best price' may perpetuate substandard yards as an unpleasant reality for the foreseeable future.[47] Unless, of course, the international public law rules that may rectify the imperfect *private law* regulation finally enter into force. It is hoped that they do. The present state of the contractual *lex mercatoria* does not suffice to ensure a real push for acceptable practices.

Notes

1 See, e.g., *International Chamber of Shipping*, 'Environmental Performance: Comparison of CO_2 Emissions by Different Modes of Transport' <www.ics-shipping.org/shipping-fact/environmental-performance-environmental-performance/> accessed 7 July 2023.

2 For example, the International Convention on Civil Liability for Oil Pollution Damage 1969 (CLC 1969) or the numerous conventions and guidelines issued under the International Convention for the Prevention of Pollution from Ships 1973 (MARPOL 1973).

3 The reader may remember, e.g., the Erica and Prestige disasters.

4 See below on the IMO and EU initiatives.

5 See, e.g., the *NGO Shipbreaking Platform* <www.shipbreakingplatform.com> accessed 7 July 2023, or the ship owner driven *Ship Recycling Transparency Initiative* <www.shiprecyclingtransparency.org/> accessed 7 July 2023.

6 Private market actors will have a varied stake in the issue, depending on their market role and their corporate social responsibility profile. See, e.g., for a rather developed approach, the *Maersk Group* <www.maersk.com/sustainability/our-esg-priorities/responsible-ship-recycling> accessed 7 July 2023, compared to the more minimalist approach displayed by e.g. *COSCO Shipping Lines Co. Ltd.* <lines.coscoshipping.com/lines_resource/pdf/quality2021en.pdf> 27 accessed 7 July 2023.

7 Commission, 'Communication from the Commission to the European Parliament, the Council, the European Economic and Social Committee and the Committee of the Regions: An EU strategy for better ship dismantling' COM (2008) 767 final, 3.

8 Commission, 'Communication from the Commission to the European Parliament, the Council, the European Economic and Social Committee and the Committee of the Regions: An EU strategy for better ship dismantling' COM (2008) 767 final, 3.

9 See an overview of this below in point 3.

10 Gabriela Argüello Moncayo, 'International law on ship recycling and its interface with EU law' [2018] 109(1) *Marine Pollution Bulletin* 301, 307 (https://doi.org/10.1016/j.marpolbul.2016.05.065).

11 The international regulation on this issue is deserving of its own study. Here, however, only the bare minimum necessary to contra-point it with the private law aspects of the issue will be presented.

12 Amendment to the Basel Convention adopted by Decision III/1 of the Parties to the Basel Convention on its third meeting, 28 November 1995.

13 The amendment did not enter into force until 5 December 2019.

14 See, e.g., the *Seatrade* case. In the case, four ships were shipped from the ports of Rotterdam and Hamburg to beaches in India, Bangladesh, and Turkey for scrapping. The vessels were considered waste in the meaning of the regulation, despite the argument made by the defense that they could have remained in service for at least another decade. All vessels contained asbestos and refrigeration units containing hydrochlorofluorocarbons, which are considered hazardous under EWSR, Annex V. Thus, the export of the vessels had taken place in violation of the export ban. The accused, the director of Seatrade Management B.V., was sentenced to a fine of €50,000 and a professional disqualification of one year for infringement of the EWSR. The *Seatrade,* District Court of Rotterdam (15 March 2018), Case No. 10/994550-15.

15 Regulation (EC) No 1013/2006 of the European Parliament and of the Council of 14 June 2006 on shipments of waste [2006] OJ L190/1, preamble para 3 and 4.

16 Industry Code of Practice on Ship Recycling, adopted by BIMCO, INTERCARGO, INTERTANKO, ICS, ITOPF, ITF and OCIMS in August 2001.

17 Technical Guideline for the Environmentally Sound Management of the Full or Partial Dismantling of Ships, published at *UNEP*, <https://www.basel.int/Implementation/ShipDismantling/TechnicalGuidelines/tabid/2767/Default.aspx#> accessed 25 September 2023.

18 ILO, 'Safety and Health in Shipbreaking' <https://www.ilo.org/public/english/standards/relm/gb/docs/gb289/pdf/meshs-1.pdf> accessed 25 September 2023.

19 IMO, 'Resolution A.964(23), IMO Guidelines on Ship Recycling' (5 December 2003).

20 IMO, 'Resolution A.964(23), IMO Guidelines on Ship Recycling' (5 December 2003) 6.

21 Report on the Conference of the Parties to the Basel Convention, seventh meeting, Geneva, October 25–29, 2004, UNEP/CHW.7/33, 63 f, point 5.

22 Gabriela Argüello Moncayo, 'International law on ship recycling and its interface with EU law' [2018] 109(1) *Marine Pollution Bulletin* 301, 302 (https://doi.org/10.1016/j.marpolbul.2016.05.065).

23 Regulation (EU) No 1257/2013 of the European Parliament and of the Council on ship recycling and amending Regulation (EC) No 1013/2006 and Directive 2009/16/EC [2013] OJ L330/1, preamble, para 4.

24 Hong Kong Convention Article 4(1).

25 Including at present 2011 Guidelines for the Development of the Ship Recycling Plan, MEPC.196(62), 2012 Guidelines for Safe and Environmentally Sound Ship Recycling, MEPC.210(63), 2012 Guidelines for the Authorization of Ship Recycling Facilities, MEPC.211(63), 2012 Guidelines for the Survey and Certification of Ships under the Hong Kong Convention, MEPC.222(64), 2012 Guidelines for the Inspection of Ships under the Hong Kong Convention, MEPC.223(64), 2015 Guidelines for the Development of the Inventory of the Hazardous Materials, MEPC.269(68). The resolutions may be accessed at <www.imo.org/en/OurWork/Environment/Pages/Ship-Recycling.aspx> accessed 7 July 2023.

26 Regulation (EU) No 1257/2013 of the European Parliament and of the Council on ship recycling and amending Regulation (EC) No 1013/2006 and Directive 2009/16/EC [2013] OJ L330/1, preamble para. 5.

27 Salescrap 87, Cl 3, Demolishcon 2004, Cl 12, Recyclecon 2012, Cl 1 and Cl 8(a).

28 See further on this in a lifecycle perspective, Jun-Ki Choi *et al*, 'Economic and environmental perspectives of end-of-life ship management' [2016] 107 *Resources, Conservation and Recycling* 82 (https://doi.org/10.1016/j.resconrec.2015.12.007).

29 Vessels for scrap are sold as specific goods and are not traded as commodities. Therefore, there are no official (international) commodity values for vessels intended for demolition. Instead, private entities, such as, e.g., shipbrokers, collect relevant data, and provide overviews of the average prices currently offered by the scrapping/recycling yards in different geographical regions. Numbers will thus have to be taken with a pinch of salt; however, generally yards that apply beaching practices have been able to offer between 50% and 70% higher prices for the vessels sold to them. See, e.g., Star Asia Shipbroking Pte. Ltd.'s weekly market report, which provides both a weekly snapshot and a five-year average. According to their data, the average price in USD pr. light displacement tonnage in May 2023 averaged ca. USD 550 in Alang, India, USD 560 in Chattogram, Bangladesh, and USD 320 in Turkey, with a specific proviso regarding the Turkish market, that if the vessel is EU-flagged, and thus covered by the stricter requirements of Regulation (EU) No. 1257/2013, the price per ton is 30–40 USD less. As per ultimo May 2023, Yards in Alang and Chattogram thus offer 70% higher prices for the vessel, <www.star-asia.com.sg/_files/ugd/4ece41_8732daf7923843909e6ee0b092d24794.pdf>, 10, accessed 7 July 2023.

30 Urs Daniel Engels, *European Ship Recycling Regulation, Entry-Into-Force Implications of the Hong Kong Convention* (Heidelberg, Springer 2013) 16 ff; Yen-Chiang Chang, Nannan Wang and Onur Sabri Durak, 'Ship recycling and marine pollution' [2010] 60(9) *Marine Pollution Bulletin* 1390 (https://doi.org/10.1016/j.marpolbul.2010.05.021).

31 In the *Tide Carrier*, the Sundhordland Court of First Instance found it to be proven that an approved demolishing yard would have been able to offer approx. NOK 39 million as opposed to the NOK 41 million which the sellers obtained by exporting the vessel to be demolished under unapproved conditions (p. 38). The court confiscated NOK 2 million, putting the sellers in the same financial position as if they had dismantled the ship in accordance with the EWSR. Sunnhordland Court of First Instance (27 November 2020) Case No. 19-183171 MED-SUHO (*The Tide Carrier*).

32 Halvor Schøyen, Umar Burki, Sunu Kurian, 'Ship-owners' stance to environmental and safety conditions in ship recycling. A case study amongst Norwegian shipping managers' [2017] 5(3) *Case Studies on Transport Policy* 499, 500 (https://doi.org/10.1016/j.cstp.2017.06.003).

33 Halvor Schøyen, Umar Burki, Sunu Kurian, 'Ship-owners' stance to environmental and safety conditions in ship recycling. A case study amongst Norwegian shipping managers' [2017] 5(3) *Case Studies on Transport Policy* 499, 503 (https://doi.org/10.1016/j.cstp.2017.06.003).

34 See, e.g., the North Sea Producer scandal. The North Sea Production Company Ltd. sold the FPSO North Sea Producer to a third party in April 2016. The vessel was not able to sail on its own steam and had been used as an offshore storage facility. Seemingly, it was presumed in the contract that the vessel should be sold for continued use. The sales contract did however contain a clause that should the vessel be dismantled, the procedure should live up to Hong Kong Convention standards. The buyer eventually abandoned the attempt to redeploy the vessel and sold it to a recycling yard in Chattogram, Bangladesh, about four months later. The original sellers – a Maersk Group subsidiary – had no contractual relationship with the ultimate buyers in Chattogram and found themselves unable to influence the processes there. Still, the Maersk Group faced scrutiny both from NGOs and journalists, and from certain authorities; see <www.maersk.com/news/articles/2019/04/04/maersk-tightens-its-ship-recycling-procedures> accessed 7 July 2023. The Bangladeshi High Court later found the circumstances surrounding the dismantling of the vessel to be illegal under domestic public administrative law; see Bangladeshi High Court (14 November 2019) WP 8466 of 2017. Further on the North Sea Producer scandal, see Ministry of Environment and Foods, file No 001-15429 of 29 December 2016, reply to question to the ministry No. 52: Exposition on the breaking of Maersk vessels in Bangladesh and India, p. 8, <www.ft.dk/samling/20161/almdel/mof/spm/52/svar/1366808/1699670.pdf> accessed 7 July 2023. See also NGO Shipbreaking Platform on the issue, <shipbreakingplatform.org/spotlight-north-sea-producer-case/#:~:text=The%20North%20Sea%20Producer%20%28ex%20Dagmar%20Maersk%29%20was,between%20Danish%20A.P.%20Moeller%20Maersk%20and%20Brazilian%20Odebrecht> accessed 7 July 2023.

35 Demolishcon 2004 and Recyclecon 2012 may be accessed in a downloadable, read-only version on BIMCO's website; see <www.bimco.org> accessed 7 July 2023. Salescrap 87 is not available on the

website as its use is no longer recommended by BIMCO. Salescrap 87, Cl 3, Demolishcon 2004, Cl 12, Recyclecon 2012, Cl 1 and Cl 8(a). Demolishcon 2004 and Recyclecon 2012 may be accessed in a downloadable, read-only version on BIMCO's website; see accessed 7 July 2023. Salescrap 87 is not available on the website as its use is no longer recommended by BIMCO.

36 BIMCO members amount to about 60% of the global merchant fleet; see <www.bimco.org/about-us-and-our-members>, accessed 7 July 2023.

37 The *ad hoc* contracts in BTB-transactions will normally be confidential and their contents are normally only displayed to the public in case they have attracted journalistic or governmental interest.

38 IMO, 'Resolution A.964(23), IMO Guidelines on Ship Recycling' (5 December 2003).

39 Industry Code of Practice on Ship Recycling, adopted by BIMCO, INTERCARGO, INTERTANKO, ICS, ITOPF, ITF and OCIMS in August 2001.

40 Jun-Ki Choi *et al*, 'Economic and environmental perspectives of end-of-life ship management' [2016] 107 *Resources, Conservation and Recycling* 82, 84 ff (https://doi.org/10.1016/j.resconrec.2015.12.007).

41 Hong Kong Convention Art. 4(1).

42 Incorporation of the Hong Kong Convention Rules is recommended, e.g., by the *Norwegian Shipowners' Union*, <rederi.no/aktuelt/2019/nederland-ratifiserer-hong-kong-konvensjonen> accessed 7 July 2023, and the *Danish Shipowners Union,* <www.danishshipping.dk/presse/nyheder/alang-er-ikke-bare-alang> accessed 7 July 2023, and Ministry of Environment and Foods, file No 001-15429 of 29 December 2016, reply to question to the ministry No. 52: Exposition on the breaking of Maersk vessels in Bangladesh and India, <www.ft.dk/samling/20161/almdel/mof/spm/52/svar/1366808/1699670.pdf> accessed 7 July 2023, question 13, p 9. Danish Shipowners Union further recommends that IMO standards are incorporated in the parties contractual relationship, see, e.g., <www.danishshipping.dk/en/press/news/sustainable-ship-recycling-is-not-a-matter-of-geography> accessed 7 July 2023. BIMCO has lobbied heavily for the ratification of the Hong Kong Convention, see, e.g., campaign launched May 2023, <www.bimco.org/trending-topics/recycling> accessed 7 July 2023.

43 Recyclecon 2012, Cl 1, point 6 and 13.

44 Recyclecon 2012, Cl 1, point 11.

45 Recyclecon 2012, Cl 1, point 12.

46 The Hong Kong Convention does not rule out beaching. However, it may prove difficult for beaching practices to be approved under the Hong Kong system.

47 See, e.g., the Starflotte scandal. Maersk Group explains the situation thus: 'The … case relates to 14 chartered-in Starflotte ships whose contracts Maersk wanted to end ahead of time in 2014. In the final agreement with the owner, Maersk incentivized recycling at the best price, which effectively means recycling at sub-standard yards, which is what the owner chose to do when the vessels were returned. … Maersk has publicly acknowledged, and regrets, that it indirectly incentivized the owner to recycle at sub-standard yards'. <www.maersk.com/news/articles/2019/04/04/maersk-tightens-its-ship-recycling-procedures> accessed 7 July 2023.

Bibliography

Chang Y-C, Wang N and Durak OS, 'Ship recycling and marine pollution' [2010] 60(9) *Marine Pollution Bulletin* 1390 (https://doi.org/10.1016/j.marpolbul.2010.05.021)

Choi J-K *et al*, 'Economic and environmental perspectives of end-of-life ship management' [2016] 107 *Resources, Conservation and Recycling* 82 (https://doi.org/10.1016/j.resconrec.2015.12.007)

Engels UD, *European Ship Recycling Regulation, Entry-Into-Force Implications of the Hong Kong Convention* (Heidelberg, Springer 2013) 16 ff

Moncayo GA, 'International law on ship recycling and its interface with EU law' [2018] 109(1) *Marine Pollution Bulletin* 301 (https://doi.org/10.1016/j.marpolbul.2016.05.065)

Schøyen H, Burki U and Kurian S, 'Ship-owners' stance to environmental and safety conditions in ship recycling. A case study amongst Norwegian shipping managers' [2017] 5(3) *Case Studies on Transport Policy* 499 (https://doi.org/10.1016/j.cstp.2017.06.003)

14

AMENDMENTS TO ARTICLES 9 AND 41 OF THE ITALIAN CONSTITUTION ON THE PROTECTION OF THE ENVIRONMENT AND INTERGENERATIONAL RESPONSIBILITY

Roberta Calvano

14.1 Preliminary Remarks

Only recently, a change occurred in the attitude of the legal culture towards the disastrous consequences of unlimited industrial economic development. Italian constitutional law studies show a widespread awareness of the problems related to a no longer sustainable production system.[1] While several constitutional texts from all over the world address the issue of sustainability and of the different topics connected with it, it must be underlined that the Constitutions of important countries such as the USA, Russia, China, and India still don't mention any of the topics connected with sustainability. This remark seems to highlight how potentially important and, in any case, culturally characterising is the choice of evoking the concept of sustainability in a constitutional text, even though in none of the Constitutions that mention sustainability it is included in the fundamental principles evoked in the preambles.[2]

Today, it is very well known in legal studies that sustainability can no longer be regarded as only 'the development that meets the needs of the present without compromising the needs of the future generations'.[3] It would be the result of a narrow perspective limiting the concept of sustainable development to the intergenerational dimension in the environmental conservation-economic growth trade-off. But, even though this is not the case for many years, as the concept of sustainable development encompasses the societal dimension as well as intragenerational equity, in the Italian public debate, a more limited vision of sustainability as a matter of environment versus economic growth is still, to some extent, widespread,[4] also due to the reform of Article 81 of the Italian Constitution,[5] introducing sustainability as limited to the field of public debt and finance.[6] A broader definition can be read in the Constitution of Belgium, which refers to durable/sustainable 'development in its social, economic and environmental aspects', attributing its care to the public

 DOI: 10.4324/9781032662046-17

authorities 'taking into account intergenerational solidarity'. The breadth of the theme, as it shows, seems to be very well illustrated by the French term 'durabilité',[7] but the amendment to the Italian Constitution approved in 2022, as will be seen in the following pages, does not seem to build much on comparative experiences.

Anyway, further development of European Union legislation, and its transposition into Italian law, will hopefully raise a deeper awareness of issues linking sustainability to labour rights, and sustainable business practices, that is to say, to private law.[8] As the framework of the European Green Deal is being put together, for instance integrating the non-financial reporting obligations imposed by the Corporate Sustainability Report Directive[9] with new obligations, the process in the Italian legal order and other Member States will hopefully follow.

Achieving sustainability involves then several legal issues, above all the central question of creating or identifying constitutional principles that, together with international documents and EU legislation, can guide the development of more effective environmental protection through private law and business practice.[10]

In fact, it is since the 1970s that the theme of environmental protection emerged progressively in the legal discourse. Modern constitutionalism, due to its deep links to the Enlightenment, has unquestioningly accepted the assumptions of this philosophic thought, and its anthropocentric perspective. This brought constitutional law to not be able to thematise the ecological question in time.[11] This was so until legal thinking and especially constitutional law scholars were confronted with the great environmental catastrophes of the 20th century and their consequences on people's lives and health.

European Constitutions after World War have set new limits to economic production that could be activated from an ecological perspective. Such a development took place, also thanks to the important evolution of international environmental law and through Constitutional Courts recognising new individual rights and principles on the basis of constitutional texts already in force. This way, an attempt has been made to make up for a delay in positive legislation by case law.

In this connection, it is worth recalling the example of the Italian legal order, in which the recognition of the environment as a legal asset worth protecting only arose in the mid-eighties. In those years, the Constitutional Court granted legal autonomy to the concept of environment in its judgement n. 210/1987, at a time when the increasing importance of the environmental issue was being made clear in EU legislation. In those years, the law establishing the Ministry of the Environment was approved in Italy (law July 15, 1986, n. 349), while a post-Chernobyl referendum took place in 1987 repealing legislation allowing the production of energy in nuclear power plants. Environmental awareness had started to spread, and the effects began to become visible in the Italian legal order.

Such goals were not reached directly, but by means of the evolving interpretation of two articles of the Constitution, Articles 9 and 32, the latter concerning the right to health as a social right, not only in individual terms but also relationally and collectively.[12] Since then, these two articles have been the centre of the constitutional debate on environmental issues, together with Article 41, slowly leading to an unexpressed idea of sustainable development.[13]

The purpose of this contribution, apart from analysing Constitutional Law No. 1 of 2022 and its contents, will therefore concern its ability to lead to an overcoming of the dichotomous environment versus economic development perspective, and to encourage economic operators to take responsibility for protecting sustainability in terms of intergenerational responsibility not only as an economic good, but also as a legal obligation.

14.2 Constitutional Law No. 1 of 2022

The constitutional law entitled 'Amendments to Articles 9 and 41 of the Constitution on the Protection of the environment' took a year and a half of parliamentary examination before its approval. It entered into force on 9 March 2022, amending Articles 9 and 41 of the Italian Constitution, respectively, dedicated to the protection of culture, research, landscape and historical and artistic heritage, and to the freedom of economic initiative.

The first of the two articles of the reform add a new third paragraph to Article 9, stating that 'the Republic protects the environment, biodiversity and ecosystems in the interest of future generations'. In declaring this, the legislator's choice was to avoid explicitly mentioning sustainability and instead to build a concept of intergenerational responsibility. The reform also states that 'the law of the State governs the ways and forms of animal protection'. Leaving aside the theme of the animals' protection for the time being, it is worth considering whether the first paragraph could be the initial step towards the future development of a different economic model or – as constitutional law scholars name it – of a new 'Economic Constitution',[14] involving sustainability as a fundamental constitutional value. The issue is very relevant on two connected levels. First, its connection with the already recalled issue concerning Article 81 of the Constitution, which develops the balanced budget principle deriving from Article 3 of the 'Fiscal compact', the Treaty on Stability, Coordination and Governance in the Economic and Monetary Union. Secondly, intergenerational responsibility introduced in the Constitution's framework potentially impacts the regulation of economic activities in private and commercial laws.

The significance of this hypothesis also appears if we consider Article 2 of the constitutional reform, which amends Article 41 second and third paragraphs of the Constitution. It adds the words 'Environment' and 'Health' to the limits already established in the second paragraph for economic activities, 'security, freedom, human dignity'. The third paragraph of Article 41, by adding the word 'environmental' (from 9 March 2022), then states that, 'The law determines the appropriate programmes and controls for public and private economic activity to be directed and coordinated for social and *environmental* purposes'.

As it shows, the reform enshrines the full recognition of the fundamental value of the environment by the constitutional legislator, amending especially Article 9 of the Constitution. This is a rule that the Constitutional Court had already referred to on several occasions in the past, in order to affirm the constitutional importance of the environment as a unitary legal asset.[15] But its impact on the Italian legal order – and especially on private law – depends mainly on the limit added to Article 41 of the Constitution, and its capacity to influence the development of private and commercial activities regulated by the civil code.

Anyway, as a result of the reform, the concept of the environment can no longer be considered to coincide with that of the right to health, in terms of a healthy environment, and more with a principle than an enforceable right. The risk that this will lead to a 'de-subjectivisation' that paradoxically weakens environmental protection, where there will not be a right to be enforced, but a principle or at most a limitation (on the freedom of economic initiative) needs to be highlighted.[16]

Furthermore, with particular regard to the limits to economic freedom, it can no longer be regarded as falling within the social benefit clause included in Article 41. This means that innovation in the Constitutional framework has occurred not only concerning the duty of public authorities to protect and promote environmental protection, but also with regard to the scope of economic freedom, setting a new limit that can potentially affect the law of contracts and the regulation of many commercial or industrial activities.[17]

Thus, from the point of view of Article 41 freedom of economic initiative, environmental protection is no longer covered by the social utility clause. This way, by potentially disconnecting the link between the protection of the environment and that of weak contracting parties in private law relationships (workers, consumers, residents in areas destined for particular activities or reclamation, clean-up, demolition, etc.) the two might come into conflict. Of course, it is a matter of imagining borderline cases, in which the continued presence of the limit of social utility, dignity, and safety may fade in the face of the new limit of environmental protection, which in its placing itself alongside – and not within – the protection of social utility.

The innovation introduced in Article 41 appears likely to have an impact on the breadth of freedom of economic initiative, placing a limit capable of influencing the regulation of contracts and commercial activities.

14.3 Issues Left Unsolved by the Reform

The reform has been considered epochal in the public debate, while it actually seems, for its main contents, rather the mere restatement of acquisitions that had already been accrued for some time through legislative and jurisprudential interventions.[18] Though, as these developments were characterised by their fragmentation, precariousness, and incompleteness stemming from their base in case law, it has been acutely observed[19] that a restatement of the principle at a constitutional level may help the Constitutional Court to safeguard the environment and more effectively connect it to new values, such as sustainability.

In fact, the elusive nature of the environmental and ecological issues for decades has left wide discretion to the legislator, as rulings by the Italian Constitutional Court show. In the ILVA case,[20] the referring courts challenged the violation by the national legislature of the fundamental rights to health and a healthy environment and the related limits to freedom of private economic initiative. In short, they asked the Constitutional Court to review the 'technical merit' of the legislative choices allegedly deemed ineffective to adequately ensure the protection of the fundamental rights mentioned. In this ruling, the Constitutional Court made it clear that

> it cannot be accepted that a judge (including this Court) considers the new legislation to be unlawful by virtue of an assessment of the merits of its inadequacy, irrespective of whether it infringes precise legislative, constitutional or ordinary parameters, overlapping their own discretionary assessments with those of the legislator and the competent authorities. Such judicial review of legislation would be possible only in the presence of a manifest unreasonableness of the new discipline.[21]

This means that a precise and analytical constitutional discipline of the environment provides legislators with the fundamental coordinates and constraints on which to build environmental policies. Consequently, the legislative and administrative choices will be less determined by political assessments, or by the uncertain choices of the judges' review on the external limits of reasonableness.

A very good example of such an analytic framework of rules is set in the EU Treaties discipline on the environment, representing a highly refined model, structured on the basis of foundations,[22] objectives,[23] methods,[24] and principles.[25] Not much room is left in the EU discipline for vagueness and imprecise subjective legal situations unfit to be legally protected. This consideration highlights the problem of the possible inadequacy especially of new Article 9, due to the very wide and generic formulation of the amended constitutional discipline.

Other critical issues are involved in the reform and offer the cue for a brief comment.

It has been pointed out that a clear and forward-looking choice has been made by the legislator from a subjective point of view. Article 9 Const., as amended, involves all subjects, bodies, and public institutions in the common task of protecting and safeguarding the assets already mentioned. At the same time, a less convincing choice in the third paragraph of Article 9 entrusts to the state legislator alone the protection of animals.[26] This provision does not seem to entail a new significant task regarding what was already guaranteed by the protection of animal health in terms of care, veterinary control, and animal prophylaxis.[27]

The objections raised by this law of constitutional amendment also arise from a procedural point of view, in terms of the general guarantee of the Constitution. In fact, for the first time in the history of the Italian Republic, an amendment to the first 12 articles of the Constitution regarding the fundamental principles has been approved, creating what has been and can be considered a dangerous precedent in terms of opening the door to possible future interventions on constitutional fundamental principles.[28] This constitutional amendment was approved by a two-thirds majority of the two chambers, thus avoiding activating the referendum procedure provided for by Article 138 Const. for cases of approval of constitutional revisions by an absolute majority.[29] The Constitutional Court case law since ruling n. 1146 of 1988 has established that supreme principles and fundamental rights cannot be the object of a constitutional amending procedure,[30] but it has never listed which among the fundamental principles of the Constitution are to be also considered 'supreme'.[31] The question remains open to interpreters. And the precedent is now there and, in some way, worrisome, as Articles 9 and 41 quite clearly concern the fundamental rights that cannot be subject to constitutional amendment. Secondarily, even though the case law of the Court is quite clear, it is claimed that by such a procedure, one may amend provisions proclaiming supreme principles in order to extend them, whereas suppressive amendments are only to be prohibited. But it is true that enlarging one right may limit another. It is not possible to analyse further the issue from a theoretical point of view, but it is clear that, as the next few pages will try to show, such an intervention can have an important impact on the form of the State.

14.4 Environment, Sustainability, and the Impact of the Limits to Economic Freedom

The word environment was added to the Italian Constitution many years after it came into force. The amendment approved in Constitutional Law No. 3/2001 (concerning Part II, Title V of the Constitution) concerning the distribution of legislative competence among the Regions and the State made the conferral of the 'protection of the environment and the ecosystem' a matter of exclusive State legislative power within Article 117 Const.

After the Title V amendment, environmental protection is structured around two pillars. On the one hand, the Regions' ownership of legislative powers over matters (territorial government, health protection, etc.) is considered of great environmental importance. On the other hand, the State was granted 'adequate and not reducible levels of protection' in environmental matters. And this is with the prominent aim of preventing the Regions from 'derogating or worsening the level of environmental protection established by the State'.[32]

Comparing the Constitution as amended with the previous one in this subject area, it can be remarked, with specific regard to the 2001 text quality, that today's legislator clarifies the concepts better, enriching the protection of the environment with the use of the terms 'biodiversity' and 'ecosystems' (now using a more correct plural form), reaching a better articulation of the discipline, by addition to the original and unchanged term 'landscape'.

The new intervention is certainly a further step in the direction of environmental protection, but it would be interesting to understand whether the constitutional amendments can be read in terms of a restatement of a long experience of legislative and judicial development, or if its contents can be considered as elements introducing a new and different perspective for the future. If both functions should be fulfilled by the reform, it will have to be analysed which balance can be found among the two.[33]

The contradictory nature of the reform is evident even in the most enthusiastic comments: 'The reform certainly does not intend to transform the economic model prefigured by the Italian Constitution, even if the Article 2 of the l. Const. 1/2022 intervenes precisely on the central norm of that model',[34] but surprisingly it is also considered that the environment would relate to a sphere other than that of economic interests. How could it change the form of the state then?

Actually, the constitutional amendment capability to establish sustainability and environmental protection as a limit to economic development, in accordance with the 2015 Paris Agreement on climate change and intergenerational equity,[35] seems weak. And the task of the constitutional legislator in order to fulfil what has been held as a responsibility of EU Member States governments and legislators – 'climate and environmental problems are the task that defines our generation' –,[36] seems to have been left unfulfilled.

The lack of deep environmental awareness and the fear of interfering with the economic interests of industry and financial groups are quite perceptible in the reform. It seems then that, incapable of a real change of perspective, the Italian legislator tried to give a greenwashing appearance to the Italian Constitution, also because environmental protection is one of the strategic objectives of development and innovation policies under the most recent international regulatory framework, as well as the Next Generation EU. Not to forget the Italian National Recovery and Resilience Plan (NRRP), in which, notably, Mission 2 is named after 'Green revolution and ecological transition', whereas the same mission is part of the UN 2030 Agenda.

However, the objective of this constitutional greenwash does not seem to have been fully achieved. Not only because of the meagre and imprecise wording, which has already been mentioned. What is also missing is an indication of the transversality of environmental protection and its necessary consideration in all public policies, in relation to Article 11 TFEU and the principle of environmental integration.

Anyway, sustainability, although surrounded by an aura of sacredness, deriving from its being formalised in a series of important international documents,[37] is now the subject of many criticisms in constitutional law[38] because it has often proved an empty clause, unable to translate into concrete and operational rules.[39]

14.5 Intergenerational Responsibility

The part of the reform involving future generations is certainly due to an idea of constitutional revision as a programme for the future. This clause, recalling that contained in Article 20a *Grundgesetz*, appears intended to build up the concept of sustainability and sustainable development. An intergenerational dimension imposes undeniably if only for reasons of effectiveness, the projection of environmental policies in the medium to long term. The reform thus tries to introduce an extremely important constraint for the legislator. The question is now to understand if the amendment to Article 9 concerning the interest of future generations is apt to introduce a parameter of validity suitable to be judicially enforced, whereas its other paragraphs are certainly rich in potential for the decision-making processes of environmental policymaking. Although Constitutional Law No. 1/2022 has not incorporated the explicit reference to sustainable development, it seems nonetheless that the overall aim of the reform was probably to introduce the idea of sustainability in the model of society prefigured in the Constitution. Its formulation though, with the generic reference to future generations, seems to lack the precision necessary to establish an effective constitutional principle, susceptible to be enacted in judicial review of legislation.[40]

The temporal dimension of the future, of the consequences in time of regulatory choices, has assumed in this debate a previously unknown importance in contemporary law, although Constitutional texts were created in this exact perspective. It must be underlined though that the construction of the concept of rights of the future generations responds to an ethical requirement, but not to legal categories.[41] The current non-existence of the holder of those rights could be overcome by resorting to the use of fiction. Not only developing new rights from the text of the constitution interpreting Article 2 of the Constitution general clause can bring to undermine those which are expressly granted,[42] but the new (next generation) rights could potentially bring a huge conflict with the present generation fundamental rights also in terms of public finance and the welfare state, as the concept of 'intergenerational equity' brought already the Constitutional Court case law on Article 81 to affirming 'the need not to place a disproportionate burden on the growth opportunities of future generations, guaranteeing them sufficient resources for a balanced development'.[43]

The solution brought by the reform to develop an idea of intergenerational responsibility, depending on how it will be interpreted, can help to smooth the conflict or even to worsen it. In addition to that, it would then be important to develop the idea and practice of constitutional duties of the present generations (including the duty to protect the environment), alongside the protection of fundamental rights. The aforementioned constitutional legislation could then play a role in preparing the ground for sustainability regulations intended to apply to companies.

This could be the case of acts such as the Corporate Sustainability Due Diligence Directive, in which the European Union legislator's objective of enabling a more effective framework for sustainable development could be pursued with the provisions in which companies are obliged to adopt rules of governance and supervision of sustainability risks along the entire value chain.

In the end, it must be considered that notwithstanding some weak points of the described constitutional framework, sustainability can be considered to encourage rational choices of all economic actors, whereas greater awareness of this issue can help promote the idea of a society in which human rights, from health to labour rights, are safeguarded throughout the whole economic process. More doubtful is whether the concept of intergenerational responsibility involved in this reform, with all the questions it leaves unsolved, will help to move faster in this direction.

Notes

1 Among the first scholars to show a deep interest in the issue was B. Caravita, *Diritto pubblico dell'ambiente* (Il Mulino, Bologna 1990).

2 As underlined in T. Groppi, 'Sostenibilità e Costituzioni. Lo Stato costituzionale alla prova del futuro' (2016) *DPCE*, n. 1, 43.

3 Our Common Future, Report of the World Commission on Environment and Development (c.d. 'Brundtland Commission'), 1987, par. 27, Annex to the UN General Assembly, A/42/427, Development and International Cooperation: Environment (2 Aug. 1987) 8.

4 M. Benvenuti, 'La revisione dell'articolo 41, commi 2 e 3, della Costituzione, i suoi destinatari e i suoi interpreti', (2023) *Rivista AIC*, n. 2, 59, available at <www.rivistaaic.it> accessed 4 July 2023.

5 Art. 81 was amended by Constitutional Law No. 1 of 2012, as a transposition of Art. 3 of the Fiscal Compact. Its paragraph 4 states: 'The content of the budget law, the fundamental rules and criteria to ensure the balance between budget revenue and expenditure and the sustainability of the debt of the general government as a whole shall be established by a law passed by an absolute majority'.

6 The use of the concept in this field can be found in the Constitutions of the following countries: Belgium, Hungary, Spain, Angola, Bolivia, Colombia, the Dominican Republic, Timor East, Nepal, Ecuador, Guinea, Kosovo, Mexico, Peru, Serbia, South Africa, Switzerland, Venezuela, and Zimbabwe.

7 The words "sustainable development" and "sustainability" appear in 53 contemporary constitutions as recalled in L. Bartolucci, 'Le generazioni future (con la tutela dell'ambiente) entrano "espressamente" in Costituzione' (2022) *Forum di Quaderni Costituzionali*, n. 2, 29, available at <www.forumcostituzionale.it> accessed 1 December 2022. See also T. Groppi, 'Sostenibilità e Costituzioni. Lo Stato costituzionale alla prova del futuro'(2016) *DPCE*, n. 1, 43.

8 An example of how this influence could be impressive can be offered by the draft directive on corporate sustainability due diligence, released by the EU Commission in February 2022.

9 Directive (EU) 2022/2464 of the European Parliament and of the Council of 14 December 2022 amending Regulation (EU) No 537/2014, Directive 2004/109/EC, Directive 2006/43/EC and Directive 2013/34/EU, as regards corporate sustainability reporting [2022] OJ L322/15. For instance, Council Directive 2002/60/EC of 27 June 2002 laying down specific provisions for the control of African swine fever and amending Directive 92/119/EEC as regards Teschen disease and African swine fever [2002] OJ L192/27.

10 The concept in a constitutional perspective is analysed in A. Jakab, 'Sustainability in European Constitutional Law' (2016) *MPIL Research Paper* 2016–16, available at <https://papers.ssrn.com/sol3/papers.cfm?abstract_id=2803304> accessed 23 December 2022.

11 A.Ciervo, 'Un altro mondo (non) è possibile. Diritto costituzionale, ambiente e Climate Justice' (2021) *Parole chiave*, n. 6, 75–77.

12 A.Ciervo, 'Un altro mondo (non) è possibile. Diritto costituzionale, ambiente e Climate Justice' (2021) *Parole chiave*, n. 6, 75.

13 A. Predieri, 'voce Paesaggio', *Enciclopedia del diritto* (xxxi, Giuffrè, Milano 1981) 507–509; earlier on, compare with M.S. Giannini, '"Ambiente": saggio sui suoi diversi aspetti giuridici' (1973) *Rivista trimestrale di diritto pubblico*, 1.

14 M. Luciani, voce 'Economia (nel diritto costituzionale)', *Digesto (discipline pubblicistiche)* (V, UTET, Torino 1990) 374–375.

15 Constitutional Court, rulings n. 196/2004, 278/2012, 9 and 85/2013, 260/2017, 210/2016, 113/2018, 7/2019. On the constitutional notion on environment see B. Caravita di Toritto and A. Morrone, 'Ambiente e Costituzione' in B. Caravita di Toritto, L. Cassetti, A. Morrone (eds), *Diritto dell'Ambiente* (Il Mulino, Bologna 2016) 30.

16 It is no coincidence that at the parliamentary hearing, someone proposed the insertion of the phrase 'requires respect for the ecosystem by all citizens' in Art. 9 in order to achieve the *Drittwirkung*: see G. Azzariti, 'Appunto per l'audizione presso la Commissione Affari costituzionali del Senato della Repubblica del 16 gennaio 2020' (2020) *Osservatorio costituzionale*, n. 1, 70–72, available at <www.osservatorio aic.it> accessed 19 December 2022.

17 E. Di Salvatore, 'Brevi osservazioni sulla revisione degli articoli 9 e 41 della Costituzione' (2022) *Costituzionalismo*, n. 1, 20, available at <www.costituzionalismo.it> accessed 19 December 2022.

18 C. De Fiores, 'Le insidie di una revisione pleonastica. Brevi note su ambiente e Costituzione' (2021) *Costituzionalismo* n. 3, 13, available at <www.Costituzionalismo.it> accessed 23 December 2022.

19 M. CECCHETTI, 'La revisione degli articoli 9 e 41 della Costituzione e il valore costituzionale dell'ambiente: tra rischi scongiurati, qualche virtuosità (anche) innovativa e molte lacune' (2021) *Forum di Quaderni Costituzionali*, n. 3, 288, available at <www.forumcostituzionale.it> accessed 19 December 2022.

20 Constitutional. Court, ruling n. 85/2013.

21 § 12.6 of ruling n. 85/2013.

22 Art. 3, par. 3 Treaty on European Union says the EU "shall work for the sustainable development of Europe based on balanced economic growth and price stability, a highly competitive social market economy, aiming at full employment and social progress, and a high level of protection and improvement of the quality of the environment".

23 Art. (191) of the Treaty on the Functioning of the European Union.

24 Arts. (192) and (193) of the Treaty on the Functioning of the European Union.

25 Charter of Fundamental Rights of the European Union, Art. (37): "A high level of environmental protection and the improvement of the quality of the environment must be integrated into the policies of the Union and ensured in accordance with the principle of sustainable development". See on this A. Lucarelli,

'Commento Art. 37', in R. Bifulco, M. Cartabia, A. Celotto (eds.), *L'Europa dei diritti* (Il Mulino, Bologna 2001) 258.

26 M. Cecchetti, 'La revisione degli articoli 9 e 41 della Costituzione e il valore costituzionale dell'ambiente: tra rischi scongiurati, qualche virtuosità (anche) innovativa e molte lacune' (2021) *Forum di Quaderni Costituzionali*, n. 3, 305, available at <www.forumcostituzionale.it> accessed 19 December 2022.

27 D'Atena, 'Dove vanno le Regioni?' (2022) *Rivista Aic*, n. 4, 198, commenting on the issue, complains about the confusion and approximation of the reform. See also F. Rescigno, 'Memoria per l'Audizione dinanzi alla Commissione affari costituzionali del Senato' (2020) *Osservatorio costituzionale*, n. 1, 49, available at <www.osservatorioaic.it> accessed 23 December 2022 concerning the discussion of draft proposals of the constitutional amendment of Art. 9.

28 Several scholars have held hearings in Parliament during the proceedings. Among others, see on this G. Azzariti, 'Appunto per l'audizione presso la Commissione Affari costituzionali del Senato della Repubblica del 16 gennaio 2020' (2020) *Osservatorio costituzionale*, n. 1, 70–72 available at <www.osservatorio aic.it> accessed 19 December 2022.

29 T.E. Frosini, 'La Costituzione in senso ambientale. Una critica' (2021) *Federalismi*, 23.6.2021, 16, 1ss, available at <www.federalismi.it> accessed 19 December 2022.

30 "The Italian Constitution contains some supreme principles that cannot be subverted or modified in their essential content even by constitutional revision laws or other constitutional laws. Such are the principles that the Constitution itself explicitly provides for as absolute limits to the power of constitutional revision, such as the republican form (Art. 139 Const.), as well as the principles that, although not expressly mentioned among those not subject to the constitutional revision procedure, belong to the essence of the supreme values on which the Italian Constitution is founded". See for a different reading of the ruling R. Bifulco, Primissime riflessioni intorno alla l. Cost. 1/2022 in materia di tutela dell'ambiente, *Federalismi, Paper*, 6.4.2022, 3 available at <www.federalismi.it> accessed 19 December 2022.

31 On this, see R. Calvano, 'Il ruolo dei principi fondamentali della Costituzione nella giurisprudenza costituzionale' in S. Staiano (ed.), *Giurisprudenza costituzionale e principi fondamentali, Atti del convegno annuale del Gruppo di Pisa* (Giappichelli, Torino 2006) 361.

32 C. De Fiores, 'Le insidie di una revisione pleonastica. Brevi note su ambiente e Costituzione' (2021) *Costituzionalismo*, n. 3, 141–142, available at <www.Costituzionalismo.it> accessed 23 December 2022.

33 M. Cecchetti, 'La revisione degli articoli 9 e 41 della Costituzione e il valore costituzionale dell'ambiente: tra rischi scongiurati, qualche virtuosità (anche) innovativa e molte lacune' (2021) *Forum di Quaderni Costituzionali*, n. 3, 305; on constitutional amendments as a balance, see G. Silvestri, 'Spunti di riflessione sulla tipologia e sui limiti della revisione costituzionale', *Studi in onore di Paolo Biscaretti di Ruffia* vol. II (Milano, 1987) 1189.

34 R. Bifulco, 'Primissime riflessioni intorno alla l. Cost. 1/2022 in materia di tutela dell'ambiente' (2022) *Federalismi*, paper, April 6 2022, n. 11, 4, available at <www.federalismi.it> accessed 19 December 2022.

35 Paris Agreement, OJ (2016), 19.10.2016, L 282/4 available at <https://eur-lex.europa.eu/legal-content/EN/TXT/PDF/?uri=CELEX:22016A1019(01)&from=IT> accessed 23 December 2022.

36 Commission, The European Green Deal, Communication from the European Commission COM (2019) 640 final.

37 See among others the already cited Paris Agreement, the document "Education 2030: Towards inclusive and equitable quality education and lifelong learning for all", by the *World Education Forum* (2015) the UN 2030 Agenda.

38 M. LUCIANI, 'Generazioni future, distribuzione temporale ella spesa pubblica e vincoli costituzionali' (2008) *Diritto e società*, n. 2, 145.

39 References in R. Bifulco, 'Primissime riflessioni intorno alla l. Cost. 1/2022 in materia di tutela dell'ambiente' (2022) *Federalismi*, paper, April 6, n. 11, 8, available at <www.federalismi.it> accessed 19 December 2022.

40 Compare I. Rivera, 'Le tonalità dell'ambiente e le generazioni future nel cammino di riforma della Costituzione' (2022) *BioLaw Journal – Rivista di BioDiritto*, n. 2, 225–229, available at <https://teseo.unitn.it/biolaw it> accessed 19 December 2022.

41 M. Luciani, 'Generazioni future, distribuzione temporale ella spesa pubblica e vincoli costituzionali' (2008) *Diritto e società*, n. 2, 145. On this see par. 38 in the Urgenda case, decided by the Dutch Supreme Court, ruling of 09.10.2018. A recent important ruling on this was taken by the BVerfG, 24.3.2021, BvR 2656/18, 99, 148.

42 A. Pace, 'Problematica delle libertà costituzionali. Parte generale', *CEDAM* (Padova 2003) 4.

43 Constitutional Court, ruling n. 88/2014. See on this C. De Fiores, 'Le insidie di una revisione pleonastica. Brevi note su ambiente e Costituzione' (2021) *Costituzionalismo* n. 3, 156, available at <www.Costituzionalismo.it> accessed 23 December 2022.

Bibliography

Azzariti G, 'Appunto per l'audizione presso la Commissione Affari costituzionali del Senato della Repubblica del 16 gennaio 2020' (2020) *Osservatorio costituzionale*, n. 1, 70–72, available at <www.osservatorio aic.it> accessed 19 December 2022

Bartolucci L, 'Le generazioni future (con la tutela dell'ambiente) entrano "espressamente" in Costituzione' (2022), *Forum di Quaderni Costituzionali*, n. 22, 29, available at <www.forumcostituzionale.it> accessed 1 December 2022

Benvenuti M, 'La revisione dell'articolo 41, commi 2 e 3, della Costituzione, i suoi destinatari e i suoi interpreti' (2023) *Rivista AIC*, n. 2, 59, available at <www.rivistaaic.it> accessed 4 July 2023

Bifulco R, 'Primissime riflessioni intorno alla l. Cost. 1/2022 in materia di tutela dell'ambiente' (2022) *Federalismi*, paper, April 6, 2022, n. 11, 4, available at <www.federalismi.it> accessed 19 December 2022

De Fiores C, 'Le insidie di una revisione pleonastica. Brevi note su ambiente e Costituzione' (2021) *Costituzionalismo*, n. 3, 141–142, accessed at <www.Costituzionalismo.it> accessed 23 December 2022

Calvano R, 'Il ruolo dei principi fondamentali della Costituzione nella giurisprudenza costituzionale' in S. Staiano (ed.), *Giurisprudenza costituzionale e principi fondamentali, Atti del convegno annuale del Gruppo di Pisa* (Giappichelli, Torino 2006) 361

Caravita B, *Diritto pubblico dell'ambiente* (Il Mulino, Bologna 1990)

Cecchetti M, 'La revisione degli articoli 9 e 41 della Costituzione e il valore costituzionale dell'ambiente: tra rischi scongiurati, qualche virtuosità (anche) innovativa e molte lacune' (2021) *Forum di Quaderni Costituzionali*, n. 3, 288, available at <www.forumcostituzionale.it> accessed 19 December 2022

Ciervo A, 'Un altro mondo (non) è possibile. Diritto costituzionale, ambiente e Climate Justice' (2021) *Parole chiave*, n. 6, 75–77

D'Atena, 'Dove vanno le Regioni?' (2022) *Rivista Aic*, n. 4, 198

Di Salvatore E, 'Brevi osservazioni sulla revisione degli articoli 9 e 41 della Costituzione' (2022) *Costituzionalismo* n. 1, 20, available at <www.costituzionalismo.it> accessed 19 December 2022

di Toritto B. Caravita, Morrone A, 'Ambiente e Costituzione' in B. Caravita di Toritto, L. Cassetti, A. Morrone (eds), *Diritto dell'Ambiente* (Il Mulino, Bologna 2016) 30

Frosini TE, 'La Costituzione in senso ambientale. Una critica' (2021) *Federalismi*, 23.6.2021, 16, 1ss, available at <www.federalismi.it> accessed 19 December 2022

Giannini MS, '"Ambiente": saggio sui suoi diversi aspetti giuridici' (1973) *Rivista trimestrale di diritto pubblico*, n. 1

Groppi T, 'Sostenibilità e Costituzioni. Lo Stato costituzionale alla prova del futuro' (2016) *DPCE*, no. 1, 43

Jakab A, 'Sustainability in European Constitutional Law' (2016) *MPIL Research Paper* 2016–16, available at <https://papers.ssrn.com/sol3/papers.cfm?abstract_id=2803304> accessed 23 December 2022

Lucarelli A, 'Commento Art. 37' in R. Bifulco, M. Cartabia, A. Celotto (eds.), *L'Europa dei diritti* (Il Mulino, Bologna 2001) 258

Luciani M, voce 'Economia (nel diritto costituzionale)', *Digesto (discipline pubblicistiche)* (V, UTET, Torino 1990) 374–375

Luciani M, 'Generazioni future, distribuzione temporale ella spesa pubblica e vincoli costituzionali' (2008) *Diritto e società*, n. 2, 145

Pace A, 'Problematica delle libertà costituzionali. Parte generale', in *CEDAM* (Padova 2003) 4

Predieri A, 'voce Paesaggio', in *Enciclopedia del diritto* (xxxi, Giuffrè, Milano 1981) 507–509

Rescigno F, 'Memoria per l'Audizione dinanzi alla Commissione affari costituzionali del Senato' (2020) *Osservatorio costituzionale*, n. 1, 49, available at <www.osservatorioaic.it> accessed 23 December 2022

Rivera I, 'Le tonalità dell'ambiente e le generazioni future nel cammino di riforma della Costituzione' (2022) *BioLaw Journal – Rivista di BioDiritto*, n. 2, 225–229, available at <https://teseo.unitn.it/biolaw it> accessed 19 December 2022

Silvestri G, 'Spunti di riflessione sulla tipologia e sui limiti della revisione costituzionale', *Studi in onore di Paolo Biscaretti di Ruffia* vol. II (Giuffrè, Milano 1987) 1189

15

THE CASE FOR A STABLE CLIMATE

The Urgent Need for a Subjective (Personality?) Right in the Portuguese Legal System

Maria Regina Redinha[1] *and Maria Raquel Guimarães*[2]

15.1 The Recognition of a 'Right to...'

Social and individual concerns with the environment and its multiple dimensions have existed for many years. But although its legal construction has some solid basis in the Portuguese legal system, particularly in constitutional law, it is difficult to identify an instrument which, in terms of macro- or micro-framework, can be traced back to the idea of the 'right to…'.[3]

The aim of this chapter is to discuss the possible recognition of a *subjective right* in the Portuguese legal system that would make it possible to work towards the protection of a stable climate.

In effect, a subjective right is the only legal formulation which makes it possible to state the prevalence of a certain interest, which may be immediately mobilised for effective protection by the legal system. To have a right is to be admitted to a 'fast-track' for the recognition of the holder's primacy in relation to other individuals and thus to have access to the preventive and repressive protection of the Law.

However, a right is not a reality unto itself, except perhaps for some extreme iusnaturalists and Natural Law theories. Rights are for what they are for like everything man does in his relationship with the outside world, and they serve precisely to protect legal interests that the Law technically transmutes into objects.[4]

Thus, if we expect the State to assume its constitutional mission, using Law as an environmental protection and reconversion instrument,[5] we must necessarily revisit the very idea of the *object of law* and discover in it the *quid* that allows us to structure a Law that can fulfil the functionality that society requires.

15.2 Stable Climate, Protection of the Environment, Quality of Life, and Fundamental Rights

There is a scientific consensus today that climate change is a real and serious threat that endangers the very particular interaction between chemical, biological, and physical processes that ensure that the Earth System works as a safe operating space for humanity (the Holocene, in geological terms).[6] Scientists are now capable of identifying (and quantifying) the planetary boundaries—among which climate change and the integrity of the biosphere have been recognised as 'core planetary boundaries'[7]—compatible with human health and welfare and, ultimately, human life on Earth.[8]

DOI: 10.4324/9781032662046-18

The essential role that the 'intangible well-functioning state of the Earth System corresponding to the geological epoch of the Holocene' plays in the preservation of humankind should consequently lead to a new object of law, legislators thus assuming their role of protecting the individual and society in general.[9]

What then makes any good a 'legal good'?[10] The emergence of a new legal good is the result of the universality and democracy of ethical-legal awareness. The interpellation by the collective, the will that it should be so in order to fulfil a purpose that goes beyond the individual, the tribe, the group, and the nation, demands that it should be enforced through the seal of legality.[11]

In the Portuguese legal system, we can even claim that this recognition of the environment as a 'legal good' can unequivocally be traced back to Article 66 of the original version of the 1976 Constitution, although still in the form of the 'legal start-up' that is the programmatic norms.[12]

In the current version of the Portuguese Constitution, Article 66 reads as follows[13]:

'Article 66
(Environment and quality of life)

1 Everyone has the right to a healthy and ecologically balanced human living environment and the duty to defend it.
2 In order to ensure the right to the environment within an overall framework of sustainable development, the State, acting via appropriate bodies and with the involvement and participation of citizens, is charged with:

  a Preventing and controlling pollution and its effects and the harmful forms of erosion;
  b Conducting and promoting town and country planning with a view to a correct location of activities, balanced social and economic development and the enhancement of the landscape;
  c Creating and developing natural and recreational reserves and parks and classifying and protecting landscapes and places, in such a way as to guarantee the conservation of nature and the preservation of cultural values and assets that are of historic or artistic interest;
  d Promoting the rational use of natural resources, while safeguarding their ability to renew themselves and ecological stability, with respect for the principle of intergenerational solidarity;
  e In cooperation with local authorities, promoting the environmental quality of rural settlements and urban life, particularly on the architectural level and as regards the protection of historic zones;
  f Promoting the integration of environmental objectives into the various policies with a sectoral scope;
  g Promoting environmental education and respect for environmental values and assets;
  h Ensuring that the fiscal policy renders development compatible with the protection of the environment and the quality of life.

In addition to that, the defence of nature and the environment is designated as one of the fundamental tasks of the State, as well as the preservation of natural resources.[14]

The recognition of a fundamental right 'to a healthy and ecologically balanced human living environment', as one of the constitutional 'social rights', does not contain an operational concept of 'environment' or of 'quality of life' for the enforcement of the standard.[15] Portuguese literature advocates a *unitarian functional vision* of the environment, including physical, chemical, and biological systems and interacting economic, social, and cultural factors that have an effect, directly or indirectly, on living beings and on the quality of human life.[16] The environment and quality of life appear as two articulated concepts, the environment being an autonomous value, insofar as it is a value for the maintenance of the quality of human life.[17]

The fundamental right to the environment involves two main aspects, one negative and the other positive, i.e., it includes the right to demand that the State and third parties refrain from conducts that endanger the environment—with a scope identical to that of 'rights, freedoms and guarantees' with direct application in private relations, as results from Constitutional Articles 17 and 18—as well as the right to State measures to ensure environmental protection, requiring positive action—political, legislative, administrative, and criminal measures—on the part of the State, whose practical operability depends on being anchored in legislation and specifically implemented.[18]

In conclusion, in the Portuguese Constitution, the environment is a legal good with a *collective* nature, protected as a fundamental *social* right, allowing action against the State and other entities, namely access to administrative and criminal judicial proceedings, in order to preserve the environment and avoid attacks from third parties.

However, the difficulty caused by the absence of legal norms that expressly recognise the rights claimed or that only define them in a programmatic way, without operational content that can be mobilised by the individual, prevents people from claiming a personal right to a stable climate and the reparation of damage caused by conducts that threaten it.[19]

The search for more 'limited' and 'precise' legal standards in the Portuguese legal system gave rise, in July 2021, to a Draft Law designed to amend the Consumer Protection Act (Law 24/96, of 31 July) in order to establish the consumer's right to environmental protection and sustainable consumption.[20] The proposal, authored by an independent Member of Parliament, and later dropped due to lack of a parliamentary majority, aimed to insert a new article into the Portuguese Consumer Act establishing that

> the producer must privilege the integration of environmental aspects in the conception of goods, taking into account their entire life cycle and aiming at a better environmental performance, namely regarding durability, reparability, reusability, recyclability and non-toxicity of the goods and their components.[21]

And, above all, Article 8-A stated that 'the producer, the supplier of goods and the provider of services are jointly liable, regardless of fault, for damage caused to consumers in breach of the provisions of this article', aiming to empower the consumer with legal standing to sue producers, sellers, and service providers.

Although it was dismissed, this proposal had the effect of opening the debate on the subject in the Portuguese Parliament, pointing towards a new pathway in the recognition of a personal right to environmental protection, with the corresponding legal guarantee of its exercise, through the mechanism of civil liability. It is noteworthy that this attempt appears in the specific context of consumer law, in defence of sustainable consumption, a path which is already beginning to be followed by the EU legislator in different legislative initiatives.[22]

This debate has also been held at the international level, leading to the proclamation of the human right to a safe, clean, and healthy environment in Resolution 48/13 of 8 October 2021 of the United Nations Human Rights Council.[23]

Despite this, a stable climate still lacks the status of a 'Common Heritage of Humanity'. The United Nations General Assembly Resolution of 6 December 1988 opted to consider climate change as a 'Common Concern of Humanity',[24] a concept which lacks precision and raises implementation problems.[25] The recognition of a Common Heritage without geographic or political borders

> should become the structural basis for the development of a regenerative economy of nature, that is, one that allows the transition from an exclusive logic of no-harm rule to a logic of production of benefits in the Common Heritage, of cleaning up and ensuring the maintenance of what belongs to all.[26]

With a pioneering approach, Portugal adopted this position and assumed the diplomatic commitment in the Climate Framework Law—Law 98/2021, of 31 December—to actively defend, in both its foreign policy and within the framework of climate diplomacy, the recognition by the United Nations of a stable climate as a Common Heritage of Humanity, in order to pave the way for the management of this Global Common Good at the international level.[27]

15.3 The Importance of the Decision Milieudefensie et al. v. Royal Dutch Shell in the Construction of a Personal Right to a Stable Climate

The debate on the possibility of individuals whose interests in maintaining a healthy environment are at stake taking a direct action against those who jeopardise it has already reached the courts in some countries. In Europe, the importance of the *Milieudefensie et al. v. Royal Dutch Shell*-judgement has been emphasised by several authors.[28]

The District Court of The Hague, on 26 May 2021, ordered RDS—Royal Dutch Shell—based in the Netherlands, as the parent company of the Shell group, to reduce the group's CO_2 emissions by at least net 45% by the end of 2030, relative to 2019 levels. The Court held that the interpretation of an 'unwritten standard of care' laid down in Book 6 Section 162 Dutch Civil Code required that, when determining the Shell group's corporate policy, 'the company should observe the due care exercised in society' and, therefore, limit the annual volumes of CO_2 emissions discharged into the atmosphere in order to protect the interests of the inhabitants of the Netherlands.[29]

This decision was widely regarded as a landmark ruling in environmental law, regardless of the final balance of interests that may be reached, taking into account that Shell filed its statement of appeal with the Dutch Court of Appeal in The Hague on 22 March 2022.

It should be noted that, for procedural reasons, the District Court did not admit the intervention in the action ('collectieve actie') of the more than 17,000 individual claimants represented by the environmental association Milieudefensie, on the grounds that the interests they intended to protect with the action were already protected by Milieudefensie and the other NGOs involved in the case.[30] It also considered that the action in question could not be based on the protection of the interests of the entire world population in relation to the dangers of climate change caused by CO_2 emissions, given that those interests would not be affected on the same level as that of the interests of the population of the Netherlands, and for that reason it also considered that the association ActionAid did not have *locus standi* to bring the action, since the focus of its action, contributing to the fight against poverty and injustice all over the world, is centred on African countries.[31]

The claim of the applicant Milieudefensie et al. for the recognition, in relation to Shell, of an 'unwritten standard of care' to contribute to the prevention of the dangers of climate change on the basis of the aforementioned Book 6, Section 162, of the Netherlands Civil Code was nevertheless satisfied, as stated *supra*. The District Court of The Hague called for the interpretation of the general clause of the 'unwritten standard of care' ('*ongeschreven zorgvuldigheidsnorm*') imposed on the defendant according to several criteria, including the policy-setting position of RDS in the Shell group, the Shell group's CO_2 emissions, the 'serious and irreversible consequences' of CO_2 emissions for the Netherlands and the Wadden region and 'the right to life and the right to respect for private and family life' of residents in the Netherlands.[32]

Account was also taken of the 'soft law' arising, in particular, from the UN Guiding Principles on Business and Human Rights (UNGP) of 2011, the need for the contribution of the various stakeholders, and not only governments, to achieve the objectives set by the Paris Agreement, and the widespread consensus that collaboration between different actors is fundamental to achieving the global objective of zero net emissions, with Shell being called upon to do its part, by ensuring that the group's emissions are reduced.[33] In the Court's view, the uncontested fact that Shell is not the only company responsible for 'tackling dangerous climate change' does not 'absolve' it from its share of responsibility in the fight against climate change, with the means at its disposal.[34] Based on the scientific reports on the issue, the Court considered that the Shell group would have to ensure the reduction of CO_2 emissions by 45%, net by the end of 2030 with reference to 2019 levels, this being an obligation of result.[35]

Climate change was seen as one of the most pressing threats to the 'rights to life and the right to respect for private and family life' of present and future generations. In the Netherlands' legal order, the Court did not recognise outright a personal right to a stable environment but admitted the negative influence of climate change on the rights of residents of the Netherlands and detected conduct contrary to the 'unwritten standard of care', that gave rise to climate change. And it recognised that the interests of those residents in the action concerned were protected by the actions of the applicant associations, justifying the associations' *locus standi*.

The turning point in *Milieudefensie et al. v. Royal Dutch Shell* was that a legal good was identified and the holders of the interests that require protection were also identified. The integration of those interests in a tool to react to their threat became closer.

The point is whether the assessment made by the Dutch court can be replicated in other legal systems, namely by Portuguese courts, where the issue has not yet been assessed.

15.4 Stable Climate as a Legal Good in the Portuguese Legal System

As awareness of the impact of climate change increases, as a consequence of the growing number of environmental disasters, climate-based litigation is also on the rise.[36] Disputes such as *Milieudefensie et al. v. Royal Dutch Shell* will certainly become more common in courts everywhere.

The protection of individuals in these disputes demands the recognition of effective legal instruments. And once the need for an operational right without recourse to indeterminate and diffuse references is acknowledged, it is necessary to delimit its object, a task which is methodologically accomplished by importing the pre-legal data provided by the best state of the science.

An important step taken in the Portuguese Climate Framework Law—Law 98/2021, of 31 December—was to integrate the notion of a stable climate—the manifestation of a stable and definable functioning pattern of the Climatic System, within the limits of natural variability that were observed after the last glaciation (Holocene period), and that resulted in a rich functional biodiversity—as was recommended by the Portuguese National Council for the Environment and Sustainable Development (CNADS).[37] A notion that passes the filter of the strictest legal technique, because, despite its intangibility, it is based on a measurable physicality that provides an objective determination and a drive towards its realisation. Furthermore, the Law has centuries of experience in dealing with intangible goods—*Et quidem naturali jure communia sunt omnium haec: aer, aqua profluens, et mare et per hoc litora maris*—[38]and is increasingly perfecting flexible instruments of adaptation, such as indeterminate concepts, general clauses, and codes of good practice to times of 'liquid modernity',[39] which undoubtedly make it possible to welcome a notion which, although complex, has the added advantage over others of scientific parameterisation.

Moreover, legal evolutionism supports this construction of subjective law, which is not unprecedented from a methodological point of view. Like organisms, legal positions, contractual relations, and rights also suffer the course of adaptation to the social, economic, and cultural, surrounding reality. At each step, new contractual models emerge forged by the creativity of private autonomy, but only some of them last with the typicality that legitimises them as instruments of generalised use and individual mobilisation.

In reality, despite the dissonance of some legal literature, there are unequivocal signs that in environmental and climate matters, times have matured for the emergence of a specific subjective right that makes up the legal sphere of individuals as such, as it seems to arise, on the legislative level, from the attribution of the 'right of action to defend subjective rights and legally protected interests' and from the recognition of the status of the subject of climate action to 'citizens, private companies and other entities governed by private law'—Articles 6(2)(a), and 8, of the Portuguese Climate Framework Law. This affirmation, moreover, follows the identical right of action contained in Article 7(2) of Law 19/2014, of 14 April, Law for Environmental Policy.

These recent legal developments are juxtaposed, therefore, with the prevision of civil liability arising from the exercise of dangerous activities *per se* or by the means of execution they require, provided for in Article 493(2) of the Civil Code.[40] Interference with the environmental or climatic balance resulting from human action or enterprise can effectively fulfil the fluid notion of danger, from which the agent's liability stems unless he/she demonstrates that his/her best effort has been used in the prevention of its harmful result.

Washing oil tanks, discharging contaminated liquids into waterways or emissions of atmospheric pollutants are 'dangerous' activities, comprised in the useful scope of application of this legal standard, a standard designed to provide compensation for the damage caused by private individuals and thus fulfilling a function of 'social control' of these activities.[41]

The construction of a subjective right thus already has secure support in the law, in terms of both its recognition and its effectiveness. A further step along this path of providing the individual with tools to directly activate their interest in a stable climate would be to recognise this subjective right as a personality right.

Indeed, a very close parallel can be found, within the scope of Labour Law, in the establishment of a true and proper right to safety at work, which today is universally recognised, both in the national legal system and internationally.[42]

Once upon a time, workers' safety was left to Dickensian chance.

In the early days of industrialisation, accidents at work followed one after another without repair, since the imperfection or malfunction of the machine was not attributable to the entrepreneurs, since there was no culpable illicit conduct on their part.

Until the social and economic weight of this damage, not softened by any welfare system, generated the objective responsibility of the employer for the use of potentially dangerous means and later the compulsory transfer of risk through the constitution of obligatory insurance. An evolution that ended up concentrating on the present obligation of security and in a correlative personality right of the employee.[43]

The axiological centrality of this right is clear today in its integration into the core standards of labour regulations, applicable to the intermediate category between employment and autonomous work, thus extending the personal scope of labour law—Article 10, Portuguese Labour Code.

In the case of climate and environmental issues, the normative incipience of protecting environmental and climate assets has also been overcome and due to the urgency of the reality, we are currently in the construction phase of positive action measures which, in the absence of regenerative effects, prevent the deterioration of the *status quo*.

Soft law resources, such as declarations and recommendations, have not been known to fully achieve this goal, so similar to what happened with occupational safety, here too the recognition of a personality right and the related obligation of non-disruption may prove to be an effective tool for pursuing the programme objectives contained in the Portuguese law on the environment and climate.

15.5 Stable Climate as a Personality Interest

A stable climate demands its recognition by the legal system as a personality interest and the efficiency of its protection through the force of the State.

If there is a legal interest, there is an interest (*inter est*), if there is a social need, there is a legal need, if there is a personal demand and a personal risk, there is the involvement of the personality, and therefore, sooner rather than later, we will have a right, a subjective right characterisable as personality right.[44]

This right is, as previously mentioned, the only way to access normative instruments that can immediately be mobilised by the individual *vis a vis* public and private entities to defend their rights as a 'person-being' and 'person-to-be'.[45] Only subjective rights can and will embody a legal prevalence and if this prevalence is required by any personal aspects, then the right has to be a personality right.

Personality rights, which are designed to protect personality interests in private relations, are instruments of civil law and offer an additional layer of protection to some legal goods already protected by fundamental rights, allowing, in these cases, multilevel protection. The complex set of tools available under Portuguese civil law as regards the protection of the personality comprises a 'general personality right', a general clause establishing the right to the free and full development of personality. This standard allows a permanent update of civil law in the evolving and dynamic context of personality rights and operates as a foundational right, where all the known and unknown, predictable and unpredictable interests of human personality may be involved, based on Article 70 of the Civil Code.[46] And it covers the right to a stable climate as a personality right, as climate change is an obstacle to the harmonious development of the personality and to quality of life.[47]

Environmental and climatic aggressions can, in fact, disturb both the physical and psychological conditions of the person at a given moment (harm their physical integrity, for example), but they can also compromise their future projection, inhibiting biological and non-biological capacities and competences (permanent damage to health, among others) or negatively conditioning behaviours and activities (inhibition of learning or pursuing certain professional careers, as well as the access to certain goods and places).[48]

The inviolability of the physical and mental person demands not only the protection of interior goods, such as privacy or honour but also requires the preservation of their living space, namely, the external, social, and environmental conditions essential for their survival and free development.[49] In addition, the human being is also part of the environment and a result of this environment. This being so, the environment does not exist as an external reality in opposition to the individual but integrates these individuals and evolves within them.[50]

The intangible well-functioning state of the Earth System, in short, a stable climate—comprising the peculiar interaction between chemical, biological, and physical processes that ensure that the

Earth System functions as a safe operating space for humanity, as mentioned *supra*—is thus a legal good protected as a personality interest, through a personality right radiating from the general right of personality.

This means that the threatened or offended person may request 'measures appropriate to the circumstances of the case' in order to prevent the consummation of the threat or to mitigate the effects of the offence already committed, as stated in Article 70(2) of the Portuguese Civil Code, regardless of the civil liability that may be incurred.[51] In other words, an injury for the purpose of Article 70(2) is an invasion of a legally protected interest, which is concrete, particularised, and actual or imminent. An injury that can be caused by climate change… *quod erat demonstrandum*.

15.6 Conclusions

The Portuguese Constitution recognises a fundamental right 'to a healthy and ecologically balanced human living environment'—Article 66—as a 'social right', with a *collective* nature, allowing action against the State and other entities and access to administrative and criminal judicial proceedings. This is, however, a programmatic norm, without an operational content that can be individually mobilised.

However, the recognition of a personal right to a *stable climate*—as the manifestation of a stable and definable functioning pattern of the Climatic System, within the limits of the Holocene period—and the consequent right to the reparation of damage caused by conducts that threaten this equilibrium, can be founded on the new Portuguese Climate Framework Law, Law 98/2021, of 31 December. Here, the legislator has included a 'right of action to defend subjective rights and legally protected interests' and conceded the status of the subject of climate action to 'citizens, private companies and other entities governed by private law'—Articles 6(2)(a), and 8 of the Portuguese Climate Framework Law. This protection is complemented by the provision of civil liability arising from the exercise of dangerous activities, provided for in Article 493(2) of the Civil Code.

This intangible stable climate can also be analysed as a personality interest, ensuring that the Earth System functions as a safe operating space for humanity, protected through a personality right radiating from the general right of personality, based on Article 70 of the Civil Code. This standard covers the right to a stable climate as a personality right, as climate change is an obstacle to the harmonious development of the personality and to quality of life. This subjective right to a stable climate is thus a personality right.

Furthermore, a parallel path in acknowledging a stable climate as a subject of personal rights has been followed at an international level, with Resolution 48/13 of 8 October 2021 of the United Nations Human Rights Council recognising the human right to a safe, clean, and healthy environment.

Notes

1 Professor of Labour Law and Civil Law at the Faculty of Law, University of Porto, Portugal; researcher at CIJ—FDUP, Centre for Interdisciplinary Research on Justice (University of Porto, Portugal).

2 Professor of Civil Law and Contract Law at the Faculty of Law, University of Porto, Portugal; researcher at CIJ—FDUP, Centre for Interdisciplinary Research on Justice (University of Porto, Portugal). This chapter presents research carried out within the framework of CIJ—Centre for Interdisciplinary Research on Justice, U.Porto, UIDB/00443/2020 (Project 'Common Home of Humanity') funded by FCT—Fundação para a Ciência e a Tecnologia.

3 As we will point out *infra*, in the text, both at a constitutional level and at an infra-constitutional level of protection of the environment, Portuguese legal literature has, until now, had difficulty in

recognising the existence of a legal instrument with an operational content that can be mobilised by the individual.

4 Quoting Roscoe Pound, 'Interests of Personality' [1915] 28(4) *Harvard Law Review* 343, 344, 'A legal system attains its end by recognizing certain interests (…), by defining the limits within which these interests shall be recognized legally and given effect through the force of the state'. But, as the author also notes, the legal system does not create those interests: it only recognises them.

5 In 1978, Jorge Figueiredo Dias, 'Sobre o papel do direito penal na protecção do ambiente' [1978] IV(1) *Revista de Direito e Economia* 4, 18–19, already pointed out the constitutional duty that the Portuguese State had to protect the environment, although limiting his analysis to the role of Criminal Law as an instrument to achieve this goal.

6 See Will Steffen, 'The Planetary Boundaries Framework: Defining a Safe Operating Space for Humanity' in Paulo Magalhães et al (eds), *The Safe Operating Space Treaty: A New Approach to Managing Our Use of the Earth System* (Cambrigde, Cambridge Scholars Publishing 2016) 24–28.

7 Will Steffen, 'The Planetary Boundaries Framework: Defining a Safe Operating Space for Humanity' in Paulo Magalhães et al (eds), *The Safe Operating Space Treaty: A New Approach to Managing Our Use of the Earth System* (Cambrigde, Cambridge Scholars Publishing 2016) 31.

8 Will Steffen, 'The Planetary Boundaries Framework: Defining a Safe Operating Space for Humanity' in Paulo Magalhães et al (eds), *The Safe Operating Space Treaty: A New Approach to Managing Our Use of the Earth System* (Cambrigde, Cambridge Scholars Publishing 2016) 32–40.

9 Paulo Magalhães, 'Climate as a Concern or a Heritage?' [2020] 21 *RED* 99, 125 (https://www.doi.org/10.24840/2182-9845_2020-0001_0006) <https://cij.up.pt//client/files/0000000001/6-artigo-paulo-magalhaes_1592.pdf> accessed 06 February 2023.

10 We use the expression 'legal good', according to its widespread use in Continental legal literature, *verbi gratia*, Fabio Roberto D'Avila and Giovanni Agostini Saavedra, 'Legal Good and Recognition: A Study of Axel Honneth's Social Theory' [2011] 2(78) *Droit et Société* 325 (https://doi.org/10.3917/drs.078.0325)—meaning, in extreme synthesis, a good warranting legal protection and axiologic support—as a direct translation, namely, of the '*Rechtsgut*', '*bene giuridico*', '*bien juridique*', '*bien jurídico*', or '*bem jurídico*' of the German, Italian, French, Spanish and Portuguese Law. About this notion, see, for all, Salvatore Pugliatti, 'Beni (teoria gen.)' in *Enciclopedia del Diritto* V (Milano, Giuffrè 1959) 164, 170 ff.

11 See Jorge Figueiredo Dias, 'Sobre o papel do direito penal na protecção do ambiente' [1978] IV(1) *Revista de Direito e Economia* 4, 8–9, referring specifically to a healthy environment as a legal good. See, also, in Criminal Law, Roland Hefendehl, 'Das Rechtsgut als materialer Angelpunkt einer Strafnorm' in Roland Hefendehl, Andrew von Hirsch and Wolfgang Wohlers (eds), *Die Rechtsgutstheorie, Legitimationsbasis des Strafrechts oder dogmatisches Glasperlenspiel?* (Baden-Baden, Nomos 2003) 119, and, before, Claus Roxin, *Strafrecht, Allgemeiner Teil* 1 (3rd edn, München, Beck 1997) §2, 34.

12 Luís Manuel Teles de Menezes Leitão, 'A responsabilidade civil por danos causados ao ambiente' [2011] 2 *Direito e Justiça* (especial), Estudos dedicados ao Professor Doutor Luís Alberto Carvalho Fernandes 381, 382 (https://doi.org/10.34632/direitoejustica.2011.11435) <https://journals.ucp.pt/index.php/direitoejustica/article/view/11435> accessed 13 July 2023; Luis Filipe Colaço Antunes, *O procedimento administrativo de avaliação de impacto ambiental* (Coimbra, Almedina 1998) 39, 45; Jorge Figueiredo Dias, 'Sobre o papel do direito penal na protecção do ambiente' [1978] IV(1) *Revista de Direito e Economia* 4, 8–9.

13 The English version of the Portuguese Constitution can be accessed in <https://www.parlamento.pt/sites/EN/Parliament/Documents/Constitution7th.pdf> accessed 06 February 2023.

14 See art. (9) (e), of the Portuguese Constitution.

15 José Joaquim Gomes Canotilho and Vital Moreira, *Constituição da República Portuguesa Anotada, Artigos 1. ° a 107. °* (II, 4th edn, Coimbra, Coimbra Editora 2007) 371–378, 814, 844.

16 José Joaquim Gomes Canotilho and Vital Moreira, *Constituição da República Portuguesa Anotada, Artigos 1. ° a 107. °* (II, 4th edn, Coimbra, Coimbra Editora 2007) 844–845. This concept was adopted by Law 19/2014, of 14 April, Law for Environmental Policy, art 5(2). See also João Menezes Leitão, 'Instrumentos de Direito Privado para Protecção do Ambiente' [1997] 7 *Revista Jurídica do Urbanismo e do Ambiente* 29, 39.

17 Joaquim Gomes Canotilho and Vital Moreira, *Constituição da República Portuguesa Anotada, Artigos 1. ° a 107. °* (II, 4th edn, Coimbra, Coimbra Editora 2007) 845, refer to an 'anthropocentric teleology' in this context.

18 Joaquim Gomes Canotilho and Vital Moreira, *Constituição da República Portuguesa Anotada, Artigos 1. ° a 107. °* (II, 4th edn, Coimbra, Coimbra Editora 2007) 371, 845–848; João Menezes Leitão, 'Instrumentos

de Direito Privado para Protecção do Ambiente' [1997] 7 *Revista Jurídica do Urbanismo e do Ambiente* 29, 55–56; Luís Carlos L Batista, *Reflexões em torno do Direito Subjectivo do Ambiente* (CJP/CIDP s.d.) 72–73 <https://www.icjp.pt/sites/default/files/media/813-1348.pdf> accessed 12 February 2023.

19 It is important to note that the problem is not unique to continental law. In the United States, the question was addressed in the U.S. Court of Appeals for the Ninth Circuit (2020) *Juliana vs. United States* (947, F.3d 1159)]: the court eluded the constitutional matter, despite having recognised the grave threat of climate change to human health and well-being. The Ninth Circuit held that ordering the government to adopt 'a comprehensive scheme to decrease fossil fuel emissions and combat climate change' would go beyond the federal court's authority, in terms of remedies, contending with the principles of separation of powers. According to the reasoning of the court, it was implicit that courts can only order injunctive relief when supported by 'limited and precise' legal standards. Differently, sustaining that a fundamental right to a clean and sustainable environment exists as a 'penumbra of the Bill of Rights' and should be recognised by the federal courts, see Robert Kemper, 'Recognizing a Fundamental Right to a Clean Environment: Why the Juliana Court Got It Wrong and How to Address the Issue Moving Forward' [2022] 16(2) *FIU Law Review* 457, 464 (https://doi.org/10.25148/lawrev.16.2.11) <https://ecollections.law.fiu.edu/lawreview/vol16/iss2/10/> accessed 06 February 2023. For further analysis of this case, see *Juliana v. United States—Ninth Circuit Holds that Developing and Supervising Plan to Mitigate Anthropogenic Climate Change Would Exceed Remedial Powers of Article III Court* [2021] 134(5) *Harvard Law Review* 1929, 1933 <https://harvardlawreview.org/wp-content/uploads/2021/03/134-Harv.-L.-Rev.-1929.pdf> accessed 06 February 2023.

20 Portuguese Draft Law 915/XIV/2.a: Amends the Consumer Protection Act to establish the right to environmental protection and sustainable consumption, 20.07.2021, <https://www.parlamento.pt/ActividadeParlamentar/Paginas/DetalheIniciativa.aspx?BID=121053> accessed 07 February 2022.

21 Art 8-A, Portuguese Draft Law 915/XIV/2.a. The proposal also established that 'goods and services intended for consumption must, whenever possible, have the least impact on the environment, preserving biodiversity and natural resources, except when public interest or technical reasons that justify it are at stake'; 'the packaging in which the goods are packed must be suitable and proportional to the respective content, with preference being given to the use of recycled, reusable and recyclable materials'; and 'provided that adequate health, hygiene and safety conditions are ensured, the consumer must not be prevented by the supplier of goods or service provider from adopting sustainable consumption habits, namely with regard to reusing packaging and using other containers'.

22 See, among others, Directive (EU) 2019/771 of the European Parliament and of the Council of 20 May 2019 on certain aspects concerning contracts for the sale of goods, amending Regulation (EU) 2017/2394 and Directive 2009/22/EC, and repealing Directive 1999/44/EC (2019) OJ L136/28, Recital no. 32: 'Ensuring longer durability of goods is important for achieving more sustainable consumption patterns and a circular economy'.

23 Resolution A/HRC/RES/48/13 United Nations General Assembly, 8 October 2021.

24 Resolution A/43/905 United Nations General Assembly, 30 November 1988.

25 See, extensively, Paulo Magalhães, 'O Clima Estável como Património Comum da Humanidade' (2022) 29 *RED* 104, 110 (https://www.doi.org/10.24840/2182-9845_2022-0003_0006) <https://cij.up.pt//client/files/0000000001/6-paulo-magalhaes_1959.pdf> accessed 07 February 2023.

26 Paulo Magalhães, 'O Clima Estável como Património Comum da Humanidade' (2022) 29 *RED* 104, 110, 126 (https://www.doi.org/10.24840/2182-9845_2022-0003_0006) <https://cij.up.pt//client/files/0000000001/6-paulo-magalhaes_1959.pdf> accessed 07 February 2023.

27 Paulo Magalhães, 'O Clima Estável como Património Comum da Humanidade' (2022) 29 *RED* 104, 110, 125 (https://www.doi.org/10.24840/2182-9845_2022-0003_0006) <https://cij.up.pt//client/files/0000000001/6-paulo-magalhaes_1959.pdf> accessed 07 February 2023 125. See art 15(1) (f), Portuguese Climate Framework Law, Law 98/2021, of 31 December.

28 Hague District Court (26 May 2021) *Vereniging Milieudefensie et al. v. Royal Dutch Shell PLC*. An English version of the decision is available at <https://uitspraken.rechtspraak.nl/#!/details?id=ECLI:NL:RBDHA:2021:5339> accessed 08 February 2022. See, among others, Chiara Macchi and Josephine van Zeben, 'Business and human rights implications of climate change litigation: Milieudefensie et al. v Royal Dutch Shell' [2021] 30(3) *RECIEL* 409 (https://doi.org/10.1111/reel.12416) <https://onlinelibrary.wiley.com/doi/epdf/10.1111/reel.12416> accessed 11 July 2023; critically, Benoit Mayer, 'The Duty of Care of Fossil-Fuel Producers for Climate Change Mitigation' [2022] 11(2) *Transnational Environmental Law* 407 (https://www.doi.org/10.1017/S2047102522000103).

29 Milieudefensie et al. v. Shell, 4.4.1–4.4.2.
30 Milieudefensie et al. v. Shell, 4.2.7.
31 Milieudefensie et al. v. Shell, 4.2.5.
32 Mlieudefensie et al. v. Shell, 4.4.2. ff.
33 Mlieudefensie et al. v. Shell, 4.4.2. ff.
34 Mlieudefensie et al. v. Shell, 4.4.37.
35 Milieudefensie et al. v. Shell, 4.4.23.
36 Robert Kemper, 'Recognizing a Fundamental Right to a Clean Environment: Why the Juliana Court Got It Wrong and How to Address the Issue Moving Forward' [2022] 16(2) *FIU Law Review* 457, 458 (https://doi.org/10.25148/lawrev.16.2.11) <https://ecollections.law.fiu.edu/lawreview/vol16/iss2/10/> accessed 06 February 2023. See also, focussing on international litigation, Benoit Mayer and Harro van Asselt, 'The rise of international climate litigation' [2023] 32(2) *RECIEL* 10 (https://doi.org/10.1111/reel.12515) <https://onlinelibrary.wiley.com/doi/full/10.1111/reel.12515> accessed 11 July 2023.
37 Art. 15(1) (f), Portuguese Climate Framework Law, Law 98/2021, of 31 December. See, also, Conselho Nacional do Ambiente e do Desenvolvimento Sustentável (CNADS), Recomendação sobre uma Lei do Clima (12th February 2021) <https://www.cnads.pt/images/documentos/2021_recomendacao–leiclima.pdf> accessed 12 February 2023: 'In a subject such as Climate, in which the good itself is difficult to define and to frame in the existing legal framework, it is fundamental to resort to the most recent scientific knowledge in order to build upon it and construct definitions that may support the legislative options. (…) The restoration of a Stable Climate implies an integrated approach to the Earth System'. For further details, Paulo Magalhães, 'O Clima Estável como Património Comum da Humanidade' [2022] 29 *RED* 104, 110, 125–126 (https://www.doi.org/10.24840/2182-9845_2022-0003_0006) <https://cij.up.pt//client/files/0000000001/6-paulo-magalhaes_1959.pdf>. The Recommendation of CNADS was followed by an unprecedented petition signed by 240 academics (<https://www.commonhomeofhumanity.org/_files/ugd/deeae3_ac50fbc916a742f98c2a7abcef67b357.pdf> accessed 24 July 2023) and then by an electronic public petition to the Portuguese Parliament.
38 *Institutas*, II, I, §I.
39 After Zygmunt Bauman, *Liquid Modernity* (Cambridge, Polity 2000).
40 A close approach can be found in João Menezes Leitão, 'Instrumentos de Direito Privado para Protecção do Ambiente' [1997] 7 *Revista Jurídica do Urbanismo e do Ambiente* 29, 51–52. For an analysis of the interaction of this article with the provisions of Law 11/87, of 7 de Abril—Environment Framework Law (repealed by Law 19/2014, of 14 April)—see 391; see also 386.
41 About this 'social control' carried out by the Civil rules on liability, protecting the environmental goods, see Guido Alpa and Renato Speciale, 'Bene culturali e ambientali' in Rodolfo Sacco (ed), *Digesto delle Discipline Privatistiche, Sezione Civile* II (4th edn, Torino, Utet 1998) 93, 106.
42 In brief, summarising this conception, cf. art 283(1) Portuguese Labour Code, Law 7/2009, of 12 February: 'The employee has the right to work in safe and healthy conditions' and the Resolution of the International Labour Organisation (ILO) on the inclusion of a safe and healthy working environment in the ILO's framework of fundamental principles and rights at work, 10 June 2022, <https://www.ilo.org/wcmsp5/groups/public/---ed_norm/---relconf/documents/meetingdocument/wcms_848632.pdf> accessed 06 February 2023.
43 For an extensive analysis of the current obligation of health and safety, Milena Silva Rouxinol, *A Obrigação de Segurança e Saúde do Empregador* (Coimbra, Coimbra Editora 2008) 91 ff.
44 Following the same path, although with some methodological differences, cf. Luís Carlos L Batista, *Reflexões em torno do Direito Subjectivo do Ambiente* (CJP/CIDP s.d.) 47–49, 83–89 <https://www.icjp.pt/sites/default/files/media/813-1348.pdf>.
45 Building on the distinction of Orlando de Carvalho, *Teoria Geral do Direito Civil* (4th edn, Coimbra, Gestlegal 2021) 205.
46 The 'general personality right' is described as a foundational right, source of all the different single-personality rights, a radical *jus in se ipsum*, by Orlando de Carvalho, *Teoria Geral do Direito Civil* (4th edn, Coimbra, Gestlegal 2021) 204–209. See also Paulo Mota Pinto, *Direitos de Personalidade e Direitos Fundamentais*, *Estudos* (Coimbra, Gestlegal 2018) 33ff, 332ff, 487 ff., and Maria Raquel Guimarães and Maria Regina Redinha, 'A Portuguese Approach to Privacy in COVID-19 Times: Through the Keyhole' in Ewoud Hondius et al. (eds), *Coronavirus and the Law in Europe* (Cambridge, Intersentia 2021) 1011 ff. However, the acknowledgement of a general right of personality is not unanimous in Portugal: cf., for a system of *numerus apertus* of single-personality rights, José de Oliveira Ascensão, *Direito Civil, Teoria Geral, Introdução, As pessoas, Os Bens* I (2nd edn, Coimbra, Coimbra Editora 2000) 86–89.

47 João Menezes Leitão, 'Instrumentos de Direito Privado para Protecção do Ambiente' [1997] 7 *Revista Jurídica do Urbanismo e do Ambiente* 29, 45, 57–59. See also Fernando Alves Correia, *O Plano Urbanístico e o Princípio da Igualdade* (Coimbra, Almedina 1989) 77; José Manuel Araújo de Barros, 'Por um Direito ao Ambiente' in CEJ (ed), *Textos: Ambiente e Consumo* 2 (Lisboa, Centro de Estudos Judiciários 1996) 228; António Menezes Cordeiro, *Tratado de Direito Civil, Parte Geral* IV (5th edn, Coimbra, Almedina 2019) 104–105. For a critical position, see, extensively, Luís Carlos L Batista, *Reflexões em torno do Direito Subjectivo do Ambiente* (CJP/CIDP s.d.) 47 ff., 55 ff <https://www.icjp.pt/sites/default/files/media/813-1348.pdf>.

48 A similar observation had already been made in the Portuguese legal literature by Luís Manuel Teles de Menezes Leitão, 'A responsabilidade civil por danos causados ao ambiente' [2011] 2 *Direito e Justiça* (especial), Estudos dedicados ao Professor Doutor Luís Alberto Carvalho Fernandes 381, 382–383. (https://doi.org/10.34632/direitoejustica.2011.11435) <https://journals.ucp.pt/index.php/direitoejustica/article/view/11435>.

49 With a similar position, see Rabindranath Capelo de Sousa, *O Direito Geral de Personalidade* (Coimbra, Coimbra Editora 1995) 184.

50 As mentioned in Luís Manuel Teles de Menezes Leitão, 'A responsabilidade civil por danos causados ao ambiente' [2011] 2 *Direito e Justiça* (especial), Estudos dedicados ao Professor Doutor Luís Alberto Carvalho Fernandes 381, 382 (https://doi.org/10.34632/direitoejustica.2011.11435) <https://journals.ucp.pt/index.php/direitoejustica/article/view/11435>, damage to the environment does not constitute damage to a legal good external to man, but rather damage to the human personality itself.

51 Also, João Menezes Leitão, 'Instrumentos de Direito Privado para Protecção do Ambiente' [1997] 7 *Revista Jurídica do Urbanismo e do Ambiente* 29, 60–61.

Bibliography

Alpa G and Speciale R, 'Bene culturali e ambientali' in Rodolfo Sacco (ed), *Digesto delle Discipline Privatistiche, Sezione Civile* II (4th edn, Torino, Utet 1998)

Antunes L F C, *O procedimento administrativo de avaliação de impacto ambiental* (Coimbra, Almedina 1998)

Ascensão J O, *Direito Civil, Teoria Geral, Introdução, As pessoas, Os Bens* I (2nd edn, Coimbra, Coimbra Editora 2000)

Barros J M A de, 'Por um Direito ao Ambiente' in CEJ (ed), *Textos: Ambiente e Consumo* 2 (Lisboa, Centro de Estudos Judiciários 1996)

Batista L C L, *Reflexões em torno do Direito Subjectivo do Ambiente* (CJP/CIDP s.d.) 72–73 <https://www.icjp.pt/sites/default/files/media/813-1348.pdf>

Bauman Z, *Liquid Modernity* (Cambridge, Polity 2000)

Canotilho J J G and Moreira V, *Constituição da República Portuguesa Anotada, Artigos 1.° a 107.°* (II, 4th edn, Coimbra, Coimbra Editora 2007)

Carvalho O, *Teoria Geral do Direito Civil* (4th edn, Coimbra, Gestlegal 2021)

Correia F A, *O Plano Urbanístico e o Princípio da Igualdade* (Coimbra, Almedina 1989)

Cordeiro A M, *Tratado de Direito Civil, Parte Geral* IV (5th edn, Coimbra, Almedina 2019)

D'Avila F R and Agostini Saavedra G, 'Legal Good and Recognition: A Study of Axel Honneth's Social Theory' [2011] 2(78) *Droit et Société* 325 (https://doi.org/10.3917/drs.078.0325)

Dias J F, 'Sobre o papel do direito penal na protecção do ambiente' [1978] IV(1) *Revista de Direito e Economia* 4

Guimarães M R and Redinha M R, 'A Portuguese Approach to Privacy in COVID-19 Times: Through the Keyhole' in Ewoud Hondius et al (eds), *Coronavirus and the Law in Europe* (Cambridge, Intersentia 2021) 1011 ff

Hefendehl R, 'Das Rechtsgut als materialer Angelpunkt einer Strafnorm' in Roland Hefendehl, Andrew von Hirsch and Wolfgang Wohlers (eds), *Die Rechtsgutstheorie, Legitimationsbasis des Strafrechts oder dogmatisches Glasperlenspiel?* (Baden-Baden, Nomos 2003)

Juliana v. United States—Ninth Circuit Holds That Developing and Supervising Plan to Mitigate Anthropogenic Climate Change Would Exceed Remedial Powers of Article III Court [2021] 134(5) *Harvard Law Review* 1929, 1933 <https://harvardlawreview.org/wp-content/uploads/2021/03/134-Harv.-L.-Rev.-1929.pdf>

Kemper R, 'Recognizing a Fundamental Right to a Clean Environment: Why the Juliana Court Got It Wrong and How to Address the Issue Moving Forward' [2022] 16(2) *FIU Law Review* 457 (https://doi.org/10.25148/lawrev.16.2.11) <https://ecollections.law.fiu.edu/lawreview/vol16/iss2/10/>

Leitão J M, 'Instrumentos de Direito Privado para Protecção do Ambiente' [1997] 7 *Revista Jurídica do Urbanismo e do Ambiente* 29

Leitão L M T M, 'A responsabilidade civil por danos causados ao ambiente' [2011] 2 *Direito e Justiça* (especial), Estudos dedicados ao Professor Doutor Luís Alberto Carvalho Fernandes 381, 382 (https://doi.org/10.34632/direitoejustica.2011.11435) <https://journals.ucp.pt/index.php/direitoejustica/article/view/11435>

Macchi C and van Zeben J, 'Business and Human Rights Implications of Climate Change Litigation: Milieudefensie et al. v Royal Dutch Shell' [2021] 30(3) *RECIEL* 409 (https://doi.org/10.1111/reel.12416) <https://onlinelibrary.wiley.com/doi/epdf/10.1111/reel.12416>

Magalhães P, 'Climate as a Concern or a Heritage?' [2020] 21 *RED* 99, 125 (https://www.doi.org/10.24840/2182-9845_2020-0001_0006) <https://cij.up.pt//client/files/0000000001/6-artigo-paulo-magalhaes_1592.pdf>

Magalhães P, 'O Clima Estável como Património Comum da Humanidade' [2022] 29 *RED* 104, 110 (https://www.doi.org/10.24840/2182-9845_2022-0003_0006) <https://cij.up.pt//client/files/0000000001/6-paulo-magalhaes_1959.pdf>

Mayer B, 'The Duty of Care of Fossil-Fuel Producers for Climate Change Mitigation' [2022] 11(2) *Transnational Environmental Law* 407 (https://www.doi.org/10.1017/S2047102522000103)

Mayer B and van Asselt H, 'The Rise of International Climate Litigation' [2023] 32(2) *RECIEL* 10 (https://doi.org/10.1111/reel.12515) <https://onlinelibrary.wiley.com/doi/full/10.1111/reel.12515>

Pinto P M, *Direitos de Personalidade e Direitos Fundamentais, Estudos* (Coimbra, Gestlegal 2018)

Pound R, 'Interests of Personality' [1915] 28(4) *Harvard Law Review* 343

Pugliatti S, 'Beni (teoria gen.)' in *Enciclopedia del Diritto* V (Milano, Giuffrè 1959)

Rouxinol M S, *A Obrigação de Segurança e Saúde do Empregador* (Coimbra, Coimbra Editora 2008)

Roxin C, *Strafrecht, Allgemeiner Teil* 1 (3rd edn, München, Beck 1997)

Steffen W, 'The Planetary Boundaries Framework: Defining a Safe Operating Space for Humanity' in Paulo Magalhães et al (eds), *The Safe Operating Space Treaty: A New Approach to Managing Our Use of the Earth System* (Cambridge, Cambridge Scholars Publishing 2016)

Sousa R C de, *O Direito Geral de Personalidade* (Coimbra, Coimbra Editora 1995)

PART 4

Conciliating Property Rights and Environmental Protection

16

PROPERTY LAW AND (MORE THAN ONE NOTION OF) SUSTAINABILITY

A New Field

Björn Hoops[1]

16.1 Introduction

Property law, more specifically private-law ownership, and sustainability have a complicated and contradictory history. In 1968, Garrett Hardin argued that nature and resources held in common were doomed to deteriorate in a 'Tragedy of the Commons' and proposed exclusive private ownership or strict state regulation as means to ensure their sustainable use.[2] Elinor Ostrom later refuted Hardin's argument, demonstrating that the Commons could be sustainably governed and that the Tragedy of the Commons would only occur under narrow conditions.[3] While Ostrom has shown that exclusive private ownership or strict state regulation is not needed for the sustainable governance of nature and resources, we have realised that property law with exclusive ownership at its core provides the legal foundation for extractive capitalism.[4] This economic system fuels global warming and the overexploitation of nature and resources, endangering biodiversity, the livelihoods of billions of people, and the stability of human societies.

Property law's contribution to threatening our very existence makes it imperative that we change our use of property law and/or adjust property law itself, to promote a sustainable economy and society. In fact, property law and its use have already started to change and contribute to more sustainable practices. For instance, New Zealand employs ownership to enhance environmental protection. New Zealand's legislature recently vested ownership of the former national park Te Urewera in a legal person embodying this natural entity, thereby introducing a new category of owners to property law.[5] Property-law scholars seek to integrate this change into the legal system and ascertain its benefits for environmental protection.[6] Beyond such legal-dogmatic research,[7] normative[8] property-law scholarship seeks change to create a more sustainable property law.[9] For example, Ugo Mattei has been advocating for a far-reaching reform of ownership to enhance the influence of local communities over local natural resources and promote the conservation of nature.[10]

This scholarship exemplifies that a new research field has been born: 'Property Law and Sustainability' (PropLS). In this field, 'sustainability' serves as a criterion for the legal-dogmatic assessment of existing law and the normative development of adjustments to the law. This contribution identifies the definitions of 'sustainability' followed by existing PropLS-scholarship,

 DOI: 10.4324/9781032662046-20

thereby examining the extent to which 'sustainability' can actually serve as a common criterion and denominator in the field. 'Sustainability' is a vague term. On the one hand, such a broad term may unite diverse scholars under a single umbrella. On the other, differences in the definition of 'sustainability' may create isolated sub-groups of scholars and undermine the exchange within the field. While, for instance, 'sustainable development', as defined by the United Nations' Brundtland Commission in 1987,[11] incorporates a social dimension that requires us to meet the basic needs of all people on Earth,[12] PropLS-scholarship tends to focus on the sustainability of our society and economy through the lens of the preservation of the natural foundations of human life (or 'ecological sustainability').[13] As this contribution argues, even within the scholarship on property law and ecological sustainability, scholars subscribe to diverging definitions of 'sustainability'. However, this contribution pleads for an inclusive field of 'Property Law and (Ecological) Sustainability' because the legal issues raised are so similar that scholars with diverging definitions of 'sustainability' will still greatly benefit from exchange and collaboration.

This contribution is structured as follows. Section 16.2 defines the two elements of PropLS: property law and sustainability. As 'sustainability' proves to be too complex for legal researchers to apply as a criterion, Section 16.3 examines more concrete criteria applied in PropLS-scholarship. As these criteria follow different definitions of 'sustainability', Section 16.4 concludes this contribution with a plea for an inclusive field of 'Property Law and Sustainability'. This contribution does not purport to deal with all PropLS-scholarship. As property law is still predominantly national law and the international debate is limited, this contribution discusses examples from Belgian and Dutch law in addition to examples from the international literature, in particular theoretical literature from common-law jurisdictions and comparative literature from Europe.

16.2 Property Law and the Many Meanings of Sustainability

Property Law and Sustainability consists of two elements: 'property law' and 'sustainability'. Property law is taken in a wide sense and is not confined to the private property law on ownership, limited property rights and condominiums. It includes the links of private property law to other branches of private law, for example, insolvency and tenancy law, as well as public property law on, for instance, constitutional property protection and land use regulation.

As for sustainability in the sense of ecological sustainability, there is not a uniform definition of sustainability, but several ones.[14] Without purporting to give a comprehensive account of the sustainability discourse, this section sketches two definitions of sustainability based on economic research (Section 16.2.1) as well as three definitions of a 'sustainable future' from sociological research (Section 16.2.2), providing the basis for examining the criteria used in PropLS-scholarship in Section 16.3.

16.2.1 Sustainability in Economics

The Brundtland Commission defined 'Sustainable development' as 'development that meets the needs of the present without compromising the ability of future generations to meet their own needs'.[15] Resources thus have to be preserved for future generations.[16] Environmental economists posit that sustainability can be equated with a constant amount of capital over time and available to succeeding generations.[17] Capital refers to natural capital, such as soils, water, and wood, and other forms of capital such as products made by human society.[18] In this context, a rough distinction is made between weak and strong sustainability, which can help categorise PropLS-scholarship.

Weak sustainability refers to an overall constant amount of capital. The composition is irrelevant, which means that natural capital can be substituted with human-made capital and *vice versa*.[19]

This type of sustainability is mainly aimed at securing the wealth of human societies through the availability of capital. However, it does not necessarily include an obligation for humans to preserve nature because human-made capital can replace natural capital.[20] Under the notion of 'weak sustainability', this substitution of natural capital, which is a technical expression for the destruction of nature, will only end where human-made capital depends upon the preservation of natural capital, for example, where the preservation of biodiversity is essential to maintaining or expanding agricultural production. Following weak sustainability involves the danger that natural capital will be depleted, eroding the natural basis of human life.[21] The discourse on green or sustainable growth, which promotes, among others, a de-coupling of economic growth from the use of fossil fuels and other natural resources, could be regarded as pursuing weak sustainability because this green or sustainable economy still requires natural resources and growth leads to the use of yet higher amounts of natural resources.[22]

The alternative concept is strong sustainability. Under this concept, natural and other forms of capital cannot be substituted and must both be maintained at current levels or increased. Human society thus takes on a responsibility to preserve nature and, if possible, to increase the amount of natural capital available.[23] The 2015 Paris Agreement, for instance, pursues strong sustainability within a certain range and certain parameters. Article 2(1) of the Agreement limits the global temperature rise to 2°C compared to pre-industrial levels. The natural capital represented by a 'healthy global climate' can thus only be substituted by human-made capital up to a certain threshold. From that point onwards, economic activities and growth would have to be entirely de-coupled from emitting harmful greenhouse gases. However, as is further explained in Subsections 16.3.2 and 16.3.3, de-coupling through energy efficiency and the use of renewable energy generally requires the destruction of other natural capital. This means that the Paris Agreement can only pursue strong sustainability to the extent that new technologies or the reduction of consumption avoid the destruction of natural capital for climate protection.

16.2.2 Sustainable Futures in Sociology

While strong and weak sustainability set goals for the sustainable preservation of resources, notions of a 'sustainable future' paint a picture of a sustainable society with a focus on the steps needed to achieve a sustainable future. The sociologists Adloff and Neckel have created a useful taxonomy for analysing these notions,[24] enabling us to categorise and contextualise further different types of PropLS-scholarship. These authors distinguish between *Modernisation*, *Transformation*, and *Control*.

'Modernisation' achieves a sustainable future through an improvement of the ecological balance of the current social order without doing away with either liberal democracy or the market economy.[25] Essential reforms include the greatest possible de-coupling of economic growth from fossil fuels and other natural resources ('green growth'). The economic notion of weak sustainability appears closely linked to this definition of a sustainable future.

By contrast, proponents of 'Transformation' argue that Modernisation is insufficient to meet the ecological challenges of our times.[26] There are a variety of approaches to Transformation, but what the proponents essentially advocate for is an economy and society no longer based on competition and growth. For instance, the degrowth movement proposes extremely restrained consumption patterns and resource conservation,[27] and the post-capitalist movement pleads for digital practices of sharing and exchanging outside market constraints.[28]

Finally, 'Control' is the darkest scenario of a sustainable future, which, as Adloff and Neckel admit, can be hardly considered sustainable. It envisions an environmental emergency, in which a technocratic government would replace liberal democracy and suspend constitutional and human rights.[29] This government would implement the measures needed to adapt to environmental disasters such as global warming. Examples of control would be plans for geo-engineering without democratic deliberations, such as the extraction of CO_2 from the atmosphere, or the deployment of the military to contain migration or social unrest.

16.3 Diverging Criteria as Operationalisation of Sustainability

The definitions of sustainability from Section 16.2 refer to aggregated amounts of resources or other goals at national or an even higher level. Even public-law scholars will have difficulty assessing law against these definitions, although they are used to examining laws and regulations solving macroeconomic or other societal problems.

Sustainability research on private property law and related branches of private law encounters an additional problem at a conceptual level. Private law is a discipline that focuses on the collaboration within groups, for example, through the rules on communities of property or in company law, or on transactions between private entities or of private entities with public entities on an equal footing, such as contracts of sale. Just as microeconomics,[30] private law generally omits to consider the aggregate consequences of all such transactions combined. Private lawyers focus on the reasonableness and fairness of single actions with or within a delineated group. Unlike public law, modern private law struggles to embrace the thought that actions that are fair and reasonable for the involved groups may, before and/or after the aggregation of these actions, result in unfair and unreasonable consequences for society as a whole.[31] For example, private lawyers tend to ignore the diffused harm done by single exercises of property rights in, and contracts of sale of, gas and oil.[32] The dilemma for private lawyers to address is that global warming and other crises undermining sustainability are exactly the result of what private lawyers perceive as perfectly reasonable and fair actions. All this suggests that weak or strong sustainability, both based on aggregates of capital accounts, or notions of a sustainable future would be all but impossible to handle for private-law researchers as a criterion of assessment.

This is not to say that public interests, such as environmental protection, cannot be pursued through private law. There is a wide range of examples in, for instance, the Netherlands alone, from the rights of municipalities under the Dutch civil code to interfere with the affairs of an association of apartment owners to safeguard construction standards[33] to tort liability of big private greenhouse gas emitters.[34] However, these private-law rules pursue specific goals that are easy to analyse and apply, such as compliance with construction standards or the reduction of the greenhouse gas emissions of one company. The vague notions of sustainability are by far less 'user-friendly'. Private-law research (and, to a lesser extent, also public-law research) thus needs to operationalise 'sustainability' by translating this abstract notion into clear criteria that mark certain individual actions as (un)fair or (un)reasonable. Note that this operationalisation does not rule out the conception of 'sustainability' as a general principle concretising rights and obligations under private law or public law. Rather, it can provide guidelines on how to interpret and apply such a principle.

This section outlines a number of criteria that operationalise sustainability in PropLS-scholarship, analyses how they align with the different notions of sustainability from Section 16.2 and gives examples of research using one of these criteria. This outline in no way purports to be comprehensive. The criteria are the limitation or ban of the extraction of a natural resource (Section 16.3.1), the promotion of energy from renewable sources (Section 16.3.2), and the promotion of energy efficiency and the circular economy (Section 16.3.3). This sub-section concludes with a few observations (Section 16.3.4).

16.3.1 Limiting or Banning the Extraction of a Natural Resource

To preserve a specific natural resource, the limitation or ban of extraction operationalises sustainability.[35] This criterion generally pursues a form of strong sustainability because it does not allow for the further degradation of a specific natural resource. However, take into account that in a globalised economy without a global authority, preserving a specific resource in one country is likely to lead to the overexploitation of a resource in another. This criterion resonates with a notion of Transformation that pleads for resource conservation. The preservation of a specific resource can be applied to a delineated situation involving a distinct group of people and is thus also accessible to private property lawyers because the criterion does not involve aggregations of actions and/or groups. Property lawyers can contribute to strong sustainability and Transformation by applying or adjusting property-law rules and doctrine so as to limit or ban extraction to the extent needed to preserve the natural resource. If all natural resources were governed in this way, property law would successfully contribute to strong sustainability and Transformation.

A tremendous contribution using this criterion is the work on the Commons by Ugo Mattei and his disciples.[36] The essence of their work is that private ownership, not adequately counterbalanced by regulation in the public interest, leads to the over-extraction of natural resources and unsustainable economies. This failure of state power leads the authors to plead for restricting private control over natural resources and the re-introduction of common ownership of natural resources and local participatory decision-making on the use of these resources. Assuming that local communities have an interest in preserving their local resources and are able to fulfil this task, this group of scholars view the Commons as a means to fulfil basic needs and foster generative and regenerative community activities.[37]

'Rights of Nature' are another approach divesting private owners of control over natural resources. Legislation or court judgments declare natural entities to be legal persons (environmental persons) with rights and liability. Countries such as Australia, Canada, Cambodia, India, New Zealand, and Spain have made forests, lakes, rivers, and national parks legal subjects.[38] Responsible and independent representatives decide on the use of the natural entity and it is, therefore, insulated from private control. Legal scholars have either advocated for such a solution[39] or have been analysing and advancing our understanding of environmental persons.[40] New Zealand has added a particular twist to the environmental persons Te Urewera and Te Awa Tupua in that the former national park Te Urewera owns itself and the river Te Awa Tupua owns its own riverbed.[41] This property right would be protected under, for instance, Article 1 of the First Protocol to the ECHR if the land were located in a signatory state.[42] It would, therefore, also insulate the natural entity to some extent from state action and democratic decision-making within the human state.

While Mattei and the Rights of Nature divest a possible private owner of their control and subject the use of a resource to a form of collective decision-making, another strain of research seeks to limit the owner's extraction and oblige them to maintain the natural resource. In the United States, 'Green Property' theorists recognised the interdependence between humans and nature and argued that to achieve a sustainable balance in the ecosystem, ownership had to be automatically limited by environmental protection.[43] Redefining property and ownership, one goal of green property theory is to ensure that human owners have a duty not to harm the environment, even in the absence of sufficient environmental regulation. Australian scholars have put forward similar arguments, for example, under the banner of 'Earth Jurisprudence'.[44] While the research on the Commons and Rights of Nature is aimed at preserving specific resources, green property theorists rely upon the reinterpretation of traditional legal concepts affecting all owners.

In Europe, with the same impetus to limit the owner's scope for manoeuvring, some scholars also rely upon open norms and concepts. They examine whether the very concept of ownership or the doctrine of the abuse of ownership could prevent owners from extracting too much or even compel them to maintain a natural entity actively.[45] Other comparative property-law research in Europe investigates the value of limited property rights to sustainability. They pose the question of whether positive obligations to maintain a forest or another natural entity could be imposed through servitudes (easements).[46] Depending on the content of such an obligation and reliable enforcement by the holder of the servitude, it may effectively protect a natural entity and thereby serve strong sustainability and Transformation.

16.3.2 Facilitating Energy from Renewable Sources

Another criterion for operationalising sustainability is the facilitation of the generation of electricity and heat from renewable sources. The replacement of fossil fuels with renewable energy sources forms part of an effort to de-couple our current level of production or even its growth from emitting greenhouse gas emissions.[47] Property-law research that uses this criterion seeks to identify and change property-law rules that deter the installation of renewable energy capacity in the form of solar panels, heat pumps, district heating systems, and other renewable energy installations.

One strain of research specifically scrutinises whether property law sufficiently facilitates the installation of, and connection to, district heating systems.[48] Among other insights, it identifies the problem that the operator of a district heating system depends on the cooperation of the landowners and needs to make agreements on how the system can be installed and maintained. Another strain of research seeks to identify and circumvent obstacles to renewable energy installations in a condominium.[49] It turns out that the majorities of apartment owners and the formalities required under the rules on apartment rights discourage them from installing renewable energy installations.[50]

A common culprit in PropLS-scholarship on renewable energy, particularly in Belgium and the Netherlands, is the doctrine of accession.[51] At the source of this debate was the insight that many households refrain from installing sources of renewable energy because of the high up-front investment. Leasing solar panels or another renewable energy installation would be an option to make renewable energy more accessible: the lessor would install, maintain, and replace the renewable energy installation for an affordable monthly fee and the lessee would save on their energy bill. The rules on accession pose an obstacle to this lease agreement. They entail that the owner of the land and house, the lessee, also becomes the owner of, for instance, the solar panels installed on the roof. The lessor thus loses their security in the form of ownership, and the solution to this problem, the creation of a right of superficies (*opstalrecht*), makes the lease more expensive and less accessible. For this reason, researchers explored and advocated for a more lenient interpretation of the rules on accession, preventing the landowner from becoming the owner of the renewable energy installation.

While a ban or restrictions to extractive ownership clearly pursue strong sustainability and Transformation, facilitating the generation of energy from renewable sources cannot be classified this easily. It is true that renewable energy replaces energy from fossil fuels, thereby protecting the climate.[52] However, solar panels, for instance, require rare metals and cause toxic waste.[53] Also, psychological and economic mechanisms tend to create an incentive to consume more and might eventually result in an increase in energy usage. Through, for instance, 'moral licencing', the awareness of using renewable energy may encourage consumers to engage in higher energy usage or other unsustainable behaviour.[54] When less costly than fossil-fuel alternatives, renewable energy may initiate an economic rebound effect, resulting in higher consumption.[55] More generally, green growth, a notion that incorporates the use of renewable energy for growing the economy, will require more resources than a zero-growth economy.[56] Depending on how these considerations play out on regeneration rates and on how different natural resources – specifically fossil fuels and rare metals – are weighed, renewable energy may still reduce the available natural capital because it consumes natural resources more quickly than they naturally regenerate. Instead of pursuing strong sustainability, this factor would only pursue weak sustainability in that case. Needless to say, in comparison to the use of fossil fuels, it does reduce the consumption of natural resources and makes human activity more 'weakly sustainable'.

Regarding notions of a sustainable future, the transition towards renewable energy appears to promote a reform of liberal democracy and a capitalistic market economy by de-coupling economic growth from fossil fuels. This criterion does not question our competitive and growth-based economic order and still heavily relies on the extraction of natural resources. Hence, it appears to be aimed at Modernisation rather than Transformation.

A final strain of PropLS-scholarship on renewable energy installations belongs to the realm of public law and may pursue yet another notion of a sustainable future to some extent. This strain investigates the extent to which the State can force owners to install renewable energy capacity in their buildings, with or without compensation.[57] The central object of investigation is the property of the owner as a constitutional and human right. On the one hand, this research acknowledges the current legal order and primarily seeks to identify ways to promote the use of renewable energy installation within the boundaries set by property rights. In this respect, this research is similar to the other strains already described. On the other hand, it contains the seed of legal coercion as a last resort to avert an environmental emergency. The more global warming progresses, the weaker property protection is likely to become and the closer this research may drag towards a notion of 'Control'.

16.3.3 Promoting Energy Efficiency and the Circular Economy

A criterion with similar strains of research and characteristics as facilitating renewable energy is the promotion of energy efficiency. For example, insulation is a means to improve the energy efficiency of buildings. Another criterion that has inspired the discussion about the rules on accession in Belgium and the Netherlands is the transition to a circular economy. In a circular economy, no materials of value are thrown away and are instead processed, upgraded (if needed), and re-used for their former or a purpose of higher value.[58] The promotion of the circular economy serves as a criterion for adjustments of the current legal rules in the work of many legal scholars.[59] The rules on accession are an example of property-law rules that potentially deter the transition to a circular economy.[60] In order for a circular economy to work, a producer needs to have access to the materials that need to be re-used. But, to ensure the producer's access to the goods, the consumer does not buy, but, instead, leases all types of products, including most of the components of their house such as the floors and non-structural walls.[61] Again, accession deprives the lessor of their ownership and thereby gives rise to an obstacle to the circular economy. PropLS-scholarship seeks to circumvent this obstacle through a more lenient interpretation of the doctrine of accession.

Energy efficiency serves sustainability because it reduces the amount of electricity and heat needed to meet housing needs. However, energy efficiency also requires insulation material made from natural resources, and more efficient housing may, through lower energy expenses and/or a good conscience, induce psychological or economic rebound effects resulting in other unsustainable behaviour.[62] Appealing as the circular economy might be, the idea of a completely circular economy should be taken with a pinch of salt. Circularity does reduce or even eliminate the extraction of a certain material, but it requires a lot of energy.[63] Depending on how these considerations play out on regeneration rates and on how different natural resources – specifically the re-used materials and the energy source – are weighed, the circular economy and energy efficiency efforts might still consume natural resources faster than they regenerate.

Even though the circular economy is an improvement compared to the current linear economy, the criteria of promoting energy efficiency and the transition to the circular economy thus do not necessarily pursue strong sustainability. However, at least they adhere to weak sustainability.

In terms of notions of a sustainable future, neither energy efficiency nor the circular economy abandons a competition- and growth-based economic order. While energy efficiency appears ideologically neutral, the circular economy makes use of the market economy by creating an economic incentive for producers, who remain the owners of their products, to manufacture more durable products than in a linear economy.[64] Although the circular economy would be essential to reducing consumption in the course of a Transformation, its dominant notion rather appears to be a Modernisation of liberal democracy and the capitalistic market economy.

16.3.4 Concluding Observations

On the bright side, the presented criteria enable property-law researchers to assess and promote sustainable property law with a focus on specific cases instead of aggregations. In this way, PropLS-research need not employ economic modelling or other techniques with which lawyers tend to be unfamiliar. PropLS-researchers should, of course, still try to take into account aggregate effects where that is possible, for instance, based on already existing insights from other disciplines. The drawback is that these criteria lead a life of their own and PropLS-research is likely not to question the extent to which they actually contribute to sustainability at a global scale. For instance, too many renewable energy installations may lead to an over-extraction of the required natural resources. To some extent, this abstraction from the 'unsustainable side' of a criterion seems necessary to make PropLS-research possible.

Moreover, the presented criteria do not all pursue the same concept of sustainability. The approaches pursuing strong sustainability and Transformation incorporate the need for restraining overall consumption and seek to adjust property law and doctrine accordingly.[65] By contrast, the approaches pursuing weak sustainability and Modernisation seek to substitute less extractive techniques for highly extractive techniques such as fossil fuels but do not necessarily foresee a reduction of consumption. Both approaches promote progress compared to the *status quo*, but only the approaches pursuing strong sustainability and Transformation are certain not to diminish the natural capital in question. Finally, one criterion relies on the coercive power of the State and may increasingly reflect a sustainable future marked by Control.

There may be various reasons for this divergence. Property-law research within a national context, such as the research on accession, is often inspired by problems encountered by companies and other actors in legal practice, a practice still dominated by growth-seeking enterprises, private ownership and other institutions that encourage extraction. Less aligned with practice is more theoretical research, such as the research on the Commons, which pursues strong sustainability and Transformation. Another reason may be the belief or non-belief that technological innovation and its diffusion can completely de-couple economic development from the unsustainable consumption of natural resources and thereby save the planet.[66]

16.4 A Plea for an Inclusive Field of Property Law and Sustainability

Property Law and Sustainability pursues a normative goal through property-law systematisation and reform, that of a sustainable economy and society. This goal is entrenched in national legal orders, for example, through constitutional clauses on a right to a healthy environment,[67] and international law.[68] However, for both the definition of sustainability and sustainability-promoting criteria and goals for its research, PropLS depends upon insights on sustainability from other disciplines such as economics and sociology. This appropriate humility on the part of legal researchers makes this field inherently interdisciplinary. However, interdisciplinarity plants the seed of divisions over what sustainability actually means, exacerbated by the fact that there is not one discipline called 'sustainability research' and that diverse disciplines contribute to our understanding of what is sustainable. Do diverging definitions of 'sustainability' have to lead to a split between different strains of PropLS-research? Can PropLS be exclusively claimed by strains of research pursuing strong sustainability and Transformation or those aiming at weak sustainability and Modernisation?

This contribution pleads that this is neither needed nor desirable. Even scholarship pursuing weak sustainability and/or Modernisation moves in the right direction and its impact on practice improves the current situation. Moreover, a sense of belonging to a certain field of research enables an exchange of thoughts and collaboration, which would otherwise be hampered by walls between different fields. Exchange and collaboration can be very fruitful because the strains of research employing different criteria often address the same or similar research questions. For instance, researchers interested in facilitating renewable energy and those working on the circular economy both encounter the doctrine of accession as a stumbling block. Also, an inclusive field allows research pursuing strong sustainability and Transformation to inspire researchers using other criteria and diffuse more easily into mainstream practice due to the stronger ties of scholars pursuing weak sustainability and Modernisation with practice.

To be sure, diverging notions of sustainability or sustainable futures are not the only stumbling block towards an inclusive field of PropLS. In civil-law jurisdictions, the division of private law and public law can hinder fruitful research due to a lack of knowledge of the respective other area and little exchange between them. An important example would be the relationship between the owners of neighbouring properties. While under private law, neighbour law and the doctrine of abuse of ownership play an important role, public law is the source of emission standards and zoning plans. Moreover, what is often unclear is the relationship between these branches of law. Evaluating and promoting the sustainability of property law in this field is thus all but impossible without an intra-disciplinary analysis of both private-law and public-law aspects. The wall between private law and public law must finally fall.

Another challenge to a unified field is the boundary between legal systems. Equipped with the functional method of comparative law,[69] PropLS-researchers already compare potential obstacles to a more sustainable economy and society in order to borrow insights and arguments that can help make property law more sustainable in one's own legal system.[70] This is fairly easy as long as jurisdictions share the same societal problems and feature similar legal concepts, such as the doctrine of accession in Belgium and the Netherlands. However, what makes comparisons in this field particularly challenging is that in each legal system, the sources on the sustainability of property law lie scattered in different branches of property law and that the tools to achieve that goal differ in nature. In some legal systems, the very notion of private ownership may be flexible enough to incorporate obligations to promote sustainability, as the Green Property theory from the United States suggests.[71] In other legal systems, PropLS-researchers may have to turn to the Constitution to impose obligations on owners, because of an inflexible concept of ownership.[72] PropLS-research thus requires dealing with sources of diverging status in each legal system. Also, the impact of these sources and the reach of the tools employed to promote sustainability vary considerably. The work on apartment rights, the doctrine of accession, or limited property rights focuses on technical property-law rules in legislation or common law. It has a significant but determinable impact within a limited area. By contrast, the reach of the concept of ownership, constitutional provisions, and the doctrine of abuse of ownership is by far greater but also more uncertain. At a conceptual level, comparative PropLS-research can identify and categorise these different layers of property law at which sustainability is promoted. *Akkermans*, for instance, distinguishes between fundamental principles of property law, ground rules, and technical rules.[73] On top of that, PropLS-research aims to make practical proposals to improve the sustainability property law in each legal system. To achieve this goal, PropLS-research must demonstrate the ability to translate and integrate the insights from foreign sources into the language and correct layers of one's own property-law system. An example of this ability is the integration of insights from the United States property theory into the civil-law doctrine of abuse of ownership in PropLS-research.[74]

While overcoming these stumbling blocks to a unified field, PropLS-scholars also need to work together to address urgent methodological issues in the field of Property Law and Sustainability. Hitherto under-researched, the following crucial issues and respective questions need to be raised and addressed in PropLS-research

1 Criteria as an operationalisation of sustainability: how from a definition of sustainability do we deduce reliable criteria for our research?
2 Law as an obstacle to a more sustainable economy and society: how do we determine whether a particular rule is detrimental to a sustainable economy and society?
3 Adjustments to the law to promote a sustainable economy and society: how do we make profound proposals to change doctrine or legislation that enhance sustainability, without an excessive impact on other public or private interests?

Answers to these questions fall outside the scope of this contribution. In Belgium and the Netherlands, some PropLS-scholars have made a modest start to answer these questions.[75] All PropLS-scholars should contribute to this methodological debate to help shape a high-quality and productive field of Property Law and Sustainability.

Notes

1 Björn Hoops holds the Chair of Private Law and Sustainability at the University of Groningen and is Marie Curie Fellow at the University of Turin from April 2022 to March 2024. This contribution seeks to give a partial theoretical underpinning for the project 'Private Law and the Energy Commons'. This project has received funding from the European Union's Horizon 2020 research and innovation programme under the Marie Skłodowska-Curie grant agreement No. 101024836. The author expresses his gratitude to the two anonymous peer reviewers for their valuable suggestions and the members of the ProSus research network for inspiring exchanges. All remaining errors are the author's alone.
2 G. Hardin, 'The Tragedy of the Commons' (1968) 162 *Science* 1243–1248.
3 E. Ostrom, *Governing the Commons: The Evolution of Institutions for Collective Action* (Cambridge University Press 2015).
4 F. Capra and U. Mattei, *The Ecology of Law: Toward a Legal System in Tune with Nature and Community* (Berrett-Koehler Publishers 2015); U. Mattei and A. Quarta, *The Turning Point in Private Law: Ecology, Technology and the Commons* (Edward Elgar Publishing 2018). Cf. P. Degens, 'Towards Sustainable

Property? Exploring the Entanglement of Ownership and Sustainability' (2021) 60 (2) *Social Science Information* 209.

5 Section 12 Te Urewera Act 2014.

6 See, e.g., B. Hoops, 'What If the Black Forest Owned Itself? A Constitutional Property Law Perspective on Rights of Nature' (2022) 11 *Transnational Environmental Law* 475 <https://doi.org/10.1017/S2047102522000322> accessed 21 September 2023; B. Hoops, 'Of Forests and Rivers Owning Themselves: Looking for a Place for Te Urewera and Te Awa Tupua in Constitutional Property Law and Theory' (2022) 7 *Journal of Law, Property, and Society* 126 <https://bit.ly/JLPS-Hoops> accessed 21 September 2023.

7 J.M. Smits, *The Mind and Method of the Legal Academic* (Edward Elgar Publishing 2012) 13–18; J.M. Smits, 'What Is Legal Doctrine? On the Aims and Methods of Legal-Dogmatic Research' in R. van Gestel, H. Micklitz and E.L. Rubin (eds), *Rethinking Legal Scholarship: A Transatlantic Dialogue* (Cambridge University Press 2017) 207–228; M. Vols, *Legal Research: One Hundred Questions and Answers* (Eleven International Publishing 2021).

8 J.M. Smits, *The Mind and Method of the Legal Academic* (Edward Elgar Publishing 2012) 35, 41–42; E.L. Rubin, 'The Practice and Discourse of Legal Scholarship' (1988) 86(8) *Michigan Law Review* 1835, 1847 and 1851.

9 For a plea to reform legal education as well, to achieve greater regulation of private entitlements to pursue sustainability, refer to N. Graham, 'Teaching Private Law in a Climate Crisis' (2021) 40(3) *University of Queensland Law Journal* 404 <https://journal.law.uq.edu.au/index.php/uqlj/article/view/6047/5029> accessed 21 September 2023.

10 F. Capra and U. Mattei, *The Ecology of Law: Toward a Legal System in Tune with Nature and Community* (Berrett-Koehler Publishers, Oakland 2015); U. Mattei and A. Quarta, *The Turning Point in Private Law: Ecology, Technology and the Commons* (Edward Elgar Publishing 2018).

11 World Commission on Environment and Development, *Report of the World Commission on Environment and Development: Our Common Future* (Oxford University Press 1987) 43.

12 F. Ekardt, *Sustainability: Transformation, Governance, Ethics, Law* (Springer 2020) 13 and 21–29.

13 A theory that could potentially unify ecological and social sustainability in property law is the theory of human flourishing. Under this theory, owners have obligations towards the community to contribute to the prerequisites for everyone's flourishing. These prerequisites not only include basic products and services, such as education, food, housing, security, and water, but also a healthy environment. See G.S. Alexander, *Property and Human Flourishing* (Oxford University Press 2018) 56 and 122.

14 See, for an overview, F. Adloff and S. Neckel, 'Futures of Sustainability as Modernization, Transformation, and Control: A Conceptual Framework' (2019) 14(4) *Sustainability Science* 1015–1025; G.R. Davies, 'Appraising Weak and Strong Sustainability: Searching for a Middle Ground' (2013) *Consilience: The Journal for Sustainable Development*, 111–124; C.J. Circo, 'Does Sustainability Require a New Theory of Property Rights?' (2009) 58(1) *Kansas Law Review* 91, 92 and 98.

15 World Commission on Environment and Development, *Report of the World Commission on Environment and Development: Our Common Future* (Oxford University Press 1987) 43.

16 G. Heal, 'Reflections—Defining and Measuring Sustainability' (2012) 6(1) *Review of Environmental Economics and Policy* 147, 154. <https://www.journals.uchicago.edu/doi/full/10.1093/reep/rer023> 21 September 2023.

17 D. Pearce, A. Markandya, E.B. Barbier and E. Barbier, *Blueprint for a Green Economy* (Earthscan 1989) 34.

18 D. Pearce, A. Markandya, E.B. Barbier and E. Barbier, *Blueprint for a Green Economy* (Earthscan 1989) 34.

19 D. Pearce, A. Markandya, E.B. Barbier and E. Barbier, *Blueprint for a Green Economy* (Earthscan 1989) 128.

20 G. Heal, 'Reflections—Defining and Measuring Sustainability' (2012) 6 (1) *Review of Environmental Economics and Policy* 147, 155–158. <https://www.journals.uchicago.edu/doi/full/10.1093/reep/rer023> 21 September 2023.

21 F. Ekardt, *Sustainability: Transformation, Governance, Ethics, Law* (Springer 2020) 22 ff.

22 K. Biely, D. Maes and S. Van Passel, 'The Idea of Weak Sustainability Is Illegitimate' (2018) 20 (1) *Environment, Development and Sustainability* 223–232 <https://link.springer.com/article/10.1007/s10668-016-9878-4> accessed 21 September 2023; F. Ekardt, *Sustainability: Transformation, Governance, Ethics, Law* (Springer 2020) 12 ff. Cf. M. Redclift, 'Sustainable Development (1987–2005): An

Oxymoron Comes of Age' (2005) 13(4) *Sustainable Development* 212–227 <https://doi.org/10.1002/sd.281> accessed 21 September 2023.

23 D. Pearce, A. Markandya, E.B. Barbier and E. Barbier, *Blueprint for a Green Economy* (Earthscan 1989) 34 and 128; G. Heal, 'Reflections—Defining and Measuring Sustainability' (2012) 6(1) *Review of Environmental Economics and Policy* 147, 158 <https://www.journals.uchicago.edu/doi/full/10.1093/reep/rer023> 21 September 2023.

24 F. Adloff and S. Neckel, 'Futures of Sustainability as Modernization, Transformation, and Control: A Conceptual Framework' (2019) 14(4) *Sustainability Science* 1015, 1016.

25 F. Adloff and S. Neckel, 'Futures of Sustainability as Modernization, Transformation, and Control: A Conceptual Framework' (2019) 14(4) *Sustainability Science* 1015, 1018–1019.

26 F. Adloff and S. Neckel, 'Futures of Sustainability as Modernization, Transformation, and Control: A Conceptual Framework' (2019) 14(4) *Sustainability Science* 1015, 1020–1021.

27 See, e.g., M. Weiss and C. Cattaneo, 'Degrowth–Taking Stock and Reviewing an Emerging Academic Paradigm' (2017) 137 *Ecological Economics* 220 <https://doi.org/10.1016/j.ecolecon.2017.01.014> accessed 21 September 2023.

28 P. Mason, *PostCapitalism. A Guide to Our Future* (Allen Lane 2015).

29 F. Adloff and S. Neckel, 'Futures of Sustainability as Modernization, Transformation, and Control: A Conceptual Framework' (2019) 14(4) *Sustainability Science* 1015, 1021–22.

30 P.A. Samuelson and W.D. Nordhaus, *Microeconomics* (17th edn, McGraw-Hill Education 2001).

31 On the development of private law thinking towards a system of distinct individuals, F. Capra and U. Mattei, *The Ecology of Law: Toward a Legal System in Tune with Nature and Community* (Berrett Koehler Publishers 2015) Ch. 3.

32 L.L. Butler, 'Property's Problem with Extremes' (2020) 55 *Wake Forest Law Review* 1, 25.

33 Art. 5:121 (4), 5:127a and 5:140a of the Dutch Civil Code (*Burgerlijk wetboek*; BW).

34 See, e.g., the *Shell* judgement of the District Court of The Hague based on Art. 6:162 Dutch Civil Code: Rechtbank Den Haag 26 May 2021, ECLI:NL: RBDHA:2021:5339.

35 P. Taylor and D.P. Grinlinton (eds), *Property Rights and Sustainability: The Evolution of Property Rights to Meet Ecological Challenges* (Brill | Nijhoff 2011) Chs. 1, 2; C.J. Circo, 'Does Sustainability Require a New Theory of Property Rights?' (2009) 58(1) *Kansas Law Review* 91, 93–94.

36 See F. Capra and U. Mattei, *The Ecology of Law: Toward a Legal System in Tune with Nature and Community* (Berrett-Koehler Publishers 2015); U. Mattei and A. Quarta, *The Turning Point in Private Law: Ecology, Technology and the Commons* (Edward Elgar Publishing 2018); U. Mattei, A. Quarta, F. Valguarnera and R. Fisher, *Property Meeting the Challenge of the Commons* (Springer 2023).

37 F. Capra and U. Mattei, *The Ecology of Law: Toward a Legal System in Tune with Nature and Community* (Berrett-Koehler Publishers 2015) Chs. 8, 9.

38 See, for a recent overview and analysis, A. Putzer, T. Lambooy, R. Jeurissen and E. Kim, 'Putting the Rights of Nature on the Map. A Quantitative Analysis of Rights of Nature Initiatives across the World' (2022) 18 *Journal of Maps* 89.

39 Starting with the seminal article of C.D. Stone, 'Should Trees Have Standing? – Towards Legal Rights for Natural Objects' (1972) 45 *Southern California Law Review* 450–501.

40 See, e.g., S. Borràs, 'New Transitions from Human Rights to the Environment to the Rights of Nature' (2016) 5(1) *Transnational Environmental Law* 113–143 <https://doi.org/10.1017/S204710251500028X> accessed 21 September 2023.

41 Section 12 Te Urewera Act 2014; Section. 14 Te Awa Tupua Act 2017.

42 See, for an analysis of the implications of Rights of Nature in constitutional property law, B. Hoops, 'What If the Black Forest Owned Itself? A Constitutional Property Law Perspective on Rights of Nature' (2022) 11 *Transnational Environmental Law* 475.

43 E.T. Freyfogle, 'Taking Property Seriously' in P. Taylor and D.P. Grinlinton (eds), *Property Rights and Sustainability: The Evolution of Property Rights to Meet Ecological Challenges* (Brill | Nijhoff 2011) 43–61; R.J. Goldstein, 'Green Wood in the Bundle of Sticks: Fitting Environmental Ethics and Ecology into Real Property Law' (1998) 25(2) *Environmental Affairs* 347; T.W. Frazier, 'The Green Alternative to Classical Liberal Property Theory' (1995) 20(2) *Vermont Law Review* 299.

44 See, e.g., P. Burdon, *Earth Jurisprudence, Private Property and the Environment* (Routledge 2015).

45 B. Akkermans, 'Sustainable Property Law: Towards a Revaluation of Our System of Property Law' in B. Akkermans and G. van Dijck (eds), *Sustainability and Private Law* (Eleven International publishing 2019) 37–58, section 4; B. Akkermans and K. van Melle, 'Regeneratieve Eigendom - naar een hernieuwde

benadering' in B. Akkermans, B. Hoops, E. Van der Sijde and B. Verheye (eds), *Privaatrecht 2050. De weg naar ecologische duurzaamheid* (Die Keure 2022) 193–208.

46 See, e.g., S. Demeyere, 'Affirmative Land Burdens in German, Dutch and Belgian Law: Possibilities, Restrictions and Workarounds' (2017) 6(2) *European Property Law Journal* 196–235 <https://doi.org/10.1515/eplj-2017-0010> accessed 21 September 2023.

47 UN-DESA, 'A Guidebook to the Green Economy Issue 1: Green Economy, Green Growth, and Low Carbon Development – History, Definitions, and a Guide to Recent Publications' (2012) <https://sustainabledevelopment.un.org/index.php?page=view&type=400&nr=634> accessed 6 April 2022.

48 M. Reynebeau, 'The Use of Private and Public Land for the Development of Heat Networks' (2019) 28(4) *European Energy and Environmental Law Review* 131–143 <https://doi.org/10.54648/eelr2019014> accessed 21 September 2023.

49 See, on upgrades to condominiums more generally, C. Van der Merwe (ed.), *European Condominium Law* (Cambridge University Press 2015) 521 ff.

50 M.C.E. van der Vleuten and I. de Jong, 'Een verbeterde regeling voor appartementsrechten – Ontwerp Wetsvoorstel' in L.C.A. Verstappen and F.J. Vonck (eds), *Vervolg 'Boek 5 BW van de toekomst, Het ontwerpwetsvoorstel appartementsrechten* (Sdu 2020) 23–74, 31 ff.

51 B. Verheye, 'Toekomst van de circulaire vastgoedeconomie' (2019) 1 *Tijdschrift voor Privaatrecht* 107–196; B. Akkermans, 'Duurzaam Goederenrecht: naar een herijking van ons goederenrechtelijk stelsel?' (2018) 55(3) *Tijdschrift voor Privaatrecht* 1437–1470; R. Koolhoven, 'Gebouwen en hun bestanddelen in een meer circulair goederenrecht. Van een wegwerpeconomie naar een kringloop van hoogwaardige, modulaire producten die worden verdienstelijkt' in R. Koolhoven, C.W. Backes, M.N. Boeve and D.R. Versteeg, *Circulair bouwen* (Instituut voor Bouwrecht 2018) 5–54; A.J. Mes, H.D. Ploeger and B.A.M. Janssen, 'Eigendom van onroerende zaken, met name natrekking (titels 1 en 3)' in L.C.A. Verstappen (ed.), *Boek 5 BW van de toekomst* (Sdu 2016) 145–217, 162.

52 F. Ekardt, *Sustainability: Transformation, Governance, Ethics, Law* (Springer 2020) 12 ff.

53 Y. Xu, J. Li, Q. Tan, A.L. Peters, and C. Yang, 'Global Status of Recycling Waste Solar Panels: A Review' (2018) *Waste Management* 450–458. Cf. F. Ekardt, *Sustainability: Transformation, Governance, Ethics, Law* (Springer 2020) 13, with respect to biogas <https://doi.org/10.1016/j.wasman.2018.01.036> accessed 21 September 2023.

54 H. Reimers, A. Jacksohn, D. Appenfeller, W. Lasarov, A. Hüttel, K. Rehdanz, I. Balderjahn and S. Hoffmann, 'Indirect Rebound Effects on the Consumer Level: A State-of-the-Art Literature Review' (2021) 3 *Cleaner and Responsible Consumption* 100032.

55 B. Alcott, 'Jevons' Paradox' (2005) 54(1) *Ecological Economics* 9–21; E. Rosenbaum, 'Rebound Effects and Green Growth – An Examination of Their Relationship in a Parsimonious Equilibrium Input-Output-Framework' (2019) *Journal of Clean Production* 121–132 <https://doi.org/10.1016/j.jclepro.2019.03.296> accessed 21 September 2023. F. Ekardt, *Sustainability: Transformation, Governance, Ethics, Law* (Springer 2020) 18.

56 K. Biely, D. Maes and S. Van Passel, 'The Idea of Weak Sustainability Is Illegitimate' (2018) 20(1) *Environment, Development and Sustainability* 223–232; F. Ekardt, *Sustainability: Transformation, Governance, Ethics, Law* (Springer 2020) 12 ff.

57 See, e.g., B. Hoops, 'Verduurzamingsverplichtingen zonder compensatie: Het krimpende gewicht van de autonomie van de eigenaar' (2020) 5 *Tijdschrift voor Bouwrecht* 445–455.

58 See, e.g., M. Braungart and W. McDonough, *Cradle to Cradle: Remaking the Way We Make Things* (Farrar, Straus and Giroux 2002).

59 See, e.g., T.J. de Römph and J.M. Cramer, 'How to Improve the EU Legal Framework in View of the Circular Economy' (2020) 38(3) *Journal of Energy & Natural Resources Law* 245–260 <https://doi.org/10.1080/02646811.2020.1770961> accessed 21 September 2023.

60 See, e.g., R. Koolhoven, 'Gebouwen en hun bestanddelen in een meer circulair goederenrecht. Van een wegwerpeconomie naar een kringloop van hoogwaardige, modulaire producten die worden verdienstelijkt' in R. Koolhoven, C.W. Bakes, M.N. Boeve and D.R. Versteeg, *Circulair bouwen* (Instituut voor Bouwrecht 2018) 5–54.

61 T. Rau and S. Oberhuber, *Material Matters* (Bertram + de Leeuw Uitgevers BV 2016).

62 L.A. Greening, D.L. Greene and C. Difiglio, 'Energy Efficiency and Consumption — The Rebound Effect — A Survey' (2000) 28(6–7) *Energy Policy* 389–401 <https://doi.org/10.1016/S0301-4215(00)00021-5> accessed 21 September 2023; H. Reimers, A. Jacksohn, D. Appenfeller, W. Lasarov, A. Hüttel, K. Rehdanz, I. Balderjahn and S. Hoffmann, 'Indirect Rebound Effects on the

Consumer Level: A State-of-the-Art Literature Review' (2021) 3 *Cleaner and Responsible Consumption* 100032 <https://doi.org/10.1016/j.clrc.2021.100032> accessed 21 September 2023.

63 M. Bianchi and M. Cordella, 'Does Circular Economy Mitigate the Extraction of Natural Resources? Empirical Evidence Based on Analysis of 28 European Economies over the Past Decade' (2023) 203 *Ecological Economics* 107607 <https://doi.org/10.1016/j.ecolecon.2022.107607> accessed 21 September 2023. On the economics of recycling, which only requires the re-use of materials but not necessarily for the same purpose or a purpose of the same or higher value: F. Grosse and G. Mainguy, 'Is Recycling "Part of the Solution"? The Role of Recycling in an Expanding Society and a World of Finite Resources' (2010) 3(1) *SAPIENS* <https://journals.openedition.org/sapiens/906> accessed 21 September 2023.

64 P. Lacy, J. Long and W. Spindler, *The Circular Economy Handbook* (Palgrave MacMillan 2020).

65 Cf. F. Ekardt, *Sustainability: Transformation, Governance, Ethics, Law* (Springer 2020) 16 ff.

66 On whether this belief is justified, F. Ekardt, *Sustainability: Transformation, Governance, Ethics, Law* (Springer 2020) 16 ff.

67 See, for instance, Section 24 of the Constitution of the Republic of South Africa, 1996.

68 Principle 1 of the 1972 UN Declaration on the Human Environment.

69 M. Siems, *Comparative Law* (Cambridge University Press, Cambridge 2014) 13 ff.

70 See, e.g., B. Verheye, 'Toekomst van de circulaire vastgoedeconomie' (2019) 1 *Tijdschrift voor Privaatrecht* 107-196; B. Akkermans, 'Duurzaam Goederenrecht: naar een herijking van ons goederenrechtelijk stelsel?' (2018) 55(3) *Tijdschrift voor Privaatrecht* 1437–1470.

71 See sub-section 16.3.1. More generally, the 'Progressive Property Theory' movement has been advocating for the notion that social values underpin the concept of ownership: G.S. Alexander, E.M. Peñalver, J.W. Singer and L.S. Underkuffler, 'A Statement of Progressive Property' (2009) 94 *Cornell Law Review* 743.

72 See, e.g., in the South African context, E. van der Sijde, 'Tenure Security for ESTA Occupiers: Building on the *Obiter* Remarks in *Baron v Claytile Limited*' (2020) 36 *South African Journal on Human Rights* 74 <https://doi.org/10.1080/02587203.2020.1773308> accessed 21 September 2023.

73 B. Akkermans, '(Normative) Models of Property Law', in B. Akkermans and A. Berlee (eds), *'Sjef-Sache'* (Eleven International Publishing 2021) 3–16.

74 B. Akkermans, 'Sustainable Property Law: Towards a Revaluation of Our System of Property Law' in B. Akkermans and G. van Dijck (eds), *Sustainability and Private Law* (Eleven International Publishing 2019) 37–58, section 4; B. Akkermans and K. van Melle, 'Regeneratieve Eigendom – naar een hernieuwde benadering' in B. Akkermans, B. Hoops, E. Van der Sijde and B. Verheye (eds), *Privaatrecht 2050. De weg naar ecologische duurzaamheid* (Die Keure 2022) 193–208.

75 B. Akkermans, B. Hoops, E. Van der Sijde and B. Verheye, 'Duurzaam privaatrecht voor de 21ste eeuw: het leven van een rechtswetenschappelijke subdiscipline' (2023) 60(1) *Tijdschrift voor Privaatrecht* 199.

Bibliography

Adloff F and Neckel S, 'Futures of Sustainability as Modernization, Transformation, and Control: A Conceptual Framework' (2019) 14(4) *Sustainability Science* 1015–1025

Akkermans B, 'Duurzaam Goederenrecht: naar een herijking van ons goederenrechtelijk stelsel?' (2018) 55(3) *Tijdschrift voor Privaatrecht* 1437–1470

Akkermans B, 'Sustainable Property Law: Towards a Revaluation of Our System of Property Law' in B. Akkermans and G. van Dijck (eds), *Sustainability and Private Law* (Eleven International Publishing 2019) 37–58

Akkermans B, '(Normative) Models of Property Law' in B. Akkermans and A. Berlee (eds), *'Sjef-Sache'* (Eleven International Publishing 2021) 3–16

Akkermans B, Hoops B, Van der Sijde E and Verheye B, 'Duurzaam privaatrecht voor de 21ste eeuw: het leven van een rechtswetenschappelijke subdiscipline' (2023) 60(1) *Tijdschrift voor Privaatrecht* 199–256

Akkermans B and Melle K, 'Regeneratieve Eigendom – naar een hernieuwde benadering' in B. Akkermans, B. Hoops, E. Van der Sijde and B. Verheye (eds), *Privaatrecht 2050. De weg naar ecologische duurzaamheid* (Die Keure 2022) 193–208

Alcott B, 'Jevons' Paradox' (2005) 54(1) *Ecological Economics* 9–21

Alexander G, *Property and Human Flourishing* (Oxford University Press 2018)

Alexander G, Peñalver E, Singer J and Underkuffler L, 'A Statement of Progressive Property' (2009) 94 *Cornell Law Review* 743–744

Bianchi M and Cordella M, 'Does Circular Economy Mitigate the Extraction of Natural Resources? Empirical Evidence Based on Analysis of 28 European Economies over the Past Decade' (2023) 203 *Ecological Economics* 107607 <https://doi.org/10.1016/j.ecolecon.2022.107607> accessed 21 September 2023

Biely K, Maes D and Van Passel S, 'The Idea of Weak Sustainability Is Illegitimate' (2018) 20(1) *Environment, Development and Sustainability* 223–232 <https://link.springer.com/article/10.1007/s10668-016-9878-4> accessed 21 September 2023

Borràs S, 'New Transitions from Human Rights to the Environment to the Rights of Nature' (2016) 5(1) *Transnational Environmental Law* 113–143 <https://doi.org/10.1017/S204710251500028X> accessed 21 September 2023

Braungart M and McDonough W, *Cradle to Cradle: Remaking the Way We Make Things* (Farrar, Straus and Giroux 2002)

Burdon P, *Earth Jurisprudence, Private Property and the Environment* (Routledge 2015)

Butler L, 'Property's Problem with Extremes' (2020) 55 *Wake Forest Law Review* 1–53

Capra F and Mattei U, *The Ecology of Law: Toward a Legal System in Tune with Nature and Community* (Berrett-Koehler Publishers 2015)

Circo C, 'Does Sustainability Require a New Theory of Property Rights?' (2009) 58(1) *Kansas Law Review* 91–159

Davies G, 'Appraising Weak and Strong Sustainability: Searching for a Middle Ground' (2013) 10 *Consilience: The Journal for Sustainable Development* 111–124

Degens P, 'Towards Sustainable Property? Exploring the Entanglement of Ownership and Sustainability' (2021) 60(2) *Social Science Information* 209–229

Demeyere S, 'Affirmative Land Burdens in German, Dutch and Belgian Law: Possibilities, Restrictions and Workarounds' (2017) 6(2) *European Property Law Journal* 196–235 <https://doi.org/10.1515/eplj-2017-0010> accessed 21 September 2023

Ekardt F, *Sustainability: Transformation, Governance, Ethics, Law* (Springer 2020) <https://doi.org/10.1016/j.wasman.2018.01.036> accessed 21 September 2023

Frazier T, 'The Green Alternative to Classical Liberal Property Theory' (1995) 20(2) *Vermont Law Review* 299–371

Freyfogle E, 'Taking Property Seriously' in P. Taylor and D.P. Grinlinton (eds), *Property Rights and Sustainability: The Evolution of Property Rights to Meet Ecological Challenges* (Brill | Nijhoff 2011) 43–61

Goldstein R, 'Green Wood in the Bundle of Sticks: Fitting Environmental Ethics and Ecology into Real Property Law' (1998) 25(2) *Environmental Affairs* 347–430

Graham N, 'Teaching Private Law in a Climate Crisis' (2021) 40(3) *University of Queensland Law Journal* 403–419 <https://journal.law.uq.edu.au/index.php/uqlj/article/view/6047/5029> accessed 21 September 2023

Greening L, Greene D, and Difiglio C, 'Energy Efficiency and Consumption — The Rebound Effect — A Survey' (2000) 28(6–7) *Energy Policy* 389–401 <https://doi.org/10.1016/S0301-4215(00)00021-5> accessed 21 September 2023

Grosse F and Mainguy G, 'Is Recycling "Part of the Solution"? The Role of Recycling in an Expanding Society and a World of Finite Resources' (2010) 3(1) *SAPIENS* <https://journals.openedition.org/sapiens/906> accessed 21 September 2023

Hardin G, 'The Tragedy of the Commons' (1968) 162 *Science* 1243–1248

Heal G, 'Reflections—Defining and Measuring Sustainability' (2012) 6(1) *Review of Environmental Economics and Policy* 147–163 <https://www.journals.uchicago.edu/doi/full/10.1093/reep/rer023> 21 September 2023

Hoops B, 'Verduurzamingsverplichtingen zonder compensatie: Het krimpende gewicht van de autonomie van de eigenaar' (2020) 5 *Tijdschrift voor Bouwrecht* 445–455

Hoops B, 'Of Forests and Rivers Owning Themselves: Looking for a Place for Te Urewera and Te Awa Tupua in Constitutional Property Law and Theory' (2022) 7 *Journal of Law, Property, and Society* 126–159 <https://bit.ly/JLPS-Hoops> accessed 21 September 2023

Hoops B, 'What If the Black Forest Owned Itself? A Constitutional Property Law Perspective on Rights of Nature' (2022) 11 *Transnational Environmental Law* 475–500 <https://doi.org/10.1017/S2047102522000322> accessed 21 September 2023

Koolhoven R, 'Gebouwen en hun bestanddelen in een meer circulair goederenrecht. Van een wegwerpeconomie naar een kringloop van hoogwaardige, modulaire producten die worden verdienstelijkt' in R. Koolhoven, C.W. Backes, M.N. Boeve and D.R. Versteeg, *Circulair bouwen* (Instituut voor Bouwrecht 2018) 5–54

Lacy P, Long J and Spindler W, *The Circular Economy Handbook* (Palgrave MacMillan 2020)

Mason, P, *PostCapitalism. A Guide to Our Future* (Allen Lane 2015)

Mattei U and Quarta A, *The Turning Point in Private Law: Ecology, Technology and the Commons* (Edward Elgar Publishing 2018)

Mattei U, Quarta A, Valguarnera F and Fisher R, *Property Meeting the Challenge of the Commons* (Springer 2023)

Mes A, Ploeger H and Janssen B, 'Eigendom van onroerende zaken, met name natrekking (titels 1 en 3)' in L.C.A. Verstappen (ed.), *Boek 5 BW van de toekomst* (Sdu 2016) 145–217

Ostrom E, *Governing the Commons: The Evolution of Institutions for Collective Action* (Cambridge University Press 2015)

Pearce D, Markandya A, Barbier E and Barbier E, *Blueprint for a Green Economy* (Earthscan 1989)

Putzer A, Lambooy T, Jeurissen R and Kim E, 'Putting the Rights of Nature on the Map. A Quantitative Analysis of Rights of Nature Initiatives across the World' (2022) 18 *Journal of Maps* 89–96

Rau T and Oberhuber S, *Material matters* (Bertram + de Leeuw Uitgevers BV 2016)

Redclift M, 'Sustainable Development (1987–2005): An Oxymoron Comes of Age' (2005) 13 (4) *Sustainable Development* 212–227 <https://doi.org/10.1002/sd.281> accessed 21 September 2023

Reimers H, Jacksohn A, Appenfeller D, Lasarov W, Hüttel A, Rehdanz K, Balderjahn I and Hoffmann S, 'Indirect Rebound Effects on the Consumer Level: A State-of-the-Art Literature Review' (2021) 3 *Cleaner and Responsible Consumption* 100032 <https://doi.org/10.1016/j.clrc.2021.100032> accessed 21 September 2023

Reynebeau M, 'The Use of Private and Public Land for the Development of Heat Networks' (2019) 28(4) *European Energy and Environmental Law Review* 131–143 <https://doi.org/10.54648/eelr2019014> accessed 21 September 2023

Römph T and Cramer J, 'How to Improve the EU Legal Framework in View of the Circular Economy' (2020) 38(3) *Journal of Energy & Natural Resources Law* 245–260 <https://doi.org/10.1080/02646811.2020.1770961> accessed 21 September 2023

Rosenbaum E, 'Rebound Effects and Green Growth – An Examination of Their Relationship in a Parsimonious Equilibrium Input-Output-Framework' (2019) *Journal of Clean Production* 121–132 <https://doi.org/10.1016/j.jclepro.2019.03.296> accessed 21 September 2023

Rubin E, 'The Practice and Discourse of Legal Scholarship' (1988) 86(8) *Michigan Law Review* 1835–1905

Samuelson P and Nordhaus W, *Microeconomics* (17th edn, McGraw-Hill Education 2001)

Siems M, *Comparative Law* (Cambridge University Press, Cambridge 2014)

Van der Sijde E, 'Tenure Security for ESTA Occupiers: Building on the *Obiter* Remarks in *Baron v Claytile Limited*', (2020) 36 *South African Journal on Human Rights* 74–92 <https://doi.org/10.1080/02587203.2020.1773308> accessed 21 September 2023

Smits J, *The Mind and Method of the Legal Academic* (Edward Elgar Publishing 2012) 13–18

Smits J, 'What Is Legal Doctrine? On the Aims and Methods of Legal-Dogmatic Research' in R. van Gestel, H. Micklitz and E.L. Rubin (eds), *Rethinking Legal Scholarship: A Transatlantic Dialogue* (Cambridge University Press 2017) 207–228

Stone C, 'Should Trees Have Standing? – Towards Legal Rights for Natural Objects' (1972) 45 *Southern California Law Review* 450–501

Taylor P and Grinlinton D (eds), *Property Rights and Sustainability: The Evolution of Property Rights to Meet Ecological Challenges* (Brill | Nijhoff 2011)

UN-DESA, 'A Guidebook to the Green Economy Issue 1: Green Economy, Green Growth, and Low Carbon Development – History, Definitions and a Guide to Recent Publications' (2012) <https://sustainabledevelopment.un.org/index.php?page=view&type=400&nr=634> accessed 6 April 2022

Van der Merwe C (ed.), *European Condominium Law* (Cambridge University Press 2015)

Verheye B, 'Toekomst van de circulaire vastgoedeconomie' (2019) 1 *Tijdschrift voor Privaatrecht* 107–196

Vleuten M and Jong I, 'Een verbeterde regeling voor appartementsrechten – Ontwerp Wetsvoorstel' in L.C.A. Verstappen and F.J. Vonck (eds), *Vervolg 'Boek 5 BW van de toekomst, Het ontwerpwetsvoorstel appartementsrechten* (Sdu 2020) 23–74

Vols M, *Legal Research: One Hundred Questions and Answers* (Eleven International Publishing 2021)

Weiss M and Cattaneo C, 'Degrowth–Taking Stock and Reviewing an Emerging Academic Paradigm' (2017) 137 *Ecological Economics* 220–230 <https://doi.org/10.1016/j.ecolecon.2017.01.014> accessed 21 September 2023

World Commission on Environment and Development, *Report of the World Commission on Environment and Development: Our Common Future* (Oxford University Press 1987)

Xu Y, Li J, Tan Q, Peters A and Yang C, 'Global Status of Recycling Waste Solar Panels: A Review' (2018) *Waste Management* 450–458

17

A NEW PARADIGM OF OWNERSHIP IN ITALIAN PRIVATE LAW?

Francesco Zecchin[1]

17.1 Reality, Ownership, and Law

It was the Greek philosophy that for the first time approached the concept of justice not as an expression of an irrational almighty god, but as a social harmony created by the king of Olympus, Zeus.[2] From this point of view, a parallel with the idea of health could be drawn, because '(...) like the physical life of the body, social life reacts with grave disturbances and diseases when its harmony is violated'.[3] Authors like Solon, after all, described justice in their poems as the health of the community,[4] and Plato was inspired by Hippocrates[5] for his conception of the relation between the nature and the function of justice in the soul.[6]

As a result, human beings were supposed to maintain this social balance or to restore it in case of alteration, and the law (*nomos*) was probably the most relevant technical instrument at their disposal to achieve this end,[7] at least ideally. It is well known that the modern concept of law was created by the Romans using Latin,[8] but it was not by a chance that in some Central Italy such as *Osco*, *iu-dex* has also been called *med-dix*. This other term has the same etymological root of *med-icus*, namely *med*, and it means rightly restoring the balance in a diseased body, in the universe, or in human affairs.[9]

Of course, over time, social harmony meant very different things, and already in the Greek '(...) law seemed to be permanently changing and to depend on changing majorities with different viewpoints'.[10] In other words, often the law was the one ruled by those who held political or military power, and its aim was not to do justice, but to maintain that power at any cost.[11]

From this point of view, the history of the 'terrible, and perhaps unnecessary right' that is ownership, as defined by Cesare Beccaria speaking about theft,[12] seems emblematic. On the one hand, its rules affect society's basic structure, especially regarding relationships between individuals.[13] On the other hand, the same property norms established from time to time, in turn, influence the social, economic, and political vision of the people.[14] To summarise, it seems especially true for ownership that there is 'a gap that constantly appears between law and reality'.[15]

This means that property law will play a crucial role in the current process towards achieving sustainability.[16] Understanding the aim of the current Italian ownership rules seems, therefore, essential, but it is not possible without an analysis of the path that led to such legislation.[17] In matters of private law – which is for instance different than economics, whose vision

DOI: 10.4324/9781032662046-21

of the relation between person and good is focused on the worth of the latter – research must turn its attention to the question of the powers that come with being an owner, since the so-called 'private enigma of ownership' revolves around this issue,[18] at least in the continental Europe.[19] As it was said, normally in this area

> (...) the word ownership is simply a verbal construction used for describing the relation between a person and a good from the law point of view, or in other words for answering to the key issue regarding the legal meaning of the possessive adjective "my".[20]

17.2 Ownership as a *Status*?

Following a diachronic perspective, it is necessary to focus briefly on the common definition of ownership used when the Italian mainland was not yet a single legal system but was divided into several states and statelets. It had a lot of influence on the property laws ruled in later times, at least as a model to be abandoned.[21]

Of course, in that period – as in the ones before and partially after – the issue was mainly related to land property. However, the rules provided for this kind of ownership were applied to movable goods, too, although the distinction between the *fundus* and other property was already known in the Age of Justinian.[22]

In the classical feudal view '(...) the sovereign (emperor, king or prince) was the owner of the entire land, but its enjoyment was given to aristocracy in exchange for their loyalty and military services'.[23] This is true in almost all of Europe, even if the French model is stricter than the Longobard or Italian one, because in the former the fief was indivisible, unalienable and devisable only to the first son, while in the latter it is in part alienable and it will be devised among the sons.[24] Anyway, this framework, on the one hand, intensified the landowner's authority, which ensured supremacy over the lower social classes; on the other one, it strengthened the solidarity among the community's members, because no one can exercise a right against other people without taking responsibility in their favour.[25]

Near the end of the Early Middle Ages, and even more during the *ius commune* period, there was a development. Despite a plurality of legal systems with different rules, in the Italian mainland the concrete relation between person and land was becoming increasingly important. It was not just the powers of the owner that were protected by law, but also the ones of who exercised an authority over the good from a practical point of view.[26] The core had become the *res* with its opportunity of use: the so-called *dominium utile*.[27] Not always but in some important cases, it resulted in forms of ownership[28] which gave actions like the *rei vindicatio*.[29]

However, ownership still meant power and exclusivity, even if it was no longer a right only for the sovereign.[30] It did not lose the importance that it had during feudalism,[31] but it was split among a limited number of people, normally of aristocratic origins or linked to the church.[32] Being an owner used to be a *status*, if it is true that in the Middle Ages and in the *ius commune* time '(...) ownership's main issue was the indissoluble bond between the situation of the person and the ones of landed property'.[33]

In other words, the Italian tenure system in the late 18th century, like in the large part of continental Europe,[34] mirrored that of feudalism time, except the presence of more than one owner: not just the sovereign, who was the owner 'in the full sense', but also those aristocrats who were entitled to exercise some kind of enjoyment's rights, like emphyteusis or superficies.[35] As a result, on the one hand, usually, the bourgeois class could not become owner of the many lands that they managed due to other rights of enjoyment; on the other hand, the social conditions of the peasantry had not changed, because the truth remained that the farther you were from the sovereign, the more you had obligations and not rights,[36] and the succession law of the time made this destiny of submission almost irreversible.[37]

To summarise, from the private law perspective it seems that in the 18th century, ownership was still especially a prerogative of noble class or clergy, who had no right to disposition of its property (because the land could not be sold and succession had strict rules) but a very wide right of enjoyment.[38] However, since the *dominium* was, anyway, granted by the sovereign, it was worsened by economic constraints, which affected the chain of actors involved with the land's management.[39]

17.3 Ownership as a Human Right

It is precisely this legacy of the *ancient régime* that led to the French Revolution, and to a new concept of ownership in almost all continental Europe during the 19th century.[40] To be more precise, the submission to the sovereign, to the noble class, and to the clergy was no longer acceptable, and this caused social, political, and economic changes that gradually required to second thought ownership, whose core moved from the good to the person.[41] The new property laws focused their attention not on absoluteness, but on exclusivity, in the sense of freedom from external constraints (*ius ceteros excludendi*).[42]

This is the spirit of Italian law, too. Just consider the point of view of the Albertine Statute of 1848 (AS) and of the first Italian Civil Code (first ICC), in force from 1865 until 1942.

The AS provided that

> All property, without exception, are inviolable. However, when it is necessary due to the legally established public interest, one may be required to surrender them in whole or in part, by means of just compensation in accordance with statutes.
>
> (Art. 29)

Following the first ICC,

> Ownership is the right to enjoy and dispose of things without any restrictions, as long as the use was no forbidden by statutes or administrative regulations.
>
> (Art. 436),

and

> No one may be constrained to surrender his property or to allow others to use them, except for a legally recognised public interest and subject to the payment of just compensation. The rules relating to expropriation for reasons of public utility are determined by special statutes.
>
> (Art. 438)

The connection with the individualistic and anthropocentric understanding expressed in the Declaration of the Rights of Man and of the Citizen of 1789 (DRMC) is clear from the above-mentioned provisions.[43] Without a revolutionary inspiration but with the aim to build a new property law in discontinuity with the *ancient régime*,[44] the Italian law adopted the vision of Art. 544 of the *Code Napoléon* (*CN*),[45] whose translation is, moreover, the text of the already mentioned Art. 436 of the first ICC.[46]

The idea was to keep the land rules that arose out of the revolutionary experience in the authoritarian regime,[47] namely, to express the concept of a land property that, thanks to its connection with the sale contract, was supposed to grant people the chance to emancipate themselves from the status in which they were socially born.

The relation between freedom and ownership was centred around the subjective right, considered as a '(…) sovereign space in which the individual exercises ownership powers, as an area of lordship that the natural order gives to human beings and that society must protect from attacks by third parties'.[48] The reference point of this model was ownership as a *human right*,[49] which theoretically has become generally available because of the contract.[50]

It was a reaction to the *ancient régime* constraints, as it is proved by the following words of Jean Baptiste Treilhard: '*Qu'étoit-ce en effect que le régime féodal, sinon l'art de faire de la propriété foncière un instrument d'asservissement?*'.[51] In other words, the liberalism view moved away from the absolutism of the sovereign to the absolutism of owners,[52] who claimed reliable guarantees against government intrusions.[53]

This is particularly well demonstrated by the already mentioned Art. 436 of the first ICC, but also and especially by its property rules. These provide for several rights of the owner – action of *vindication* (Art. 439 of the first ICC), right to the space above, on and under the ground (Art. 440 of the first ICC), right to surround the land (Art. 442 of the first ICC), and right of enjoyment (Art. 443 of the first ICC) – and no limits, unless the ones which are needed in the key to socio-economic perspective, like legal servitudes (Art. 533 ff. of the ICC).[54]

Nevertheless, the idea of overcoming classism between people and the corresponding system of feudalism's enjoyment rights *in rem* through the combination of property and contract[55] – or in the light of the transition 'from status to contract', using Henry Maine's famous words[56] – failed on a practical level.[57]

On the one hand, both the *CN* and the first ICC hid a regulation of land relation that complied with 'a social typicality derived essentially from a recognition or acknowledgement of how they formally looked'.[58] On the other hand, deep down, the glorification of the individual will 'ultimately satisfied only the owner's interest'[59] to the detriment of who depends on him.[60]

In other words, the idea that ownership was able to satisfy the needs of people and free them did not succeed. Over the next few years it became clear that the opportunity to enjoy land was guaranteed to new owners thanks to the labour of who were still in an absolute inferior social position, and whose interests were not protected by the brand new rules.[61] This is the reason why in many countries agriculture started to be regarded as something similar to feudalism's enjoyment rights *in rem*, like the emphyteusis in France.[62] However, these attempts were essentially pointless, because they were overwhelmed by the dominant principles of *laissez faire, laissez passer*.[63]

To sum up, the right not to be interfered by the State ended up being an ultimate power over a property, which was reflected on the people who worked the land, because their survival depended actually on such a job.[64] Moreover, this situation quickly spread beyond the agricultural production's borders: due to the industrial revolution, the workers' freedom was increasingly at the mercy of the 'owner' of the fabric.[65]

It was not by chance that not all the Italian scholars of that time shared this vision. The medieval approach to the ownership issue had not been entirely dismissed.[66] Its positive aftermaths, somehow lost with the new rules, were pointed out by influential authors, like Gian Domenico Romagnosi and Antonio Rosmini,[67] and this led other ones to less strict and tempered reading of the owner's powers on account of a social point of view.[68]

17.4 Property Law and Social Function

It is thanks to this debate too, that Italian law, like many other legal systems,[69] moved towards property rules which do not express an idea of ultimate power, but try to find a balance between the owners' rights and their living in a society, that is, the so-called 'social function'.

The current Italian Civil Code (current ICC) is inspired by this idea.[70] In general, there is a reshaping of ownership compared to other legal arrangements, such as obligations, which are no longer ruled within the framework of property, but have an independent regulation.[71] More particularly, the wording of Art. 832 ICC, on the one hand, drops the reference to 'without any restrictions' in favour of a more balanced 'fully and exclusively' regarding the owner's power to enjoy and dispose of a good. On the other hand, in relation to the limits of this power, the current norm replaces the specific reference to use that was 'not forbidden by statutes or administrative regulations' with a more generic one to 'within the bounds of and observing the duties established by the law'.[72]

Nevertheless, the attempt to adopt expressly the idea that the power of enjoyment and disposition should, nevertheless, be exercised '(...) in accordance with the social function of the right itself' failed.[73] It would be accepted only in the Italian Constitution of 1948 (IC), probably due to a greater recognition of social changes that took place in the 19th century.[74] Art. 42(2) IC, once clarified that 'private ownership is recognised and guaranteed by statute', specifies that still statute 'provides for the manner of acquisition and enjoyment and its limitation, in order to ensure its social function and make it accessible to all'. This means that while in the current ICC the social function is an external factor limiting the owner's power, in the IC it turns into a real intrinsic feature, an essential element which show the very substance of ownership.[75]

The opposition to the ownership individualistic conception in favour of the social function, moreover, is supported by the whole structure of Art. 42 IC. Especially where it states that '(...) in cases provided for by statute, and subject to compensation', expropriation is allowed if there are 'grounds of general interest' (para. 3) and that 'the statute lays down the rules and limits of legitimate and testamentary inheritance' (para. 4).

But there is another factor that cannot be overlooked in the idea of ownership as expressed in the IC, and it is still connected with the social function: the 'fragmentation' of property law since the second half of the 19th century has started been pointed out with Filippo Vassalli.[76] There is no longer a single idea of ownership, but instead of a plurality of laws, each with its own rules. As a result, we are experiencing a somewhat revival of the medieval concept of property; not in the sense that more than one subject can be owner to the same good – as we have seen happened in that age – but rather in the sense there is a different regulation according to the value that the legal system assigns to the relation between the property and its owner.[77] This is the reason why, on the one hand, the Italian Constitutional Court[78] rejects the idea that, apart from the specific cases where the connection between the person and the good is particularly close,[79] ownership can be included among human rights.[80] On the other hand, Luigi Mengoni suggests including ownership among fundamental rights, since the first part of Art. 42(2) provides that statute regulates ownership, but cannot remove it from the law.[81]

In matters of private law, all this means that due to the social function, whenever the relation between owners and their goods does not show there is a human right, it is not just possible losing the property through expropriation and compensation, but also seeing the power of enjoyment or disposition restricted without compensation. In other words, the social function in the Italian legal system means both compulsory purchase and the conformation of owner's powers to social purposes. But while with reference to compulsory purchase compensation is provided for, this is not the case for the conformation of the owner's powers. This is because through the so-called 'functionalization' property is not taken away, but only limited. From this point of view, however, there is a need not to mask behind this tool a *de facto* eminent domain, as when the restrictions would be so strong they would prevent the owner from exercising his rights in practice (the so-called 'non-ablative', or 'anomalous' expropriations,[82] whose effects have to be compensated for by an indemnity).[83] In other words, it is necessary not to undermine the essential or minimal core content of what it means being an owner.[84]

In any case, it seems undeniable that ownership is increasingly becoming an economic tool, subject to an assessment of social compatibility, rather than a classical individual right.[85] Following the Italian Constitution of 1948,

> (...) it is no longer a human right, marking out a personal sphere free from the intrusion of any political power, but a right of contribution to the development of the Country. The owner is protected in this participative role, and therefore the aim of the Constitution is not so much their personality as the functionality of the socio-economic system.[86]

After all, the current ICC has already specific rules designed to fulfil the social function of a good, like in cases of rural estate (Artt. 846 ff. current ICC), of construction property (Artt. 869 ff. current ICC), and land ownership (Artt. 840 ff. and 873 ff. current ICC). The social function's boundaries depend on the type of good, of course.[87] And such boundaries take three different forms: limitation of the good's power to enjoy or to dispose, like in the case of building's cubage; imposition of specific conditions for using these powers, like in rentals for residential use contract law; or duty to exercise them in specific procedures, like if one good has artistical or cultural value.[88]

17.5 Ownership and Sustainability?

At this point, the question that perhaps most naturally comes up is whether the social function is able to accept and direct property law to the paradigm that is also re-focusing the Italian legal system, i.e. sustainability.[89] And to answer this question, it is useful to point out that in the Italian legal system there are already some rules that might be considered adopted in the light of sustainability.

For instance, Art. 873 of the current ICC might be read in the spirit of improving people's quality of life from environmental and, perhaps, social sustainability. This Article reads that 'constructions on neighbouring land, if not joined or adjoining, must be kept at a distance of not less than three metres', and it is justified, precisely, with the need of avoiding negative effects on the livability of buildings, and the health of those who live in them.[90]

But also the rules on urban planning that limits the building activity can be seen as pursuing social and environmental protection in their core. Territorial planning – whose multiple levels now involve the whole territory and no longer only city areas[91] – imposes on owners not just limits *ab externo*, but internal constraints of a negative nature relating to the *jus aedificandi* (no building activity, no building activity except for residential use, etc.).[92] In other words, these limits neutralise individual preferences in the now-recognised direction of a social and ecological balance that is defined as 'urban sustainability'.[93]

Moreover, in light of environmental and economic sustainability, the Italian regulations on waste disposal, contained in Legislative Decree No 152 of 3 April 2006, should also be mentioned. On the one hand, these regulations relate to movable goods while the 'functionalisation' of ownership has been traditionally carried out in relation to immovable property.[94] On the other hand, the used notion of 'refusal' shows that it is not the exercise of the power of enjoyment that is at stake, but that of disposition. To be more precise, waste shall mean a material or object, resulting from human activity or natural processes, that the holder of which discards or has decided or is required to discard (Art. 183 Leg. Dec. 152/06). And it is ruled out that owners can neither exercise ancient dereliction, i.e. relinquishment of movable property by abandonment, nor sell such property to someone who is not identified by the legislation as those who can dispose of it.[95] Additionally, the regulation sets out a series of priority criteria for waste management (Art. 179 Leg. Dec. 152/06), consisting of prevention, preparation for reuse, recycling, other types of recovery, and, finally, disposal. Lastly, Art. 184 Leg. Dec. 152/06 makes differentiation of waste between urban (internal and external) and special (dangerous and non-dangerous). According to the type of waste, there are different duties and costs for disposal.

Now, the main *ratio* behind these norms is clear: to focus the waste management system first and foremost on environmental sustainability. The *ratio* appears also to be economic since there is a tendency to reduce waste production to a minimum and seek a way to reuse it as a resource, moving from linear models (extraction-production-consumption-disposal) to circular models (extraction-production-use-reuse-remake-recycle). To summarise, the legislation naturally has an ecological purpose, but it seems to involve items aimed at achieving better eco-efficiency.[96]

All these property-conforming limits are, fundamentally, constitutionally lawful precisely because of the social function that the Italian Constitution requires owners to fulfil while exercising their rights.[97] The above-mentioned rules, in other words, could be seen as legitimate restrictions on the owner's powers because they lie within the social function expressed in Art. 42(2) IC.

After all, the ownership's history is also a history of direction of power's use towards a particular purpose.[98] From this point of view, the rural property rules seem very meaningful since law has always attempted to direct ownership towards productivity.[99]

Perhaps not in early Roman law, but almost surely already in the Classical Period, the individualistic element that characterises ownership was strongly attenuated.[100] It has been proved that the definition of property as *ius utendi et abutendi* does not actually appear in the Roman sources, and in any case it should not be read in the sense of arbitrary right.[101]

In Justinian law too it seemed possible to recognise the promotion of the productive approach in the *agri deserti* rule: the *dominus* who had neglected cultivation lost ownership of their land in favour of those who had made it productive through their work. The rule, in addition to making tax collection easier, no doubt responded to the wish to promote and protect rural work,[102] so much so that according to Vittorio Scialoja, it was inspired by a 'a higher social principle'.[103]

But a similar approach also marked the following ages. Just think that in Longobard Italy, the imposition in the law of the 'massaro' – i.e. the peasant 'who holds the only small farm grown by his family'[104] – certainly responded to the need to neutralise the Roman ruling class, but at the same time emphasised the '(…) factual relation between land and farmers according to the conception of the rights in rem proper to the German tradition'.[105]

The protective purpose towards those who make the land fruitful, that is, who yield it productive, began to have a private emphasis with the already mentioned medieval concept of *dominium utile*, which strongly marked '(…) a work conquest, an intensification of the ties with the land, for the future guaranteed by temporary contractual relationships'.[106] In spite of the fact that the opportunity to use land was still linked to a

> (...) personal bond of subordination, a licensee who gives due services and pays fees more or less quickly, feels a right to maintain for himself and his descendants the land he possesses and farms, and to keep it always on the same conditions.[107]

Once the 19th-century liberalism view was over, the 1948 IC – within the economic and social situation framework of the time, of course – made owners accountable with respect to the production capacity of goods, and of land particularly, precisely in light of their social function.[108] The preparatory papers, after all, assumed the power of the legislator to reallocate land and promote specific forms of use[109] and focused their attention to the range of powers granted by the IC.[110] Art. 44 IC, more specifically, 'in order to achieve the rational development of land and to set out fair social relationships', empowered the legislator to impose 'duties and constraints on private land ownership'. This means that it is required making specific regulatory provisions to enhance the land's production capacity.[111] So much that according to one legal scholar, in contravention to Art. 42 IC, an owner of a rural plot would not even have 'freedom to choose the use of property, to choose or not to choose an economic end: instead of freedom of economic enterprise, there is a duty to use the land for production'.[112]

Coming back to the issue of sustainability and Italian property law, it is possible, therefore, to follow a well-trodden path by such a legal system, although with a slightly different destination: not productivity *per se*, which has shown all its weaknesses,[113] but sustainable development. After all,

> (...) the setting of the ownership borders designed to fulfil its social function is itself a setting of the right's content, which consequently changes its meaning depending on the evolution of socio-economic conditions as conceived in the culture of a people.[114]

And perhaps the imitation could also involve the approach that was frequently used in the second half of the last century to increase the level of land productivity, namely the so-called 'promotional view'.[115] That is to say, to give preference to rules that guide owners towards sustainable use of goods by offering support services for those who join the politically recognised best practice, as already happens to some extent in the waste disposal law (Art. 179 ff. Leg. Dec. 152/06).

The issue would deserve further consideration, but as a contribution for further discussion, which is newly open, it seems, in short, that the social function has the capacity to justify norms that *de iure condito* – or even *de iure condendo*, perhaps by changing the rules originally conceived in the rural productivity view – steer the exercise of the owner's powers – not only of material goods but also of immaterial ones[116] – in the transition from the linear to the sustainable development model. And that is not because the social function does not have its own meaning, and hence the legislator can choose it from time to time,[117] but because the Italian Constitution, whose provisions give meaning to the social function,[118] not only does not support the idea of linear development without limits[119] but begins to explicitly welcome sustainability, as it is shown by the most recent constitutional reform of both Arts. 9 and 41.[120]

On the one hand, the perspective would then be that of what is expected of the owner as such, with the result that it is no longer be necessary to argue for the idea of rights for people who do not yet exist, such as future generations,[121] whose reference in the Italian Constitution seems to be rather rhetorical.[122] On the other hand, the goal could not be achieved without the intervention of the legislator – normally applicable by analogy and specified in administrative acts[123] – which is the only entity to whom Art. 42(2) gives the task of conforming the owners' powers.[124] It would be perhaps a modern way to fulfil the task that – as we have referred at the beginning of this writing – has typified law since the Greek philosophy, that is, restoring and maintaining the balance of the world we live in.

Notes

1 Università Cattolica del Sacro Cuore (Sede di Piacenza-Cremona), Via Emiliana Parmense, 84 – 29122 Piacenza PC, Italy.

2 Werner Jaeger, 'Praise of Law: The Origin of Legal Philosophy and the Greeks' in Paul Sayre (ed), *Interpretations of Modern Legal Philosophies: Essays in Honor of Roscoe Pound* (New York, Oxford University Press 1947) 352, 355.

3 Werner Jaeger, 'Praise of Law: The Origin of Legal Philosophy and the Greeks' in Paul Sayre (ed), *Interpretations of Modern Legal Philosophies: Essays in Honor of Roscoe Pound* (New York, Oxford University Press 1947) 352, 356.

4 Werner Jaeger, *Paideia. The Ideals of Greek Culture* (vol. I, New York, Oxford University Press 1939) 139.

5 See Jacques Jouanna, *Ippocrate* (Torino, Società Editrice Internazionale 1994), and Hans Georg Gadamer, *The Enigma of Health. The Art of Healing in a Scientific Age* (Redwood City, Stanford University Press 1996) 31 ff.

6 Werner Jaeger, *Paideia. The Ideals of Greek Culture* (vol. II, New York, Oxford University Press 1939) 129–131.

7 Werner Jaeger, 'Praise of Law: The Origin of Legal Philosophy and the Greeks' in Paul Sayre (ed), *Interpretations of Modern Legal Philosophies: Essays in Honor of Roscoe Pound* (New York, Oxford University Press 1947) 352, 360 ff, for whom it was Sophocles that more than other poets praised '(…) the triumph of the human civilization as it has risen victoriously from the darkness of the primitive ages and enforced on nature the rule of creative reason. (…) He develops language and thought and finally emerges as a city builder and founder of states' (p. 362). For the Plato's concept of law, see Werner Jaeger, *Paideia. The Ideals of Greek Culture* (vol. III, New York, Oxford University Press 1939) 245 ff.

8 Aldo Schiavone, *Ius. L'invenzione del diritto in occidente* (Torino, Einaudi 2017) 5 ff.

9 Émile Benveniste, *Il vocabolario delle istituzioni indoeuropee* (vol. II, Torino, Einaudi 2001) 376–383.

10 Werner Jaeger, 'Praise of Law: The Origin of Legal Philosophy and the Greeks' in Paul Sayre (ed), *Interpretations of Modern Legal Philosophies: Essays in Honor of Roscoe Pound* (New York, Oxford University Press 1947) 352, 364.

11 Werner Jaeger, 'Praise of Law: The Origin of Legal Philosophy and the Greeks' in Paul Sayre (ed), *Interpretations of Modern Legal Philosophies: Essays in Honor of Roscoe Pound* (New York, Oxford University Press 1947) 352, 365.

12 Cesare Beccaria, 'Dei delitti e delle pene' in Luigi Firpo and Gianni Franconi (eds), *Edizione nazionale delle opere di Cesare Beccaria* (vol. I, Milano, Mediobanca 1984) 75, although endnote 3 shows a change in Beccaria's opinion on ownership, because in the first draft and the *editio princeps* of the *Delitti* the sentence reads '(…) terrible but perhaps necessary right'. However, this variation could be ascribed to the Beccaria's thought: the rejection of the idea of private property was so explicit in him as to trigger the irony of a letter from Pietro to Alessandro Verri, at the time of Beccaria's planned trip to Russia: 'Put him in front of a Moscow minister, with a task for which he does not have the resources, and you will see what a poor figure he will be. He will begin the Code by removing the right of ownership: he will make you a novel' [letter dated 16.09.1767, in Emanuele Greppi and Alessandro Giulini, *Carteggio di Pietro e Alessandro Verri* (vol. I, Milano, L.F. Cogliati 1923) 63].

13 See Paolo Grossi, *L'inaugurazione della proprietà moderna* (Napoli, Guida editori 1980) 17–18, 'To the constant issue that arises from the necessary interaction between individual, community and ownership, every historically written legal system has given different answers depending on the geo-political conditions and economic structure, with the distinctions that the link with a certain ideology always tends to mark and exaggerate. Sometimes by absolutizing and intensifying, sometimes by attenuating

the proper content of the ownership; sometimes by identifying it in a unique and exclusive relation - politically very significant - in the strictest discrimination between privileged owner and socially unprotected non-owner; sometimes by relativising it in the head of a plurality of subjects. The history of ownership is thus correctly configured as the history of these answers, each of them has its individual autonomy and typicality within the specific boundaries of the cultural setting that influences but also justifies it. The historical line runs formally as a unitary one, but it is substantially marked by moments endowed with relevant specificity', and Paolo Grossi, 'Proprietà (dir. interm.)' in *Enc. Dir.* (XXXVII, Milano, Guiffré 1988) 226, 230 'Ownership will never be a technical rule but it will be always an answer to neverending question of the connection between human being and things, of the tension between the world of subjects and the world of phenomena, and the one who sets out to trace its history must always attempt to place it within a mentality and a land system with an eminently interpretative purpose'.

14 Luigi Mengoni, 'Problema e sistema nella controversia sul metodo giuridico' [1976] 3 Jus 40, who applies the so-called hermeneutic circle in relation to law, as well.

15 Luigi Mengoni, 'Diritto e politica nella dottrina giuridica' in Carlo Castronovo, Antonio Albanese and Andrea Nicolussi (eds), *Scritti I. Metodo e teoria giuridica* (Milano, Giuffré 2011) 149, 152.

16 See, *ex multis*, Bram Akkermans, 'Sustainable Property Law - Towards a Revaluation of Our System of Property Law' in Bram Akkermans and Gijs van Dijck (eds), *Sustainabilty and Private Law* (The Hague, Eleven International Publishing 2020) 37, 38 ff.

17 Compare Vicenzo Visocchi, 'La proprietà privata nel mondo agricolo' [1951] 6–7 Rivista di economia agraria 352, 356 ff.

18 Stefano Rodotà, *Il terribile diritto. Studi sulla proprietà privata* (Bologna, Il Mulino 1990) 19 ff.

19 Paolo Grossi, 'Proprietà (dir. interm.)' in *Enc. Dir.* (XXXVII, Milano, Guiffré 1988) 226, 232.

20 Paolo Grossi, 'Proprietà (dir. interm.)' in *Enc. Dir.* (XXXVII, Milano, Guiffré 1988) 226, 232, 226.

21 Paolo Grossi, 'Tradizioni e modelli nella sistemazione post-unitaria della proprietà' [1977] 5–6 Quad. fior. 201, 209 ff.

22 Paul B Matthews, 'La proprietà: storia e comparazione' in Carlo Castronovo and Salvatore Mazzamuto (eds), *Manuale di diritto privato europeo* (T. II, Milano, Giuffré 2007) 21, 31.

23 Paul B Matthews, 'La proprietà: storia e comparazione' in Carlo Castronovo and Salvatore Mazzamuto (eds), *Manuale di diritto privato europeo* (T. II, Milano, Giuffré 2007) 21, 28–29.

24 Vicenzo Visocchi, 'La proprietà privata nel mondo agricolo' [1951] 6–7 Rivista di economia agraria 352, 363–364.

25 Cesare Salvi, *L'invenzione della proprietà. La destinazione universale dei beni e i suoi nemici* (Venice, Marsilio Editori 2021) 88.

26 Paolo Grossi, 'Proprietà (dir. interm.)' in *Enc. Dir.* (XXXVII, Milano, Guiffré 1988) 226, 238.

27 Paolo Grossi, 'Proprietà (dir. interm.)' in *Enc. Dir.* (XXXVII, Milano, Guiffré 1988) 226, 239.

28 Paolo Grossi, 'Proprietà (dir. interm.)' in *Enc. Dir.* (XXXVII, Milano, Guiffré 1988) 226, 242 f.

29 Paolo Grossi, 'Proprietà (dir. interm.)' in *Enc. Dir.* (XXXVII, Milano, Guiffré 1988) 226, 244.

30 Paolo Grossi, 'Proprietà (dir. interm.)' in *Enc. Dir.* (XXXVII, Milano, Guiffré 1988) 226, 246.

31 Paolo Grossi, 'Proprietà (dir. interm.)' in *Enc. Dir.* (XXXVII, Milano, Guiffré 1988) 226, 238.

32 Vicenzo Visocchi, 'La proprietà privata nel mondo agricolo' [1951] 6–7 Rivista di economia agraria 352, 364–365.

33 Vicenzo Visocchi, 'La proprietà privata nel mondo agricolo' [1951] 6–7 Rivista di economia agraria 352, 365.

34 Like Germany, Prussia, and Austria, but surprisingly not in France, where the peasant was often landholder, there is always another person who could 'interfere with his enjoyment, impede his work, consume his produce; and when he has done with these, other, dressed in black, make their appearance, and sweep off the clearest part of his harvest' [Alexis de Tocqueville, *The Old Regime and the French Revolution* (Mineola, Dover Publications 2012) 49]. See also Marc Bloch, *I caratteri originali della storia rurale francese* (Torino, Einaudi 1973) 148, and Adhémar Esmein, *Cours èlèmentaire d'histoire du droit français a l'usage des étudiants de première année* (Paris, Librairie de la Société du Recueil Général des Lois et des Arrêts et du Journal du Palais 1898) 687 ff.

35 Paolo Grossi, 'Proprietà (dir. interm.)' in *Enc. Dir.* (XXXVII, Milano, Guiffré 1988) 226, 243.

36 In relation to Italy, especially to the Po valley, Salvatore Pugliese, *Condizioni Economiche e Finanziarie della Lombardia nella prima metà del secolo XVIII* (Tornio, Fratelli Bocca Editore 1924) 46–47, says that 'Enriched merchants, curials, bankers, tax collectors, stewards of noble houses and other people of similar professions began (…) to turn their slowly accumulated capital to the farming industry, taking the

risk of renting landed estates belonging to the nobility or the church, which they then had cultivated by wage-earners, or more generally sublet or sharecropped in small plots. These speculators, mostly greedy and merciless, oppress their dependent peasants in a thousand ways, in order to squeeze the maximum profit out of them, and make the name of tenant farmers odious and sow discontent in the countryside'. See also Pier Silverio Leicht, 'I tipi di azienda agraria in Italia nell'alto medioevo' in *Atti del Congresso Nazionale di Diritto Agrari* (Firenze, Ricci 1935) 47, who shows that already in the Early Middle Ages the king's or bishop's vassal '(...) rarely pursues aims of economic improvement, but exploits the agricultural peoples of the lands submitted to him as rudest as possibile. Therefore, feudal baronies represents a regression, from an agricultural point of view'.

37 Natalino Irti, *Dal diritto civile al diritto agrario (Momenti di storia giuridica francese)* (Milano, Giuffré 1962) 6 ff.

38 Paolo Grossi, 'Proprietà (dir. interm.)' in *Enc. Dir.* (XXXVII, Milano, Guiffré 1988) 226, 246, 247.

39 Compare Vicenzo Visocchi, 'La proprietà privata nel mondo agricolo' [1951] 6–7 Rivista di economia agraria 352, 364 ff.

40 Alexis de Tocqueville, *The Old Regime and the French Revolution* (Mineola, Dover Publications 2012) 42, and Natalino Irti, *Dal diritto civile al diritto agrario (Momenti di storia giuridica francese)* (Milano, Giuffré 1962) 6.

41 Paolo Grossi, 'Proprietà (dir. interm.)' in *Enc. Dir.* (XXXVII, Milano, Guiffré 1988) 226, 246 ff.

42 Cesare Salvi, *Le immissioni industriali: rapporti di vicinato e tutela dell'ambiente* (Milano, Giuffré 1979) 55 ff, and Tomaso Scozzafava Oberdan, *I beni e le forme giuridiche di appartenenza* (Milano, Giuffré 1982) 197 ff. See also John H Merryman, 'Ownership and Estate (Variations on a Theme by Lawson)' [1974] 48 Tul. Law Rev. 916, 924 ff.

43 Art. 2 – 'Le but de toute association politique est la conservation des droits naturels et imprescriptibles de l'homme. Ces droits sont la liberté, la propriété, la sûreté et la résistance à l'oppression', and art. 17 – 'La propriété étant un droit inviolable et sacré, nul ne peut en être privé, si ce n'est lorsque la nécessité publique, légalement constatée, l'exige évidemment, et sous la condition d'une juste et préalable indemnité'.

44 Marco Comporti, 'Ideologia e norma nel diritto di proprietà' [1974] I Riv. dir. civ. 285, 297.

45 'La propriété est le droit de jouir et disposer des choses de la manière la plus absolue, pourvu qu'on n'en fasse pas un usage prohibé par les lois ou par les règlement'. Stefano Rodotà, *Il terribile diritto. Studi sulla proprietà privata* (Bologna, Il Mulino 1990) 76 ff analyses in depth the steps that led from the DRMC to art. 544 of the *CN* and highlights some differences precisely with regard to the political background of the two dispositions, but ends by stating that '(...) the structure of ownership, as highlighted by art. 544, decisively contributed to imposing an individualistic interpretation of it' (p. 106). Regarding the Lockian origin of the provisions, see Paolo Grossi, *La proprietà e le proprietà nell'officina dello storico* (Napoli, ESI 2006) 23 ff, and Cesare Salvi, *L'invenzione della proprietà. La destinazione universale dei beni e i suoi nemici* (Venice, Marsilio Editori 2021) 94 ff.

46 On the derivation of art. 436 of the first ICC from the *CN* compare Paolo Grossi, 'Tradizioni e modelli nella sistemazione post-unitaria della proprietà' [1977] 5–6 Quad. fior. 201, 227 ff; Marco Comporti, 'Ideologia e norma nel diritto di proprietà' [1974] I Riv. dir. civ. 285, 330 ff, and Anna Moscarini, *Proprietà privata e tradizioni costituzionali comuni* (Milano, Giuffré 2006) 61 ff.

47 Stefano Rodotà, *Il terribile diritto. Studi sulla proprietà privata* (Bologna, Il Mulino 1990) 91 ff.

48 Natalino Irti, *Dal diritto civile al diritto agrario (Momenti di storia giuridica francese)* (Milano, Giuffré 1962) 16.

49 Stefano Rodotà, *Il terribile diritto. Studi sulla proprietà privata* (Bologna, Il Mulino 1990) 75 ff. See also Francesco S Bianchi, *Corso di codice civile italiano* (IX, 1, Torino, Unione tipografica 1890) 592, and Jacopo Mattei, *Il codice civile italiano* (II, Venezia, Naratovich 1873) 43.

50 After all, both into the *CN* and into the first ICC contract, like many others juridical issue, did not have specific rules, but its regulation was included in the property modes of acquisition. Compare Adolfo di Majo, *Obbligazioni in generale* (Bologna, Cedam 1985) 23 ff. Ownership was the point of view of the whole legal system: Pietro Rescigno, *Enc. Dir.* (XXXVII, Milano, Guiffré 1988) 254, 256.

51 Jean Baptiste Treilhard, 'Discours sur le même objet' in *Code civil français. Discours et exposé des motifs* (Brussels, G. Huyghe 1804) 557.

52 Stefano Rodotà, *Il terribile diritto. Studi sulla proprietà privata* (Bologna, Il Mulino 1990) 105, and Paul B Matthews, 'La proprietà: storia e comparazione' in Carlo Castronovo and Salvatore Mazzamuto (eds), *Manuale di diritto privato europeo* (T. II, Milano, Giuffré 2007) 21, 24 ff.

53 Marco Comporti, 'Ideologia e norma nel diritto di proprietà' [1974] I Riv. dir. civ. 285, 292.

54 Marco Comporti, 'Ideologia e norma nel diritto di proprietà' [1974] I Riv. dir. civ. 285, footnote 11.
55 Summarised in these terms the revolutionary attempt Rosalba Alessi, *Autonomia privata e rapporti agrari* (Napoli, Jovene 1982) 22.
56 Henry JS Maine, *Ancient Law Its Connection with the Early History of Society and Its Relation to Modern Ideas* (London, John Murray 1908) 151 (italics in the text).
57 See Sjef van Erp and Bram Akkermans, *Cases, Materials and Text on Property Law* (London, Bloomsbury Publishing 2012) 59 ff.
58 Giovanni Galloni, 'Agricoltura (diritto dell'). Quali prospettive per gli anni '80' in Antonio Carrozza (ed), *Diritto agrario* (Milano, Giuffré 1993) 1–12.
59 Natalino Irti, *Dal diritto civile al diritto agrario (Momenti di storia giuridica francese)* (Milano, Giuffré 1962) 102.
60 Compare Marc Bloch, *I caratteri originali della storia rurale francese* (Torino, Einaudi 1973) 228 ff.
61 Marc Bloch, *I caratteri originali della storia rurale francese* (Torino, Einaudi 1973) 271 ff, and Gian Gastone Bolla, 'Della proprietà fondiaria agraria come situazione soggettiva e come istituzione tipica' [1952] I Riv. dir. agr. 512, 512 ff.
62 Rosalba Alessi, *Autonomia privata e rapporti agrari* (Napoli, Jovene 1982) 33 ff.
63 Gian Gastone Bolla, 'Della proprietà fondiaria agraria come situazione soggettiva e come istituzione tipica' [1952] I Riv. dir. agr. 512.
64 Michele Graziadei, 'The Structure of Property Ownership and the Common Law/Civil Law Divide' in Lionel Smith and Michele Graziadei (eds), *Comparative Property Law. Global Perspectives* (Cheltenham, Edward Elgar Publishing Limited 2017) 90 ff.
65 Cesare Salvi, *Le immissioni industriali: rapporti di vicinato e tutela dell'ambiente* (Milano, Giuffré 1979) 20 ff.
66 Paolo Grossi, 'Tradizioni e modelli nella sistemazione post-unitaria della proprietà' [1977] 5–6 Quad. fior. 201, 214 ff.
67 Paolo Grossi, 'Tradizioni e modelli nella sistemazione post-unitaria della proprietà' [1977] 5–6 Quad. fior. 201, 233 ff.
68 See *ex pluribus*, Giovanni Lomonaco, *Istituzioni di diritto civile italiano* (II, Napoli, Jovene 1884) 31 ff; Francesco De Filippis, *Corso completo di diritto civile comparato* (II, Milano, Vallardi 1887) 135 ff, and F. Degni, 'I limiti del diritto di proprietà' (1901) *Il Filangieri*, 721 ff.
69 Cesare Salvi, *Il contenuto del diritto di proprietà* (Milano, Giuffrè Francis Lefebvre 2019) 37 ff. See John H Merryman, 'Ownership and Estate (Variations on a Theme by Lawson)' [1974] 48 Tul. Law Rev. 916, whose opinion is that 'Italian property law is not exactly like that of any other civil law nation, but it probably comes as close to a 'typical' civil law property system as any other and closer than most. Much of what is said here is accordingly applicable to the land law in other West European nations and throughout Latin America' (918).
70 See Antonio Gambaro, *Il diritto di proprietà* (Milano, Giuffrè 1995) 135–136, and from an European perspective Christian von Bar, *Gemeineuropäisches Sachenrecht* (vol. I, München, Beck 2015) 31 ff.
71 Michele Giorgianni, 'Obbligazione (diritto privato)' in *Noviss. dig. it.* (XI, Torino, Utet 1965) 581, 582 ff.
72 About the transition from an individualistic approach to one that also takes the social point of view into account compare Salvatore Pugliatti, *La proprietà nel nuovo diritto* (Milano, Giuffrè 1954) 123 ff; Marco Comporti, 'Ideologia e norma nel diritto di proprietà' [1974] I Riv. dir. civ. 285, 312 ff, and Antonio Gambaro, *Il diritto di proprietà* (Milano, Giuffrè 1995) 136 ff. Compare also the Ministro Guardasigilli, *Relazione del Ministro Guardasigilli al Codice Civile* (Roma, Istituto poligrafico e zecca dello Stato: Libreria dello Stato 2010) 83: 'Today, the configuration of ownership shifts to another level: it is no longer a primigenial or natural right of a person, to which specific limitations are imposed in correlation with certain social needs or the need to coordinate with other rights. The extent of this right is organically determined by the reason why the power is recognised and thus changes in relation to the different goods susceptible to appropriation'.
73 Salvatore Pugliatti, *La proprietà nel nuovo diritto* (Milano, Giuffrè 1954) 140 ff, and Pietro Rescigno, 'Per uno studio sulla proprietà' [1972] I Riv. dir. civ. 1, 33 ff.
74 Luigi Mengoni, 'Proprietà e libertà' in Carlo Castronovo, Antonio Albanese and Andrea Nicolussi (eds), *Scritti I. Metodo e teoria giuridica* (Milano, Giuffré 2011) 73, 78.
75 Salvatore Pugliatti, *La proprietà nel nuovo diritto* (Milano, Giuffrè 1954) 270 ff; Pietro Rescigno, 'Per uno studio sulla proprietà' [1972] I Riv. dir. civ. 1, 45 ff, and Luigi Mengoni, 'Proprietà e libertà' in

Carlo Castronovo, Antonio Albanese and Andrea Nicolussi (eds), *Scritti I. Metodo e teoria giuridica* (Milano, Giuffré 2011) 73, 84 ff. See also the Ministro Guardasigilli, *Relazione del Ministro Guardasigilli al Codice Civile* (Roma, Istituto poligrafico e zecca dello Stato: Libreria dello Stato 2010) 83, '(...) limitations on property have always been set up as something simply outside the right, such that they do not affect the unlimited content of the right itself'. A different opinion is expressed by Francesco Santoro-Passarelli, 'Proprietà privata e Costituzione' [1972] XXVI Riv. trim. dir. e proc. civ. 953, 960 ff.

76 Filippo Vassalli, 'Per una definizione legislativa del diritto di proprietà, Relazione alla prima riunione del Comitato per le relazioni giuridiche italo-tedesche (Roma giugno 1938)' in *Studi giuridici* (II, Milano, Giuffrè 1960) 329 ff. Compare also Salvatore Pugliatti, *La proprietà nel nuovo diritto* (Milano, Giuffrè 1954) 145 ff; P Pietro Rescigno, 'Per uno studio sulla proprietà' [1972] I Riv. dir. civ. 1, 20 ff; Marco Comporti, 'Ideologia e norma nel diritto di proprietà' [1974] I Riv. dir. civ. 285, 310 ff; Luigi Mengoni, 'Proprietà e libertà' in Carlo Castronovo, Antonio Albanese and Andrea Nicolussi (eds), *Scritti I. Metodo e teoria giuridica* (Milano, Giuffré 2011) 73, 79; Paolo Grossi, *La proprietà e le proprietà nell'officina dello storico* (Napoli, ESI 2006) 23; Antonio Gambaro, *Il diritto di proprietà* (Milano, Giuffrè 1995) 295.

77 Salvatore Pugliatti, *La proprietà nel nuovo diritto* (Milano, Giuffrè 1954), *passim*, e Pietro Rescigno, 'Per uno studio sulla proprietà' [1972] I Riv. dir. civ. 1, 20 ff.

78 Corte costituzionale (9 May 1968) n. 55, and Corte costituzionale (11 February 1971) n. 22.

79 Carlo Castronovo, *Eclissi del diritto civile* (Milano, Giuffré 2015) 64 ff.

80 See Francesco Santoro-Passarelli, 'Proprietà privata e Costituzione' [1972] XXVI Riv. trim. dir. e proc. civ. 953, 967 ff; Pietro Rescigno, 'Per uno studio sulla proprietà' [1972] I Riv. dir. civ. 1, 1 ff; Marco Comporti, 'Ideologia e norma nel diritto di proprietà' [1974] I Riv. dir. civ. 285, 321 ff; Luigi Mengoni, 'Proprietà e libertà' in Carlo Castronovo, Antonio Albanese and Andrea Nicolussi (eds), *Scritti I. Metodo e teoria giuridica* (Milano, Giuffré 2011) 73, 73 ff; Antonio Baldassarre, 'Diritti inviolabili' [1989] XI Enc. giur. Treccani 1, 26 ff; Stefano Rodotà, *Il terribile diritto. Studi sulla proprietà privata* (Bologna, Il Mulino 1990); Cesare Salvi, 'La proprietà privata e l'Europa. Diritto di libertà o funzione sociale' (2009) Riv. crit. dir. priv. 409, 409 ff; Francesco D Busnelli, 'La faticosa evoluzione dei principi europei tra scienza e giurisprudenza' [2009] II Riv. dir. civ. 292, 292 ff; Antonio Gambaro, *Il diritto di proprietà* (Milano, Giuffrè 1995) 56 ff, has reservations. *Contra* more recently Francesco Galgano, 'Danno non patrimoniale e diritti dell'uomo' [2009] 25(4–5) Contr. e impr. 883, 891 ff, and Rita Rolli, 'La proprietà come diritto dell'uomo?' [2011] 27(4–5) Contr. e impr. 1014, 1014 ff.

81 Luigi Mengoni, 'Proprietà e libertà' in Carlo Castronovo, Antonio Albanese and Andrea Nicolussi (eds), *Scritti I. Metodo e teoria giuridica* (Milano, Giuffré 2011) 73, 84 ff, Francesco Santoro-Passarelli, 'Proprietà privata e Costituzione' [1972] XXVI Riv. trim. dir. e proc. civ. 953, 958, says that a statute which formally or substantially abolished ownership would be uncostitutional. Regarding the difference between human rights and fundamental ones, see also Massimo S Giannini, 'Basi costituzionali della proprietà privata' [1971] 4–5 Pol. dir. 443, 466, fn. 32.

82 Corte costituzionale (20 January 1966) n. 6, and Corte costituzionale (29 May 1968) n. 55.

83 Luigi Mengoni, 'Proprietà e libertà' in Carlo Castronovo, Antonio Albanese and Andrea Nicolussi (eds), *Scritti I. Metodo e teoria giuridica* (Milano, Giuffré 2011) 73, 89, '(...) in the view of functional theory, the relation between freedom of ownership and social function is not presented as an opposition, which can lead to a compression of freedom to the point of destroying it', and p. 92, where he speaks about just compensation for such a case, too. Compare Antonio Gambaro, *La proprietà. Beni, proprietà, possesso* (Milano, Giuffré 2017) 179 ff, and Cesare Salvi, *Il contenuto del diritto di proprietà* (Milano, Giuffrè Francis Lefebvre 2019) 89 ff.

84 Marco Comporti, 'Ideologia e norma nel diritto di proprietà' [1974] I Riv. dir. civ. 285, 325 ff, and Stefano Rodotà, *Il terribile diritto. Studi sulla proprietà privata* (Bologna, Il Mulino 1990) 399 ff.

85 Cesare Salvi, *Il contenuto del diritto di proprietà* (Milano, Giuffrè Francis Lefebvre 2019) 86. Rodolfo Sacco, *La proprietà* (Torino, Giappichelli 1968) 79, speaks about social function of the ownership's boundaries than of ownership itself.

86 L Luigi Mengoni, 'Proprietà e libertà' in Carlo Castronovo, Antonio Albanese and Andrea Nicolussi (eds), *Scritti I. Metodo e teoria giuridica* (Milano, Giuffré 2011) 73, 90, and Carlo Castronovo, *Eclissi del diritto civile* (Milano, Giuffré 2015) 26 ff.

87 Michele Costantino, *Contributo alla teoria della proprietà* (Napoli, Jovene 1967) 97 ff.

88 Stefano Rodotà, *Il terribile diritto. Studi sulla proprietà privata* (Bologna, Il Mulino 1990) 420 ff.

89 See Enrico Caterini, *Sostenibilità e ordinamento civile. Per una riproposizione della questione sociale* (Napoli, ESI 2018), and Ugo Mattei and Alessandra Quarta, *Punto di svolta. Ecologia, tecnologia e diritto privato. Dal capitale ai beni comuni* (San Sepolcro, Aboco 2018).

90 Corte di Cassazzione (17 December 2012) n. 23189, and Antonio Gambaro, *La proprietà. Beni, proprietà, possesso* (Milano, Giuffré 2017) 263.

91 Antonio Gambaro, *La proprietà. Beni, proprietà, possesso* (Milano, Giuffré 2017) 304 ff.

92 Feliciano Benvenuti, 'I limiti dello *ius aedificandi* e la natura giuridica della licenza edilizia' in *Atti del congresso internazionale sulla pianificazione regionale e provinciale* (Trento, Camera di commercio di Trento 1956) 15, 35 ff. Compare in the key of comparative property law Christian von Bar, *Gemeineuropäisches Sachenrecht* (vol. I, München, Beck 2015) 577 ff.

93 See Lorenzo Casini, *L'equilibrio degli interessi nel governo del territorio* (Milano, Giuffré 2005) 244 ff, and Giulia Dimitrio, 'Le regole di destinazione e di uso edilizio dei suoli agricoli' in Franco G Scoca, Paolo Stella Richter and Paolo Urbani (eds), *Trattato di diritto del territorio* (I, Torino, Giappichelli 2018) 220, 263.

94 Luigi Capogrossi Colognesi, 'Proprietà (dir. rom)' in *Enc. Dir.* (XXXVII, Milano, Guiffré 1988) 160, 196.

95 Antonio Gambaro, *La proprietà. Beni, proprietà, possesso* (Milano, Giuffré 2017) 378.

96 Marco Di Lullo, 'La nozione e la disciplina (pubblicistica) dei "rifiuti": beni da valorizzare?' [2020] 3 Il diritto dell'economia 555, 572 ff.

97 Luigi Mengoni, 'Proprietà e libertà' in Carlo Castronovo, Antonio Albanese and Andrea Nicolussi (eds), *Scritti I. Metodo e teoria giuridica* (Milano, Giuffré 2011) 73, 91 ff.

98 Luca Nivarra, 'La funzione sociale della proprietà: dalla strategia alla tattica' [2013] 31(4) Rivista critica del diritto privato 503, 510 ff.

99 Antonio Gambaro, *La proprietà. Beni, proprietà, possesso* (Milano, Giuffré 2017) 178 ff.

100 See Vicenzo Visocchi, 'La proprietà privata nel mondo agricolo' [1951] 6–7 Rivista di economia agraria 352, 363, and Fernand de Visscher, 'Individualismo ed evoluzione della proprietà nella Roma repubblicana' [1957] 23 Studia et documenta historiae et iuris 26, 26 ff.

101 Stefano Rodotà, *Il terribile diritto. Studi sulla proprietà privata* (Bologna, Il Mulino 1990) 79, and footnote 8.

102 Biondo Biondi, 'Influssi cristiani sulla legislazione agraria romana' in *Dopo il primo convegno internazionale di diritto agrario. Valutazioni e prospettive in un incontro di giuristi italiani* (Milano, Giuffré 1958) 106 ff.

103 Vittorio Scialoja, *Teoria della proprietà nel diritto romano* (II, Roma, Sampaolesi 1931) 46.

104 Gian P Bognetti, 'La proprietà della terra nel passaggio dal mondo antico al medio evo occidentale' in *Dopo il primo convegno internazionale di diritto agrario. Valutazioni e prospettive in un incontro di giuristi italiani* (Milano, Giuffré 1958) 137.

105 Gian P Bognetti, 'La proprietà della terra nel passaggio dal mondo antico al medio evo occidentale' in *Dopo il primo convegno internazionale di diritto agrario. Valutazioni e prospettive in un incontro di giuristi italiani* (Milano, Giuffré 1958) 141.

106 Natalino Irti, *Dal diritto civile al diritto agrario (Momenti di storia giuridica francese)* (Milano, Giuffré 1962) 5.

107 Guido Astuti, 'Aspetti del regime giuridico medioevale della proprietà fondiaria e dei contratti agrari' in *Dopo il primo convegno internazionale di diritto agrario. Valutazioni e prospettive in un incontro di giuristi italiani* (Milano, Giuffré 1958) 79.

108 Compare Salvatore Pugliatti, *La proprietà nel nuovo diritto* (Milano, Giuffrè 1954) 176: '(…) today the ownership's concept is essentially represented by land ownership, which is basically ownership of productive assets (and not merely consumer goods), thus land ownership is typified as functional property'. As pointed out by Luigi Mengoni ('Strumenti giuridici e progresso dell'agricoltura' in *Agricoltura sviluppo agricolo e reddito del lavoro agricolo. Atti del sesto convegno di studi di economia e politica del lavoro. Roma - Palazzo Barberini - 20-21 maggio 1965* (Roma, Arti Grafiche Wanzer 1966) 140, 149), '(…) the social function of ownership does not tend so far as to force the owner to work the land directly, but they keep all his owner powers only when such a situation arises; separated from this activity, ownership becomes anaemic, it is bound to surrender to land entrepreneur tenant or to share with associated entrepreneur some of its essential features'.

109 Salvatore Rodotà, 'Art. 42, Rapporti economici, II' in Giuseppe Branca (ed), *Commentario della Costituzione* (Bologna, Zanichelli 1982) 177 ff.

110 See, for instance, Franco Carresi, 'La proprietà terriera privata', in Piero Calamandrei and Alessandro Levi (dir), *Commentario sistematico alla Costituzione italiana* (I, Firenze, Barbera 1950) 388 ff, but with some doubts Costantino Mortati, 'La costituzione e la proprietà terriera' in *Atti del terzo congresso nazionale di diritto agrario, Palermo - 19-23 ottobre 1952* (Milano, Giuffré 1954) 263, 267 ff.

111 Salvatore Rodotà, 'Art. 42, Rapporti economici, II' in Giuseppe Branca (ed), *Commentario della Costituzione* (Bologna, Zanichelli 1982) 241 ff, and more broadly Salvatore Pugliatti, *La proprietà nel nuovo diritto* (Milano, Giuffrè 1954) 181 ff.

112 Carlo Esposito, 'Note esegetiche sull'art. 44 della Costituzione' [1949] I Riv. dir. agr. 169, 169 ff, and Natalino Irti, 'Profili della programmazione agricola' [1972] I Riv. dir. agr. 394, 394 ff.

113 Antonio Gambaro, *La proprietà. Beni, proprietà, possesso* (Milano, Giuffré 2017) 180, and Mauro Pennasilico, 'Contratto ecologico e conformazione dell'autonomia negoziale' [2017] 4 Giust. civ. 809, 833 ff.

114 Luigi Mengoni, 'Proprietà e libertà' in Carlo Castronovo, Antonio Albanese and Andrea Nicolussi (eds), *Scritti I. Metodo e teoria giuridica* (Milano, Giuffré 2011) 73, 90.

115 Sul punto Norberto Bobbio, 'Sulla funzione promozionale del diritto' [1969] XXIII Riv. trim. dir. e proc. civ. 1312, 1312 ff.

116 On the process of extending property law to immaterial goods, as a result of technological development, too, compare Cesare Salvi, *Il contenuto del diritto di proprietà* (Milano, Giuffrè Francis Lefebvre 2019) 8 ff.

117 Antonio Gambaro, *La proprietà. Beni, proprietà, possesso* (Milano, Giuffré 2017) 194 ff.

118 Stefano Rodotà, *Il terribile diritto. Studi sulla proprietà privata* (Bologna, Il Mulino 1990) 418 ff, and Cesare Salvi, *Il contenuto del diritto di proprietà* (Milano, Giuffrè Francis Lefebvre 2019) 44 f.

119 Stefano Rodotà, *Il terribile diritto. Studi sulla proprietà privata* (Bologna, Il Mulino 1990) 419 ff, points out that the interpretation of the Italian Costitution in the meaning of production's maximisation was determined by the economicism that marked the culture of the 1950s and 1960s, to such an extent that it influenced legal studies, too.

120 Gabriella Marcatajo, 'La riforma degli articoli 9 e 41 della Costituzione e la valorizzazione dell'ambiente' [2022] 2 ambientediritto.it 118, 118 ff.

121 See Michael W Monterossi, *L'orizzonte intergenerazionale del diritto civile. Tutela, soggettività, azione* (Pisa, Edizioni ETS 2020) 259 ff, and more prudently Antonio D'Aloia, 'Generazioni future (diritto costituzionale)' [2016] Annali IX *Enc. Dir.* 331, 366 ff. Raffaele Bifulco, *Diritto e generazioni future. Problemi giuridici della responsabilità intergenerazionale* (Milano, Franco Angeli 2008), *passim*, seems, however, inclined towards the idea of present generations' duty rather than the future generations' rights.

122 Antonio D'Aloia, 'Generazioni future (diritto costituzionale)' [2016] Annali IX *Enc. Dir.* 331, 366.

123 Antonio Gambaro, *La proprietà. Beni, proprietà, possesso* (Milano, Giuffré 2017) 213 ff.

124 Aldo Mazzimi Sandulli, 'Profili costituzionali della proprietà privata' [1972] XXVI Riv. trim. dir. proc. civ. 465, 473 ff, and Luigi Mengoni, 'Proprietà e libertà' in Carlo Castronovo, Antonio Albanese and Andrea Nicolussi (eds), *Scritti I. Metodo e teoria giuridica* (Milano, Giuffré 2011) 73, 89. Along this line, regarding the problematic issue of the so called 'common goods' (Ugo Mattei and Alessandra Quarta, *Punto di svolta. Ecologia, tecnologia e diritto privato. Dal capitale ai beni comuni* (San Sepolcro, Aboco 2018)), compare Carlo Castronovo, *Eclissi del diritto civile* (Milano, Giuffré 2015) 52 ff and 178 ff.

Bibliography

Akkermans A, 'Sustainable Property Law – Towards a Revaluation of Our System of Property Law' in Bram Akkermans and Gijs van Dijck (eds), *Sustainabilty and Private Law* (The Hague, Eleven International Publishing 2020) 37

Alessi R, *Autonomia privata e rapporti agrari* (Napoli, Jovene 1982)

Astuti G, 'Aspetti del regime giuridico medioevale della proprietà fondiaria e dei contratti agrari' in *Dopo il primo convegno internazionale di diritto agrario. Valutazioni e prospettive in un incontro di giuristi italiani* (Milano, Giuffré 1958)

Baldassarre A, 'Diritti inviolabili' [1989] XI Enc. giur. Treccani 1

v. Bar C, Gemeineuropäisches Sachenrecht (vol. I, München, Beck 2015) Beccaria C, 'Dei delitti e delle pene' in Luigi Firpo and Gianni Franconi (eds), *Edizione nazionale delle opere di Cesare Beccaria* (vol. I, Milano, Mediobanca 1984) 75
Benveniste E, *Il vocabolario delle istituzioni indoeuropee* (vol. II, Torino, Einaudi 2001)
Benvenuti F, 'I limiti dello *ius aedificandi* e la natura giuridica della licenza edilizia' in *Atti del congresso internazionale sulla pianificazione regionale e provinciale* (Trento, Camera di commercio di Trento 1956) 15
Bianchi FS, *Corso di codice civile italiano* (IX, 1, Torino, Unione tipografica 1890)
Bifulco R, *Diritto e generazioni future. Problemi giuridici della responsabilità intergenerazionale* (Milano, Franco Angeli 2008)
Biondi B, 'Influssi cristiani sulla legislazione agraria romana' in *Dopo il primo convegno internazionale di diritto agrario. Valutazioni e prospettive in un incontro di giuristi italiani* (Milano, Giuffré 1958) 106 ff
Bloch M, *I caratteri originali della storia rurale francese* (Torino, Einaudi 1973)
Bobbio N, 'Sulla funzione promozionale del diritto' [1969] XXIII Riv. trim. dir. e proc. civ. 1312
Bognetti GP, 'La proprietà della terra nel passaggio dal mondo antico al medio evo occidentale' in *Dopo il primo convegno internazionale di diritto agrario. Valutazioni e prospettive in un incontro di giuristi italiani* (Milano, Giuffré 1958) 137 ff
Bolla GG, 'Della proprietà fondiaria agraria come situazione soggettiva e come istituzione tipica' [1952] I Riv. dir. agr. 512
Busnelli FD, 'La faticosa evoluzione dei principi europei tra scienza e giurisprudenza' [2009] II Riv. dir. civ. 292, 292 ff
Capogrossi Colognesi L, 'Proprietà (dir. rom)' in *Enc. Dir.* (XXXVII, Milano, Guiffré 1988) 160
Carresi F, 'La proprietà terriera privata', in Piero Calamandrei and Alessandro Levi (dir), *Commentario sistematico alla Costituzione italiana* (I, Firenze, Barbera 1950) 388 ff
Casini L, *L'equilibrio degli interessi nel governo del territorio* (Milano, Giuffré 2005)
Castronovo C, *Eclissi del diritto civile* (Milano, Giuffré 2015)
Caterini E, *Sostenibilità e ordinamento civile. Per una riproposizione della questione sociale* (Napoli, ESI 2018)
Comporti M, 'Ideologia e norma nel diritto di proprietà' [1974] I Riv. dir. civ. 285
Costantino M, *Contributo alla teoria della proprietà* (Napoli, Jovene 1967)
D'Aloia A, 'Generazioni future (diritto costituzionale)' [2016] Annali IX Enc. Dir. 331, 366 ff
De Filippis F, *Corso completo di diritto civile comparato* (II, Milano, Vallardi 1887)
de Tocqueville A, *The Old Regime and the French Revolution* (Mineola, Dover Publications 2012)
de Visscher F, 'Individualismo ed evoluzione della proprietà nella Roma repubblicana' [1957] 23 Studia et documenta historiae et iuris 26
Di Lullo M, 'La nozione e la disciplina (pubblicistica) dei "rifiuti": beni da valorizzare?' [2020] 3 Il diritto dell'economia 555, 572 ff
di Majo A, *Obbligazioni in generale* (Bologna, Cedam 1985)
Dimitrio G, 'Le regole di destinazione e di uso edilizio dei suoli agricoli' in Franco G Scoca, Paolo Stella Richter and Paolo Urbani (eds), *Trattato di diritto del territorio* (I, Torino, Giappichelli 2018) 220
v. Erp S and Akkermans B, Cases, Materials and Text on Property Law (London, Bloomsbury Publishing 2012) Esmein A, *Cours èlèmentaire d'histoire du droit français a l'usage des étudiants de première année* (Paris, Librairie de la Société du Recueil Général des Lois et des Arrêts et du Journal du Palais 1898)
Esposito C, 'Note esegetiche sull'art. 44 della Costituzione' [1949] I Riv. dir. agr. 169, 169 ff
Gadamer HG, *The Enigma of Health. The Art of Healing in a Scientific Age* (Redwood City, Stanford University Press 1996)
Galgano F, 'Danno non patrimoniale e diritti dell'uomo' [2009] 25(4–5) Contr. e impr. 883, 891 ff
Galloni G, 'Agricoltura (diritto dell'). Quali prospettive per gli anni '80' in Antonio Carrozza (eds), *Diritto agrario* (Milano, Giuffré 1993) 1–12
Gambaro A, *Il diritto di proprietà* (Milano, Giuffrè 1995)
Gambaro A, *La proprietà. Beni, proprietà, possesso* (Milano, Giuffré 2017)
Giannini MS, 'Basi costituzionali della proprietà privata' [1971] 4–5 Pol. dir. 443
Giorgianni M, 'Obbligazione (diritto privato)' in *Noviss. dig. it.* (XI, Torino, Utet 1965) 581
Graziadei M, 'The Structure of Property Ownership and the Common Law/Civil Law Divide' in Lionel Smith and Michele Graziadei (eds), *Comparative Property Law. Global Perspectives* (Cheltenham, Edward Elgar Publishing Limited 2017) 90 ff

Greppi E and Giulini E, *Carteggio di Pietro e Alessandro Verri* (vol. I, Milano, L.F. Cogliati 1923)
Grossi P, 'Tradizioni e modelli nella sistemazione post-unitaria della proprietà' [1977] 5–6 Quad. fior. 201
Grossi P, *L'inaugurazione della proprietà moderna* (Napoli, Guida editori 1980)
Grossi P, 'Proprietà (dir. interm.)' in *Enc. Dir.* (XXXVII, Milano, Guiffré 1988) 226
Grossi P, *La proprietà e le proprietà nell'officina dello storico* (Napoli, ESI 2006)
Guardasigilli, *Relazione del Ministro Guardasigilli al Codice Civile* (Roma, Istituto poligrafico e zecca dello Stato: Libreria dello Stato 2010)
Irti N, *Dal diritto civile al diritto agrario (Momenti di storia giuridica francese)* (Milano, Giuffré 1962)
Irti N, 'Profili della programmazione agricola' [1972] I Riv. dir. agr. 394
Jaeger W, *Paideia. The Ideals of Greek Culture* (vol. I, New York, Oxford University Press 1939)
Jaeger W, 'Praise of Law: The Origin of Legal Philosophy and the Greeks' in Paul Sayre (ed), *Interpretations of Modern Legal Philosophies: Essays in Honor of Roscoe Pound* (New York, Oxford University Press 1947) 352
Jouanna J, *Ippocrate* (Torino, Società Editrice Internazionale 1994)
Leicht PS, 'I tipi di azienda agraria in Italia nell'alto medioevo' in *Atti del Congresso Nazionale di Diritto Agrari* (Firenze, Ricci 1935) 47
Lomonaco G, *Istituzioni di diritto civile italiano* (II, Napoli, Jovene 1884)
Maine HJS, *Ancient Law Its Connection with the Early History of Society and Its Relation to Modern Ideas* (London, John Murray 1908)
Marcatajo G, 'La riforma degli articoli 9 e 41 della Costituzione e la valorizzazione dell'ambiente' [2022] 2 ambientediritto.it 118, 118 ff
Mattei J, *Il codice civile italiano* (II, Venezia, Naratovich 1873)
Mattei U and Quarta A, *Punto di svolta. Ecologia, tecnologia e diritto privato. Dal capitale ai beni comuni* (San Sepolcro, Aboco 2018)
Matthews PB, 'La proprietà: storia e comparazione' in Carlo Castronovo and Salvatore Mazzamuto (eds), *Manuale di diritto privato europeo* (T. II, Milano, Giuffré 2007) 21
Mengoni L, 'Strumenti giuridici e progresso dell'agricoltura' in *Agricoltura sviluppo agricolo e reddito del lavoro agricolo. Atti del sesto convegno di studi di economia e politica del lavoro. Roma - Palazzo Barberini - 20-21 maggio 1965* (Roma, Arti Grafiche Wanzer 1966) 140
Mengoni L, 'Problema e sistema nella controversia sul metodo giuridico' [1976] 3 Jus 40
Mengoni L, 'Proprietà e libertà' in Carlo Castronovo, Antonio Albanese and Andrea Nicolussi (eds), *Scritti I. Metodo e teoria giuridica* (Milano, Giuffré 2011) 73
Mengoni L, 'Diritto e politica nella dottrina giuridica' in Carlo Castronovo, Antonio Albanese and Andrea Nicolussi (eds), *Scritti I. Metodo e teoria giuridica* (Milano, Giuffré 2011) 149
Merryman JH, 'Ownership and Estate (Variations on a Theme by Lawson)' [1974] 48 Tul. Law Rev. 916
Monterossi MW, *L'orizzonte intergenerazionale del diritto civile. Tutela, soggettività, azione* (Pisa, Edizioni ETS 2020)
Mortati C, 'La costituzione e la proprietà terriera' in *Atti del terzo congresso nazionale di diritto agrario, Palermo – 19-23 ottobre 1952* (Milano, Giuffré 1954) 267 ff
Moscarini A, *Proprietà privata e tradizioni costituzionali comuni* (Milano, Giuffré 2006)
Nivarra L, 'La funzione sociale della proprietà: dalla strategia alla tattica' [2013] 31(4) Rivista critica del diritto privato 503, 510 ff
Oberdan TS, *I beni e le forme giuridiche di appartenenza* (Milano, Giuffré 1982)
Pennasilico M, 'Contratto ecologico e conformazione dell'autonomia negoziale' [2017] 4 Giust. civ. 809, 833 ff
Pugliatti S, *La proprietà nel nuovo diritto* (Milano, Giuffrè 1954)
Pugliese S, *Condizioni Economiche e Finanziarie della Lombardia nella prima metà del secolo XVIII* (Tornio, Fratelli Bocca Editore 1924)
Rescigno P, 'Per uno studio sulla proprietà' [1972] I Riv. dir. civ. 1
Rescigno P, *Enc. Dir.* (XXXVII, Milano, Guiffré 1988) 254
Rodotà S, *Il terribile diritto. Studi sulla proprietà privata* (Bologna, Il Mulino 1990)
Rolli R, 'La proprietà come diritto dell'uomo?' [2011] 27(4–5) Contr. e impr. 1014, 1014 ff
Sacco R, *La proprietà* (Torino, Giappichelli 1968)
Salvatore Rodotà, 'Art. 42, Rapporti economici, II' in Giuseppe Branca (ed), *Commentario della Costituzione* (Bologna, Zanichelli 1982) 177 ff

Salvi C, *Le immissioni industriali: rapporti di vicinato e tutela dell'ambiente* (Milano, Giuffré 1979)
Salvi C, 'La proprietà privata e l'Europa. Diritto di libertà o funzione sociale' [2009] 3 Riv. crit. dir. priv. 409
Salvi C, *Il contenuto del diritto di proprietà* (Milano, Giuffrè Francis Lefebvre 2019)
Salvi C, *L'invenzione della proprietà. La destinazione universale dei beni e i suoi nemici* (Venice, Marsilio Editori 2021)
Sandulli AM, 'Profili costituzionali della proprietà privata' [1972] XXVI Riv. trim. dir. proc. civ. 465
Santoro-Passarelli F, 'Proprietà privata e Costituzione' [1972] XXVI Riv. trim. dir. e proc. civ. 953
Schiavone A, *Ius. L'invenzione del diritto in occidente* (Torino, Einaudi 2017)
Scialoja V, *Teoria della proprietà nel diritto romano* (II, Roma, Sampaolesi 1931)
Treilhard JB, 'Discours sur le mème objet' in *Code civil français. Discours et exposé des motifs* (Brussels, G. Huyghe 1804) 557
Vassalli F, 'Per una definizione legislativa del diritto di proprietà, Relazione alla prima riunione del Comitato per le relazioni giuridiche italo-tedesche (Roma giugno 1938)' in *Studi giuridici* (II, Milano, Giuffrè 1960) 329 ff
Visocchi V, 'La proprietà privata nel mondo agricolo' [1951] 6–7 Rivista di economia agraria 352

18

USING TOKENISATION IN SUPPORT OF A 'SUPERFICIES SUSTAINABLE'

A Dutch Case Study on LegalTech to Support Sustainability Goals

Jasper Verstappen and Tim van Zuijlen[1]

18.1 Introduction

This contribution will focus on a set of unique challenges that are created in the context of the law on the registration of property rights.[2] More specifically, the focus will be on the right of superficies employed in the context of refurbishments to the immovable property to create a more sustainable immovable property. The question at the heart of this contribution is whether technology can be leveraged to support an effective response to the challenges faced by land registries as a result of sustainability refurbishments to immovable properties.

The right of superficies is a limited property right that separates the ownership of a building or construction from the ownership of land or a piece of immovable property.[3] It therefore enables its holder to own or acquire buildings, constructions, or vegetation in, on, or above an immovable property that is owned by someone else. Without this right, the mentioned attachments will become the property of the owner of the land or building through accession, in line with the general principle of *superficies solo cedit*. In cases of sustainability refurbishments, the right of superficies can be used to separate the ownership of, for example, the solar panels on the roof of a house from the ownership of the house itself. Literature has noted that the registry of these sustainability refurbishments raises issues of inefficiency and uncertainty.[4] A recent Dutch proposal addresses these issues by adding a 'superficies sustainable' variant to the right of superficies.[5] This proposal provides an excellent illustration of the challenges that refurbishments of immovable property to increase the sustainable character of that property create for land registries, and how to address these.

Because of this proposal, this contribution uses the Dutch legal system as an example of how to address these issues with sustainable refurbishments. Equivalent rights of superficies exist throughout Europe, including Germany and France, and this Dutch example might be of value to the discussion there.[6] A comparative analysis does not fit within the scope of this contribution, but this would be a valuable next step.

The purpose of this contribution is to illustrate how legal measures, specifically the 'superficies sustainable', can be combined with technological innovations to efficiently and effectively face the

DOI: 10.4324/9781032662046-22

challenges that the registration of sustainability refurbishments provides. Blockchain technology has oftentimes been proposed as an emerging technology that could greatly benefit land registries. This contribution will show that a very nuanced, very precise, and very limited application of the technology is sufficient and could serve as support in responding to those challenges by, for example, introducing a 'superficies sustainable'.

This contribution will take three steps. First, Section 18.2 will explain what 'superficies sustainable' is. This section will highlight the challenges created by sustainability refurbishments of immovable property and will describe how the proposal for a 'superficies sustainable' aims to resolve these challenges. Section 18.3 will provide an overview of the way blockchain technology functions in practice using five technical aspects. This approach allows us to explain the general functioning of the technology without engaging in a lengthy exploration of its technological underpinnings. Subsequently, the third section will sketch past proposals for implementing this technology as a land registration system. It will show that such ideas tended to be overambitious and that the practical functioning of land registration systems is too complex to simply supplant with a distributed ledger. Lastly, Section 18.4 will show that supplanting a land registry is not necessary. Instead, the technology can be used to create a rights management system to support a right like the 'superficies sustainable' and assist land registries in facing the challenges created through refurbishments to immovable properties that aim to make such properties more sustainable.

18.2 The Proposed 'Superficies Sustainable'

18.2.1 Goal and Purpose

In early 2022, a collaboration of the Netherlands' Cadastre, the Royal Dutch Association of Civil-law notaries, university researchers and notarial practitioners proposed an updated variation of the right of superficies.[7] This 'superficies sustainable' is designed to address issues that are unique to improvements made to immovable properties in order to increase the sustainability of that property, as well as the financing of such projects. Examples of such projects include the construction of solar panels on an existing property, the installation of a heat pump, or a ground-coupled heat exchanger.[8]

The drafters of the proposal observe that several variations of the right of superficies exist within the Netherlands. These include the standard right of superficies as well as a variation of superficies that is specific to utility services: the so-called 'superficies utilities'.[9] The advantage offered by 'superficies utilities' is found in Article 6(3) sub a no. 4 *Kadasterbesluit*. This article holds that registration of the 'superficies utilities' is possible without cadastral measurements being taken. Hence, this is a more cost-effective and more time-efficient alternative to registering a classic right of superficies for utilities. However, the drafters note that the 'superficies utilities' create an obstacle to employing it for sustainable improvements to properties: if a 'superficies utilities' is used for such sustainability improvements, the object of the right of superficies is most often not sufficiently identified in the contract of sale and in the legal acts relevant for the financing.[10] This is caused by the fact that, historically, the 'superficies utilities' was used for objects of minimal value, such as pipes, small pumps, and measuring devices.[11] The drafters have provided a proposal in which they strive to create a 'superficies sustainable' that combines the advantages of the 'superficies utilities' with the transparency and specificity of the standard right of superficies. With this, the drafters attempt to create a variation of the right of superficies in which a cadastral measurement is not necessary, whilst the specificity of the object is guaranteed.[12] This would enable a cost-effective and time-efficient solution, whilst at the same time ensuring legal certainty.

18.2.2 Design

The proposed 'superficies sustainable' is limited to 'installations and attachments to immovable property that generate, store, or convert electricity, heat or cold from sustainable sources, installations and attachments to immovable property that provides that property with electricity, heat or cold from a sustainable source, or improve the energy-efficiency of immovable property'.[13]

By registering these objects as a 'superficies sustainable' rather than a 'superficies utility', the recognisability of the object would be more pronounced, and the right would give a better indication of the value of the object as compared to the 'superficies utilities'. According to the drafters, this prevents uncertainty and the need for additional research into the value of the objects and their quality as security.[14] Whether or not a specific installation or attachment fulfils the criteria of the 'superficies sustainable' must be determined by the parties themselves as well as the civil-law notary drafting the deed.[15]

The distinguishing aspect of the 'superficies utilities' as compared to the standard right of superficies lies in the fact that the 'superficies utilities' does not oblige the registering party to commit to cadastral measurements.[16] The challenge in creating a variation of the right of superficies specific for the sustainability context lies in overcoming the financial burden that cadastral measurements impose, whilst not sacrificing the legal certainty that the context within which such a right operates requires. It is for this reason that the drafters have opted to provide parties registering a 'superficies sustainable' with the option to opt in to a measurement. The proposal submits that article 6(2) sub b of the *Kadasterbesluit* should stipulate that measurement will not be done, unless:

> the registration explicitly requests the registrar to perform a measurement and the parcel with a provisional or the portion of the parcel is in the opinion of the registrar suited for measurement.[17]

The drafters point out that, depending on the facts and circumstances surrounding the object of the right of superficies, a measurement might or might not be appropriate. A measurement, for example, might be appropriate if it concerns a sustainable system situated outside a property, such as the outdoor unit of a heat pump. Other situations, in which the object might more closely resemble the object of the 'superficies utilities', might not merit the additional costs of the measurement. Hence, the drafters propose that measures will only take place upon the explicit request of the person making the registration.[18] The drafters point out that their proposal includes a similar system for the 'superficies utilities'. This system, in their view, limits – although not fully prevents – 'contamination issues' (*besmettingsproblematiek*) in which, when relevant, an object of a right of superficies rather than being specifically described is registered to the parcel as a whole.[19]

18.3 Tokenisation Platforms

In the context of the practical functioning of blockchain and tokenisation technology, it is helpful to differentiate between two types of promises made by the technology: promises with regard to technological capabilities and promises with regard to its usefulness in specific use cases. This distinction is essential in order to make a prediction on whether the technology could reasonably be expected to be successful in a specific context.

Section 18.3.1 will touch upon the first promise: its technological capabilities. A baseline understanding of these technological capabilities, and the technical underpinnings thereof, is essential as the novel technological elements dictate the extent to which the use of this technology provides added value.

Sections 18.3.2 and 18.4 deal with the second promise: its usefulness in the context of a specific use case. The relevant use case referred to here is the application of tokenisation technology in the context of the registration of land and limited property rights.[20]

18.3.1 Technical Aspects

Promises by the technology regarding its technical capabilities relate to its asserted unique technical outcomes. This section, rather than engaging in an in-depth analysis of the actual functioning of the technology, will elaborate on these asserted unique technical aspects. This contribution refers to these technical outcomes as 'technical aspects'. These aspects discern the technology from more commonplace equivalents. They describe how a blockchain platform must, for example, be differentiated from an online marketplace and how smart contracts must be differentiated from classic software applications. This approach, focussing on the technical outcomes rather than the technical underpinnings, will ensure a certain baseline in understanding of the technology in question, without having to perform an in-depth technological deep dive. For the sake of brevity and in order not to distract from the legal and sustainability focus of this contribution, a short description of these technical aspects is the most suitable approach. The technology in question commits to five technical aspects: immutability, pseudonymity, transparency, automatic execution, and automatic enforcement. The following sections will shortly touch upon these aspects.[21]

18.3.1.1 Immutability

A database built on a blockchain is immutable: no single party can alter the state of the information as recorded in the ledger. This immutability works on a transaction level and on a record level. The former is guaranteed through public-key cryptography. In the context of digital currencies, which is the context in which this technology found its first practical application, this public-key cryptography was first proposed as a method through which secure communication is possible between the parties. Both parties are in possession of a key pair consisting of a public key and a private key. The public key is known to everyone, whilst the private key is only known to the person who holds it. This system enables one person to send a secret message over an unprotected network. Imagine the following situation:

- Alice wants to send a secret message to Bob.
- Alice looks up Bob's public key and encrypts her message using Bob's public key.
- Alice now transfers the message to Bob. Only the person with the key that corresponds to Bob's public key (i.e., Bob's private key) can decrypt the message.
- Bob receives the message and uses his private key to decrypt the message. He can now read the message.

Additionally, if a third party were to alter the encrypted message in transit, the recipient would be unable to decrypt the communication in a manner that would render a legible message. The effective end result is a situation in which persons can send transactions and are able to trust that no person has altered the information, whilst both sender and recipient are unable to claim that the transaction was not sent.[22]

Additionally, the technology guarantees immutability on a record level through a consensus mechanism. Whilst a multitude of different mechanisms exists, they all ensure that all participants and users of the database agree on whether the object described in the transaction they receive is authentic, the person making the transaction has a right to make that transaction, and that no person other than the recipient can claim an entitlement to the object described by the transaction.[23] The key aspect of immutability is important in the context of employing this technology in the context of the registration of land and limited property rights. One important purpose of land registration systems is the creation of legal certainty with regard to long-lasting property relations.[24] If applied properly, this technology might be able to leverage the aspects of immutability in order to support this purpose.

18.3.1.2 Transparency and Pseudonymity

The previously mentioned consensus mechanism plays an essential role in guaranteeing the immutability of the database. In order to do so, the consensus mechanism requires a certain involvement of participants and users. It requires, in other words, for some users to check the validity of transactions.[25] After all, if these checks cannot be performed, the integrity of the database cannot be guaranteed. At the very least, those performing the checks must know the sender, the recipient, the time of initiation, and the object of the transaction. It is for this reason that the system is inherently transparent: without transparency, users are unable to perform the necessary checks. The key aspect of transparency aligns with requirements that the relevant legal framework imposes: a fundamental principle that underpins the classical model of property law is the principle of transparency.[26] This technological aspect of transparency could be leveraged in order to align implementation of this technology in the area of property registration with the legal requirements. More specifically, it could be used to support legal interventions aimed at tackling the shortcoming, as identified by the drafters of the proposal, that the object of a right of superficies in the context of sustainability refurbishments to immovable property is often not sufficiently identified in a contract of sale.[27]

It should be noted that, as a result of public-key encryption, the information in the system cannot automatically be attributed to a natural person. Public-key encryption makes that person, rather than operating on the database by their own name, operate through a pseudonym. That pseudonym is their own public key. To conclude: the consensus mechanism requires the system to operate with a certain degree of transparency, whilst the public-key encryption that operates as a foundation under the database, ensures that pseudonymity is guaranteed.[28]

18.3.1.3 Automatic Execution and Automatic Enforcement

Immutability, transparency, and pseudonymity are a result of the fact that the databases under consideration here are blockchain-based databases. Within the larger category of blockchain databases, several different variations exist. The types of databases that are relevant for this contribution, i.e., the databases that form the foundation for token platforms, all allow for programming upon that database. Programs and software that run on such a database are known as 'smart contracts'. Hence, from a definitional point of view, smart contracts are nothing more and nothing less than software. Smart contracts differentiate themselves from other software, i.e., software contexts different from a blockchain (e.g., a program that runs on a laptop), through two additional technical aspects. The first of which is the automatic execution ensured through smart contracts and the latter of which is the automatic enforcement that they create.[29] It should be noted that the term 'smart contract' does not refer to any legal qualification. A smart contract is not a legal concept nor a legal agreement. The question of whether a smart contract is a binding legal agreement is a separate question and is outside the scope of this contribution.[30]

The software operates on these platforms much like a person would. An account can be held by a person (an externally owned account, or EOA) or by a contract. An account held by a contract is known as a 'contract account'. This contract account, after being deployed on the platform, cannot be accessed by persons anymore. The contract account contains the code that describes the software. This software can be interacted with by interacting with the contract account.[31] An example of a simple smart contract is an escrow-like construction: Alice transfers an object to the smart contract for safekeeping. The smart contract will observe the holdings of Alice. Thanks to the transparent nature of the platform, the smart contract is able to do so. The code contained within that contract account describes that, on the condition of a person having transferred a certain sum to Alice's account, the object will be transferred to that person. Hence, the condition is a transfer to Alice, and the result is a transfer by the smart contract to the sender of the transaction to Alice. If, at a certain point, Bob transfers that sum to Alice, the smart contract will observe this, determine the predefined condition to be fulfilled, and execute the stipulated result by transferring the object to Bob. Bob has triggered the condition contained within the smart contract with his transfer, and the conditions contained therein are executed automatically. Thanks to the fact that the smart contract is able to interact with the assets of a person on that platform, any conditions specified in its code will be enforced by way of that contract.

Within that contract account, the smart contract code is contained. Individuals can interact with the smart contract, through the contract account. The transparent nature of the platform ensures that all are able to inform themselves of the conditions for execution of the smart contract, as well as the stipulated results that will be effectuated upon fulfilment of those conditions. The execution of the smart contract happens irrevocably once the conditions are fulfilled. Neither the designer or programmer of the smart contract nor the person interacting with the smart contract account, or its beneficiary is able to prevent the execution and the enforcement of the smart contract at that point.[32] Smart contracts, in the context of this contribution, are therefore viewed as pieces of software. These pieces of software, enabled by its automatic execution and automatic enforcement, can be employed for more general applications, such as the creation of a system that supports the achievement of sustainability goals in the context of land and property rights registration.

18.3.1.4 Tokenisation Platforms

The platforms that are at the heart of this contribution were initially designed as transaction platforms. The option to run software on top of such a platform, provided that the programming language used allows for sufficient flexibility,[33] makes it possible to create applications with a wide range of functionalities. In this manner, the platform becomes a generalised programming layer upon which software applications can be created. These include, for example, supply chain applications, Internet-of-things applications, and financial services applications.[34] For the purposes of this contribution, one possibility created by smart contracts is especially relevant: the possibility of giving rise to tokenisation platforms. These tokenisation platforms allow for the creation and transaction of tokens. These tokens are metadata files and function as units that represent a record.[35] One potential use of these tokenisation platforms that has generated a considerable amount of attention in recent years are platforms that allow artists to generate a token from their art and sell that token. This use-case concerns so-called 'non-fungible tokens' (NFTs) and the platforms are NFT platforms. In this example, the token is a record that describes a connection with a piece of art.[36]

The connection described within that token is not limited to art alone. Programmers, fintech start-ups, and researchers have proposed a range of potential applications that include, for example, financial products, cargo in the area of logistics, and physical properties. This contribution is concerned with this latter category and places this in the context of a more sustainable right of superficies. The following sections focus on the manner in which legal scholars have attempted to find potential uses for the technology in the area of law under investigation here. Section 18.3.2 shows how the grand ideas on generalised blockchain platforms are overly ambitious and argues that a more nuanced position is appropriate, especially in matters concerning sustainability. Such a nuanced position will be proposed in Section 18.4, which connects the aim and purpose of the Dutch proposal for a 'superficies sustainable' to tokenisation platforms. This section then shows how a tool could be developed that supports the cadastral authorities in tackling the challenges that the drafters of the proposal for a 'superficies sustainable' have highlighted. This tool must be limited in scope and application and does not supplant any land registration system. Instead, the purpose of such a tool is to leverage the transparency and immutability offered by the technology and generate object-specificity and recognisability. Hence, rather than supplanting current registries, such a tool would provide support to registrars and notaries in facing the unique challenges that improving the sustainability of properties creates from a registration perspective.

18.3.2 Blockchain-Based Land Registration

Since its very inception, blockchain technology has been the basis of many unkept promises. These promises include retail applications, voting solutions, decentralised autonomous investment vehicles, new avenues for artists to exploit their work, and the potential to create legally binding agreements in a distributed environment.[37] One of the most prominent of such promises was, and the relevant context to a tokenisation-based system to support a sustainable variation of the right to superficies, the idea of transferring land registration systems to a blockchain-based platform. This idea can be traced back to Nick Szabo, who is widely credited for being the originator of the idea of smart contracts.[38] In a 1998 paper that describes the outlines of a technical solution, based on a distributed database, meant to serve as a property title registry, he proposes 'a secure, distributed title database to prevent [forgery and destruction of public records] in the future'.[39]

Whilst it seems, at least to us, that the technology in question has promised the world, but delivered no more than an atlas, the promise reflected in that 1998 statement has been echoed in different shapes and forms by computer scientists, programmers, lawyers, and legal scholars.[40] These statements, upon which will be reflected below, form the background against which we would like to propose a less ambitious and more realistic application of the technology in question: an implementation of the technology to support sustainability goals that supports the existing registration systems.

In order to ensure a certain degree of conciseness in this contribution, these echoes are not repeated in full here. Instead, some observations on the potential effectiveness of the proposal made by one of the registrars of the Netherlands Cadastre during the annual International Property Registries Association and International Centre of Registration Law Conference (Ipra-Cinder) will suffice. Here it was argued that land registration systems, in general, have a common basic tripartite structure, consisting of an object (spatial unit), a right (right *in rem*) and a subject (the holder of the right that is related to the object).[41] Situations in which that basic structure can become relatively complicated relate to circumstances in which two or all three of these layers are complex and/or are used in unusual or extraordinary situations. An example of such a situation is as follows:

> a case where there are multiple persons, each entitled to different shares in various rights (e.g.: a right of bare ownership, encumbered with the right of usufruct and a building right),

> with a mortgage right or seizure on the right of ownership or the building right, with regard to a building on a plot of land (parcel), which building has been divided into apartment rights. Will this fit in a land administration based on blockchain technology?[42]

In a follow-up publication, several particular challenges are highlighted which render a solution based on the technology under investigation here not the obvious choice for complex land registration purposes. First and foremost, a land registration system can only be successful if all parties involved, including property owners, banks, and notaries, have trust in the system. For a blockchain-based solution, this means that a successful solution cannot be based on technological aspects alone. Organisational and institutional aspects must be considered as well.[43] A second observation relates to the governance of such a system. In this regard, the question must be posed who would oversee designing and maintaining the system, as well as keeping the data. By extension thereof, a pivotal question is whether such a blockchain-based system must be on a private blockchain, a public blockchain, or a hybrid blockchain.[44] If a system other than a public system is chosen, a valid question is whether the added value of a blockchain platform would outweigh the benefits of a system provided by a central public institution charged with cadastral duties. Thirdly, a challenge with regard to the creation of such a system, especially if the proposed system is built on a public ledger, lies in guaranteeing the accuracy and completeness of the information used as initial input.[45] The immutable character of the platform will render it very difficult, if not impossible, to correct information with retroactive effect.[46]

Regardless of the many promises and predictions made since the 1998 article by Nick Szabo, these challenges have not been overcome. It seems to be that, in order to employ some of the advantages that are offered by the technology, the only way in which the technology can be used is in a supporting yet subservient role to existing cadastral authorities, thereby potentially serving the interests of the parties involved. Regarding the balance between old-fashioned cadastral authorities and blockchain-based land registration databases, to us, it seems that with the passage of time, the statement that 'a future reality is somewhere in the middle'[47] is increasingly likely to be true. A use case in which a tokenisation-based tool, subservient to the cadastral system, might be employed in order to serve a very specific purpose, i.e., the remediation of the challenges illustrated by the drafters of the proposal, could be an example of such a middle-ground.

18.4 A Tokenisation-Based Tool to Support the 'Superficies Sustainable'

Section 18.3.2 has shown the overambitious nature of the many ideas and initiatives focussed on using the technology in the context of land registration systems. It seems that these ideas reflect either an overestimation of the possibilities of the functioning and functionality of the technology, an underestimation of the complexity of land registration systems, or a misconception of both. The context within which land registration operates, as well as its inherent complexity, renders it impossible to be supplanted with a blockchain-based solution. However, the 'superficies sustainable' provides a clearly defined and limited context within which this technology might be employed in a role supportive of, and subservient to, a land registration system. We suggest exploring the possibilities of employing this technology in a supporting role to cadastral authorities in the implementation and operation of the 'superficies sustainable', as well as the tackling of the specific challenges created through refurbishments to immovable properties made in view of increasing the properties' sustainable character. The inherent characteristics of this technology (or its 'technical aspects', see Section 18.3.1) align well with the challenges that the drafters of the 'superficies sustainable' have identified. Previously we quoted Jacques Vos, a registrar with the Dutch cadastral authorities, who took a nuanced position when he concludes that 'a future reality is somewhere in the middle'.[48] This section will show that a limited application of the technology is possible in a limited context, with limited goals. The application, first of all, is limited because it concerns a role that is merely and exclusively supporting. Secondly, the context is limited because it concerns only the newly proposed 'superficies sustainable' in Dutch law. Lastly, the goals are limited since the purpose of such a software solution would be restricted to providing an instrument to cadastral authorities in support of achieving the particular sustainability challenges identified by the drafters of the proposal: object-specificity, recognisability, and cost-efficiency.[49]

18.4.1 Token-Based 'Rights Management' in Support of Sustainability Goals

A tokenisation-based solution makes it easier to create a database of rights. Hence, the difference between many of the ideas and proposals discussed in Section 18.3.2 lies in the fact that rather than registering land or parcels, the registration of rights becomes the standard. By taking rights registration as a starting point, a more carved-out, and therefore limited, solution becomes possible. For example, as this section will show, a software solution that is limited only to the right of superficies, and potentially even one variation of the right of superficies such as the 'superficies sustainable', is possible. Moreover, applying the technology in such a limited context enables the implementation of the software in support of land registration. By doing so, the technology can be used to create a 'legal tech' solution to assist cadastral authorities in achieving the sustainability challenges they are confronted with. The following paragraphs will sketch an outline of what such a software tool could look like.

Earlier Section 18.3.1.4 defined tokens as 'metadata files that function as units that represent a record'.[50] A record can describe a great variety of things. Amongst the objects that could potentially be recorded in a token, is a right. For example, the Monetary Authority of Singapore, its central bank, has defined a token as being a cryptographically secured representation of a token holder's rights to receive a benefit or to perform specified functions.[51]

The country of Lichtenstein has enacted its 2019 Law on Tokens and TT Service Providers. In the translation provided by the Lichtenstein government, a token is defined as a piece of information on a TT System which (1) can represent claims of rights of memberships against a person, right to property of other absolute or relative rights, and (2) is assigned to one or more TT Identifiers.[52]

These definitions make clear that from a technical point of view, these tokens can represent rights. Moreover, these definitions also show that legal recognition of the rights represented by tokens is not without precedent.

It has been suggested that such tokens might be used to transfer ownership of assets. This creates potential issues, especially if applied in a cross-border context, related to the identification of the parties to prove ownership, plurality of owners, and extinction of ownership.[53] A focussed and limited solution would be based on using smart contracts to articulate limited rights, rather than the full right of ownership. Such a solution might be inspired by a rights management system, developed in the ambit of information science. Such a system is based on any combination of three properties: the ability to read code, the ability to write code and the ability to execute code. Such a combination of properties has been compared to the Hohfeldian 'bundle of sticks' analogy.[54] Using these properties, it becomes possible to program limited rights by interpreting the ability to read as the right to possess, the ability to write as the right to dispose, and the ability to execute as the right to use.[55] More complicated proprietary entitlements can be contained within a token through natural language, a comprehensive rights management system on the basis of the tripartite classification illustrated above and, if necessary, be supported by smart contracts.

For example, in the right of usufruct, the payments the holder of the usufruct must make may be automatised; or in the right to build, the smart contract could consult directly with the local Administration to verify whether the land can be developed, or the periodic payment to which the owner of the ground is entitled may be paid automatically, with non-payment leading to the automatic resolution of the right.[56]

Rather than creating a blockchain-based land registration system, a system that employs the technology in a manner as illustrated above would create a rights management system. Such a system, as made clear through the example given above, can be employed with respect to land rights. However, many of the obstacles illustrated in Section 18.3.2 would apply to rights management systems *mutatis mutandis*.[57] Moreover, for the purposes of the sustainability challenges created by the classic variation of the right of superficies, no rights management register is necessary. After all, the challenges that the 'superficies sustainable' aims to tackle are concerned with object--specificity, transparency, and cost efficiency. For these challenges, no management of rights is necessary. A more elementary implementation of the technology, one that is limited to the identifiability of the object rather than the management of the right, will suffice.

This system could be created by designing a platform layer that exists parallel to, in support of, and subservient to, a land registration system. On this platform, tokens could be created that represent the 'superficies sustainable'. A design choice could be made to include tokens that represent the classical right of superficies and or the 'superficies utilities'. A simple earmark would be enough to make this difference clear. Through the metadata within that token, a link could be made to the registration of the relevant right within the land registration itself. A smart contract could be programmed and interfaced with the land registration of the object or parcel to which that right relates. In a system like this, the token reflects the right as contained in the land registration. It does not replace the land registration system. Whilst the token does not, in and of itself, have any legal significance (such is retained, by force of law, by the land registry), it does generate a certain degree of certainty. This is a direct result of this system leveraging the immutability, automatic execution, and automatic enforcement of the underlying database. Through careful programming, combined with solid assurance, the technology can guarantee that the reflection provided by the token mirrors the land registry. Moreover, this system leverages the transparency offered by the underlying technology. By doing so in a manner that clearly reflects the right as well as the object to which it relates, it allows for greater recognisability and object-specificity. Lastly, a system such as this can be built upon existing infrastructure and, in light of its limited character, can be done in a relatively inexpensive manner.

18.5 Conclusion

In this contribution, we presented a limited approach to a solution in which the technology can be used to support the implementation of the 'superficies sustainable' and enable cadastral authorities to meet the unique demands created by the move towards a more sustainable world. We argue that the endeavour of the 'superficies sustainable' could be supported by a legal tech solution, created by a limited implementation of the emerging technology of tokenisation platforms. This limited implementation of a tokenisation platform in which the 'superficies sustainable' is reflected, provided it is linked to the land registration, leverages the immutability, transparency, automatic execution, and automatic enforcement of the technology in question, without unhinging the registry itself. The technology can enhance the recognisability and the object-specificity, without creating an additional financial burden on users. A platform like this supports a land registry in achieving solutions in response to challenges created by the move towards sustainable immovable property.

The basis of this approach lies in the Dutch proposal on a 'superficies sustainable', which presents a concrete and clearly defined set of challenges created by a movement towards more sustainable properties. The drafters of this proposal have identified that these particular challenges are best tackled through a reassessment of existing legal instruments. Their solution of a 'superficies sustainable' seems to be appropriate and proportional. The next step would be to analyse whether such a proposal, perhaps combined with a tokenisation-based tool as suggested in this contribution, might add to solutions in other legal systems that encounter similar issues with the registration of sustainability refurbishments.

Notes

1 The authors wish to thank Dario Hug and the anonymous peer reviewers for their helpful suggestions and insightful comments. Any oversights that remain are ours alone.

2 For a broader context, see Bram Akkermans, 'Sustainable Property Law – Towards a Revaluation of Our System of Property Law' in Bram Akkermans and Gijs van Dijck (eds), *Sustainability and Private Law* (The Hague, Eleven International Publishing 2020) 37–58.

3 Sjef van Erp and Bram Akkermans, *Cases, Materials and Text on National, Supranational and International Property Law* (Oxford, Hart Publishing 2012) 280.

4 Björn Hoops and Loes AM Meijerink, 'Het Opstalrecht Duurzaam' [2022] 7353 *Weekblad voor Privaatrecht Notariaat en Registratie* 1.

5 Voorstel tot wijziging van Kadasterbesluit ter invoering van het opstalrecht duurzaam in de bijwerking van de basisregistratie kadaster en gerelateerde zaken <https://www.kadaster.nl/documents/1953498/3276179/Wijziging+van+het+Kadasterbesluit+ter+invoering+van+het+Opstalrecht+duurzaam+in+de+bijwerking+van+de+Basisregistratie+Kadaster-.pdf/759cb366-d38a-0fde-ddd3-08934f3bfb92?t=1641383878702> accessed 15 July 2023. Hereinafter: the proposal.

6 See, for example, on the German *Erbaurecht* in the *Erbaurechtgesetz* Klaus Vieweg and Sigrid Lorz, *Sachenrecht* (Munich, Vahlen 2022) 599 ff; Christian Larroumet, *Les Biens, Droits Réels Principaux* (6th edn, Paris, Economica 2019) 472 ff.

7 Voorstel tot wijziging van Kadasterbesluit ter invoering van het opstalrecht duurzaam in de bijwerking van de basisregistratie kadaster en gerelateerde zaken <https://www.kadaster.nl/documents/1953498/3276179/Wijziging+van+het+Kadasterbesluit+ter+invoering+van+het+Opstalrecht+duurzaam+in+de+bijwerking+van+de+Basisregistratie+Kadaster-.pdf/759cb366-d38a-0fde-ddd3-08934f3bfb92?t=1641383878702> accessed 15 July 2023.

8 Björn Hoops and Loes AM Meijerink, 'Het Opstalrecht Duurzaam' [2022] 7353 *Weekblad voor Privaatrecht Notariaat en Registratie* 1.

9 Jacques Vos and Reuben Roes, 'Onduidelijkheden Rondom Inschrijving En Registratie' [2018] 7180 *Weekblad voor Privaatrecht Notariaat en Registratie* 115, 120–121; Björn Hoops and Loes AM Meijerink, 'Het Opstalrecht Duurzaam' [2022] 7353 *Weekblad voor Privaatrecht Notariaat en Registratie* 1.

10 Björn Hoops and Loes AM Meijerink, 'Het Opstalrecht Duurzaam' [2022] 7353 *Weekblad voor Privaatrecht Notariaat en Registratie* 1.

11 Hendrik Ploeger and Hafida Bounjouh, *Erfpacht En Opstal* (Deventer, Kluwer 2019) 98.

12 Björn Hoops and Loes AM Meijerink, 'Het Opstalrecht Duurzaam' [2022] 7353 *Weekblad voor Privaatrecht Notariaat en Registratie* 1.

13 See sub C of the proposal for the newly proposed art. 6(4) Kadasterbesluit: 'Een opstalrecht duurzaam betreft uitsluitend het aanleggen en houden van een installatie en aanhorigheden in, op of boven de onroerende zaak van een ander, van welke aard dan ook, die hetzij elektriciteit, hetzij warmte of koude uit hernieuwbare bronnen opwekt, opslaat of omvormt, een gebouw op de onroerende zaak van elektriciteit, warmte of koude uit hernieuwbare bronnen voorziet of de energiezuinigheid van een gebouw op de onroerende zaak verbetert'.

14 Björn Hoops and Loes AM Meijerink, 'Het Opstalrecht Duurzaam' [2022] 7353 *Weekblad voor Privaatrecht Notariaat en Registratie* 1, 2; see, also, the explanatory memorandum to art. 6(4) of the Proposal.

15 See Explanation to Artikel I.C – artikel 6, vierde lid, van het Kadasterbesluit to the proposal.

16 Barbara AM Janssen, *Wie Heeft de Leiding?* (Deventer, Kluwer 2010) 234–235.

17 Art. 1 A of the Proposal.

18 Björn Hoops and Loes AM Meijerink, 'Het Opstalrecht Duurzaam' [2022] 7353 *Weekblad voor Privaatrecht Notariaat en Registratie* 1, 3.

19 Björn Hoops and Loes AM Meijerink, 'Het Opstalrecht Duurzaam' [2022] 7353 *Weekblad voor Privaatrecht Notariaat en Registratie* 1, 3; Björn Hoops and Loes AM Meijerink, 'Naschrift' [2022] 7358 *Weekblad voor Privaatrecht, Notariaat, en Registratie* 103; see also, for cadastral contamination in a more general context Martin Daverschot, 'Voorkoming Kadastrale Besmetting in Notariële Praktijk' [2013] 6984 *Weekblad voor Privaatrecht Notariaat en Registratie* 612.

20 An alternative – and quite promising – case study for the application of tokenisation technology as LegalTech in a sustainability context is the creation of an emission trading platform; see Adam J Sulkowski, 'The Tao of DAO: Hardcoding Business Ethics on Blockchain' [2020] 3(2) *Business & Finance Law Review* 156; See more broadly Özgür Arslan-Ayadin, Prabal Shrestha and James Thewissen, 'Blockchain as a Technology Backbone for an Open Energy Market' in André Dorsman, Özgür Arslan-Ayadin and James Thewissen (eds), *Regulations in the Energy Industry: Financial Economic and Legal Implications* (Cham, Springer 2020).

21 See also, see Jasper Verstappen, 'Comparative Law and Legal Issues Arising From Distributed Ledgers and Smart Contracts' in Bram Akkermans and Anna Berlee (eds), *'Sjef-Sache': Essays in Honour of Prof. mr. dr. J.H.M. (Sjef) van Erp on the Occasion of his Retirement* (The Hague, Eleven International Publishing 2021) 485, 490–496.

22 David Chaum, 'Blind Signatures for Untraceable Payments' in David Chaum, Ronald L Rivest and Alan T Sherman (eds), *Advances in Cryptology* (Cham, Springer 1982) 199; David Chaum and Hans van Antwerpen, 'Undeniable Signatures' in Gilles Brassard (ed), *Advances in Cryptology – CRYPTO '89 Proceedings* (Cham, Springer 1990) 212 and Jasper Verstappen, *Legal Agreements on Smart Contract Platforms in European Systems of Private Law* (Cham, Springer 2023) 15–50.

23 Andreas Antonopoulos, *Mastering Bitcoin: Programming the Open Blockchain* (2nd edn, Sebastopol, O'Reilly 2017) 3.

24 See for a broader discussion Sjef van Erp, 'Land Registration and 'Disruptive' (or 'Trustworthy'?) Technologies: Tokenisation of Immovables Property' in Anabel Fraga, Elena Ioriatti and Sjef van Erp (eds), *IMOLA II Project: The European Land Register Document (ELRD): A Common Semantic Model for Land Registers Interconnection* (Brussels, European Land Registry Association 2019) 169–172.

25 Andreas Antonopoulos, *Mastering Bitcoin: Programming the Open Blockchain* (2nd edn, Sebastopol, O'Reilly 2017) 196 and 225 ff.

26 Sjef van Erp, *European and National Property Law: Osmosis or Growing Antagonism?* (Groningen, Europa Law Publishers 2006) 14–16; see for an overview of the publicity in the context of land registration systems Anna Berlee, *Access to Personal Data in Public Land Registers: Balancing Publicity of Property Rights With the Right to Privacy and Data Protection* (The Hague, Eleven International Publishing 2018) 17–58.

27 Björn Hoops and Loes AM Meijerink, 'Het Opstalrecht Duurzaam' [2022] 7353 *Weekblad voor Privaatrecht Notariaat en Registratie* 1.

28 See, for more practical context, Primavera De Filippi, Morshed Mannan and Wessel Reijers, 'The Alegality of Blockchain Technology' [2022] 41(3) *Policy and Society* 358, 358–360 (https://doi.org/10.1093/polsoc/puac006).

29 It should be noted that this contribution takes the Ethereum platform as an example of how smart contracts work. This choice has been made because this Ethereum is the most widely used and most mature solution currently available. It should be noted that alternative solutions exist. These solutions might have implemented the aspects of automatic execution and automatic enforcement slightly differently. However, to ensure the conciseness of this contribution, such alternative implementations will not be explored further here.

30 See, on this, Mateja Durovic and André Janssen, 'The Formation of Blockchain-Based Smart Contracts in the Light of Contract Law' [2018] 26(6) *European Review of Private Law* 753 (https://doi.org/10.54648/erpl2018053).

31 See, for a detailed technical explanation, Gavin Wood, 'Ethereum: A Secure and Decentralised Generalised Transactions Ledger' (*Berlin Version*, 25 August 2023) <https://ethereum.github.io/yellowpaper/paper.pdf> accessed 15 July 2023; Vitalik Buterin, 'Ethereum White Paper – A Next Generation Smart Contract & Decentralized Application Platform' (*Ethereum*, 2014) <https://ethereum.org/en/whitepaper/> accessed 15 July 2023.

32 Andreas Antonopoulos, *Mastering Bitcoin: Programming the Open Blockchain* (2nd edn, Sebastopol, O'Reilly 2017) 129; Andreas Antonopoulos and Gavin Wood, *Mastering Ethereum: Building Smart Contracts and DApps* (Sebastopol, O'Reilly 2018) 128–129.

33 A programming language in this context must be Turing-complete; see Andreas Antonopoulos and Gavin Wood, *Mastering Ethereum: Building Smart Contracts and DApps* (Sebastopol, O'Reilly 2018) 8.

34 Gurtu Amulya and Jestin Johny, 'Potential of Blockchain Technology in Supply Chain Management: A Literature Review' [2019] 49 *International Journal of Physical Distribution & Logistics Management* 881 (https://doi.org/10.1108/IJPDLM-11-2018-0371); Tiago M Fernández-Caramés and Paula Fraga-Lamas, 'A Review on the Use of Blockchain for the Internet of Things' [2018] 6 *IEEE Access* 32979 (https://doi.org/10.1109/ACCESS.2018.2842685); Alexander Shelkovnikov, 'Blockchain Applications in Banking' (*Deloitte*, 2016) <https://www2.deloitte.com/content/dam/Deloitte/uk/Documents/Innovation/deloitte-uk-blockchain-app-in-banking.pdf> accessed 15 July 2023.

35 Oleskii Konashevych, 'General Concept of Real Estate Tokenization on Blockchain: The Right to Choose' [2020] 9(1) *European Property Law Journal* 21, 31 (https://doi.org/10.1515/eplj-2020-0003).

36 For an in-depth functional description, see Andres Guadamuz, 'The Treachery of Images: Non-Fungible Tokens and Copyright' [2021] 16(12) *Journal of Intellectual Property Law & Practice* 1367 (https://doi.org/10.1093/jiplp/jpab152).

37 See, for example, Andreas Antonopoulos, *Mastering Bitcoin: Programming the Open Blockchain* (2nd edn, Sebastopol, O'Reilly 2017) 5; Xin Sun et al, 'A Simple Voting Protocol on Quantum Blockchain' [2019] 58 *International Journal of Theoretical Physics* 275 (https://doi.org/10.1007/s10773-018-3929-6); Quinn DuPont, 'Experiments in Algorithmic Governance: A History and Ethnography of "The DAO" a Failed Decentralized Autonomous Organization' in Malcolm Campbell-Verduyn (ed), *Bitcoin and Beyond* (Abingdon, Routledge 2017); Andres Guadamuz, 'The Treachery of Images: Non-Fungible Tokens and Copyright' [2021] 16(12) *Journal of Intellectual Property Law & Practice* 1367 (https://doi.org/10.1093/jiplp/jpab152); Kevin Werbach and Nicolas Cornell, 'Contracts: Ex Machina' in Marcelo C Compagnucci, Mark Fenwick and Stefan Wrbka (eds), *Smart Contracts: Technological, Business and Legal Perspectives* (Oxford, Hart 2021); Mateja Durovic and André Janssen, 'The Formation of Blockchain-Based Smart Contracts in the Light of Contract Law' [2018] 26(6) *European Review of Private Law* 753 (https://doi.org/10.54648/erpl2018053); Joshua AT Fairfield, 'Tokenized: The Law of Non-Fungible Tokens and Unique Digital Property' [2022] 97 *Indiana Law Journal* 1261.

38 Nick Szabo, 'Smart Contracts: Building Blocks for Digital Free Markets' [1996] 8(1) *Extropy: Journal of Transhumanist Thought* 50; Kevin Werbach and Nicolas Cornell, 'Contracts: Ex Machina' in Marcelo C Compagnucci, Mark Fenwick and Stefan Wrbka (eds), *Smart Contracts: Technological, Business and Legal Perspectives* (Oxford, Hart 2021); Primavera de Filippi, Chris Wray and Giovanni Sileno, 'Smart Contracts' [2021] 10(2) *Internet Policy Review* 2 (https://doi.org/10.14763/2021.2.1549).

39 Nick Szabo, *Secure Property Titles with Owner Authority* (Satoshi Nakamoto Institute, 1998) <nakamotoinstitute.org/secure-property-titles> accessed 15 July 2023.

40 Jan Vreuger, 'Dutch Blockchain, Real Estate and Land Registration' [2020] 12(2) *Journal of Property, Planning, and Environmental Law* 93 (https://doi.org/10.1108/JPPEL-11-2019-0053); Amnon Lehavi, 'The Future of Property Rights: Digital Technology in the Real World' in Amnon Lehavi and Ronit Levine-Schnur (eds), *Disruptive Technology, Legal Innovation, and the Future of Real Estate* (New York, Springer International Publishing 2020); Benjamin Verheye, 'Real Estate Publicity in a Blockchain World: A Critical Assessment' [2017] 6(3) *European Property Law Journal* 441 (https://doi.org/10.1515/eplj-2017-0020); Georg von Wangenheim, 'Blockchain-Based Land Registers: A Law-and-Economics Perspective' in Amnon Lehavi and Ronit Levine-Schnur (eds), *Disruptive Technology, Legal Innovation, and the Future of Real Estate* (New York, Springer International Publishing 2020); Benjamin Verheye, 'Land Registration in the Twenty-First Century: Blockchain Land Registers from a Civil Law Perspective' in Amnon Lehavi and Ronit Levine-Schnur (eds), *Disruptive Technology, Legal Innovation, and the Future of Real Estate* (New York, Springer International Publishing 2020); Maria Kaczorowska, 'Blockchain-Based Land Registration: Possibilities and Challenges' [2019] 13(2) *Masaryk University Journal of Law and Technology* 339 (https://doi.org/10.5817/MUJLT2019-2-8); Oleksii Konashevych, 'Constraints and Benefits of the Blockchain Use for Real Estate and Property Rights' [2020] 12(2) *Journal of Property, Planning, and Environmental Law* 109 (https://doi.org/10.1108/JPPEL-12-2019-0061); Nicolás Nogueroles Peiró and Eduardo J Martinez García, 'Blockchain and Land Registration Systems' [2017] 6(3) *European Property Law Journal* 296 (https://doi.org/10.1515/eplj-2017-0017).

41 Jacques Vos, *Blockchain-Based Land Administration: Panacea, Illusion or Something in between? Legal Interference of Registrars in the e-Conveyancing Process* (European Land Registry Association, 2017) <https://www.elra.eu/wp-content/uploads/2017/02/10.-Jacques-Vos-Blockchain-based-Land-Registry.pdf> accessed 15 July 2023, 1.

42 Jacques Vos, *Blockchain-Based Land Administration: Panacea, Illusion or Something in between? Legal Interference of Registrars in the e-Conveyancing Process* (European Land Registry Association, 2017) <https://www.elra.eu/wp-content/uploads/2017/02/10.-Jacques-Vos-Blockchain-based-Land-Registry.pdf> accessed 15 July 2023, 2.

43 Jacques Vos, 'The Impact of "Disruptive" IT and the Registrar's Role in Future e-Conveyancing' [2017] 1 *International Review IPRA/CINDER* 68, 69.

44 Jacques Vos, 'The Impact of "Disruptive" IT and the Registrar's Role in Future e-Conveyancing' [2017] 1 *International Review IPRA/CINDER* 68, 70.

45 Jacques Vos, 'The Impact of "Disruptive" IT and the Registrar's Role in Future e-Conveyancing' [2017] 1 *International Review IPRA/CINDER* 68, 71.

46 See, for a more in-depth exploration of existing obstacles to a successful implementation of a blockchain-based land registration system, Rod Thomas, 'Blockchain's Incompatibility for Use as a Land Registry: Issues of Definition, Feasibility and Risk' [2017] 6(3) *European Property Law Journal* 361 (https://doi.org/10.1515/eplj-2017-0021).

47 Jacques Vos, 'Blockchain and Land Administration: A Happy Marriage?' [2017] 6(3) *European Property Law Journal* 293, 295 (https://doi.org/10.1515/eplj-2017-0018).

48 Jacques Vos, 'Blockchain and Land Administration: A Happy Marriage?' [2017] 6(3) *European Property Law Journal* 293, 295 (https://doi.org/10.1515/eplj-2017-0018).

49 Björn Hoops and Loes AM Meijerink, 'Het Opstalrecht Duurzaam' [2022] 7353 *Weekblad voor Privaatrecht Notariaat en Registratie* 1.

50 Oleskii Konashevych, 'General Concept of Real Estate Tokenization on Blockchain: The Right to Choose' [2020] 9(1) *European Property Law Journal* 21, 31 (https://doi.org/10.1515/eplj-2020-0003).

51 Monetary Authority of Singapore, 'MAS Clarifies Regulatory Position on the Offer of Digital Tokens in Singapore' <https://www.mas.gov.sg/news/media-releases/2017/mas-clarifies-regulatory-position-on-the-offer-of-digital-tokens-in-singapore#:~:text=The%20Monetary%20Authority%20of%20Singapore,289)%20(SFA)> accessed 15 July 2023; see for more elaborate also Monetary Authority of Singapore, 'A Guide to Digital Token Offerings' <https://www.mas.gov.sg/-/media/MAS/Sectors/Guidance/Guide-to-Digital-Token-Offerings-26-May-2020.pdf> accessed 15 July 2023; Philipp Hacker and Chris Thomale, 'Crypto-Securities Regulation: ICOs, Token Sales and Cryptocurrencies under EU Financial Law' [2018] 15(4) *European Company and Financial Law Review* 645, 647 (https://doi.org/10.1515/ecfr-2018-0021).

52 See art. 2(1) sub c of the Law of 3 October on Tokens and TT Service Providers, nr 2019.301, available via: <https://www.regierung.li/files/medienarchiv/950-6-01-09-2021-en.pdf> (last visited and found active on 15 July 2023). For the official, and legally binding, German version, see art. 2(1) sub c of Gesetz vom 3. Oktober 2019 über Token und VT-Dienstleister, which defines a token as '*eine Information auf einem VT-System, die: 1. Forderungs- oder Mitgliedschaftsrechte gegenüber einer Person, Rechte an Sachen oder andere absolute oder relative Rechte repräsentieren kann; und 2. einem oder mehreren VT-Identifikatoren zugeordnet wird*', available via: http://www.gesetze.li/konso/pdf/2019.301 (last visited and found active on 15 July 2023).

53 Rosa M Garcia-Teruel and Héctor Simón-Moreno, 'The Digital Tokenization of Property Rights. A Comparative Perspective' [2021] 41 *Computer Law & Security Review* 105543, 8–11 (https://doi.org/10.1016/j.clsr.2021.105543).

54 Oleskii Konashevych, 'General Concept of Real Estate Tokenization on Blockchain: The Right to Choose' [2020] 9(1) *European Property Law Journal* 21, 45 (https://doi.org/10.1515/eplj-2020-0003); Joshua AT Fairfield, 'Property as the Law of Virtual Things' [2022] 7 *Frontiers in Research Metrics and Analytics* 981964 (https://doi.org/10.3389/frma.2022.981964); For more on the bundle of sticks analogy see James E Penner, *Property Rights: A Re-Examination* (Oxford, Oxford University Press 2020).

55 Oleskii Konashevych, 'General Concept of Real Estate Tokenization on Blockchain: The Right to Choose' [2020] 9(1) *European Property Law Journal* 21, 45 (https://doi.org/10.1515/eplj-2020-0003).

56 Rosa M Garcia-Teruel and Héctor Simón-Moreno, 'The Digital Tokenization of Property Rights. A Comparative Perspective' [2021] 41 *Computer Law & Security Review* 105543, 11 (https://doi.org/10.1016/j.clsr.2021.105543).

57 See for more on blockchain-based rights management systems in the ambit of property law John Quinn and Barry Connolly, 'Distributed Ledger Technology and Property Registers: Displacement or Status Quo' [2021] 13(2) *Law, Innovation and Technology* 377 (https://doi.org/10.1080/17579961.2021.1977223).

Bibliography

Akkermans B, 'Sustainable Property Law – Towards a Revaluation of Our System of Property Law' in Bram Akkermans and Gijs van Dijck (eds), *Sustainability and Private Law* (The Hague, Eleven International Publishing 2020) 37–58

Amulya G and Johny J, 'Potential of Blockchain Technology in Supply Chain Management: A Literature Review' [2019] 49 *International Journal of Physical Distribution & Logistics Management* 881 (https://doi.org/10.1108/IJPDLM-11-2018-0371)

Antonopoulos A, *Mastering Bitcoin: Programming the Open Blockchain* (2nd edn, Sebastopol, O'Reilly 2017)

Antonopoulos A and Wood G, *Mastering Ethereum: Building Smart Contracts and DApps* (Sebastopol, O'Reilly 2018)

Arslan-Ayadin Ö, Shrestha P and Thewissen J, 'Blockchain as a Technology Backbone for an Open Energy Market' in André Dorsman, Özgür Arslan-Ayadin and James Thewissen (eds), *Regulations in the Energy Industry: Financial Economic and Legal Implications* (Cham, Springer 2020) 65–85

Berlee A, *Access to Personal Data in Public Land Registers: Balancing Publicity of Property Rights with the Right to Privacy and Data Protection* (The Hague, Eleven International Publishing 2018)

Buterin V, 'Ethereum White Paper – A Next Generation Smart Contract & Decentralized Application Platform' (*Ethereum*, 2014) <https://ethereum.org/en/whitepaper/>

Chaum D, 'Blind Signatures for Untraceable Payments' in David Chaum, Ronald L Rivest and Alan T Sherman (eds), *Advances in Cryptology* (Cham, Springer 1982) 199–203

Chaum D and van Antwerpen H, 'Undeniable Signatures' in Gilles Brassard (ed), *Advances in Cryptology – CRYPTO '89 Proceedings* (Cham, Springer 1990) 212–216

Daverschot M, 'Voorkoming Kadastrale Besmetting in Notariële Praktijk' [2013] 6984 *Weekblad voor Privaatrecht Notariaat en Registratie* 612

De Filippi M, Mannan M and Reijers W, 'The Alegality of Blockchain Technology' [2022] 41(3) *Policy and Society* 358 (https://doi.org/10.1093/polsoc/puac006)

De Filippi P, Wray C and Sileno G, 'Smart Contracts' [2021] 10(2) *Internet Policy Review* 2 (https://doi.org/10.14763/2021.2.1549)

DuPont Q, 'Experiments in Algorithmic Governance: A History and Ethnography of "The DAO" a Failed Decentralized Autonomous Organization' in Malcolm Campbell-Verduyn (ed), *Bitcoin and Beyond* (Abingdon, Routledge 2017) 157–177

Durovic M and Janssen A, 'The Formation of Blockchain-Based Smart Contracts in the Light of Contract Law' [2018] 26(6) *European Review of Private Law* 753 (https://doi.org/10.54648/erpl2018053)

Fairfield JAT, 'Property as the Law of Virtual Things' [2022] 7 *Frontiers in Research Metrics and Analytics* 981964 (https://doi.org/10.3389/frma.2022.981964)

Fairfield JAT, 'Tokenized: The Law of Non-Fungible Tokens and Unique Digital Property' [2022] 97 *Indiana Law Journal* 1261

Fernández-Caramés TM and Fraga-Lamas P, 'A Review on the Use of Blockchain for the Internet of Things' [2018] 6 *IEEE Access* 32979 (https://doi.org/10.1109/ACCESS.2018.2842685)

Garcia-Teruel RM and Simón-Moreno H, 'The Digital Tokenization of Property Rights. A Comparative Perspective' [2021] 41 *Computer Law & Security Review* 105543, 8–11 (https://doi.org/10.1016/j.clsr.2021.105543)

Guadamuz A, 'The Treachery of Images: Non-Fungible Tokens and Copyright' [2021] 16(12) *Journal of Intellectual Property Law & Practice* 1367 (https://doi.org/10.1093/jiplp/jpab152)

Hacker P and Thomale C, 'Crypto-Securities Regulation: ICOs, Token Sales and Cryptocurrencies under EU Financial Law' [2018] 15(4) *European Company and Financial Law Review* 645 (https://doi.org/10.1515/ecfr-2018-0021)

Hoops B and Heijerink LAM, 'Het Opstalrecht Duurzaam' [2022] 7353 *Weekblad voor Privaatrecht Notariaat en Registratie* 1

Hoops B and Meijerink LAM, 'Naschrift' [2022] 7358 *Weekblad voor Privaatrecht, Notariaat, en Registratie* 103

Janssen BAM, *Wie Heeft de Leiding?* (Deventer, Kluwer 2010)

Kaczorowska M, 'Blockchain-Based Land Registration: Possibilities and Challenges' [2019] 13(2) *Masaryk University Journal of Law and Technology* 339 (https://doi.org/10.5817/MUJLT2019-2-8)

Konashevych O, 'Constraints and Benefits of the Blockchain Use for Real Estate and Property Rights' [2020] 12(2) *Journal of Property, Planning, and Environmental Law* 109 (https://doi.org/10.1108/JPPEL-12-2019-0061)

Konashevych O, 'General Concept of Real Estate Tokenization on Blockchain: The Right to Choose' [2020] 9(1) *European Property Law Journal* 21 (https://doi.org/10.1515/eplj-2020-0003)

Larroumet C, *Les Biens, Droits Réels Principaux* (6th edn, Paris, Economica 2019)

Lehavi A, 'The Future of Property Rights: Digital Technology in the Real World' in Amnon Lehavi and Ronit Levine-Schnur (eds), *Disruptive Technology, Legal Innovation, and the Future of Real Estate* (New York, Springer International Publishing 2020) 59–79

Nogueroles Peiró N and Martinez García EJ, 'Blockchain and Land Registration Systems' [2017] 6(3) *European Property Law Journal* 296 (https://doi.org/10.1515/eplj-2017-0017)

Penner JE, *Property Rights: A Re-Examination* (Oxford, Oxford University Press 2020)

Ploeger H and Bounjouh H, *Erfpacht En Opstal* (Deventer, Kluwer 2019)

Quinn J and Connolly B, 'Distributed Ledger Technology and Property Registers: Displacement or Status Quo' [2021] 13(2) *Law, Innovation and Technology* 377 (https://doi.org/10.1080/17579961.2021.1977223)

Shelkovnikov A, 'Blockchain Applications in Banking' (*Deloitte*, 2016) <https://www2.deloitte.com/content/dam/Deloitte/uk/Documents/Innovation/deloitte-uk-blockchain-app-in-banking.pdf>

Sulkowski AJ, 'The Tao of DAO: Hardcoding Business Ethics on Blockchain' [2020] 3(2) *Business & Finance Law Review* 156

Sun X et al, 'A Simple Voting Protocol on Quantum Blockchain' [2019] 58 *International Journal of Theoretical Physics* 275 (https://doi.org/10.1007/s10773-018-3929-6)

Szabo N, 'Smart Contracts: Building Blocks for Digital Free Markets' [1996] 8(1) *Extropy: Journal of Transhumanist Thought* 50

Szabo N, *Secure Property Titles with Owner Authority* (Satoshi Nakamoto Institute, 1998) <nakamotoinstitute.org/secure-property-titles>

Thomas R, 'Blockchain's Incompatibility for Use as a Land Registry: Issues of Definition, Feasibility and Risk' [2017] 6(3) *European Property Law Journal* 361 (https://doi.org/10.1515/eplj-2017-0021)

van Erp S, *European and National Property Law: Osmosis or Growing Antagonism?* (Groningen, Europa Law Publishers 2006)

van Erp S, 'Land Registration and 'Disruptive' (or 'Trustworthy'?) Technologies: Tokenisation of Immovables Property' in Anabel Fraga, Elena Ioriatti and Sjef van Erp, *IMOLA II Project: The European Land Register Document (ELRD): A Common Semantic Model for Land Registers Interconnection* (Brussels, European Land Registry Association 2019)

van Erp S and Akkermans B, *Cases, Materials and Text on National, Supranational and International Property Law* (Oxford, Hart Publishing 2012)

Verheye B, 'Real Estate Publicity in a Blockchain World: A Critical Assessment' [2017] 6(3) *European Property Law Journal* 441 (https://doi.org/10.1515/eplj-2017-0020)

Verheye B, 'Land Registration in the Twenty-First Century: Blockchain Land Registers from a Civil Law Perspective' in Amnon Lehavi and Ronit Levine-Schnur (eds), *Disruptive Technology, Legal Innovation, and the Future of Real Estate* (New York, Springer International Publishing 2020)

Verstappen J, 'Comparative Law and Legal Issues Arising From Distributed Ledgers and Smart Contracts' in Bram Akkermans and Anna Berlee (eds), *'Sjef-Sache': Essays in Honour of Prof. mr. dr. J.H.M. (Sjef) van Erp on the Occasion of his Retirement* (The Hague, Eleven International Publishing 2021) 485

Verstappen J, *Legal Agreements on Smart Contract Platforms in European Systems of Private Law* (Cham, Springer 2023) 15

Vieweg K and Lorz S, *Sachenrecht* (Munich, Vahlen 2022)

von Wangenheim G, 'Blockchain-Based Land Registers: A Law-and-Economics Perspective' in Amnon Lehavi and Ronit Levine-Schnur (eds), *Disruptive Technology, Legal Innovation, and the Future of Real Estate* (New York, Springer International Publishing 2020)

Vos J, 'Blockchain and Land Administration: A Happy Marriage?' [2017] 6(3) *European Property Law Journal* 293 (https://doi.org/10.1515/eplj-2017-0018)

Vos J, *Blockchain-Based Land Administration: Panacea, Illusion or Something in between? Legal Interference of Registrars in the e-Conveyancing Process* (European Land Registry Association 2017) <https://www.elra.eu/wp-content/uploads/2017/02/10.-Jacques-Vos-Blockchain-based-Land-Registry.pdf>

Vos J, 'The Impact of "Disruptive" IT and the Registrar's Role in Future e-Conveyancing' [2017] 1 *International Review IPRA/CINDER* 68

Vos J and Roes R, 'Onduidelijkheden Rondom Inschrijving En Registratie' [2018] 7180 *Weekblad voor Privaatrecht Notariaat en Registratie* 115

Vreuger J, 'Dutch Blockchain, Real Estate and Land Registration' [2020] 12(2) *Journal of Property, Planning, and Environmental Law* 93 (https://doi.org/10.1108/JPPEL-11-2019-0053)

Werbach K and Cornell N, 'Contracts: Ex Machina' in Marcelo C Compagnucci, Mark Fenwick and Stefan Wrbka (eds), *Smart Contracts: Technological, Business and Legal Perspectives* (Oxford, Hart 2021) 313–382

Wood G, 'Ethereum: A Secure and Decentralised Generalised Transactions Ledger' (*Berlin Version*, 25 August 2023) <https://ethereum.github.io/yellowpaper/paper.pdf>

19
THE ECOLOGICAL FUNCTION OF PROPERTY IN THE CONSTITUTION OF SLOVENIA

Miha Juhart[1]

19.1 Introduction

Since the 1992 United Nations Conference on Environment and Development in Rio de Janeiro, the terms sustainable development and sustainable use have become commonplace in international environmental documents. The terms encompass the environment, economic development, and social rights, all of which are interconnected and crucial to the growth of society.[2] As a result of countries' acceptance of international commitments for the provision of sustainable development, it is now their duty and responsibility to include measures aimed at accomplishing these objectives into their domestic legal framework. The manner in which land is utilised is one area of regulation that is vital for achieving these objectives. As space is a limited resource, sustainable development goals can only be attained via the efficient use of land. When enabling the use of space, it is therefore impossible to simply consider the private interest of the landowner, but also the broader general interest of the community or the public interest. This creates a tension between the private interest of the landowner, who views property as the largest right that ensures any sort of use, and the public interest, which aims to achieve generally acceptable goals, to which sustainable development increasingly belongs.

Land use restrictions for grounds of public interest have existed for a very long time. Here, however, the question of whether the public interest is substantial enough to sustain property restrictions may arise. It is crucial, therefore, that property was included among fundamental rights and liberties during the founding of the legal system. This also allows for an abstract evaluation of the constitutionality of public law restrictions on the limiting of private property. Due to its (really its holders and objects) engagement in social and economic relationships, it is necessary to impose (external) limitations to this power, which denote the limits of property's consequences.[3]

The Republic of Slovenia's recent Constitution[4] *expressis verbis* mentions the ecological function of the property when it regulates the conditions for limitation of the right to property in the public interest. This contribution is not intended to give a comparative legal analysis of the grounds for the limitation of the right to property in order to ensure the values associated with the environment and its protection.[5] In the chapter, I would like to present where are limits of the restriction of property rights in the legal system, in which the purpose of environmental protection is explicitly emphasised. The case law of the Constitutional Court in this field is relatively modest. So far, it is

 DOI: 10.4324/9781032662046-23

mostly related to property rights in the forests. The Constitutional Court in the argumentation of its decisions does not refer specifically to the ecological function of property but refers in general terms to the public interest. Therefore, it is not yet possible to assess, whether the explicit mention of the ecological function of property contributes more to interest in environmental protection and sustainable development.

19.2 The Ecological Function of Property in the Constitution of the Republic of Slovenia

The Constitution includes the right to property among fundamental rights and liberties. The Constitution protects private property and inheritance rights (Article 33 of the Constitution). Article 67 of the Constitution specifies that the manner in which property is acquired and enjoyed shall be governed by law in order to ensure its economic, social, and environmental function. When we discuss the economic, social, and ecological functions of property, we are referring primarily to the owner's responsibilities and limitations in acquiring and enjoying the property.[6] These responsibilities and limitations are outlined in the Constitution, but they should be governed in greater detail by statute. Specifically, environmental protection, community interests (ensuring efficient use of land, the possibility of expropriation, etc.), protection of public goods, natural resources, and land, as well as restrictions due to neighbourly relations and the prohibition of economic activities contrary to the public interest, should be considered. In this view, the social purpose of property must be understood. The second aspect of this social function is the provision of funds for the state's social functions (such as social insurance) and cash for state financing. Particularly interesting about the new Slovenian Constitution is its emphasis on the ecological function of property.[7]

The ecological content of property encompasses nature with its components, forces, connections, transformations, and laws, which serve as the basis of life for all living things, such as animals and plants, and are also a part of these things. The idea of nature comprises the biosphere, while the aspects of the environment consist of nature management, natural resource conservation, and landscape preservation. Air, water, and soil are elements. The primary objective is human behaviour that ensures the preservation of human life's foundations and recreational opportunities. The forms of protection are to protect, nurture and develop. The goals are also defined as (1) keeping as many areas as possible unbuilt; protecting natural resources in terms of their ability to exist; (2) rational use of goods, limitation in particular if it is a question of rare goods; (3) special protection of assets that are at risk (extinction of animal and plant species); (4) fertility of the land; (5) landscape image; (6) vegetation, especially free-living flora and fauna; (7) water, air, peace, climate, and recreational conditions.[8] Due to their interdependence and interconnectedness, the listed goods cannot be safeguarded individually. The necessity for comprehensive environmental protection that considers interdependencies is becoming increasingly apparent. To achieve these goals, the state must include in its laws directives and prohibitions requiring individuals to do, permit, or refrain from doing something. Such obligations may apply to everyone (prohibitions on the handling of certain hazardous substances and emissions); they may be regarded as specially protected areas (natural parks, reserves, monuments) for specific species of flora and fauna; a strict regime applies to air and water (emissions); a special regime applies to forests.[9] Traffic, industrial, and residential emissions, as well as leisure-time population mobility, place a rising burden on the environment. This results in treatments that are detrimental to soil, water, and air, which are all essential to life. It also pertains to the preservation of flora and fauna, as the ecological balance must be preserved.[10] Therefore, state interventions are permitted to defend life's basics, even if they curtail fundamental rights and liberties, subject to a rigorous proportionality test.[11] The relationship between property and economic, social, and ecological functions conflicts with the fundamental constitutional notion of property freedom. The result of weighing both interests must be harmonisation. This harmonisation is implemented by the proportionality principle.[12]

When a legislator interferes with the constitutionally protected rights of an individual, such interference shall be subject to additional scrutiny to determine if it is constitutionally permissible. The proportionality test prohibits excessive interference by the law with the rights of persons and demands a proper assessment of whether the statutory measures are consistent with its intent. The measure must be justified by the objective, namely in a manner that has the least possible impact on the rights and interests of the affected people. 'The measures must be appropriate for the fulfilment of the legislator's objectives, necessary for their implementation in accordance with the objective interests of citizens and must not be disproportionate to the social or political significance of these objectives.'[13] The aforementioned clause of Article 67 of the Constitution should guide this weighing. Consequently, this clause grants the legislature the ability to regulate the process of gaining and enjoying the property, considering the three uses of the property described above. The legislature is not required to specify which function the restriction is meant to serve. All three functions must be viewed as interdependent and interconnected. The legislative ability to encroach on the property is limited. If the legislature exceeds its authority, it is no more a matter of regulating the method of property enjoyment, but of infringing on the right to private property. The location of this boundary depends not only on the nature of the property but also on the legal obligation imposed on the owner in determining the way of property enjoyment.[14]

In Slovenia, the property law is governed by the Law of Property Code (*Stvarnopravni zakonik* – SPZ), which was adopted in 2002 and has been in force since 1 January 2003.[15] Article 37 specifies the definition of property and its content. Property is the right to possess, use, and enjoy it to the greatest extent feasible, and to dispose of it. Only the law can impose restrictions on use, enjoyment, and disposal, thereby interfering with the essence of property rights. The most extensive use is relative, as the owner must adhere to legal limits on its application, even if these are against his or her will, interests, economic necessities, or the purpose for which the property right was acquired. Regarding public law constraints, the sole recourse the owner has is the constitutional guarantee of the protection of private property.[16]

Environmental Protection Act (*Zakonu o varstvu okolja*) also stipulates the general limitation of the substance of the property right due to the ecological function of the property.[17] Article 17 provides that in order to consider ecological function in the use of property, it is required to ensure the maintenance and enhancement of environmental quality, the protection of natural values, and biodiversity.

19.3 Effecting the Ecological Function of Property – The Example of Forests

Exploiting forests is also directly tied to the requirement for sustainable development and natural resource utilisation. The preservation of tree species, habitat for living organisms, and the living conditions of some indigenous peoples are all fundamentally important, but so are the economic activities connected to the use of wood and the control of greenhouse gas emissions into the atmosphere.[18] Forests, as a natural resource, combine the essential elements of sustainable development, making them an outstanding example for demonstrating a method of restricting property rights to achieve this goal. Because the measures may seriously infringe on individuals' rights, they are motivated to initiate procedures to protect their individual interests, including a request to the Constitutional Court for an assessment of the compliance of the prescribed restrictions with the Constitution.

Forests Act (*Zakon o gozdovih* – ZG)[19] is a fundamental instrument that comprehensively regulates the legal rules applying to the forest as an object of property. It governs the protection, cultivation, exploitation, and use of forests, as well as the disposal of forests as a natural resource, with the objective of ensuring sustainable co-natural and multifunctional management in accordance with the principles of environmental protection and natural values, the permanent and optimal functioning of forests as an ecosystem, and the realisation of their functions.[20] The connection between the purpose of the law and the constitutional requirement to ensure the ecological function of property is crystal evident. The law makes explicit reference to environmental preservation and certain aspects of sustainable development, which might considerably affect the constitutional assessment of the restrictions. The unique significance of the forest to society as a whole is a justification for greater interference with private property rights.

Article 5 of the ZG specifies the limitations on the forest owner's property rights. In the same provision, the Act also refers to Article 67 of the Constitution and the functions of property, which exceed a merely individualistic conception. The ecological, social, and productive functions of forests are safeguarded through the exercise of property rights. The forest owner must (1) allow free access to his/her forest, excluding cases of for-profit tourism or recreational activities; (2) permit beekeeping, hunting, and recreational gathering of fruits, herbaceous plants, mushrooms, and wild animals in accordance with the regulations; and (3) manage the forest in accordance with management plans and administrative acts. It is an extremely stringent restriction on property rights, based on both the social and ecological functions of property. The forest owner must accept these restrictions, and his/her property rights may be exercised within the prescribed boundaries. If the property is unrestricted, the owner of the forest must refrain from particular behaviours that are typical of the owner's activity. The law goes a step further. In some parts, the forest owner is assigned responsibilities that he/she is obliged to carry out in his/her forest to maintain its general benefit and function.

19.3.1 General Right of Access

Article 5 of the ZG stipulates that the first restraint on property rights is the possibility of free access. Free access means that individuals can freely wander through the forest and have other access to the forest. It is a universal right that everyone may exercise. The right to free access is comparable to the usage of a public road. Whoever possesses the right of free access shall exercise it in a manner that places the least burden on the forest owner. Any type of damage is prohibited, and the user must compensate the owner. If a landowner dedicates a forest to a particular economic activity, the right of free access may be restricted only in extraordinary circumstances. In this instance, however, the owner must first acquire a public authority decision authorising the activity and restricting access. In such a case, the forest owner may install a fence in extraordinary circumstances. Also banned is the erection of a fence in one's own forest. Unless a special exemption is allowed, the owner may not restrict the right of access to the forest for walking or recreation.[21]

19.3.2 Limiting Hunting and Allowing Fruit Picking

Increased encroachment on the forest owner's property rights implies a general right for anyone to practise beekeeping, hunting, and recreational gathering of fruits, herbs, and mushrooms in a foreign forest. This general right has a direct impact on the forest owner's economic position. True, the benefits mentioned are minor in comparison to the yield provided by the wood. However, because it is an economic function with its own value, the owner's benefit is limited. This is especially true for hunting and fruit-gathering regulations. According to Slovenian law, the Republic of

Slovenia is responsible for managing wild game.[22] The right to hunt is also part of the wild game management framework. The hunting right is not connected to the property right on the land, i.e., it is not a part of it but rather belongs to the state. The state has the authority to transfer hunting rights to certain legal entities under public or private law that perform specific tasks related to wild games and also assume liability for damage caused by the game. These are primarily hunting organisations that individuals with a hunting interest can join. This means that hunting can only be done by individuals who have obtained a hunting permit from the hunting organisation. Authorised individuals do not have any real estate rights in the forest, but they can carry out their activity in a foreign forest subject to the restrictions set out in Article 5 of the ZG. Such hunting regulation, which deprives the forest owner of any economic benefit, was subject to the Constitutional Court's decision. The Constitutional Court concluded in two decisions that the legal system is consistent with the Constitution. The Constitutional Court emphasised in the first decision that the legislator did not violate the constitutional guarantee of private property when it completely removed the benefits of hunting from the forest owner.[23] By regulating the right to hunt as a public law right separate from the property right on the land, which does not originate from the property right on the land, it moved within the limits of determining the way the property is used. The Constitutional Court specifically stated in this regard that the regulation in which the right to hunt is an independent right afforded by public law that does not originate from the property right on land falls within the limits of determining the way of enjoying the property.[24] Although the property right on the land does not include entitlements related to the exercise of the hunting right, the forest owners retain all of the essential elements of the property right. In another case, the Constitutional Court ruled on the forest owner's obligation to allow movement on his or her land for hunting purposes.[25] The Constitutional Court determined that the legislator's goal in encroaching on the right to private property was constitutionally consistent. As a result, it applied a strict test to determine the admissibility of property restrictions and concluded that the obligations of landowners to allow qualified persons to use their land for the exercise of the right to hunt and the restrictions on owners in the use of land and forests for the exercise of the right to hunt are a necessary and appropriate measure to achieve a constitutionally compatible goal. The Constitutional Court determined that the challenged ZG provision meets the proportionality test in the narrower sense. It stated that the protection of wild game should be prioritised and that the benefits of exercising the right to hunt in a manner determined by law, which ensures the protection of the natural asset, outweigh the weight of the interference with the rights of land and forest owners.

The Constitutional Court's decision also addressed the limitation of property rights, stating that the owner of the forest must allow anyone to pick mushrooms and fruits. The petition was denied by the Constitutional Court based on arguments related to the ecological function of property. At the same time, the content of property as a legal institute is determined by the functions bestowed upon it by the legal order. The fact that an individual's property rights are not unlimited but must consider the interests of other members of the community and the community as a whole was already a part of the notion of property rights. The property right is protected in democratic countries' constitutions as a special form of material protection of human freedom, which, along with other constitutional rights, should ensure the possibility of a free life, meeting life's needs, and personal development in society. At the same time, these legal systems recognise that the individual's freedom in the property field cannot be unrestricted, binding only to the individual owner, but that the interests of other members of the community must also be considered in the exercise of this right. According to Article 67 of the Constitution, the way property is acquired and enjoyed must be established by law to ensure its economic, social, and environmental function. The same article also affords the legislator the authority to determine the substance of a property right in the context of its functions. The statute regulated forest owner entitlements based on the purpose and use of the forests so that it also considered the benefit of non-owners of the forests in accordance with the social and ecological function of the property. The contested provision grants them free access and movement in the forest, as well as the right to beekeeping, hunting, and recreational gathering of fruits, herbs, mushrooms, and wild animals. In cases where the forest owner also grows trees for the sake of the fruits, the forest owner has the option of having the competent body of the local community issue a ban on the collection of such fruits at his request.[26]

19.3.3 Forest Management

When managing forests, the forest owner must consider the content of forest management plans. The forest economic plan is adopted by the Slovenian Forest Service (TSF). The TSF is a public entity founded by the Republic of Slovenia that provides public forestry services in all privately owned and publicly owned Slovenian woods. It is a part of the public administration. Its tasks and activities related to all fields of forest management at the national, regional, and local levels: forest management planning, monitoring of forests, silviculture and forest protection, use of forest technologies, construction, and maintenance of forest roads, monitoring of wild animal populations, hunting, forestry extension service for forest owners, rural development activities, raising awareness, and educating forest owners.[27]

TSF issues an individual decision to each forest owner, with which the more precise manner of forest management is defined for a period of five years. The decision primarily describes the amount and structure of trees to be felled to the greatest extent possible, as well as forest restoration cultivation work. The forest owner may only utilise the forest within the parameters of the issued decision, and he/she is required to carry out all specified tasks for the restoration and protection of the forest, even if the costs exceed the benefits. In addition, the institution may, by special decision, require the forest owner to carry out at his or her own expense all measures essential for the conservation of the forest, such as sanitary felling, harvesting, and removal of sick or insect-infested, badly damaged, or fallen trees. Nonetheless, the manner in which forests are managed, as prescribed by law, violates the right to private property, and no petition has been filed with the Constitutional Court to determine whether these restrictions comply with the Constitution.

19.4 Conclusion

Forest property rights have long been subject to extremely stringent restrictions. These restrictions pertain to various aspects of the substance of property rights. It is not just about restrictions that the owner must enable everyone to utilise the forest to a certain extent. The constraints on the extraction and use of the economic benefit offered by forest ownership are substantially more delicate in terms of their substance. The most severe intrusion on private property is unquestionably the obligation of the forest owner to carry out restoration and conservation efforts at his or her own expense, even if the expenditures exceed the owner's benefit. The Constitutional Court typically accepts such constraints on property rights as consistent with the provision of Article 67 of the Constitution, which provides that property rights might be limited by law to achieve their economic, social, and ecological functions. In order to fulfil the aims of sustainable development, the legislator has a variety of choices for limiting property rights, as illustrated by the example of the forest. We can reasonably anticipate that additional exceptional measures, which would be established by the legislator as a limitation of the property right on other places and things, could also pass the assessment of compliance with the Constitution.

Notes

1 PhD, Professor of Law.
2 Maša Kovič Dine, *Mednarodne obveznosti držav za zagotavljanje trajnostnega razvoja gozdov*, vol 73 (Zbornik znanstvenih razprav, 2013) 163.
3 Jan Zobec, *Komentar Ustave Republike Slovenije* (2011) 67 člen, Art. 67, 1 <http://e-kurs.si> accessed 26 May 2023.
4 Official Gazette of the Republic of Slovenia, Nos. 33/91-I, 42/97 – UZS68, 66/00 – UZ80, 24/03 – UZ3a, 47, 68, 69/04 – UZ14, 69/04 – UZ43, 69/04 – UZ50, 68/06 – UZ121,140,143, 47/13 – UZ148, 47/13 – UZ90,97,99, 75/16 – UZ70a and 92/21 – UZ62a.
5 For a comparative legal analysis, see in more detail M Pajvančić, N Petrušić, *The Social Function of Property Ownership in the Context of Sustainable Development* (Зборник радова Правног факултета у Нишу, 2018) Број 79. The analysis shows that the purpose of environmental protection is ensured through the general conditions for the limitation of the right to property in the public interest. Very detailed on environmental protection as an element of public interest allowing for interference with the right to property N. Bryner, 'Public Interests and Private Land: The Ecological Function of Property in Brazil' (2016) 34 Virginia Environmental Law Journal 1.
6 Lojze Ude, Razvoj lastninske zakonodaje, Podjetje in delo (No. 5-6/1994) 739.
7 Lojze Ude, Ustava Republike Slovenije, Pravna praksa (No. 8/1992) 2.
8 Janez Šinkovec, Ustavno zagotavljanje lastnine, Podjetje in delo (No. 6/1992) 569.
9 Janez Šinkovec, Ustavno zagotavljanje lastnine, Podjetje in delo (No. 6/1992) 569.
10 Janez Šinkovec, Nekatere spremembe v sistematiki lastninskih razmerij: Ustavna načela, svoboščine in lastnina, Podjetje in delo (No. 6-7/2001) 908–914.
11 For more on the proportionality test, see Lovro Šturm, Matej Avbelj, *Komentar Ustave Republike Slovenije*, (2019) 2 člen <http://e-kurs.si> accessed 26 May 2023.
12 Andrej Berden, *SPZ s komentarjem* (ed. Juhart, Tratnik, Vrenčur) (GV Založba 2004, Ljubljana) 187.
13 Decision of the Constitutional Court No. U-I-77/93.
14 Decision of the Constitutional Court No. U-I-70/04.
15 Official Gazette of the RS, Nos. 87/02, 91/13 and 23/20.
16 Renato Vrenčur, *SPZ s komentarjem* (Juhart, Tratnik, Vrenčur eds, Uradni list 2016, Ljubljana) 218.
17 ZVO-2 – Official Gazette of the Republic of Slovenia, No. 44/22.
18 Maša Kovič Dine, *Mednarodne obveznosti držav za zagotavljanje trajnostnega razvoja gozdov*, vol. 73 (Zbornik znanstvenih razprav, 2013) 163.
19 Official Gazette of the Republic of Slovenia, No. 30/93, 56/99 – ZON, 67/02, 110/02 – ZGO-1, 115/06 – ORZG40, 110/07, 106/10, 63/13, 101/13 – ZDavNepr, 17/14, 22/14 – CC Dec., 24/15, 9/16 – ZGGLRS and 77/16.
20 ZG, Art. 1.
21 Nana Weber, Smetenje v gozdu in dolžnosti lastnika gozdnega zemljišča, Pravna praksa (No. 1-2, 2019) 16.
22 Game and Hunting Act (Official Gazette of the Republic of Slovenia, No. 16/04, 120/06 – CC decision, 17/08, 46/14 – ZON-C, 31/18, 65/20, 97/20 – corr., 44/22 and 158/22).
23 Decision of the Constitutional Court No. U-I-98/04.
24 Jan Zobec, Komentar Ustave Republike Slovenije (2011) 67 člen, Art. 67, 12 <http://e-kurs.si> accessed 26 May 2023.
25 Decision of the Constitutional Court No. U-I-40/06.
26 Decision of the Constitutional Court of the Republic of Slovenia in case U-I-51/95.
27 Slovenia Forest Service < http://www.zgs.si/eng/homepage/index.html > accessed 26 May 2023.

Bibliography

Berden A, *SPZ s komentarjem* (ed. Juhart, Tratnik, Vrenčur) (GV Založba 2004, Ljubljana) 187
Bryner N, 'Public Interests and Private Land: The Ecological Function of Property in Brazil' (2016) 34 Virginia Environmental Law Journal 1
Dine M, *Mednarodne obveznosti držav za zagotavljanje trajnostnega razvoja gozdov*, vol 73 (Zbornik znanstvenih razprav, 2013) 163
Pajvančić M, Petrušić N, *The Social Function of Property Ownership in the Context of Sustainable Development* (Зборник радова Правног факултета у Нишу, 2018) Број 79

Šinkovec J, *Ustavno zagotavljanje lastnine, Podjetje in delo* (No. 6/1992) 569
Šinkovec J, *Nekatere spremembe v sistematiki lastninskih razmerij: Ustavna načela, svoboščine in lastnina, Podjetje in delo* (No. 6-7/2001) 908–914
Šturm L, Avbelj M, *Komentar Ustave Republike Slovenije* (2019) 2 člen <http://e-kurs.si> accessed 26 May 2023
Ude L, *Razvoj lastninske zakonodaje, Podjetje in delo* (No. 5-6/1994) 739
Ude L, *Ustava Republike Slovenije, Pravna praksa* (No. 8/1992) 2
Vrenčur R, *SPZ s komentarjem* (Juhart, Tratnik, Vrenčur eds, Uradni list 2016, Ljubljana) 218
Weber N, *Smetenje v gozdu in dolžnosti lastnika gozdnega zemljišča, Pravna praksa* (No. 1-2, 2019) 16
Zobec J, *Komentar Ustave Republike Slovenije* (2011) 67 člen, Art. 67, 1 <http://e-kurs.si> accessed 26 May 2023

20

ELECTROMOBILITY RECONSTRUCTION OF BUILDINGS AND SUSTAINABLE PROPERTY MANAGEMENT

Tatjana Josipović[1]

20.1 Introduction

The green transition to climate neutrality by 2050 is one of the most important strategic goals of the European Union. Europe should become the first climate-neutral continent in the world. The European Commission launched the European Green Deal[2] in December 2019 as the new EU's growth strategy to reach the 2050 goal. The main target is to ensure a fair and prosperous society with a modern resource-efficient and competitive economy with no net emissions of greenhouse gases. This must be achieved by cutting emissions, investing in green technologies, and protecting the natural environment. The overall strategy is aimed at the protection, conservation, and enhancement of the EU's natural capital, the protection of the health and well-being of citizens from any environment-related risks and impacts. It is emphasised that the green transition must be effective, just, and inclusive for citizens, national, regional, and local authorities, as well as for civil society and industry.[3]

Making Europe climate-neutral by 2050 is a very challenging and a complex process requiring an efficient cooperation between EU Institutions and Member States. In June 2021, Regulation (EU) 2021/1119 was adopted establishing the framework for achieving climate neutrality (the European Climate Law).[4] In the new Regulation, long-term goals for climate neutrality by 2050, set forth in the European Green Deal, are laid down as legally binding targets for both EU Institutions and Member States.[5] The Regulation also defines the Intermediate Union's climate targets which both the EU and its Member States are bound to achieve by 2030. They have committed themselves to reducing, by 2030, net greenhouse gas emissions by at least 55% compared to the levels of 1990.[6] The measures for the reduction of emissions by 2030 are adopted within the 'Fit for 55' package (hereinafter: 'Fit for 55'). The European Commission has already adopted a series of proposals to revise and update the EU legislation and has put in place the current regulations, which are of extreme importance for the EU's intermediate climate goals to be achieved by 2030.

A green and inclusive transition can be successful only if it is based on a coordinated, comprehensive and cross-sectoral approach to mutually linked EU policies that are crucial for the green transition. Within the 'Fit for 55', a reform of many EU policies is planned that is connected with land and forest use, renewable energy, energy efficiency, alternative fuels infrastructure, energy taxation, construction and renovation of buildings, and transport etc. It is particularly highlighted

 DOI: 10.4324/9781032662046-24

that a modernisation of the EU's building stock and the green transformation of transport are crucial sectors for achieving the climate objectives for 2030 and 2050. Constructions and renovations of buildings must be carried out more efficiently, by considering energy-saving and resource-saving methods.[7] The green transformation of transport calls for a reduction of transport emissions by 2050. It is necessary to accelerate the shift to sustainable and smart mobility through the use of zero- and low-emission vehicles.[8]

Transformation of the construction and building sectors is very important for any green transition because most of the existing buildings (75%) are not energy-efficient. In fact, they are responsible for 40% of energy consumption and for 36% of greenhouse gas emissions from energy.[9] The 'Fit for 55' package envisages a comprehensive revision of the Energy Performance of Buildings Directive (hereinafter: EPBD).[10,11] The main target is to increase the renovation rates of both public and private buildings, to modernise and decarbonise buildings, to build new zero-emission buildings by 2030 and to transform the existing buildings to zero-emission buildings by 2050.[12]

In addition, the green transformation of transport is essential because transport is responsible for 25% of total emissions.[13] Today, the number of low- and zero-emission vehicles in the vehicle fleet is very low. To achieve climate neutrality, a reduction of 90% in transport emissions by 2050 is necessary. A shift to sustainable and smart mobility must be based on a transition to zero-emission vehicles. It is planned that at least 30 million zero-emission vehicles will be in operation on the roads of Europe by 2030. By 2050, nearly all cars should be zero-emission ones.[14] The implementation of these plans has already started in the form of agreements on the new CO_2 standards for any new cars. In October 2022, within the 'Fit for 55' package, the first agreement was reached between the European Parliament and the Council, making sure that all new cars and vans, registered in Europe by 2035, would be zero-emission ones.[15] However, such a gradual transition to zero-emission vehicles must be accompanied by the construction of charging and refuelling infrastructure. It is expected that by 2025, about 1 million public charging facilities will be needed for 13 million zero- and low-emission vehicles. Indeed, by 2030, as many as 3 million public charging points will be installed.[16,17]

20.2 Private Charging Points in Buildings

Sustainable mobility after the transition to electric vehicles[18] involves, besides the construction of a large number of public charging stations,[19] increasingly intensive development of the so-called private charging in residential and non-residential private buildings.[20] The situation presses for the installation of the necessary mobility infrastructure in, or next to, private residential and commercial buildings.[21] Empirical research shows that a shortage of charging points for electric cars in private residential and commercial buildings, where the owners of electrically powered vehicles live or work, may have a negative impact on people's interest to buy such vehicles. Most owners of electric vehicles charge their vehicles at home or at work, i.e., during long parking times. Besides the price, the possibility of access to private charging in the buildings where the owners live, or work is for most consumers a decisive factor in a decision on whether to buy an electrically powered vehicle.[22]

Therefore, an efficient shift to low- and zero-emission mobility is inseparably linked with the development of sustainable construction, renovation, and housing. Measures for the energy performance of buildings, among other things, also include various installations to enhance electromobility in private buildings. In the EPBD, and particularly after the amendments of 2018, emphasis has already been placed on the importance of the installation of cable and charging infrastructure with charging points for electric vehicles when constructing or renovating private buildings. All Member States must ensure that, by their national rules on the requirements for legal construction or renovation, installation of a sufficient number of charging points and ducting infrastructures is provided.[23] It is held that ' building codes can be effectively used to introduce targeted requirements to support the deployment of recharging infrastructure in car parks of residential and non-residential buildings.[24] In the 'Fit for 55' package, there is a proposal to enhance the promotion of electric mobility through building policies. Indeed, in the Proposal for a revision of the EPBD of 2021, the number of mandatory charging points in residential and non-residential buildings has increased.[25]

The European rules on the deployment of charging points for electrically powered cars in private buildings lay down the obligations of Member States to provide, in their national laws, the installation of charging points as the responsibility of investors or owners of buildings in the course of construction or major renovation of buildings. The Member States' obligation to plan the installation of a particular number of charging points is a matter of public law and it is regulated within the framework of the construction rules on the requirements for the issuance of construction or major renovation permits, i.e., the conditions for legal constructions or renovations. The provisions of the EPBD on charging points in private buildings do not lay down the obligations of Member States to introduce measures to ensure the installation of charging facilities in the already existing buildings not undergoing major renovations. There is no obligation of Member States to directly intervene in private law relations between owners/landlords and tenants regarding the decisions on the installation of the charging infrastructure in private buildings. The only obligation of Member States is to provide measures to simplify the deployment of charging points in the new and existing buildings and to address possible regulatory barriers, including the permission and approval procedures.[26] However, it is expressly emphasised that the Member States' obligations are '(..) without prejudice to the property and tenancy law of the Member States'.[27]

The proposal for a revision of the EPBD of 2021, adopted within the 'Fit for 55' package, regarding the Member States' obligations in connection with the installation of charging infrastructure in private buildings, brings about a significant change. Apart from the already existing obligations of Member States, it is proposed to expressly provide for their obligation to remove barriers to the installations of charging facilities in residential buildings with parking spaces, '(..) in particular the need to obtain consent from the landlord or co-owners for a private recharging point for own use'.[28,29]

It arises from the current proposals for a revision of the EPBD that the European legislator is of the opinion that the barriers existing when installing charging infrastructure in private buildings can no longer be efficiently accomplished only by way of a building policy. Clearly, the European legislator's position is that the traditional private law regulation of lease/rent agreements, and of the property management of buildings involving any construction changes, may also be barriers to the installation of infrastructure aimed at electromobility in existing buildings. Namely, the installation of charging points in private buildings requires that, in order to carry out any major renovation, construction works or financing, they must previously be decided by the owners of the building. These decisions are rendered in accordance with the national property law regulations on the management of residential and non-residential buildings, private ownership/co-ownership/condominium, or in accordance with the regulations which, through lease/rent agreements provide for the relations between landlords and lessees/leaseholders. Indeed, the owners' decisions on undertaking construction works for the installation of charging points are, as a rule, the necessary condition for the initiation of administrative procedures to issue building permits for constructions/renovations. In a large number of cases, these are national rules on property management of buildings based on unanimous decision-making processes involving major renovations. However, such rules are no longer adequate for effective sustainable management and electromobility. To adjust buildings to electromobility, new regulation of private law relations among building owners, co-owners, landlords, and tenants is necessary when the management of buildings is concerned. It must be more efficient to enable speedy renovation and introduction of structural changes in buildings before the installation of the charging infrastructure.

The new EU proposals for the enhancement of energy efficiency in private buildings indirectly provide for the obligation of Member States to reform their national property law on the management of buildings, as well as tenancy law, in order to remove any barriers to private charging points ensuing from the rules on the management of buildings or flats. From the sphere of public law, we must step into the sphere of private law, particularly in the sphere of property law and tenancy law. These new trends in the European regulations to remove the barriers in the development of sustainable mobility in private buildings bring about numerous challenges for both the European legislator and the Member States. On the one hand, the European Union is facing a major challenge of how to explain its competencies in regard to Article 345 TFEU laying down that the Treaties are without prejudice to the rules of Member States governing the system of property ownership. On the other hand, a particularly challenging task is the one placed by the European Union its Member States: to establish new rules on property management to ensure efficient electromobility reconstructions of buildings that requires a completely new approach to the organisation of private law relations among the co-owners of buildings, or between landlords and tenants, as well as the introduction of some new subjective rights of flat owners and tenants. An extremely challenging task is how to organise the financing and the process of making decisions on the structural changes regarding the common parts of buildings and the rented flats in order to achieve efficient electromobility in existing buildings. It is also necessary to take into account not only the necessity of ensuring a just balance in mutual relations among flat owners, and between landlords and tenants, but also that possible restrictions on ownership/co-ownership imposed for the sake of electromobility must be proportional and justified in the context of the protection of fundamental rights.

20.3 Removing Barriers to Electromobility by Reforms of Property Management

20.3.1 General

Due to an increasingly intensive green transition in existing buildings, national property law rules on the management and decisions on how to upgrade multi-owned buildings are becoming more and more serious regulatory barriers to the deployment of charging infrastructure and private charging points.[30] This became obvious when amending the EPBD of 2018. Member States were faced with an obligation to address possible regulatory barriers to the deployment of charging points in buildings, including the necessary permission and approval procedures.[31] Such regulatory barriers were even more implicitly identified in the Proposal for a new EPBD (recast) of 2021 aiming at putting more effort to promote electromobility in private buildings.[32] The barriers to electromobility reconstructions in existing buildings mostly arise from the primary focus of the traditional concept of property management of multi-owned buildings on the protection of individual interests of owners and landlords for which the consent/agreement of owners/landlords is required when installing a charging infrastructure.

The traditional property management of multi-owned buildings is based on an individualistic concept of ownership as an absolute and exclusive *right in rem*,[33] on private autonomy and freedom of contracting. With the help of these concepts, national legislators resort to various ways of solving the conflicts arising from the necessity to decide on the management of buildings being often faced with conflicting interests of the owners. The traditional approach gives preference to the protection of individual owners' rights and interests within a community of owners. This is essentially the main reason why, for any upgrades of buildings, it is necessary to have the consent or agreement of the owners. Preference is given to a consensual approach to management within owners' community/association, to make it possible for them to regulate the management of the building in conformity with their individual ownership rights and interests. Indeed, the owners' mutual relations, concerning their decisions on the management of their building, its maintenance, repairs, upgrades of common parts, management costs, and the like, are, as a rule, taken care of by the owners' agreement, management agreement or by individual decisions on management.[34] Therefore, any legal relations among the owners regarding the management of their building are regulated by default property law rules applying only if there is no other agreement made by the owners. At the same time, the freedom of contracting various property management issues, primarily because of the protection of individual ownership rights of the owners, is sometimes limited by strict property law rules. The traditional rules on property management of multi-owned buildings sometimes require unanimous decisions by all owners, particularly when major renovations or extraordinary management must be decided. The main goal is to ensure that every owner freely and autonomously participates in any decision processes on renovations or upgrades that may violate his/her rights, may have a negative impact on the exercise of his/her property rights on a flat or a building, or may cause unexpected costs for upgrades. Complex decision-making procedures are then envisaged to ensure the decision-making freedom of every owner. In addition, in order to protect the individual property rights of the owners, there are some restrictions on upgrades in flats. Namely, owners must make sure that improvements will not cause any damage to the building, encroach upon any common part of the building or other flats, i.e., that they will not jeopardise other owner's interests. Within the framework of the traditional tenancy law rights, the focus is also on the protection of individual ownership rights of landlords. The rights of tenants to any upgrades in rented flats are limited. Priority is also given to the protection of landlords (flat owners) whose prior consent is a prerequisite for any changes in the flat or in common parts of the building.

Removing barriers to the development of electromobility in multi-owned buildings requires a shift from an individualistic property management approach to a new concept by which, because of the interests of the community of flat owners in sustainable electromobility, their individual property rights are limited.[35] When deciding on upgrades to achieve electromobility, priority should in some way be given to owners' interests in installing private charging points, to the collective interests of the community of owners to modernise the building for electromobility reconstruction and to the protection of public interests to achieve climate neutrality through electromobility.[36] These goals can be achieved by various approaches to management that are crucial for the electromobility construction of the building. Any of these approaches brings new challenges to national legislators when dealing with the regulation of the management of multi-owned buildings.

The first approach ought to be to extend the general rules on ordinary property management to decision-making on electromobility. For example, it can be expressly prescribed that the installation of private charging points, although they are considered as upgrades, can be considered as ordinary maintenance on which the owners normally decide by simple majority.[37] However, the extension of the catalogue of ordinary management measures to electromobility upgrades would also require interventions into other provisions on property management. Sometimes, electromobility reconstruction measures require serious constructional improvements resulting in high costs of reconstruction. Therefore, it would be useful to expressly prescribe who bears the costs of electromobility upgrades and possibly also special court protection of owners who were outvoted when electromobility upgrades were decided.

Moreover, electromobility reconstruction can also be regulated by special rules within separate laws on enhancing electromobility. Such rules, as *lex specialis,* would have precedence over the general rules on property management set forth in general condominium law. For example, it is possible to lay down that in order to make decisions on electromobility upgrades, the specific majority is required depending on the number of owners who are deciding on upgrades and/or on the size of their co-owned parts. General rules of condominium law on property management would then apply only in a subsidiary manner. Therefore, such an approach would require a detailed analysis of the adequacy of general rules on property management which would be implemented in the absence of any special rules on electromobility upgrades (e.g., on the participation in the costs of the electromobility reconstruction, on court protection of outvoted owners, on the decisions on upgrades when there is no majority owners' decision and the like). If the subsidiary application of general rules on property management fails to contribute to the efficient installation of private charging points, it would be necessary to regulate, by a separate law (*lex specialis*), also other aspects of electromobility reconstructions.

An option is to expressly regulate that owners and tenants have special right within their property/tenancy rights to install private charging points (the so-called right to plug) which may be exercised even without express consent obtained from other owners and landlords. This would constitute the most radical change of property management requiring a redefinition of the content of the subjective right of ownership in multi-owned buildings and the subjective rights of tenants and landlords. Therefore, it would be necessary to provide also efficient mechanisms for the protection of owners' and landlords' rights and interests which may be seriously infringed by the installation of private charging points.

The aforementioned approaches result in bigger or smaller changes of the rules on the consent/agreement/permission of owners regarding electromobility upgrades. Basically, they each bring some 'moderation' of the rules on unanimous decisions and the rules on express and formal decision-making as the most serious barriers to the efficient installation of private charging points. Clearly, because of electromobility, each of these possible approaches may lead to some new limitations on ownership of individual owners/landlords, although the traditional concept of ownership in multi-unit buildings, already deeply rooted and organised as co-ownership or condominium, does not change.

A certain number of countries have already introduced some of these models in their condominium laws to remove the barriers to the installation of private charging points in private parking places or garages located in buildings, as well as the installation of ducting infrastructures in the common parts of buildings. In some Member States, a trend in the reforms of management of multi-owned buildings has already existed for several decades. Ineffective traditional property management was obvious even before any requests for energy performance and electromobility in existing buildings. The main motive of reforms is to make the decision-making process for upgrades a lot easier. Reforms are carried out by a gradual transition from unanimous decisions to simple or absolute majority decisions for improvements or constructional measures. For example, German *WEG*, since its adoption in 1951, was first amended in 2007, and then in 2020. The last amendments to the rules aimed at alleviating the decision-making process regarding upgrades due to climate changes, protection of environment, energy efficiency, and promotion of electromobility.[38] However, in some Member States, such reforms are only in the planning phase.[39] The ongoing reforms very often include other aspects of property management to alleviate and speed up any decision-making processes to achieve electromobility (e.g., costs of upgrades, protection of outvoted owners, formal rules on the owners' meetings and decision-making processes, renovation fund, dispute resolution, structure of the owners' association, and the like). Indeed, it has become clear that the reforms to remove barriers to electromobility in existing buildings cannot be reduced to mere and express organisation of alternative decision-making processes on electromobility upgrades but that many other aspects of legal relations among owners, tenants and landlords must also be reformed.

20.3.2 Alternative Decision-Making Processes for Electromobility Reconstructions

20.3.2.1 Privileged Status of Owners When Installing Charging Facilities

Alternative methods on making decisions on electromobility reconstructions are based on the concept that the owner who, as a member of the community of owners, has initiated the procedure to decide on the installation of private charging points, has a privileged position in relation to other owners of the building. His/her privileged position is ensured in various ways, and with different intensities, depending on the way and the extent to which national legislator has decided to 'depart' from the traditional rules on property management which are based on giving express consent to reconstruction.

20.3.2.2 Obligation of Giving Consent to Electromobility Reconstruction

The owner's privileged position may be based on separate rules which do not exclude the owners' consent to electromobility reconstruction but they do bind all other owners to give consent to reconstruction of common parts of the building in order to install private charging points. These are the rules prescribing express and active consent given by the owners for upgrading common parts of the building. A major difference in relation to the traditional concept of decision-making for upgrades lies in the fact that the law expressly binds the owners to give consent for electromobility reconstruction of their common parts of the building if all legally prescribed prerequisites have previously been met. Therefore, the consent of all other owners may not depend on their subjective assessment of the justification and opportunism of the proposed reconstruction and they may not refuse it without foundation. Refusing the consent is justified only if, for the interventions in common parts of the building, to install private charging points, not all the listed prerequisites established by law are fulfilled.

For example, the Austrian reform of condominium law (having entered into force on 1 January 2022) is based on the concept of obligatory consent to electromobility reconstruction.[40] By the reform, the catalogue is extended of the so-called 'privileged reconstructions' in common parts of the buildings to which owners must agree if all legal prerequisites are met. It is expressly prescribed that owners may not refuse their consent to the works carried out in their common parts of the building for the installation of private charging points, including the necessary infrastructure according to technical standards. They must give their consent to the changes in common parts of the building if the charging facility cannot be hooked up to the existing installations, or if this would not be a reasonable idea.[41] In such a case, it is indisputably presumed that the changes in common parts of the building are in the owner's serious interest and he/she does not have to specifically prove his/her interest in the installation of the charging point.[42] If any of the owners still refuses, without justified reason, his/her consent for electromobility reconstructions in any common part of the building will be replaced by a court decision.[43] In the end, the installation of a private charging point is carried out on the basis of express consent given by all co-owners, or on the basis of court decisions substituting refusals.[44] This is the circumstance that significantly determines the privileged position of the owner who wants to install a private charging point, as well as the efficiency of the approach. To obtain all consents, it is still necessary to organise an alternative decision-making process, and when the consent is refused, court proceedings must be conducted to replace it with a court decision. Naturally, this may significantly slow down the decision-making process. There is a possibility that some owners do not take part in the decision-making process, and by being passive, they block the adoption of a consensual decision on electromobility upgrades.[45]

A balance between the privileged position of the owner who seeks the installation of a private charging point, and all other owners who are obliged to give consent, is achieved by several legal instruments. First of all, the protection of the interest of the owners' community is ensured by a list of prerequisites for the installation. The fulfilment of these prerequisites is a condition for the obligation of giving consent to reconstruction (e.g., inadequacy of the existing installations in the common parts of the building). The protection of individual ownership rights and interests in flats is exercised on the basis of the rules prohibiting the electromobility reconstruction in common parts of the building causing major and permanent damage to other owners' flats.[46] Finally, giving consent to electromobility reconstruction does not mean that other owners have the obligation to participate in the costs of the reconstruction. The owner who requested the installation of a charging point must bear all the costs of the reconstruction in common parts of the building, as well as any other costs arising from the maintenance resulting from the reconstruction.[47]

20.3.2.3 Legal Fiction of Consent to Electromobility Reconstruction

A privileged position of the owner, when installing a private recharging point, is based on the obligation of other owners to expressly agree on electromobility reconstruction and it is primarily manifested in the exclusion of the possibility that any of the owners, for any subjective or objective reason, refuses his/her consent. Consent may justifiably be refused only in the cases specified by law (e.g., if the hookup possibility to the existing infrastructure in common parts of buildings already exists). However, a prerequisite for electromobility reconstruction continues to be an express owner's consent to upgrades, whereby a refused consent without foundation is substituted by a court decision. This is why all owners must actively take part in the decision-making process on electromobility reconstruction and they must expressly declare whether they give or refuse consent to upgrades.

The rule that all owners must actively participate in the decision-making process on electromobility reconstruction and express their will and whether they agree on upgrades can, in practice, constitute a serious regulatory barrier to the installation of charging points, particularly when involving residential buildings with a large number of flats. Problems may arise because of the absence of any observations made by the owners on whether they agree or do not agree on a particular electromobility reconstruction. Positive or negative observations are sometimes missing for various reasons (e.g., failure to be present at the owners' meeting, lack of any interest in property management, absence because of living in another location, and the like). If an owner fails to declare his/her position on electromobility reconstruction, such kind of passivity cannot be replaced by a court decision. In that case, removing the barriers to electromobility requires a different mechanism to ensure the fulfilment of the requirement that all owners must agree on electromobility reconstructions. The Austrian legislator has solved this problem of the absence of any observations made by owners by a separate rule on the so-called legal fiction of consent by passive owners. It is presumpted that there is a consent of the owner, who fails to expressly declare himself or herself regarding the reconstruction but had previously been orderly informed about the planned changes in common parts of the building and still failed to complain within two months following the delivery of the information. Such passivity of owner is then considered as tacit consent to carry out the reconstruction.[48] However, such legal fiction can be effective only if the owner was orderly and fully informed about the planned works and if he/she was indeed in the position to make an informed decision on whether to give or to refuse the consent.[49]

Legal fiction of consent to electromobility reconstruction can remove the barriers to the installation of charging points resulting from passive and indifferent owners who do not actively take part in property management. Then, a passive owner, like all other owners who have expressly agreed to upgrade the common parts of the building, must put up with the fact that constructional measures are taken to install a charging point. Indeed, regardless of the owner's passivity, he/she must be protected just like all other owners who actively participated in the decision-making process. The same rules on the protection of ownership rights to common parts and to flats, and on the protection of financial interests by not participating in the costs for electromobility reconstruction, must apply to a passive owner. As a matter of fact, legal fiction of consent strengthens the privileged position of the owners who want to install private charging points. Nevertheless, his/her position is to some extent weakened by the fact that reconstruction measures must be postponed until the prerequisites for the effects of the legal fiction of consent will have been fulfilled.

20.3.2.4 The Owner's Request for a Decision on the Method of Carrying Out Electromobility Reconstruction

A privileged position of an owner regarding electromobility reconstruction of common parts of the building can also be ensured by recognising the owner's individual subjective right to request the installation of a private charging point. In that direction, German legislator has modernised condominium law by amending the Law on Apartment Ownership of 2020.[50] The reform has introduced the flat owner's right to request from the community of owners to take adequate construction measures for installing charging facilities and installations in common parts of an immovable. It is an individual, subjective right widening the catalogue of traditional ownership powers normally recognised to owners of the multi-owned building.[51] The owner's privileged position is strengthened because undertaking electromobility reconstruction does not require the consent of all other owners. They must put up with the reconstruction of common parts aimed at electromobility upgrade. However, the owner is not absolutely free in terms of how to carry out the constructional measures. Other owners decide on the model of the execution of upgrades in common parts of their building. The decision is made in accordance with the rules on regular management and by simple majority of votes of all owners whereby every owner has one vote.[52] In brief, other owners do not decide whether any upgrades are allowed but rather, how they will be carried out. Unless a owners' decision on the way upgrades will be carried out in their common parts, or in case there is a dispute among owners, the owner requesting the installation of the charging facility can obtain a decision in court proceedings.[53]

The regulation of the owner's privileged position for electromobility reconstruction by recognising the request for reconstruction, carried out without the remaining owners' consent, largely contributes to the removal of barriers to electromobility in a multi-owned building. On the one hand, the barrier connected with an express owners' consent to electromobility reconstruction is removed. On the other hand, the active participation of other owners in the process of deciding on upgrades is required. Despite the fact that other owners do not have to give their consent to reconstruction, they must render an express and active decision on how the relevant upgrade will be carried out. By deciding on the model of carrying out constructional changes to install a charging point, other flat owners are brought into the position to align the execution of the upgrade with the interests of the owners' community and with their individual ownership rights to flats.[54] The decision binds the owner, who requested electromobility reconstruction, to carry out the upgrade by taking into account other owners' interests. Besides, the costs of reconstruction are borne by the owner who requested the installation of the charging facility. He/she is then also entitled to all the benefits of the reconstruction.[55] Finally, the protection of other owners is ensured by express prescription of the limitations to request electromobility reconstruction of common parts of buildings. Constructional changes may not be requested if they will fundamentally change existing flats, if they are carried out without the owners' consent, or if his/her position would thus be violated in relation to other owners.[56] By the exclusion of the rule under which owners bring the decision on the permission to carry out electromobility upgrades, the privileged position of the owner requesting the upgrade is strengthened. However, a request that other owners decide on the way an upgrade will be carried out, may also constitute a barrier to electromobility, particularly in cases where there is a dispute among them how to carry it out, or when they refuse to make such a decision. The circumstance that court proceedings must then be initiated to render a decision on the way the upgrade will be carried out, may significantly slow down the practical implementation of the electromobility upgrade.[57]

20.3.2.5 Notification to the Owners' Community of the Intention to Carry Out Electromobility Reconstruction

The privileged position of the owner who wants to install a private charging point at his/her own expense will become significantly stronger if the consent for upgrades, and the method of carrying out the works, is substituted by a notification to the owners' community regarding the intention of installing a private charging point and the necessary installation. By the notification, instead of the owners' consent, the barriers to the installation arising from the owners' consent to upgrade (or to the method of its execution) are removed. In some Member States, the rules on property management in multi-owned buildings have already been reformed accordingly,[58] and in others, such reforms have so far only been proposed.[59]

Notifications regarding the intended installation of a charging point, addressed to the owners' community, as a substitute for any owner's decision allowing reconstruction, undoubtedly contribute to efficient transition to electromobility in multi-owned buildings. However, such an approach also requires the owner's obligation that the constructional measures will be carried out at his/her own expense, in conformity with the description of the works in the notification and in construction- and other regulations. Special instruments for the protection of the owners' community and the owner's individual interests are also required. In the first place, it is important that the process of notifying, the form and the content of the notification on the installation of private charging points are prescribed in separate rules (e.g., a detailed technical intervention plan, a description and the scope of constructional measures, distribution of the costs of the installation, the time limit within which, before the works begin, a notification must be sent, and similar details). Moreover, it is necessary to lay down effective remedies for the owners' community by which, in the cases established by law, and within a specified time limit upon the notification, the owners can justifiably oppose the installation of the private charging point and the court protection of owners in disputes.[60] For example, in the law of Portugal, it is expressly provided that within legal time, other owners are entitled to oppose the installation. It is stated, that justified grounds for opposing may be the existence of a commonly used charging point of the same capacity, a decision made by the majority of owners (2/3 of the value of the building), that within 90 days a charging point for their common use will be installed and that the majority owners' decision on the installation infrastructure for the charging facility constitutes a major risk for persons or property, or for the architectural structure of the building.[61]

20.3.2.6 Common Charging Points Priority

A privileged position of the owner who has installed a private charging facility in his/her parking place, or a garage located in the building, for the exclusive use should not be a disadvantage, or prevent the installation of common charging points to secure better and common charging of electrically powered cars of all owners. Priority is given to the protection of common interests of the owners' community regarding shared charging of electric vehicles over the exclusive and individual use of private charging points. The aim is to give priority to efficient exercise of common interests of a larger number of flat owners for charging electric vehicles by using common charging points. For example, Austrian law sets forth that the owner who installs a private recharging point at the request of other owners, must stop using it if the owners, under general rules on upgrades in common parts of the building,[62] decide to install common charging points. Such a request made by other owners is allowed upon the expiry of five years following the installation of the individual charging point.[63] For example, in Portuguese law, it is expressly provided that the owners' community may justifiably oppose the construction of electric infrastructure in common parts of buildings after having received an owner's notification of his/her intention to install the necessary infrastructure. If it is decided by the co-owners' majority vote that within 90 days they will install a common charging point for their common use, the owner is not allowed to carry out any upgrades in common parts of the building for his/her individual use of the charging facility.[64]

20.3.2.7 Erga omnes *Effects of the Privileged Owner's Position When Using a Private Charging Point*

An important measure to remove barriers to electromobility in existing buildings is the protection of the owner who, in conformity with the decision of the owners' community, has installed a charging point to be used in the future. It is particularly important to protect the owner in the cases when the owners of some flats change. Decisions on electromobility reconstruction may have positive effects in terms of removing barriers to electromobility transition in the existing buildings only if they have *erga omnes* effects, i.e., if they have positive effects on future owners of flats. In this way, the owner's privileged position is ensured to continuously use a private charging point. Therefore, it is important to ensure, by separate rules, *erga omnes* effects of the decisions on the management and use of common parts of buildings for charging electric vehicles. In Austrian law, the rule on *erga omnes* effects of the decisions of the owners' community is prescribed. The rules on the use of common parts adopted in agreement reached by all owners, or by the court, remain valid even if there is a change of flat owners. Every new flat owner is *ex lege* bound by the decisions made earlier by the community of owners.[65] The same rule is also in force in German law.[66]

20.4 Conclusion

A just stipulation of legal relations among the owners in multi-owned buildings has been an everlasting challenge for national legislators and this is why national condominium laws are constantly being reformed and modernised. The major challenge have always been the rules on property management and the community of owners, meant to ensure efficient maintenance of buildings, the participation of all owners in the decision-making process on the management of their building and the appropriate division of the management costs among them.[67] These rules impose specific ownership right limitations, as well as additional obligations on individual owners, in the interest of other owners and of the community of owners.

The green transition places new challenges before national legislators to regulate legal relations among the owners in multi-owned buildings. The establishment of a just balance between sometimes opposed, individual rights and interests of owners regarding property management is no longer sufficient. Property management must be regulated in a way to ensure a transition to sustainable housing and business adjusted to public interests, climate change, energy efficiency, and energy performance of buildings. The new approach to property management requires a completely new balancing between individual ownership rights of flat owners on the one hand, and public interests aimed at the green transition on the other hand. Due to the green transition, the social function of ownership, i.e., the rule that ownership obliges, and that owners must contribute to common good, has become even more prominent in modern regulation of property management.

The development of property management for the sake of sustainable electromobility in existing buildings is also based on the social function of ownership and on the limitations on ownership in public interest connected with electromobility as one of the important elements of the green transition. This is precisely why, in national condominium laws, where reforms of property management, because of electromobility, have already been carried out in various ways and with different intensities, the position of owners who intend to install charging facilities is privileged. The new rules on electromobility reconstructions introduce new limitations on individual ownership rights of other owners. They range from the owners' obligation to agree with electromobility renovations, to their duty to take part in the decision-making process regarding the method of renovation, all the way to their obligation to put up with any renovations if they have been informed about them and if all legal prerequisites for them have been fulfilled. All these reforms, although they are still not expressly referred to as such, are directed to the gradual introduction of the new and special 'right to plug' of an owner or tenant which must be respected by other owners/landlords and will limit their individual ownership rights.[68] These reforms have contributed to the modernisation of the overall property management of multi-owned buildings. The existing management problems, because of sustainable mobility and energy efficiency, have been a motive for extensive reforms of decision-making processes in multi-owned buildings. The reforms of national condominium laws confirm that the EU initiatives connected with climate changes, even when the EU does not have any direct authority for private law harmonisation, can have an indirect impact on the modernisation of national private law bodies (in this case of property law and tenancy law).

The proposed EU rules in the Proposal for a revision of the EPBD of 2021 are also based on the concept of social component of ownership and its limitation in public interest. The new EU rules on electro-performance of buildings will bind Member States to remove barriers to the installation of charging points in their national legislations and in particular the obligation to obtain consent from landlords and owners to a private charging point for the owner's own use.[69] If the offered proposal for an EPBD revision is accepted, Member States will be obliged to harmonise their national condominium laws with the European requests for electromobility in existing buildings. Because this obligation will be prescribed by an EU directive, Member States, when removing barriers to electromobility arising from property management of multi-owned buildings, will be able to choose legal instruments which best correspond to their national regulation of ownership, condominium, private autonomy and freedom of contracting. Finally, it will result in a situation where the Member States' national property law will contribute to the European climate law and it will become more focused on sustainable property law.

The reforms of national condominium laws that have already been carried out show that the interventions in condominium law to remove barriers to electromobility are possible even without a substantial change of the basic principles on which the regulation of ownership in multi-owned buildings is based. The removal of barriers to electromobility in existing buildings can be accomplished by the introduction of new statutory limitations on owners in the decision-making process involving electromobility renovations aimed at fulfilling the request for public interest in the green transition. However, it is important to take into account the fact that the new statutory limitations on ownership for electromobility reconstructions must be justified and proportional and that they must respect the very essence of ownership as a fundamental right.[70] There is no option by which the 'right to plug', directly or indirectly recognised to an individual owner or a tenant in multi-owned buildings, may lead to a situation where it always has absolute and unlimited priority over the ownership rights of other owners and landlords only because this contributes to public interest connected with climate changes. Because of ownership rights and interests of other co-owners, the 'right to plug' must also be limited. In some specific cases, the rights and interests of other owners/landlords to use common parts and their flats must be observed and they will thus exclude electromobility reconstruction. Adequate decision-making processes must be ensured to make it possible for owners to make informed decisions on electromobility upgrades. Efficient court protection of their ownership rights during electromobility reconstruction must be provided. Indeed, it is to be expected that the reforms of condominium law, because of sustainable electromobility in existing buildings, will probably not remove all barriers to electromobility arising from decision-making processes in property management. These barriers can be removed to the extent to which it is really justified to give priority to public interests in the way which lays down only proportional limitations on ownership rights of other flat owners. Only such an approach can ensure the green transition proclaimed in the European Green Deal.

Notes

1 Faculty of Law, University of Zagreb, Croatia.

2 Commission, 'The European Green Deal' (Communication) COM (2019) 640 final. <https://eur-lex.europa.eu/legal-content/EN/TXT/?qid=1576150542719&uri=COM%3A2019%3A640%3AFIN> accessed 29.01.2023.

3 See the Commission, 'The European Green Deal' (Communication)COM (2019) 640 final, p. 2.

4 Council Regulation (EC) 2021/1119 of 30 June 2021 establishing the framework for achieving climate neutrality and amending Regulations (EC) No. 401/2009 and (EU) 2018/1999 (2021) OJ L 243/1.

5 The European Climate Law provides for the framework for irreversible and gradual reduction of net greenhouse gas emissions, intermediate Union's climate targets and Member States' obligations for adaptation to climate changes. See Arts. 1, 2, 4, and 5 of the European Climate Law.

6 See Art. 4.1 of the European Climate Law.

7 At the time when the Commission launched the European Green Deal (2019), the annual renovation rate of the buildings varied from 0.4% to 1.2% in the Member States. See the European Green Deal, pp. 9–10.

8 Ibid, pp. 10–11.

9 Ibid, p. 9. See the Commission, 'A Renovation Wave for Europe – greening our buildings, creating jobs, improving lives' Communication COM (2020) 662 final, p. 1, <https://eur-lex.europa.eu/legal-content/EN/TXT/?uri=CELEX%3A52020DC0662> accessed on 29.01.2023.

10 Council Directive 2010/31/EU of 19 May 2010 on the energy performance of buildings [2010] OJ L 153/13, amended in Council Directive (EU) 2018/844 of 30 May 2018 amending Directive 2010/31/EU on the energy performance of buildings and Directive 2012/27/EU on energy efficiency [2018] OJ L 156/75.

11 See the Commission, 'Proposal for a Directive of the European Parliament and of the Council on the energy performance of buildings (recast)' COM (2021) 802 final <https://eur-lex.europa.eu/legal-content/EN/TXT/?uri=CELEX%3A52021PC0802&qid=1641802763889> accessed 29.01.2023.

12 See the European Green Deal, p. 9. For more, see the Renovation Wave for Europe where the key principles are defined for building renovation towards 2030 and 2050 to ensure faster and deeper renovations and a transition to green buildings for the future. There is a warning in the Document that most of the existing buildings are not energy-efficient. This piece of information is particularly worrying because 85% of the EU's building stock was built before 2001 (220 million building units), i.e., because of the fact that a large share of buildings was built without the energy performance requirement. In addition, there is a warning that 85–95% of the existing buildings will still exist in 2050. See European Green Deal, p. 1.

13 See the Commission, 'What is the European Green Deal?' (Factsheet) FS/19/6714 <https//ec.europa.eu/commission/presscorner/detail/en/fs_19_6714> accessed 22.01.2023.

14 See the Commission, 'Sustainable and Smart Mobility Strategy – putting European transport on track for the future' COM (2020) 789 final, p. 3 <https://eur-lex.europa.eu/legal-content/EN/TXT/?uri=COM:2020:789:FIN> accessed 29.01.2023.

15 See the Commission, 'Press release: Zero emission vehicles: first "Fit for 55" deal will end the sale of new CO_2 emitting cars in Europe by 2035' (Press Release) IP/22/6462 <https://ec.europa.eu/commission/presscorner/detail/en/ip_22_6462> accessed 22.01.2023.

16 See the European Green Deal p.11; Sustainable and Smart Mobility Strategy, p. 5.

17 The Member States' obligation to ensure an appropriate number of charging points accessible to the public is regulated by Council Directive 2014/94/EU of 22 October 2014 on the deployment of alternative fuels infrastructure [2014] OJ L 307/1. Within the framework of the 'Fit for 55' package, Commission, 'Proposal for a Regulation of the European Parliament and of the Council on the deployment of alternative fuels infrastructure repealing Directive 2014/94/EU of the European Parliament and of the Council' COM (2021) 559 final <https://eur-lex.europa.eu/legal-content/EN/TXT/?uri=CELEX%3A52021PC0559> accessed 29.01.2023.

18 In EU law, 'electric vehicle' is defined as '(..) a motor vehicle equipped with a powertrain containing at least one non-peripheral electric machine as energy converter with an electric rechargeable energy storage system, which can be recharged externally'. See Art. 2 (2) Directive 2014/94/EU on the deployment of alternative fuels infrastructure.

19 In EU law, 'charging point' is defined as '(..) an interface that is capable of charging one electric vehicle at a time or exchanging a battery of one electric vehicle at a time'. See Art. 2 (3) of Directive 2014/94/EU on the deployment of alternative fuels infrastructure.

20 In EU law, 'charging point' is defined as '(..) an interface that is capable of charging one electric vehicle at a time or exchanging a battery of one electric vehicle at a time'. See Art. 2 (3) of Directive 2014/94/EU on the deployment of alternative fuels infrastructure.

21 See the Renovation Wave for Europe, p. 2; Sustainable and Smart Mobility Strategy, p. 6.
See Commission, 'Report from the Commission to the European Parliement and the Council. Promotion of e-mobility through buildings policy' COM (2023) 76 final, p. 1 <https://eur-lex.europa.eu/legal-content/EN/TXT/PDF/?uri=CELEX:52023DC0076> accessed 16.7.2023

22 See ElementEnergy, 'Electric mobility: Inevitable, or not? Final report' [2022], pp. 35–39 <https://www.platformelectromobility.eu/wp-content/uploads/2022/01/20220110_InevitableEV_Final.pdf> accessed 29.01.2023.

23 See Arts 8.2-7. EPBD (a new version, after the original Art. 8 EPBD/2010 was replaced by a new Art. 8 based on Art. 1.5 of Directive 2018/844).

24 Taken from Rec. 23 Directive 2018/844.

25 See Art. 12 of the Proposal for a Directive on the energy performance of buildings (recast).

26 See Art. 8.7 EPBD.

27 Taken from Art. 8.7 EPBD.

28 Taken from Art. 12.8 of the Proposal for a Directive on the energy performance of buildings (recast).
See also Commission, 'Report from the Commission to the European Parliement and the Council. Promotion of e-mobility through buildings policy' COM (2023) 76 final, p. 8.

29 The need to remove the barriers to electromobility in buildings is particularly obvious because a large portion of the EU population live in flats, i.e., in buildings and in various forms of condominium or rented flats. According to the Eurostat data, in 2021, 46% of the EU population lived in a flat. However, 30% of

the EU population lived as tenants in rented housing. See Eurostat 'Housing in Europe. 2022 interactive edition' [2022] <https://ec.europa.eu/eurostat/cache/digpub/housing/bloc-1a.html> accessed 24.09.2023.

30 See the European Commission, Directorate-General for Energy, 'Promotion of e-mobility through buildings policy: final report' Publications Office of the European Union [2022] pp. 11, 90, 92 <https://data.europa.eu/doi/10.2833/57848> accessed 04.02.2023.

31 See Art. 8/7 EPBD replaced by Art. 1/5 Directive (EU) 2018/844. In the Recital of Directive (EU) 2018/844, in this context, it is particularly emphasised that it is necessary to remove any '(..) barriers such as split incentives and administrative complications which individual owners encounter when trying to install a recharging point on their parking space' (taken from p. 23, Recital).

32 See the Explanatory Memorandum of Proposal for a Directive on energy performance of buildings (recast), p. 15. See points 36–40, Recital of the Proposal; Art. 12 of the Proposal (Infrastructure for Sustainable Mobility).

33 See Bram Akkermans, 'Sustainable property law: towards a revaluation of our system of property law' [2020] <https://ssrn.com/abstract=3645983> accessed 23.09.2023.

34 See, for example, para. 10 of the German Act on Apartment Ownership (Wohnungseigentumsgesetz in der Fassung der Bekanntmachung vom 12. Januar 2021 (BGBl. I S. 34), das durch Artikel 7 des Gesetzes vom 7. November 2022 (BGBl. I S. 1982) geändert worden ist, hereinafter: German WEG); para. 17 and para. 26 of the Austrian Act on Apartment Ownership (Bundesgesetz über das Wohnungseigentum (Wohnungseigentumsgesetz 2002 – WEG 2002) StF: BGBl. I Nr. 70/2002 idF BGBl. I Nr. 114/2002 (DFB), hereinafter: Austrian WEG); Art. 5: 127 Dutch Civil Code (BURGERLIJK WETBOEK, Staatsblad 2010, 494 <https://zoek.officielebekendmakingen.nl/stb-2010-494.html> accessed 24.09.2023); Art. 85/4, 375 of the Croatian Property Act (Zakon o vlasništvu I drugim stvarnim pravima, Narodne novine, br. 91/96 NN 81/2015 <https://narodne-novine.nn.hr/clanci/sluzbeni/2015_07_81_1548.html> accessed 24.09.2023).

35 In that regard, some authors state that property management of multi-owned buildings (in the context of removing the barriers to achieve energy efficiency) must be based on collectivist understanding of ownership, co-dependency, and cooperation of owners. David Weatherall, Frankie McCarthy and Susan Bright, 'Property law as a barrier to energy upgrades in multi-owned properties: insights from a study of England and Scotland' [2018] 11 Energy Efficiency pp. 1652–1653. In the context of the protection of environment, other authors highlight the need for the creation of a sustainable property law. See Bram Akkermans, 'Sustainable property law: towards a revaluation of our system of property law' [2020] <https://ssrn.com/abstract=3645983> accessed 23.09.2023.

36 In that regard, see also David Weatherall, Frankie McCarthy and Susan Bright, 'Property law as a barrier to energy upgrades in multi-owned properties: insights from a study of England and Scotland' [2018] 11 Energy Efficiency pp. 1652–1653.

37 For example, in French law of 2021, in a separate Act on the fight against climate changes (Loi n° 2021–1104 du 22 août 2021 portant lutte contre le dérèglement climatique et renforcement de la résilience face à ses effets) by Art. 111, the amended general rules providing for property management by the owners' community of multi-owned buildings. There is separate rule that on the installation of charging points for electric or plug-in hybrid vehicles simple majority of owners present at a general meeting is necessary and that thereby, general rule of Art. 25 is derogated (pursuant to which decisions are rendered by the majority votes of all owners). See Art. 24 (5) (1) Loi n° 65-557 du 10 juillet 1965 fixant le statut de la copropriété des immeubles bâtis.

38 See Entwurf eines Gesetzes zur Förderung der Elektromobilität und zur Modernisierung des Wohnungseigentumsgesetzes und zur Änderung von kosten- und grundbuchrechtlichen Vorschriften (Wohnungseigentumsmodernisierungsgesetz/ WEMoG), 23.03.2020, pp. 25–27, pp. 71–73 (hereinafter: WEMoG 2020 Begründung) <https://www.bmj.de/SharedDocs/Downloads/DE/Gesetzgebung/RegE/RegE_WEMoG.pdf?__blob=publicationFile&v=3> accessed 23.09.2023.

See Marco Scharmer, 'IWD - Das deutsche Wohnungseigentumsmodernisierungsgesetz (WEMoG): Eine Reform, die den Namen auch verdient!' [2020] 33 Wohnrechtliche Blätter: WOBL, p. 408; Andreas Kappus, 'Das Wohnungeigentumsmodernisierungsgesetz' (2020) 50 Neue Juristische Wochenschrift p. 3617; Holger Pauly, 'Bauliche Veränderungen des Gemeinschaftseigentum nach In-Kraft-Treten des Wohnungsmodernsiereungsgesetz (WRModG)' [2021] 3 Zeitschrift für deutsches und internationales Bau- und Vergaberecht p. 215; Johannes Hogenschurz, 'Komm. Par.1. WEG' in Johannes Hogenschurz (ed.) *Wohnungseigentumsgesetz*, C.H. Beck, München [2021] p. 4; Jan Wilhelm, *Sachenrecht*, De Gruyter, Berlin/Boston, MA [2021] p. 1202. Similar development is also obvious in Austrian law. The Austrian condominium law was first provided for in the Act on Apartment Ownership (Wohnungseigenutmsgesetz) of 1948. This Act was later replaced by the Act on Apartment Ownership of 1975 where the rules on deciding on construction improvements were amended twice, to simplify the decision-making process (1981 and 1993). The new Act on Apartment Ownership, with even simpler rules on making decisions on upgrades, was adopted in 2002 (Wohnungseigentumsgesetz/hereinafter: WEG 2002). The last amendments to this Act entered into force in 2022 (WEG-Novelle, 2022). The final changes to the rule on upgrade decisions, among other things, also included the promotion of electromobility in multi-owned buildings. See Regierungsvorlage: Bundesgesetz, mit dem das Wohnungseigentumsgesetz 2002 geändert wird (WEG-Novelle 2022), 18.11.2021, p. 1 (hereinafter: WEG-Novelle 2022 – Erläuterungen) <https://www.parlament.gv.at/dokument/XXVII/I/1174/fname_1011941.pdf> accessed on 4.02.2023.

39 For an overview of a good regulator and/or policy practice, see European Commission, Directorate-General for Energy, 'Promotion of e-mobility through buildings policy: final report' Publications Office of the European Union [2022] <https://data.europa.eu/doi/10.2833/57848> accessed 4.02.2023, pp. 96–99.

40 WEG-Novelle 2022, BGBl. I Nr 222/2021.

41 See para. 16/2/2 of the Austrian WEG.

42 See WEG-Novelle 2022 – Erläuterungen p. 9.

43 Para. 52/1/2 of the Austrian WEG.
See WEG-Novelle 2022 – Erläuterungen p. 11.

44 See WEG-Novelle 2022 – Erläuterungen p. 8.

45 For more see under 3.2.3.

46 Para. 16/5 of the Austrian WEG.
See WEG-Novelle 2022 – Erläuterungen p. 3.

47 Para. 16/6 of the Austrian WEG.

48 Art. 16/5 of the Austrian WEG. See WEG-Novelle 2022 – Erläuterungen p. 10; Marco Scharmer, 'IWD-Die WEG-Novelle 2021', [2020] 11 Wohnrechtliche Blätter: WOBL, p. 363. This is an exception from the general principle of civil law according to which silence is not considered as consent. The exception is justified by the necessity to make the consent to the installation of private charging points easier in order to achieve climate targets. See WEG-Novelle 2022 – Erläuterungen pp. 1, 3.

49 In that regard, paras 16/5 and 24/5 of the Austrian WEG expressly provide how information about the planned upgrade is passed and what it must contain (e.g., a description of the planned upgrade, the consequences of passivity/not delivering the any observation). See WEG-Novelle 2022 – Erläuterungen p. 11.

50 Gesetz zur Förderung der Elektromobilität und zur Modernisierung des Wohnungseigentumsgesetzes und zur Änderung von kosten- und grundbuchrechtlichen Vorschriften (Wohnungseigentumsmodernisierungsgesetz/WEMoG) 16.10.2020, BGBl I Nr. 47, S. 2187.

51 See Andreas Kappus, 'Das Wohnungeigentumsmodernisierungsgesetz' [2020] 50 Neue Juristische Wochenschrift p. 3620; Holger Pauly, 'Bauliche Veränderungen des Gemeinschaftseigentum nach In-Kraft-Treten des Wohnungsmodernsiereungsgesetz (WRModG)' [2021] 3 Zeitschrift für deutsches und internationales Bau- und Vergaberecht p. 215.

52 See para. 20/2/2 and para. 25 of the German WEG. See Oliver Elzer, 'Komm. para. 20. WEG' in Christian Grüneberg (ed.), *Bürgerliches Gesetzbuch*, C.H. Beck, München [2022] p. 3083.

53 See para. 43, para. 44/1, German WEG. See WEMoG 2020 Begründung pp. 70, 71; Oliver Elzer, 'Komm. Par. 20. WEG' in Johannes Hogenschurtz (ed.), *Wohnungseigentumsgesetz*, C.H. Beck, München [2021] p. 566.

54 See Holger Pauly, 'Bauliche Veränderungen des Gemeinschaftseigentum nach In-Kraft-Treten des Wohnungsmodernsiereungsgesetz (WRModG)' [2021] 3 Zeitschrift für deutsches und internationales Bau- und Vergaberecht p. 215; Marco Scharmer, 'IWD - Das deutsche Wohnungseigentumsmodernisierungsgesetz (WEMoG): Eine Reform, die den Namen auch verdient!' [2020] 33 Wohnrechtliche Blätter p. 410.

55 See para. 21/1 of the German WEG, see WEMoG 2020 Begründung p. 71; Oliver Elzer, 'Komm. para. 20. WEG', in Johannes Hogenschurtz (ed.) *Wohnungseigentumsgesetz*, C.H. Beck, München [2021] p. 564.

56 Para. 20/4, German WEG. See Oliver Elzer, 'Komm. para. 20. WEG', in Johannes Hogenschurtz (ed.) *Wohnungseigentumsgesetz*, C.H. Beck, München [2021] p. 563.

57 In German law, similar rules also exist for tenants. They are strict rules whose application may not be excluded by a contract. By Art. 2 of WEMoG, para. 554 of the German Civil Code is replaced (Bürgerliches Gesetzbuch/BGB). Tenants are recognised the right to request consent, from their landlords, for constructional measures on the rented object (e.g., parking place and garage) in order to install private charging points. Landlords are obliged to give their consent except in the cases specified in the law.

58 For example, Portugal; see Art. 29 Decree-Law No. 39/2010 (Decreto-Lei n.º 39/2010) on charging points in existing buildings.

59 For example, the Netherlands; see the bill notification scheme of charging points for associations of owners (Wet notificatieregeling oplaadpunten verenigingen van eigenaars) of 2021 by which the Civil Code is proposed to be amended by Art. 118b < https://www.internetconsultatie.nl/notificatieregeling_oplaadpunten_vve> accessed 5.02.2023.

60 The concept of notification of the intention to carry out the works for the installation of electric cables for charging point has existed in French law since 2021 for tenants, as well. A tenant is obligated to notify the owners' association, of his/her intention to install the infrastructure for the charging point in the parking space for his/her private use. The owners' association have a deadline to oppose such works but only if having a serious and legitimate reason and unless they themselves decide to carry out the upgrades to equip the parking space with the necessary installations in due time. See Art. R113-6 to Art. 113-10 of the amended Code of Construction and Housing (Code de la construction et de l'habitation) by Decree No. 2021-872 of 30 June 2021 <https://www.legifrance.gouv.fr/loda/id/LEGIARTI000043816034/2021-07-01/> accessed 5.02.2023. The concept of notification which in the Portuguese law is valid for owners in multi-owned buildings applies accordingly when a tenant, at his/her own expense, intends to install electric cables for a charging point. See Art. 29/1, Decree-Law No. 39/2010.

61 See Art. 29/3 of the Portuguese Decree-Law No. 39/2010.

62 According to the reformed rules of the Austrian WEG, making decisions on the installation of common charging points is considered to be extraordinary management whereby the owners decide by a majority vote calculated on the basis of their co-ownership shares (para. 29/1). The decision is made if the owners, who together possess more than half of the immovable, have voted. By the WEG Novelle 2022, an alternative method of calculating the majority was introduced in order to favour the owners who actively participate in the management and to make the process of making decisions on upgrades easier. The majority also exists when the majority of owners, who took part in the decision-making process, voted on the upgrade and together possess 2/3 of the immovable (Art. 24/4 WEG). This alternative rule has been valid since 1 July 2022. See WEG-Novelle 2022 – Erläuterungen pp.4,5,16-18; Marco Scharmer 'IWD-Die WEG-Novelle 2021' [2020] 33 Wohnrechtliche Blätter p. 364.

63 See Art. 16/3 of the Austrian WEG Novelle 2022. See WEG-Novelle 2022 – Erläuterungen pp. 13, 14.

64 See Art. 29/3/b of the Portuguese Decree-Law No. 39/2010.

65 See para. 17/3 of the Austrian WEG. Decisions made by the flat owners' community have erga omnes effects even regardless of not being entered in land register. See Martin Schauer, 'Die Eigentümergemeinschaft (§§ 18 ff WEG 2002)' [2002] 4, 5 Wohnrechtliche Blätter p.138.

66 See para. 10/3 of the German WEG. Decisions made by the community of flat owners also have erga omnes effects without being entered in the land register. See Fridolin Bartholome, 'Komm. para. 23. WEG' in Johannes Hogenschurz (ed.) *Wohnungseigentumsgesetz*, C.H. Beck, München [2021] pp. 600–601.

67 On condominium law reforms in different national legislations see Cornelius Van Der Merwe, 'Comparative survey of jurisdictions represented' in Cornelius Van Der Merwe (ed.), *European Condominium Law*, Cambridge University Press, Cambridge [2015] pp. 25–54.

68 See WEG-Novelle 2022 – Erläuterungen p. 10.

69 Charter of Fundamental Rights of the European Union [2012] OJ C 326/391 in Art. 17/1 lays down that property is a fundamental right, and it expressly lists the prerequisites for the limitations on the use of property rights because of general public interest.

70 See, for example, Art. 52 of the Charter of Fundamental Rights of the European Union.

Bibliography

Akkermans B, 'Sustainable property law: towards a revaluation of our system of property law' [2020] <https://ssrn.com/abstract=3645983>

Bartholome F, 'Komm. para. 23. WEG' in Hogenschurz J (ed), *Wohnungseigentumsgesetz* (München, C.H. Beck 2021)

Elzer O, 'Komm. Par. 20. WEG' in Hogenschurtz J (ed), *Wohnungseigentumsgesetz* (München, C.H. Beck 2021)

Elzer O, 'Komm. para. 20. WEG' in Grüneberg C (ed), *Bürgerliches Gesetzbuch* (München, C.H. Beck 2022)

European Commission, Directorate-General for Energy, 'Promotion of e-mobility through buildings policy: final report' Publications Office of the European Union [2022] <https://data.europa.eu/doi/10.2833/57848>

Hogenschurz J, 'Komm. Par.1. WEG' in Hogenschurz J (ed) *Wohnungseigentumsgesetz* (München, C.H. Beck 2021)

Kappus A, 'Das Wohnungeigentumsmodernisierungsgesetz' [2020] 50 *Neue Juristische Wochenschrift*: 3617–3623

Pauly H, 'Bauliche Veränderungen des Gemeinschaftseigentum nach In-Kraft-Treten des Wohnungsmodernsiereungsgesetz (WRModG)' [2021] 3 *Zeitschrift für deutsches und internationales Bau- und Vergaberecht*: 214–218

Scharmer M, 'IWD - Das deutsche Wohnungseigentumsmodernisierungsgesetz (WEMoG): Eine Reform, die den Namen auch verdient!' [2020] 33 *Wohnrechtliche Blätter: WOBL*: 408–411

Scharmer M, 'IWD-Die WEG-Novelle 2021' [2020] 11 *Wohnrechtliche Blätter: WOBL*: 363 – 364

Schauer M, 'Die Eigentümergemeinschaft (§§ 18 ff WEG 2002)' [2002] 4, 5 Wohnrechtliche Blätter: 135–143

Van Der Merwe C, 'Comparative survey of jurisdictions represented' in Van Der Merwe C (ed), *European Condominium Law* (Cambridge, Cambridge University Press 2015).

Weatherall D, McCarthy F and Bright S, 'Property law as a barrier to energy upgrades in multi-owned properties: insights from a study of England and Scotland' [2018] 11 *Energy Efficiency,* pp 1641–1655

Wilhelm J, *Sachenrecht* (Berlin/Boston, MA, De Gruyter 2021).

21
DISTRIBUTIONAL EFFECTS OF DISASTER MANAGEMENT

Willemijn van Doorn-Hoekveld and Marleen van Rijswick

21.1 Introduction

One of the Sustainable Development Goals (SDGs) is related to climate change. The first target of SDG 13 is to 'strengthen resilience and adaptive capacity to climate-related hazards and natural disasters in all countries'. Indicators to achieve these targets are – among others – the adoption and implementation of national disaster risk reduction strategies in line with the UN Sendai Framework for Disaster Risk Reduction 2015–2030.

The Sendai Framework[1] outlines priorities for action to prevent and reduce existing disaster risks by understanding disaster risk; strengthening disaster risk governance in its institutional and participatory aspects; investing in resilience, with an emphasis on the need to build critical infrastructure better from the start; and being prepared to 'Build Back Better'.[2]

In this contribution, we will address the distributional effects of one specific branch of disaster risk management, flood risk management (FRM). We focus on a specific area of private law, i.e., property law. We show that property law is an essential factor for inequality in FRM.

The moment a government implements disaster risk management often involves a restriction – or, in some cases, even the deprivation – of (the use of) property. Thereby the public interest and public law affect private law. Such restrictions involve a redistribution of rights. This topic cannot be separated from assessing whether such a (re)distribution is just. The question of what a 'just' distribution is, depends on the theory of justice, as well as the context of society. In this chapter, we will briefly address the theory of distributive, environmental and climate justice.

21.2 Distributional Effects of Flood Risk Management

Distribution is the way in which an available supply of something is shared among people or spread over an area. The subject of the distribution varies. It can be on income, species, natural resources, or the effects of climate change, for example.[3] In many works on distribution, the focus is on efficiency, which can be seen in several studies of the distributional effects of FRM, which emphasise cost-benefit analysis.[4]

In this chapter, we concentrate on the distributional effects of FRM. In this policy field, there is a constant (re)distribution of burdens and benefits. (Flood) risks are not equally divided over an

DOI: 10.4324/9781032662046-25

area and by taking preventive measures, such as the erection of flood defence structures; the risks are, instead, re-arranged. People living in areas with high flood risks benefit more from FRM than people living in relatively 'risk-free' areas. This does not alter the fact that everyone in society derives some degree of benefit from flood protection, given the disruptive nature that a flood can have on a society. Burdens are limited to specific personal losses or adverse consequences of FRM: a loss of income, devaluation of property, or even deprivation of property or expropriation.[5]

21.3 Theories of Justice

In this paragraph, we will discuss four theories of justice that are relevant to assess the distributional effects of disaster risk reduction or FRM, which have been introduced in Section 21.2 and will be further elaborated upon in Section 21.4.3. We confine ourselves to *adaptation* and not to *mitigation*, although – which is especially true for environmental and climate justice – mitigating measures impact a more equitable redistribution of the effects of climate change and thus the reduction of climate change-driven disaster risks, such as flooding, wildfires, and severe droughts.

21.3.1 Distributive Justice

The starting point to define distributive justice is Aristotle, who defined justice as [the] 'manifestation of complete virtue in relation to others'.[6] Distributive justice is, according to this philosopher,

> the distribution of honour, money or any of the others things divisible among those who share a regime (for in these things it is possible for one person to have a share that is either equal or unequal to another's).[7]

Equality is relevant in this regard. This is explained by Wright: 'All individuals in the political community are measured against some distributive standard (e.g., merit or need), and goods and advantages are allocated to different individuals in the same proportion as their respective measurements'.[8] This is considered a fair or just distribution and differs per society and it depends as well on which philosophical tradition is adhered to. For utilitarians, maximising the total potential happiness of society as a whole through the aggregation of individual happiness leads to a just society.[9] According to egalitarianism, institutions must ensure equality of opportunity in the distribution of resources across society and the treatment of individuals.[10] For liberals, individual authority is the key value, for which not only protection of rights by the government is necessary, but also the provision of adequate healthcare, good education and a minimum of social security.[11] The core value of 'self-ownership' connected to free-market mechanisms and a government as small as possible is important for libertarians.[12] And for Rawls,[13] basic liberties and fair equality of opportunity must be guaranteed. Inequalities are only justified where they are to the greatest benefit of the least advantaged.[14]

21.3.2 Environmental Justice

Environmental justice is seen as a precursor to climate justice. Environmental justice is a grassroots movement that started in the 1970s in North America. Defining the environment is relevant to grasp the main goals of the movement. Instead of considering the environment as wilderness or nature that is disconnected from everyday life,[15] it should be considered as the place where 'people live, work and play'.[16] It also includes the indigenous views of the relationship between human beings and nonhuman nature.[17] The main denunciation of the movement was the unequal distribution of environmentally harmful health impacts (e.g., pollution and waste facilities) and risks that were disproportionally borne by poor black or ethnic minority Americans.[18] The criticism focuses on this unequal distribution of risks and benefits and efforts to ensure the fair and meaningful participation of affected communities in the decision-making process, thus involving both procedural and distributive justice.[19]

21.3.3 Climate Justice

Climate justice emphasises systemic causes and unequal, inequitable and unjust environmental impacts on the climate, with a particular focus on climate change.[20] It focuses especially on the claim that the impacts of climate change would fall disproportionally on already vulnerable groups.[21] The movement endeavours to bring about the social, ecological, and economic justice for everyone, who contributed the least in causing global warming, but are likely to be the most affected by it.[22] This includes the 'equal treatment in policymaking, inclusivity of all vulnerable groups and fair distribution of resources that correspond to the impact of climate change'.[23]

When addressing the distributional effects of climate change one can take a view of intergenerational justice, which 'encompasses the distribution of benefits among individuals that exist across time (i.e., in different generations), including the continuous and overlapping generations of the present and the future'. Another view is that of intragenerational justice, 'which addresses spatially dispersed individuals in the present'.[24] Although questions regarding intergenerational justice are of great importance for FRM, in this chapter, we confine ourselves to intragenerational inequalities.

21.4 Flood Risk Management and Property Rights

21.4.1 Flood Risk Management Strategies

Although states have their own disaster and FRM frameworks, their flood risk and disaster management can be categorised into different strategies. Being by and large:

Prevention: proactive spatial planning or 'keeping the people away from the water'.

Protection: also known as 'defence'. It includes hard infrastructural measures, such as flood defences, or 'keeping the water away from the people'.

Mitigation: reducing the effects of flooding by taking mitigating measures.

Preparedness: informing the public about risks, and drafting disaster and evacuation plans.

Response: acting during a flood, trying to prevent as much damage and casualties as possible.

Recovery: bouncing back to the situation as it was before the disaster.[25]

The first five strategies are designed to prevent or reduce the adverse consequences of flooding. The *recovery* strategy aims to return to a 'normal', liveable situation as soon as possible. Although the first four strategies aim to prevent damage caused by flooding, the measures carried out as part of these strategies cause loss as well.

In the *prevention* strategy, the loss will result from the imposition of restrictions on land use or zoning. An example is the rezoning of areas with high flood risks. This type of measure may lead to devaluation of property or a loss of income, e.g., when a prohibition is imposed to grow a certain type of high-value crop.

In the *protection* strategy, the designation of land as a water retention area that may be employed during times of high-water discharges of rivers or the erection of flood defences or the creation of river by-passes is well-known examples. This can lead to the devaluation of property or in some cases, the property must be taken to realise the flood defence works.

Mitigation measures include building regulations and codes to create flood-proof buildings. This can lead to an increase in building costs.

Preparedness measures involve the public about flood risks, e.g., by designing risk maps and by drafting evacuation plans. This could theoretically lead to devaluation of property, if it becomes clear to potential buyers that the property is in an area with high flood risks or in case insurance companies refuse to give insurance to properties located in flood risk areas or only do so against high costs.

Recovery consists of emergency management (during or right after the disaster), restoration (restoration of repairable essentials of life), reconstruction (the reconstruction of infrastructure and housing for the post-disaster population) and – desirably – commemorative betterment reconstruction (normal activities at pre-disaster level, but with improvements and developments).[26] The damage is the one caused by the disaster during the surge.

21.4.2 Restrictions/Limitations of Property Rights

Measures carried out in FRM cause damage that relates to property. Property is protected in many constitutions and in most cases is considered absolute or unlimited. Every person is entitled to natural use and undisturbed enjoyment of his own property.[27] This does not alter the fact that limitations to this absolute right are also possible, which is also provided for by the constitution.[28]

An example of how property rights are protected and may be infringed can be found in Article 1 of the First Protocol of the European Convention on Human Rights (hereafter FP ECHR) states:

1. Every natural or legal person is entitled to the peaceful enjoyment of his possessions. No one shall be deprived of his possessions except in the public interest and subject to the conditions provided for by law and by the general principles of international law.
2. The preceding provisions shall not, however, in any way impair the right of a State to enforce such laws as it deems necessary to control the use of property in accordance with the general interest or to secure the payment of taxes or other contributions or penalties'.

According to this provision, as well as many other (constitutional) provisions, the limitation or deprivation can exist in different forms and to different extents. What all have in common is the fact that the deprivation must be in the public or general interest and must be provided for by law and entail – in most cases – a just compensation. These compensation mechanisms differ per state. One can identify different ranges of compensation regimes.

In the Netherlands, for example, the compensation regime of no-fault liability[29] is based on the French principle of *égalité devant les charges publiques* (equality before the public burdens). This leads to a well-developed compensation scheme for damage in the field of spatial planning and zoning. Damage that forms a disproportional burden in comparison to others who are in a more-or-less similar situation will be compensated for the disproportionally large burden. In Flanders, a similar regime exists for damage resulting from spatial planning.

In common law countries, the 'regulatory takings doctrine' is relevant in this regard. When governmental regulations limit the use of private property in such a way that the landowner is effectively deprived of all economically reasonable use or value of his property, the government is required to pay just compensation for these takings.

21.4.3 Inequalities

As already stated above, the distribution of risks is not equally divided in society. Vulnerable groups are more likely to be affected by flooding than other groups. In addition, this group is also impacted even more, when it comes to the consequences of FRM measures. Property rights are an important element of inequality, inter alia because most strategies aim at homeowners. In this paragraph, we address inequality in the context of FRM.

21.4.3.1 (Un)level Playing Field

The basis of both the environmental and the climate justice movements is that – although environmental and climate changes are objective developments – the *impacts* of these developments are not equal or fairly distributed at all. Starting with the baseline. A policy that led to racial inequality in the U.S.A. is the 'G.I. Bill', enacted after World War II, although its aim was to improve the socio-economic conditions of veterans.[30] It provided education and training for returned war veterans as well as the possibility to borrow money with little or even no advance payment and 30-year repayment periods to purchase land, houses, etc.[31] Black veterans were kept from enjoying these advantages and thus this Bill created a gap between homeownership between white and black Americans. Black residents resided in less-developed neighbourhoods. Different studies show the inequality of homeownership still exists. For example, data showed that in the second quarter of 2022, 75% of white Americans owned their house, against only 45% of black Americans, and 48% for Hispanic households. The white-black gap in homeownership changed just slightly in the last three decades.[32] This ethnic gap in homeownership is also shown in other countries. Data show that Dutch Moroccans in the Netherlands are 47% less often homeowners than native Dutch.[33]

21.4.3.2 Inequality in the Prevention Strategy

The first inequalities in the preventive strategy are traceable to the unlevel playing field. For example, in New Orleans, the areas with the highest flood risks were hit the hardest by Katrina, and those areas were disproportionally inhabited by Afro-Americans. This can be traced back to the Civil War of 1861–1865 after which white people selected these regions, with high disadvantages such as air, noise and water pollution, and flooding, for the Afro-Americans to build their communities.[34] Other historical housing policies in the U.S.A. led to inequality too. The 'red lining' is an example of drafting geographical boundaries (red lines) of areas in a city with predominantly black or Latino residents that were considered to be at high risk of 'default and for which no mortgage lending would occur'. These areas were typically the areas with a higher risk of extreme weather.[35] And these discussed policies made homeownership and socio-economic development more difficult or even nearly impossible in these areas.

For areas in which the risks are acknowledged and proactive spatial planning is carried out, the compensation regimes would provide the compensation for the devaluation of property. However, general regulations for renters are not at place and show the unequal treatment between (home) owners and those who own no immovable property.

21.4.3.3 Inequality in the Protection Strategy

In a sense, structural measures do not favour or disadvantage certain groups of people. However, because these very costly measures are financed by general means, in many cases, there are approaches at hand to assess which areas should be protected to a specific level. An example is the Dutch risk approach. Here the norms for primary flood defences are based on the probability of flooding. The probability is determined by the potential consequences of a flood incident in a certain area.[36] Areas that are highly (economically) developed, are therefore better protected than less-developed areas. Well-protected areas also face another effect, the so-called 'levee effect'. Because of a (false) sense of safety in dike-protected areas, development takes place in these areas.[37] And that development leads to more consequences of flooding and thus leads to the necessity of better protection.

21.4.3.4 Inequality in the Mitigation Strategy

Mitigating measures are not likely to be carried out without direction or coercion from other parties. Studies have shown that there is a very promising connection between insurance schemes and mitigation.[38] This promising connection is linked to ownership. For tenants it is more difficult to take mitigating measures; they do not always have the means to do so, and they need permission from their landlords.[39]

21.4.3.5 Inequality in the Preparation Strategy

In this strategy, we can identify instruments that are available for all inhabitants of a society, such as risk maps that inform the public about the flood risks in a specific area.[40] It also includes informing the public by informing buyers or tenants about the risks of the area in which the building is located.[41] In this strategy, information is, in general, present for all inhabitants and therefore does not create or increase inequality.

21.4.3.6 Inequality in the Recovery Strategy

Studies showed that after Katrina homeowners were more likely to return to their homes a month after the storm (75%), than renters (with a return rate of less than 25%).[42] This can be explained by the fact that homeowners have a stronger connection to their property, but also because renters have less power to decide and finance the restoration of housing.

The Intergovernmental Panel on Climate Change (IPCC) considers insurance against natural disasters, such as floods, an adaptation response.[43] It is a form of risk-sharing and works if enough people participate.[44] Completely private systems can increase inequality, because households or persons who cannot afford the premiums are excluded from protection.[45] The other side of the coin is that public-private insurance schemes can be affordable for less prosperous groups as well when the government is financially contributing.

21.5 Discussion and Conclusion

The first target of SDG 13 is 'to strengthen resilience and adaptive capacity to climate-related hazards and natural disasters in all countries'. The priorities set out in the Sendai Framework aim to prevent and reduce existing disaster risk by strengthening disaster risk governance in its institutional and participatory aspects, investing in resilience, with an emphasis on the need to build critical infrastructure better from the start and preparedness to 'Build Back Better'. We conclude that despite the overall aim of disaster risk reduction and FRM: to protect society as a whole and to protect specific groups more than others, the realisation of disaster risk reduction or FRM does increase – already existing – inequalities.

It all starts with the unlevel playing field of homeowners, renters and areas that face high risks. The consequences of floods are often more severe in practice for people who do not own property. They are less protected in most preventive and protective FRM approaches in different countries. Connected to this is, in the preparation strategy, the communication about what people themselves can (and in the case of tenants *may*) do to protect themselves and prepare for disasters. Also here the focus is mainly on homeowners, whereby apartment owners, for example, are again worse off than 'ordinary' homeowners because they need permission from everyone in the owners' association for investments.[46]

When a disaster does occur, homeowners still appear to be better off: either they have insurance, or the compensation from governments is particularly aimed at them (recovery).

The protection of property is part of private law, which is not the cause of the problem of injustice, but it is currently not part of the solution either. The unfair distribution of risks and adverse consequences of disaster risk reduction measures must be borne in mind in order to design a successful, effective and just flood risk reduction approach. This applies not only to the reduction of flood risks but also to other disasters, such as droughts, water availability, and wildfires. The role of property law, tenancy law and the obligations and rights of the rightsholders is currently not sufficiently on the minds of policymakers and legislators in the field of disaster management. Property law can or even *should* be used to create new ways of looking at disaster risk reduction. The separation between private and public laws should therefore be eliminated to make the distribution of the burdens and benefits of the FRM fairer and thus, instead of letting the most vulnerable groups bear the impacts of climate change, protect these groups the most.

Notes

1 A non-legally binding agreement, adopted by the Third UN World Conference on Disaster Risk Reduction on March 18, 2015, and endorsed by consensus by the UN General Assembly with Resolution A/RES/69/283 on 3 June 2015.

2 UNISDR, *Reading the Sendai Framework for Disaster Risk Reduction 2015–2030*, UNISDR, Geneva, 2016, para 3.h.

3 W.J. van Doorn-Hoekveld, 'Equal distribution of burdens in flood risk management. The application of the 'égalité principle' in the compensation regimes of the Netherlands, Flanders, and France' (2017) 10(1) *Review of European Administrative Law* pp. 79–108.

4 See, e.g., C.W. Sanchirico, 'Taxes versus legal rules as instruments for equity: a more equitable view' (2000) 29(2) *The Journal of Legal Studies* pp. 797–820; R. Cooter & T. Ulen, *Law and Economics*, 6th ed., Pearson Education, Boston, MA 2012; E. Penning-Rowsell & J. Pardoe, 'Who benefits and who loses from flood risk reduction?' (2012) 30 *Environment and Planning C: Government and Policy* pp. 448–466.

5 W.J. van Doorn-Hoekveld, 'Distributional effects of EU flood risk management and the law. The Netherlands, Flanders, and France as case studies' (DPhil thesis, Utrecht University 2017) p. 18.

6 Aristotle, *Politics*, cited in: R.W. Wright, 'Substantive corrective justice' (1992) 77 *Iowa Law Review* pp. 625–711, 685.

7 Aristotle, *Nicomachaen Ethics* (translated by R.C. Bartlett & S.D. Collins), The Chicago University Press, Chicago, IL 2011, p. 94.

8 R.W. Wright, 'Substantive corrective justice' (1992) 77 *Iowa Law Review* pp. 625–711.

9 J.S. Mill, *Utilitarianism*, Dent, London 1863.

10 A. Senn, *Inequality Re-Examined*, Clarendon Press, Oxford 1992.

11 J. Anderson & A. Honneth, 'Autonomy, vulnerability, recognition, and justice' in J. Christman & J. Anderson (eds.), *Autonomy and the Challenges to Liberalism*, Cambridge University Press, Cambridge 2005, pp. 127–149; see also: A.M. Keessen, J.M. Hamer, H.F.M.W. van Rijswick & M. Wiering, 'The concept of resilience from a normative perspective: examples from Dutch adaptation strategies' (2013) 18 *Ecology and Society* p. 45.

12 F.A. Hayek, *The Road to Serfdom*, Routledge, London 1944; R. Nozik, *Anarchy, State and Utopia*, Blackwell, Oxford 1974.

13 J. Rawls, *A Theory of Justice*, Harvard University Press, Cambridge, MA 1971; J. Rawls, *Justice as Fairness: A Restatement*, Harvard University Press, Cambridge, MA 2001.

14 C. Johnson, E. Penning-Rowsell & D. Parker, 'Natural and imposed injustices: the challenges in implementing 'fair' flood risk management policy in England' (2007) 173 *The Geographical Journal* pp. 374–390, 376.

15 D. Schlosberg & L.B. Collins, 'From environmental to climate justice: climate change and the discourse of environmental justice' (2014) 5 *WIREs Climate Change* pp. 359–374.

16 P. Novotny, Where we live, work and play: the environmental justice movement and the struggle for a new environmentalism (2001) 8(1) *Journal of Political Ecology* p. 26.

17 D. Schlosberg & L.B. Collins, 'From environmental to climate justice: climate change and the discourse of environmental justice' (2014) 5 *WIREs Climate Change* pp. 359–374.

18 W.R. Williams, 'Environmental injustice in America and its politics of scale' (1999) 18 *Political Geography* pp. 49–73; K. Jenkins, 'Setting energy justice apart from the crowd: Lessons from environmental and climate justice' (2018) 39 *Energy Research & Social Science* pp. 117–121; G. Evans & L. Phelan, 'Transition to post-carbon society: linking environmental justice and just transitions' (2016) 99 *Energy Policy* pp. 329–339.

19 G. Evans & L. Phelan, 'Transition to post-carbon society: linking environmental justice and just transitions' (2016) 99 *Energy Policy* pp. 329–339, 331.

20 G. Evans & L. Phelan, 'Transition to post-carbon society: linking environmental justice and just transitions' (2016) 99 *Energy Policy* pp. 329–339, 331.

21 D. Schlosberg & L.B. Collins, 'From environmental to climate justice: climate change and the discourse of environmental justice' (2014) 5 *WIREs Climate Change* pp. 359–374.

22 A. Aylett, 'Conflict, collaboration and climate change: participatory democracy and urban environmental struggles in Durban, South Africa' (2010) 34(3) *International Journal of Urban and Regional Research*, pp. 478–495.

23 C. Saraswat & P. Kumar, 'Climate justice in lieu of climate change: a sustainable approach to respond to the climate change injustice and an awakening of the environmental movement' (2016) 1 *Energy, Ecology and Environment* pp. 67–74, 67 <https://doi.org/10.1007/s40974-015-0001-8> accessed 18.09. 2023.

24 J.D. Teodoro, N. Doorn, J. Kwakkel & T. Comes, 'Flexibility for intergenerational justice in climate resilience decision-making: an application on sea-level rise in the Netherlands' (2022) 18 *Sustainability Science*, pp. 1355–1365 <https://doi.org/10.1007/s11625-022-01233-9> accessed 18.09.2023.

25 These strategies are also visible in the Floods Directive and accompanying communication of the European Commission.

26 W.J. van Doorn-Hoekveld & H.F.M.W. van Rijswick, 'Property, property rights and natural hazards and beyond' in T. Thaler, T. Hartmann, L. Slavíková & B. Tempels (eds), *Homeowners and the Resilient City. Climate-Driven Natural Hazards and Private Land*, Macmillan Cham, Palgrave 2022, pp. 35–51, 44 <https://doi.org/10.1007/978-3-031-17763-7> accessed 18.09.2023.

27 M. De Villiers, 'Nuisances in Roman law' (1897) 13(4) *Law Quarterly Review* pp. 387–394.

28 M.R. Cohen, 'Property and sovereignty' (1927) 13(1) *Cornell Law Review* pp. 8–30, 22; J.R. Nolon, 'Takings and property rights legislation' (1996) 2(2) *Albany Law Environmental Outlook* pp. 61–69; A. Kostruba, 'The place and role of right depriving legal facts in the legal regulation mechanism of civil property relations' (2018) 23(82) *Utopía y Praxis Latinoamericano* pp. 171–183.

29 The competent authority acts lawfully, yet causes damage by doing so, e.g., causing devaluation of property by restricting land use.

30 D.H. Onkst, '"First a negro… incidentally a veteran": Black World War Two veterans and the G.I. Bill of Rights in the Deep South, 1944–1948' (1998) 31(3) *Journal of Social History* pp. 517–543, 518.

31 L.A. Dundon & J.S. Camp, 'Climate justice and home-buyout programs: renters as forgotten population in managed retreat actions' (2021) 11 *Journal of Environmental Studies and Sciences* pp. 420–433, 421 <https://doi.org/10.1007/s13412-021-00691-4>.

32 E. Derenoncourt, Chi Hyun Kim, M. Kuhn & M. Schularick, 'Wealth of two nations: the U.S. Racial Wealth Gap, 1860–2020' (2022) *National Bureau of Economic Research Working Paper* 30101, <https://www.nber.org/papers/w30101>; See also: L. Sullivan, T. Meschede, L. Dietrich, T. Shapiro, A. Traub, C. Ruetschlin & T. Draut, *The Racial Wealth Gap: Why Policy Matters*, Demos.org, 2015, p. 9 <https://www.demos.org/research/racial-wealth-gap-why-policy-matters> accessed 18.09.2023.

33 W. Uunk, 'Does the ethnic gap in homeownership vary by income? An analysis on Dutch survey data' (2017) 32(1) *Housing Studies* pp. 95–114 <https://doi.org/10.1080/02673037.2016.1181718>.
34 F.O Adeala & J.S. Picou, 'Hurricane Katrina-linked environmental injustice: race, class, and place differentials in attitudes' (2017) 41(2) *Disasters* pp. 228–257, 234 <https://doi.org/10.1111/disa.12204>.
35 L.A. Dundon & J.S. Camp, 'Climate justice and home-buyout programs: renters as forgotten population in managed retreat actions' (2021) 11 *Journal of Environmental Studies and Sciences* pp. 420–433, 426 <https://doi.org/10.1007/s13412-021-00691-4>.
36 W.J. van Doorn-Hoekveld, H.K. Gilissen, F.A.G. Groothuijse & H.F.M.W. van Rijswick, 'Adaptation to climate change in dutch flood risk management: innovative approaches and related challenges' (2022) 18(2) *Utrecht Law Review* pp. 51–69 <https://doi.org/10.36633/ulr.860>.
37 R.A. Collenteur, H. de Moel, B. Jongman & G. Di Baldassarre, 'The failed-levee effect: do societies learn from flood disasters?' (2015) 76 *Natural Hazards* pp. 373–388; G. Di Baldassarre, A. Castellarin & A. Brath 'Analysis on the effects of levee heightening on flood propagation: some thoughts on the River Po' (2009) 54(6) *Hydrological Sciences Journal* pp. 1007–1017; S.N. Lane, C., Landström & S.J. Whatmore, 'Imagining flood futures: risk assessment and management in practice' (2011) 369 *Philosophical Transactions if the Royal Society A* pp. 1784–1806.
38 C.B.R. Suykens, D. Tarlock, S.J. Priest, W.J. van Doorn-Hoekveld & H.F.M.W. van Rijswick, 'Sticks and carrots for reducing property-level risks from floods: an EU–US comparative perspective' (2019) 44(5) *Water International* pp. 622–639 <https://doi.org/10.1080/02508060.2019.1640957>; C.B.R. Suykens, S.J. Priest, W.J. van Doorn-Hoekveld, T. Thuillier & M. van Rijswick, 'Dealing with flood damages: will prevention, mitigation, and ex post compensation provide for a resilient triangle?' (2016) 21(4) *Ecology and Society* p. 1 <http://dx.doi.org/10.5751/ES-08592-210401>.
39 This is related to the fact that tenants have low insurance penetration. See, e.g., P. Sayers, E.C. Penning-Rowsell & M. Horritt, 'Flood vulnerability, risk and social disadvantage: current and future patterns in the UK' (2018) 18(2) *Regional Environmental Change* pp. 339–352 <https://doi.org/10.1007/s10113-017-1252-z>.
40 Examples are found in the Netherlands <www.overstroomik.nl> or in Germany <floodlabel.com> accessed 21 September 2023.
41 For a rental period of nine years or more: art.1.3.3.3.2 Decree Integrated Water Policy (*Waterwetboek*).
42 J.R. Elliott & J. Pais, 'Race, class, and Hurricane Katrina: social differences in human responses to disaster' (2006) 35(2) *Social Science Research* pp. 295–321, 309 <https://doi.org/10.1016/j.ssresearch.2006.02.003>.
43 IPCC, 'Climate change: impacts, adaptation, and vulnerability' (2014); L.A. Nurse, R.F. McLean, J. Agard, L.P. Briguglio, V. Duvat-Magnan, N. Pelesikoti, E. Tompkins, A. Webb (eds), Cambridge University Press, Cambridge 2014 <https://report.ipcc.ch/ar6/wg2/IPCC_AR6_WGII_FullReport.pdf>.
44 N. Doorn, L. Brackel & S. Vermeulen, 'Distributing responsibilities for climate adaptation: examples from the water domain' (2021) 13(7) *Sustainability* p. 3676, 10 <https://doi.org/10.3390/su13073676>.
45 S.J. Priest, M.J. Clark & E.J. Treby, 'Flood insurance: the challenge of the uninsured' (2005) 37(3) *Area* pp. 295–302 <https://doi.org/10.1111/j.1475-4762.2005.00633.x>; E.C. Penning-Rowsell, 'Flood insurance in the UK: a critical perspective' (2015) 2(6) *Wires Water* pp. 601–608 <https://doi.org/10.1002/wat2.1104>.
46 J. Bürkert, H.K. Gilissen & H.F.M.W. van Rijswick, 'Increasing climate resilience in urbanized areas: practical recommendations for improving stakeholder participation from the Netherlands', in M. Forsberg, A. Nilsson, & C. Zetterberg (eds), *Festskrift till Jan Darpö*, Iustus Förlag AB, Uppsala 2022, pp. 97–123; H.K. Gilissen, D. Hegger, H. Mees, M. van den Ende & H.F.M.W. van Rijswick, *Op weg naar een handreiking voor de implementatie van klimaatadaptatie: Lessen uit het project 'Klimaatbestendige woonwijken regio Utrecht'*, UCWOSL Utrecht University, Utrecht 2021.

Bibliography

Adeala FO & Picou JS, 'Hurricane Katrina-linked environmental injustice: race, class, and place differentials in attitudes' (2017) 41(2) *Disasters* pp. 228–257 <https://doi.org/10.1111/disa.12204>.

Anderson J & Honneth A, 'Autonomy, vulnerability, recognition, and justice' in J. Christman & J. Anderson (eds), *Autonomy and the Challenges to Liberalism*, Cambridge University Press, Cambridge 2005, pp. 127–149.

Aristotle, *Politics*, cited in: R.W. Wright, 'Substantive corrective justice' (1992) 77 *Iowa Law Review* pp. 625–711.

Aristotle, *Nicomachaen Ethics* (translated by R.C. Bartlett & S.D. Collins), The Chicago University Press, Chicago, IL 2011.

Aylett A, 'Conflict, collaboration and climate change: participatory democracy and urban environmental struggles in Durban, South Africa' (2010) 34(3) *International Journal of Urban and Regional Research* pp. 478–495.

Bürkert J, Gilissen HK & van Rijswick HFMW, 'Increasing climate resilience in urbanized areas: practical recommendations for improving stakeholder participation from the Netherlands' in M. Forsberg, A. Nilsson, & C. Zetterberg (eds), *Festskrift till Jan Darpö*, Iustus Förlag AB, Uppsala 2022, pp. 97–123.

Cohen MR, 'Property and sovereignty' (1927) 13(1) *Cornell Law Review* pp. 8–30.

Collenteur RA, de Moel H, Jongman B & Di Baldassarre G, 'The failed-levee effect: do societies learn from flood disasters?' (2015) 76 *Natural Hazards* pp. 373–388.

Cooter R & Ulen T, *Law and Economics*, 6th ed., Pearson Education, Boston, MA 2012.

De Villiers M, 'Nuisances in Roman law' (1897) 13(4) *Law Quarterly Review* pp. 387–394.

Derenoncourt E, Chi Hyun Kim, Kuhn M & Schularick M, 'Wealth of two nations: the U.S. Racial Wealth Gap, 1860–2020' (2022) *National Bureau of Economic Research Working Paper* 30101 <https://www.nber.org/papers/w30101>.

Di Baldassarre G, Castellarin A & Brath A, 'Analysis on the effects of levee heightening on flood propagation: some thoughts on the River Po' (2009) 54(6) *Hydrological Sciences Journal* pp. 1007–1017.

Doorn N, Brackel L & Vermeulen S, 'Distributing responsibilities for climate adaptation: examples from the water domain' (2021) 13(7) *Sustainability* p. 3676 <https://doi.org/10.3390/su13073676>.

Dundon LA & Camp JS, 'Climate justice and home-buyout programs: renters as forgotten population in managed retreat actions' (2021) 11 *Journal of Environmental Studies and Sciences* pp. 420–433 < https://doi.org/10.1007/s13412-021-00691-4>.

Elliott JR & Pais J, 'Race, class, and Hurricane Katrina: social differences in human responses to disaster' (2006) 35(2) *Social Science Research* pp. 295–321 <https://doi.org/10.1016/j.ssresearch.2006.02.003>.

Evans G & Phelan L, 'Transition to post-carbon society: linking environmental justice and just transitions' (2016) 99 *Energy Policy* pp. 329–339.

Gilissen HK, Hegger D, Mees H, van den Ende M & van Rijswick HFMW, *Op weg naar een handreiking voor de implementatie van klimaatadaptatie: Lessen uit het project 'Klimaatbestendige woonwijken regio Utrecht'*, UCWOSL Utrecht University, Utrecht 2021.

Hayek FA, *The Road to Serfdom*, Routledge, London 1944

IPCC, 'Climate change: impacts, adaptation, and vulnerability' (2014), LA. Nurse, RF. McLean, J. Agard, LP. Briguglio, V. Duvat-Magnan, N. Pelesikoti, E. Tompkins, A. Webb (eds) Cambridge University Press, Cambridge 2014 <https://report.ipcc.ch/ar6/wg2/IPCC_AR6_WGII_FullReport.pdf>.

Jenkins K, 'Setting energy justice apart from the crowd: lessons from environmental and climate justice' (2018) 39 *Energy Research & Social Science* pp. 117–121.

Johnson C, Penning-Rowsell E & Parker D. 'Natural and imposed injustices: the challenges in implementing 'fair' flood risk management policy in England' (2007) 173 *The Geographical Journal* pp. 374–390.

Keessen AM, Hamer JM, van Rijswick HFMW & Wiering M, 'The concept of resilience from a normative perspective: examples from Dutch adaptation strategies' (2013) 18 *Ecology and Society* p. 45.

Kostruba A, 'The place and role of right depriving legal facts in the legal regulation mechanism of civil property relations' (2018) 23(82) *Utopía y Praxis Latinoamericano* pp. 171–183.

Lane SN, Landström C & Whatmore SJ, 'Imagining flood futures: risk assessment and management in practice' (2011) 369 *Philosophical Transactions if the Royal Society A* pp. 1784–1806.

Mill JS, *Utilitarianism*, Dent, London 1863.

Nolon JR, 'Takings and property rights legislation' (1996) 2(2) *Albany Law Environmental Outlook* pp. 61–69.

Novotny P, 'Where we live, work and play: the environmental justice movement and the struggle for a new environmentalism' (2001) 8(1) *Journal of Political Ecology* p. 26.

Nozik R, *Anarchy, State and Utopia*, Blackwell, Oxford 1974.

Onkst DH, '"First a negro… incidentally a veteran": Black World War Two veterans and the G.I. Bill of Rights in the Deep South, 1944–1948' (1998) 31(3) *Journal of Social History* pp. 517–543.

Penning-Rowsell EC, 'Flood insurance in the UK: a critical perspective' (2015) 2(6) *Wires Water* pp. 601–608 <https://doi.org/10.1002/wat2.1104>.

Penning-Rowsell R & Pardoe J, 'Who benefits and who loses from flood risk reduction?' (2012) 30 *Environment and Planning C: Government and Policy* pp. 448–466.

Priest SJ, Clark MJ & Treby EJ, 'Flood insurance: the challenge of the uninsured' (2005) 37(3) *Area* pp. 295–302 <https://doi.org/10.1111/j.1475-4762.2005.00633.x>.

Rawls J, *A Theory of Justice*, Harvard University Press, Cambridge, MA 1971.

Rawls J, *Justice as Fairness: A Restatement*, Harvard University Press, Cambridge, MA 2001.

Sanchirico CW, 'Taxes versus legal rules as instruments for equity: a more equitable view' (2000) 29(2) *The Journal of Legal Studies* pp. 797–820.

Saraswat C & Kumar P, 'Climate justice in lieu of climate change: a sustainable approach to respond to the climate change injustice and an awakening of the environmental movement' (2016) 1 *Energy, Ecology and Environment* pp. 67–74 <https://doi.org/10.1007/s40974-015-0001-8> accessed 18.09.2023.

Sayers P, Penning-Rowsell EC & Horritt M, 'Flood vulnerability, risk and social disadvantage: current and future patterns in the UK' (2018) 18(2) *Regional Environmental Change* pp. 339–352 <https://doi.org/10.1007/s10113-017-1252-z>.

Schlosberg D & Collins LB, 'From environmental to climate justice: climate change and the discourse of environmental justice' (2014) 5 *WIREs Climate Change* pp. 359–374.

Senn A, *Inequality Re-Examined*, Clarendon Press, Oxford 1992.

Sullivan L, Meschede T, Dietrich L, Shapiro T, Traub A, Ruetschlin C & Draut T, *The Racial Wealth Gap: Why Policy Matters*, Demos.org, 2015 <https://www.demos.org/research/racial-wealth-gap-why-policy-matters> accessed 18.09.2023.

Suykens CBR, Priest SJ, van Doorn-Hoekveld WJ, Thuillier T & van Rijswick M, 'Dealing with flood damages: will prevention, mitigation, and ex post compensation provide for a resilient triangle?' (2016) 21(4) *Ecology and Society* p. 1 <http://dx.doi.org/10.5751/ES-08592-210401>.

Suykens CBR, Tarlock D, Priest SJ, van Doorn-Hoekveld WJ & van Rijswick HFMW, 'Sticks and carrots for reducing property-level risks from floods: an EU–US comparative perspective' (2019) 44(5) *Water International* pp. 622–639 <https://doi.org/10.1080/02508060.2019.1640957>.

Teodoro JD, Doorn N, Kwakkel J & Comes T, 'Flexibility for intergenerational justice in climate resilience decision-making: an application on sea-level rise in the Netherlands' (2022) 18 *Sustainability Science* pp. 1355–1365 <https://doi.org/10.1007/s11625-022-01233-9> accessed 18.09.2023.

UNISDR, *Reading the Sendai Framework for Disaster Risk Reduction 2015–2030*, UNISDR, Geneva 2016.

Uunk W, 'Does the ethnic gap in homeownership vary by income? An analysis on Dutch survey data' (2017) 32(1) *Housing Studies* pp. 95–114 <https://doi.org/10.1080/02673037.2016.1181718>.

van Doorn-Hoekveld WJ, 'Distributional effects of EU flood risk management and the law. The Netherlands, Flanders, and France as case studies' (DPhil thesis, Utrecht University 2017).

van Doorn-Hoekveld WJ, 'Equal distribution of burdens in flood risk management. The application of the 'égalité principle' in the compensation regimes of the Netherlands, Flanders, and France' (2017) 10(1) *Review of European Administrative Law* pp. 79–108.

van Doorn-Hoekveld WJ, Gilissen HK, Groothuijse FAG & van Rijswick HFMW, 'Adaptation to climate change in Dutch flood risk management: innovative approaches and related challenges' (2022) 18(2) *Utrecht Law Review* pp. 51–69 <https://doi.org/10.36633/ulr.860>.

van Doorn-Hoekveld WJ & van Rijswick HFMW, 'Property, property rights and natural hazards and beyond' in T. Thaler, T. Hartmann, L. Slavíková & B. Tempels (eds), *Homeowners and the Resilient City. Climate-Driven Natural Hazards and Private Land*, Macmillan Cham, Palgrave 2022 pp. 35–51 <https://doi.org/10.1007/978-3-031-17763-7> accessed 18.09.2023.

Williams WR, 'Environmental injustice in America and its politics of scale' (1999) 18 *Political Geography* pp. 49–73.

Wright RW, 'Substantive corrective justice' (1992) 77 *Iowa Law Review* pp. 625–711.

PART 5

Green Economy Transition

Sustainable Consumption and Production

22
THE INTERPLAY BETWEEN ECODESIGN AND CONSUMER SALES LAW

Victor Mehnert and Hans-W. Micklitz

22.1 Introduction

With the European Green Deal[1] and the new Circular Economy Action Plan,[2] the European Commission created a framework to actively involve consumers in the transition to ecological sustainability.[3] The central building blocks are the extension of the Ecodesign Directive 2009/125/EC[4] (Ecodesign Directive) to establish sustainable product designs and revise consumer law incorporating the circular economy concept. In 2022, the European Commission published a proposal for an ecodesign regulation[5] extending the ecodesign approach to all physical goods and expands the catalogue of ecodesign requirements to include aspects such as availability of spare parts, repairability, reusability and durability.[6] In consumer law, sustainability has long had a shadowy existence. The Green Deal aims to change this *status quo*. Consumers will be empowered to choose sustainable products and assert their rights to such products.[7] Complementarily, the European Commission published a proposal for a directive to promote the repair of goods (RD-P) in Spring 2023.[8] In fact, if properly shaped, right to repair laws would allow consumers to sue manufacturers directly for the right to repair products as required by ecodesign implementing regulations, explicitly linking consumer and ecodesign laws. During the ongoing consultation, the European Commission discussed far-reaching adaptations of the Sale of Goods Directive[9] (SGD) and emphasised its crucial role in circular economies.[10] However, it must be shown that the two proposals on ecodesign and the sales law remain half-hearted. The minor amendments to the SGD focus on the right to repair products outside their legal guarantee periods, without building clear linkages. This is a missed opportunity.[11] The central thesis of this chapter is that sustainability product standards set by the Ecodesign Directive can and should be made enforceable through individual claims granted to consumers by the SGD. Compliance with ecodesign requirements must be a prerequisite for conformity under the SGD. Integrating ecodesign standards into the sales contract is, in principle, possible under the 2019 version of the SGD. However, an explicit linkage of the two regulatory instruments is urgently required. The regulatory mechanism in the RD-P could serve as a starter. More is needed, however. According to the authors, the legal guarantee period in Article 10(1) SGD as well as the presumption period in Article 11(1) SGD should be prolonged to enable consumers to effectively enforce sustainability standards through contract law.

DOI: 10.4324/9781032662046-27

22.2 Recent Developments: Consumer Law and Sustainability

There have been increasing attempts to strengthen sustainable consumption outside EU law by promoting repair. Individual EU Member States have already taken initiatives. In France, there has been a repair index since 2021, through which consumers can identify a product's reparability and the availability of spare parts and repair services when making a purchase decision.[12] Austria and Germany introduced a 'repair voucher', enticing consumers to request a repair instead of buying a new product by subsidising repairs.[13] In the U.S., several states have enacted right to repair laws intended to give consumers access to repairs, spare parts and repair instructions.[14] While the sustainability-promoting effects of the right to repair products are certainly recognised,[15] the focus is more on strengthening consumers' property rights, enabling them to repair their own appliances.[16]

The planned EU right to repair forms an integral part of the greening of the economy.[17] European consumer law is built on the premise that consumers contribute to promoting the internal market by choice.[18] Indeed, freedom of choice lies at the core of the internal market policy.[19] The direction set by the (new) Circular Economy Action Plan and the Green Deal is not meant to undermine consumer choice, but rather to provide the necessary incentives to move away from growth through ever-greater consumption towards a model of green growth.[20] A package of linked initiatives is intended to create a coherent framework promoting the manufacture and consumption of sustainable products in the first place. The consumer position shall be strengthened outside and beyond the rules on ecodesign and the right to repair through reliable and relevant information about products at the point of sale, *inter alia* on their life span, the availability of repair services, spare parts and repair instructions.[21] Protecting against unfair business practices such as premature obsolescence and greenwashing shall be improved.[22] The proposal to amend the Consumer Rights Directive (CRD) would extend the current CRD Articles 5(1) and 6(1)'s information obligations,[23] and the envisaged amendment of the Unfair Commercial Practices Directive (UCPD) would provide for ten additional blacklisted business practices related primarily to preventing greenwashing.[24] Finally, the proposed 'Green Claims Directive' complements the UCPD.[25] In the current chapter, we decided to place the right to repair centre-stage.

The RD-P focuses on promoting repairs of products outside their legal guarantee periods. Typically, the largest share of defects occurs after the expiry of the legal guarantee. According to the European Commission, it is here where the potential for promoting repair is highest.[26] At the heart of the proposal lies the producer's repair obligation, under which, at consumers' request, producers must repair products free of charge or against consideration, purchased goods subject to repairability requirements set out in Annex II RD-P.[27] This will only include products with reparability requirements from 2021 onwards, explicitly linking the RD-P to the Ecodesign Directive.[28] The link insinuates that consumers may refer to individual claims against manufacturers on the repairability requirements of the Ecodesign Directive.[29] On closer inspection, however, the current proposal leaves one key problem unsolved: the pricing of repairs. Article 5(1) does not impose pricing requirements on manufacturers who opt to charge a price for a repair. Rather, manufacturers are merely obliged to repair products listed in Annex II at the consumer's request. The manufacturers determine the conditions under which the product will be repaired.

22.3 Strengthening the Linkage between the Ecodesign Directive and the SGD

During the consultation of the RD-P far-reaching adaptations of the SGD were discussed, such as extending legal guarantee periods or prioritising repair over replacement, let alone the tricky issue of the price of the repair.[30] None of them survived the initial debates. Instead, Article 13(2) SGD, that restricts the consumer's freedom of choice,[31] is based on economic rather than ecological considerations: the seller must repair the good where the costs for replacement are equal to or greater than the costs for repair. So far, the proposed amendments to the SGD have played a subordinate role in implementing the Green Deal. However, the sustainability potential of consumer sales law has attracted ever stronger attention in the academic environment.[32]

The following section goes beyond the current proposal to show that consumer sales law is suitable for ensuring compliance with public-law sustainability standards by way of private enforcement, thereby contributing to achieving the goals of the Green Deal. Fully using the SGD's sustainability potential requires a holistic perspective, which integrates the SGD with existing initiatives to promote sustainable consumption, particularly with the Ecodesign Directive.

22.3.1 *The Problem of Deficient Market Surveillance*

Legal rules can only be effective if they are complied with or properly enforced. Article 3 of the Ecodesign Directive confirms that the Member States are responsible for enforcing compliance. In 2016, the European Commission had already stated that 10%–25% of the products on the internal market could not be controlled and that better market surveillance was needed.[33] Supervisory authorities often lack the human and financial resources to carry out comprehensive controls. Intra-authority decision-making processes and pursuing self-interest may stand in the way of effective enforcement.[34] In the proposed Ecodesign Regulation, the European Commission emphasises that effective enforcement of ecodesign requirements is crucial for achieving its objective that market surveillance must be increased and that Member States should take the necessary measures.[35] However, one might have expected more substantive proposals on how to overcome well-known deficits and what the contribution of EU law could be.

One conceivable solution to reduce the problem of incomplete market surveillance is to enforce ecodesign product standards via consumer sales law. This presupposes that public and private enforcement mechanisms can be combined. Before the Green Deal, the Ecodesign Directive and consumer sales law had no significant points of policy overlap; however, they are now united under the Circular Economy Action Plan and the promotion of sustainable consumption. It must be recalled that the SGD 2019 was not designed to achieve sustainability goals but rather to fully harmonise the internal market in the face of the challenges posed by the increasing digitalisation of the economy as part of the Digital Single Market Strategy.[36] Nevertheless, in its current format, the SGD 2019 offers consumers the possibility of asserting their rights related to the lack of sustainability of their purchased goods and promoting sustainable consumption overall.[37] Indeed, if one assumes that compliance with ecodesign product standards is a prerequisite for the (objective) conformity of goods with contracts, Article 7(1) SGD grants consumers the possibility of enforcing conformity through individual claims. This requires further explanation.

22.3.2 *Starting Point: Objective Requirements for Conformity, Article 7(1) SGD*

The gateway for enforcing ecodesign standards through consumer sales law is the concept of defects in the SGD. In contrast to the Consumer Sales Directive (CSD),[38] the SGD 2019 requires that goods must meet both subjective and objective requirements without deviation (Articles 6 and 7 SGD). According to Article 7(5) SGD 2019, a deviation from the objective requirements by way of a party agreement is only possible under strict conditions. Thus, objective requirements are an important factor in assessing the conformity of goods with contracts. It could be argued that purchased goods only meet the objective requirements of conformity with the contract if they comply with ecodesign implementation rules, provided they exist. This would allow consumers to assert individual rights in the case of lack of conformity. The reference to environmental standards such as the ecodesign regulations can be methodologically anchored in the SGD.

22.3.2.1 Reference to Union Law and Product-Specific Regulation

The first connecting factor is found in the text of the SGD itself. In the context of defects, the law on the sale of goods is receptive to specifications from other, product-specific regulations when determining the objective conformity of the product with the contract. According to Article 7(1)(a) and Recital 32 SGD, goods must be fit for the purposes for which goods of the same type would normally be used, taking into account, where applicable, any existing Union and national laws, technical standards or, in the absence of such technical standards, applicable sector-specific industry codes of conduct.[39] Thus, it is important that the product is suitable for the usual purpose and can be used according to applicable law.[40] In the much-debated 'Diesel Cases', the German Federal Court of Justice held that illegal defeat devices had been installed in motor vehicles, according to Article 5(2) Regulation (EC) No. 715/2007,[41] which reduced the effect of emission control systems. The infringement of Regulation (EC) No. 715/2007 meant that operating these motor vehicles could be restricted or prohibited under German law, thereby limiting their suitability for normal use.[42] Therefore, a violation of ecodesign regulations could be considered a defect under Article 7(1)(a) SGD. However, this judgement is of limited importance regarding everyday consumer goods. Only a few cases can be imagined in, which a lack of reparability or recyclability affects normal use. In fact, these cases are likely the exception to the rule. Therefore, the ruling is not apt to serve as a blueprint for submitting the normal use of a product to a compliance test under potential public regulations.

In contrast, Article 7(1)(d) SGD appears more relevant. This article addresses the physical and the 'digital' properties of goods. Regarding Article 2(2)(d) CSD, the CJEU had ruled that the consumer may reasonably expect compliance with public-law requirements, even in the absence of specific contractual clauses.[43] Therefore, a product which does not comply with these requirements does not show the quality and performance which are normal in goods of the same type and which the consumer can reasonably expect, given the nature of the goods.[44] A reference to public-law regulations concerning the sustainability of products, especially the Ecodesign Directive, therefore appears possible under Article 7(1)(d) SGD. The SGD includes a specific reference to product durability, which is critical for promoting more sustainable consumer habits. Durability in the SGD refers to the ability of the goods to maintain their required functions and performance through normal use.[45] Recital 32 SGD states that product-specific Union legislation is the most appropriate means of introducing durability and other product-related requirements concerning specific types or groups of products. The SGD should complement the objectives pursued in such Union product-specific legislation. Two conclusions can be drawn: first, product-specific Union legislation such as the Ecodesign Directive should be decisive for determining conformity with the contract; and second, not only the durability of goods, but other properties, such as reparability, should also be based on specific Union law applicable to them. However, instead of establishing an explicit link between legitimate consumer expectations and Union-wide applicable product-specific legislation, both the ecodesign regulation and the right to repair are silent on the issue. The linkage is hidden in Recital 32 of the SGD under, which all relevant matters, such as the nature of the goods, the price and the frequency and intensity of use are to be considered for assessing conformity with the contract. Furthermore, product-specific Union legislation is the most appropriate means to introduce durability and other product-related requirements for specific types or groups of products, using adapted criteria and the SGD should therefore be complementary to the objectives pursued under such legislation.

The same is true concerning the wording of Article 7(1)(d) SGD. According to Recital 32, a case-by-case approach is needed, in which all relevant circumstances must be considered. Nevertheless, such a reading is somewhat at odds with the objective formulated in Recital 24 SGD:

> In order to provide clarity as to what a consumer can expect from the goods and what the seller would be liable for in the event of failure to deliver what is expected, it is essential to fully harmonise rules for determining whether goods are in conformity.

Instead of following a case-by-case approach, referring to existing product-specific Union requirements would provide legal certainty.

To find a balance between those two poles, Article 7(1)(d) SGD can and should be interpreted as follows: product-specific union legislation establishes a minimum standard that must be met. In addition, other circumstances such as price may influence what can reasonably be expected over and above the minimum level. Price should not influence whether ecodesign standards are to be considered *at all*. That way, sellers could only sell products that meet ecodesign requirements.

22.3.2.2 Article 35 CISG as an Interpretation Aid

The role of standards in the conformity assessment has raised attention within CISG, which applies to B2B only, but is of major importance for interpreting EU sales law.[46] The concept of defects and the remedies available to the buyer under the CISG already served as a model for the CSD.[47] Due to its exemplary position, the CISG contributes to developing a better understanding of supranational sales law.[48] However, the question is how far the parallel reaches. The CISG Advisory Council published an opinion, which draws together examples from various legal orders and highlights the opportunities and limits of using public-law standards as a benchmark for private relations. The opinion discusses in detail whether public-law product standards can be used as benchmarks to assess the objective contractual conformity of performance in the private-law relationship between buyer and seller, especially when the parties have not explicitly included these standards in the contract.[49] However, it takes a rather traditional stand by concluding that it cannot assume the primary task of private-law sales contract to enforce public-law standards without or even against, the parties' will.[50] The interpretation must be read in the context of B2B relations where there might be more leeway between the parties for individual agreements on product quality. In consumer law, private and public laws are strongly intermingled, not least due to the mandatory character of consumer law. However, a deeper comparison between the two reveals that the SGD might start from a different premise, but that the CISG and the SDG may nevertheless be aligned, at least to some extent.

Just as with Article 35 CISG and Article 2(1) CSD, SGD Article 7(1)(a) determines conformity with the contract, inter alia, according to whether the goods are suitable for the usual purpose. The SGD deviates from the primacy of the party agreement and establishes that conformity with the contract entails fulfilling objective requirements; in this respect, it no longer corresponds to the concept of the CISG and the CSD. Public standards (enshrined in particular public-law rules) and private standards (for instance voluntary technical standards elaborated by standardisation bodies) are often concerned with technical, ethical, environmental and health and safety considerations and/or the process of designing, manufacturing or producing the goods.[51] Under Article 35 CISG existing product regulations must be considered. Whether public or private standards are incorporated in a specific case depends on different factors.[52] For example, a standard (such as an ecodesign requirement) may become relevant for assessing conformity if the seller explicitly referred to the standard in the negotiations with the buyer before the conclusion of the contract or publicly expressed the intention of complying with the standard. This idea is also reflected in Article 7(1)(d) SGD: when assessing objective conformity with the contract, it must be considered what buyers could reasonably expect based on public statements made by the seller, particularly in advertising or labelling. If compliance is not explicitly stated, the degree of awareness of the standard and its discoverability are decisive. If it is a prominent standard whose requirement can be easily found and viewed, this could indicate that it must be complied with.[53] Overall, it looks as if the parallel between the CISG and the SGD reaches further than the CISG-Opinion indicates.

22.3.3 Proposal of the Authors: Explicit Reference to the Ecodesign Directive in Article 7(1)(d) SGD

Thus far, the reference to product-specific legal provisions has only been half-heartedly anchored in the SGD. The decisive factor for assessing the objective contractual conformity of goods is the buyer's legitimate expectations, which must be assessed based on individual case circumstances.[54] Specifications in public product law are becoming an essential part of an overall sustainable product policy and the European Commission stresses the importance of a coherent and comprehensive approach to creating a circular economy.[55] If ecodesign standards are to be enforced through the SGD, their importance within the SGD must be strengthened. In the RD-P, the European Commission has explicitly linked the obligation of manufacturers to repair to the implementation of regulations to be taken under the ecodesign rules. Such ecodesign product standards should ensure that products are technically repairable and that repair claims can be enforced accordingly. Naturally, manufacturers may charge a fee for carrying out repairs. The regulatory approach should be transferred to the SGD, so that a product would necessarily fail to meet the objective requirements of conformity with the contract if it did not comply with the requirements of product-specific legal provisions of Union law that apply to it.

An annex with a list of Union laws that must be complied with should therefore be added to the SGD and Article 7(1)(d) SGD should be amended as follows:

> [...] be of the quantity and possess the qualities and other features, including in relation to durability, functionality, compatibility and security normal for goods of the same type and, which the consumer may reasonably expect given the nature of the goods and taking into account any public statement made by or on behalf of the seller or other persons in previous links of the chain of transactions, including the producer, particularly in advertising or on labelling. *The consumer may at least expect that goods meet product specific environmental standards, which are provided by Union legal acts as listed in Annex (...).*

In this way, the legitimate expectations of consumers could be explicitly linked to compliance with sustainability standards; for example, buyers could demand free[56] repairs in the event of a lack of conformity, according to Article 13(1) SGD. In addition, to ecodesign implementing regulations, other Union laws could also be included in the list, such as the Battery Regulation,[57] which aims to ensure that appliance batteries are durable and replaceable. For reasons of clarity and similar to the RD-P, there should be an exhaustive, but flexible list of environmental regulations that must be complied with and that are linked to Article 7(1)(d) SGD.

22.3.4 Further Adjustments to the SGD Needed

In the following, we refer back to proposals made before the publication of the two EU measures on reparability and ecodesign, hoping that they are taken up in the legislative debates.

22.3.4.1 Legal Guarantee Periods

The reference to ecodesign standards related to product defects alone is insufficient to promote sustainable consumption. Currently, there is a maximum legal guarantee period of two years, after which the seller is no longer liable for any product defects and the consumer cannot file a claim for remedies such as repairs, Article 10(1) SGD. Therefore, longer legal guarantee periods could promote longer product lifespans.[58] An extension of the legal guarantee was considered during the consultation,[59] but ultimately not realised. However, the question arises as to how such an extension of the legal guarantee period would have to be structured. Article 10(3) SGD allows member states to introduce longer legal guarantee periods into national law. Some countries have already used this possibility, whereby several different approaches have been identified[60]: in Sweden, the legal guarantee period has been extended uniformly to three years. In Norway and Iceland, the legal guarantee is two to five years for goods that have a life span longer than two years. Finally, Finland and the Netherlands have product-specific legal guarantees based on the expected life span of the goods. For many consumer goods, guidelines have been developed by consumer organisations and industry regarding expected life spans.[61] The latter model offers the possibility of transferring the product-specific approach of the Ecodesign Directive to the legal guarantee approach.[62] Although ecodesign regulations do not always directly specify a life span, the expected life span could also be indirectly derived from how long spare parts must be provided. If, for example, certain spare parts have to be provided for refrigerators for at least seven years, the legal guarantee period could be adapted accordingly. Admittedly, prolonging the legal guarantee period bears the risk of price increases.[63] However, it was noted that longer legal guarantee periods do not necessarily lead to systematic price increases[64] and that prolonging the legal guarantee improves consumer protection.[65]

Still, regulatory measures that promote sustainability should not place an undue burden on industry and trade, which, in the end, is paid by consumers. The goal of promoting sustainable products must always be the priority. Therefore, legal guarantee periods should only be extended to the extent that doing so enables the sustainable use of goods. Thus, seller liability during the period of legal guarantee could be 'graduated': for example, if an ecodesign regulation specifies a minimum life of seven years for a certain product, seven years could be the maximum legal guarantee period to ensure the product's sustainability, but a distinction could be made between coverage during the standard legal guarantee period of two years and that during the extended legal guarantee period. During the standard legal guarantee period, buyers could exercise all remedies as usual, but after its expiry, the buyer rights could be limited to the repair of the product and the buyer would no longer be entitled to a replacement. The seller, meanwhile, should no longer be able to refuse to repair a product on the grounds of disproportionate cost (Article 13(3) SGD), at least concerning the (hypothetical) cost of a replacement product (Article 13(2) SGD). This leads to high compliance requirements and an increased incentive to market or distribute only easily repairable products (Figure 22.1).

However, a blanket extension of the legal guarantee period to cover all possible remedies could harm sustainability efforts, especially if the burden of proof is also reversed.[66] Consumers could be induced to consume goods up to their breaking point and then have them replaced every few years, increasing emissions and resource consumption, which is the opposite of sustainable product use. If consumer rights are limited to a repair claim after the usual legal guarantee period of two years,

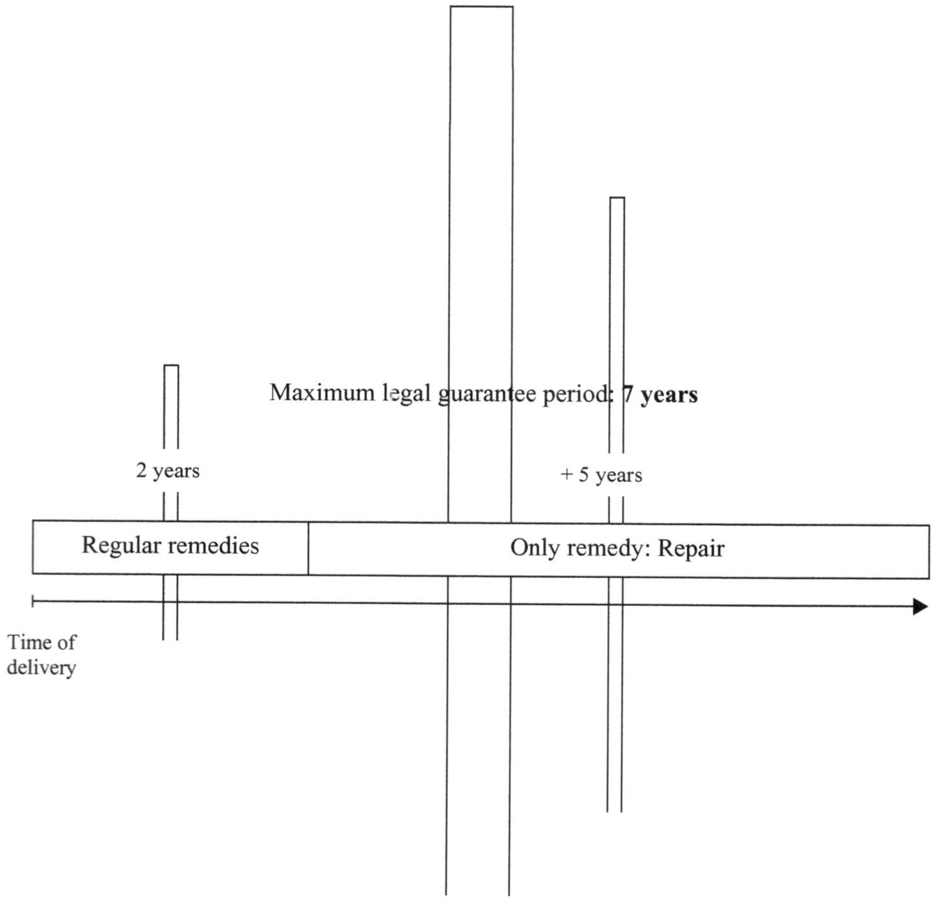

Figure 22.1 Timeline showing the subdivision of an exemplary maximum legal guarantee period of seven years. The first bracket covers the first two years after delivery, during which all remedies are available to the consumer. The second bracket covers the years 3–7, in which the consumer can only demand repair.

Example: Regulation (EU) 2019/2019 Annex II Nr. 3 (a):

'*From 1 March 2021, refrigerating appliances shall meet the following requirements*: (a) Availability of spare parts: (1) manufacturers, importers or authorised representatives of refrigerating appliances shall make available to professional repairers at least the following spare parts: thermostats, temperature sensors, printed circuit boards and light sources, for a minimum period of *seven years* after placing the last unit of the model on the market'.

sellers will likely face much lower overall costs, particularly if guaranteeing repairs encourages them to sell more durable products.

22.3.4.2 Presumption Periods

The extension of the legal guarantee periods must be closely connected with another problem. The longer from the date of purchase, the more difficult it becomes for consumers to prove that the defect existed at the delivery time.[67] Thus, according to Article 11(1) SGD, any lack of conformity, which becomes apparent within one year from the time a produce was delivered shall be presumed to have existed at the time of delivery, unless proven otherwise or unless this presumption is incompatible with the nature of the goods or with the nature of the lack of conformity. This is a considerable relief for the consumer. The presumption period had already been doubled to one year under the SGD, but because promoting sustainable product consumption through consumer sales law presupposes that buyers must still be able to assert their rights years later, a one-year period seems too short. Therefore, extending the legal guarantee period would have to be accompanied by extending the presumption period.

Some member states, such as France, already provide for a longer period to reverse the burden of proof (two years).[68] Thus, it is possible that the presumption period should also be adapted to align with the specific product's legal guarantee period. However, the already mentioned problem of 'moral hazard' must be considered here. Consumers could be induced by a long presumption period not to treat products with due consideration and then simply demand a new product.[69] However, if buyer rights under an extended legal guarantee period are limited to the right to repair from the third year onwards, as suggested here, this problem might not arise. Because repairing products often takes a considerable amount of time and contacting the seller and sending in the product is inconvenient for the consumer, consumers may be discouraged from mistreating their products. Of course, the seller would still be free to provide evidence to the contrary.

22.3.4.3 Deviating from Objective Requirements, Article 7(5) SGD

A foreseen in Article 7(5) of the SGD, the possibility of allowing deviations from the objective requirements for conformity by explicit and separate consent of the consumer is potentially a double-edged sword. Sellers could use this as an opportunity to induce consumers to purchase products that do not comply with applicable ecodesign standards by offering discounts or price promotions elsewhere. Thus, it is worth asking whether contractual derogation should be permitted at all.[70] However, the deviation could also serve a legitimate purpose from a sustainability perspective because it could possibly foster sales of used goods, fulfilling a function in the circular economy.[71] For example, an unconditional civil law obligation to guarantee ecodesign standards without the possibility of deviation could become an unmanageable challenge for some sellers and even prevent sustainable consumption. In the case of SMEs, for example those are not companies with large turnover volumes but nevertheless are considered sellers under the SGD and, thus, subject to the laws on the sale of consumer goods, it is hardly feasible to arrange with manufacturers to continuously provide software updates to comply with the objective requirement for conformity as established in Article 7(3) SGD. Mandatory compliance with ecodesign standards in every sales contract would hinder trade in second-hand goods and, preventing a culture of further use and reuse. However, exemption from Article 7(5) SGD should be limited to SMEs. Promoting competition in the spare parts and repair market might contribute to overcoming the downside of the Article 7(5) exemption.[72]

22.3.5 *Positive Effects of Linking Ecodesign and Consumer Sales Law*

The explicit linking of the SGD and the Ecodesign Directive would positively affect both consumers and the environment. If noncompliance with ecodesign requirements means a defect under the SGD, sellers must expect complaints from consumers if they sell products that are not ecodesign compliant. This could lead, for example, to consumers wanting to exchange non-repairable or non-recyclable products. The resulting extra work for sellers is likely to encourage them to sell only ecodesign-compliant products in the first place. An ecodesign-friendly orientation of the SGD could further promote repairs under consumer warranties. The incentives described above to sell only repairable products make repairs possible in the first place. By extending the legal guarantee period for repair claims, consumers can still benefit from a free repair several years after purchasing the goods; this is particularly helpful if existing defects do not show up within the first two years. This raises other questions outside this article's scope, e.g., if a repair should always be the primary remedy.[73]

Ecodesign rules are also suitable for concretising the duration of the updating obligations anchored in the SGD. These are relevant from a sustainability perspective but have thus far not been properly addressed. According to Article 7(3) SGD, one of the objective requirements for conformity with a contract is that the seller must ensure that consumers of digital goods regularly receive updates needed to maintain conformity for the reasonably expected duration of the product's use. However, the article does not define the duration of this updating obligation. SGD Recital 31 assumes that consumers would normally expect to receive updates at least during the period of legal guarantee. However, in our opinion, consumers could reasonably expect updates beyond that period if an ecodesign regulation stipulates that manufacturers must provide updates for a product for a certain period of time.[74]

22.4 Results

The enforcement of ecodesign standards through the sales contract is already possible under the SGD and serves the objectives of the Green Deal. Article 7(1) SGD is to be interpreted to the effect that public-law product standards, such as ecodesign regulations, must be considered when assessing the objective conformity of goods. However, product-specific regulations are not yet given the necessary prominence in the SGD. Their importance should be strengthened by adding to Article 7(1)(d) SGD that buyers may expect compliance with environmental regulations as a minimum standard. According to the RD-P, the Union regulations to be complied with should be listed exhaustively in an annex to the SGD and it should be possible to supplement them flexibly.

The legal guarantee period must be extended on a product-specific basis. The specifications of the ecodesign implementation regulations on the duration of the provision of spare parts serve as a guideline. To achieve more sustainable consumption by extending the legal guarantee period, the buyer's rights in the case of defects should be limited to claims for repairs from the third year onwards. A parallel extension of the presumption period is also necessary. This way, consumers can enforce their rights individually without depending solely on overburdened and under-resourced national supervisory authorities. Linking the SGD and Ecodesign law could bring consumers closer to a 'right' to repair, which deserves its name. However, even if it is possible to integrate ecodesign law into consumer sales law through targeted amendments, the question remains of whether incorporating ecological sustainability into consumer law might actually call for a structural realignment of consumer law, which goes beyond the scope of claimed linkage between ecodesign and sales law. Here, even in the academic environment, the debate has just begun.

Notes

1 Commission, Communication from the Commission to the European Parliament, the European Council, the Council, the European Economic and Social Committee and the Committee of the Regions – The European green deal, COM(2019) 640 final.

2 Commission, Communication from the Commission to the European Parliament, the Council, the European Economic and Social Committee and the Committee of the Regions – A new circular economy action plan – For a cleaner and more competitive Europe, COM(2020) 98 final.

3 Commission, Communication from the Commission to the European Parliament, the European Council, the Council, the European Economic and Social Committee and the Committee of the Regions – The European green deal, COM(2019) 640 final 9.

4 Council Directive 2009/125/EC of the European Parliament and of the Council of 21 October 2009 establishing a framework for the setting of ecodesign requirements for energy-related products (Ecodesign Directive) [2009] OJ 285/10.

5 Commission, Proposal for a Regulation of the European Parliament and of the Council establishing a framework for setting ecodesign requirements for sustainable products and repealing Directive 2009/125/EC, COM(2022) 142 final.

6 Commission, Communication from the Commission to the European Parliament, the Council, the European Economic and Social Committee and the Committee of the Regions – A new circular economy action plan – For a cleaner and more competitive Europe, COM(2020) 98 final 4.

7 Commission, Communication from the Commission to the European Parliament, the European Council, the Council, the European Economic and Social Committee and the Committee of the Regions – The European Green Deal, COM(2019) 640 final 9.

8 Commission, Proposal for a Directive of the European Parliament and of the Council on common rules promoting the repair of goods and amending Regulation (EU) 2017/2394, Directives (EU) 2019/771 and (EU) 2020/1828, COM(2023) 155 final.

9 Council Directive (EU) 2019/771 of the European Parliament and the Council of 20 May 2019 on certain aspects concerning contracts for the sale of goods, amending Regulation (EU) 2017/2394 and Directive 2009/22/EC, and repealing Directive 1999/44/EC (Sale of Goods Directive) [2019] OJ L 136/28.

10 Commission, Call for evidence for an impact assessment, Ref. Ares (2022)175084-11/01/2022.

11 Susanne Augenhofer and others, European Commission's Proposal for a Directive on Common Rules Promoting the Repair of Goods – Feedback of the European Law Institute (European Law Institute 2023) 15.

12 See Halte á l'obsolescence programée, 'The French repairability index: A first assessment – one year after its implementation' (*HOP*, 1 February 2022) <https://www.halteobsolescence.org/wp-content/uploads/2022/02/Rapport-indice-de-reparabilite.pdf> accessed 4 July 2023.

13 Bundesministerium Klimaschutz, Umwelt, Energie, Mobilität, Innovation und Technologie, 'Förderaktion Reparaturbonus' (*Reparaturbonus*) <https://www.reparaturbonus.at/> accessed 4 July 2023; Verbraucherzentrale Thüringen, Reparaturbonus 3.0 <https://www.reparaturbonus-thueringen.de/> accessed 4 July 2023.

14 Leah-Chan Grinvald and Ofer Tur-Sinai, 'Intellectual Property Law and the Right to Repair' [2019] SSRN Journal 64, 76; Daniel Moore, 'You Gotta Fight for Your Right to Repair: The Digital Millennium Copyright Act's Effect on Right-to-Repair Legislation' [2019] Texas A&M Law Review 509, 516.

15 Joshua Turiel, 'Consumer Electronic Right to Repair Laws: Focusing on an Environmental Foundation' [2021] William & Mary Environmental Law and Policy Review 579; Aaron Perzanowski, *The Right to Repair – Reclaiming the Things We Own* (Cambridge University Press 2022) 27.

16 Leah-Chan Grinvald and Ofer Tur-Sinai, 'The Right to Repair: Perspectives from the United States' [2020] Australian Intellectual Property Journal 98, 99.

17 Commission, Communication from the Commission to the European Parliament, the Council, the European Economic and Social Committee and the Committee of the Regions – A new circular economy action plan – For a cleaner and more competitive Europe, COM(2020) 98 final 6.

18 Stephen Weatherill, *Contract Law of the Internal Market* (Intersentia 2016) 26.

19 Norbert Reich and Hans-W. Micklitz, 'Chapter 1 Economic Law, Consumer Interests, and EU Integration in Norbert Reich' in Hans-W. Micklitz, Peter Rott and Klaus Tonner (eds.), *European Consumer Law* (3rd edn Intersentia 2014) 45.

20 This is not the place to discuss the feasibility and the implications of green growth.

21 Commission, Communication from the Commission to the European Parliament, the Council, the European Economic and Social Committee and the Committee of the Regions – A new circular economy action plan – For a cleaner and more competitive Europe, COM(2020) 98 final 6.
22 Commission, Communication from the Commission to the European Parliament, the Council, the European Economic and Social Committee and the Committee of the Regions – A new circular economy action plan – For a cleaner and more competitive Europe, COM(2020) 98 final 6.
23 Proposal for a Directive of the European Parliament and of the Council – amending Directives 2005/29/EC and 2011/83/EU as regards empowering consumers for the green transition through better protection against unfair practices and better information, COM 2022 143 final 16.
24 Proposal for a Directive of the European Parliament and of the Council – amending Directives 2005/29/EC and 2011/83/EU as regards empowering consumers for the green transition through better protection against unfair practices and better information, COM 2022 143 final 17.
25 Commission, Proposal for a Directive of the European Parliament and of the Council on substantiation and communication of explicit environmental claims (Green Claims Directive), COM(2023) 166 final; see Ignacio Carreño, 'To Address 'Greenwashing' and Misleading Environmental Claims, the European Commission Publishes a Proposal on 'Green Claims' and Their Substantiation' [2023] EJRR 1.
26 Commission, Proposal for a Directive of the European Parliament and of the Council on common rules promoting the repair of goods and amending Regulation (EU) 2017/2394, Directives (EU) 2019/771 and (EU) 2020/1828, COM(2023) 155 final 7.
27 Article 5(1) RD-P.
28 Commission, Proposal for a Directive of the European Parliament and of the Council on common rules promoting the repair of goods and amending Regulation (EU) 2017/2394, Directives (EU) 2019/771 and (EU) 2020/1828, COM(2023) 155 final 7.
29 Hans-W. Micklitz and others, *Right to Repair* (Publications of the Advisory Council for Consumer Affairs 2022) 57.
30 Commission, Call for evidence for an impact assessment, Ref. Ares (2022)175084-11/01/2022.
31 Susanne Augenhofer and others, *European Commission's Proposal for a Directive on Common Rules Promoting the Repair of Goods – Feedback of the European Law Institute* (European Law Institute 2023) 19.
32 Alberto De Franceschi, 'Planned Obsolescence Challenging the Effectiveness of Consumer Law and the Achievement of a Sustainable Economy' [2018] EuCML 217; Bert Keirsbilck and others, *Sustainable Consumption and Consumer Protection Legislation* (European Parliament IMCO Committee In-depth Analysis 2020); Vanessa Mak and Evelyne Terryn, 'Circular Economy and Consumer Protection: The Consumer as a Citizen and the Limits of Empowerment Through Consumer Law' [2020] Journal of Consumer Policy 227; Elias Van Gool and Anaïs Michel, 'The New Consumer Sales Directive 2019/771 and Sustainable Consumption: A Critical Analysis' [2021] EuCML 136.
33 European Commission, 'New Energy Efficiency Labels Explained' (Press corner, 11 March 2019) <https://ec.europa.eu/commission/presscorner/detail/en/MEMO_19_1596> accessed 4 July 2023; Commission, Ecodesign Work Programme 2016–2019, COM (2016) 773 final, 11; see European Parliament, 'Report on the Implementation of the Ecodesign Directive (2009/125/EC)' (Committee on the Environment, Public Health and Food Safety, 7 May 2018) <https://www.europarl.europa.eu/doceo/document/A-8-2018-0165_EN.html?redirect> accessed 4 July 2023; European Court of Auditors, 'EU action on Ecodesign and Energy Labelling: important contribution to greater energy efficiency reduced by significant delays and non-compliance' (Publication Office of the European Union, 2020) <https://www.eca.europa.eu/Lists/ECADocuments/SR20_01/SR_Ecodesign_and_energy_labels_EN.pdf> accessed 4 July 2023.
34 Franziska Weber and Michael Faure, 'The Interplay between Public and Private Enforcement in European Private Law: Law and Economics Perspective' [2015] European Review of Private Law 540; Hans-W. Micklitz and others, *Right to Repair* (Publications of the Advisory Council for Consumer Affairs 2022), 51.
35 Commission, Proposal for a Regulation of the European Parliament and of the Council establishing a framework for setting ecodesign requirements for sustainable products and repealing Directive 2009/125/EC, COM (2022) 142 final Recitals 88, 92.
36 Council Directive (EU) 2019/771 of the European Parliament and the Council of 20 May 2019 on certain aspects concerning contracts for the sale of goods, amending Regulation (EU) 2017/2394 and Directive 2009/22/EC, and repealing Directive 1999/44/EC (Sale of Goods Directive), [2019] OJ L 136/28 Recital 1.
37 See Eléonore Maitre-Ekern/Carl Dalhammar, 'Towards a Hierarchy of Consumption Behaviour in the Circular Economy' [2019] Maastricht Journal of European and Comparative Law 394.

38 Council Directive 1999/44/EC of the European Parliament and of the Council of 25 May 1999 on certain aspects of the sale of consumer goods and associated guarantees [1999] OJ L 171/12.
39 Elias Van Gool and Anaïs Michel, 'The New Consumer Sales Directive 2019/771 and Sustainable Consumption: A Critical Analysis' [2021] EuCML 136, 147
40 Fryderyk Zoll and others, 'Academic Proposal for Amending the Directive (EU) 2019/771 of the European Parliament and of the Council of 20 May 2019 on Certain Aspects Concerning Contracts for the Sale of Goods' [2022] Transformacje Prawa Prywatnego 109, 128.
41 Council Regulation (EC) No 715/2007 of the European Parliament and the Council of 20 June 2007 on type approval of motor vehicles regarding the emissions from light passenger and commercial vehicles (Euro 5 and Euro 6) and on access to vehicle repair and maintenance information [2007] OJ L 171/1.
42 BGH NJW 2019, 1133.
43 Case 145/20 *Porsche Inter Auto and Volkswagen* [2022] ECLI:EU:C:2022:572, para 54.
44 Case 145/20 *Porsche Inter Auto and Volkswagen* [2022] ECLI:EU:C:2022:572, para 55.
45 Council Directive (EU) 2019/771 of the European Parliament and the Council of 20 May 2019 on certain aspects concerning contracts for the sale of goods, amending Regulation (EU) 2017/2394 and Directive 2009/22/EC, and repealing Directive 1999/44/EC (Sale of Goods Directive), [2019] OJ L 136/28 Recital 32.
46 Reinhard Zimmermann, 'The Convention on the International Sale of Goods and Its Application in Comparative Perspective' [2004] RabelsZ 427, 428.
47 Michael Joachim Bonell, 'The CISG, European Contract Law and the Development of a World Contract Law' [2008] The American Journal of Comparative Law 7.
48 Case 247/16 *Heike Schottelius v Falk Seifert* [2017] ECLI:EU:C:2017:638, paras 39, 43.
49 CISG-AC Opinion No. 19, 'Standards and Conformity of the Goods under Article 35 CISG', 5.
50 CISG-AC Opinion No. 19, 'Standards and Conformity of the Goods under Article 35 CISG', 4.
51 CISG-AC Opinion No. 19, 'Standards and Conformity of the Goods under Article 35 CISG', 5.
52 CISG-AC Opinion No. 19, 'Standards and Conformity of the Goods under Article 35 CISG', 16.
53 CISG-AC Opinion No. 19, 'Standards and Conformity of the Goods under Article 35 CISG', 15.
54 On legitimate expectations see Hans-W. Micklitz, 'Principles of Social Justice in European Private Law' in Piet Eeckhout and Takis Tridimas (eds.), *Yearbook of European Law: Volume 19: 1999/2000*, 167–204.
55 Commission, Proposal for a Directive of the European Parliament and of the Council on common rules promoting the repair of goods and amending Regulation (EU) 2017/2394, Directives (EU) 2019/771 and (EU) 2020/1828, COM(2023) 155 final 6.
56 'Free', in this context, means that the consumer does not have to spend any additional fee for a repair service; however, the repair claim is not entirely free of cost, see on the insurance function of legal guarantees: Francesco Parisi, 'The Harmonization of Legal Warranties in European Sales Law: An Economic Analysis' [2004] The American Journal of Comparative Law 403, 408.
57 Commission, Proposal for a Regulation of the European Parliament and of the Council concerning batteries and waste batteries, repealing Directive 2006/66/EC and amending Regulation (EU) No 2019/1020, COM(2020) 798 final.
58 Klaus Tonner and Rosalind Malcom, *How an EU Lifespan Guarantee Model Could Be Implemented Across the European Union* (Study for the European Parliament, Directorate General for Internal Policies 2017) 54; Evelyne Terryn, 'A Right to Repair? Towards Sustainable Remedies in Consumer Law' [2019] ERPL 851, 859; Vanessa Mak and Enna Lujinovic, 'Towards a Circular Economy in EU Consumer Markets – Legal Possibilities and Legal Challenges and the Dutch Example' [2019] EuCML 8.
59 Commission, Call for evidence for an impact assessment, Ref. Ares (2022)175084-11/01/2022.
60 Klaus Tonner and Rosalind Malcom, *How an EU Lifespan Guarantee Model Could Be Implemented Across the European Union* (Study for the European Parliament, Directorate General for Internal Policies 2017) 19; ICF, Study on the costs and benefits of extending certain rights under the Consumer Sales and Guarantees Directive 1999/94/EC (European Commission Publications Office 2017) 13; Elias Van Gool and Anaïs Michel, 'The New Consumer Sales Directive 2019/771 and Sustainable Consumption: A Critical Analysis' [2021] EuCML 136, 141.
61 Consumentenbond, 'UNETO-VNI Tabel met gemiddelde gebruiksduurverwachtingen' (*Techniek Nederland*, 2014) <https://www.technieknederland.nl/cms/streambin.aspx?documentid=55615> accessed 4 July 2023.
62 Bert Keirsbilck and others, *Sustainable Consumption and Consumer Protection Legislation* (European Parliament IMCO Committee In-depth Analysis 2020) 20; Carlos Montalvo and others, *A Longer Lifetime*

for Products: Benefits for Consumers and Companies (Study for the European Parliament's Committee on Internal Market and Consumer Protection (IMCO) 2016) 55.

63 ICF, Study on the costs and benefits of extending certain rights under the Consumer Sales and Guarantees Directive 1999/94/EC (European Commission Publications Office 2017) 29; Vanessa Mak and Enna Lujinovic, 'Towards a Circular Economy in EU Consumer Markets – Legal Possibilities and Legal Challenges and the Dutch Example' [2019] EuCML 8.

64 Kilian Bizer and others, *Die ökonomischen Auswirkungen einer Verbesserung des deutschen Gewährleistungsrechts* (Studie im Auftrag des Verbraucherzentrale Bundesverbands e.V. 2016) 38.

65 ICF, Study on the costs and benefits of extending certain rights under the Consumer Sales and Guarantees Directive 1999/94/EC (European Commission Publications Office 2017) 61.

66 See below Section 22.3.4.2.

67 Case 497/13 *Froukje Faber v Autobedrijf Hazet Ochten BV* [2015] ECLI:EU:C:2015:357, para 54.

68 See Juliette Senechal, 'The Implementation of the EU Directives 2019/770 and 2019/771 in France' [2021] EuCML 266, 267.

69 Elias Van Gool and Anaïs Michel, 'The New Consumer Sales Directive 2019/771 and Sustainable Consumption: A Critical Analysis' [2021] EuCML 136, 144.

70 Fryderyk Zoll and others, 'Academic Proposal for Amending the Directive (EU) 2019/771 of the European Parliament and of the Council of 20 May 2019 on Certain Aspects Concerning Contracts for the Sale of Goods' [2022] Transformacje Prawa Prywatnego 109, 131.

71 This is discussed below, Section 22.3.4.3.

72 See, e.g., Dana Beldiman and others, 'Spare Parts and Design Protection – Different Approaches to a Common Problem Recent Developments from the EU and US Perspective' [2020] GRUR Int 673; Victor Mehnert, 'Reparaturen für alle? – Rechtliche Perspektiven des 'Right to Repair'' [2023] ZRP 9.

73 General priority of repair over replacement was also discussed during the Consultation of the RD-P, see Commission, Commission Staff Working Document, Impact Assessment Report Accompanying the document Proposal for a Directive of the European Parliament and of the Council on common rules promoting the repair of goods and amending Regulation (EU) 2017/2394, Directives (EU) 2019/771 and (EU) 2020/1828, SWD (2023) 59 final, 39; see also Evelyne Terryn, 'A Right to Repair? Towards Sustainable Remedies in Consumer Law' [2019] ERPL 851, 859; Elias Van Gool and Anaïs Michel, 'The New Consumer Sales Directive 2019/771 and Sustainable Consumption: A Critical Analysis' [2021] EuCML 136, 144.

74 Louisa Specht-Riemenschneider and Victor Mehnert, 'Updates und das Recht auf Reparatur – Zum Zusammenspiel von vertragsrechtlichen Aktualisierungspflichten und der Ökodesign-Richtlinie' [2022] ZfDR 313.

Bibliography

Augenhofer S and others, *European Commission's Proposal for a Directive on Common Rules Promoting the Repair of Goods – Feedback of the European Law Institute* (European Law Institute 2023) 15

Beldiman D and others, 'Spare Parts and Design Protection – Different Approaches to a Common Problem Recent Developments from the EU and US Perspective' [2020] GRUR International Vol. 69, 673

Bizer K and others, *Die ökonomischen Auswirkungen einer Verbesserung des deutschen Gewährleistungsrechts* (Studie im Auftrag des Verbraucherzentrale Bundesverbands e.V. 2016) 38

Bonell M, 'The CISG, European Contract Law and the Development of a World Contract Law' [2008] The American Journal of Comparative Law Vol. 56, 7

Carreño I, 'To Address 'Greenwashing' and Misleading Environmental Claims, the European Commission Publishes a Proposal on 'Green Claims' and Their Substantiation' [2023] EJRR Vol. 14, 1

De Franceschi A, 'Planned Obsolescence Challenging the Effectiveness of Consumer Law and the Achievement of a Sustainable Economy' [2018] EuCML Vol. 7, 217

Grinvald L-C and Tur-Sinai O, 'Intellectual Property Law and the Right to Repair' [2019] SSRN Journal 64, 76

Grinvald L-C and Tur-Sinai O, 'The Right to Repair: Perspectives from the United States' [2020] Australian Intellectual Property Journal Vol. 31, 98, 99

Keirsbilck B and others, *Sustainable Consumption and Consumer Protection Legislation* (European Parliament IMCO Committee In-depth Analysis 2020) 20

Maitre-Ekern E and Dalhammar C, 'Towards a Hierarchy of Consumption Behaviour in the Circular Economy' [2019] Maastricht Journal of European and Comparative Law Vol. 26, 394
Mak V and Lujinovic E, 'Towards a Circular Economy in EU Consumer Markets – Legal Possibilities and Legal Challenges and the Dutch Example' [2019] EuCML Vol. 8, 8
Mak V and Terryn E, 'Circular Economy and Consumer Protection: The Consumer as a Citizen and the Limits of Empowerment through Consumer Law' [2020] Journal of Consumer Policy Vol. 43, 227
Mehnert V, 'Reparaturen für alle? – Rechtliche Perspektiven des 'Right to Repair' [2023] ZRP Vol. 56, 9
Micklitz H-W, 'Principles of Social Justice in European Private Law' in Piet Eeckhout and Takis Tridimis (eds.), *Yearbook of European Law: Volume 19: 1999/2000* (Oxford University Press 2000), 167–204
Micklitz H-W and others, *Right to Repair* (Publications of the Advisory Council for Consumer Affairs 2022), 51
Montalvo C and others, *A Longer Lifetime for Products: Benefits for Consumers and Companies* (Study for the European Parliament's Committee on Internal Market and Consumer Protection (IMCO) (2016) 55
Moore D, 'You Gotta Fight for Your Right to Repair: The Digital Millennium Copyright Act's Effect on Right-to-Repair Legislation' [2019] Texas A&M Law Review Vol. 6, 509, 516
Parisi F, 'The Harmonization of Legal Warranties in European Sales Law: An Economic Analysis' [2004] The American Journal of Comparative Law Vol. 52, 403, 408
Perzanowski A, *The Right to Repair – Reclaiming the Things We Own* (Cambridge University Press 2022) 27
Reich N and Micklitz H-W, 'Chapter 1 Economic Law, Consumer Interests, and EU Integration' in Norbert Reich, Hans-W. Micklitz, Peter Rott and Klaus Tonner (eds.), *European Consumer Law* (3rd edn, Intersentia 2014) 45
Senechal J, 'The Implementation of the EU Directives 2019/770 and 2019/771 in France' [2021] EuCML Vol. 10, 266, 267
Specht-Riemenschneider L and Mehnert V, 'Updates und das Recht auf Reparatur – Zum Zusammenspiel von vertragsrechtlichen Aktualisierungspflichten und der Ökodesign-Richtlinie' [2022] ZfDR Vol. 2, 313
Terryn E, 'A Right to Repair? Towards Sustainable Remedies in Consumer Law' [2019] ERPL Vol. 27, 851, 859
Tonner K and Malcom R, *How an EU Lifespan Guarantee Model Could Be Implemented Across the European Union* (Study for the European Parliament, Directorate General for Internal Policies 2017) 19
Turiel J, 'Consumer Electronic Right to Repair Laws: Focusing on an Environmental Foundation' [2021] William & Mary Environmental Law and Policy Review Vol. 45, 579
Van Gool E and Michel A, 'The New Consumer Sales Directive 2019/771 and Sustainable Consumption: A Critical Analysis' [2021] EuCML Vol. 10, 136, 147
Weatherill S, *Contract Law of the Internal Market* (Intersentia 2016) 26
Weber F and Faure M, 'The Interplay between Public and Private Enforcement in European Private Law: Law and Economics Perspective' [2015] European Review of Private Law Vol. 23, 540
Zimmermann R, 'The Convention on the International Sale of Goods and Its Application in Comparative Perspective' [2004] RabelsZ Vol. 68, 427, 428
Zoll F and others, 'Academic Proposal for Amending the Directive (EU) 2019/771 of the European Parliament and of the Council of 20 May 2019 on Certain Aspects Concerning Contracts for the Sale of Goods' [2022] Transformacje Prawa Prywatnego Vol. 3, 109, 131

23

COOLING-OFF HOT DEALS

A Plea for Green Sludge in Distance Sales Contracts

Marta Santos Silva[1] *and Tomàs Gabriel García-Micó*[2]

23.1 Introduction

Modern consumption patterns have profound environmental ramifications.[3] According to the words of former President of the European Commission, Jean-Claude Juncker, 'We would need at least four planets to maintain our present way of life, production, and consumption'.[4] Should developing countries start following the rates at which the developed countries are consuming,[5] we will need 8 or 12 planets to store the accumulated waste we will create.

The mainstream narrative of environmentally minded academics and stakeholders refers to the need of preserving the planet for *future* generations.

Additionally, many Europeans perceive climate change as a distant and geographically removed phenomenon; however, recent disasters within Europe – exemplified by the floods in Slovenia[6] and the torrential rainfall in Italy[7] during the summer of 2023, marking the warmest recorded period in history[8] – underscore the fallacy of this perception. The impacts of climate change are increasingly evident within the continent's borders. Recognising the connection between these events and the broader global climate crisis is crucial for fostering a sense of urgency and collective responsibility to address the environmental challenges that current generations in Europe and the world are facing.[9]

Despite the scepticism from some political sectors,[10] scientific evidence shows the need of acting upon this environmental emergency. Legal academics bear a societal responsibility to address environmental challenges by critically examining the efficacy and adequacy of legal frameworks in mitigating such challenges, and in proposing solutions, which may – as is the case of the present contribution – be informed by the insights of other academic disciplines, such as psychology and environmental economics.

Building into the work initiated in the U.S.A. by Sunstein and Thaler[11] and brought to Europe particularly through the coordinated efforts of Alemanno and Sibony,[12] we argue that the right of withdrawal from distance contracts, that has been portrayed as a vital safeguard of consumers against the information asymmetry[13] vis-à-vis sellers, can act as a 'grey nudge', or as a subtle and possibly unintended form of influence[14] that steers individuals towards abuse of this right and, thereby, towards environmental degradation caused by, among others, excessive packaging and transport of goods. By encouraging consumers to acquire goods[15] with the confidence[16] that, if they no longer wish to keep such products, they may return them within a certain period without

DOI: 10.4324/9781032662046-28

needing to invoke a motive, the chronic exercise of the right of withdrawal by some consumers contributes to resource depletion and increased carbon emissions.

One of the scholars who made 'nudge' famous, Richard Thaler, identifies 'sludge' with 'nudging for evil', one whose removal would 'make the world a better place.[17] While the other scholar, Cass Sunstein, considers 'helpful to reserve the term "sludge" for impositions that have a negative balance',[18] he admits that sludge can also promote legitimate goals.[19]

We contend Thaler's understanding and assert that sludge can have positive implications when used as a deterrent – or 'counternudge'[20] – for individuals to engage in actions that may be detrimental to themselves or society.

We thus understand 'sludge' as a strategy aimed at influencing behaviours by intentionally introducing obstacles and complexities. This deliberate increase in 'friction' makes decision processes and prosocial activities more cumbersome. By eliminate dealing with the problem of chronic returns, the model we propose helps mitigating social, economic and environmental adverse consequences associated with the excessive production, consumption and circulation of goods.

Just as nudge, sludge emerges as a non-coercive, choice-preserving instrument of the environmental regulatory toolkit. The suggestion we propose of imposing sludge on chronic returners is *green-by-design* and builds into previous works that demonstrate the utility of defaults and how inertia and the status quo bias shape consumer behaviour.

In this contribution, we posit that it is incumbent upon the State to mandate private companies selling goods online to green sludge chronic returners. For the purpose of this contribution, these are customers who purchase items and return their purchases significantly more often than average customers.[21] This imposition by the State on companies serves as a countermeasure to the 'traction' exerted by persuasive marketing strategies and the instrumentalization of a connivent EU Consumer Law.

In this contribution, our aim is to investigate the extent to which mandated private green sludge targeting the exercise of the right of withdrawal by chronic returners can be implemented in a way that is respectful of consumers rights, the economic freedom of businesses but also the responsibility we all have for the environment.

For such a purpose, we start by introducing the problem statement, as well as our research question. In Chapter 2, we briefly discuss the issue of consumerism and how the EU has been pursuing the sustainability transition. In Chapter 3, we discuss the concepts of nudge and sludge, and we refer, on the one hand, to nudges and sludges that steer ecologically benign behaviour ('green') and, on the other hand, to those that contribute to ecological depletion ('grey'). As a follow-up, in Chapter 4, we refer to the adverse environmental consequences associated with product returns, while also examining scholarly suggestions for viable alternatives aimed at mitigating them. Subsequently, in Chapter 5, we introduce our recommendation of mandated green sludge, namely a pre-contractual cooling-off period. This proposal reflects our understanding that policymaking should be behaviourally informed in order to better protect the environment. Finally, we offer our concluding remarks.

Through this contribution, our objective is to enhance the understanding of the right of withdrawal paradox: on the one hand, this right was designed to address information asymmetry within B2C relationships; on the other, it has been opportunistically wielded by some of the consumers it aims at protecting. Indeed, in addition to returners who withdraw from a contract in good faith due to a lack of information about the good or its functionality, there are those who premeditate return at the time of purchase and those who use returns to repair the 'damage' done by their shopping impulsive and/or compulsive urges. Excessive returns put the economic viability of businesses at stake and exacerbate environmental strain.

We propose a solution designed to act as a discerning filter for the exercise of the right of withdrawal, making it difficult for consumers to carry on with their chronic exercise of the right of withdrawal. The overarching aim of such a solution is to ensure a fairer distribution of the unwarranted returns' costs and to curb the environmental repercussions associated with these.

To our knowledge, at the time of writing,[22] there are no other papers that explore the importance of green sludge in distance sales contracts.[23] Our intent is thus to contribute to both academic discourse and political deliberations, fostering an informed dialogue on the environmental consequences of the right of withdrawal as prescribed in the Consumer Rights Directive and advocating a solution which helps ensuring this right will be exercised for the aim of fighting information asymmetry, as it was originally prescribed for.

23.2 Navigating Consumerism and Its Implications within the Framework of EU Consumer Law

As Frank Trentmann thoroughly described in his famous *opus*, we live in an 'empire of things'.[24] Despite the advent of new-age consumption trends (sustainable[25] and ethical consumption,[26] 'green',[27] 'conscious',[28] or 'minimalist' consumerism'[29] or even 'environmentally significant consumption'[30]), the promises by the circular economy[31] and the dissemination of post-capitalistic-aligned theories (degrowth,[32] sufficiency,[33] voluntary simplicity[34]) our contemporary lifestyle is underscored by a culture of abundance, perpetuated by a robust industry heavily reliant on advertising.

The firm focus on profit has led to practices of premature obsolescence[35] and, most recently, to what has been called 'dark patterns'.[36] These have been defined as 'a proliferating species of sludge [...] or market manipulation [...]', 'user interfaces whose designers knowingly confuse users, make it difficult for users to express their actual preferences, or manipulate users into taking certain actions'.[37]

The surge in overconsumption is propelled by industries, but it is also influenced by peers and social media.[38] There is a tendency to identify ownership with identity ('own in order to become', reflecting conspicuous consumption) or happiness. More concerningly, research indicates that up to 40% of online consumer spending can be attributed to spontaneous or impulsive purchases.[39]

The culture of affluence raises significant environmental sustainability concerns: it leads to climate change, biodiversity loss, depletion of resources and an unmanageable amount of waste.[40]

The concerns with sustainable development have been first formally addressed in the famous Brundtland Report by the United Nations' World Commission on Environment and Development (WCED) in 1987. This report famously defined sustainability as 'meeting the needs of the present without compromising the ability of future generations to meet their own needs'.[41]

The European Union incorporates sustainable development into its primary objectives and Treaties.[42] It has recently set the objective to become the first climate-neutral continent by 2050[43] by decoupling economic growth from resource use and adding to a competitive, resource-efficient European Union. The EU Green Deal is integral to the Commission's strategy to implement the United Nations 2030 Agenda and its Sustainable Development Goals (SDGs).[44] Under its umbrella, the Circular Economy Action Plan 2020[45] seeks to enhance the EU Green Deal by scaling up the transition, initiated in 2015, from a linear to a circular economy.[46]

In the past years, the EU has taken several actions and initiatives to follow the principles outlined in the Brundtland report, promoting sustainability and integrating sustainable development into its policies and strategies.

Besides the European Green Deal, another key strategy within the EU was the New Consumer Agenda. It specifically places emphasis on the rule of consumers in the green and digital transition, focusing on areas such as product information, sustainable consumption and digitalization. In this context, the initiative 'Empowering Consumers for the Green Transition'[47] imposed an obligation to provide information on reparability ahead of purchase, and protection against unfair practices linked to early obsolescence; the 'Sustainable Products Initiative'[48] promoted the adoption of ecodesign beyond energy appliances, while the 'Substantiating Green Claims'[49] initiative focused on enhancing transparency and consumer trust. Simultaneously, the 'Right to Repair'[50] initiative aims to facilitate the repair of goods post-purchase.

Even though the European Union has demonstrated a robust commitment to the environment, the translation of such a commitment into concrete, legally binding measures is a complex and time-consuming process, particularly due to the need to align diverse stakeholders' interests. While long-term legal frameworks play a vital role, diversifying the environmental policy toolset is crucial to addressing the pressing challenges posed by climate change.

In the forthcoming section, we introduce nudges and sludges as prompt, behavioural-informed interventions and explore how useful these may be in fostering sustainable purchasing behaviours.

23.3 The Green Potential of Nudges and Sludges

23.3.1 Definition and Classifications of Nudge, and Its Ethical Implications

In 2008, Richard H. Thaler and Cass R. Sunstein defined nudge as 'any aspect of the choice architecture that alters people's behavior in a predictable way without forbidding any options or significantly changing their economic incentives'.[51]

This concept, subject of considerable scholarly debate,[52] is associated with libertarian paternalism.[53] Instead of enforcing specific behaviours through sanctions or legal obligations, the 'choice architects' (or 'social planners'[54]) guide citizens in a direction they deem, themselves, beneficial, operating on a principle of self-judgement (the *paternalistic* component of 'libertarian paternalism'). Citizens retain the freedom to decide whether to adhere to the suggested course of action or not,[55] thereby the process is, simultaneously, *libertarian.*[56]

Ethical concerns surrounding nudges have garnered attention,[57] particularly in what regards transparency[58] and the manipulation critique. The lack of transparency is often viewed[59] through the lens that nudges 'work best in the dark',[60] even though recent empirical evidence suggests that even when people are aware of nudges, their effect remains.[61] The opacity of some nudges raises concerns about bypassing reflective or deliberative processes, creating a potential risk of manipulation.

Criticism also extends to the impact of nudges on autonomy and agency. Diverse perspectives on autonomy exist within academic literature, ranging from J. Stuart Mill's assertion in 'On Liberty'[62] that individuals have the right to demand the freedom to make their own choices, to more nuanced views like Buss's, who defines autonomy as acting in accordance with one's character or minimal human flourishing.[63]

There are several classifications of nudge in the scholarship.[64] While acknowledging these are useful,[65] this contribution will specifically concentrate on delineating distinctions between grey and green nudges, on the one hand, and nudge and sludge, on the other.

23.3.2 Grey and Green Nudge and Sludge

In the context of this discussion, a 'grey' nudge refers to any aspect of a decision-making context ('choice architecture')[66] that encourages choices harmful to the environment.[67] Several prevalent methods used on social networks for increasing sales may be classified as 'grey nudges'. These methods include the 'buy-now' buttons on e-commerce platforms, showcasing a company's 'likes' or ratings, highlighting products based on users' previous posts and preferences[68] and incorporating augmented reality.[69] These practices foster impulsive[70] and sometimes unintended[71] buying behaviour[72] that contributes to the accumulation of products and subsequently exerts an adverse impact on the environment throughout the entire life cycle of the purchased items.

The concept of 'grey nudge' stands in opposition to the concept of 'green nudge', a term more prevalent in environmental policy and economics[73] than in legal scholarship.[74] Green nudges involve alterations in choice architecture aimed at improving people's behaviour towards the environment.[75] More broadly, green nudging is part of research on Pro-Environmental Behaviour (PEB) exploring interventions that aim to reduce the adverse environmental ramifications of human behaviours.[76]

In the context of environmental conservation, a notable example of a nudge involves default settings for electronic devices or appliances that encourage energy efficiency. Governments work with manufacturers to set default preferences on devices that prioritize energy-saving settings, thereby nudging consumers towards more environmentally friendly. This approach aligns with the concept of choice architecture, where the default option helps achieving the intended effect (energy-saving mode) due to the inertia of habit.[77]

Presently, there is an increased urgency for administering[78] green nudges for the sake of protecting the environment. Although there are challenges in making them readily applicable for environmental policy design and implementation,[79] they are progressively conquering a place within the environmental regulatory toolbox.[80] Operating alongside regulations, liability rules and market-based incentives, green nudges purport to tackle the limitations inherent in traditional environmental regulatory instruments.[81]

Despite potential limitations regarding long-lasting effects[82] when considered in isolation, recent studies show that green nudges, especially when complemented by digital tools,[83] can emerge as potent policy instruments for shaping future environmental policies.[84] Nudges present competitive advantages in relation to conventional tools that often involve a costly,[85] intricate[86] and resistance-prone[87] approval process. Furthermore, the effectiveness of traditional instruments – sometimes referred to as 'shoves' in opposition to 'nudges'[88] – when used in isolation[89] tends to be also limited.[90]

In traditional economics, the prevailing notion is that people's primary motivation is monetary gain, suggesting that market-based incentives are the most effective. However, Behavioural Law and Economics challenged this perspective by showing that, when they make decisions, individuals deviate from rational behaviour (embodied by the concept of *homo economicus*[91]) and use methods of thought called heuristics.[92]

Considering that private consumption accounts for more than a quarter of all greenhouse gas emissions,[93] using nudges as a complement of legislative measures may increase the likelihood of greening consumer behaviours.[94] A good reflection of the effectiveness of nudges is the establishment of behavioural insights[95] units in more than 35 OECD member countries and international organizations such as the World Bank and the World Health Organization.[96]

Nudges are particularly effective[97] in improving energy efficiency and managing waste, not only for individuals less attuned to environmental concerns but also for those who are environmentally conscious, effectively bridging the so-called 'intention-action gap'.[98]

Green nudges may be grouped into three types: those appealing to individuals' self-image or self-identity, those appealing to social conformism, and those involving the modification of defaults.[99] The first type simplifies product information presentation, enhances salience, or taps

Table 23.1 Fundamental difference between grey nudge, grey sludge, green nudge, and green sludge.

	Nudge	*Sludge*
Grey	Steers environmentally detrimental behaviours	Increases friction to environmentally friendly behaviours
Green	Steers environmentally friendly behaviours	Increases friction to environmentally detrimental behaviours

into individuals' social identity. The second utilizes peer comparison[100] fostering social status competition by encouraging the display of green behaviour.[101] The third capitalizes on biases, specifically the inertia effect and the suggestion effect, through default modifications.[102]

As we explore the importance of behavioural interventions within decision-making, it becomes imperative to expand our focus beyond nudges. Within the landscape of choice architecture, another contrasting phenomenon emerges known as 'sludge', thus far still notably absent in European legal literature.[103]

Nudge and sludge are alterations in choice architecture aimed at influencing behaviours. Nudges strive to achieve this by enhancing *traction*, subtly guiding individuals towards preferred choices. In contrast, sludge aims at influencing behaviours by intentionally introducing *friction*,[104] that is to say, obstacles and complexities in the decision process.[105]

The ethical assessment of sludge centres on the level of manipulativeness embedded in its design. For example, if a business implements a somewhat convoluted process for consumers to redeem a shopping voucher, it might be categorized as a 'bad' nudge, as it is practically known that, in most cases, it will not be utilized.

Of particular relevance to this discussion is the distinction between green and grey sludges. Grey sludges involve manipulating the choice environment by exploiting cognitive biases to make behaviour change more challenging[106] (e.g., selling new printers at a lower price than new cartridges). Conversely, green sludges entail introducing obstacles and delays that make unsustainable behaviours more difficult or less likely, such as the proposal we introduce in this contribution. In both grey and green sludges, the choice architect capitalizes on inertia, but only in the latter is this executed in a manner that is positive and benevolent for the environment and thus for society as a whole (Table 23.1).[107]

In this analysis, we posit that a fundamental element of consumer protection in the EU, the right of withdrawal, aligns with characteristics of a grey nudge. We thus propose a remedy to mitigate its detrimental impact on the environment – a mandated private green sludge.

23.4 The Grey Right of Withdrawal in Distance Sales of Physical Goods

23.4.1 The Need for Reimagining EU Consumer Law through a Green Lens

The trajectory of EU Consumer Law, initially designed to safeguard and promote the Common and then Internal Market, was centred on enabling seamless transnational transactions and ensuring legal certainty for economic agents operating across borders. In the 1980s, consumer policy and law sought to establish a uniform level of protection for consumers across Member States.[108]

However, the adequacy of EU Consumer Law in meeting the requisites for sustainable development,[109] and environmental sustainability in particular,[110] has recently been questioned.

Despite the global shift towards circular economic models, EU Consumer Law appears to perpetuate a mindset that predominantly facilitates consumption. While acknowledging its importance for both consumers and businesses, there is a growing recognition that the environmental impact of consumption needs to be addressed. The traditional concept of the Internal Market must be reshaped. This involves critically reconsidering the objectives and strategies underpinning Consumer Law[111] to ensure it aligns with contemporary challenges, particularly those related to environmental sustainability and *responsibility*. The call is for a more holistic and forward-looking approach that balances commercial interests and the need of preserving the environment.

European consumers find themselves navigating spaces that have been intentionally designed to encourage consumer spending[112]. Since its inception, the advertising industry has strategically leveraged behavioural insights in order to achieve this goal.[113] However, this amplified culture of consumption comes with social costs, namely consumer indebtedness[114] and environmental harm. Within the framework of EU Consumer Law, several rules and policies do not foster environmentally sustainable behaviours, thus classifying as 'grey nudges'. To illustrate, in the subsequent section we will delve into the example of the right of withdrawal.

23.4.2 The Right of Withdrawal and the Unsustainable Impact of Returns

Articles 9–16 of the Consumer Rights Directive (CRD)[115] regulate the right of withdrawal, a unilateral, short-term right for a consumer to reconsider and revoke their decision to enter into a contract.[116] Alongside other key measures such as the EU mandatory disclosure rules and the oversight of general contract terms,[117] the right of withdrawal has been acknowledged by the European legislator[118] as a fundamental tenet of consumer protection. It provides to the consumer a 'cooling-off period'[119] or a 'period for sober reflection',[120] during which they may reconsider the contract they have entered into.

The obligation to uphold the right of withdrawal extends to both distance contracts and door-to-door traders.[121] Member States are precluded from mandating a right of withdrawal for brick-and-mortar contracts,[122] in accordance with the maximum harmonisation principle established in Article 4 of the CRD. However, Article 3(6) gives traders the option to offer consumers contractual arrangements that surpass the protection specified in this Directive. Consequently, this right has been voluntarily extended by traders to brick-and-mortar customers, serving as a voluntary indicator of product or service quality.[123] Evidence shows that product return policies enhance trust[124] and increase transactions,[125] especially online.[126] The right of withdrawal consists of a 'regret mechanism'[127] that allows consumers to return most products without needing to provide a justification for the return (*ad nutum*).[128] The primary consequence of a consumer exercising the right of withdrawal is the termination of the contractual relationship between the trader and the consumer, as stipulated in Article 12 of the CRD. Additionally, any related contracts, such as a consumer credit agreement used to finance a distance or doorstep selling contract (as per Article 15 CRD), are automatically terminated. Upon the consumer's withdrawal, the trader is obligated to promptly refund all payments received from the consumer, ensuring completion within 14 days from the receipt of the termination notice (as per Article 13, paragraph 1, sentence 1 CRD).[129]

The right of withdrawal was incorporated into European Consumer Law as an exception to the overarching contractual principle that once a contract is concluded, it establishes a binding commitment between the parties involved (*pacta sunt servanda*).[130]

This rule aims to protect consumers from potentially exploitative marketing tactics and it addresses information and bargaining power asymmetries inherent in business-to-consumer (B2C) relationships.[131] In contrast to commercial premises where consumers can physically perceive a product, in distance contracts, the goods or services obtained are regarded in economy theory as 'experience goods', as their characteristics are only ascertainable upon inspection and consumption.[132] Consequently, when making online purchases, consumers place trust in the product provider that the online representation accurately reflects the product and that the price demanded for the product corresponds to their quality and, ultimately, their value.[133] The possibility to withdraw from the distance contract should give consumers an incentive to conclude it, as it lowers the risks of the transaction and of post-decision regret.[134] The right of withdrawal serves thus as a recourse for consumers when the received product differs from their expectations.[135]

The right of withdrawal, as stipulated in the CRD, has a broad scope and its interpretation has been notably expansive.

First, by allowing consumers to test the product, this right surpasses the entitlements of consumers in brick-and-mortar stores.[136]

Second, according to Article 14(2) of the CRD, consumers are only held liable for depreciation costs that do not result from activities necessary 'to establish the nature, characteristics, and functioning of the goods'. It resulted from the CJEU's ruling in the *Messner* case[137] that consumers bear no responsibility for any depreciation resulting from reasonable use of the product in most instances. While Article 14(1), paragraph 2 requires customers to return goods at their own costs, many traders and online platforms voluntarily adopt 'free returns' policies for the purpose of competing with their peers.[138] Remarkably, this policy now extends to cases where returned goods are damaged,[139] with prominent platforms offering free returns and reimbursement for damaged items even up to the 30th day after product receipt.[140]

Even though anecdotal evidence shows that actual withdrawal from contracts is rare,[141] from a business standpoint, the exercise of the right of withdrawal poses significant challenges.

Shipping, processing, and refurbishment are so costly[142] that large retailers such as Amazon[143] and Walmart allow customers to keep the item, because it often costs more to ship and process the returned product than the product is worth.[144] Additionally, companies often grapple with the difficulty of reselling returned products in the second-hand market, leading to potential product wastage.

The financial ramifications of free returns, encompassing expenses for processing and executing withdrawals, repairing or replacing damaged returned goods, as well as legal costs, are distributed across all consumers. This means that even individuals who do not utilize this right[145] either because they are familiar with the purchased product or they are averse to it namely for sustainability reasons[146] bear a share of the associated costs. The right of withdrawal is thus potentially detrimental to consumers collectively.[147]

By encouraging a rash[148] 'buy-first, reflect-later' approach to shopping,[149] free returns work as a grey nudge,[150] because they contribute to the depletion of natural resources and the increase of carbon emissions.

In light of the aforementioned considerations, there is a need for a balanced approach that considers consumer convenience on the one hand and the broader environmental impact of consumption patterns, on the other. Against this context, scholars have been questioning whether the right of withdrawal should be kept in its existing configuration,[151] and exploring avenues for more sustainable alternatives.

23.4.3 Sustainable Alternatives to the Right of Withdrawal

European scholarship has recognized the environmental challenges linked to product returns, which affect both e-commerce and traditional supply chains. While conventional supply chains register greater energy consumption in transportation, primarily from passenger transport, the e-commerce sector tends to have higher energy consumption in packaging due to additional protective materials like corrugated board.[152]

In response to these environmental impacts, proposals have emerged that resonate with the concerns of green consumers while also protecting the economic interests of smaller e-commerce traders.[153] These suggestions encompass:

- reducing the period for exercising the right of withdrawal;
- extending the exceptions to the exercise of the right of withdrawal in Article 16 CRD[154];
- imposing a fee prior to exercising the right of withdrawal,[155] particularly on abusive consumers[156];
- blacklisting and penalising abusive consumers.[157]

Despite their thoughtful intent, we hold the view that these measures are not sufficiently convincing.

The reduction of the period for exercising the right of withdrawal is problematic for two primary reasons. Firstly, it may imply that a well-intentioned consumer does not have enough time to thoroughly experience the product, potentially leading to their exclusion from warranted protection. Secondly, studies show that consumers are more inclined to exercise the right of withdrawal when they are given less time to do so.[158] This tendency could stem from our inclination towards procrastination and the optimism bias, which leads consumers to believe they will eventually return the product.

It could be valuable to add two additional categories of goods to the existing exceptions to the right of withdrawal: those that cannot be resold as new[159] and single-use products not prohibited by the Single-Use Plastics Directive.[160] However, extending the list of exceptions much further might not be advisable: the current list is already comprehensive and, because it affects the *pacta sunt servanda* principle, the right of withdrawal should remain the exception, not the rule.[161]

We are also hesitant to endorse imposing a fee for exercising the right of withdrawal or for using the product without prejudice to the exercise of the right of withdrawal. Offering the consumer a choice[162] between a contract with and without a right to withdraw would force the consumer to decide whether such a right would be worth the price of a withdrawal (which would function as an *insurance premium* against the risk of having to or wishing to return the product[163]). Such regulations could adversely impact those the right of withdrawal seeks to protect, namely consumers acting in good faith who may be unable[164] or unwilling to pay for such privileges due to factors like loss aversion bias.[165]

The effectiveness of blacklisting and penalising abusive consumers hinges on effective control and enforcement measures. We acknowledge the complexities involved in ensuring such control. Moreover, the best customers based on Customer Lifetime Value are the customers that buy frequently, return frequently (even if abusively) and buy again, who are also more likely to advocate for a brand and spreading electronic word-of-mouth (e-WOM).[166] This may make traders and platforms hesitant when deciding to blacklist them.[167]

In conclusion, we do not support the elimination or substantial reform of the right of withdrawal as stipulated in EU consumer legislation.[168] The right of withdrawal consists of a cornerstone of consumer protection that acts as a safety net for consumers dealing with products they cannot inspect before purchase,[169] thereby bolstering consumer confidence. Maintaining a high level of consumer protection should remain[170] a priority objective of EU policies.[171]

Recognizing that consumer needs evolve, the right of withdrawal serves as a crucial safeguard against potentially unfair practices, such as deceptive marketing strategies (dark patterns) that could prompt consumers to make impulsive decisions. This right plays a crucial role in preventing the execution of inefficient contracts.

Its implementation not only enhances consumer welfare but also safeguards the consumer's right to self-determination.[172] On the side of the trader, the existence of the right of withdrawal incentivizes businesses to uphold elevated standards in product descriptions, customer service, and overall transparency. This, in turn, fosters a more responsible and ethical business environment.

In our contribution, we propose an alternative approach that leaves the right of withdrawal intact but avoids that it is abused by those that the right does not aim to protect, namely chronic returners.

23.5 Greening E-Commerce by Sludging Chronic Returners

23.5.1 The Detrimental Consequences of Chronic Returns in Distance Contracts

Scholarly discourse is divided on whether the exercise of the right of withdrawal should be bound by the principle of good faith.[173] Those who endorse the view it should may agree to the classification that distinguished between legitimate, opportunistic and fraudulent exercise of the right of withdrawal.[174]

Legitimate withdrawal arises in response to situations characterized by information asymmetry, a primary rationale for instituting the right of withdrawal in distance contracts. Various factors, including the desire for uniqueness, product compatibility, perceived risk and social influence, contribute to fostering legitimate withdrawal behaviours.[175]

While recognizing that consumers deviate from purely rational economic behaviour, cases where withdrawal results from impulsive consumer actions illustrate opportunistic withdrawals that aim at mitigating regret and financial losses.[176]

Fraudulent returns occur when consumers return empty boxes, damaged goods or stolen items,[177] thereby breaching return policies. Multiple factors contribute to these behaviours, among which are thrill-search, past experiences, knowledge of regulations, social norms, consumer anomia (social isolation), attitude towards complaining and the perceived impact of fraudulent returns.[178]

Internally, immoral consumers[179] are more prone to cheating during product returns.[180] In the context of a liberal return policy with specified timeframes, packaging methods, and refund agreements, consumers may view these as rights, potentially further weakening their moral perspectives.

This contribution is dedicated to mitigating chronic returns, which typically encompasses the opportunistic exercise of the right of withdrawal. While we understand, and explained above, that no limits should be set to legitimate withdrawals (which, by their own nature, should happen only occasionally) we hold the stance that fraudulent returns must be unequivocally denied and enforced.

Empirical findings show that right of withdrawal fosters opportunistic[181] returns by chronic returners, sometimes also termed 'returnaholics'.[182] One common practice is *fittingrooming*, where consumers intentionally order multiple sizes (referred to as size sampling) and colours of the same product, intending to return all but one item. Another example is *wardrobing*, wherein a product is purchased for a specific event, such as a wedding or Halloween party,[183] or for a social media publication,[184] with a premeditated plan to return it for a refund immediately after use.[185]

In the absence of legislative clarity on whether good faith is a requisite for the exercise of the right of withdrawal,[186] or perhaps in tandem with it due to associated challenges in proof and enforcement, exploring potential solutions becomes imperative.

Studies indicate that it is primarily the same consumers exercising their right of withdrawal across various purchased goods.[187] It is also observed that immorality in these instances appears to be more situational than inherent, thus being more prone to reversion.

Retailers are grappling with ways to address the escalating operational costs linked to returns, and in this context, we consider that mandated green sludge is an alternative worth exploring.

23.5.2 Mandated Green Sludge: A Primer

The imposition of the duty on traders and platforms to persuade consumers to behave in a certain way has been termed 'mandated private nudge'. Mandated private nudge involves the imposition of a duty (*mandated*) on businesses towards individuals (*private*) with a specific profile, being that the latter remain free to act contrarily to the direction proposed (*nudge*). Mandated private (green) sludge consists, conversely, on imposing on businesses the obligation of fostering more sustainable consumer behaviours.

The mandated green sludge advocated in this contribution would consist of the need of consumers whose rate of returns is above average ('chronic returners') to confirm their intention of purchasing a product after a certain time has elapsed. This period would work as a *pre*-cooling-off period, as consumers would have to endure it *before* being able to conclude the contract. During this time the ordered product would be placed on hold for the customer free of charge.[188] Price and all remaining conditions would remain unaltered for the whole length of such a period.

The length of time the consumer would need to endure could depend on their return record: the more returns the purchaser had under their belt, the longer the pre-cooling-off period would be. The length of time could also depend on other product-related characteristics that would have an impact on the environment, such as the distance the product would need to travel until the hands of the consumer, as well as the size and weight of the product returned .The reasons for imposing such a 'pre-cooling-off period' as well as how such a time would be calculated would have to be clearly disclosed to the consumer at the time where they first manifested the intention of purchasing the good.

The pre-cooling-off period would only be imposed for durable consumer goods[189] and products considered non-urgent. Medicine and other healthcare-related products, hygiene products and food would be excluded, not only for public order reasons but also because they are excluded from the products in relation to which the right of withdrawal can be exercised.

Information on shopping behaviour, namely of how many times a consumer has returned a purchased product, and the value of such returns is necessary to discriminate the consumers who may be classified as 'chronic returners'. While in practice technology makes this information easily available,[190] this information may be considered 'personal data' under Article 4(1) of the GDPR, and particularly certain forms of processing such information will fall within the category of profiling through automated processing mechanisms (Article 4(4) of the GDPR). Any processing to be done by an economic operator must necessarily comply with the principles of lawfulness, fairness and transparency, purpose limitation, data minimization, accuracy, storage limitation, integrity and confidentiality, and accountability, provided for in Article 5(1) and (2) of the GDPR. Any treatment of the consumer's financial data would thus need to be clear in purpose, and the data subject will have to expressly consent on such a treatment.[191]

While consent cannot be imposed, the lack of consent may lead to the impossibility of buying the products for which a confirmation would be needed and which, as stated above, do not include first-need products, such as medicine or food.

The proposed method is based upon what we know about consumer behaviour, most particularly the power of green defaults and inertia and the status quo bias.

The implementation of this mandated sludge means that, after manifestation of the intention of concluding the contract, the contract is yet to be confirmed, but the consumer must endure a time until they can do so.

By changing the choice architecture in which now the 'still-not-concluded' contract is the default, we increase the probability that the contract will not be concluded. This is because inertia and procrastination, sometimes described as 'effort' or an 'effort tax',[192] are a force compelling individuals to stick to the current state of affairs. The status quo bias explains consumers' exaggerated tendency to maintain the current state of affairs.[193]

In the cases where the consumer gives up the prospect of confirming the transaction, this would most likely have materialized in a product return, should the contract had been concluded in the usual terms.

Conversely, in the cases where the consumer fights back the sludge this is likely to be indicative of a real intention of acquiring (and possibly keeping) the good. Managing to endure the pre-cooling-off period and purchasing the product in the end is a real exercise of self-determination that Consumer Law wishes to protect and should continue protecting.

23.6 Conclusion

By incentivizing consumers to acquire goods with the confidence that they may return them within a certain period and without need of invoking a motive if they change their minds or no longer need them, the well-justified existence of a right of withdrawal has had a detrimental environmental impact, because it increases the amount of objects produced, transportation-related emissions, repackaging, causes waste and feeds into a spiral of consumerism that is detrimental also for citizen's financial health and well-being.

In this contribution, we argued that the State should impose green sludge on private companies selling goods online to counteract the grey nudge[194] that the mandatory right of withdrawal in distance contracts represents. Such green sludge could decrease returns by making e-purchasing more difficult for those that chronically engage in the exercise of the right of withdrawal.

Our proposal builds into previous works demonstrating the importance of green defaults, as well as the power of inertia, and the status quo and optimism biases.

While restricting e-commerce transactions to chronic returners could affect the freedom of economic initiative of both businesses and consumers, mandated green sludges are not only more likely to pass the proportionality test, but they are also choice-preserving: contrarily to what would happen if they were subject to pure mandates, consumers are *still* given the possibility of opting purchasing goods despite the sludge.

In conclusion, embracing green sludge emerges as a promising strategy, not only to curtail the environmental impact of e-commerce chronic returns but also to foster a collective responsibility for sustainable trade in the evolving landscape of online retail.

Notes

1 Postdoctoral Researcher at the Research Center for Justice and Governance (JusGov) at the University of Minho. This work is funded by the contract-program 'Stimulus to Scientific Employment – Institutional' CEECINST/00157/2018 by the *Fundação para a Ciência e a Tecnologia.*

2 Postdoctoral Researcher with the Private Law Department (Universitat de Barcelona) and Visiting Postdoctoral Researcher at the Research Center for Justice and Governance (JusGov) at the University of Minho. This work is part of the activities of the *Grup de recerca consolidat en Dret civil català UB* (Consolidated Research Group in Catalan Private Law UB) 2021 SGR 00347. The author participates as

research member in the project PID2021-126354OB-I00/MICIN/AEI/10.13039/501100011033/FEDER, UE, about 'Responsabilidad contractual y extracontractual de las plataformas en línea' (*Contract and tort liability of online platforms*) funded by the Spanish Ministry of Science and Innovation (*Ministerio de Ciencia e Innovación*), the Spanish State Agency of Research (*Agencia Estatal de Investigación*) and the European Fund of Regional Development.

3 See, e.g., Michael Redclift, *Wasted: Counting the Costs of Global Consumption* (London, Routledge 1996), Paul C Stern et al, *Environmentally Significant Consumption* (Washington, DC, National Academies Press 1997), Richard Wilk, 'Consumption, Human Needs, and Global Environmental Change' [2022] 12(1) *Global Enviromental Change*.

4 European Commission, 'EU Invests in the Planet – Ten Initiatives for a Modern and Clean Economy – The Juncker Commission's Contribution to the One Planet Summit' (Brussels, 2018), <https://op.europa.eu/en/publication-detail/-/publication/4c7ab013-1dcd-11e8-ac73-01aa75ed71a1/language-en> (last accessed 30 November 2023).

5 United Nations, 'Changing Consumption and Production Patterns in Developed and Developing Countries Discussed in Commission on Sustainable Development' (New York, 23 April 1999) <https://www.google.es/url?sa=t&rct=j&q=&esrc=s&source=web&cd=&cad=rja&uact=8&ved=2ahUKEwjL87Tk0-2CAxXZTaQEHS3LDhkQFnoECBkQAQ&url=https%3A%2F%2Fpress.un.org%2Fen%2F1999%2F19990423.endev509.html&usg=AOvVaw1KAoQQZnUZcTWhIfjytbpj&opi=89978449>.

6 Reuters reported, along with other agencies, that this was considered the worst natural disaster in the history of Slovenia (see <https://www.reuters.com/world/europe/nature-fights-back-slovenias-worst-floods-kill-six-2023-08-07/>, last accessed 11 September 2023). See also <https://confluence.ecmwf.int/display/FCST/202308+-+Rainfall+-+Slovenia#:~:text=According%20to%20Slovenia%27s%20environmental%20agency,whole%20of%20August%2C%20ARSO%20said>.

7 The rainfalls in Emilia-Romagna this Summer involved that, in a 36-hour frame, the region received more than 50% of its mean annual rainfall (<https://www.theguardian.com/environment/2023/may/19/weather-tracker-italy-floods-exacerbated-months-drought>, last accessed 11 September 2023), causing significant damage in the region (<https://www.reuters.com/world/europe/northern-italy-floods-caused-huge-damage-says-governor-2023-05-18/>, last accessed 11 September 2023).

8 July 2023 made a record in terms of temperatures, being considered the highest on record (of any month) since 1940. See <https://climate.copernicus.eu/july-2023-warmest-month-earths-recent-history> (last accessed 11 September 2023). August 2023 was also the warmest since there are records, and the second warmest month in history, just after July 2023 (<https://climate.copernicus.eu/august-2023-second-warmest-month-closes-warmest-summer>, last accessed 11 September 2023).

9 Merja Elo et al, *Interdisciplinary Perspectives on Planetary Well-Being* (London, Routledge 2024) 3, and others.

10 Some studies show that the beliefs and worries related to climate change are more prominent in left-wing voters and liberals, rather than in right-wing voters. See Thea Gregersen et al, 'Political Orientation Moderates the Relationship between Climate Change Beliefs and Worry about Climate Change' [2020] 11(1) *Frontiers in Psychology* (https://doi.org/10.3389/fpsyg.2020.01573) and Jakob Schwörer, 'Mainstream Parties and Global Warming: What Determines Parties' Engagement in Climate Protection' [2023] *European Journal of Political Research* (https://doi.org/10.1111/1475-6765.12602).

11 See Cass Sunstein and Richard H Thaler, *Nudge. Improving Decisions about Health, Wealth and Happiness* (London, Penguin Books 2008).

12 See Anne-Lise Sibony and Alberto Alemanno, *Nudge and the Law* (London, Hart Publishing 2015).

13 Patricia Sánchez Abril, Francisco Oliva Blázquez, and Joan Martínez Evora, 'The Right of Withdrawal in Consumer Contracts: A Comparative Analysis of American and European Law' [2018] 3 *InDret* 2; P Rekaiti and Roger Van den Bergh, 'Cooling-Off Periods in the Consumer Laws of the EC Member States. A Comparative Law and Economics Approach' [2000] 23(4) *Journal of Consumer Policy* 371, 379 (https://doi.org/10.1023/A:1007203426046Rekaiti); Jan M Smits, 'The Right to Change Your Mind? Rethinking the Usefulness of Mandatory Rights of Withdrawal in Consumer Contract Law' (2011) Maastricht European Private Law Institute Working Paper 2011/01, 6 <https://papers.ssrn.com/sol3/papers.cfm?abstract_id=1719104>; Evelyne Terryn and Elias Van Gool, 'The Role of European Consumer Regulation in Shaping the Environmental Impact of E-Commerce' [2021] 3 EuCML 89, 96.

14 Cass R Sunstein, *The Ethics of Influence. Government in the Age of Behavioural Science* (Cambridge, Cambridge University Press 2016) 78 ff.

15 A right of withdrawal exists also for off-premises sales for distance and off-premises service contracts (arts 2(5)–(8) and (9) Consumer Rights Directive). See also Evelyne Terryn and Elias Van Gool, 'The Role of European Consumer Regulation in Shaping the Environmental Impact of E-Commerce' [2021] 3 *EuCML* 89, 94, fn 82.
16 Horst GM Eidenmüller, 'Why Withdrawal Rights?' (2010) Social Sciences Research Network Working Paper, 5 (http://doi.org/10.2139/ssrn.1660535).
17 Richard Thaler, 'Nudge, Not Sludge' [2018] 361(641) *Science* 431 (https://doi.org/10.1126/science.aau9241).
18 Cass R Sunstein, 'Sludge Audits' [2022] 6(4) *Behavioural Public Policy* 654, 660 (https://doi.org/10.1017/bpp.2019.32). We thank one of the anonymous peer-reviewers for this reference.
19 Cass R Sunstein, 'Sludge and Ordeals' [2019] 68 *Duke Law Journal* 1843, 1865 ff.
20 See TM Wilkinson, 'Counter-Manipulation and Health Promotion' [2017] 10(3) *Public Health Ethics* 257 (https://doi.org/10.1093/phe/phw044).
21 David Maschitto, 5 Ways to Control Chronic Returners, *Tecsys* (25 January 2022) <https://www.tecsys.com/blog/2022/01/5-ways-to-control-chronic-returners> accessed 29 November 2023.
22 September 2023.
23 We thank one of the anonymous peer-reviewers for pointing this out.
24 Frank Trentmann, *Empire of Things: How We Became a World of Consumers, from the Fifteenth Century to the Twenty-First* (New York, Harper Collins 2016).
25 Lucia Reisch and John Thøgersen, *Handbook of Research on Sustainable Consumption* (Cheltenham, Edward Elgar 2015); Oksana Mont, *A Research Agenda for Sustainable Consumption Governance* (Cheltenham, Edward Elgar 2019).
26 Lieske Voget-Kleschin, Christian Baatz, and Konrad Ott, 'Ethics and Sustainable Consumption', in Lucia Reisch and John Thøgersen (eds), *Handbook of Research on Sustainable Consumption* (Cheltenham, Edward Elgar 2015) 121.
27 Lewis Akenji, 'Consumer Scapegoatism and Limits to Green Consumerism' [2014] 63 *Journal of Cleaner Production* 13 (https://doi.org/10.1016/j.jclepro.2013.05.022); Anna Jakubczak and Małgorzata Gotowska, 'Green Consumerism vs. Greenwashing' [2020] XXIII(2) *European Research Studies Journal* (https://doi.org/10.35808/ersj/1816); Sachdeva et al list some conducts that may be subsumed within the tag of 'green consumerism', such as purchasing appliances with energy star labels or buying organic products (Sonya Sachdeva, Jennifer Jordan and Nina Mazar, 'Green Consumerism: Moral Motivations to a Sustainable Future' [2015] 6 *Current Opinions in Psychology* 60 (https://doi.org/10.1016/j.copsyc.2015.03.029)).
28 Talat S Genc and Pietro De Giovanni, 'Dynamic Pricing and Green Investments under Conscious, Emotional, and Rational Consumers' [2021] 2 *Cleaner and Responsible Consumption* (https://doi.org/10.1016/j.clrc.2021.100007); Valérie Guillard, Emmanuelle Le Nagard and Gisele Campos Ribeiro, 'A Typology of Consumers Regarding Perceived Obsolescence: The Paradox of Eco-Conscious Consumers' [2023] 412 *Journal of Cleaner Production* (https://doi.org/10.1016/j.jclepro.2023.137202).
29 Krzysztof Blonski and Jolanta Witek, 'Minimalism in Consumption' [2019] LIII(2) *Annales Universitatis Mariae-Curie-Sklodowska, Sectio H – Oeconomia* 7; Jiyun Kang, Cosette M Joyner Martinez and Catherine Johnson, 'Minimalism as a Sustainable Lifestyle: Its Behavioral Representations and Contributions to Emotional Well-Being' [2021] 27 *Sustainable Production and Consumption* 802 (https://doi.org/10.1016/j.spc.2021.02.001).
30 Paul C Stern et al, *Environmentally Significant Consumption* (Washington, DC, National Academies Press 1997), Richard Wilk, 'Consumption, Human Needs, and Global Environmental Change' [2002] 12(1) *Global Enviromental Change*; F Duchin, 'Consumption, Environmentally Significant' in Niel J Smelser and Paul B Bates (eds), *International Encyclopedia of the Social & Behavioral Sciences* (London, Elsevier 2001).
31 Rebecca KM Clube and Mike Tennant, 'What Would a Human-Centred "Social" Circular Economy Look Like? Drawing from Max-Neef's Human-Scale Development Proposal' [2023] 383 *Journal of Cleaner Production* 1 (https://doi.org/10.1016/j.jclepro.2022.135455).
32 See, generally, Giacomo D'Alise, Federico Demaria, and Giorgos Kellis, *Degrowth. A Vocabulary for a New Era* (London, Routledge 2015).
33 Nancy MP Bocken, Laura Niessen and Samuel W Short, 'The Sufficiency-Based Circular Economy – An Analysis of 150 Companies' [2022] 3 *Frontiers in Sustainability* (https://doi.org/10.3389/frsus.2022.899289).

34 Mary Grigsby, *Buying Time and Getting by: The Voluntary Simplicity Movement* (New York, State University of New York Press 2004); Shuji Ohira and Akiko Masuda, 'Voluntary Simplicity Consumption in Japan. Alternative Consumption and Backward Consumption among Minimalists' [2021] 49 *Advances in Consumer Research* 246.

35 See generally Anaïs Michel, *Premature Obsolescence: In Search of an Improved Legal Framework* (Cambridge, Intersentia 2023) and also Alberto De Franceschi, 'Planned Obsolescence Challenging the Effectiveness of Consumer Law and the Achievement of a Sustainable Economy: The Apple and Samsung Cases' [2018] 7(6) *EuCML* 217.

36 First reference has been made by Expert User Experience (UX) Harry Brignull, 'Dark Patterns: Dirty Tricks Designers Use to Make People Do Stuff', *90 Percent of Everything* (8 July 2010) <https://90percentofeverything.com/2010/07/08/dark-patterns-dirty-tricks-designers-use-to-make-people-do-stuff/index.html> accessed 30 November 2023.

37 See Jamie Luguri and Lior Jacob Strahilevitz, 'Shining a Light on Dark Patterns' [2021] 13 *Journal of Legal Analysis* 44 (https://doi.org/10.1093/jla/laaa006) and also Martin Brenncke, 'Regulating Dark Patterns' [2024] *Notre Dame Journal International & Comparative Law* (forthcoming). We argue that dark patterns should not be identified with 'sludge' (see Chapter 3.2 below) and instead with what Karen Yeung (Karen Yeung, 'Hypernudge': Big Data as a Mode of Regulation by Design' [2017] 20(1) *Information, Communication & Society* (https://doi.org/10.1080/1369118X.2016.1186713)) calls 'hypernudging' (see also Marjolein Lanzing, '"Strongly Recommended": Revisiting Decisional Privacy to Judge Hypernudging in Self-Tracking Technologies' [2019] 32 *Philosophy Technology* 549 (https://doi.org/10.1007/s13347-018-0316-4) and Viktorija Morozovaite, 'Hypernudging in the Changing European Regulatory Landscape for Digital Markets' [2023] 15(78) *Policy Internet* (https://doi.org/10.1002/poi3.329) or with 'digital nudging' (see Markus Weinmann, Christoph Schneider and Jan vom Brocke, 'Digital Nudging' [2016] 58(6) *Business & Information Systems Engineering* 433 (https://doi.org/10.1007/s12599-016-0453-1). as they share a commonality – they both occur in the online domain. A legal definition of 'dark patterns' in the specific context of online interfaces of online platforms may be found in Recital 67 of the Digital Services Act, and they are prohibited in art. 25(1) of the same act. Dark patterns can be also subsumed in other legal instruments in current EU law, such as the Unfair Commercial Practices Directive. See European Commission, Directorate-General for Justice and Consumers, Francisco Lupiáñez-Villanueva et al, *Behavioural Study on Unfair Commercial Practices in the Digital Environment – Dark Patterns and Manipulative Personalisation – Final Report* (Brussels, Publications Office of the European Union 2022) <https://data.europa.eu/doi/10.2838/859030> accessed 30 November 2023.

38 Robin J Tanner et al, 'Of Chameleons and Consumption: The Impact of Mimicry on Choice and Preferences' [2008] 34(6) *Journal of Consumer Research* 754 (https://psycnet.apa.org/doi/10.1086/522322).

39 This is the conclusion reached by a literature review of more than 50 articles published in the US and Europe on impulsive buying Tommy KH Chan, Christy MK Cheung and Zach WY Lee, 'The State of Online Impulsive-Buying Research: A Literature Analysis' [2017] 54(2) *Information & Management* 204 (https://doi.org/10.1016/j.im.2016.06.001). Additional studies have found similar figures. See, e.g., Maleku Alemu and Shimels Zewdie, 'Consumers' Impulsive Buying Behavior: Structured Systematic Literature Review' [2021] 8(10) *International Journal of Management, Accounting and Economics* (https://doi.org/10.5281/zenodo.5975984).

40 See Joan Wallet et al, Growing a Circular Economy: Ending the Throwaway Society (House of Commons 2014) 6 and Carl Dalhammar et al, *Moving Away from the Throwaway Society. Five Policy Instruments for Extending the Life of Consumer Durables* (Mistra Sustainable Consumption 2022) 7.

41 Report of the World Commission on Environment and Development: Our Common Future.

42 The Treaty on European Union (TEU), art. 3(3), outlines the EU's goals, emphasizing the promotion of economic, social and territorial cohesion, along with a commitment to environmental sustainability. Additionally, the Treaty on the Functioning of the European Union (TFEU) incorporates provisions addressing environmental protection, sustainable agriculture and fisheries.

43 Commission, 'Communication from the Commission to the European Parliament, the European Council, the Council, the European Economic and Social Committee and the Committee and the Committee of the Regions, The European Green Deal' COM/2019/640 final.

44 United Nations, Sustainable Development Goals Report 2023: Special Edition (New York, United Nations 2023).

45 Commission, 'Communication from the Commission to the European Parliament, the Council, the European Economic and Social Committee and the Committee of the Regions. A new Circular Economy Action Plan. For a cleaner and more competitive Europe' COM(2020) 98 final.

46 The first Circular Economy Action dated from 2015. Commission, 'Communication from the Commission to the European Parliament, the Council, the European Economic and Social Committee and the Committee of the Regions. Closing the loop – An EU action plan for the Circular Economy' COM(2015) 614 final.

47 Commission, 'Proposal for a Directive on empowering consumers for the green transition through better protection against unfair practices and better information' COM/2022/143 final.

48 Commission, 'Proposal for a Regulation of the European Parliament and of the Council establishing a framework for setting ecodesign requirements for sustainable products and repealing Directive 2009/125/EC' COM(2022) 142 final.

49 Commission, 'Proposal for a Directive of the European Parliament and of the Council on substantiation and communication of explicit environmental claims (Green Claims Directive)' COM(2023) 166 final.

50 Commission, 'Proposal for a Directive of the European Parliament and of the Council on common rules promoting the repair of goods and amending Regulation (EU) 2017/2394, Directives (EU) 2019/771 and (EU) 2020/1828' COM(2023) 155 final.

51 Cass Sunstein and Richard H Thaler, *Nudge. Improving Decisions about Health, Wealth and Happiness* (London, Penguin Books 2008) 7.

52 Vincent Berthet and Benjamin Ouvrard, 'Nudge towards a Consensus View' [2019] 5(1) *Psychology and Cognitive Sciences Open Journal* 1 (https://www.doi.org/10.17140/PCSOJ-5-143), consider that the concept of nudge is 'ill-defined'. Scholarship has been abundantly discussing this concept. See Daniel M Hausman and Brynn Welch, 'Debate: To Nudge or Not to Nudge' [2010] 18 *Journal of Political Philosophy* 123 (https://doi.org/10.1111/j.1467-9760.2009.00351.x); Yashar Saghai, 'Salvaging the Concept of Nudge' [2013] 29 *Journal of Medical Ethics* 487; Pelle Guldborg Hansen, 'The Definition of Nudge and Libertarian Paternalism: Does the Hand Fit the Glove?' [2016] 7(1) *EJRR* and Philippe Mongin and Mikaeel Cozic, 'Rethinking Nudge: Not One but Three Concepts' [2018] 2(1) *Behavioural Public Policy* 107.

53 Cass Sunstein and Richard H Thaler, *Nudge. Improving Decisions about Health, Wealth and Happiness* (London, Penguin Books 2008) 7. Some call it, instead 'soft paternalism'. See Daniel M Hausman and Brynn Welch, 'Debate: To Nudge or Not to Nudge' [2010] 18 *Journal of Political Philosophy* 123 (https://doi.org/10.1111/j.1467-9760.2009.00351.x).

54 Antonios Karampatzos, *Private Law, Nudging and Behavioural Economic Analysis. The Mandated-Choice Model* (London, Routledge 2020) 4.

55 Cass R Sunstein, 'Nudging: A Very Short Guide' [2014] 37 *Journal of Consumer Policy* 583, 584 (https://doi.org/10.1007/s10603-014-9273-1); Péter Cserne, 'Making Sense of Nudge-Scepticism: Three Challenges to EU Law's Learning from Behavioural Sciences' in Anne-Lise Sibony and Alberto Alemanno (eds), *Nudge and the Law: A European Perspective* (London, Hart Publishing 2015) 279.

56 Cass Sunstein and Richard H Thaler, *Nudge. Improving Decisions about Health, Wealth and Happiness* (London, Penguin Books 2008) 6.

57 Luc Bovens, 'The Ethics of Nudge' in Till Grüne-Yanoff and Sven Ove Hansson (eds), *Preference Change: Approaches from Philosophy, Economics and Psychology* (London, Springer 2009) 207; Daniel M Hausman and Brynn Welch, 'Debate: To Nudge or Not to Nudge' [2010] 18 *Journal of Political Philosophy* 123 (https://doi.org/10.1111/j.1467-9760.2009.00351.x); Sarah Conly, *Against Autonomy. Justifying Coercive Paternalism* (Cambridge, Cambridge University Press 2013) 30; Christian Schubert, 'Green Nudges: Do the Work? Are They Ethical?' [2017] 132 *Ecological Economics* 329 (https://doi.org/10.1016/j.ecolecon.2016.11.009).

58 This does not affect the fact that some transparent nudges might be unethical (Andreas T Schmidt and Bart Engelen, 'The Ethics of Nudging: An Overview' [2020] 7 *Philosophy Compass* (https://doi.org/10.1111/phc3.12658). We thank one anonymous reviewer for this remark and the reference.

59 Samuli Reijula and Ralph Hertwig, 'Self-Nudging and the Citizen Choice Architect' [2020] 6 *Behavioural Public Policy* 119.

60 Luc Bovens, 'The Ethics of Nudge' in Till Grüne-Yanoff and Sven Ove Hansson (eds), *Preference Change: Approaches from Philosophy, Economics and Psychology* (London, Springer 2009) 217.

61 LC Van Gestel, MA Adriaanse and DTD De Ridder, 'Do Nudges Make Use of Automatic Processing? Unraveling the Effects of a Default Nudge under Type 1 and Type 2 Processing' [2021] 5(1–3) *Comprehensive Results in Social Psychology* 4 (https://doi.org/10.1080/23743603.2020.1808456); Denise de Ridder, Floor Kroese and Laurens van Gestel, 'Nudgeability: Mapping Conditions of Susceptibility to Nudge Influence' [2022] 17(2) *Perspectives on Psychological Science* 346 (https://doi.org/10.1177/1745691621995183). We thank one of the anonymous reviewers for this remark and references.

62 John Stuart Mill, *On Liberty* (Batoche Books 2001).

63 Sarah Buss, 'Autonomous Action: Self-Determination in the Passive Mode' [2012] 122 *Ethics* 647, 685, fn. 89 (https://doi.org/10.1086/666328).

64 According to Robert Baldwin, 'From Regulation to Behaviour Change: Giving Nudge the Third Degree' [2014] 77(6) *The Modern Law Review* 831 (https://doi.org/10.1111/1468-2230.12094)., there are first-, second- and third-degree nudges.

65 See, particularly, the one which distinguishes between paternalistic and non-paternalistic nudges Anne van Aaken, 'Judge the Nudge: In Search of the Legal Limits of Paternalistic Nudging in the EU' in Anne-Lise Sibony and Alberto Alemanno (eds), *Nudge and the Law: A European Perspective* (London, Hart Publishing 2015) 83. See also Marta Santos Silva, 'Nudging and Other Behaviourally Based Policies as Enablers for Environmental Sustainability' [2022] 11(1) *Laws* (https://doi.org/10.3390/laws11010009).

66 It is believed that the expression was coined by Cass Sunstein and Richard H Thaler, *Nudge. Improving Decisions about Health, Wealth and Happiness* (London, Penguin Books 2008) 81 ff.

67 This does not relate to the fact they would operate in a 'grey' zone (namely between legitimate and illegitimate methods of influencing consumer behaviour), but more as opposed to 'green nudges' as will be explained in this section, in a clear allusion to the colour of smoke from exhaust pipes and indication of fire running out of materials to burn.

68 Ming Chung Han, 'Checkout Button and Online Consumer Impulse-Buying Behavior in Social Commerce: A Trust Transfer Perspective' [2023] 74 *Journal of Retailing and Consumer Services* (https://doi.org/10.1016/j.jretconser.2023.103431).

69 Wen-Chin Hsu, Mu-Heng Lee, and Kai-Wen Zheng, 'From Virtual to Reality: The Power of Augmented Reality in Triggering Impulsive Purchases' [2024] 76 *Journal of Retailing and Consumer Services* 76 (https://doi.org/10.1016/j.jretconser.2023.103604).

70 Joasia A Luzak, 'To Withdraw or Not to Withdraw or Not to Withdraw? Evaluation of the Mandatory Right of Withdrawal in Consumer Distance Selling Contracts Taking into Account Its Behavioural Effects on Consumers', [2013] Centre for the Study of European Contract Law Working Paper, Amsterdam Law School Legal Studies Research Paper, 4.

71 This is a situation leading to the so-called 'buyer's remorse'. Patricia Sánchez Abril, Francisco Oliva Blázquez, and Joan Martínez Evora, 'The Right of Withdrawal in Consumer Contracts: A Comparative Analysis of American and European Law' [2018] 3 *InDret* 5.

72 Ming Chung Han, 'Checkout Button and Online Consumer Impulse-Buying Behavior in Social Commerce: A Trust Transfer Perspective' [2023] 74 *Journal of Retailing and Consumer Services* (https://doi.org/10.1016/j.jretconser.2023.103431).

73 Mevlude Akbulut-Yuksel and Catherine Boulatoff, 'The Effects of a Green Nudge on Municipal Solid Waste: Evidence from a Clear Bag Policy' [2021] 106 *Journal of Environmental Economics and Management* (https://doi.org/10.1016/j.jeem.2020.102404); Matthias Buchholz and Oliver Musshoff, 'Tax or Green Nudge? An Experimental Analysis of Pesticide Policies in Germany' [2021] 48(4) *European Review of Agricultural Economics* 940 (https://doi.org/10.1093/erae/jbab019); Christian Schubert, 'Green Nudges: Do the Work? Are They Ethical?' [2017] 132 *Ecological Economics* 329 (https://doi.org/10.1016/j.ecolecon.2016.11.009); Clément Carrel, Marie-Laure Gavard-Perret and Cindy Caldara, 'Factors of Effectiveness of Green Nudges for More Eco-Responsible Behaviour – Systematic Review and Research Directions' [2023] 38(3) *Recherche et Applications En Marketing* 1 (https://doi.org/10.1177/20515707231177814).

74 Frederik Carlsson et al, 'The Use of Green Nudges as an Environmental Policy Instrument' [2021] 15(2) *Review of Environmental Economics and Policy* 216; Marta Santos Silva, 'Nudging and Other Behaviourally Based Policies as Enablers for Environmental Sustainability' [2022] 11(1) *Laws* (https://doi.org/10.3390/laws11010009).

75 See also Christian Schubert, 'Green Nudges: Do the Work? Are They Ethical?' [2017] 132 *Ecological Economics* 329, 331 (https://doi.org/10.1016/j.ecolecon.2016.11.009). Compare with Frederik Carlsson et al, 'The Use of Green Nudges as an Environmental Policy Instrument' [2021] 15(2) *Review of Environmental Economics and Policy* 216, 217, who define it as 'behavioural interventions aimed at reducing negative externalities'.

76 Katherine White, Rishad Habib, and David J Hardisty, 'How to SHIFT Consumer Behaviors to Be More Sustainable: A Literature Review and Guiding Framework' [2019] 83(3) *Journal of Marketing* (https://doi.org/10.1177/0022242919825649).

77 Jimena González-Ramírez, 'The Effectiveness of a Green Default Nudge in Achieving Resource Conservation', Social Sciences Research Network Working Paper, 1 November 2022 (http://dx.doi.org/10.2139/ssrn.4264945, last accessed 30 November 2023).

78 The expression 'administration of nudges' underscores the purposeful and strategic nature involved in implementing these subtle encouragements. See Moritz von Zahn et al, 'The Smart Green Nudge: Reducing Product Returns through Enriched Digital Footprints & Causal Machine Learning', SAFE Working Paper, October 2022, 7.

79 Hilary Byerly et al, 'Nudging Pro-Environmental Behavior: Evidence and Opportunities' [2018] 16(3) *Frontiers in Ecology and the Environment* (https://doi.org/10.1002/fee.1777) 159, 167.

80 Despoina Alempaki, Andrea Isoni and Daniel Read, 'Tainted Nudge' [2023] 176 *Organizational Behavior and Human Decision Processes* (https://doi.org/10.1016/j.obhdp.2023.104244). Michael G Faure and Roy A Partain, *Environmental Law and Economics. Theory and Practice* (Cambridge, Cambridge University Press 2019) 32, call nudge 'suasive mechanism'.

81 The criticism to command-and-control instruments is often collectively referred to as 'New Governance'. See further On Amir and Orly Lobel, 'Liberalism and Lifestyle: Informing Regulatory Governments with Behavioural Research' [2012] 1 *European Journal of Risk Regulation* 17.

82 Francesca Cellina et al, 'Significant but Transient: The Impact of an Energy Saving App Targeting Swiss Households' [2024] 335 *Applied Energy* (https://doi.org/10.1016/j.apenergy.2023.122280).

83 Moritz von Zahn et al, 'The Smart Green Nudge: Reducing Product Returns through Enriched Digital Footprints & Causal Machine Learning', SAFE Working Paper, October 2022; Stefan Stieglitz et al, 'The Potential of Digital Nudging to Bridge the Gap between Environmental Attitude and Behavior in the Usage of Smart Home Applications' [2023] 72 *International Journal of Information Management* (https://doi.org/10.1016/j.ijinfomgt.2023.102665).

84 Mevlude Akbulut-Yuksel and Catherine Boulatoff, 'The Effects of a Green Nudge on Municipal Solid Waste: Evidence from a Clear Bag Policy' [2021] 106 *Journal of Environmental Economics and Management* (https://doi.org/10.1016/j.jeem.2020.102404); Siaw-Chui Wee, Weng-Wai Choong and Sheau-Ting Low, 'Can "Nudging " Play a Role to Promote Pro-Environmental Behaviour?' [2021] 5(4) *Environmental Challenges* (https://doi.org/10.1016/j.envc.2021.100364).

85 Steve Brooker and Philip Cullum, 'Five Steps to Consumer-Friendly Better Regulation' [2008] 18(1) *Consumer Policy Review* 4; Alberto Alemanno, *What Can EU Health Law Learn from Behavioural Sciences? The Case of EU Lifestyle Regulation* (London, Hart Publishing 2015) 235–254, 342 ff; Alberto Alemanno and Alessandro Spina, 'Nudging Legally. On the Checks and Balances of Behavioural Regulation' [2014] 12(2) *International Journal of Constitutional Law* (https://doi.org/10.1093/icon/mou033).

86 Alberto Alemanno and Alessandro Spina, 'Nudging Legally. On the Checks and Balances of Behavioural Regulation' [2014] 12(2) *International Journal of Constitutional Law* 20 (https://doi.org/10.1093/icon/mou033).

87 Alberto Alemanno, *What Can EU Health Law Learn from Behavioural Sciences? The Case of EU Lifestyle Regulation* (London, Hart Publishing 2015) 246. There seems to be more resistance to more coercive measures. See Alberto Alemanno and Alessandro Spina, 'Nudging Legally. On the Checks and Balances of Behavioural Regulation' [2014] 12(2) *International Journal of Constitutional Law* 16 (https://doi.org/10.1093/icon/mou033).

88 Cass Sunstein, 'Nudges vs Shoves: The Benefits of Preserving Choice' [2014] 127(6) *Harvard Law Review*.

89 Marius Alt et al, 'Synergies of Interventions to Tackle Climate Change – A Meta-Analysis' (Ispra, Italy/Brussels, Belgium/Petten, The Netherlands: European Commission, Joint Research Centre, December 2022).

90 Felix Ekardt, *Sustainability: Transformation, Governance, Ethics, Law* (London, Springer 2020).

91 Patricia Sánchez Abril, Francisco Oliva Blázquez, and Joan Martínez Evora, 'The Right of Withdrawal in Consumer Contracts: A Comparative Analysis of American and European Law' [2018] 3 *InDret* 30.

92 Sometimes, heuristics quickly return approximate answers that are better than deliberative thinking, viewed from the perspective of ecological rationality (we thank one of the anonymous for this remark). Other times, heuristics give rise to systematic errors called *biases* (Eliezer Yudkowsky, 'Cognitive Biases Potentially Affecting Judgment of Global Risks' in Milan Ćirković and Nick Bostrom (eds), *Global Catastrophic Risks* (Oxford, Oxford University Press 2011) 91.

93 Christian Thorun et al, *Nudge-Ansätze Beim Nachhaltigen Konsum: Ermittlung Und Entwicklung von Maßnahmen Zum ‚Anstoßen' Nachhaltiger Konsummuster* (Bundesministeriums für Umwelt, Naturschutz, Bau und Reaktorsicherheit 2016) 16.

94 Marta Santos Silva, 'Nudging and Other Behaviourally Based Policies as Enablers for Environmental Sustainability' [2022] 11(1) *Laws* (https://doi.org/10.3390/laws11010009) and Marta Santos Silva and Luísa Cortat Simonetti Gonçalves, 'Nudging Consumers towards Sustainable Practices Regarding Plastics in a Post-COVID-19 Europe' in Ewoud Hondius et al (eds), *Coronavirus and the Law in Europe* (Cambridge, Intersentia 2021) 947.

95 Behavioural insights refers here to the range of literature about human behaviour from behavioural economics, social psychology, neuroscience and sociology (David Halpern and Rory Gallagher, 'Can "Nudging" Change Behaviour? Using "Behavioural Insights" to Improve Program Redesign' in John Wanna, Hsu-Ann Lee and Sophie Yates (eds), *Managing under Austerity, Delivering under Pressure. Performance and Productivity in Public Service* (Sydney, ANU Press 2015) 165 f.

96 Alberto Alemanno, *What Can EU Health Law Learn from Behavioural Sciences? The Case of EU Lifestyle Regulation* (London, Hart Publishing 2015) 342 f.

97 See, on the effect of nudge, Anne Nielsen et al, *Nudging and Pro-Environmental Behaviour* (Nordic Council of Ministers 2016) 18 f; Hendrik Bruns et al, 'Can Nudges Be Transparent and yet Effective?' [2018] 65 *Journal of Economic Psychology* 41 (https://doi.org/10.1016/j.joep.2018.02.002); Christian Schubert, 'Green Nudges: Do the Work? Are They Ethical?' [2017] 132 *Ecological Economics* 329 (https://doi.org/10.1016/j.ecolecon.2016.11.009).

98 Mark White, *The Manipulation of Choice. Ethics and Libertarian Paternalism* (London, Palgrave Macmillan 2013) 17 ff., even well-minded consumer lawyers admittedly fall back into their comfort zones. See Hans-W Micklitz, 'Squaring the Circle? Reconciling Consumer Law and the Circular Economy' [2019] 6 *EuCML* 229, 230.

99 Christian Schubert, 'Green Nudges: Do the Work? Are They Ethical?' [2017] 132 *Ecological Economics* 329 (https://doi.org/10.1016/j.ecolecon.2016.11.009). See also Moritz von Zahn et al, 'The Smart Green Nudge: Reducing Product Returns through Enriched Digital Footprints & Causal Machine Learning', SAFE Working Paper, October 2022, 6.

100 NJ Goldstein, RB Cialdini and V Griskevicius, 'A Room with a Viewpoint: Using Social Norms to Motivate Environmental Conservation in Hotels' [2008] 35(3) *Journal of Consumer Research* 472 (https://doi.org/10.1086/586910).

101 Christian Schubert, 'Green Nudges: Do the Work? Are They Ethical?' [2017] 132 *Ecological Economics* 329 (https://doi.org/10.1016/j.ecolecon.2016.11.009) 332 ff.

102 Cass Sunstein, 'Green Defaults Can Combat Climate Change' [2021] 5 *Nature Human Behaviour*.

103 Certain U.S. publications on the subject are accessible, often found in disciplines outside of law. See Yu Luo et al, 'A Meta-Analytic Cognitive Framework of Nudge and Sludge' [2023] 10(11) *Royal Society Open Science* (https://doi.org/10.1098/rsos.230053). See, however, the publications authored by legal scholar Cass Sunstein (Cass R Sunstein, 'Sludge and Ordeals' [2019] 68 *Duke Law Journal* 1843; Cass R Sunstein, 'Sludge Audits' [2022] 6(4) *Behavioural Public Policy* 654).

104 Cass R Sunstein, 'Sludge and Ordeals' [2019] 68 *Duke Law Journal* 1843, 1850 and Cass R Sunstein and Julien L Gosset, 'Optimal Sludge? The Price of Program Integrity' [2020] 70 *Duke Law Journal* 74, 75.

105 According to Thaler (Richard Thaler, 'Nudge, Not Sludge' [2018] 361(641) *Science* 431 (https://doi.org/10.1126/science.aau9241)), sludge 'just mucks things up'. It usually involves the activation of what Kanneman calls 'system 2' thinking. See Daniel Kahneman, *Thinking, Fast and Slow* (Farrar, Straus and Giroux 2011).

106 Mark Petticrew et al, 'Dark Nudges and Sludge in Big Alcohol: Behavioral Economics, Cognitive Biases, and Alcohol Industry Corporate Social Responsibility' [2020] 4 *The Milkbank Quarterly* 1290, 1296 (https://doi.org/10.1111/1468-0009.12475).

107 Antonios Karampatzos, *Private Law, Nudging and Behavioural Economic Analysis. The Mandated-Choice Model* (London, Routledge 2020) ix.
108 Hans-W Micklitz, 'Squaring the Circle? Reconciling Consumer Law and the Circular Economy' [2019] 6 *EuCML* 229.
109 Klaus Tonner, 'Consumer Protection and Environmental Protection: Contradictions and Suggested Steps Towards Integration' [2000] 23(1) *Journal of Consumer Policy* 63, 64 (https://doi.org/10.1023/A:1006307709249).
110 Hans-W Micklitz, 'Squaring the Circle? Reconciling Consumer Law and the Circular Economy' [2019] 6 *EuCML* 229.
111 Hans-W Micklitz, 'Squaring the Circle? Reconciling Consumer Law and the Circular Economy' [2019] 6 *EuCML* 229, 234.
112 Arguing the law electively "codes" certain assets, endowing them with the capacity to protect and produce private wealth, see Katharina Pistor, *The Code of Capital. How the Law Creates Wealth and Inequality* (Princeton University Press 2019).
113 Gerhard Wagner and Horst Eidenmüller, 'Down by Algorithms? Siphoning Rents, Exploiting Biases, and Shaping Preferences: Regulating the Dark Side of Personalized Transactions' [2018] 20 *The University of Chicago Law Review* 581.
114 According to the Eurostat, in 2022, 9.2% of inhabitants of the EU27 were in arrears in relation to their payments of mortgage or rent, utility bills or hire purchase (<https://ec.europa.eu/eurostat/databrowser/view/ILC_MDES05$DV_422/default/table?lang=en> accessed 14 September 2023. In the USE, credit card usage reached a new milestone in summer 2023, with balances totalling over $1 trillion (https://www.gao.gov/blog/american-credit-card-debt-hits-new-record-whats-changed-post-pandemic, last accessed 30 November 2023).
115 Directive 2011/83/EU of the European Parliament and of the Council of 25 October 2011 on consumer rights (2011) OJ L 304/260.
116 Marco BM Loos, 'Rights of Withdrawal' in Geraint Howells and Reiner Schulze (eds), *Modernising and Harmonising Consumer Contract Law* (Sellier 2009) 241.
117 Antonios Karampatzos, *Private Law, Nudging and Behavioural Economic Analysis. The Mandated-Choice Model* (London, Routledge 2020) 86.
118 The right of withdrawal was first regulated in Europe by Council Directive 85/577/EEC of 20 December 1985.
119 Alberto Alemanno, 'Nudge and the European Union' in Holger Straßheim and Silke Beck (eds), *Handbook of Behavioural Change and Public Policy* (Cheltenham, Edward Elgar 2019) 271.
120 Antonios Karampatzos, *Private Law, Nudging and Behavioural Economic Analysis. The Mandated-Choice Model* (London, Routledge 2020) 36.
121 The *ratio* of the right in the context of door-to-door sales consists of the risk that the consumer might have concluded the contract under psychological pressure from the salesperson. See Antonios Karampatzos, *Private Law, Nudging and Behavioural Economic Analysis. The Mandated-Choice Model* (London, Routledge 2020) 36.
122 This may justify the fact that returns are greater in e-commerce. See Henrik Pålsson, Fredrik Pettersson and Lena Winslott Hiselius, 'Energy Consumption in E-Commerce versus Conventional Trade Channels-Insights into Packaging, the Last Mile, Unsold Products and Product Returns' [2017] 164 *Journal of Cleaner Production* 765, 776 (https://doi.org/10.1016/j.jclepro.2017.06.242).
123 Antonios Karampatzos, *Private Law, Nudging and Behavioural Economic Analysis. The Mandated-Choice Model* (London, Routledge 2020) 94. Studies show that when a voluntary provision is perceived as friendly, buyers tend to reciprocate by not exploiting sellers too heavily. G Borges and B Ilrenbusch, 'Fairness Crowded Out by Law: An Experimental Study on Withdrawal Rights' [2007] 163 *Journal of Institutional and Theoretical Economics* 84.
124 Horst GM Eidenmüller, 'Why Withdrawal Rights?' (2010) Social Sciences Research Network Working Paper (http://doi.org/10.2139/ssrn.1660535) 8; Jan M Smits, 'The Right to Change Your Mind? Rethinking the Usefulness of Mandatory Rights of Withdrawal in Consumer Contract Law' [2011] Maastricht European Private Law Institute Working Paper no. 2011/01, 9; Joasia A Luzak, 'To Withdraw or Not to Withdraw or Not to Withdraw? Evaluation of the Mandatory Right of Withdrawal in Consumer Distance Selling Contracts Taking into Account Its Behavioural Effects on Consumers' [2013] Centre for the Study of European Contract Law Working Paper, Amsterdam Law School Legal Studies Research Paper, 15 ff.

125 Tomas Suwelack, Jens Hogreve, and Wayne D Hoyer, 'Understanding Money-Back Guarantees: Cognitive, Affective, and Behavioral Outcomes' [2011] 87(4) *Journal of Retailing* 462–478, 462. See also Patricia Sánchez Abril, Francisco Oliva Blázquez and Joan Martínez Evora, 'The Right of Withdrawal in Consumer Contracts: A Comparative Analysis of American and European Law' [2018] 3 *InDret* 3, 27.

126 Marco BM Loos, 'Rights of Withdrawal' in Geraint Howells and Reiner Schulze (eds), *Modernising and Harmonising Consumer Contract Law* (Sellier 2009) 241, 245–246; Omri Ben-Shahar and Eric A Posner, 'The Right to Withdraw in Contract Law' [2011] 40(1) *The Journal of Legal Studies* 115–148, 119. In the pursuit of strengthening and harmonizing the European internal market, proponents argue that European consumers should engage in more cross-border transactions (GM Horst Eidenmüller, 'Why Withdrawal Rights?' 17 August 2010, http://doi.org/10.2139/ssrn.1660535, p. 6); European Commission (September 2006). Consumer protection in the internal market. Special Eurobarometer 252, p. 2; European Commission (October 2008). Consumer protection in the internal market. Special Eurobarometer 298, p. 2; European Commission (2 October 2008). Great potential of cross-border shopping going unfulfilled, says EU report. Press release, IP/08/1447, p. 1. A significant portion of these transactions occurs remotely, with a growing trend of online transactions (Joasia A Luzak, 'To Withdraw or Not to Withdraw or Not to Withdraw? Evaluation of the Mandatory Right of Withdrawal in Consumer Distance Selling Contracts Taking into Account Its Behavioural Effects on Consumers' [2013] Centre for the Study of European Contract Law Working Paper, Amsterdam Law School Legal Studies Research Paper, p. 4).

127 Antonios Karampatzos, *Private Law, Nudging and Behavioural Economic Analysis. The Mandated-Choice Model* (London, Routledge 2020) 37.

128 Reiner Schulze and Jonathan Morgan, 'Right of Withdrawal' in Gerhard Dannemann and Stefan Vogenauer (eds), *The Common European Sales Law in Context* (Oxford, Oxford University Press 2013) 294.

129 Joasia A Luzak and Vanessa Mak, 'The Consumer Rights Directive' Centre for the Study of European Contract Law Working Paper Series No. 2013-01, 13.

130 Joasia A Luzak, 'To Withdraw or Not to Withdraw or Not to Withdraw? Evaluation of the Mandatory Right of Withdrawal in Consumer Distance Selling Contracts Taking into Account Its Behavioural Effects on Consumers' [2013] Centre for the Study of European Contract Law Working Paper, Amsterdam Law School Legal Studies Research Paper, 1; Patricia Sánchez Abril, Francisco Oliva Blázquez, and Joan Martínez Evora, 'The Right of Withdrawal in Consumer Contracts: A Comparative Analysis of American and European Law' [2018] 3 *InDret* 31.

131 Pamaria Rekati and Roger Van den Bergh, 'Cooling-Off Periods in the Consumer Laws of the EC Member States: A Comparative Law and Economics Approach' [2000] 23(4) *Journal of Consumer Policy* 371, 391–392 (https://doi.org/10.1023/A:1007203426046).

132 Horst GM Eidenmüller, 'Why Withdrawal Rights?' (2010) Social Sciences Research Network Working Paper (http://doi.org/10.2139/ssrn.1660535) 7; Christian Twigg-Flesner and Reiner Schulze, 'Protecting National Choice Information and the Right of Withdrawal' in Geraint Howells, Iain Ramsay and Thomas Wilhelmsson, *Handbook of Research on International Consumer Law* (Cheltenham, Edward Elgar 2010) 145; Patricia Sánchez Abril, Francisco Oliva Blázquez, and Joan Martínez Evora, 'The Right of Withdrawal in Consumer Contracts: A Comparative Analysis of American and European Law' [2018] 3 *InDret* 32; Antonios Karampatzos, *Private Law, Nudging and Behavioural Economic Analysis. The Mandated-Choice Model* (London, Routledge 2020) 36. See also Recital 14 of Directive 97/7/EC of 20 May 1997 on the protection of consumers in respect of distance contracts and paragraph 37 of the Consumer Rights Directive (1997) OJ L144/19.

133 See, however, Joasia A Luzak, 'To Withdraw or Not to Withdraw or Not to Withdraw? Evaluation of the Mandatory Right of Withdrawal in Consumer Distance Selling Contracts Taking into Account Its Behavioural Effects on Consumers' [2013] Centre for the Study of European Contract Law Working Paper, Amsterdam Law School Legal Studies Research Paper, 7 who claims that consumers, despite the right of withdrawal, still not be able to properly assess their value.

134 Pamaria Rekati and Roger Van den Bergh, 'Cooling-Off Periods in the Consumer Laws of the EC Member States: A Comparative Law and Economics Approach' [2000] 23(4) *Journal of Consumer Policy* 371 (https://doi.org/10.1023/A:1007203426046) 381; Joasia A Luzak, 'To Withdraw or Not to Withdraw or Not to Withdraw? Evaluation of the Mandatory Right of Withdrawal in Consumer Distance Selling Contracts Taking into Account Its Behavioural Effects on Consumers' [2013] Centre for the Study of European Contract Law Working Paper, Amsterdam Law School Legal Studies Research Paper, 6.

135 An alternative solution to address the need for the right of withdrawal is emerging through technology, as referred to in the subsequent section. See Commission, 'New Consumer Agenda: Strengthening consumer resilience for sustainable recovery' COM(2020) 696, 8–9.

136 As Joasia A Luzak, 'To Withdraw or Not to Withdraw or Not to Withdraw? Evaluation of the Mandatory Right of Withdrawal in Consumer Distance Selling Contracts Taking into Account Its Behavioural Effects on Consumers' [2013] Centre for the Study of European Contract Law Working Paper, Amsterdam Law School Legal Studies Research Paper, 6 rightly points out, 'a consumer is not allowed to take a pillow back home from a shop and test it for a night to check whether he is satisfied with its firmness before he would purchase it'.

137 CJEU C-489/07 *Pia Messner* [2009]. See also Jonathon Watson, 'Withdrawal Rights' in Christian Twigg-Flesner (ed), *Research Handbook on EU Consumer and Contract Law* (Cheltenham, Edward Elgar 2016).

138 Evelyne Terryn and Elias Van Gool, 'The Role of European Consumer Regulation in Shaping the Environmental Impact of E-Commerce' [2021] 3 *EuCML* 89, 95. Free returns rank as the second most crucial factor, following only free shipping, in the considerations of consumers when making online purchases. See Power Reviews, 'Consumer Survey: Returns in Retail in 2021' (Power Reviews, 2021) <https://www.powerreviews.com/insights/consumer-survey-retail-returns-2021/> accessed 12 June 2023. The big retail businesses, such as Amazon, Walmart and others offer by themselves a right to return goods, usually within 30 days after the delivery of the good. Antonios Karampatzos, *Private Law, Nudging and Behavioural Economic Analysis. The Mandated-Choice Model* (London, Routledge 2020) 94, fn. 37.

139 Antonios Karampatzos, *Private Law, Nudging and Behavioural Economic Analysis. The Mandated-Choice Model* (London, Routledge 2020) 89, 95. A more balanced proposition is put forth by Ben-Shahar and Posner, who categorically state that when a product has moderate depreciation, consumers should bear the depreciation costs (Omri Ben-Shahar and Eric A Posner, 'The Right to Withdraw in Contract Law' [2011] 40(1) *The Journal of Legal Studies* 115, 118).

140 See the terms of conditions of reimbursement of Amazon Spain <https://www.amazon.es/gp/help/customer/display.html/ref=orc_hp_s_retest?ie=UTF8&nodeId=201149940> accessed 19 June 2023, and Amazon International <https://www.amazon.com/gp/help/customer/display.html?nodeId=GKQNFKFK5CF3C54B&ref_=hp_GNW5VKFXMF72FFMR_Refunds> accessed 17 September 2023.

141 Péter Cserne, *Freedom of Contract and Paternalism* (London, Springer 2012) 114.

142 Daria Dzyabura et al, 'Leveraging the Power of Images in Managing Product Return Rates' [2019] 42(6) *Marketing Science* 3 (https://doi.org/10.1287/mksc.2023.1451).

143 According to the lastly available data, Amazon is the most visited online marketplace in Europe, with more than 1,070 million visits per month Statista, 'Leading Online Marketplaces in Europe as of July 2022, Based on Number of Monthly Visits', 2022 <https://www.statista.com/statistics/288056/leading-retail-websites-in-europe-based-on-unique-visitors/>. quite far away from its direct competitor, eBay, with less than the half of monthly visits (534.6 millions). The importance of Amazon may also be seen in terms of annual revenue. Amazon, in 2021, had a yearly revenue of 51,322.34 million euros [Statista, 'Total revenue of Amazon Europe from 2011 to 2021' (*Amazon*, 2022) <https://www.statista.com/statistics/934963/revenue-of-amazon-europe/> accessed 19 June 2023] while eBay had closely an 80% less than Amazon (9,541.59 million euros) [Statista, 'Annual net revenue of eBay from 2013 to 2022' (*eBay*, 2023) <https://www.statista.com/statistics/507881/ebays-annual-net-revenue/> accessed 19 June 2023].

144 Wall Street Journal (2022) Retailers' many unhappy returns. <https://www.wsj.com/articles/retailers-many-unhappy-returns-11641387605>, last accessed 17 of November 2023.

145 Oren Bar-Gill and Omri Ben-Shahar, 'Regulatory Techniques in Consumer Protection. A Critique of European Consumer Contract Law' [2013] 50 *CMLRev* 109, 121 (https://doi.org/10.54648/cola2013039); Pamaria Rekati and Roger Van den Bergh, 'Cooling-Off Periods in the Consumer Laws of the EC Member States: A Comparative Law and Economics Approach' [2000] 23(4) *Journal of Consumer Policy* 371, 383 (https://doi.org/10.1023/A:1007203426046).

146 This crosssubsidization is referred to by Oren Bar-Gill and Omri Ben-Shahar, 'Regulatory Techniques in Consumer Protection. A Critique of European Consumer Contract Law' [2013] 50 *CMLRev* 109, 121 (https://doi.org/10.54648/cola2013039).

147 Pamaria Rekati and Roger Van den Bergh, 'Cooling-Off Periods in the Consumer Laws of the EC Member States: A Comparative Law and Economics Approach' [2000] 23(4) *Journal of Consumer Policy* 371, 381 (https://doi.org/10.1023/A:1007203426046); Marco BM Loos, 'The Case for a Uniform and

Efficient Right of Withdrawal from Consumer Contracts in European Contract Law' [2007] 15(1) *Zeitschrift Für Europaeisches Privatrecht* 5, 10; Horst GM Eidenmüller, 'Why Withdrawal Rights?' (2010) Social Sciences Research Network Working Paper (http://doi.org/10.2139/ssrn.1660535).

148 Marco BM Loos, 'The Case for a Uniform and Efficient Right of Withdrawal from Consumer Contracts in European Contract Law' [2007] 15(1) *Zeitschrift Für Europaeisches Privatrecht* 5.

149 We reject the notion of viewing the right of withdrawal as a remedy for instances of 'temporary madness' Pamaria Rekati and Roger Van den Bergh, 'Cooling-Off Periods in the Consumer Laws of the EC Member States: A Comparative Law and Economics Approach' [2000] 23(4) *Journal of Consumer Policy* 371, 372 and 375 ff (https://doi.org/10.1023/A:1007203426046); Patricia Sánchez Abril, Francisco Oliva Blázquez, and Joan Martínez Evora, 'The Right of Withdrawal in Consumer Contracts: A Comparative Analysis of American and European Law' [2018] 3 *InDret* 7. Instead, we align with perspectives such as Watson, 'Withdrawal Rights'., asserting that impulsive or unreflective behaviour should not be regarded as an impairment of the will, as suggested by certain authors like Horst GM Eidenmüller, 'Why Withdrawal Rights?' (2010) Social Sciences Research Network Working Paper (http://doi.org/10.2139/ssrn.1660535) 5. If transactions driven by impulse were deemed as instances of impaired will, resulting in void contracts, businesses, particularly those reliant on impulsive buying patterns like e-commerce, could face substantial financial losses due to cancelled or contested transactions. Furthermore, if cooling-off periods exclusively served the economic function of enabling withdrawal from contracts made under cognitive or emotional distress, the necessity for the actual transfer and utilization of the goods would be negated. In such a scenario, the right of withdrawal would prove inadequate in addressing the situation. See Péter Cserne, *Freedom of Contract and Paternalism* (London, Springer 2012).

150 While consumers possess a compulsory entitlement to a withdrawal right (as stipulated in art. 12(1) of the Distance Marketing of Financial Services Directive; art. 25 of the Consumer Credit Directive; art. 12(1) of the Timeshare Directive; art. 25 CRD), it is crucial to note that the actual exercise of the right of withdrawal, akin to all other consumer rights (Hans-W Micklitz, 'Squaring the Circle? Reconciling Consumer Law and the Circular Economy' [2019] 6 *EuCML* 229, 235), is discretionary rather than obligatory for consumers.

151 Horst GM Eidenmüller, 'Why Withdrawal Rights?' (2010) Social Sciences Research Network Working Paper (http://doi.org/10.2139/ssrn.1660535) 5; Jan M Smits, 'The Right to Change Your Mind? Rethinking the Usefulness of Mandatory Rights of Withdrawal in Consumer Contract Law' (2011) Maastricht European Private Law Institute Working Paper 2011/01 <https://papers.ssrn.com/sol3/papers.cfm?abstract_id=1719104>; Joasia A Luzak, 'To Withdraw or Not to Withdraw or Not to Withdraw? Evaluation of the Mandatory Right of Withdrawal in Consumer Distance Selling Contracts Taking into Account Its Behavioural Effects on Consumers' [2013] Centre for the Study of European Contract Law Working Paper, Amsterdam Law School Legal Studies Research Paper; Evelyne Terryn and Elias Van Gool, 'The Role of European Consumer Regulation in Shaping the Environmental Impact of E-Commerce' [2021] 3 *EuCML* 89.

152 Henrik Pålsson, Fredrik Pettersson and Lena Winslott Hiselius, 'Energy Consumption in E-Commerce versus Conventional Trade Channels-Insights into Packaging, the Last Mile, Unsold Products and Product Returns' [2017] 164 *Journal of Cleaner Production* 765, 776 (https://doi.org/10.1016/j.jclepro.2017.06.242).

153 Evelyne Terryn and Elias Van Gool, 'The Role of European Consumer Regulation in Shaping the Environmental Impact of E-Commerce' [2021] 3 *EuCML* 89, 95–100; Pålsson, Pettersson, and Hiselius, 'Energy Consumption in E-Commerce versus Conventional Trade Channels-Insights into Packaging, the Last Mile, Unsold Products and Product Returns'.

154 These exceptions are aimed at considering the interests of traders as well (Evelyne Terryn and Elias Van Gool, 'The Role of European Consumer Regulation in Shaping the Environmental Impact of E-Commerce' [2021] 3 *EuCML* 89, 97). See also Recital 4 CRD and CJEU C-681/17, *Slewo* (2019) ECLI:EU: C:2019:255, para 39. While a withdrawal right is mandatory and cannot be relinquished by contractual agreement in the *conclusion* of the contract, it can therefore be relinquished during the *performance* stage of the contract. Jonathon Watson, 'Withdrawal Rights' in Christian Twigg-Flesner (ed), *Research Handbook on EU Consumer and Contract Law* (Cheltenham, Edward Elgar 2016) 256.

155 Research has shown that the spending rates of customers who had to pay a fee to returning a product were from 75% to 100% lower than when they did not have to make such payment (Bower and Maxham 2012) *apud* Antonios Karampatzos, *Private Law, Nudging and Behavioural Economic Analysis. The Mandated-Choice Model* (London, Routledge 2020).

156 Evelyne Terryn and Elias Van Gool, 'The Role of European Consumer Regulation in Shaping the Environmental Impact of E-Commerce' [2021] 3 *EuCML* 89.
157 Evelyne Terryn and Elias Van Gool, 'The Role of European Consumer Regulation in Shaping the Environmental Impact of E-Commerce' [2021] 3 *EuCML* 89, 98.
158 See studies cited in Evelyne Terryn and Elias Van Gool, 'The Role of European Consumer Regulation in Shaping the Environmental Impact of E-Commerce' [2021] 3 *EuCML* 89, 96.
159 Evelyne Terryn and Elias Van Gool, 'The Role of European Consumer Regulation in Shaping the Environmental Impact of E-Commerce' [2021] 3 *EuCML* 89, 97. This standpoint is aligned with Ben Shahar and Posner's, who consider that if the depreciation of a good following consumer's use is substantial, no right of withdrawal should be permitted (Omri Ben-Shahar and Eric A Posner, 'The Right to Withdraw in Contract Law' [2011] 40(1) *The Journal of Legal Studies* 115, 118).
160 Directive (EU) 2019/904 of the European Parliament and of the Council of 5 June 2019 on the reduction of the impact of certain plastic products on the environment (2019) OJ L155/1. For an overview of the environmental impact of the latter, see Marta Santos Silva and Luísa Cortat Simonetti Gonçalves, 'Nudging Consumers towards Sustainable Practices Regarding Plastics in a Post-COVID-19 Europe' in Ewoud Hondius et al (eds), *Coronavirus and the Law in Europe* (Cambridge, Intersentia 2021) 947.
161 Jonathon Watson, 'Withdrawal Rights' in Christian Twigg-Flesner (ed), *Research Handbook on EU Consumer and Contract Law* (Cheltenham, Edward Elgar 2016) 253.
162 Joasia A Luzak, 'To Withdraw or Not to Withdraw or Not to Withdraw? Evaluation of the Mandatory Right of Withdrawal in Consumer Distance Selling Contracts Taking into Account Its Behavioural Effects on Consumers' [2013] Centre for the Study of European Contract Law Working Paper, Amsterdam Law School Legal Studies Research Paper, 32 ff.
163 Horst GM Eidenmüller, 'Why Withdrawal Rights?' (2010) Social Sciences Research Network Working Paper (http://doi.org/10.2139/ssrn.1660535) 9.
164 Evelyne Terryn and Elias Van Gool, 'The Role of European Consumer Regulation in Shaping the Environmental Impact of E-Commerce' [2021] 3 *EuCML* 89, 99.
165 This bias suggests that the disutility of giving up an object is greater than the utility associated with acquiring it. Daniel Kahneman, Jack L. Knetsch and Richard H. Thaler, 'Anomalies: The Endowment Effect, Loss Aversion, and Status Quo Bias' [1991] 5(1) *Journal of Economic Perspectives* 193, 194.
166 Navjit Bhasin, 'The Four Personas of a Serial Returner' *Women's Wear Daily* (4 October 2019) <https://wwd.com/feature/personas-of-a-serial-returner-1203333197/>.
167 This blacklisting brings some concerns. See Horst GM Eidenmüller, 'Why Withdrawal Rights?' (2010) Social Sciences Research Network Working Paper (http://doi.org/10.2139/ssrn.1660535) 21.
168 Jan M Smits, 'The Right to Change Your Mind? Rethinking the Usefulness of Mandatory Rights of Withdrawal in Consumer Contract Law' (2011) Maastricht European Private Law Institute Working Paper 2011/01 <https://papers.ssrn.com/sol3/papers.cfm?abstract_id=1719104> 11, signalizes the counter-argument 'that a mandatory withdrawal right may lead to a crowding out effect: individual retailers will no longer be able to show their willingness towards consumers to give them an extra right, thus seducing them to purchase their products'. See this discussion in Section 4.2.
169 Technological advancements such as augmented reality diminish the rationale supporting the argument for the necessity of the right of withdrawal in scenarios where consumers cannot physically preview the product beforehand. On technological solutions for the problems caused by returns, see Daria Dzyabura et al, 'Leveraging the Power of Images in Managing Product Return Rates' [2019] 42(6) *Marketing Science* (https://doi.org/10.1287/mksc.2023.1451), Priya Ambilkar et al, 'Product Returns Management: A Comprehensive Review and Future Research Agenda' [2022] 60(12) *International Journal of Production Research* 3920 (https://doi.org/10.1080/00207543.2021.1933645).
170 Jonathon Watson, 'Withdrawal Rights' in Christian Twigg-Flesner (ed), *Research Handbook on EU Consumer and Contract Law* (Cheltenham, Edward Elgar 2016) 264–265.
171 See art. 114 TFEU and art. 38 Charter of Fundamental Rights of the European Union.
172 Horst GM Eidenmüller, 'Why Withdrawal Rights?' (2010) Social Sciences Research Network Working Paper (http://doi.org/10.2139/ssrn.1660535) 11.
173 In the same sense, see Evelyne Terryn and Elias Van Gool, 'The Role of European Consumer Regulation in Shaping the Environmental Impact of E-Commerce' [2021] 3 *EuCML* 89, 99; CJEU C-489/07, *Messner* [2009] ECLI:EU:C:2009:502, paras 25–27. In the contrary sense that there is no requirement to exercise the right of withdrawal in good faith see Cour de Cassation (FR)7/3/2006 (2006) JCP (G)737

II, 10056 and Marco BM Loos, 'Rights of Withdrawal' in Geraint Howells and Reiner Schulze (eds), *Modernising and Harmonising Consumer Contract Law* (Sellier 2009) 241.

174 Lloyd C Harris, 'Fraudulent Return Proclivity: An Empirical Analysis' [2008] 84(8) *Journal of Retailing* 461 (https://doi.org/10.1016/j.jretai.2008.09.003).

175 Zhi Pei and Audhesh Paswan, 'Consumers' Legitimate and Opportunistic Product Return Behaviors in Online Shopping' [2018] 19(4) *Journal of Electronic Commerce Research* 301.

176 Hsiu-Hua Chang and Ting-Shan Yang, 'Consumer Rights or Unethical Behaviors: Exploring the Impacts of Retailer Return Policies' [2022] 64 *Journal of Retailing and Consumer Services* (https://doi.org/10.1016/j.jretconser.2021.102779).

177 Mengxia Zhang and Lan Luo, 'Can Consumer-Posted Photos Serve as a Leading Indicator of Restaurant Survival? Evidence from Yelp' [2022] 69(1) *Management Science* (https://doi.org/10.1287/mnsc.2022.4359).

178 Lloyd C Harris, 'Fraudulent Return Proclivity: An Empirical Analysis' [2008] 84(8) *Journal of Retailing* 461 (https://doi.org/10.1016/j.jretai.2008.09.003) 465.

179 Zhi Pei and Audhesh Paswan, 'Consumers' Legitimate and Opportunistic Product Return Behaviors in Online Shopping' [2018] 19(4) *Journal of Electronic Commerce Research* 301.

180 Jochen Wirtz and Doreen Kum, 'Consumer Cheating on Service Guarantees' [2004] 32(2) *Journal of the Academy of Marketing Science* 159 (https://doi.org/10.1177/0092070303261416); Zhi Pei and Audhesh Paswan, 'Consumers' Legitimate and Opportunistic Product Return Behaviors in Online Shopping' [2018] 19(4) *Journal of Electronic Commerce Research* 301, 319.

181 Pamaria Rekati and Roger Van den Bergh, 'Cooling-Off Periods in the Consumer Laws of the EC Member States: A Comparative Law and Economics Approach' [2000] 23(4) *Journal of Consumer Policy* 371 (https://doi.org/10.1023/A:1007203426046) 381; Jan M Smits, 'The Right to Change Your Mind? Rethinking the Usefulness of Mandatory Rights of Withdrawal in Consumer Contract Law' (2011) Maastricht European Private Law Institute Working Paper 2011/01, <https://papers.ssrn.com/sol3/papers.cfm?abstract_id=1719104>, 9; Antonios Karampatzos, *Private Law, Nudging and Behavioural Economic Analysis. The Mandated-Choice Model* (London, Routledge 2020) 103.

182 Antonios Karampatzos, *Private Law, Nudging and Behavioural Economic Analysis. The Mandated-Choice Model* (London, Routledge 2020) 91; Pamaria Rekati and Roger Van den Bergh, 'Cooling-Off Periods in the Consumer Laws of the EC Member States: A Comparative Law and Economics Approach' [2000] 23(4) *Journal of Consumer Policy* 371 (https://doi.org/10.1023/A:1007203426046) 381.

183 Klas Hjort and Bjorn Lantz, '(R)e-Tail Borrowing of Party Dresses: An Experimental Study' [2012] 40 *International Journal of Retail & Distribution Management* 997.

184 That is the example of the Instagram page #outfitoftheday or #ootd, which has more than half a million followers.

185 Evelyne Terryn and Elias Van Gool, 'The Role of European Consumer Regulation in Shaping the Environmental Impact of E-Commerce' [2021] 3 *EuCML* 89, 98.

186 Evelyne Terryn and Elias Van Gool, 'The Role of European Consumer Regulation in Shaping the Environmental Impact of E-Commerce' [2021] 3 *EuCML* 89, 99.

187 Georg Borges and Bernd Ilrenbusch, 'Fairness Crowded Out by Law: An Experimental Study on Withdrawal Rights' [2007] 163(1) *Journal of Institutional and Theoretical Economics* 84.

188 The practice of allowing the consumer to rethink their intention of purchasing a flight is widespread in international companies that allow periods ranging from six hours up to 14 days for thinking. However, in all these cases, the consumer must pay to think, which due to the loss aversion bias will usually lead to this possibility being less opted for by prospective flyers. For such a reason, this would classify as a grey, rather than as a green nudge.

189 Durable products are those '[t]hat may be used repeatedly or continuously over a period of more than a year' (UN, 2009).

190 See, e.g., the Behamics, a cloud-based software as a service that optimizes conversion and product return rates in e-commerce by combining Behavioural Science with AI (<https://xpreneurs.io/news/behamics>, last retrieved 30 of November 2023).

191 For the sake of space, the applicable data protection considerations cannot be developed in this contribution.

192 Eric J Johnson and Daniel G Goldstein, 'Decisions by Default' [2013] *The Behavioral Foundations of Policy* 417.

193 Daniel Kahneman, Jack L Knetsch and Richard H Thaler, 'Anomalies: The Endowment Effect, Loss Aversion, and Status Quo Bias' [1991] 5(1) *Journal of Economic Perspectives* 193, 194.

194 They act as 'counternudge'. See Anne-Lise Sibony and Alberto Alemanno, *Nudge and the Law* (London, Hart Publishing 2015) 328.

Bibliography

Akbulut-Yuksel M and Boulatoff C, 'The Effects of a Green Nudge on Municipal Solid Waste: Evidence from a Clear Bag Policy' [2021] 106 *Journal of Environmental Economics and Management* 102404 (https://doi.org/10.1016/j.jeem.2020.102404)

Akenji L, 'Consumer Scapegoatism and Limits to Green Consumerism' [2014] 63 *Journal of Cleaner Production* 13 (https://doi.org/10.1016/j.jclepro.2013.05.022)

Alemanno A, *What Can EU Health Law Learn from Behavioural Sciences? The Case of EU Lifestyle Regulation* (London, Hart Publishing 2015)

———, 'Nudge and the European Union' in Holger Straßheim and Silke Beck, *Handbook of Behavioural Change and Public Policy* (Cheltenham, Edward Elgar 2019) 271

Alemanno A and Spina A, 'Nudging Legally. On the Checks and Balances of Behavioural Regulation' [2014] 12(2) *International Journal of Constitutional Law* 429 (https://doi.org/10.1093/icon/mou033)

Alempaki D, Isoni A and Read D, 'Tainted Nudge' [2023] 176 *Organizational Behavior and Human Decision Processes* 104244 (https://doi.org/10.1016/j.obhdp.2023.104244)

Alemu M and Zewdie S, 'Consumers' Impulsive Buying Behavior: Structured Systematic Literature Review' [2021] 8(10) *International Journal of Management, Accounting and Economics* 756 (https://doi.org/10.5281/zenodo.5975984)

Alt M et al, 'Synergies of Interventions to Tackle Climate Change – A Meta-Analysis' (Ispra, Italy/Brussels, Belgium/Petten, The Netherlands: European Commission, Joint Research Centre, December 2022)

Ambilkar P et al, 'Product Returns Management: A Comprehensive Review and Future Research Agenda' [2022] 60(12) *International Journal of Production Research* 3920 (https://doi.org/10.1080/00207543.2021.1933645)

Amir O and Lobel O, 'Liberalism and Lifestyle: Informing Regulatory Governments with Behavioural Research' [2012] 1 *European Journal of Risk Regulation* 17 (https://doi.org/10.1017/S1867299X00001768)

Baldwin R, 'From Regulation to Behaviour Change: Giving Nudge the Third Degree' [2014] 77(6) *The Modern Law Review* 831 (https://doi.org/10.1111/1468-2230.12094)

Bar-Gill O and Ben-Shahar O, 'Regulatory Techniques in Consumer Protection. A Critique of European Consumer Contract Law' [2013] 50 *CMLRev* 109 (https://doi.org/10.54648/cola2013039)

Ben-Shahar O and Posner EA, 'The Right to Withdraw in Contract Law' [2011] 40(1) *The Journal of Legal Studies* 115

Berthet V and Ouvrard B, 'Nudge towards a Consensus View' [2019] 5(1) *Psychology and Cognitive Sciences Open Journal* 1 (https://www.doi.org/10.17140/PCSOJ-5-143)

Bhasin N, 'The Four Personas of a Serial Returner', *Women's Wear Daily* (4 October 2019) <https://wwd.com/feature/personas-of-a-serial-returner-1203333197/>

Blonski K and Witek J, 'Minimalism in Consumption' [2019] LIII(2) *Annales Universitatis Mariae-Curie-Sklodowska, Sectio H – Oeconomia* 7

Bocken MP, Niessen L and Short SW, 'The Sufficiency-Based Circular Economy – An Analysis of 150 Companies' [2022] 3 *Frontiers in Sustainability* 1 (https://doi.org/10.3389/frsus.2022.899289)

Borges G and Ilrenbusch B, 'Fairness Crowded Out by Law: An Experimental Study on Withdrawal Rights' [2007] 163(1) *Journal of Institutional and Theoretical Economics* 84

Bovens L, 'The Ethics of Nudge' in Grüne-Yanoff T and Hansson SO (eds), *Preference Change: Approaches from Philosophy, Economics and Psychology* (London, Springer 2009) 207

Brenncke M, 'Regulating Dark Patterns' [2024] *Notre Dame Journal International & Comparative Law* (https://papers.ssrn.com/sol3/papers.cfm?abstract_id=4588652)

Brignull H, 'Dark Patterns: Dirty Tricks Designers Use to Make People Do Stuff', *90 Percent of Everything* (8 July 2010) <https://90percentofeverything.com/2010/07/08/dark-patterns-dirty-tricks-designers-use-to-make-people-do-stuff/index.html>

Brooker S and Cullum P, 'Five Steps to Consumer-Friendly Better Regulation' [2008] 18(1) *Consumer Policy Review* 4

Bruns H et al, ‘Can Nudges Be Transparent and Yet Effective?’ [2018] 65 *Journal of Economic Psychology* 41 (https://doi.org/10.1016/j.joep.2018.02.002)

Buchholz M and Musshoff O, ‘Tax or Green Nudge? An Experimental Analysis of Pesticide Policies in Germany’ [2021] 48(4) *European Review of Agricultural Economics* 940 (https://doi.org/10.1093/erae/jbab019)

Buss S, ‘Autonomous Action: Self-Determination in the Passive Mode’ [2012] 122 *Ethics* 647 (https://doi.org/10.1086/666328)

Byerly H et al, ‘Nudging Pro-Environmental Behavior: Evidence and Opportunities’ [2018] 16(3) *Frontiers in Ecology and the Environment* 159 (https://doi.org/10.1002/fee.1777)

Carlsson F et al, ‘The Use of Green Nudges as an Environmental Policy Instrument’ [2021] 15(2) *Review of Environmental Economics and Policy* 216

Carrel C, Gavard-Perret ML and Caldara C, ‘Factors of Effectiveness of Green Nudges for More Eco-Responsible Behaviour – Systematic Review and Research Directions’ [2023] 38(3) *Recherche et Applications En Marketing* 1 (https://doi.org/10.1177/20515707231177814)

Cellina F et al, ‘Significant but Transient: The Impact of an Energy Saving App Targeting Swiss Households’ [2024] 335 *Applied Energy* 122280 (https://doi.org/10.1016/j.apenergy.2023.122280)

Chan TCH, Cheung CMK and Lee ZWY, ‘The State of Online Impulsive-Buying Research: A Literature Analysis’ [2017] 54(2) *Information & Management* 204 (https://doi.org/10.1016/j.im.2016.06.001)

Chang HH and Yang TS, ‘Consumer Rights or Unethical Behaviors: Exploring the Impacts of Retailer Return Policies’ [2022] 64 *Journal of Retailing and Consumer Services* 102779 (https://doi.org/10.1016/j.jretconser.2021.102779)

Chin Hsu W, Lee MH and Zheng KW, ‘From Virtual to Reality: The Power of Augmented Reality in Triggering Impulsive Purchases’ [2024] 76 *Journal of Retailing and Consumer Services* 76 (https://doi.org/10.1016/j.jretconser.2023.103604)

Chung Han M, ‘Checkout Button and Online Consumer Impulse-Buying Behavior in Social Commerce: A Trust Transfer Perspective’ [2023] 74 *Journal of Retailing and Consumer Services* 103431 (https://doi.org/10.1016/j.jretconser.2023.103431)

Clube RKM and Tennant M, ‘What Would a Human-Centred “Social” Circular Economy Look like? Drawing from Max-Neef’s Human-Scale Development Proposal’ [2023] 383 *Journal of Cleaner Production* 1 (https://doi.org/10.1016/j.jclepro.2022.135455)

Conly S, *Against Autonomy. Justifying Coercive Paternalism* (Cambridge, Cambridge University Press 2013)

Cserne P, *Freedom of Contract and Paternalism* (London, Springer 2012)

———, ‘Making Sense of Nudge-Scepticism: Three Challenges to EU Law’s Learning from Behavioural Sciences’ in Sibony AL and Alemanno A (eds), *Nudge and the Law: A European Perspective* (London, Hart Publishing 2015) 279

D’Alise G, Demaria F and Kellis G, *Degrowth. A Vocabulary for a New Era* (London, Routledge 2015)

Dalhammar C et al, *Moving Away from the Throwaway Society. Five Policy Instruments for Extending the Life of Consumer Durables* (Mistra Sustainable Consumption 2022)

De Franceschi A, ‘Planned Obsolescence Challenging the Effectiveness of Consumer Law and the Achievement of a Sustainable Economy: The Apple and Samsung Cases’ [2018] 7(6) *EuCML* 217

de Ridder D, Kroese F and van Gestel L, ‘Nudgeability: Mapping Conditions of Susceptibility to Nudge Influence’ [2022] 17(2) *Perspectives on Psychological Science* 346 (https://doi.org/10.1177/1745691621995183)

Duchin F, ‘Consumption, Environmentally Significant’ in Niel J Smelser and Paul B Bates (eds), *International Encyclopedia of the Social & Behavioral Sciences* (London, Elsevier 2001) 2678

Dzyabura D et al, ‘Leveraging the Power of Images in Managing Product Return Rates’ [2019] 42(6) *Marketing Science* 1125 (https://doi.org/10.1287/mksc.2023.1451)

Eidenmüller HGM, ‘Why Withdrawal Rights?’ (2010) Social Sciences Research Network Working Paper (http://doi.org/10.2139/ssrn.1660535)

Ekardt F, *Sustainability: Transformation, Governance, Ethics, Law* (London, Springer 2020).

Elo M et al, *Interdisciplinary Perspectives on Planetary Well-Being* (London, Routledge 2024)

Faure MG and Partain RA, *Environmental Law and Economics. Theory and Practice* (Cambridge, Cambridge University Press 2019)

Genc TS and De Giovanni P, ‘Dynamic Pricing and Green Investments under Conscious, Emotional, and Rational Consumers’ [2021] 2 *Cleaner and Responsible Consumption* 100007 (https://doi.org/10.1016/j.clrc.2021.100007)

Goldstein NJ, Cialdini RB and Griskevicius V, 'A Room with a Viewpoint: Using Social Norms to Motivate Environmental Conservation in Hotels' [2008] 35(3) *Journal of Consumer Research* 472 (https://doi.org/10.1086/586910)

González-Ramírez J, 'The Effectiveness of a Green Default Nudge in Achieving Resource Conservation', Social Sciences Research Network Working Paper, 1 November 2022 (http://doi.org/10.2139/ssrn.4264945)

Gregersen T et al, 'Political Orientation Moderates the Relationship between Climate Change Beliefs and Worry about Climate Change' [2020] 11(1) *Frontiers in Psychology* 1 (https://doi.org/10.3389/fpsyg.2020.01573)

Grigsby M, *Buying Time and Getting by: The Voluntary Simplicity Movement* (New York, State University of New York Press 2004)

Guillard V, Le Nagard E and Campos Ribeiro G, 'A Typology of Consumers Regarding Perceived Obsolescence: The Paradox of Eco-Conscious Consumers' [2023] 412 *Journal of Cleaner Production* 137202 (https://doi.org/10.1016/j.jclepro.2023.137202)

Guldborg Hansen P, 'The Definition of Nudge and Libertarian Paternalism: Does the Hand Fit the Glove?' [2016] 7(1) *EJRR* 155

Halpern D and Gallagher R, 'Can "Nudging" Change Behaviour? Using "Behavioural Insights" to Improve Program Redesign' in Wanna J, Lee HA and Yates S (eds), *Managing under Austerity, Delivering under Pressure. Performance and Productivity in Public Service* (Sydney, ANU Press 2015) 165

Harris LC, 'Fraudulent Return Proclivity: An Empirical Analysis' [2008] 84(8) *Journal of Retailing* 461 (https://doi.org/10.1016/j.jretai.2008.09.003).

Hausman DM and Welch B, 'Debate: To Nudge or Not to Nudge' [2010] 18 *Journal of Political Philosophy* 123 (https://doi.org/10.1111/j.1467-9760.2009.00351.x)

Hjort K and Lantz B, '(R)e-Tail Borrowing of Party Dresses: An Experimental Study' [2012] 40 *International Journal of Retail & Distribution Management* 997.

Jakubczak A and Gotowska M, 'Green Consumerism vs. Greenwashing' [2020] XXIII(2) *European Research Studies Journal* 167 (https://doi.org/10.35808/ersj/1816)

Johnson EJ and Goldstein DG, 'Decisions by Default' [2013] *The Behavioral Foundations of Policy* 417.

Kang J, Joyner Martinez CM and Johnson C, 'Minimalism as a Sustainable Lifestyle: Its Behavioral Representations and Contributions to Emotional Well-Being' [2021] 27 *Sustainable Production and Consumption* 802 (https://doi.org/10.1016/j.spc.2021.02.001)

Kahneman D, *Thinking, Fast and Slow* (Farrar, Straus and Giroux 2011)

Kahneman D, Knetsch JL and Thaler RH, 'Anomalies: The Endowment Effect, Loss Aversion, and Status Quo Bias' [1991] 5(1) *Journal of Economic Perspectives* 193

Karampatzos A, *Private Law, Nudging and Behavioural Economic Analysis. The Mandated-Choice Model* (London, Routledge 2020)

Lanzing M, '"Strongly Recommended": Revisiting Decisional Privacy to Judge Hypernudging in Self-Tracking Technologies' [2019] 32 *Philosophy & Technology* 549 (https://doi.org/10.1007/s13347-018-0316-4)

Loos MBM, 'The Case for a Uniform and Efficient Right of Withdrawal from Consumer Contracts in European Contract Law' [2007] 15(1) *Zeitschrift Für Europaeisches Privatrecht* 5

———, 'Rights of Withdrawal' in Howells G and Schulze R (eds), *Modernising and Harmonising Consumer Contract Law* (Sellier 2009) 241

Luguri J and Strahilevitz LJ, 'Shining a Light on Dark Patterns' [2021] 13 *Journal of Legal Analysis* 43 (https://doi.org/10.1093/jla/laaa006)

Luo Y et al, 'A Meta-Analytic Cognitive Framework of Nudge and Sludge' [2023] 10(11) *Royal Society Open Science* 1 (https://doi.org/10.1098/rsos.230053)

Lupiáñez-Villanueva F et al, *Behavioural Study on Unfair Commercial Practices in the Digital Environment – Dark Patterns and Manipulative Personalisation – Final Report* (Brussels, Publications Office of the European Union 2022) (https://data.europa.eu/doi/10.2838/859030)

Luzak JA, 'To Withdraw or Not to Withdraw or Not to Withdraw? Evaluation of the Mandatory Right of Withdrawal in Consumer Distance Selling Contracts Taking into Account Its Behavioural Effects on Consumers' [2013] Centre for the Study of European Contract Law Working Paper, Amsterdam Law School Legal Studies Research Paper

Luzak JA and Mak V, 'The Consumer Rights Directive' [2013] Centre for the Study of European Contract Law Working Paper Series No. 2013-01

Maschitto D, 5 Ways to Control Chronic Returners, *Tecsys* (25 January 2022) <https://www.tecsys.com/blog/2022/01/5-ways-to-control-chronic-returners>

Michel A, *Premature Obsolescence: In Search of an Improved Legal Framework* (Cambridge, Intersentia 2023)

Micklitz HW, 'Squaring the Circle? Reconciling Consumer Law and the Circular Economy' [2019] 6 *EuCML* 229

Mill JS, *On Liberty* (Batoche Books 2001)

Mongin P and Cozic M, 'Rethinking Nudge: Not One but Three Concepts' [2018] 2(1) *Behavioural Public Policy* 107

Mont O, *A Research Agenda for Sustainable Consumption Governance* (Cheltenham, Edward Elgar 2019)

Morozovaite V, 'Hypernudging in the Changing European Regulatory Landscape for Digital Markets' [2023] 15(1) *Policy Internet* 78 (https://doi.org/10.1002/poi3.329)

Nielsen A et al, *Nudging and Pro-Environmental Behaviour* (Nordic Council of Ministers 2016)

Ohira S and Masuda A, 'Voluntary Simplicity Consumption in Japan. Alternative Consumption and Backward Consumption Among Minimalists' [2021] 49 *Advances in Consumer Research* 246

Pålsson H, Pettersson F and Winslott Hiselius L, 'Energy Consumption in E-Commerce versus Conventional Trade Channels-Insights into Packaging, the Last Mile, Unsold Products and Product Returns' [2017] 164 *Journal of Cleaner Production* 765 (https://doi.org/10.1016/j.jclepro.2017.06.242)

Pei Z and Paswan A, 'Consumers' Legitimate and Opportunistic Product Return Behaviors in Online Shopping' [2018] 19(4) *Journal of Electronic Commerce Research* 301

Petticrew M et al, 'Dark Nudges and Sludge in Big Alcohol: Behavioral Economics, Cognitive Biases, and Alcohol Industry Corporate Social Responsibility' [2020] 4 *The Milkbank Quarterly* 1290 (https://doi.org/10.1111/1468-0009.12475)

Pistor K, *The Code of Capital. How the Law Creates Wealth and Inequality* (Princeton University Press 2019)

Redclift M, *Wasted: Counting the Costs of Global Consumption* (London, Routledge 1996)

Reijula S and Hertwig R, 'Self-Nudging and the Citizen Choice Architect' [2020] 6 *Behavioural Public Policy* 119

Reisch L and Thøgersen J, *Handbook of Research on Sustainable Consumption* (Cheltenham, Edward Elgar 2015)

Rekati P and Van den Bergh R, 'Cooling-Off Periods in the Consumer Laws of the EC Member States: A Comparative Law and Economics Approach' [2000] 23(4) *Journal of Consumer Policy* 371 (https://doi.org/10.1023/A:1007203426046)

Sachdeva S, Jordan J and Mazar N, 'Green Consumerism: Moral Motivations to a Sustainable Future' [2015] 6 *Current Opinions in Psychology* 60 (https://doi.org/10.1016/j.copsyc.2015.03.029)

Saghai Y, 'Salvaging the Concept of Nudge' [2013] 29 *Journal of Medical Ethics* 487

Sánchez Abril P, Oliva Blázquez F and Martínez Evora J, 'The Right of Withdrawal in Consumer Contracts: A Comparative Analysis of American and European Law' [2018] 3 *InDret* 27

Santos Silva M, 'Nudging and Other Behaviourally Based Policies as Enablers for Environmental Sustainability' [2022] 11(1) *Laws* 9 (https://doi.org/10.3390/laws11010009)

Santos Silva M and Cortat Simonetti Gonçalves L, 'Nudging Consumers towards Sustainable Practices Regarding Plastics in a Post-COVID-19 Europe' in Hondius E et al (eds), *Coronavirus and the Law in Europe* (Cambridge, Intersentia 2021) 947

Schmidt AT and Engelen B, 'The Ethics of Nudging: An Overview' [2020] 7 *Philosophy Compass* 1 (https://doi.org/10.1111/phc3.12658)

Schubert C, 'Green Nudges: Do the Work? Are They Ethical?' [2017] 132 *Ecological Economics* 329 (https://doi.org/10.1016/j.ecolecon.2016.11.009)

Schulze R and Morgan J, 'Right of Withdrawal' in Dannemann G and Vogenauer S (eds) *The Common European Sales Law in Context* (Oxford, Oxford University Press 2013) 294

Schwörer J, 'Mainstream Parties and Global Warming: What Determines Parties' Engagement in Climate Protection' [2023] 63(1) *European Journal of Political Research* 303 (https://doi.org/10.1111/1475-6765.12602)

Sibony AL and Alemanno A, *Nudge and the Law* (London, Hart Publishing 2015)

Smits JM, 'The Right to Change Your Mind? Rethinking the Usefulness of Mandatory Rights of Withdrawal in Consumer Contract Law' (2011) Maastricht European Private Law Institute Working Paper 2011/01, <https://papers.ssrn.com/sol3/papers.cfm?abstract_id=1719104>

Stern PC et al, *Environmentally Significant Consumption* (Washington DC, National Academies Press 1997)

Stieglitz S et al, 'The Potential of Digital Nudging to Bridge the Gap between Environmental Attitude and Behavior in the Usage of Smart Home Applications' [2023] 72 *International Journal of Information Management* 102665 (https://doi.org/10.1016/j.ijinfomgt.2023.102665)

Sunstein CR, 'Nudges vs Shoves: The Benefits of Preserving Choice' [2014] 127(6) *Harvard Law Review* 210

———, 'Nudging: A Very Short Guide' [2014] 37 *Journal of Consumer Policy* 583 (https://doi.org/10.1007/s10603-014-9273-1)

———, *The Ethics of Influence. Government in the Age of Behavioural Science* (Cambridge, Cambridge University Press 2016)

———, 'Sludge and Ordeals' [2019] 68 *Duke Law Journal* 1843

———, 'Green Defaults Can Combat Climate Change' [2021] 5 *Nature Human Behaviour* 548

———, 'Sludge Audits' [2022] 6(4) *Behavioural Public Policy* 654 (https://doi.org/10.1017/bpp.2019.32)

Sunstein CR and Gosset JL, 'Optimal Sludge? The Price of Program Integrity' [2020] 70 *Duke Law Journal* 74

Sunstein CR and Thaler RH, *Nudge. Improving Decisions about Health, Wealth and Happiness* (London, Penguin Books 2008)

Tanner RJ et al, 'Of Chameleons and Consumption: The Impact of Mimicry on Choice and Preferences' [2008] 34(6) *Journal of Consumer Research* 754 (https://psycnet.apa.org/doi/10.1086/522322)

Terryn E and Van Gool E, 'The Role of European Consumer Regulation in Shaping the Environmental Impact of E-Commerce' [2021] 3 *EuCML* 89

Thaler R, 'Nudge, Not Sludge' [2018] 361(641) *Science* 431 (https://doi.org/10.1126/science.aau9241)

Thorun C et al, *Nudge-Ansätze Beim Nachhaltigen Konsum: Ermittlung Und Entwicklung von Maßnahmen Zum 'Anstoßen' Nachhaltiger Konsummuster* (Bundesministeriums für Umwelt, Naturschutz, Bau und Reaktorsicherheit 2016)

Tonner K, 'Consumer Protection and Environmental Protection: Contradictions and Suggested Steps Towards Integration' [2000] 23(1) *Journal of Consumer Policy* 63 (https://doi.org/10.1023/A:1006307709249)

Trentmann F, *Empire of Things: How We Became a World of Consumers, from the Fifteenth Century to the Twenty-First* (New York, Harper Collins 2016)

Twigg-Flesner C and Schulze R, 'Protecting National Choice Information and the Right of Withdrawal' in Howells G, Ramsay I and Wilhelmsson T (eds), *Handbook of Research on International Consumer Law* (Cheltenham, Edward Elgar 2010) 130

van Aaken A, 'Judge the Nudge: In Search of the Legal Limits of Paternalistic Nudging in the EU' in Sibony AL and Alemanno A (eds), *Nudge and the Law: A European Perspective* (London, Hart Publishing 2015) 83

Van Gestel LC, Adriaanse MA and De Ridder DTD, 'Do Nudges Make Use of Automatic Processing? Unraveling the Effects of a Default Nudge under Type 1 and Type 2 Processing' [2021] 5(1–3) *Comprehensive Results in Social Psychology* 4 (https://doi.org/10.1080/23743603.2020.1808456)

Voget-Kleschin L, Baatz C, and Ott K, 'Ethics and Sustainable Consumption', in Reisch L and Thøgersen J (eds), *Handbook of Research on Sustainable Consumption* (Cheltenham, Edward Elgar 2015) 121

von Zahn M et al, 'The Smart Green Nudge: Reducing Product Returns through Enriched Digital Footprints & Causal Machine Learning', SAFE Working Paper, October 2022

Wagner G and Eidenmüller H, 'Down by Algorithms? Siphoning Rents, Exploiting Biases, and Shaping Preferences: Regulating the Dark Side of Personalized Transactions' [2018] 20 *The University of Chicago Law Review* 581

Wallet J et al, *Growing a Circular Economy: Ending the Throwaway Society* (House of Commons 2014)

Watson J, 'Withdrawal Rights' in Twigg-Flesner C (ed), *Research Handbook on EU Consumer and Contract Law* (Cheltenham, Edward Elgar 2016) 256

Wee SC, Choong WW and Low ST, 'Can "Nudging" Play a Role to Promote Pro-Environmental Behaviour?' [2021] 5(4) *Environmental Challenges* (https://doi.org/10.1016/j.envc.2021.100364)

Weinmann M, Schneider C and vom Brocke J, 'Digital Nudging' [2016] 58(6) *Business & Information Systems Engineering* 433 (https://doi.org/10.1007/s12599-016-0453-1).

White K, Habib R, and Hardisty DJ, 'How to SHIFT Consumer Behaviors to Be More Sustainable: A Literature Review and Guiding Framework' [2019] 83(3) *Journal of Marketing* 22 (https://doi.org/10.1177/0022242919825649)

White M, *The Manipulation of Choice. Ethics and Libertarian Paternalism* (London, Palgrave Macmillan 2013)

Wilk R, 'Consumption, Human Needs, and Global Environmental Change' [2002] 12(1) *Global Environmental Change* 5
Wilkinson TM, 'Counter-Manipulation and Health Promotion' [2017] 10(3) *Public Health Ethics* 257 (https://doi.org/10.1093/phe/phw044)
Wirtz J and Kum D, 'Consumer Cheating on Service Guarantees' [2004] 32(2) *Journal of the Academy of Marketing Science* 159 (https://doi.org/10.1177/0092070303261416)
Yeung K, 'Hypernudge': Big Data as a Mode of Regulation by Design' [2017] 20(1) *Information, Communication & Society* 118 (https://doi.org/10.1080/1369118X.2016.1186713)
Yudkowsky E, 'Cognitive Biases Potentially Affecting Judgment of Global Risks' in Milan Ćirković and Nick Bostrom (eds), *Global Catastrophic Risks* (Oxford, Oxford University Press 2011) 91
Zhang M and Luo L, 'Can Consumer-Posted Photos Serve as a Leading Indicator of Restaurant Survival? Evidence from Yelp' [2022] 69(1) *Management Science* 25 (https://doi.org/10.1287/mnsc.2022.4359)

24

HOW CAN WE PERSUADE CONSUMERS TO PURCHASE MORE SUSTAINABLE PRODUCTS? A REVIEW OF EUROPEAN LEGAL DEVELOPMENTS

Carl Dalhammar and Eléonore Maitre-Ekern

24.1 Introduction

The role of products, and their environmental impacts in different life cycle phases, has received more and more attention from policymakers in the last decade, in particular in Europe.[1] At the European Union (EU) level, the Integrated Product Policy (IPP)[2] was the first initiative that addressed products and their life cycle impacts in a more holistic way. In the last 30 years, the EU has also adopted several laws that aim at improving the environmental performance of products. These include laws that remove toxic substances from products (e.g., the REACH Regulation[3] and the RoHS Directive[4]), laws that enforce the collection and recycling of used products as a responsibility for producers (the Waste Framework Directive[5] and other vertical instruments, most prominently the WEEE Directive for electronics[6]), and rules that set mandatory standards for the energy efficiency performance of products (The Ecodesign Directive[7]) and mandates labelling of product energy performance (The Energy Labelling Regulation[8]).

These mandatory rules have been complemented with voluntary policies, including ecolabels and sustainability criteria in public procurement, which provide additional incentives for manufacturers to design more sustainable products.[9] Such policies are adopted both at the EU level and among EU Member States.

While many of the laws were initially resisted by industries, over time many corporations have realized that the rules can also boost the competitiveness of EU industries.[10] Poor-performing products, that have poor functionality or are non-compliant with rules, can no longer be sold on the EU Internal Market. On the global stage, the EU is generally considered a 'green leader', and this is even more obvious in the case of product regulation. European laws on products and chemicals, and related standards, are also influencing manufacturers – and legal developments – in other jurisdictions, implying that EU policies have external effects, also known as the 'Brussels Effect'.[11] The European Commission has also indicated a clear willingness to 'lead' the developments. This

DOI: 10.4324/9781032662046-29

is also obvious in some recent policy documents, such as the latest standardization strategy, which states a clear objective to strengthen the global role of the European standardization system.[12]

With the emergence of the circular economy (CE) as a distinct policy field, there is an increasing interest in regulating new aspects of the product lifecycle. Most notably, several product policies and laws are aimed at prolonging the lifetime[13] of products and components. These include rules that more directly intend to increase lifetime through product design, but also policies that aim to increase the number of repairs. For instance, the EU has, through the EU Ecodesign scheme, regulated the lifetime of two products groups (light bulbs and vacuum cleaners), and set requirements that manufacturers and producers must issue spare parts to independent repairers and consumers for a number of product groups (these examples will be discussed later in this chapter).

There is also leadership at the national level, and especially France has emerged as a frontrunner for CE product policies, for instance by criminalizing planned obsolescence, adopting policies to mitigate the destruction of unsold goods, and setting up a repair fund under extended producer responsibility schemes (these policies will be discussed below). Several EU Member States have gone further than EU rules, and increased consumer protection notably through extended legal guarantee periods, which is a way of inciting producers and consumers to extend product lifetime respectively through better design and prolonged use. There is also leadership at the local level. Examples include the City of Vienna, which has set up a local repair network and pays for part of the costs for consumer repairs through a voucher scheme,[14] and the city of Eskilstuna which has the world's first shopping mall for re-used and recycled goods.[15]

In March 2022, The European Commission presented its 'Sustainable products initiative', with a *chapeau* communication entitled 'On making sustainable products the norm'.[16] The initiative contains several legislative and policy proposals, including a proposal for a new Ecodesign Regulation that will replace the existing Ecodesign Directive as well as a legislative proposal to Empower Consumers in the Green Transition.[17,18] The new Ecodesign Regulation will enable the Commission to adopt requirements for a much broader range of products than now (where it is limited to energy-related products), including important product groups such as textiles and furniture. Moreover, the Regulation clearly states the aim to regulate product lifetime and repairability, whereas the focus has until now been almost exclusively on energy efficiency. Another important innovation in the Regulation is the introduction of digital product passports to track products throughout their life cycles. Subsequently, the Regulation also aims to address the problem of destruction of unsold products and e-returns. As for the Commission's proposal to amend two existing consumer directives to empower consumers, a strong focus is placed on providing consumers with information on durability and avoiding unsubstantiated green claims. Finally, it should be noted that a highly anticipated legislative proposal on 'right-to-repair' (R2R) is expected in 2023.

Thus, we are seeing the emergence of a new 'landscape' of CE laws aiming for the first time at addressing the issue of product lifetime in a more comprehensive way. In this chapter, we discuss both existing and proposed law and policies, at the EU level as well as the national level through selected examples from EU Member States. Some general trends are appearing, which we analyse under five specific headings, representing five different stages of 'consumption intervention points'; cf. Figure 24.1.

Choice editing by regulators[19] implies a process where the legal sustainability requirements on products are constantly becoming more stringent in the EU Internal Market, meaning that consumers can only choose products that have an increasingly improved sustainability performance. Research indicates that it is hard to change consumer behaviour, and it is therefore crucial that all products on the market have a decent sustainability performance.

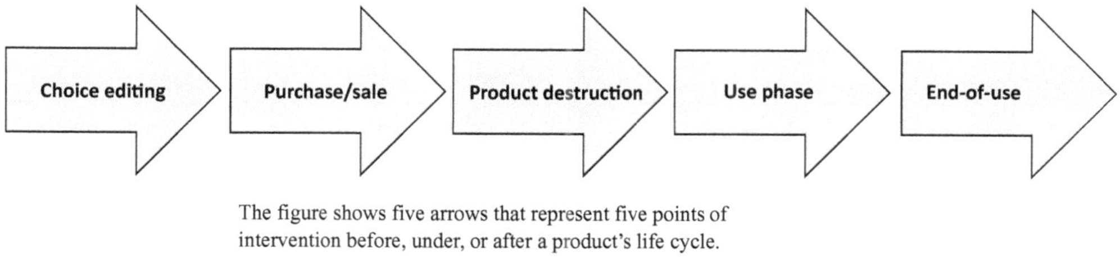

Figure 24.1 Five different stages of 'consumption intervention points'.

The next intervention point is the point of *purchase/sale*: here, we analyse existing and emerging policies that aims to incentivize consumers to purchase more sustainable products at the point of sale.

Product destruction is an under-researched and under-regulated policy area.[20] With this term we refer to products that are produced but never sold (overproduction), or that are sold but never used, either because they are e-commerce returns or other customer returns, or because they are defective/recalled/expired products. There is now a lot of interest in Europe for how we can adopt policies to address the problem of product destruction.

In the *use phase*, the consumer can choose to maintain and repair a product to extend its lifetime. Here we look at policies that promote such behaviour, focusing on policies that support consumer repairs.

Finally, at the *end-of-use* phase, we look at policies that aims to promote product re-use, remanufacturing, and reconditioning, rather than recycling/energy recovery/landfilling.

We base our analysis on legal documents and relevant literature. We primarily look at consumer-oriented policies and laws, and less at public procurement, though some examples of the latter are provided. We discuss and analyse existing and proposed EU policies, and also bring up relevant national policies.

Finally, we reflect upon these policies and some key issues for future European policy developments.

24.2 Regulatory Choice Editing

The safest way to ensure that consumers do not choose unsustainable products is to remove the worst-performing ones from the market. As it is hard to change consumer behaviour, regulating what is available on the market is a safer way to ensure that consumers choose more sustainable products. In the last 30 years, we have seen an ever-increasing number of laws that edit out unsustainable products.

24.2.1 EU Laws

Some key EU legal interventions are outlined in Table 24.1. These all place requirements on manufacturers and producers that put products on the Internal Market.

Regulations on *chemicals in products* have been applied for a long time, but they are constantly updated, and new substance restrictions are added. In the last 15 years we have also set *minimum energy efficiency standards* for an increasing number of energy-using product groups – e.g., white goods, electronics, and electric motors – under the Ecodesign Directive. *Extended producer responsibility* (EPR) schemes are also set for a growing number of product groups; EPR schemes

Table 24.1 **Key EU legal interventions**

Type of intervention	*Examples of laws*
Removing toxic substances from products	REACH, RoHS Directive, Toy Safety Directive, Products Safety Directive, Battery Directive
Ensuring mandatory minimum energy performance standards for products	Ecodesign Directive
Mandating extended producer responsibility: producers pay for collection and recycling systems; e.g., for packaging, vehicles, batteries, electronics, and fishing gear	Directive on Packaging and Packaging Waste, ELV Directive, WEEE Directive, Battery Directive, Directive on the reduction of the impact of certain plastic products on the environment
Setting minimum requirements on durability/lifetime of products; exists for vacuum cleaner and lighting products	Ecodesign Directive
Mandating producers to supply spare parts and repair manuals to professional repairers and consumers	Ecodesign Directive

are not strong drivers for changing product design but contain market access requirements for producers.[21]

Under the Ecodesign Directive, lifetime/durability requirements have been set for two product groups, namely, and as previously mentioned, vacuum cleaners, and lighting products.[22] Since 2019, a number of R2R requirements have been set for several product groups, including white goods and electronic displays. These typically require[23] various aspects of repairability, including (1) that certain spare parts are made available for professional repairers, and that some parts should also be available to consumers; (2) that spare parts must be delivered within a number of days; (3) that professional repairers should have the necessary software access; and (4) that dismantling of key components should be possible with commonly available tools. As an example, the proposed Ecodesign requirements for cell phones state that spare parts must be delivered within a specified number of working days, obliges manufacturers, importers, or authorized representatives to indicate an expected maximum pre-tax price for spare parts, and has rules that promote easier disassembly of phones.[24]

The European Commission's proposal for a new Ecodesign Regulation[25] is expected to further strengthen requirements on manufacturers in several ways: (1) ecodesign rules can be set for all types of products, including furniture and textiles[26]; (2) there is a strong focus on various new types of ecodesign requirements, such as product lifetime and repairability, and microplastics[27]; (3) new types of requirements on manufacturers are expected, such as mandating that new products put on the market must contain 'recycled content'.[28]

The proposed Regulation also foresees the introduction of Digital product passports,[29] which can be used to communicate data on a product and its supply chain and share it across entire value chains in order for various actors, including consumers, to have a better understanding of the materials and products they use and their embodied environmental impact. This will introduce yet another requirement for producers.

24.2.2 *National Laws and Policies*

Rules with product specifications can mainly be introduced at the EU level, as national laws would distort the functioning of the Internal Market. There are however other ways in which EU Member

States can contribute to incentivizing the design of more sustainable products or to removing the worst performers from the market. For instance, several states have introduced longer mandatory guarantees in consumer laws. As an example, Norway has, already since 2007, an extended five-year guarantee for all products that are expected to last longer than two years.[30] Sweden recently introduced a three-year guarantee for most consumer goods.[31] In Finland, the rule regarding the burden of proof has been amended in 2022 such that for a one-year period the defect will be presumed to have existed at the time of purchase.[32] Such national rules, which go beyond the minimum two-year guarantee period or six-month reversal of the burden of proof provided by EU law,[33] do not only contribute to protecting consumer rights but can also incentivize longer product lifetimes. If consumers can make a claim for a defect for an extended period, then it places a corresponding duty on producers to provide a remedy for such defect for a longer period. The most forward and cheaper remedy would be to increase the durability of the products, while other remedies could include easy repair, such as dismantling or spare parts availability.

As part of its broad and ambitious Anti-Waste and Circular Economy Law from 2020,[34] France set out to introduce a repair fund, where producers will have to pay some of the costs for repairing broken products that are not falling under the legal guarantee.[35] Such schemes can provide additional incentives to design long-lived products. Initially planned for 1 January 2022, the fund was eventually introduced on 15 December 2022 and provides, in particular, a rebate – between Euros 10 (for a coffee machine) and 45 (for a computer) – on the price of repair. The rebate originally aimed to amount to about 20% of the price of repair, and the French Agency for Ecological Transition (ADEME) thus announced an expected 28% increase in repair activities by 2027 compared to 2022.[36] However, a decree from December 2021 amended the law and replaced the 20% rebate with a 10% one,[37] provoking the ire of civil society, who argue that it would not be enough to shift consumer behaviours.[38]

With regard to rules aiming at removing the most unsustainable products from the market, France also innovated in 2015 by introducing a rule criminalizing the act of planned obsolescence.[39] Planned obsolescence is defined as 'all the techniques by which the company placing on the market aims to deliberately reduce the lifespan of a product in order to increase its replacement rate'. The law was adopted in particular in reaction to public outrage at the exposure that some manufacturers intentionally limited the lifespan of their products to ensure repetitive consumption.[40] Although the provision is hard to enforce in practice, in particular because it requires proving an intention, it sends a clear signal to the market that some of the most unsustainable and outrageous commercial practices will no longer be tolerated.

More recently, France also adopted new information requirements, which because of their mandatory character could constitute a potential market entry barrier. A new law mandates producers of various product groups – such as electrical and electronic products, packaging, textiles – to label their products for several criteria, including repairability, sustainability, recycled materials, micro plastics, etc.[41] For large companies (i.e., with an annual turnover of more than 50 million euros and placing at least 25,000 products on the national market each year), many rules will apply already from 1 January 2023, and then progressively in 2024 and 2025 for smaller companies. According to our conversations with representatives of the industry, it will be a challenge to comply also for many larger corporations.

24.3 Purchase/Sale

Regulatory choice editing means that only products with a decent sustainability performance are allowed on the market. The next policy objective is to persuade consumers to choose the most sustainable products put on the market. We then need policies that can underpin such decision-making.

24.3.1 EU Level Laws and Policies

The mandatory EU energy label – regulated under the Energy Label Regulation[42] – can support consumers to purchase the most energy-efficient products on the market. For some product groups, the label also communicates information on water consumption and noise levels. The label is applied for a number of product groups, including TVs and white goods. The EU also has a voluntary ecolabelling scheme ('The Flower'),[43] covering a number of product groups.

In the new proposed Ecodesign Regulation,[44] a new type of labelling is proposed (see especially Article 14). It is expected that new types of environmental aspects – most notably related to product lifetime and repairability – can be communicated through a mandatory label. This is similar in nature to the existing and planned French labelling (respectively on repair and durability, see below). In fact, the EU labelling scheme can be seen as an attempt to pre-empt new national labelling initiatives and harmonize existing practices within the EU.[45] The proposed Regulation also states that an alternative approach is to communicate product issues related to the CE (lifetime, repairability, etc.) under the existing Energy Label, when appropriate.

Recently, the European Commission proposed a legislative reform on consumer law for empowering consumers for the green transition.[46] More specifically, it proposed to amend two existing consumer directives,[47] to promote *inter alia* the following information and support to consumers: information on the existence and length of a producer's commercial guarantee of durability for all types of goods, or the absence of such guarantee in case of energy-using goods; information on the availability of free software updates for all goods with digital elements, digital content and digital services; information on the reparability of products, through a reparability score or other relevant repair information, where available, for all types of goods; ensuring that traders do not mislead consumers about environmental and social impacts, durability, and reparability of products; and a ban of certain practices related to early obsolescence of goods.

The proposal is quite complicated, and we will not go into details into all proposed amendments in this chapter but discuss some of them only. The information about guarantees for durability (or the lack thereof in the case of energy-using goods)[48] is an important aspect for ensuring that consumers are able to make enlightened choices upon sale. The question is how the information will be delivered to consumers, in particular in view of the possible development of a durability index in line with the proposed Ecodesign Regulation. It is not clear at this point whether those two 'durability information' requirements could be achieved together to avoid overwhelming consumers or if they would even be compatible. Another concern about the proposed information about guarantees for durability is that it does not mandate the provision of a guarantee for durability. The information is only to be provided if producers offer such a guarantee. A similar provision was developed in France in 2014 pertaining to information on the availability of spare parts. The fact that only certain products had the information attached to it clearly limited the impact of the provision (see more about this below).[49]

Further, a Commission proposal on laws to promote R2R was expected in late 2022 but has been delayed until 2023. It is expected that it will deal with the obligations of producers to repair a product during – or after – the legal guarantee period. Another issue considered is whether a product that has been repaired should receive a new legal guarantee as it if it was a 'new' product put on the market.

24.3.2 National Initiatives

France launched a mandatory repair index[50] on 1 January 2021. The index initially applied to five categories of electronic devices – namely, smartphones, laptops, televisions, washing machines and lawnmowers – and was recently extended to four additional product categories – namely vacuum cleaners, dishwashers, top-loading washing machines and pressure washers. The index score is placed next to the price tag in stores or on internet platforms and takes the form of a score from 1 to 10, in addition to using colour codes (green, yellow, and red). An assessment of the repair index after the first year found that most French consumers know about the Index and find it useful for purchasing decisions.[51]

As provided for in its Anti-Waste Law, France will turn its repair index into a durability index by 2024.[52] The new index shall include criteria related notably to *reliability* (or robustness), which is linked to the intrinsic performance of the product, whether mechanical or electrical, and may include aspects such as durability/hardiness, correct maintenance of the product and related consumer information, and; *upgradability*, which relates to the need for products to be usable over time even technological changes may occur and thus may include considerations of compatibility of software and hardware upgrades.

Spain announced in March 2021 plans to introduce a repairability index inspired by France's model.[53] It is reported that the Spanish Ministry of Consumer Affairs is currently preparing the draft law.[54] Similarly, the Belgian Ministry of Environment put forward, in September 2022, a proposal for the adoption of a repair index. This follows the government's stated intention, in its Federal Action Plan for the CE 2021–2024, to adopt an index.[55]

In 2014, France was also a pioneer in adopting a requirement for professional sellers to provide information about the availability of spare parts provided that such information was given to them by the producer.[56,57] The French provision was amended in 2021 and now requires that when the information is not provided to the professional seller, the spare parts are deemed to be unavailable, and it shall be communicated as such to consumers.[58]

Another way to stimulate sustainable behaviour at the point of purchase/sale is to persuade consumers to consider alternatives to buying new products, e.g., by leasing, sharing, or buying second-hand products.[59] Many cities support such activities by, for example, supporting tool libraries, various re-use activities, and specific re-use streams (e.g., construction products). There is also support for the second-hand sector, e.g., by letting it access material streams for re-use/recycling parks. One example of a more advanced scheme is the already mentioned ReTuna in Eskilstuna, the world's first re-use shopping mall, where private re-use companies receive various support activities from the city.[60]

Further, many national consumers agencies support changed consumer habits through information about options for purchasing new products. Many public actors also sell used ICT and furniture to re-users, supporting them in this way, or buying re-used products through public procurement.[61] There are also additional proposals in the literature on how to support the re-use sector, such as supporting quality labels for re-used products.[62]

24.4 Product Destruction

Many items that are produced are never actually used. They are unsold goods or customer returns, which go directly to disposal.[63] This could be considered the most extreme expression of our 'linear' economy. Among the product groups where this practice is common, we find textiles, foodstuff, and consumer electronics. According to Rödig et al., the sum of textiles and electronics products being destroyed in the EU alone is expected to reach €21.74 billion in 2022.[64]

There is limited research on these issues, and it was not until lately that policymakers have started to address the issue.

24.4.1 EU Level Policies

In its proposal for a new Ecodesign Regulation, Article 20 first establishes a general obligation of transparency for economic operators who discard unsold consumer products. It further provides for the possibility to adopt delegated acts on how this information shall be disclosed and to prohibit economic operators from destroying unsold consumer products. It is believed that the European Commission will start with the textiles sector. In the first step, transparency will be required. Later, there may be a banning of certain destruction practices.

24.4.2 Member State's Policies

Also on this topic, France can be considered a leader, as it has introduced a ban on the destruction of unsold goods as part of the French Anti-Waste Law. However, due to a lack of enforcement, and various deviations allowed, the law does not seem to be very effective in stopping the destruction of unsold products.[65] Germany has introduced a 'duty of care' law[66] that aims at stopping the destruction of waste, but the legal obligations are vague and the effects of the law have not been evaluated.[67] Belgium has reduced the value-added tax (VAT) on donated goods.[68] This VAT reduction may reduce the number of goods destroyed by allowing companies to make use of donation as alternatives. However, charities can most likely not take care of large, donated volumes.[69]

There are additional proposals in the literature on how to address the issue.[70] For instance, one option is to legislate that free e-commerce returns are not allowed, as cost-free consumer returns may incentivize unsustainable behaviour. Thus, we are likely to see more policies in the near future.

24.5 Use Phase

In the use phase, we want consumers to take care of their products and maintain them, and to repair them when broken rather than discarding them and buying new products.

A first key issue is that we have indeed convinced consumers to buy a more expensive, high-quality product to start with, as consumers are more likely to take good care of such products. Further, since many countries have legislated on longer consumer guarantees, it is crucial that consumers know of their rights, save receipts, etc., so that they actually enforce their consumer rights. Thus, we need to educate consumers in these respects. Further, we need to educate people on how to maintain their products.

When it comes to the issue of consumer repairs, the willingness to repair is influenced by (1) policies that make its legally possible to repair stuff, (2) policies that make it more economically attractive to repair stuff, and (3) policies that promote a 'repair culture'.[71] Both some States of the USA, and the EU and its Member States, have started to address the issue of corporations using intellectual property rights (IPRs) and consumer contracts to restrict repairs.[72] Due to limitations of space, we do not further discuss these issues here.

24.5.1 EU Policies

Further, as outlined above, the EU has started to mandate producers – under the Ecodesign Directive – to provide spare parts and tools, and repair manuals to independent repairers, and also to consumers. This will secure that spare parts are available, and increase competition in repair markets, which can reduce the cost of repairs. These rules are also important as they override producers' attempts to use IPRs to block the provision of spare parts.

As was also discussed above, the European Commission plans to introduce a R2R legislative package in 2023, which will further strengthen consumer rights in relation to repairs.

24.5.2 National Policies

Also here, France can be considered a leader. Its repairability index includes aspects such as availability of spare parts, which can influence consumers' willingness to repair their products. Further, the repair fund presented above, and organized through producer responsibility schemes, requires producers to pay some of the costs for repairing broken products, after the legal guarantee period has expired.[73] Consumers thus benefit from a reduction in the cost of repairing out-of-warranty goods. Although this is meant to incentivize consumers to seek repair instead of replacement – which is often seen as cheaper and easier than repair –, the fact that the reduction only amounts to 10% of the repair costs makes environmental associations doubt its effect on truly mainstreaming repair.

Extended legal guarantee periods and the postponed reversal of the burden of proof discussed above with regard to choice editing are also relevant at this stage.[74] Indeed, legal guarantees can provide a safeguard to consumers who decide to opt for the more durable product and enable them to use that product for a longer period of time even if a defect occurs.[75] However, consumers are too often faced with expired guarantees or are unable to prove the defect is not their doing. Strict rules on guarantees thus tend to limit consumer willingness to seek remedies and may thus lead to new purchases when a repair may have been possible. Faced with such difficulties, consumers may also think twice about choosing (again) the more expensive durable alternatives.

Other countries have adopted supporting policies. For instance, Sweden has lowered the VAT for some repair services and also offers tax rebates for repairs at home.[76] Some cities are also leading the way: Vienna and Graz have introduced local repair networks and use vouchers to subsidize consumer repairs at these repairers.[77] Some cities also support 'repair cafés' and other activities that promote a repair culture.

24.6 End-of-Life

Eventually, all products will reach and end-of-life (EOL) stage, where the consumer stops using the product. Then, it is crucial that the lifetime of products or components is increased through re-use, repair, remanufacturing or reconditioning/refurbishment. This can be supported in various ways. As outlined above, public agencies and companies can purchase reconditioned products or sell their used products to reconditioners. Reconditioners also propose some policies to support their businesses. For instance, a lot of high-quality furniture is destroyed when government agencies and corporations move offices, though they could be reconditioned and re-sold; a ban on throwing away products with value could support the reconditioners.[78]

For private consumers, it is important that they can support re-use. That is why there should be re-use options at recycling stations, and cities can also support the re-use sector in other ways (see also above), e.g., by supporting them with materials or logistical solutions. Several European cities are setting up re-use centres for specific waste streams where there is large potential for re-use, such as construction and demolition waste.

24.7 Concluding Discussion

In this contribution, we have outlined the legal and policy developments in consumption for the CE. We were able to identify several trends, in particular: (1) new aspects of the CE are being regulated, notably with a focus on the promotion of product repairs and longer products lifetimes; (2) new issues are being addressed; most notably, the destruction of unsold product; (3) new policies are emerging at the EU level, but also some Member States are also adopting policies, often quite innovative ones. And even at the level of cities, new approaches are being tested.

One important issue going forward concerns the harmonization of national measures at the EU level. Countries like France act as important 'policy innovators', but issues like mandatory product labels should be addressed at the EU level: it is unreasonable but also quite ineffective that manufacturers and producers should have to comply with many different national labelling schemes. A second issue pertains to the need to bring consumers to choose high-quality products with longer lifetimes: 'regulatory choice editing' is certainly the most effective way to ensure the most unsustainable products are not allowed in the EU, but these are minimum measures. However, in order to make more progress towards a CE we must also convince consumers to buy more sustainable products.

The issue of destruction on new products – unsold products, consumer returns, etc. – also highlights how long the road ahead still is: many current business models and business practices are fundamentally unsustainable: overproduction of low-cost, low-quality products; on the one hand, ease and economy of throwing away old products and replacing by new ones and, on the other hand, burden of extending product lifetime and finding new users. Thus, we need to fundamentally question some current practices in the market.

Acknowledgements

This research was funded by the Swedish FORMAS Forskarråd, through the project 'Mapping out and overcoming barriers for circular products: the policy context for corporations that want to "go circular"' (grant no. 2022–02045).

Notes

1 For an overview, see Carl Dalhammar, 'The application of 'life cycle thinking' in European environmental law: theory and practice' [2015] 12 *Journal for European Environmental and Planning Law* 97 (https://doi.org/10.1163/18760104-01202002); Cándido Garcia Molyneux, 'The dilemma of EU waste policy and circular economy: deconstructing the environmental regulation of an electronic tablet' in Amandine Orsini and Elena Kavvatha (eds), *EU environmental governance. Current and future challenges* (New York, Routledge 2021); Michael G. Faure and Carl Dalhammar, 'Principles for the design of a policy framework to address product life cycle impacts' in Eléonore Maitre-Ekern, Carl Dalhammar and Hans Christian Bugge (eds), *Preventing Environmental Damage from Products: An Analysis of the Policy and Regulatory Framework in Europe* (Cambridge, Cambridge University Press 2018); Leonidas Milios, 'Advancing to a circular economy: three essential ingredients for a comprehensive policy mix' [2018] 13 *Sustainability Science* 861 (https://doi.org/10.1007/s11625-017-0502-9); Eléonore Maitre-Ekern, 'Towards an integrated product regulatory framework based on life-cycle thinking' in Beate Sjåfjell and Anja Wiesbrock (eds), *The Greening of European Business under EU Law* (New York, Routledge 2019).

2 Commission, 'Green Paper on Integrated Product Policy' COM(2001) 68 final; Commission 'Communication from the Commission to the Council and the European Parliament – Integrated Product Policy – Building on Environmental Life-Cycle Thinking' COM(2003) 0302 final.
3 Council Regulation (EC) 1907/2006 of 18 December 2006 concerning the Registration, Evaluation, Authorization and Restriction of Chemicals (REACH) [2006] OJ L396/1.
4 Council Directive 2011/65/EU of 8 June 2011 on the restriction of the use of hazardous substances in electrical and electronic equipment [2011] OJ L174/88.
5 Council Directive 2008/98/EC of 19 November 2008 on waste and repealing certain Directives [2018] OJ L312/3.
6 Council Directive 2012/19/EU of 4 July 2012 on electronic waste and electronic equipment (WEEE) [2012] OJ L197/38.
7 Council Directive 2009/125/EC of 21 October 2009 establishing a framework for the setting of ecodesign requirements for energy-related products [2009] OJ L285/10.
8 Council Regulation (EU) 2017/1369 of 4 July 2017 setting a framework for energy labelling and repealing Directive 2010/30/EU [2017] OJ L198/1.
9 Michael G. Faure and Carl Dalhammar, 'Principles for the design of a policy framework to address product life cycle impacts' in Eléonore Maitre-Ekern, Carl Dalhammar and Hans Christian Bugge (eds), *Preventing Environmental Damage from Products: An Analysis of the Policy and Regulatory Framework in Europe*, (Cambridge, Cambridge University Press 2018).
10 See, e.g., Carl Dalhammar, 'Industry attitudes towards ecodesign standards for improved resource efficiency' [2016] 123 *Journal of Cleaner Production* 155 (https://doi.org/10.1016/j.jclepro.2015.12.035).
11 When similar policies and laws as those in the EU are adopted outside the EU, this is a 'de jure' Brussels effect. When corporations outside the EU voluntarily adopt EU standards – primarily to ensure that their products can be exported to the Internal Market – this constitutes a 'de facto' Brussels effect. See Anu Bradford, *The Brussel Effect: How the European Union Rules the World* (Oxford, Oxford University Press 2020).
12 Commission, 'An EU Strategy on Standardisation. Setting global standards in support of a resilient, green and digital EU single market' COM(2022) 31 final.
13 In this chapter, we do not get into details on how we can define concepts like 'lifetime' – which is far from straightforward – here. For such discussions, see Carl Dalhammar, Leonidas Milios and Jessika L. Richter, 'Increasing the lifetime of products – policies and consumer perspectives' (2021) Swedish Energy Agency, ER 2021: 25 <https://energimyndigheten.a-w2m.se/Home.mvc?ResourceId=201400> accessed 12 June 2023.
14 Carl Dalhammar et al, *Moving Away from the Throwaway Society. Five Policy Instruments for Extending the Life of Consumer Durables* (Götingen, Chalmers University of Technology 2022).
15 For more information about the mall ReTuna, see <https://www.retuna.se/english/> accessed 12 June 2023.
16 Commission, 'Communication from the Commission on making sustainable products the norm' COM(2022) 140 final.
17 Commission, 'Proposal for a Directive amending Directives 2005/29/EC and 2011/83/EU as regards empowering consumers for the green transition through better protection against unfair practices and better information' COM(2022) 143 final.
18 Policy proposals include in particular: Commission, 'Communication from the Commission Ecodesign and Energy Labelling Working Plan 2022–2024' 2022/C 182/01; and Commission, 'Communication of the Commission, EU Strategy for Sustainable and Circular Textiles' COM(2022) 141 final.
19 Regulatory choice editing should not be confused with the choice editing methods used in behavioural interventions ('nudging'). See the chapter by Marta Santos Silva and Tomàs Gabriel García-Micò in this book.
20 Hedda Roberts et al, 'Product destruction: exploring unsustainable production-consumption systems and appropriate policy responses' [2023] 35 *Sustainable Production and Consumption* 300 (https://doi.org/10.1016/j.spc.2022.11.009).
21 Note that a 'producer' can be the manufacturer, the importer, or another party that fulfils the obligations of the EPR laws.
22 For an overview and analysis, see Eléonore Maitre-Ekern and Carl Dalhammar, 'Regulating planned obsolescence: a review of legal approaches to increase product durability and reparability in Europe' [2016] 26 *RECIEL* 378, 383–385 (https://doi.org/10.1111/reel.12182).

23 For an overview, see Carl Dalhammar, Leonidas Milios and Jessika L. Richter, 'Increasing the lifetime of products – policies and consumer perspectives' (2021) Swedish Energy Agency, ER 2021: 25 <https://energimyndigheten.a-w2m.se/Home.mvc?ResourceId=201400> accessed 12 June 2023, 53–54 and Annex IV.

24 Commission, 'Proposal for a Commission Regulation laying down ecodesign requirements for mobile phones, cordless phones and slate tablets pursuant to Directive 2009/125/EC of the European Parliament and of the Council and Regulation (EU) 2023/826' C/2023/3538 final, Annex II.

25 Commission, 'Proposal for a Regulation of the European Parliament and of the Council establishing a framework for setting ecodesign requirements for sustainable products and repealing Directive 2009/125/EC' COM(2022) 142 final.

26 Commission, 'Proposal for a Regulation of the European Parliament and of the Council establishing a framework for setting ecodesign requirements for sustainable products and repealing Directive 2009/125/EC' COM(2022) 142 final, Art. 1.

27 Commission, 'Proposal for a Regulation of the European Parliament and of the Council establishing a framework for setting ecodesign requirements for sustainable products and repealing Directive 2009/125/EC' COM(2022) 142 final, Annex I.

28 Commission, 'Proposal for a Regulation of the European Parliament and of the Council establishing a framework for setting ecodesign requirements for sustainable products and repealing Directive 2009/125/EC' COM(2022) 142 final, Annex I.

29 Commission, 'Proposal for a Regulation of the European Parliament and of the Council establishing a framework for setting ecodesign requirements for sustainable products and repealing Directive 2009/125/EC' COM(2022) 142 final, Art. 2(29); Article 7–13.

30 Norwegian Act of 21 June 2002, no. 34 relating to Consumer Purchases [Forbrukerkjøpsloven], Section 27(2).

31 Swedish Consumer Sales Act [Konsmentköplag 2022:260] Chapter 4, Section 14.

32 Finnish Consumer Protection Act (38/1978) [Kuluttajansuojalaki], Section 15, as amended.

33 Council Directive (EU) 2019/771 of 20 May 2019 on certain aspects concerning contracts for the sale of goods [2019] OJ L136/28.

34 French Law n° 2020-105 of 10 February 2020 on combating waste for the circular economy, published in JORF n°0035 of 11 February 2020.

35 Article L. 541-10-4 and 5 of the French Environmental Code.

36 ADEME, In Extenso Innovation Croissance (Benoît Tinetti, Beatriz Berthoux, Arthur Robin, Nathan Setayesh) and Mathieu Hestin (independant expert), 'Fonds réparation de la filière des Equipements Electriques et Electroniques ménagers, étude préalable' [2021] <https://www.actu-environnement.com/media/pdf/news-38076-etude-ademe-fonds-reparation.pdf>, 12–13 accessed 12 June 2023.

37 French Decree No. 2021-1904 of December 30, 2021 on various adaptation provisions relating to extended producer responsibility, published in JORF n° 0304 of 31 December 2021, text n° 6, Article 1 modifying Article R. 541-147 of French Environmental Code.

38 See <https://www.halteobsolescence.org/appel-a-mobilisation-citoyenne-pour-sauver-le-fonds-reparation/> accessed 12 June 2023.

39 French Law n° 2015-992 of 17 August 2015, published in JORF n° 0189 of 18 August 2015, p. 1426, n° 1.

40 That is, Light bulbs and printers; see Markus Krajewski, 'The Great Lightbulb Conspiracy' (*IEEE Spectrum*, 24 September 2014) < https://spectrum.ieee.org/the-great-lightbulb-conspiracy#toggle-gdpr> accessed 12 June 2023.

41 French Decree No. 2022-748 of 29 April 2022, relating to consumer information on the environmental qualities and characteristics of waste-generating products, published in JORF n° 0101 of 30 April 2022, text n° 4.

42 Council Regulation (EU) 2017/1369 of 4 July 2017 setting a framework for energy labelling and repealing Directive 2010/30/EU [2017] OJ L198/1.

43 Council Regulation (EC) No 66/2010 of 25 November 2009 on the EU Ecolabel [2010] OJ L27/1.

44 Commission, 'Proposal for a Regulation of the European Parliament and of the Council establishing a framework for setting ecodesign requirements for sustainable products and repealing Directive 2009/125/EC' COM(2022) 142 final.

45 Many corporations that we have talked to are very critical of national mandatory labelling schemes and view them as a barrier for trade on the Internal Market; the proposed Ecodesign Regulation aims to harmonize practices.

46 Commission, 'Proposal for a directive of the European Parliament and of the Council amending Directives 2005/29/EC and 2011/83/EU as regards empowering consumers for the green transition through better protection against unfair practices and better information' COM(2022) 143 final.

47 That is, Council Directive 2005/29/EC of 11 May 2005 concerning unfair business-to-consumer commercial practices in the internal market and amending Council Directive 84/450/EEC, Directive 97/7/EC, 98/27/EC and 2002/65/EC and Regulation (EC) No. 2006/2004 ('Unfair Commercial Practice Directive') [2005] OJ L149/22; Council Directive 2011/83/EU of 25 October 2011 on consumer rights, amending Council Directive 93/13/EEC and Directive 1999/44/EC and repealing Council Directive 85/577/EEC and Directive 97/7/EC [2011] OJ L304/64.

48 Commission, 'Proposal for a directive of the European Parliament and of the Council amending Directives 2005/29/EC and 2011/83/EU as regards empowering consumers for the green transition through better protection against unfair practices and better information' COM(2022) 143 final; Proposed amendments to Article 5, paragraph 1.

49 See <https://www.quechoisir.org/decryptage-disponibilite-des-pieces-detachees-l-affichage-se-fait-mal-n8371/> accessed 12 June 2023.

50 French Decree No. 2020-1757 of 29 December 2020 relating to the repairability index for electrical and electronic equipment, JORF n°0316.

51 HOP, 'The French repairability index, a first assessment – one year after its implementation' (*Halte à l'Obsolescence Programmée*, 1 March 2022) <https://www.halteobsolescence.org/lindice-de-reparabilite-tient-il-ses-promesses/> accessed 12 June 2023.

52 See also discussion in Carl Dalhammar et al, *Moving away from the Throwaway Society. Five Policy Instruments for Extending the Life of Consumer Durables* (Götingen, Chalmers University of Technology 2022).

53 See <https://www.lamoncloa.gob.es/lang/en/gobierno/news/Paginas/2021/20210315reparability-label.aspx> accessed 12 June 2023.

54 Arantxa Herranz, 'Derecho a reparar, el gran reto pendiente del consumo' (*ABC Antropía*, 17 March 2022) <https://www.abc.es/antropia/abci-derecho-reparar-aparatos-electronicos-dia-consumidor-20220315160936_noticia.html> accessed 12 June 2023.

55 Service Public Fédéral and Economie, 'Plan d´action fédéral pour une économie circulaire 2021-2024' (2021) <https://www.health.belgium.be/fr/plan-daction-federal-pour-une-economie-circulaire-2021-2024#anchor-41031> accessed 12 June 2023.

56 See article L. 111-of the French Consumer Code.

57 See Eléonore Maitre-Ekern and Carl Dalhammar, 'Regulating planned obsolescence: a review of legal approaches to increase product durability and reparability in Europe' [2016] 26 *RECIEL* 378, 383–385 (https://doi.org/10.1111/reel.12182).

58 The article was modified by Article 18 of the French Law n° 2021-1485 aimed at reducing the environmental footprint of digital technology in France, JORF n° 0266 of 16 November 2021.

59 Eléonore Maitre-Ekern and Carl Dalhammar, 'Towards a hierarchy of consumption behaviour in the circular economy' [2019] 26(3) *Maastricht Journal of European and Comparative Law* 394 (https://doi.org/10.1177/1023263X19840943).

60 See <https://www.retuna.se/english/> accessed 12 June 2023.

61 Katarina Crafoord, Carl Dalhammar and Leonidas Milios, 'The use of public procurement to incentivize longer lifetime and remanufacturing of computers' [2018] 73 *Procedia CIRP* 137 (https://doi.org/10.1016/j.procir.2018.03.316); Matilda Öhgren et al, 'Public procurement of reconditioned furniture and the potential transition to product service system solutions' [2019] 83 *Procedia CIRP* 151 (https://doi.org/10.1016/j.procir.2019.02.134).

62 Ida Gåvertsson, Leonidas Milios and Carl Dalhammar, 'Quality labelling for re-used ICT equipment to support consumer choice in the circular economy' [2020] 43 *Journal of Consumer Policy* 353 (https://doi.org/10.1007/s10603-018-9397-9).

63 Hedda Roberts et al, 'Product destruction: exploring unsustainable production-consumption systems and appropriate policy responses' [2023] 35 *Sustainable Production and Consumption* (https://doi.org/10.1016/j.spc.2022.11.009).

64 Lisa Rödig et al, 'Policy brief on prohibiting the destruction of unsold goods European environmental bureau' (2021) *European Environmental Bureau* <https://eeb.org/wp-content/uploads/2021/10/Prohibiting-the-destruction-of-unsold-goods-Policy-brief-2021.pdf> accessed 12 June 2023.

65 Hedda Roberts et al, 'Product destruction: exploring unsustainable production-consumption systems and appropriate policy responses' [2023] 35 *Sustainable Production and Consumption* 300, 304 (https://doi.org/10.1016/j.spc.2022.11.009).

66 German Act Reorganising the Law on Closed Cycle Management and Waste of 24 February 2012 BGBl (*Kreislaufwirtschaftsgesetz*). I S. 212, 1474 (Nr. 10).

67 Hedda Roberts et al, 'Product destruction: exploring unsustainable production-consumption systems and appropriate policy responses' [2023] 35 *Sustainable Production and Consumption* 300, 304 (https://doi.org/10.1016/j.spc.2022.11.009).

68 Belgian Law amending the Value Added Tax Code with a view to exempting from VAT donations of non-food items to the most deprived, M.B. of 6 May 2019.

69 Hedda Roberts et al, 'Product destruction: exploring unsustainable production-consumption systems and appropriate policy responses' [2023] 35 *Sustainable Production and Consumption* 300, 304 (https://doi.org/10.1016/j.spc.2022.11.009).

70 Hedda Roberts et al, 'Product destruction: exploring unsustainable production-consumption systems and appropriate policy responses' [2023] 35 *Sustainable Production and Consumption* 300, 307–309 (https://doi.org/10.1016/j.spc.2022.11.009).

71 Sarah Svensson-Hoglund et al, 'Barriers, enablers and market governance: a review of the policy landscape for repair of consumer electronics in the EU and the U.S.' [2021] 288(125488) *Journal of Cleaner Production* (https://doi.org/10.1016/j.jclepro.2020.125488).

72 Sarah Svensson-Hoglund et al, 'Barriers, enablers and market governance: a review of the policy landscape for repair of consumer electronics in the EU and the U.S.' [2021] 288(125488) *Journal of Cleaner Production* (https://doi.org/10.1016/j.jclepro.2020.125488).

73 Hannah Thompson, 'France to introduce out-of-warranty repair fund for electronics' (*The Connexion*, 25 April 2022) <https://www.connexionfrance.com/article/French-news/France-to-introduce-out-of-warranty-repair-fund-for-electronics> accessed 12 June 2023.

74 Eléonore Maitre-Ekern and Carl Dalhammar, 'Towards a hierarchy of consumption behaviour in the circular economy' [2019] 26(3) *Maastricht Journal of European and Comparative Law* (https://doi.org/10.1177/1023263X19840943).

75 Vanessa Mak and Enna Lujinovic, 'Towards a circular economy in EU consumer markets – legal possibilities and legal challenges and the Dutch example' [2019] 8(1) *Journal of Consumer and Market Law* 4.

76 Carl Dalhammar, 'Sustainability, the circular economy and consumer law in Sweden' [2020] 9(3) *Journal of European Consumer and Market Law* 125, 127.

77 For an overview see Carl Dalhammar et al, *Moving away from the Throwaway Society. Five Policy Instruments for Extending the Life of Consumer Durables* (Götingen, Chalmers University of Technology 2022).

78 Matilda Öhgren et al, 'Public procurement of reconditioned furniture and the potential transition to product service system solutions' [2019] 83 *Procedia CIRP* (https://doi.org/10.1016/j.procir.2019.02.134).

Bibliography

ADEME, In Extenso Innovation Croissance (Benoît Tinetti, Beatriz Berthoux, Arthur Robin, Nathan Setayesh) and Mathieu Hestin (independant expert), 'Fonds réparation de la filière des Equipements Electriques et Electroniques ménagers, étude préalable' [2021] <https://www.actu-environnement.com/media/pdf/news-38076-etude-ademe-fonds-reparation.pdf>

Bradford A, *The Brussel Effect: How the European Union Rules the World* (Oxford, Oxford University Press 2020)

Crafoord K, Dalhammar C and Milios L, 'The use of public procurement to incentivize longer lifetime and remanufacturing of computers' [2018] 73 *Procedia CIRP* (https://doi.org/10.1016/j.procir.2018.03.316)

Dalhammar C, 'The application of 'life cycle thinking' in European environmental law: theory and practice' [2015] 12 *Journal for European Environmental and Planning Law* (https://doi.org/10.1163/18760104-01202002)

Dalhammar C, 'Industry attitudes towards ecodesign standards for improved resource efficiency' [2016] 123 *Journal of Cleaner Production* 155 (https://doi.org/10.1016/j.jclepro.2015.12.035)

Dalhammar C, 'Sustainability, the circular economy and consumer law in Sweden' [2020] 9(3) *Journal of European Consumer and Market Law*, 125–128

Dalhammar C, Milios L and Richter J L, 'Increasing the lifetime of products – policies and consumer perspectives' (2021) Swedish Energy Agency, ER 2021: 25 <https://energimyndigheten.a-w2m.se/Home.mvc?ResourceId=201400>

Dalhammar C et al, *Moving Away from the Throwaway Society. Five Policy Instruments for Extending the Life of Consumer Durables* (Götingen, Chalmers University of Technology 2022)

Faure M G and Dalhammar C, 'Principles for the design of a policy framework to address product life cycle impacts' in Eléonore Maitre-Ekern, Carl Dalhammar and Hans Christian Bugge (eds), *Preventing Environmental Damage from Products: An Analysis of the Policy and Regulatory Framework in Europe* (Cambridge, Cambridge University Press 2018) 57–86

Garcia Molyneux C, 'The dilemma of EU waste policy and circular economy: deconstructing the environmental regulation of an electronic tablet' in Amandine Orsini and Elena Kavvatha (eds), *EU Environmental Governance. Current and Future Challenges* (New York, Routledge 2021) 57–78

Gåvertsson I, Milios L and Dalhammar C, 'Quality labelling for re-used ICT equipment to support consumer choice in the circular economy' [2020] 43 *Journal of Consumer Policy* (https://doi.org/10.1007/s10603-018-9397-9).

Herranz A, 'Derecho a reparar, el gran reto pendiente del consumo' (*ABC Antropía*, 17 March 2022) <https://www.abc.es/antropia/abci-derecho-reparar-aparatos-electronicos-dia-consumidor-20220315160936_noticia.html>

HOP, 'The French repairability index, a first assessment – one year after its implementation' (*Halte à l'Obsolescence Programmée*, 1 March 2022) <https://www.halteobsolescence.org/lindice-de-reparabilite-tient-il-ses-promesses/>

Krajewski M, 'The great lightbulb conspiracy' (*IEEE Spectrum*, 24 September 2014) <https://spectrum.ieee.org/the-great-lightbulb-conspiracy#toggle-gdpr>

Maitre-Ekern E, 'Towards an integrated product regulatory framework based on life-cycle thinking' in Beate Sjåfjell and Anja Wiesbrock (eds), *The Greening of European Business under EU Law* (New York, Routledge 2019) 144–162

Maitre-Ekern E and Dalhammar C, 'Regulating planned obsolescence: a review of legal approaches to increase product durability and reparability in Europe' [2016] 26 *RECIEL* (https://doi.org/10.1111/reel.12182)

Maitre-Ekern E and Dalhammar C, 'Towards a hierarchy of consumption behaviour in the circular economy' [2019] 26(3) *Maastricht Journal of European and Comparative Law* (https://doi.org/10.1177/1023263X19840943)

Mak V and Lujinovic E, 'Towards a circular economy in EU consumer markets – legal possibilities and legal challenges and the dutch example' [2019] 8(1) *Journal of Consumer and Market Law*, 4–12

Milios L, 'Advancing to a circular economy: three essential ingredients for a comprehensive policy mix' [2018] 13 *Sustainability Science* (https://doi.org/10.1007/s11625-017-0502-9)

Öhgren M et al, 'Public procurement of reconditioned furniture and the potential transition to product service system solutions' [2019] 83 *Procedia CIRP* (https://doi.org/10.1016/j.procir.2019.02.134)

Roberts H et al, 'Product destruction: exploring unsustainable production-consumption systems and appropriate policy responses' [2023] 35 *Sustainable Production and Consumption* (https://doi.org/10.1016/j.spc.2022.11.009)

Rödig L et al, 'Policy brief on prohibiting the destruction of unsold goods European Environmental Bureau' (2021) *European Environmental Bureau* <https://eeb.org/wp-content/uploads/2021/10/Prohibiting-the-destruction-of-unsold-goods-Policy-brief-2021.pdf>

Svensson-Hoglund S et al, 'Barriers, enablers and market governance: a review of the policy landscape for repair of consumer electronics in the EU and the U.S.' [2021] 288(125488) *Journal of Cleaner Production* (https://doi.org/10.1016/j.jclepro.2020.125488)

Thompson H, 'France to introduce out-of-warranty repair fund for electronics' (*The Connexion*, 25 April 2022) <https://www.connexionfrance.com/article/French-news/France-to-introduce-out-of-warranty-repair-fund-for-electronics>

25

THE BATTLE AGAINST PLANNED OBSOLESCENCE – LEGAL REMEDIES

Denis Philippe[1]

25.1 Introduction

In 1932, Bernard London wrote that consumption is good for the economy and that planned obsolescence, facilitated by the new more efficient machinery in the industry, could end the depression.[2] Already in 1924, the Pheobus cartel was formed by several manufacturers of electric light bulbs, who agreed to produce light bulbs that were more fragile and thus reduced their lifespan to 1,000 hours instead of 2,500 hours.[3] At the time of the European Green Deal, this view has been abandoned.[4] The growing frustration among consumers regarding the fact that our consumer goods do not last as long as they should and the changing attitude of consumers and legislators towards the environment explain why nowadays, 'planned' and 'premature obsolescence' are the new buzzwords.[5]

To explain this phenomenon, some people refer to the concept of programmed obsolescence, according to which product lifespan is 'programmed' by the manufacturers.[6] Reducing the lifecycle is a convenient way for manufacturers to encourage consumers to buy a new device sooner, generating more profits, as Bernard London explained. This voluntary reduction of the lifespan of consumer goods has several important climatic and financial implications. The phenomenon not only increases the level of production and provokes the depletion of natural resources as a result of land destruction.[7] It also has the effect that products that fail earlier than expected produce waste and put increasing pressure on natural resources that are becoming scarcer. The fight against planned obsolescence also contributes to restoring consumer's trust in businesses. This contribution examines the current legal framework and legislative proposals regarding planned obsolescence. We will begin providing definitions and typologies of planned obsolescence (Section 25.2); we will continue by analysing the current status of the law (Section 25.3), then we will analyse some national initiatives (Section 25.4), and we will be concluding with European initiatives (Section 25.5).

25.2 Definitions and Typologies

25.2.1 The Concepts of Planned and Premature Obsolescence

Planned obsolescence is a business strategy by which a product's lifetime is deliberately shortened to increase its replacement rate.[8] Given the controversy surrounding the concept of 'programmed

DOI: 10.4324/9781032662046-30

obsolescence' and the fact that the responsibility for products' lifespan is shared between the producer and the consumer, it seems preferable to refer to 'premature obsolescence'. Indeed, compared to the notion of 'programmed obsolescence', the concept of 'premature obsolescence' has the advantage of being neutral and general[9]; some consumers like the change of products and are not aware of the important contribution of durability and other factors, like the evolution of the technology, to the obsolescence of the product.[10] Premature obsolescence covers all practices that reduce product lifetime so that the product does not reach its optimal lifetime (from an environmental point of view) without being the result of normal wear and tear.[11]

25.2.2 The Different Types of Premature Obsolescence[12]

Different types of premature obsolescence can shorten the life of products. Based on existing doctrine and practice, it is possible to identify five distinct categories of premature obsolescence.[13]

First, there is *material* obsolescence, which affects the physical quality of the product. This category includes products designed with an electronic device that blocks the device's functioning after a certain period of use, those manufactured with a more fragile component, or the productions where the defective parts are quasi-inaccessible.[14] A product may become technologically obsolete through incompatibility or specific manufacturing techniques which make the repair more difficult or wear parts which create a functional defect after a short period of time; in the Phoebus case, the light bulbs were very fragile and broke after a short period of time.[15]

Second, there is *functional* obsolescence, namely the reduction of an object's usefulness or desirability caused by an outdated design feature that cannot be easily changed or updated. Third, we have *indirect* obsolescence, which makes the use of a product impossible because the items essential to the product's functioning are no longer available, such as in the case of lack of spare parts.[16]

Fourth, when the price of repair is too high in comparison with the sales price of a new product, the product is *economically* obsolete.[17]

Last, a final case of so-called *psychological* obsolescence corresponds to products that have become unattractive and unsatisfactory in the eyes of consumers (by way of example, fashion). This subdivision of obsolescence into different categories is purely theoretical. In practice, an object's usefulness or desirability decreases because of a combination of several categories, making the phenomenon more complex to understand.[18]

25.3 Current Legal Framework

In this section, the current legal framework will be set out. We will first elaborate on the Belgian consumer law and civil regulation which offers a solution to consumers to fight against premature obsolescence. In the second sub-section, product regulations will be discussed.

25.3.1 Consumer Law

Article 1,134, 3° of the (old) Belgian Civil Code – now Article 5.73 of the new Civil Code – provides, in some instances, an obligation on the seller to provide information. However, it is up to the buyer to prove that the conditions for this obligation have been met. The buyer would have to prove that the lifetime of the product would be longer, and that the lifetime of the product was part of the contract. This obligation of good faith is also present in most of the European legal systems. Additionally, Article VI.2 of the Code of Economic Law[19] imposes an express obligation on the business[20] to provide information to the consumer. This is followed by a list of information the seller must expressly communicate to the consumer. However, no information about the duration of the product was required. The duration of the product is arguably an essential characteristic of the good, which must be disclosed in any event by the business. Under the general provisions of Belgian law, the seller must deliver a conforming good. Conformity is assessed at the time of conclusion of the contract; however, programmed obsolescence will be difficult for the purchaser to detect at the time of conclusion of the contract since the product appears to conform at that time; the purchaser's remedy will therefore lie more in the guarantee against hidden defects. We believe that limiting a product's durability constitutes a hidden defect. However, one must remember that the burden of proof also lies with the buyer, and that is not easy to assess *in abstracto* the lifespan of a product.

In Consumer Law, the guarantee of conformity and the guarantee of hidden defects are combined in a single concept of conformity. The guarantee will apply if the product does not meet the usual requirements that the consumer may expect and durability. We will see later that the notion of durability has been introduced in the Belgian Civil Code thanks to the European directives. In conclusion, the remedies of general contract law do exist, but greater protection and awareness for the buyer would require that the durability of the product, together with the ease of repair and/or availability of spare parts, be expressly included in the legal characteristics of the product, what is since recently the case.[21]

25.3.2 Product Regulations: The Ecodesign Directive 2009/125/EC

The EU Ecodesign Directive,[22] adopted in 2009 by the European Parliament and the Council, establishes a European-wide framework to set mandatory environmental requirements for 'energy-related products'[23] sold in all Member States. These products comprise energy-using products that use, generate, transfer or measure energy (e.g., boilers and computers) and energy-impacting products that only affect energy consumption (e.g., insulation and showerheads).[24] Its scope currently covers more than 30 product groups. The implementation of measures is based on EU internal market rules governing which products may be placed on the market.[25] The ultimate aim of the Ecodesign Directive is that manufacturers of energy-using products will be required, at the design stage, to reduce the energy consumption and other environmental impacts of products. While the Directive's primary aim is to reduce energy use, it also enforces other environmental considerations, including materials used, water use, polluting emissions, waste production, end-of-life management, and recyclability.[26] The Ecodesign Directive set product-specific minimum environmental requirements. Furthermore, the Directive imposes requirements on energy efficiency, but also more and more[27] requirements on the minimum lifespan of some products or components, on the design for facilitating the repairability, on the availability and delivery of spare parts, on the provision of information, on the availability of updates and secure data deletion.[28] Therefore, at least currently, the European Union does not have in place general sustainability or circularity requirements for all products on the internal market.

25.4 Experiences in Other National Laws

25.4.1 The French Legislation and Case Law on 'Planned Obsolescence'. The Repairability Index

France was the first country to pass legislation on programmed obsolescence worldwide, having a definition of the phenomenon as early as 2015.[29] Programmed obsolescence has been made a criminal offence by Articles L 441-2 and L 454-6 of the French Consumer Code. In its current wording, Article L. 441-2 of the French Consumer Code reads as follows

> The practice of planned obsolescence, which is defined as the set of techniques, including software-related ones, by which the person responsible for placing the product on the market seeks to deliberately reduce the lifetime of this product, is prohibited.

Furthermore, one must add it is a criminal offence under Article L. 454-6 §1-2 of the French Consumer Code. According to this Article, offenders risk two years of incarceration and a 300.000€ fine or up to 5% of the company's average annual turnover. This Article foresees also additional sanctions against the individuals involved. This Article on individuals finds almost no application because deliberate action must be proved.

As far as case law is concerned, since the French law on planned obsolescence has entered into force, a small number of procedures have been initiated based on this text, namely *HOP v. X (including Epson), HOP v. Apple* and *UFC-Que Choisir v. Nintendo.*

Regarding HOP v. X (including Epson), in 2017, the French consumer group '*Halte à l'Obsolescence Programmée*' ('HOP') filed a complaint against several printer companies, including Epson, a Japanese cartridge manufacturer, for programmed obsolescence. In its complaint, HOP specifically accuses Epson of using two techniques to shorten the lifetime of its printers or the related ink cartridges. HOP claims that Epson signal that the cartridges are empty while they sometimes still contain between 20% and 40% of ink. In addition, the consumer may be forced to change all the cartridges in order to be able to continue to use all the printer's functions.[30] HOP also underlines that this phenomenon has been observed among other cartridge manufacturers and would indicate a possible anticompetitive agreement, a practice that is criminally punishable under the French Commercial Code. According to HOP, the practices can be considered criminal under the provision quoted earlier. Following filing a complaint on 18 September 2017, the Public Prosecutor opened an investigation on 28 December of the same year. More than three years after HOP's complaint against Epson for programmed obsolescence and deception, the case has still not been trialled.[31]

The second case on planned obsolescence was the HOP v. Apple case. Apple strongly encouraged its customers of the iPhone 6S and 7 to update their smartphones with iOS 11.2, reducing the smartphone performance. This prompted many users to replace their devices prematurely. In 2017, many iPhone customers started complaining about their devices slowing down significantly after updating their phones. HOP alerted the DGCCRF,[32] France's watchdog on competition and consumer rights. A few months earlier, Apple had admitted to voluntarily slowing down the performance of its iPhones after a certain period of use to 'extend their lifespan'. Eventually, the DGCCRF and Apple settled in early 2020. Apple was fined 25 million euros for 'deceptive commercial practices by omission'. A new penal claim has been introduced against Apple by HOP on 7 December 2022. Serialization (also known as 'pairing'), consists in associating the serial numbers of a product's components and peripherals with that of the iPhone, in particular via microchips. This practice has recently been extended to the parts most frequently subject to failure (screens, battery, camera, etc.).[33] It enables the manufacturer to limit repair possibilities, particularly for unauthorized repairers.

HOP has launched two forms to collect testimonials from users who have encountered a problem.

Another case that caused a stir is the *UFC-Que Choisir v. Nintendo* case. In this case, no fewer than 25,000 consumers filed a complaint against Nintendo for their defective joysticks (the Joy-con

Drift: due to a defect, the joystick moves without even being touched by the player, rendering them unusable). UFC indicates that 65% of consumers were affected within a year of purchasing the console, even if they played less than five hours a week. In 25% of cases, the breakdown occurred in the first six months after purchase. The obsolescence of this product means that consumers often must buy a new set of game controllers after a short time, also because of the disproportionate costs and the practical burdens that consumers would face when trying to exercise repairs. This situation prompted the European Consumers' Organisation (BEUC) and the UFC-Que Choisir to file a complaint with the European Commission and the national consumer protection bodies on premature obsolescence and misleading action, misleading omission, as laid forward in Articles 7 and 8 of the UCPD.[34] 9.100 testimonials have been gathered in Europe.[35]

Another relevant aspect in France is the repairability index. Since 1 January 2021, in France, electric and electronic products put on the market must be accompanied by a repairability index.[36] The index measures the documentation, the possibility of disassembling, and the price and availability of spare parts.[37]

As demonstrated, the index is very clear and straightforward. On the one hand, the product receives a score out of 10. The higher the score, the more repairable a product is. On the other hand, it works with a colour code, which goes from bright red for non-repairable products to dark green for easily repairable products, this in order to make the index even easier to read.[38] Thanks to the mandatory display of the repairability index, consumers can immediately get a good idea of the repairability and thus the durability of the product at a glance. This enables consumers to compare products more easily and make an informed choice between several products. It is definitely an innovation that should be welcomed, especially in the context of the fight against premature obsolescence.[39]

25.4.1.1 The Italian Cases against Apple and Samsung

Nearly a century after the Phoebus cartel, the Italian Competition Authority (AGCM) has issued two landmark decisions against various companies belonging to the Apple[40] and Samsung[41] groups. These two decisions issued on 25 September 2018 are the first sanctions taken against manufacturers for having used planned obsolescence. The AGCM found that both were guilty of unfair commercial practices by releasing firmware updates for their mobile phones, which caused serious malfunctions and significantly reduced their performance, speeding up their replacement with more recent products.

The AGCM found that Apple deceived consumers in two ways. Firstly, consumers got the insistent proposal to install iOS 10. Subsequently, customers were imposed updates whose characteristics impacted on the smartphone's performance, without offering any means of restoring the original functionality. This, according to the AGCM, was considered as an omissive and deceptive practice. Secondly, the AGCM found that Apple did not adequately inform their customers about the characteristics of the lithium batteries, such as their average duration and deterioration factors. Consequently, Apple was fined to pay 10 million euros based on unfair, deceptive and aggressive commercial practices as stipulated in Articles 20–22 and 24 of the Italian Consumer Code (5 million for each of the two contested practices). The maximum fine was deemed appropriate due to the misconduct's size and seriousness.

In addition to the Apple's case, there is the Samsung case. Consumers having purchased Samsung Note 4 received the insistent proposal of firmware updates that slowed the performance of their smartphones and whose characteristics and impact were described in an omissive and deceptive manner, limiting the possibility to restore the original functionality of the appliance to smartphones still covered by the legal guarantee of conformity and without considering alternative solutions. Samsung was fined to pay 5 million euros due to the misconduct's size and seriousness, on the grounds of unfair, deceptive, and aggressive commercial practices.

25.4.1.2 Other Countries

In Belgium, the consumer association *Test Achats* has filed a class action against Apple,[42] because, according to complain the company has wilfully put on the market updates software for iPhone 6 models which slowed down the functioning of these iPhones, so that the consumer has no other choice than to buy new models.[43] *Test Achats* has created a weblink to collect testimonials from consumer complaining about the shortened lifespan of products. The Portuguese consumer organization *Deco Proteste* introduced a class action for 115,000 iPhone users; it claims that 'Apple deliberately manipulated, and without informing its users, the performance of its most popular devices, the iPhone 6, 6 Plus, 6S, and 6S Plus'.[44] In Norway, the State extends the consumer guarantee for products.[45] We can find also initiatives in numerous countries.[46] For example, consumer associations in five countries (Belgium, Spain, Italy, Portugal and Brazil) issued a formal notice to the printer manufacturer Hewlett-Packard (HP). The organizations accuse HP of deliberately blocking the use of ink cartridges other than its own via software updates.[47]

25.5 European Initiatives

As part of the European Green Deal, several initiatives have been taken at the European level to combat planned obsolescence.[48] The European Green Deal is a package of policy initiatives from the European Commission to support the green transition in the European Union Europe aims to reduce CO_2 emissions by 55% by 2030 compared to 1990 and with the end goal of climate neutrality by 2050. The European Green Deal was launched at the beginning by the Von der Leyen Commission. All relevant policy areas must contribute to the ultimate climate goal to achieve these ambitious goals. The package includes initiatives around climate, environment, energy, transport, industry, agriculture, and sustainable finance, which are closely intertwined. In mid-2021, the European Commission presented a set of measures called 'Fit for 55', to enact the ambitions of the Green Deal into legislation. The climate package consists of many legislative proposals to achieve the climate goals set out in the Green Deal.

25.5.1 Empowering Consumers through Better Information

The European Union has launched numerous initiatives to fight against planned obsolescence and to improve consumer awareness.[49] On 30 March 2022, the European Commission presented a new proposal for a directive to empower consumers to act in favour of the ecological transition through better protection against unfair practices and better information on greenwashing and planned obsolescence. This proposal amends Directive 2005/29/EC on Unfair Commercial Practices Directive (UCPD) and Directive 2011/83/EU on consumer rights. With the new proposal of the Consumer Rights Directive, the Commission intends to strengthen the pre-contractual information obligation of companies on the durability of products and the possible existence of a commercial guarantee, on the one hand and, on the means of repairing products (repair index or an instruction manual, information on the availability of spare parts, etc.) on the other hand. The means of fulfilling this information obligation should, however, remain at the discretion of companies which may choose to provide this information on the packaging, product description, or website. With the UCPD amendment, the Commission wants to prohibit companies from misleading consumers about the environmental characteristics of their products, by extending the products' characteristics. It also adds new infringements to the 'blacklist' of commercial practices, such as linking a product to vague and generic commercial characteristics on its performance or labelling products with a quality label not approved by public authorities. The Commission's new legislative proposal is examined by the European Parliament and the Council under the ordinary legislative co-decision procedure, with the Parliament deciding on this text on an equal footing with the Council.

25.5.1.1 Unfair Commercial Practices Directive (UPCD) 2019/2161 (So-Called 'Omnibus Directive')

The EU directive on unfair commercial practices from 2005s objective is to foster consumer awareness and make it easier for businesses to trade across borders. This Directive was recently amended by the EU Directive 2019/2161. The new Directive introduced the obligation for enterprises to inform consumers about the product's durability. The Commission Staff Working Document clarifies that from the UCPD point of view, planned obsolescence is not unfair *per se*. However, under Article 7 of the UPCD '*A trader who* fails to inform [emphasis added] *the consumer that a product has been designed with a limited lifetime might, according to the specific circumstances of the individual case, be considered to have omitted to provide material information*'.[50] After a case-by-case assessment, the amendments also introduced new practices regarded as misleading, such as making an environmental claim related to future environmental performance without clear, objective, and verifiable commitments and targets and without an independent monitoring system. Lastly, new practices were added to the blacklist (e.g., *not informing about features introduced to limit durability*).

The aim of the amendments to the UPCD is threefold. First, it aims to provide legal certainty for traders. Second, it facilitates enforcement of cases, and last, it aims for enforcement by the consumer through collective redress.[51]

25.5.1.2 Consumer Rights Directive (Directive (EU) 2019/2161)

The proposal for the consumer rights directive also introduced rules regarding premature obsolescence. The proposal introduced a general obligation of the trader to provide the consumer with some specific items of pre-contractual information, among which information on the durability and reparability of products can be found, information that the trader must provide to the consumer before the purchase, regardless of whether the purchase takes place online or in a shop.[52] The proposal stipulates that the trader shall give pre-contractual information to the consumer on the following six items

1 the existence and length of a producer's commercial guarantee of durability for all types of goods, when this information is made available by the producer;
2 the mention that no information has been provided by the producer about the existence of a producer's guarantee of durability for energy-using goods;
3 the existence and length of the period during which the producer commits to providing software updates for goods with digital elements;
4 the existence and length of the period during which the provider commits to providing software updates for digital content and digital services;

5 the reparability score of the good as applicable under Union law;
6 other repair information, should no reparability score be available at the Union level, such as information on the availability of spare parts and a repair manual.

These six additions to the required pre-contractual information are welcomed by the authors[53]: thanks to these provisions, the trader can no longer withhold certain information that the seller knows to be relevant to the consumer. The availability of such information may also influence the consumer's decision to purchase or repair a good, and above all, this allows consumers to prove more easily the existence of a lack of conformity.[54] Consumers are faced with sustainability labels which are not transparent or credible. The EU aims to empower the consumer. Traders must guarantee that their environmental claim is based on recognized scientific evidence and the technological state of the art.[55]

25.5.2 Encouraging Repair in Case of Defect[56]

25.5.2.1 Sales of Goods Directive 2019/771

Another initiative of European law in this regard is Directive 2019/771.[57] Recital 32 of the Directive provides that conformity implies that the good meets the durability criteria.[58] The average duration of the good is considered as the legitimate expectations of the consumer.[59] At the same time, recital 48 of the Directive states that consumers should enjoy a choice between repair and replacement. Enabling consumers to require repair should encourage sustainable consumption and contribute to greater product durability. As a result of this European Directive, Article 1649-quater, §3,4° was inserted into the Belgian old Civil Code and in Article L 212-4 of the consumer code in Luxemburg. This new legislation provides that, in order to be compliant, the good must meet the following requirements, and more specifically:

> *4° être en quantité et présenter les qualités et d'autres caractéristiques, y compris en termes de durabilité...*

The change to the law makes durability an element of conformity.

In addition, Directive 2019/771 also amended the existing legal guarantee[60] for consumer purchases. The legal guarantee gives consumers the right to have defective products repaired during the legal guarantee. The rule regarding the legal guarantee in consumer purchases before the reform by Directive 2019/771 was confusing for consumers. The legal guarantee was valid for two years but was split into two systems. For the first six months after delivery, the Directive provided for a legal presumption that the conformity defect was attributable to the seller. After the expiry of these initial six months, it was up to the consumer to prove that the defect was due to the seller's responsibility. This meant that, in practice, the consumer could only benefit from an effective legal guarantee of six months. The Directive introduces an important innovation, namely the presumption (and thus the burden of proof) of the existence of a conformity defect for a duration of two years; the liability lies thus exclusively with the seller for the entire duration of the legal guarantee. This change represents an important development in the fight against planned obsolescence since this proposal will increase transparency for consumers as regards commercial guarantees and will thereby also incentivize producers to offer commercial guarantees of durability longer than two years.

A few comments on the guarantees issue should be made.

Durability is currently mentioned in the sales of goods regulation but imposes no specific sanction on the manufacturer. The new rules introduce adequately the possibility to compare the availability of software updates and the duration of this availability and concerning the commercial guarantee of durability makes the producer (not only the seller) directly liable to the consumer during the entire period of that guarantee for repair or replacement of the goods.

We also welcome the new presumption in the existence of the conformity request during the guarantee period. But the problem is that liability defects frequently occur after the two years guarantee period. Finland, for instance, has aligned the guarantee period with the average duration of the product.[61]

Concerning the remedies in case of defects, we think consumers must be more encouraged to choose repair instead of replacement. Must repair be the sole remedy for consumers? We do not think that we must go as far as this stage: we have to assess the efficiency of the new ecodesign provisions. Nonetheless, it is the seller and not the producer the first responsible for these defects. Yet, durability is more the producer's responsibility than the seller's and the mere contractual liability between seller and consumer is insufficient to address the durability problem. Maybe the producer and the seller could be jointly liable.

25.5.2.2 Right to Repair

On 7 April 2022, the European Parliament adopted resolution on the right to repair,[62] calling on the Commission to legislate in this field. This is no surprise since the European Parliament has been a strong advocate for improving repair options for consumers for over a decade and has recently called for additional measures in different resolutions.[63]

The right to repair is intended to give consumers the practical means to self-repair their products, for instance by a better availability of spare parts or a manual explaining how to repair the good[64] or choose a third-party service provider instead of going through the manufacturer. Nowadays, several obstacles prevent consumers from opting for repair, including unavailability of information, lack of access to spare parts, lack of standardization and interoperability or other technical barriers, and the cost of repair.[65]

Ecodesign can be very useful in this regard. For example, designing a computer or a smartphone that can have memory added post-purchase benefits. Besides forcing new designs, reparability also means continuing the production of spare parts for a certain number of years after discontinuing the production of a given good.

The European Parliament stresses in its Resolution of 7 April 2022 on the right to repair, in particular, that those software updates must be made available for a minimum period in line with the Digital Content Directive. Furthermore, the Parliament asks the Commission to propose in its initiative, among other things, harmonized rules for such consumer information, including, among other information, repair scores, information on estimated lifespan, and information on spare parts. This information should be made available at the time of purchase. Furthermore, the Parliament calls on the Commission and Member States to develop financial incentives for repair services so to make repairs convenient and attractive to consumers and to require manufacturers to design their products in such a manner that they last longer, can be safely repaired and their parts can be easily accessed and removed. Likewise, the Parliament requests the Commission to empower the consumer to make better decisions through better information. In addition, the Parliament would like a voluntary EU ecolabel encouraging the industry to uptake of labelling policies that relay crucial information to consumers on the lifespan of goods.

Also, the Parliament supports the extension of the legal guarantee beyond two years, as this might provide an incentive to choose repair over replacement. Lastly, the European Parliament calls on the Commission to propose measures that reward traders for providing repair possibilities for remanufactured or refurbished goods to increase consumer trust.[66]

The Commission on 22 March 2023 submitted a Directive proposal on the right to repair.[67] (hereinafter the *Proposal*).

The text proposes, among other things, the creation of an online platform at the national level, matchmaking consumers with repairers and promoting refurbished goods[68] and that repair is prioritized when costs are cheaper than replacement.

The text introduces, also, an obligation on repairers to issue, upon request, a price quote and conditions for repair in a standardized form and a voluntary EU easy repair standard (European Standard for repair services).[69]

This proposal has been criticized because it did not address the costs of financial accessibility of the repair. Besides, they did not take sufficiently into account the function of independent repairers which are not in the official retail network.[70]

25.6 Leveraging Product Regulations to Extend Product Regulations for Extending Product Lifetime. The 2022 Proposal for a Sustainable Products Regulation

On 30 March 2022, the European Commission launched a proposal for Regulation[71] establishing a general framework for setting ecodesign requirements for sustainable products (hereinafter the SPR proposal) This SPR proposal would repeal the current applicable Ecodesign Directive discussed earlier in this contribution.[72]

The most significant changes to the existing regulation would be a broadening of the scope of the ecodesign legislation to all products on the internal market; furthermore, companies that discard unsold consumer products would be subject to transparency requirements, new obligations for online marketplaces concerning market surveillance.[73]

The proposed regulation sets ecodesign requirements with a focus on sustainability aspects and is applicable to the broadest possible range of products.

The most significant proposed changes to the existing regulation will be briefly discussed hereinafter.

First, the proposal looks to broaden the scope of the ecodesign legislation to all products on the internal market. In contrast, the current legislation only focuses on energy-related products.[74] The goal of the regulation is not only energy efficiency but durability, and the possibility of repair is placed on the same level of importance. The requirements will be specified for each kind of product by delegated acts of the Commission.

A second significant change is the introduction of a digital product passport. The passport could contain, for example, information related to the traceability of the product and information on the performance (durability, reparability…). The proposal also clarifies that the information shall be connected through a data carrier.[75]

A third proposed change is an information requirement regarding the destruction of unsold goods. Companies which discard unsold consumer products will be forced to publicly disclose certain information, such as the reasons for throwing away the goods (type of products) and the quantity of such goods. This measure will increase enterprise and consumer awareness.[76] Environmental information must also be communicated to the consumer when he buys the product.[77]

The proposal also looks to introduce new obligations for online marketplaces concerning market surveillance. The proposal stipulates the obligation for online marketplaces to cooperate with the surveillance authorities and to inform the market surveillance authorities of any action taken.[78]

The proposal thus introduces information, performance and ecodesign requirements, market surveillance, and rules on circumvention prevention.[79] It has been discussed by the EU ministers on 29 September 2022, a discussion at which some Member States raised some concerns.[80]

25.7 Ecolabel[81]

Labelling has first been introduced for energy products.[82] Since 2019, ecolabel is required before selling energy products in the EU market.[83] The products and their environmental characteristics must be registered in a European database,[84] and updates must also be registered.[85] The voluntary label has existed also for other specific products since 1992.[86] The Ecolabel must apply to a wider range of products and be enlarged to a sustainability label.[87]

25.8 Concluding Remarks

This chapter analysed planned obsolescence, its different types, current legal framework, and experiments in national legal laboratories. Planned obsolescence is a complex topic because it is linked to technical considerations. It should be evident that (planned) obsolescence is a standard business practice nowadays in almost every industry. Overall, close interaction between environmental and economic laws is badly needed.

We have seen that, thankfully, more and more the European Union and the Member States – and specifically French policymakers – are increasingly starting to recognize the problems caused by planned obsolescence and acting against this behaviour, by implementing regulations supporting sustainability, reparability, reusability, and rules to empower consumers to combat planned obsolescence. France even went as far as to make planned obsolescence a criminal offence.

One of the problems lies in the fact that the durability of a product is not easy to determine with precision, and it is difficult to determine the liability referring to an imprecise element.[88]

Furthermore, not only the seller must be accountable for the planned obsolescence but also the producer and consumer. Consumer awareness has been promoted through better information but 'the legislator does not yet give the "consumer-actor" the means to fully play his role'.[89] Psychological obsolescence will also be challenging to tackle, and consumer awareness will be key.[90]

Last but not least, we consider that regulation of the market of spare parts could also be more developed[91] to tackle properly and sustainability the issue of premature obsolescence.

Notes

1 Denis Philippe, Professor at the UCLouvain and Paris Ouest, Lawyer at the Luxemburg and Brussels Bar. Cochairman of the Belgian Luxemburg Branch of the European Law Institute.

2 Bernard London, 'Ending the Depression through Planned Obsolescence' (*Wikimedia*, 1932) <https://upload.wikimedia.org/wikipedia/commons/2/27/London_%281932%29_Ending_the_depression_through_planned_obsolescence.pdf> accessed 25 April 2023; Giles Slade, *Made to Break: Technology and Obsolescence in America* (Harvard, Harvard University Press 2009) 72–75, 290; Samuel Becher and Anne-Lise Sibony, 'Confronting Product Obsolescence' [2021] 27 *Columbia Journal of European Law* 97, Eléonore Maitre-Ekern and Carl Dalhammar, 'Regulating Planned Obsolescence: A Review

of Legal Approaches to Increase Product Durability and Reparability in Europe' [2016] 25(3) *Review of European Community & International Environmental Law* 378 (https://doi.org/10.1111/reel.12182); Carl Dalhammar, 'The Application of 'Life Cycle Thinking' in European Environmental Law: Theory and Practice' [2015] 12(2) *Journal for European Environmental & Planning Law* 97 (https://doi.org/10.1163/18760104-01202002).

3 Coralie Schaub, 'La vie gâchée des objets' *Libération* (Paris, 28 October 2012) <https://www.liberation.fr/futurs/2012/10/28/la-vie-gachee-des-objets_856572/> accessed 24 September 2023.

4 *A European Green Deal. Striving to be the First Climate-Neutral Continent* <https://commission.europa.eu/strategy-and-policy/priorities-2019-2024/european-green-deal_en> accessed 16 April 2023.

5 Anaïs Michel, 'La directive 1999/44/CE sur la garantie des biens de consommation: un remède efficace contre l'obsolescence programmée' [2016] 2 *Revue européenne de droit de la consommation* 207, 208; Anaïs Michel, 'Premature Obsolescence, in Search of an Improved Legal Framework' (PhD Law thesis, Katholieke Universiteit Leuven 2022).

6 Anaïs Michel, 'L'obsolescence prématurée de nos produits de consommation: Un débat à remodeler' [2021] 3 *Le Climatoscope* 20, 21.

7 Céline Senay, 'La lutte contre l'obsolescence programmée une législation bientôt en vue ? Une étude comparée avec le régime français' [2021] 46 *R.D.T.I.*

8 Anaïs Michel, 'Premature Obsolescence, in Search of an Improved Legal Framework' (PhD Law thesis, Katholieke Universiteit Leuven 2022) 43; Samuel Becher and Anne-Lise Sibony, 'Confronting Product Obsolescence' [2021] 27 *Columbia Journal of European Law* 97, 102.

9 Anaïs Michel, 'Premature Obsolescence, in Search of an Improved Legal Framework' (PhD Law thesis, Katholieke Universiteit Leuven 2022) 44.

10 Anaïs Michel, 'Premature Obsolescence, in Search of an Improved Legal Framework' (PhD Law thesis, Katholieke Universiteit Leuven 2022) 46.

11 Anaïs Michel, 'Premature Obsolescence, in Search of an Improved Legal Framework' (PhD Law thesis, Katholieke Universiteit Leuven 2022) 51.

12 Anaïs Michel, 'Premature Obsolescence, in Search of an Improved Legal Framework' (PhD Law thesis, Katholieke Universiteit Leuven 2022) 52, 98.

13 See on this classification, Anaïs Michel, 'Premature Obsolescence, in Search of an Improved Legal Framework' (PhD Law thesis, Katholieke Universiteit Leuven 2022) 52–64.

14 Anaïs Michel, 'L'obsolescence prématurée de nos produits de consommation: Un débat à remodeler' [2021] 3 *Le Climatoscope* 20, 21.

15 Anaïs Michel, 'Premature Obsolescence, in Search of an Improved Legal Framework' (PhD Law thesis, Katholieke Universiteit Leuven 2022) 54.

16 Céline Senay, 'La lutte contre l'obsolescence programmée une législation bientôt en vue? Une étude comparée avec le régime français' [2021] 46 *R.D.T.I* 9.

17 Anaïs Michel, 'L'obsolescence prématurée de nos produits de consommation: Un débat à remodeler' [2021] 3 *Le Climatoscope* 20, 21.

18 Anaïs Michel, 'L'obsolescence prématurée de nos produits de consommation: Un débat à remodeler' [2021] 3 *Le Climatoscope* 20, 21–23.

19 Code of economic law [2013] Belgian Official Journal 29 March 2013, 19975; Act of 21 December 2013 concerning the implementation of Book VI 'Market practices and consumer protection' in the Code of Economic Law and concerning the implementation of the definitions specific for Book VI and the provisions for the enforcement, specific for Book VI in Books I [2013] Belgian Official Journal 30 December 2013, 103506.

20 'Business', 'company', or 'enterprise' is used with the same meaning in this chapter.

21 Denis Philippe and Anne Jonlet, 'L'Europe s'attaque à l'obsolescence programmée!' *La tribune* (Brussels, June 2022).

22 Directive 2009/125/EC of the European Parliament and of the Council of 21 October 2009 establishing a framework for the setting of ecodesign requirements for energy-related products ('Ecodesign Directive') [2009] OJ L285/172.

23 European Parliament, 'Resolution of 31 May 2018 on the implementation of the Ecodesign Directive (2009/125/EC) (2017/2087(INI))' [2020] OJ C76/192, recital U; Liesbeth Van Acker, *The New Eco-Design Package: An Important Step towards a Circular Economy* (Nijmegen, Ars Aequi libri 2020) 797.

24 Ecodesign Directive, art. 2. The Ecodesign Directive clarifies that it also covers the parts that are meant to be incorporated into energy-related products (e.g., ventilation units in electric motors).

25 See Olivier Mouton, 'L'obsolescence prématurée à l'épreuve du droit à la réparabilité européen' (PhD Law thesis, Katholieke Universiteit Leuven 2022) 39.
26 Commission, 'Commission Staff Working Document accompanying the Proposal for a Directive of the European Parliament and of the Council establishing a framework for the setting of eco-design requirements for energy of eco-design requirements for energy-related products' COM(2008) 399 final 15.
27 Olivier Mouton, 'L'obsolescence prématurée à l'épreuve du droit à la réparabilité européen' (PhD Law thesis, Katholieke Universiteit Leuven 2022) 45.
28 See also Energy Transition Act. Loi de transition énergétique pour la croissance verte. See, for example, Commission Regulation (EU) 2019/2023 of 1 October 2019 laying down ecodesign requirements for household washing machines and household washer-dryers pursuant to Directive 2009/125/EC of the European Parliament and of the Council, amending Commission Regulation (EC) No 1275/2008 and repealing Commission Regulation (EU) No 1015/2010 [2019] OJ L315/285.
29 Art. 99 de la loi n° 2015-992 du 17 août 2015 relative à la transition énergétique pour la croissance verte.
30 Namely, the scanner and the ability to print in both colour and black.
31 *La plainte contre Epson déposée par HOP va-t-elle devenir obsolète?* <https://www.halteobsolescence.org/la-plainte-contre-epson-deposee-par-hop-va-t-elle-devenir-obsolete/> accessed 5 July 2023.
32 Direction générale de la concurrence, de la consommation et de la répression des fraudes.
33 *Nouvelle plainte de HOP contre Apple pour obsolescence programmée et entraves à la réparation* <https://www.halteobsolescence.org/nouvelle-plainte-de-hop-contre-apple-pour-obsolescence-programmee-et-entraves-a-la-reparation/> accessed 2 July 2023.
34 *Action européenne coordonnée contre la panne du Joy-Con Drift* <https://www.quechoisir.org/action-ufc-que-choisir-nintendo-switch-action-europeenne-coordonnee-contre-la-panne-du-joy-con-drift-n87699/> accessed 2 July 2023; Kevin Dachez, 'Nintendo Switch: comment faire réparer gratuitement vos Joy-Con' *Phonandroid* (25 May 2022) accessed 2 July 2023.
35 *Action européenne coordonnée contre la panne du Joy-Con Drift* <https://www.quechoisir.org/action-ufc-que-choisir-nintendo-switch-action-europeenne-coordonnee-contre-la-panne-du-joy-con-drift-n87699/> accessed 2 July 2023.
36 L. n° 2020-105, February 10, 2020, art. 16 and 130; D. n° 2020-1757, December 29, 2020, on the reparability index for electrical and electronic equipment, NOR: TRED2023604DL. nr. 2020-105, 10 February 2020, art. 16 et 130; D. n° 2020-1757, 29 December 2020, relatif à l'indice de réparabilité des équipements électriques et électroniques, NOR: TRED2023604D.
37 See Anaïs Michel, 'Premature Obsolescence, in Search of an Improved Legal Framework' (PhD Law thesis, Katholieke Universiteit Leuven 2022) 296, 552–562.
38 See, for a critical opinion on this index, Olivier Mouton, 'L'obsolescence prématurée à l'épreuve du droit à la réparabilité européen' (PhD Law thesis, Katholieke Universiteit Leuven 2022) 53.
39 *Indice de réparabilité* <https://www.ecologie.gouv.fr/indice-reparabilite> accessed 5 July 2023. See Anaïs Michel, 'Premature Obsolescence, in Search of an Improved Legal Framework' (PhD Law thesis, Katholieke Universiteit Leuven 2022) 297.
40 Autorità Garante della Concorrenza e del Mercato, 'Provvedimento n. 27365 dd.' (25 September 2018) 176–226.
41 Autorità Garante della Concurrenza e del Mercato, 'Provvedimento n. 27363 dd.' (25 September 2018).
42 See De Standaard, 12 October 2020, 19.
43 See also *Test Achats* <https//www.test-achats.be/plainte/entreprises/apple/100000177> accessed 15 October 2022.
44 See *Apple hit with new lawsuit over alleged iPhone 'planned obsolescence', this time in Portugal* <https://9to5mac.com/2021/03/01/apple-lawsuit-portugal-planned-obsolescence/> accessed 3 July 2023.
45 Eléonore Maitre-Ekern and Carl Dalhammar, 'Regulating Planned Obsolescence: A Review of Legal Approaches to Increase Product Durability and Reparability in Europe', [2016] 25(3) *Review of European Community & International Environmental Law* 378, 379 (https://doi.org/10.1111/reel.12182).
46 See Anaïs Michel, 'Premature Obsolescence, in Search of an Improved Legal Framework' (PhD Law thesis, Katholieke Universiteit Leuven 2022) 42.
47 *Test Achats* <https://www.test-achats.be/hightech/imprimantes/news/hp-mis-en-demeure> accessed 4 July 2023.
48 *A European Green Deal. Striving to be the First Climate-Neutral Continent* <https://commission.europa.eu/strategy-and-policy/priorities-2019-2024/european-green-deal_en> accessed 2 July 2023.

49 Anaïs Michel, 'Premature Obsolescence, in Search of an Improved Legal Framework' (PhD Law thesis, Katholieke Universiteit Leuven 2022) 34.

50 Commission, 'Commission Staff Working Document, Guidance on the implementation/application of Directive 2005/29/EC on Unfair Commercial Practices, accompanying the document communication from the Commission to the European Parliament, the Council, the European Economic and Social Committee and the Committee of the Regions. A Comprehensive Approach to stimulating cross-border e-Commerce for Europe's citizens and businesses' COM(2016) 320 final).

51 The rule was transposed in Belgian law on the 8 May 2022 Moniteur belge, 2 June 2022 and in Luxembourgish law on 8 December 2021 Journal Officiel, 9 December 2021.

52 Art. 5 to 8 of the Directive 2011/83/EU of the European Parliament and of the Council of 25 October 2011 on consumer rights, amending Council Directive 93/13/EEC and Directive 1999/44/EC of the European Parliament and of the Council and repealing Council Directive 85/577/EEC and Directive 97/7/EC of the European Parliament and of the Council [2011] OJ L304/64.

53 Anaïs Michel, 'La directive 1999/44/CE sur la garantie des biens de consommation: un remède efficace contre l'obsolescence programmée' [2016] 2 *Revue européenne de droit de la consommation* 207, 233–234.

54 See also Final Compromise amendments on the Draft Report on the Commission, 'Proposal for a directive of the European Parliament and of the Council on amending Directives 2005/29/EC and 2011/83/EU as regards empowering consumers for the green transition through better protection against unfair practices and better information' COM(2022) 143.

55 See art. 4 of the Proposal for a Directive of the European Parliament and of the Council on substantiation and communication of explicit environmental claims [2023] COM(2023)166 final.

56 See on the right to repair, Samuel Becher and Anne-Lise Sibony, 'Confronting Product Obsolescence' [2021] 27 *Columbia Journal of European Law* 97, 126.

57 Directive (EU) 2019/771 of the European Parliament and of the Council of 20 May 2019 on certain aspects concerning contracts for the sale of goods, amending Regulation (EU) 2017/2394 and Directive 2009/22/EC, and repealing Directive 1999/44/EC [2019] OJ L136/28.

58 It was transposed in Luxemburg by the Law of 8 December 2021, Journal Officiel, 2021.

59 Olivier Mouton, 'L'obsolescence prématurée à l'épreuve du droit à la réparabilité européen' (PhD Law thesis, Katholieke Universiteit Leuven 2022) 60.

60 Hervé Jacquemin and Pauline Limbrée, 'La garantie légale applicable aux biens, aux contenus numériques et aux services numériques' [2022] 2 *R.D.T.I.* 51.

61 Consumer Protection Act, 38/1978, Sections 12 and 15; Olivier Mouton, 'L'obsolescence prématurée à l'épreuve du droit à la réparabilité européen' (PhD Law thesis, Katholieke Universiteit Leuven 2022) 63.

62 European Parliament, 'European Parliament resolution of 7 April 2022 on the right to repair' 2022/2515 (RSP).

63 See the resolutions mentioned in the recitals of the resolution of 7 April 2022.

64 Anaïs Michel, 'Premature Obsolescence, in Search of an Improved Legal Framework' (PhD Law thesis, Katholieke Universiteit Leuven 2022) 370.

65 European Parliament, 'European Parliament resolution of 7 April 2022 on the right to repair' 2022/2515 (RSP).

66 European Parliament, 'European Parliament resolution of 7 April 2022 on the right to repair' 2022/2515 (RSP).

67 Proposal for a Directive of the European Parliament and of the Council on common rules promoting the repair of goods and amending Regulation (EU) 2017/2394, Directives (EU) 2019/771 and (EU) 2020/1828 [2023] COM(2023) 155 final.

68 Art. 7 of the Proposal.

69 Art. 4 of the Proposal.

70 *La nouvelle proposition européenne sur le droit à la réparation n'est pas suffisante* <https://www.halteobsolescence.org/la-nouvelle-proposition-europeenne-sur-le-droit-a-la-reparation-nest-pas-suffisante/> accessed 15 April 2023.

71 Proposal for a Regulation of the European Parliament and of the Council establishing a Framework for setting Ecodesign Requirements for Sustainable Products and Repealing Directive 2009/125/EC [2022] COM(2022) 142 final.

72 This proposal is based on Art. 114 of the TFEU, which gives competence to the Union in environmental matters.

73 Commission, 'Briefing EU legislation in progress: ecodesign for sustainable products' <https://www.europarl.europa.eu/RegData/etudes/BRIE/2022/733524/EPRS_BRI(2022)733524_EN.pdf> 5–6.
74 Art. 1(2) of the SPR proposal.
75 Art. 9 of the SPR proposal. See Olivier Mouton, 'L'obsolescence prématurée à l'épreuve du droit à la réparabilité européen' (PhD Law thesis, Katholieke Universiteit Leuven 2022) 50.
76 Art. 20 of the SPR proposal.
77 Adèle Chasson, Laetitia Vasseur and Alice Papillon, 'Durable and repairable products: 20 steps to a sustainable Europe. HOP's public policy guide to end premature obsolescence in the European Union' (2020) HOP White Paper <https://www.halteobsolescence.org/wp-content/uploads/2020/11/Livre-Blanc-europeen.pdf> accessed 1 September 2022. Art 20 of the SPR proposal.
78 Art. 29 of the SPR proposal.
79 Commission, 'Briefing EU legislation in progress: ecodesign for sustainable products' <https://www.europarl.europa.eu/RegData/etudes/BRIE/2022/733524/EPRS_BRI(2022)733524_EN.pdf>, 5–6.
80 Valentina Romano, 'EU ministers challenge Commission's 'green products' regulation' *Euractiv* (4 October 2022).
81 See<https://commission.europa.eu/energy-climate-change-environment/standards-tools-and-labels/products-labelling-rules-and-requirements/energy-label-and-ecodesign/about_en> accessed 5 July 2023.
82 See Regulation (EC) No 66/2010 of the European Parliament and of the Council of 25 November 2009 on the EU Ecolabel [2010] OJ L27/184.
83 See Anaïs Michel, 'Premature Obsolescence, in Search of an Improved Legal Framework' (PhD Law thesis, Katholieke Universiteit Leuven 2022) 222. See Regulation (EU) 2017/1369 of the European Parliament and of the Council of 4 July 2017 setting a framework for energy labelling and repealing Directive 2010/30/EU [2017] OJ L198/1.
84 Anaïs Michel, 'Premature Obsolescence, in Search of an Improved Legal Framework' (PhD Law thesis, Katholieke Universiteit Leuven 2022) nr. 570.
85 See Anaïs Michel, 'Premature Obsolescence, in Search of an Improved Legal Framework' (PhD Law thesis, Katholieke Universiteit Leuven 2022) nr. 571.
86 Council Regulation (EEC) No 880/92 of 23 March 1992 on a Community eco-label award scheme [1992] OJ L99/1 Regulation (EU) 2017/1369 of the European Parliament and of the Council of 4 July 2017 setting a framework for energy labelling and repealing Directive 2010/30/EU [2017] OJ L198/1.
87 See Anaïs Michel, 'Premature Obsolescence, in Search of an Improved Legal Framework' (PhD Law thesis, Katholieke Universiteit Leuven 2022) nrs 583 and 584.
88 Criteria are developed to assess the optimal lifetime of a product. See Anaïs Michel, 'Premature Obsolescence, in Search of an Improved Legal Framework' (PhD Law thesis, Katholieke Universiteit Leuven 2022) nr 93.
89 Natacha Sauphanor-Brouillaud, 'Le consommateur, levier de la transition écologique' in Aude-Solveig Epstein and Marie Nioche (dirs), *Le droit économique, levier de la transition écologique?* (Brussels, Bruylant 2022) 137.
90 Natacha Sauphanor-Brouillaud, 'Le consommateur, levier de la transition écologique' in Aude-Solveig Epstein and Marie Nioche (dirs), *Le droit économique, levier de la transition écologique?* (Brussels, Bruylant 2022) 137.
91 See, Olivier Mouton, 'L'obsolescence prématurée à l'épreuve du droit à la réparabilité européen' (PhD Law thesis, Katholieke Universiteit Leuven 2022); see, on the French regulation, Anaïs Michel, 'Premature Obsolescence, in Search of an Improved Legal Framework' (PhD Law thesis, Katholieke Universiteit Leuven 2022) 252.

Bibliography

Becher S and Sibony A-L, 'Confronting Product Obsolescence' [2021] 27 *Columbia Journal of European Law* 97

Chasson A, Vasseur L and Papillon A, 'Durable and Repairable Products: 20 Steps to a Sustainable Europe. HOP's Public Policy Guide to End Premature Obsolescence in the European Union' (2020) HOP White Paper <https://www.halteobsolescence.org/wp-content/uploads/2020/11/Livre-Blanc-europeen.pdf>

Dachez K, 'Nintendo Switch: comment faire réparer gratuitement vos Joy-Con' *Phonandroid* (25 May 2022) <https://www.phonandroid.com/nintendo-switch-comment-faire-reparer-gratuitement-vos-joy-con.html> last accessed 25 February 2024

Dalhammar C, 'The Application of 'Life Cycle Thinking' in European Environmental Law: Theory and Practice' [2015] 12(2) *Journal for European Environmental & Planning Law* 97 (https://doi.org/10.1163/18760104-01202002)

Jacquemin H and Limbrée P, 'La garantie légale applicable aux biens, aux contenus numériques et aux services numériques' [2022] 2 *R.D.T.I.* 51

London B, 'Ending the Depression through Planned Obsolescence' (*Wikimedia*, 1932) <https://upload.wikimedia.org/wikipedia/commons/2/27/London_%281932%29_Ending_the_depression_through_planned_obsolescence.pdf>

Maitre-Ekern E and Dalhammar C, 'Regulating Planned Obsolescence: A Review of Legal Approaches to Increase Product Durability and Reparability in Europe' [2016] 25(3) *Review of European Community & International Environmental Law* 378 (https://doi.org/10.1111/reel.12182)

Michel A, 'La directive 1999/44/CE sur la garantie des biens de consommation: un remède efficace contre l'obsolescence programmée' [2016] 2 *Revue européenne de droit de la consommation* 207

Michel A, 'L'obsolescence prématurée de nos produits de consommation: Un débat à remodeler' [2021] 3 *Le Climatoscope* 20

Michel A, 'Premature Obsolescence, In Search of an Improved Legal Framework' (PhD Law thesis, Katholieke Universiteit Leuven 2022)

Mouton O, 'L'obsolescence prématurée à l'épreuve du droit à la réparabilité européen' (PhD Law thesis, Katholieke Universiteit Leuven 2022)

Philippe D and Jonlet A, 'L'Europe s'attaque à l'obsolescence programmée!', *La tribune* (Brussels, June 2022) <https://latribune.avocats.be/fr/l-europe-s-attaque-a-l-obsolescence-programmee> last accessed 25 February 2024

Sauphanor-Brouillaud N, Le consommateur, levier de la transition écologique' in Aude-Solveig Epstein and Marie Nioche (dirs), *Le droit économique, levier de la transition écologique?* (Brussels, Bruylant 2022) 137

Schaub C, 'La vie gâchée des objets' *Libération* (Paris, 28 October 2012) <https://www.liberation.fr/futurs/2012/10/28/la-vie-gachee-des-objets_856572/>

Senay C, 'La lutte contre l'obsolescence programmée une législation bientôt en vue? Une étude comparée avec le régime français' [2021] 46 (4) *R.D.T.I.* 5

Slade G, *Made to Break: Technology and Obsolescence in America* (Harvard, Harvard University Press 2009)

Van Acker L, *The New Eco-Design Package: An Important Step towards a Circular Economy* (Nijmegen, Ars Aequi libri 2020)

26

LEVERAGING REGULATORY MEASURES TO ENHANCE THE SUSTAINABILITY POTENTIAL OF SERVITISATION IN THE EU

Janja Hojnik[1]

26.1 Introduction

Servitisation has been very generously accepted in the domain of environmental studies. It is broadly acknowledged that since present trends in economic and population growth continue, the natural environment is increasingly being stressed.[2] Servitisation business models could have an important impact in this respect, even though it is equally important not to oversimplify the relationship between servitisation and sustainability.

For these reasons, servitisation and a similar concept of Product Service Systems became a popular subject for researchers engaged with sustainability and business from the mid-1990s onwards. Sustainability researchers argued that if one were to focus on the service, rather than the product, it would become much easier to design systems with radically lower impacts.[3] In product-oriented business models, firms have the incentive to maximise the number of products sold as the principal method of boosting turnover, increasing market share and generating profits. In service-oriented business models, scholars claimed, firms have an incentive to prolong the service life of products to ensure they are used as intensively as possible.[4] Authors argued that servitisation could play a major role in a more circular economy in the future and result in resource savings because if products are provided per use rather than sold, vendors and manufacturers will have an interest in prolonging the life of the product for as long as possible to get the maximum use from it.[5] Legal researchers started observing servitisation only after their colleagues in other research fields, such as business studies, IT or ecology, but this topic has now increased focus.[6]

This chapter examines how regulatory institutions, in particular at the EU level, encourage the potential of servitisation business models for ecologically sustainable growth.[7] The chapter finds that while there are a number of legally binding and soft law frameworks, mechanisms and initiatives related to sustainability, their impact on the servitised industry remains to be fully discovered. It is submitted that in the future, it will be important to move from incentives for resource efficiency and recycling towards mandatory standards, as increasingly pursued by the 2020 European Green Deal[8] and the subsequent legal instruments.

DOI: 10.4324/9781032662046-31

26.2 Servitisation and the New EU Industrial Impetus

Right at the outbreak of the COVID epidemic in Europe, on 10 March 2020, the European Commission adopted the EU Industrial Strategy[9] to lead the twin transitions towards climate neutrality and digital leadership. In 2021 it was updated to include some COVID lessons.[10] The Strategy outlines three drivers for industrial transformation: global competition, climate neutrality, and a digital future. The EU Industrial Strategy reflects a new paradigm for the European Union, i.e., the European industrial sovereignty, namely in its strategic autonomy in key technologies and access to raw materials.

The European industry needed a new strategy considering that manufacturing in developed economies is under massive pressure.[11] The story of deindustrialisation of developed economies started in the 1950s and the value added by manufacturing as a percentage of GDP is now less than 15% in most OECD countries.[12] Levitt's statement of 1972 that 'everybody is in service'[13] is thus becoming progressively true. Nevertheless, the financial crisis of 2008 and the recession that followed have led many people and companies to seek alternative sources of employment and income and have headed the developed economies in general to recognise the danger of over-reliance on financial services. Supply cuts during the pandemic have further convinced European leaders that it is important to bring the industry back to Europe. Finally, economic recovery after every big crisis proved more difficult in countries with a weak industrial sector. Europe thus needs to 'rebalance' the economy, with a particular focus on manufacturing.[14] It is believed that 'an industrial renaissance' or 'reindustrialisation' can bring jobs and growth back to Europe.[15]

To survive in developed economies providing products alone is, however, insufficient and it is being suggested that manufacturing companies have to move up the value chain, innovating and creating ever more sophisticated products and services, so they do not have to compete on the basis of cost alone.[16] Despite the conventional separation between manufacturing and services, it is increasingly realistic for manufacturers to offer services; in fact, they increasingly base entire competitive strategies on service innovation.[17] Additionally, as the complexity and variety of business activities grow and digitalisation is spreading, the boundary between services and manufacturing becomes increasingly elusive.[18] To describe these processes, the term '*servitization*' has burgeoned, reflecting the fact that we are in the midst of a redefinition of what is manufacturing.

Vandermerwe and Rada describe that companies initially considered themselves to be in goods *or* services and then moved to offer goods *combined* with closely related services, and finally to a position where companies offer 'bundles' consisting of customer-focused combinations of goods *and* services.[19] Other business literature explains that traditionally the tendency has been for managers to view services as unavoidable in the context of marketing strategies – here, the main part of total value creation was considered to stem from physical goods, and services were assumed purely as an add-on to products.[20] From this beginning, it is now believed that a dramatic change has occurred in the way services are produced and marketed by manufacturing companies. The provision of services has now turned into 'a conscious and explicit strategy' with services becoming a main differentiating factor in a totally integrated products and service offering.[21] In this situation, the services are considered as fundamental value-added activities and the product to be just a part of the offering.[22] This continuum from traditional manufacturer where companies merely offer services as an add-on to their products, through to service providers where companies have services as the main part of their value creation is called 'product-service continuum'[23] and is often used as a starting point for the categorisation of different types of servitisation in management research.[24]

26.3 A Myriad of Different Servitisation Business Models

Servitisation is nowadays considered omnipresent in manufacturing companies in developed economies.[25] While some (traditional) services support the supplier's product (e.g., maintenance), other (more advanced) services support the client's action in relation to the supplier's product (e.g., IT consulting). While the former ensures the proper functioning of the product, the latter enables the supplier to explore how services support particular clients' initiatives and advance the operation of the customer's company.[26] Accordingly, the most commonly provided service in practice is still the delivery of products, followed by the provision of spare parts and consumables, a customer helpline or support desk, and product or systems training.[27]

Moreover, servitisation transactions may include renting cars instead of buying them; contracting services of irrigation instead of acquiring irrigation systems; or securing server capacity instead of procuring computers.[28] One of the archetypal examples of a successful strategy of mixing the supply of goods and services is Rolls-Royce. The Economist reported in 2009 that

> instead of selling airlines first engines and then parts and service, Rolls-Royce has convinced its customers to pay a fee for every hour that an engine runs. Rolls-Royce in turn promises to maintain it and replace it if it breaks down. They aren't selling engines; they are selling hot air out the back of an engine. (…) It is sometimes necessary to be good at making things to sell the services connected with them. At Rolls-Royce it is difficult to see where one begins and the other ends.[29]

In addition to this model of renting aeroplanes (called *Power-by-the-Hour*), Rolls Royce has also adopted sensors that are able to monitor 24/7 the aeroplane engine status (*TotalCare* programme) that is considerably simplifying the maintenance process.[30] Rolls Royce has thus succeeded in transition from being a pure manufacturer to being an integrated solutions provider.[31] Moreover, Volvo is essentially manufacturing cars, but they are today in the entire scale of activities associated with the automobile transportation from insurance to gas stations and roadside assistance networks.[32]

Although servitisation often involves large multinationals supporting high-value capital equipment, there is a rising number of cases where servitisation is not limited to expensive capital goods. For example, AB Electrolux installs a washing machine in a customer's home, maintains and repairs it when necessary and charges customers by the laundry load.[33] It is now also common for car dealers to offer loans or leasing services to their customers, so they are not just selling products but also services – much the same as Amazon is now offering easier and more comfortable services around books and IBM has become, with the growth of its software and consultancy, from a manufacturer of computer hardware to a service company.[34]

The servitisation philosophy is also the basis of access-based consumption, where instead of buying and owning things, consumers want access to goods and prefer to pay for the experience of temporarily accessing them. Ownership is no longer the ultimate expression of consumer desire.[35] In this way consumers are able to access goods which they could otherwise not afford or which they choose not to own due to concerns such as space limitations or the environment, thereby paying for usage rather than ownership.[36] While public access to goods, such as books in public libraries or public transportation, has been known for centuries, the Internet has fuelled new business models of access-based consumption.[37] Indeed, we have seen a proliferation of consumption models in which access is enabled through sharing or pooling of resources/products/services redefined through technology and peer communities. Examples of access models vary from car- or bike-sharing programs (Zipcar, Hubway) to online borrowing programs for DVDs, bags, fashion, or jewellery (Netflix, Bag Borrow or Steal, Rent the Runway, Borrowed Bling).[38] Moreover, companies such as Uber and Lyft connect car owners and those in need of transport over an online platform (i.e., ridesharing). Internet platform services for ridesharing or home sharing have caused great turmoil across sectors and it is being envisioned that what is ahead is 'a shift in the dominant business model, one in which all consumer goods will be available as a service and all consumer services will be available on demand'.[39]

From a sustainability point of view, it needs to be pointed out that an important concept among the servitisation substitutes is the 'product-service system' (PSS). PSS is a concept that originated in the north of Europe at the end of the 1990s that describes companies offering solutions aimed at increasing market share and consumer satisfaction, but also with a parallel objective of reducing the consumption of products through alternative scenarios of product use instead of acquiring it.[40] The first scholarly authority on PSS is Goedkoop, who defined it in 1999 as 'a combination of products and services in a system that provides functionality for consumers and reduces environmental impact'.[41] Correspondingly, Mont highlights that PSS offers a system of integrated products and services that are intended to reduce the environmental impact through alternative scenarios of product use.[42] The PSS embodies a transition from 'well-being based on the product' to 'well-being based on access to the product'.[43] The difference between servitisation and PSS thus arises in the motivation of the process, considering that the PSS is closely coupled to the debates on sustainability and emphasises the use of the function of a product without necessarily owning it with the aim of reducing the environmental impact.[44] While Pigosso et al. argue that a PSS can only be considered a 'true' PSS if it was designed to minimise environmental and social impacts throughout its whole cycle,[45] most authors today adopt a broader approach to PSS, where PSS is seen as a synonym of servitisation[46] and it does not necessarily reduce environmental impact. Although servitisation and PSS have emerged from differing perspectives on the world, they are converging towards a common conclusion that manufacturing companies should be focusing on selling integrated solutions.[47]

26.4 Sustainable Servitisation of European Industry

Political prioritisation of circular economy is enhancing those new business models that can offer interesting solutions to the consumers, while at the same time offering ecological modernisation of the economy. Albeit traditional non-digital services, such as aftersales repair, remain important, digitalisation of products and services increases effectiveness of the business. Entrepreneurs and consumers alike are looking for digital solutions. If they offer greater sustainability, we are dealing with a winning combination that helps preserve natural resources.

26.4.1 Servitisation That Bolsters Ecological Modernisation

In the domain of environmental law it is broadly acknowledged that since present trends in economic and population growth continue, the natural environment is increasingly being stressed.[48] As already evident from the previous sections, it is one of the cornerstones of servitisation that in addition to strengthening alliances among the parties involved in the business it may also aim to bolster sustainability.[49] It can provide an incentive for suppliers to increase the durability of the products by changing their business models and redesigning products. Practical examples include business models such as bike and car sharing, ride sharing, systems for renting cars, aircrafts, machines, or irrigation systems instead of buying them. Consequently, servitisation has a strong potential to reduce environmental impacts associated with linear production-consumption systems based on product sales, ownership, and disposal models and thereby enabling businesses to compete in a relatively regulated market.[50]

Circularity and servitisation are also seen as very correlated by the Council of the EU. In this respect, its Recommendation on climate neutrality of 2022 sets forth:

> Circular value retention activities, including repair, re-use, remanufacturing and servitisation business models, can foster affordable and sustainable access to goods and services. They also generate jobs and opportunities at various skill levels, including for women, persons with disabilities and groups in vulnerable situations, through social economy entities active in those areas. By promoting product circularity, they reduce carbon emissions vastly, while the jobs created are in proximity to the products that need to be maintained, refurbished, or shared.[51]

Servitisation could thus play a major role in a more circular economy in the future and result in resource savings because if products are provided per use rather than sold, the vendors and manufacturers have an interest in prolonging the life of the product as long as possible to get the maximum use from it.[52] However, consumers are provided with the economic incentive to use products in a more efficient way[53] and manufacturers adopting the servitisation strategy may increase consumer's interest in environmental issues.[54]

Moreover, many sustainable solutions improving the quality of the global environment are offered by the digitising industry, where the expansion of smart objects enables the introduction of smart cities with smart infrastructure, where sensors observe continuing safety of a bridge by monitoring data on vibration and pressure; smart public lighting that can allegedly cut a city's energy costs up to 80%; smart waste management with sensors on garbage collection; or even smart farms with smart irrigation (that supposed to lead to 60% decrease of water consumption).[55] Private customer-oriented solutions include smart homes with energy-efficient ovens, thermostats and refrigerators that adjust power consumption to cheaper rates of electricity during the night.[56] This way, servitisation can help minimise the consumption of scarce resources and environmental degradation.[57]

26.4.2 What If Environmental Improvement of Servitisation Is Marginal Only?

Notwithstanding optimistic expectations by a long line of researchers, some scholars recently realised that the correlation between servitisation and dematerialisation is considerably more complex than it appeared at first. As pointed out by Tukker in a review of eight different types of Product Service Systems (PSS), the majority will result in marginal environmental improvements at best, while some (such as product lease) could even lead to less responsible user behaviour and hence to increased environmental impacts, because PSS makes users less responsible for careful use of the product.[58]

While Vasques and Ono found that services for shared usage of washing machines and dryers among neighbours seem to be better accepted when they promote convenience and comfort for low price, instead of taking care of the environment,[59] Möhlmann, Moeller, and Wittkowski even found no effect of environmentalism on preferring renting instead of owning good, when surveying accommodation, car sharing and online peer-to-peer network.[60] It is moreover hard to see, why is Uber that is performing transportation services using diesel machine cars more environmentally sustainable than for example conventional taxis running on bio-gas. Consequently, there is no irrefutable evidence regarding the link between environmental motivations and participation in the sharing economy. There are also increasingly problematic social sustainability dilemmas related to this business model that concern the danger of broad precarisation of the labour market.[61] The services economy, lease economy and sharing economy are thus not per se environmentally sustainable.

26.4.3 Regulatory Incentives for Sustainable Servitisation

Notwithstanding all the above considerations, servitisation is still primarily economically motivated and its sustainability does not come automatically. Services providers and consumers alike will act sustainably only if regulation supports such orientation of servitisation business models. The Internet enables advanced services. Yet, regulation is needed for this advancement to be both environmentally and socially sustainable. Servitisation can provide an incentive for suppliers to increase the durability of products by changing their business models and redesigning products.[62]

Consequently, it is important for policymakers and regulators to promote these positive aspects of servitisation – be it on a voluntary or mandatory basis. This is more so, considering that the shift from products to services cannot be assumed to be eco-efficient and there are a number of potential rebound effects.[63] Particularly powerful players in the market may thus oppose servitisation solutions because their specific competitive advantage lies in mass production.[64] Policymakers have essentially two forms of enhancing sustainable manufacturing.

On the one hand, there are direct policy instruments addressed to the manufacturers, who are encouraged through regulatory or other policy instruments to incorporate mechanisms of sustainability into their production and service system. On the other hand, however, there are more indirect instruments that are addressed to the consumer to achieve the aim of sustainable consumption and thus a reduction in the usage of scarce natural resources. This refers to binding and non-binding rules, practices, and awareness-raising with the purpose of reducing the use of natural resources and toxic materials as well as the releases of waste over the life cycle of the service or product that consumers are purchasing and using, which then influence the conduct of manufacturers.[65]

When framing the EU response to the servitisation trend the EU institutions first of all need to respect the EU multi-level governance system, thereby respecting the competences of various levels of governance.[66] EU industrial policy is horizontal in nature and aims at securing framework conditions favourable to industrial competitiveness.[67] It falls among those policies, where the EU has the competence to carry out actions to support the actions of the Member States (Article 6 TFEU). Consequently, it is the Member States that are the holders of their respective industrial policies, thereby adopting corresponding national strategies for the digitalisation of manufacturing.[68] The EU's response, revealed in the Commission's action plan on the digitising industry,[69] is in this respect focused on coordination between national and EU-level initiatives and in developing policy actions, such as investments in digital innovations and infrastructure, accelerating the development of ICT standards, exploring regulatory conditions and adaptation of the workforce, including up-skilling. Moreover, EU industrial policy is well integrated into a number of other EU policies such as those relating to competition, consumers, trade, the internal market, research and innovation, employment, environmental protection and public health. This interleaving between industry and other policies logically affects the servitisation trend as well. The majority of these fields are, however, not within the exclusive competence of the EU institutions and even though regulation at the EU level is in most instances crucial so as to prevent a myriad of different national approaches that would create chaos and partition of the internal market, authorities at the national and local levels, in line with the principles of subsidiarity and proportionality, need to get involved. Moreover, other constitutional and institutional settings must be respected in the regulatory process, including industry involvement and self-regulation in line with the 'New Approach' so that the market itself defines the technical solutions while public authorities only set the general regulatory requirements.[70]

Considering these constitutional restrictions, the EU regulatory response progresses from various soft law documents towards binding and generally applicable EU legislation. Servitisation is consistent with the goals defined in the EU Commission's Communication entitled 'The Roadmap to a Resource Efficient Europe'[71] which outlines how we can transform Europe's economy into a sustainable one by 2050. It proposes ways to increase resource productivity and decouple economic growth from resource use and its environmental impact. Although the vision is an ambitious one the practical tools to directly steer this development are currently fairly limited.[72] While there is no specific policy focus on supporting servitisation by the EU Commission, several EU strategies are targeting the aims of sustainable consumption and production as well as resource and energy efficiency,[73] such as the 2020 Circular Economy Action Plan[74] and the 2021 update of the EU Industrial Strategy.[75] In this respect, servitisation is occasionally mentioned. In the New Legislative Framework (NLF) of 2022,[76] the Commission admits the wide-spread occurrence of servitisation business models but states that the NLF does not define new aspects in the value chain, such as the increased role of services within products and what this means from a regulatory perspective at the point in time of placing on the market. In contrast, the NLF focuses on ensuring product safety in respect of post-market placement, as services are typically provided once the product is in operation.

The service business model is concretely visible in the recent Energy Efficiency Directive,[77] which calls for a common framework that

> should give energy utilities the option of offering energy services to all final customers, not only to those to whom they sell energy. This increases competition in the energy market because energy utilities can differentiate their product by providing complementary energy services.[78]

Energy services that include a variety of activities, such as energy analysis and audits, energy management, maintenance and operation, monitoring and evaluation of savings,[79] are thus a response to the constantly increasing criticality of energy-related issues that lead to the situation, where a mere fuelling is no longer considered as equivalent to energy supply.[80] The EU Energy Efficiency Directive may hence be perceived as direct instrument in support of servitisation and it is expected that this 'support' will further increase under the recent Commission's proposal to recast this Directive, as part of the 'Fit for 55' package.[81]

Additionally, the EU has a broad policy on waste management. This policy is based on the so-called *principle of extended producer responsibility (EPR)* requiring manufacturers of products containing toxic or environmentally unsustainable materials[82] to take responsibility for management throughout key parts of their lifecycle, especially for management of post-consumer waste.[83] According to Thomas Lindhqvist, who first introduced EPR,[84] it refers to

> an environmental protection strategy to reach an environmental objective of a decreased total environmental impact from a product, by making the manufacturer of the product responsible for the entire life-cycle of the product and especially for the take-back, recycling and final disposal of the product.[85]

The concept thus implies that responsibilities, which were traditionally assigned to consumers and authorities responsible for waste management, are to be shifted to the producer of the products and are thus consistent with the polluter pays principle. The environmental benefits deriving from the widespread application of EPR potentially include more efficient use of resources, cleaner products and technologies, more efficient manufacturing, increased recycling and greener consumption.[86] The Packaging Directive[87] provides in the preamble that

> it is essential that all those involved in the production, use, import and distribution of packaging and packaged products become more aware of the extent to which packaging becomes waste, and that in accordance with the polluter-pays principle they accept responsibility for such waste.

The recent proposal of 30 November 2022 for a Packaging Regulation[88] that will replace and upgrade the current Packaging Directive further strengthens this by aiming to harmonise monitoring and reporting obligations, including producer reporting obligations under EPR schemes.[89]

Additionally, the Directive on Waste Electric Equipment (WEEE Directive)[90] provides for the creation of collection schemes where consumers return their WEEE free of charge, however, 'in order to give maximum effect to the concept of producer responsibility', each producer is responsible for financing the management of the waste from his own products.[91] The current Batteries Directive[92] is providing that financing schemes for the management of waste batteries and accumulators should 'help to achieve high collection and recycling rates and to give effect to the principle of producer responsibility'.[93] It is expected that the recently proposed Batteries Regulation of October 2020[94] that should repeal the Directive will further increase EPR schemes and bring the playing field for EPR schemes for EV and industrial batteries.

Moreover, the Euratom Radioactive Waste Directive[95] provides in *Article 7* that 'Member States shall ensure that the prime responsibility for the safety of spent fuel and radioactive waste management facilities and/or activities rest with the licence holder'. Although critics say that the implementation of EPR is not always successful in practice[96] or that the collection targets stipulated by the directives are not always as high as they could be,[97] the directives encourage manufacturers to add corresponding services to the products. Although they are offering recycling services due to the legal requirements for waste handling, these recycling services have become for many companies a source of income that has increased customer loyalties by taking care of waste for customers.[98]

The central EU legislation in the field of chemicals is the so-called REACH Regulation,[99] which aims to improve the protection of human health and the environment through the better and earlier identification of the intrinsic properties of chemical substances. This is done by the four processes of REACH, namely the registration, evaluation, authorisation, and restriction of chemicals. REACH also aims to enhance innovation and competitiveness of the EU chemicals industry. The REACH Regulation places responsibility on the industry to manage the risks from chemicals and to provide safety information on the substances. Manufacturers and importers are required to gather information on the properties of their chemical substances, which will allow their safe handling, and to register the information in a central database in the European Chemicals Agency (ECHA) in Helsinki. The Regulation also calls for the progressive substitution of the most dangerous chemicals (referred to as 'substances of very high concern') when suitable alternatives have been identified.[100]

Moreover, the Eco-design Directive is setting eco-design requirements for energy-using products.[101] The proposal for a *new Ecodesign for Sustainable Products Regulation*,[102] published on 30

March 2022, builds upon this directive, while not being limited to energy-related products. The proposed regulation is called 'the cornerstone of the Commission's approach to more environmentally sustainable and circular products' and establishes a framework to set ecodesign requirements for specific product groups to significantly improve their circularity, energy performance and other environmental sustainability aspects. The Commission has stated it would start legislating for high-impact industries with gaps in sustainability rules, such as textiles, furniture, mattresses, tyres, detergents, paints, and lubricants, as well as iron, steel, and aluminium.[103] In this respect, e.g., the recent *EU Strategy for Sustainable and Circular Textile*,[104] specifically addresses the need to implement extended producer responsibility in the field of fast fashion, by boosting reuse and recycling of textile waste. Moreover, it is worth noting at this point, that improved recycling efficiencies and higher material recovery for packaging materials, as aimed by the mentioned Proposal for a Regulation on packaging and packaging waste also include textiles.[105]

26.5 Conclusion

Europe has done a long road from the times when people were hiring pineapples to the times when many Europeans own a smartphone and robots are used in factories. As found by the EU High-Level Group on Business Services servitisation 'offers opportunities to revolutionize the manufacturing sector'.[106] It is expected that in the future billions or even trillions of everyday physical things will be enhanced by a small electronic device to provide local intelligence and connectivity to the Internet, thus bridging the gap between the physical world and the information world.[107]

The European Commission has made a multidimensional response to this economic development by adopting or announcing to adopt a series of action plans, guidelines or legislative proposals related to servitisation. While some commentators claim that the EU may not adopt the legislation concerning digitising industry too fast, since it is not yet known, how exactly smart industry will develop, others call for immediate response to avoid distinct legislative activities by the Member States, thereby putting the EU in an inferior position towards other global superpowers, such as US and China that have also developed their own manufacturing strategies. Servitisation in many aspects makes sense and it is thus reasonable that it gets regulatory support; however, this does not mean that it is always necessary to rush into new regulation when amending existing legislation would suffice.[108] At the same time, it is also important that regulators leave enough space for dynamism while ensuring fair competition and developing the framework conditions prudently.[109]

Business models that are covered by servitisation, PSS and sharing economy are already considered an economic megatrend. They are increasingly overwhelming and to align them with goals of environmental sustainability could have significant effects for the state of the limited natural resources. EU Commission's response to systematic challenges in this respect (bringing together industrial, ecological, social, public health and other challenges) is thorough and realistic.

When connecting servitisation and sustainability EU regulators have so far particularly focused on extending product responsibility. This way EU is trying to promote environmental improvements of products and product systems, thereby implying that responsibilities, which were traditionally assigned to consumers and authorities responsible for waste management, are to be shifted to the producer of the products, so as to make the manufacturer of the product responsible for the entire life-cycle of the product and especially for the take-back, recycling and final disposal of the product.[110] Thus, for manufacturers, this means an obligation of pollution prevention applied across the product lifecycle, extending beyond the factory gates.[111] Although this policy does not mean differentiation and image (i.e., to make the purchase of a product attractive to the customer), this increasingly comes as a side-effect, considering that extension of products will often constitute physical products as well as the associated accessories or services, like maintenance, engineering, and software.[112] European test will in this respect be in particular on the ability of the Member States to agree on the environmental perspective of their industries and economy and not to use environmental standards as a basis for regulatory competition.

Notes

1 EU Law Professor, University of Maribor, Slovenia.

2 Oksana Mont, 'Editorial for the Special Issue of the Journal of Cleaner Production on Product Service Systems' [2003] 11 *Journal of Cleaner Production* 815 (https://doi.org/10.1016/S0959-6526(02)00163-4). Bocken and colleagues developed an approach, through the categorisation of eight business model archetypes, that might contribute to building up the business model for sustainability: Nancy MP Bocken et al, 'A Literature and Practice Review to Develop Sustainable Business Model Archetypes' [2014] 65 *Journal of Cleaner Production* 42 (https://doi.org/10.1016/j.jclepro.2013.11.039).

3 See, e.g., Arnold Tukker, 'Product Services for a Resource-Efficient and Circular Economy: A Review' [2015] 97 *Journal of Cleaner Production* 76 (https://doi.org/10.1016/j.jclepro.2013.11.049).

4 Walter R Stahel, 'Product Life as a Variable: The Notion of Utilization' [1986] 13 *Science and Public Policy* 185 (https://doi.org/10.1093/spp/13.4.185); Friedrich Schmidt-Bleek, *Wieviel Umwelt Braucht Der Mensch?: MIPS – das Mass Für ökologisches Wirtschaften* (Basel, Springer-Verlag 2013).

5 Frank Tietze and Erik Hansen, 'To Own or to Use? How Product Service Systems Facilitate Eco-Innovation Behavior' (2013) Centre for Technology Management working paper series 4 <https://www.ifm.eng.cam.ac.uk/uploads/Working_paper/16_04_Tietze_Hansen.pdf> accessed 28 September 2023.

6 See, e.g., Bert Keirsbilck and Evelyne Terryn (eds), *Circular Economy and Consumer Protection* (Cambridge, Intersentia 2019); Bert Keirsbilck et al, *Servitization and Circular Economy: Economic and Legal Challenges* (Cambridge, Intersentia 2023).

7 Due to the limited space in this chapter please also refer to my previous work on this topic, Janja Hojnik, 'The Servitization of Industry: EU Law Implications and Challenges' [2016] 53 *Common Market Law Review* 1575 (https://doi.org/10.54648/cola2016143); Janja Hojnik, 'Ecological Modernization through Servitization: EU Regulatory Support for Sustainable Product–Service Systems' [2018] 27(2) *Review of European, Comparative & International Environmental Law* 162 (https://doi.org/10.1111/reel.12228).

8 Commission, 'European Green Deal' COM(2019) 640 final.

9 Commission, 'Identifying and addressing barriers to the Single Market' COM(2020) 93 final.

10 2021 update of the Commission, 'EU Industrial Strategy' COM(2021) 350 final.

11 Andy Neely, 'The Servitization of Manufacturing: An Analysis of Global Trends' (14th European Operations Management Association Conference, Ankara, January 2007); Lena Johansson, 'Servicification of Swedish Manufacturing' [2010] *The National Board of Trade* <https://www.kommerskollegium.se/globalassets/publikationer/rapporter/2016-och-aldre/report-2010-1-servicification-of-swedish-manufacturing.pdf>.

12 Matthieu Crozet and Emmanuel Milet, 'The Servitization of French Manufacturing Firms' (2014) CEPII Working Paper 2014-10 <http://www.cepii.fr/pdf_pub/wp/2014/wp2014-10.pdf>. Gintare Kemekliene et al, 'Services Employment in Europe, Now and in the Future' (Background paper for UNI-Europa Conference, Athens, April 2007); Ronald Schettkat and Lara Yocarini, 'The Shift to Services Employment: A Review of the Literature' [2006] 17(2) *Structural Change and Economic Dynamics* 127 (https://doi.org/10.1016/j.strueco.2005.04.002).

13 Theodore Levitt, 'Production-Line Approach to Service' [1972] 50 *Harvard Business Review* 20. See also Victor R Fuchs, *The Growing Importance of the Service Industries* (New York, National Bureau of Economic Research 1965).

14 Finbarr Livesay, 'The Need for a New Understanding of Manufacturing and Industrial Policy in Leading Economies' [2012] 7(3) *Innovations: Technology, Governance, Globalization* 193. Bruce Tether and Elif Bascavusoglu-Moreau, 'Servitization: The Extent and Motivations for Service Provision amongst UK Manufacturers' (2012) IMP Innovation, Strategy and Sustainability Working Paper <https://research.manchester.ac.uk/en/publications/servitization-the-extent-and-motivations-for-service-provision-am>; Finbarr Livesay, 'The Need for a New Understanding of Manufacturing and Industrial Policy in Leading Economies' [2012] 7(3) *Innovations: Technology, Governance, Globalization* 193. Howard Lightfoot,

Tim Baines and Palie Smart, 'The Servitization of Manufacturing: A Systematic Literature Review of Interdependent Trends' [2013] 33(11/12) *International Journal of Operations & Production Management* 1408 (https://doi.org/10.1108/IJOPM-07-2010-0196); Louis Brennan et al, 'Manufacturing in the World: Where Next?' [2015] 35(9) *International Journal of Operations & Production Management* 1253 (https://doi.org/10.1108/IJOPM-03-2015-0135). See also Commission, 'Towards Knowledge Driven Reindustrialisation, European Competitiveness Report 2013' SWD(2013) 347 final, 3.

15 David Bailey and Karl Aiginger, 'The High Road: Europe Must Seize the Potential of Advanced Manufacturing' (*EurActiv*, 26 January 2016) <https://www.euractiv.com/section/innovation-industry/opinion/the-high-road-europe-must-seize-the-potential-of-advanced-manufacturing/> accessed 28 September 2023; Elżbieta Bieńkowska, 'Reindustrialisation of Europe: Industry 4.0 – Innovation, Growth and Jobs, Speech' (The Reindustrialisation of Europe: Industry 4.0 – Innovation, jobs and growth. A high-level one-day conference, Brussels, June 2015).

16 Michael E Porter and Christian HM Ketels, 'UK Competitiveness: Moving to the Next Stage' (2003) DTI Economics Paper n. 3 <https://www.hbs.edu/ris/Publication%20Files/file14771_83b42e5a-7e88-49be-9d33-2fc7585a87d9.pdf> accessed 28 September 2023; Andy Neely, 'Exploring the Financial Consequences of the Servitization of Manufacturing' [2009] 1 *Operations Management Research* 103 (https://doi.org/10.1007/s12063-009-0015-5).

17 Tim Baines, 'Exploring Service Innovation and the Servitization of the Manufacturing Firm' [2015] 58 *Research Technology Management* 9 (https://doi.org/10.5437/08956308X5805002).

18 Matthieu Crozet and Emmanuel Milet, 'The Servitization of French Manufacturing Firms' (2014) CEPII Working Paper 2014-10 <http://www.cepii.fr/pdf_pub/wp/2014/wp2014-10.pdf> accessed 28 September 2023. Gintare Kemekliene et al, 'Services Employment in Europe, Now and in the Future' (Background paper for UNI-Europa Conference, Athens, April 2007); Joel Goldhar and Daniel Berg, 'Blurring the Boundary: Convergence of Factory and Service Processes' [2010] 21(3) *Journal of Manufacturing Technology Management* 341 (https://doi.org/10.1108/17410381011024322).

19 Sandra Vandermerwe and Juan Rada, 'Servitization of Business: Adding Value by Adding Services' [1988] 6(4) *European Management Journal* 314 (https://doi.org/10.1016/0263-2373(88)90033-3).

20 Heiko Gebauer and Thomas Friedli, 'Behavioral Implications of the Transition Process from Products to Services' [2005] 20(2) *Journal of Business & Industrial Marketing* 70 (https://doi.org/10.1108/08858620510583669).

21 Tim S Baines et al, 'The Servitization of Manufacturing: A Review of Literature and Reflection on Future Challenges' [2009] 20(5) *Journal of Manufacturing Technology Management* 547, 556 (https://doi.org/10.1108/17410380910960984).

22 Rogelio Oliva and Robert Kallenberg, 'Managing the Transition from Products to Services' [2003] 14(2) *International Journal of Service Industry Management* 160 (https://doi.org/10.1108/09564230310474138).

23 Rogelio Oliva and Robert Kallenberg, 'Managing the Transition from Products to Services' [2003] 14(2) *International Journal of Service Industry Management* 160 (https://doi.org/10.1108/09564230310474138) Heiko Gebauer and Thomas Friedli, 'Behavioral Implications of the Transition Process from Products to Services' [2005] 20(2) *Journal of Business & Industrial Marketing* 70 (https://doi.org/10.1108/08858620510583669); Heiko Gebauer, 'Identifying Service Strategies in Product Manufacturing Companies by Exploring Environment–Strategy Configurations' [2008] 37(3) *Industrial Marketing Management* 278 (https://doi.org/10.1016/j.indmarman.2007.05.018).

24 Accordingly, Tukker has developed a model of three main forms of business models: product-oriented services, use-oriented services and result-oriented services – Arnold Tukker, 'Eight Types of Product–Service System: Eight Ways to Sustainability? Experiences from SusProNet' [2004] 13(4) *Business Strategy and the Environment* 246 (https://doi.org/10.1002/bse.414). See also Laura Smith, Roger Maull and Irene C.L. Ng, 'Servitization and Operations Management: A Service Dominant-Logic Approach' [2014] 34(2) *International Journal of Operations & Production Management* 242 (https://doi.org/10.1108/IJOPM-02-2011-0053).

25 David Opresnik and Marco Taisch, 'The Value of Big Data in Servitization' [2015] 165 *International Journal of Production Economics* 174, 174 (https://doi.org/10.1016/j.ijpe.2014.12.036).

26 Valérie Mathieu, 'Product Services: From a Service Supporting the Product to a Service Supporting the Client' [2001] 16(1) *Journal of Business & Industrial Marketing* 39, 40 (https://doi.org/10.1108/08858620110364873).

27 Bruce Tether and Elif Bascavusoglu-Moreau, 'Servitization: The Extent and Motivations for Service Provision amongst UK Manufacturers' (2012) IMP Innovation, Strategy and Sustainability Working

Paper <https://research.manchester.ac.uk/en/publications/servitization-the-extent-and-motivations-for-service-provision-am>, 17.

28 Commission, 'Communication from the Commission to the European Parliament, the Council, the European Economic and Social Committee and the Committee of the Regions. Innovation for a sustainable future – The Eco-innovation Action Plan (Eco-AP)' COM(2011) 899 final (*Eco-innovation Action Plan*).

29 Editor's Note, 'Britain's Lonely High-Flier' (*The Economist*, 8 January 2009) <https://www.economist.com/briefing/2009/01/08/britains-lonely-high-flier> accessed 28 September 2023.

30 See Marco Ardolino, Nicola Saccani and Marco Perona, 'The Impact of Digital Technologies and Ecosystems on the Servitization of Companies: A Preliminary Analysis' (Spring Servitization Conference, Aston, May 2015), 51–58.

31 Charlotta Windahl and Nicolette Lakemond, 'Integrated Solutions from a Service-Centered Perspective: Applicability and Limitations in the Capital Goods Industry' [2010] 39(8) *Industrial Marketing Management* 1278 (https://doi.org/10.1016/j.indmarman.2010.03.001); Tim Brady, Andrew Davies and David M Gann, 'Creating Value by Delivering Integrated Solutions' [2005] 23(5) *International Journal of Project Management* 360 (https://doi.org/10.1016/j.ijproman.2005.01.001). Similarly, Hoyer is offering trucks provided by MAN on a pay-as-you-go basis, Xerox developed advanced document management solutions and Alstom so-called Train-Life services - Tim Baines, 'Exploring Service Innovation and the Servitization of the Manufacturing Firm' [2015] 58 *Research Technology Management* 9 (https://doi.org/10.5437/08956308X5805002).

32 Sandra Vandermerwe and Juan Rada, 'Servitization of Business: Adding Value by Adding Services' [1988] 6(4) *European Management Journal* (https://doi.org/10.1016/0263-2373(88)90033-3). More in Björn Remneland Wikhamn, Jan Ljungberg and Alexander Styhre, 'Enacting Hard and Soft Product Offerings in Mature Industries: Moving towards Servitisation in Volvo' [2013] 17(4) *International Journal of Innovation Management* 1 (https://doi.org/10.1142/S136391961350014X).

33 Michael W Toffel, 'Contracting for Servicizing' (2008) Harvard Business School Technology & Operations Mgt. Unit Research Paper 08-063 <https://papers.ssrn.com/sol3/papers.cfm?abstract_id=1090237> accessed 28 September 2023.

34 Louis V Gerstner Jr, *Who Says Elephants Can't Dance?: Leading an Enterprise through Dramatic Change* (New York, Harper Business 2002). For a number of other examples across various industry sectors see Gunter Lay, *Servitization in Industry* (Cham, Springer 2014).

35 Jeffrey F. Durgee and Gina Colarelli O'Connor, 'An Exploration into Renting as Consumption Behavior' [1995] 12(2) *Psychology and Marketing* 89 (https://doi.org/10.1002/mar.4220120202); Christopher Lovelock and Evert Gummesson, 'Whither Services Marketing? In Search of a New Paradigm and Fresh Perspectives' [2004] 7(1) *Journal of Service Research* 20 (https://doi.org/10.1177/1094670504266131).

36 Fleura Bardhi and Giana M. Eckhardt, 'Access-Based Consumption: The Case of Car Sharing' [2012] 39(4) *Journal of Consumer Research* 881 (https://doi.org/10.1086/666376); Walker Smith, 'The Uber-All Economy of the Future' [2016] 20 *Independent Review* 383, 385.

37 It coincided with the global economic crisis during which consumers reconsider their values and spending habits as well as with urbanisation and high population concentration that create a 'critical mass' of supply and demand and facilitate better matches – see Vassilis Hatzopoulos and Sofia Roma, 'Caring for Sharing? The Collaborative Economy under EU Law' [2017] 54(1) *Common Market Law Review* 81 (https://doi.org/10.54648/cola2017004).

38 Fleura Bardhi and Giana M. Eckhardt, 'Access-Based Consumption: The Case of Car Sharing' [2012] 39(4) *Journal of Consumer Research* 881 (https://doi.org/10.1086/666376); Walker Smith, 'The Uber-All Economy of the Future' [2016] 20 *Independent Review* 383, 385.

39 Walker Smith, 'The Uber-All Economy of the Future' [2016] 20 *Independent Review* 383, 385.

40 Fernanda Hänsch Beuren, Marcelo Gitirana Gomes Ferreira and Paulo A. Cauchick Miguel, 'Product-Service Systems: A Literature Review on Integrated Products and Services' [2013] 47 *Journal of Cleaner Production* 222 (https://doi.org/10.1016/j.jclepro.2012.12.028); Alessandro Annarelli, Cinzia Battistella and Fabio Nonino, 'Product Service System: A Conceptual Framework from a Systematic Review' [2016] 139 *Journal of Cleaner Production* 1011 (https://doi.org/10.1016/j.jclepro.2016.08.061).

41 Mark J Goedkoop et al, 'Product Service Systems, Ecological and Economic Basics' (1999) Ministry of Housing, Spatial Planning and the Environment, Communications Directorate 1999-36 <https://scirp.org/reference/referencespapers.aspx?referenceid=1520975> accessed 28 September 2023.

42 Oksana Mont, 'Clarifying the Concept of Product–Service System' [2002] 10(3) *Journal of Cleaner Production* 237 (https://doi.org/10.1016/S0959-6526(01)00039-7). See also Ezio Manzini and Carlo Vezzoli,

'Product-Service Systems and Sustainability: Opportunities for Sustainable Solutions' (2002) UNEP <https://wedocs.unep.org/bitstream/handle/20.500.11822/8123/-Product-Service%20Systems%20and%20Sustainability_%20Opportunities%20for%20Sustainable%20Solutions-20021192.pdf?sequence=2&isAllowed=y> accessed 28 September 2023.

43 Ezio Manzini, Carlo Vezzoli and Garrette Clark, 'Product-Service Systems. Using an Existing Concept as a New Approach to Sustainability' [2001] 1(2) *Journal of Design Research* (https://doi.org/10.1504/JDR.2001.009811).

44 Fernanda Hänsch Beuren, Marcelo Gitirana Gomes Ferreira and Paulo A. Cauchick Miguel, 'Product-Service Systems: A Literature Review on Integrated Products and Services' [2013] 47 *Journal of Cleaner Production* 224 (https://doi.org/10.1016/j.jclepro.2012.12.028); Arnold Tukker, 'Eight Types of Product–Service System: Eight Ways to Sustainability? Experiences from SusProNet' [2004] 13(4) *Business Strategy and the Environment* (https://doi.org/10.1002/bse.414); Arnold Tukker and Ursula Tischner, *New Business for Old Europe: Product-Service Development, Competitiveness and Sustainability* (Sheffield, Greenleaf 2006); Tim S Baines et al, 'The Servitization of Manufacturing: A Review of Literature and Reflection on Future Challenges' [2009] 20(5) *Journal of Manufacturing Technology Management* (https://doi.org/10.1108/17410380910960984)

45 Daniela C A Pigosso et al, 'Is the Industrial Product-Service System Really Sustainable?' in [editors], *Proceedings of the 2nd CIRP IPS2 Conference 2010* (Linköping, Linköping University Electronic Press 2012), 59–65.

46 Tim S Baines et al, 'State-of-the-Art in Product-Service Systems' [2007] 221(10) *Proceedings of the Institution of Mechanical Engineers, Part B: Journal of Engineering Manufacture* 1543 (https://doi.org/10.1243/09544054JEM858).

47 Tim S Baines et al, 'The Servitization of Manufacturing: A Review of Literature and Reflection on Future Challenges' [2009] 20(5) *Journal of Manufacturing Technology Management* (https://doi.org/10.1108/17410380910960984); Arnold Tukker and Ursula Tischner, *New Business for Old Europe: Product-Service Development, Competitiveness and Sustainability* (Sheffield, Greenleaf 2006).

48 Oksana Mont, 'Editorial for the Special Issue of the Journal of Cleaner Production on Product Service Systems' [2003] 11 *Journal of Cleaner Production* (https://doi.org/10.1016/S0959-6526(02)00163-4).

49 Craig A Hart, *Climate Change and the Private Sector: Scaling Up Private Sector Response to Climate Change* (London, Routledge 2013); Fernanda Hänsch Beuren, Marcelo Gitirana Gomes Ferreira and Paulo A. Cauchick Miguel, 'Product-Service Systems: A Literature Review on Integrated Products and Services' [2013] 47 *Journal of Cleaner Production* 225 (https://doi.org/10.1016/j.jclepro.2012.12.028).

50 Andrius Plepys, Eva Heiskanen and Oksana Mont, 'European Policy Approaches to Promote Servicizing' [2015] 97 *Journal of Cleaner Production* 117, 118 (https://doi.org/10.1016/j.jclepro.2014.04.029).

51 Council, 'Recommendation of 16 June 2022 on ensuring a fair transition towards climate neutrality' OJ 2022/C 243/04, para. 12.

52 Commission, 'Communication from the Commission to the European Parliament, the Council, the European Economic and Social Committee and the Committee of the Regions. Innovation for a sustainable future – The Eco-innovation Action Plan (Eco-AP)' COM(2011) 899 final (*Eco-innovation Action Plan*); Frank Tietze and Erik Hansen, 'To Own or to Use? How Product Service Systems Facilitate Eco-Innovation Behavior' (2013) Centre for Technology Management working paper series 4 <https://www.ifm.eng.cam.ac.uk/uploads/Working_paper/16_04_Tietze_Hansen.pdf>.

53 Commission, 'European Green Deal', COM(2019) 640 final. See also Commission, 'Green Deal: New proposals to make sustainable products the norm and boost Europe's resource independence' (*European Commission*, 30 March 2022) <https://ec.europa.eu/commission/presscorner/detail/en/ip_22_2013> accessed 28 September 2023.

54 Ezio Manzini, Carlo Vezzoli and Garrette Clark, 'Product-Service Systems. Using an Existing Concept as a New Approach to Sustainability' [2001] 1(2) *Journal of Design Research* (https://doi.org/10.1504/JDR.2001.009811).

55 Michael Miller, *The Internet of Things: How Smart TVs, Smart Cars, Smart Homes, and Smart Cities Are Changing the World* (Indianapolis, IN, Que 2015), 263–280.

56 Michael Miller, *The Internet of Things: How Smart TVs, Smart Cars, Smart Homes, and Smart Cities Are Changing the World* (Indianapolis, IN, Que 2015), 263–280. Additionally, Miller (at p. 290) mentions CO_2 sensors that monitor automobile emissions, pollution from factories and even toxic gases generated on farms; water sensors that monitor water quality in oceans, rivers, and lakes, and determine whether water is suitable for fish and plant life; radiation sensors that monitor radiation levels etc.

57 Tim S Baines et al, 'State-of-the-Art in Product-Service Systems' [2007] 221(10) *Proceedings of the Institution of Mechanical Engineers, Part B: Journal of Engineering Manufacture* 1543 (https://doi.org/10.1243/09544054JEM858).

58 Arnold Tukker, 'Eight Types of Product–Service System: Eight Ways to Sustainability? Experiences from SusProNet' [2004] 13(4) *Business Strategy and the Environment* 246 (https://doi.org/10.1002/bse.414).

59 Rosana Aparecida Vasques and Maristela Mitsuko Ono, 'When Sharing Is (Almost and/or Possibly) Better Than Owning: A Case Study on a "Full Service" Collective Laundry' [2016] 14 *Revista GestãoOrganizacional* 97 (http://dx.doi.org/10.21714/1679-18272016v14Esp.p97-105).

60 Mareike Möhlmann, 'Collaborative Consumption: Determinants of Satisfaction and the Likelihood of Using a Sharing Economy Option Again' [2015] 14(3) *Journal of Consumer Behaviour* 193 (https://doi.org/10.1002/cb.1512); Sabine Moeller and Kristina Wittkowski, 'The Burdens of Ownership: Reasons for Preferring Renting' [2010] 20(2) *Managing Service Quality: An International Journal* 176 (https://doi.org/10.1108/09604521011027598).

61 Cristiano Codagnone, Fabienne Abadie and Federico Biagi, 'The Future of Work in the Sharing Economy: Market Efficiency and Equitable Opportunities or Unfair Precarisation?' [2016] *JCR Science for Policy Report. European Commission* <https://publications.jrc.ec.europa.eu/repository/handle/JRC101280> accessed 28 September 2023.

62 Andrius Plepys, Eva Heiskanen and Oksana Mont, 'European Policy Approaches to Promote Servicizing' [2015] 97 *Journal of Cleaner Production* 117, 118 (https://doi.org/10.1016/j.jclepro.2014.04.029).

63 Matteo Bartolomeo et al, 'Eco-Efficient Producer Services—What Are They, How Do They Benefit Customers and the Environment and How Likely Are They to Develop and Be Extensively Utilised?' [2003] 11(8) *Journal of Cleaner Production* 829 (https://doi.org/10.1016/S0959-6526(02)00157-9).

64 P Andrius Plepys, Eva Heiskanen and Oksana Mont, 'European Policy Approaches to Promote Servicizing' [2015] 97 *Journal of Cleaner Production* 117, 119 (https://doi.org/10.1016/j.jclepro.2014.04.029); Kai Hockerts and Rolf Wüstenhagen, 'Greening Goliaths versus Emerging Davids—Theorizing about the Role of Incumbents and New Entrants in Sustainable Entrepreneurship' [2010] 25(5) *Journal of Business Venturing* 481 (https://doi.org/10.1016/j.jbusvent.2009.07.005).

65 More on this in Lucia A Reisch and John Thøgersen, *Handbook of Research on Sustainable Consumption* (Cheltenham Glos, Edward Elgar Publishing 2017).

66 Gary Marks, Liesbet Hooghe and Kermit Blank, 'European Integration from the 1980s: State-Centric v. Multi-Level Governance' [1996] 34(3) *JCMS: Journal of Common Market Studies* 341 (https://doi.org/10.1111/j.1468-5965.1996.tb00577.x).

67 Commission, 'Industrial Policy in an Enlarged Europe' COM(2002) 714 final.

68 In Germany the development of manufacturers equipping their products and machines with intelligent digital systems is referred to as '*Industrie 4.0*', the French term for the same is '*Industrie du Futur*', in the Netherlands and in Slovakia strategists talk about '*Smart Industry*', in the UK about *'Catapult'* (High Value Manufacturing), in Spain about '*Industria Conectada 4.0*', and in Italy, it is referred to as '*Fabbrica Intelligente*'. More in Bauernhansl, 'Industry 4.0: Challenges and Limitations in the Production. Keynote' [2013] ATKearney, Factory of the year; Anne Dujin, Cornelia Geissler and Dirk Horstkötter, 'INDUSTRy 4.0: The New Industrial Revolution' [2014] Roland Berger Strategy Consultants, Munich <https://bdi.eu/media/themenfelder/digitalisierung/downloads/20151117_Digitalisierte_Wirtschaft_Industrie_40_Gutachten_der_Noerr_LLP.pdf>.

69 Commission, 'Digitising European Industry Reaping the full benefits of a Digital Single Market' COM(2016) 180 final. This communication is accompanied by three further communications, i.e., on a European Cloud Initiative, on Priorities for ICT Standardisation, the E-Government action plan, and the Staff Working Document on the Internet of Things.

70 See Commission, 'A vision for the internal market for industrial product' COM(2014) 25 final, 5 and Commission, 'Staff Working Document, Evaluation of the New Legislative Framework' SWD/2022/0364 final. See also Thomas Klindt, Susanne Wende and Steffen Burrer, 'Standardisierung und Produktkonformität' [2015] *Digitalisierte Wirtschaft/Industrie 4.0* 100, 106; Rolf H Weber and Romana Weber, *Internet of Things: Legal Perspectives* (Zürich, Springer Science & Business Media 2010) 23.

71 Commission, 'Roadmap to a Resource Efficient Europe' COM(2011) 571 final.

72 Susanne Fischer et al, 'Leasing Society' (2012) *Directorate General for Internal Policies* <https://www.europarl.europa.eu/RegData/etudes/etudes/join/2012/492460/IPOL-ENVI_ET(2012)492460_EN.pdf> accessed 28 September 2023.

73 Andrius Plepys, Eva Heiskanen and Oksana Mont, 'European Policy Approaches to Promote Servicizing' [2015] 97 *Journal of Cleaner Production* 117, 118 (https://doi.org/10.1016/j.jclepro.2014.04.029).
74 Commission, 'A new Circular Economy Action Plan' COM(2020) 98 final.
75 Commission, 'Updating the 2020 New Industrial Strategy: Building a stronger Single Market for Europe's recovery' COM(2021) 350 final.
76 Commission, 'Staff Working Document, Evaluation of the New Legislative Framework' SWD/2022/0364 final.
77 Directive 2012/27/EU on energy efficiency [2012] OJ L315/1.
78 Directive 2012/27/EU on energy efficiency [2012] OJ L315/1, preamble, rec. 20. Additionally, the Directive obliges the Member States to set up national energy efficiency obligation schemes for energy utilities or other alternative policy measures that achieve the same amount of energy savings.
79 Paolo Bertoldi, Silvia Rezessy and Edward Vine, 'Energy Service Companies in European Countries: Current Status and a Strategy to Foster Their Development' [2006] 34(14) *Energy Policy* 1818 (https://doi.org/10.1016/j.enpol.2005.01.010). More on energy services in: Thorsten Helms, 'Asset Transformation and the Challenges to Servitize a Utility Business Model' [2015] 91 *Energy Policy* 98 (https://doi.org/10.1016/j.enpol.2015.12.046); Miriam Benedetti et al, 'A Proposal for Energy Services' Classification Including a Product Service Systems Perspective' [2015] 30 *Procedia CIRP* 251 (https://doi.org/10.1016/j.procir.2015.02.121); Yudi Fernando and Sofri Yahya, 'Challenges in Implementing Renewable Energy Supply Chain in Service Economy Era' [2015] 4 *Procedia Manufacturing* 454 (https://doi.org/10.1016/j.promfg.2015.11.062).
80 Miriam Benedetti et al, 'A Proposal for Energy Services' Classification Including a Product Service Systems Perspective' [2015] 30 *Procedia CIRP* 251, 252 (https://doi.org/10.1016/j.procir.2015.02.121). For public procurement aspects of energy services see Colin Nolden, Steve Sorrell and Friedemann Polzin, 'Innovative Procurement Frameworks for Energy Performance Contracting in the UK Public Sector' (2015) SPRU Working Paper Series 2015-31 <https://www.sussex.ac.uk/webteam/gateway/file.php?name=2015-31-swps-nolden-et-al.pdf&site=25> accessed 28 September 2023.
81 Proposal for a Directive on energy efficiency (recast), COM(2021) 558.
82 E.g., paint, batteries, beverage containers, pesticide containers, electronics, packaging, cell phones, sharps, radioactive devices, motor oils, plastic bags, and smoke detectors.
83 Clifton Curtis et al, 'Extended Producer Responsibility and Product Stewardship for Tobacco Product Waste' [2014] 4(3) *International Journal of Waste Resources* 2, 2 (https://doi.org/10.4172%2F2252-5211.1000157).
84 Thomas Lindhqvist, *Extended Producer Responsibility in Cleaner Production: Policy Principle to Promote Environmental Improvements of Product Systems*, vol 2000 (Lund, Lund University 2000) ii.
85 The definition was published in English for the first time in: Thomas Lindhqvist, *Extended Producer Responsibility in Cleaner Production: Policy Principle to Promote Environmental Improvements of Product Systems*, vol 2000 (Lund, Lund University 2000).
86 Allen L White, Mark Stoughton and Linda Feng, 'Servicizing: The Quiet Transition to Extended Product Responsibility' [1999] *US Environmental Protection Agency* 20 <https://www.tellus.org/pub/Servicizing%20-%20The%20Quiet%20Transition%20to%20Extended%20Product%20Responsibility.pdf> accessed 28 September 2023. See also Dorothy Maxwell and Rita Van der Vorst, 'Developing Sustainable Products and Services' [2003] 11(8) *Journal of Cleaner Production* 883 (https://doi.org/10.1016/S0959-6526(02)00164-6).
87 Directive 94/62/EC of 20 December 1994 on packaging and packaging waste [1994] OJ L365/10.
88 Proposal for a Regulation on Packaging and packaging waste, COM(2022) 677.
89 Proposal for a Regulation on Packaging and packaging waste, COM(2022) 677, 3.
90 Directive 2002/96/EC on waste electrical and electronic equipment (WEEE) [2003] OJ L37/24.
91 Directive 2002/96/EC on waste electrical and electronic equipment (WEEE) [2003] OJ L37/24, rec. 20. More on this in: Meike Altvater and Christina Brandmann, 'Extended Producer Responsibility: The EU WEEE Directive Goes Global – Strict Law and Order Required or Self-Regulating Market Power a Promising Alternative?' [2012] *Electronics Goes Green 2012+, ECG 2012 – Joint International Conference and Exhibition, Proceedings* <https://search.ricest.ac.ir/dl/search/defaultta.aspx?DTC=49&DC=578711>; Grit Walther et al, 'Implementation of the WEEE-Directive — Economic Effects and Improvement Potentials for Reuse and Recycling in Germany' [2010] 47 *International Journal of Advanced Manufacturing Technology* 461 (https://doi.org/10.1007/s00170-009-2243-0). In autumn 2022 the Commission started the evaluation of the progress made under the Directive – see here <https://ec.europa.eu/info/law/

better-regulation/have-your-say/initiatives/13420-Waste-from-electrical-and-electronic-equipment-evaluating-the-EU-rules_en> last accessed 7 July 2023.

92 Directive 2006/66/EC on batteries and accumulators and waste batteries and accumulators [2006] OJ L266/1.

93 Directive 2006/66/EC on batteries and accumulators and waste batteries and accumulators [2006] OJ L266/1, rec. 19. More on this in Thomas Lindhqvist, 'Policies for Waste Batteries' [2010] 14(4) *Journal of Industrial Ecology* 537 (https://doi.org/10.1111/j.1530-9290.2010.00266.x).

94 Proposal for a Regulation concerning batteries and waste batteries, COM(2020) 798 final.

95 Council Directive 2011/70/Euratom establishing a Community framework for the responsible and safe management of spent fuel and radioactive waste [2011] OJ L199/48.

96 Kieren Mayers and others, 'Implementing Individual Producer Responsibility for Waste Electrical and Electronic Equipment through Improved Financing' [2013] 17(2) *Journal of Industrial Ecology* 186 (https://doi.org/10.1111/j.1530-9290.2012.00528.x); Noah Sachs, 'Planning the Funeral at the Birth: Extended Producer Responsibility in the European Union and the United States' [2006] 30 *Harvard Environmental Law Review* 51.

97 Thomas Lindhqvist, *Extended Producer Responsibility in Cleaner Production: Policy Principle to Promote Environmental Improvements of Product Systems*, vol 2000 (Lund, Lund University 2000).

98 Emilie Aner and Magnus Rentzhog, 'Everybody Is in Services: Servicification and Trade Policy' (*Servicification – Why free trade in services matter to manufacturing*, ECIPE, 25 April 2013) <https://dokumen.tips/documents/-everybody-is-in-services-servicification-and-trade-policy.html?page=1>

99 Regulation (EC) No 1907/2006 of 18 December 2006 concerning the Registration, Evaluation, Authorisation and Restriction of Chemicals (REACH) [2006] OJ L396/1.

100 See more in: John Goodier, 'EU Chemicals Regulation: New Governance, Hybridity and REACH' [2016] 30(2) *Reference Reviews* 35 (https://www.doi.org/10.1108/RR-11-2015-0259).

101 Directive 2005/32/EC of 6 July 2005 establishing a framework for the setting of ecodesign requirements for energy-using products [2005] OJ L191/29. Recently proposed to be amended – Ecodesign for Sustainable Products Regulation, COM(2022) 142. More on this Liesbet van Acker, 'The New Ecodesign Package: An Important Step towards a Circular Economy' [2020] 69 *Ars Aequi: Juridisch Studentenblad* 793.

102 Proposal for a Regulation establishing a framework for setting ecodesign requirements for sustainable products and repealing Directive 2009/125/EC, COM(2022) 142.

103 On making sustainable products the norm Commission, 'On making sustainable products the norm' COM(2022) 140 final, 9.

104 Commission, 'EU Strategy for Sustainable and Circular Textiles' COM(2022) 141 final.

105 According to the Proposal for a Regulation on packaging and packaging waste, COM(2022) 677, all packaging will be fully recyclable by 2030, Extended Responsibility Schemes fees will be properly modulated and mandatory targets for recycled content for plastic packaging will be set – see page 101.

106 High-Level Group on Business Services, Final Report, April 2014, 28.

107 Bruce Sterling, *Shaping Things* (The MIT Press 2005); Hakima Chaouchi, *The Internet of Things: Connecting Objects* (London, John Wiley & Sons 2013); Fawzi Behmann and Kwok Wu, *Collaborative Internet of Things (C-IoT): For Future Smart Connected Life and Business* (London, John Wiley & Sons 2015).

108 WDMA European Office, '"Industrie 4.0: Mastering the Transition", 10 Key Recommendations for a European Framework for the Successful Digital Transition in Industry' [2016] *VDMA European Office* 10 <https://www.pac.gr/bcm/uploads/industrie-4-0-mastering-the-transition-december-2015.pdf> accessed 28 September 2023.

109 WDMA European Office, '"Industrie 4.0: Mastering the Transition", 10 Key Recommendations for a European Framework for the Successful Digital Transition in Industry' [2016] *VDMA European Office* 6 <https://www.pac.gr/bcm/uploads/industrie-4-0-mastering-the-transition-december-2015.pdf> accessed 28 September 2023.

110 Thomas Lindhqvist, *Extended Producer Responsibility in Cleaner Production: Policy Principle to Promote Environmental Improvements of Product Systems*, vol 2000 (Lund, Lund University 2000).

111 Allen L White, Mark Stoughton and Linda Feng, 'Servicizing: The Quiet Transition to Extended Product Responsibility' [1999] *US Environmental Protection Agency* 19 <https://www.tellus.org/pub/Servicizing%20-%20The%20Quiet%20Transition%20to%20Extended%20Product%20Responsibility.pdf>.

112 Klaus-Dieter Thoben, Jens Eschenbächer and Harinder Jagdev, 'Extended Products: Evolving Traditional Product Concepts' (7th International Conference on Concurrent Enterprising, Bremen, June 2001) 437; Klara Demyttenaere, Ivo Dewit and Alexis Jacoby, 'The Influence of Ownership on the Sustainable Use of Product-Service Systems – A Literature Review' [2016] 47 *Procedia CIRP* 180 (https://doi.org/10.1016/j.procir.2016.03.071).

Bibliography

Altvater M and Brandmann C, 'Extended Producer Responsibility: The EU WEEE Directive Goes Global – Strict Law and Order Required or Self-Regulating Market Power a Promising Alternative?' [2012] *Electronics Goes Green 2012+, ECG 2012 – Joint International Conference and Exhibition, Proceedings* <https://search.ricest.ac.ir/dl/search/defaultta.aspx?DTC=49&DC=578711>

Annarelli A, Battistella C and Nonino F, 'Product Service System: A Conceptual Framework from a Systematic Review' [2016] 139 *Journal of Cleaner Production* 1011 (https://doi.org/10.1016/j.jclepro.2016.08.061)

Aparecida Vasques R and Mitsuko Ono M, 'When Sharing Is (Almost and/or Possibly) Better than Owning: A Case Study on a "Full Service" Collective Laundry' [2016] 14 *Revista Gestão Organizacional* (http://dx.doi.org/10.21714/1679-18272016v14Esp.p97-105)

Ardolino M, Saccani N and Perona M, 'The Impact of Digital Technologies and Ecosystems on the Servitization of Companies: A Preliminary Analysis' (Spring Servitization Conference, Aston, May 2015)

Bailey D and Aiginger K, 'The High Road: Europe Must Seize the Potential of Advanced Manufacturing' (*EurActiv*, 26 January 2016) <https://www.euractiv.com/section/innovation-industry/opinion/the-high-road-europe-must-seize-the-potential-of-advanced-manufacturing/>

Baines T S, 'Exploring Service Innovation and the Servitization of the Manufacturing Firm' [2015] 58 *Research Technology Management* (https://doi.org/10.5437/08956308X5805002)

Baines T S et al, 'State-of-the-Art in Product-Service Systems' [2007] 221(10) *Proceedings of the Institution of Mechanical Engineers, Part B: Journal of Engineering Manufacture* (https://doi.org/10.1243/09544054JEM858)

Baines T S et al, 'The Servitization of Manufacturing: A Review of Literature and Reflection on Future Challenges' [2009] 20(5) *Journal of Manufacturing Technology Management* (https://doi.org/10.1108/17410380910960984)

Bardhi F and Eckhardt G M, 'Access-Based Consumption: The Case of Car Sharing' [2012] 39(4) *Journal of Consumer Research* 881 (https://doi.org/10.1086/666376)

Bartolomeo M et al, 'Eco-Efficient Producer Services—What Are They, How Do They Benefit Customers and the Environment and How Likely Are They to Develop and Be Extensively Utilised?' [2003] 11(8) *Journal of Cleaner Production* (https://doi.org/10.1016/S0959-6526(02)00157-9)

Behmann F and Wu K, *Collaborative Internet of Things (C-IoT): For Future Smart Connected Life and Business* (London, John Wiley & Sons 2015)

Benedetti M et al, 'A Proposal for Energy Services' Classification Including a Product Service Systems Perspective' [2015] 30 *Procedia CIRP* 251 (https://doi.org/10.1016/j.procir.2015.02.121)

Bertoldi P, Rezessy S and Vine E, 'Energy Service Companies in European Countries: Current Status and a Strategy to Foster Their Development' [2006] 34(14) *Energy Policy* (https://doi.org/10.1016/j.enpol.2005.01.010)

Bieńkowska E, 'Reindustrialisation of Europe: Industry 4.0 – Innovation, Growth and Jobs, Speech' (The Reindustrialisation of Europe: Industry 4.0 – Innovation, Jobs and Growth. A High-Level One-Day Conference, Brussels, June 2015)

Bocken N M P et al, 'A Literature and Practice Review to Develop Sustainable Business Model Archetypes' [2014] 65 *Journal of Cleaner Production* (https://doi.org/10.1016/j.jclepro.2013.11.039)

Brady T, Davies A and Gann D M, 'Creating Value by Delivering Integrated Solutions' [2005] 23(5) *International Journal of Project Management* (https://doi.org/10.1016/j.ijproman.2005.01.001)

Brennan L et al, 'Manufacturing in the World: Where Next?' [2015] 35(9) *International Journal of Operations & Production Management* 1253 (https://doi.org/10.1108/IJOPM-03-2015-0135)

Chaouchi H, *The Internet of Things: Connecting Objects* (London, John Wiley & Sons 2013)

Codagnone C, Abadie F and Biagi F, 'The Future of Work in the Sharing Economy: Market Efficiency and Equitable Opportunities or Unfair Precarisation?' [2016] *JCR Science for Policy Report. European Commission* <https://publications.jrc.ec.europa.eu/repository/handle/JRC101280>

Commission, 'Green Deal: New Proposals to Make Sustainable Products the Norm and Boost Europe's Resource Independence' (*European Commission*, 30 March 2022) <https://ec.europa.eu/commission/presscorner/detail/en/ip_22_2013>

Crozet M and Milet E, 'The Servitization of French Manufacturing Firms' (2014) CEPII Working Paper 2014-10 <http://www.cepii.fr/pdf_pub/wp/2014/wp2014-10.pdf>

Curtis C et al, 'Extended Producer Responsibility and Product Stewardship for Tobacco Product Waste' [2014] 4(3) *International Journal of Waste Resources* (https://doi.org/10.4172%2F2252-5211.1000157)

Demyttenaere K, Dewit I and Jacoby A, 'The Influence of Ownership on the Sustainable Use of Product-Service Systems – A Literature Review' [2016] 47 *Procedia CIRP* 180 (https://doi.org/10.1016/j.procir.2016.03.071)

Durgee J F and Colarelli O'Connor G, 'An Exploration into Renting as Consumption Behavior' [1995] 12(2) *Psychology and Marketing* (https://doi.org/10.1002/mar.4220120202)

Editor's Note, 'Britain's Lonely High-Flier' (*The Economist*, 8 January 2009) <https://www.economist.com/briefing/2009/01/08/britains-lonely-high-flier>

Fernando Y and Yahya S, 'Challenges in Implementing Renewable Energy Supply Chain in Service Economy Era' [2015] 4 *Procedia Manufacturing* (https://doi.org/10.1016/j.promfg.2015.11.062)

Fischer S et al, 'Leasing Society' (2012) *Directorate General for Internal Policies* <https://www.europarl.europa.eu/RegData/etudes/etudes/join/2012/492460/IPOL-ENVI_ET(2012)492460_EN.pdf>

Fuchs V R, *The Growing Importance of the Service Industries* (New York, National Bureau of Economic Research 1965)

Gebauer H, 'Identifying Service Strategies in Product Manufacturing Companies by Exploring Environment–Strategy Configurations' [2008] 37(3) *Industrial Marketing Management* (https://doi.org/10.1016/j.indmarman.2007.05.018)

Gebauer H and Friedli T, 'Behavioral Implications of the Transition Process from Products to Services' [2005] 20(2) *Journal of Business & Industrial Marketing* 70 (https://doi.org/10.1108/08858620510583669)

Gerstner Jr L V, *Who Says Elephants Can't Dance?: Leading an Enterprise through Dramatic Change* (London, Harper Business 2002)

Goedkoop M J et al, 'Product Service Systems, Ecological and Economic Basics' (1999) Ministry of Housing, Spatial Planning and the Environment, Communications Directorate 1999-36 <https://scirp.org/reference/referencespapers.aspx?referenceid=1520975>

Goldhar J and Berg D, 'Blurring the Boundary: Convergence of Factory and Service Processes' [2010] 21(3) *Journal of Manufacturing Technology Management* 341 (https://doi.org/10.1108/17410381011024322)

Hänsch Beuren F, Gitirana Gomes Ferreira M and Cauchick Miguel P A, 'Product-Service Systems: A Literature Review on Integrated Products and Services' [2013] 47 *Journal of Cleaner Production* (https://doi.org/10.1016/j.jclepro.2012.12.028)

Hart C A, *Climate Change and the Private Sector: Scaling Up Private Sector Response to Climate Change* (London, Routledge 2013)

Hatzopoulos V and Roma S, 'Caring for Sharing? The Collaborative Economy under EU Law' [2017] 54(1) *Common Market Law Review* (https://doi.org/10.54648/cola2017004)

Helms T, 'Asset Transformation and the Challenges to Servitize a Utility Business Model' [2015] 91 *Energy Policy* 98 (https://doi.org/10.1016/j.enpol.2015.12.046)

Hockerts K and Wüstenhagen R, 'Greening Goliaths versus Emerging Davids—Theorizing about the Role of Incumbents and New Entrants in Sustainable Entrepreneurship' [2010] 25(5) *Journal of Business Venturing* (https://doi.org/10.1016/j.jbusvent.2009.07.005).

Hojnik J, 'The Servitization of Industry: EU Law Implications and Challenges' [2016] 53 *Common Market Law Review* (https://doi.org/10.54648/cola2016143)

Hojnik J, 'Ecological Modernization through Servitization: EU Regulatory Support for Sustainable Product–Service Systems' [2018] 27(2) *Review of European, Comparative & International Environmental Law* (https://doi.org/10.1111/reel.12228)

Johansson L, 'Servicification of Swedish Manufacturing' [2010] *The National Board of Trade* <https://www.kommerskollegium.se/globalassets/publikationer/rapporter/2016-och-aldre/report-2010-1-servicification-of-swedish-manufacturing.pdf>

Keirsbilck B and Terryn E (eds), *Circular Economy and Consumer Protection* (Cambridge, Intersentia 2019)

Keirsbilck B et al, *Servitization and Circular Economy: Economic and Legal Challenges* (Cambridge, Intersentia 2023)

Kemekliene G et al, 'Services Employment in Europe, Now and in the Future' (Background paper for UNI-Europa Conference, Athens, April 2007)
Lay G, *Servitization in Industry* (Cham, Springer 2014)
Levitt T, 'Production-Line Approach to Service' [1972] 50 *Harvard Business Review* 41–52
Lightfoot H, Tim Baines and Palie Smart, 'The Servitization of Manufacturing: A Systematic Literature Review of Interdependent Trends' [2013] 33(11/12) *International Journal of Operations & Production Management* (https://doi.org/10.1108/IJOPM-07-2010-0196)
Lindhqvist T, *Extended Producer Responsibility in Cleaner Production: Policy Principle to Promote Environmental Improvements of Product Systems*, vol 2000 (Lund, Lund University 2000)
Lindhqvist T, 'Policies for Waste Batteries' [2010] 14(4) *Journal of Industrial Ecology* 537 (https://doi.org/10.1111/j.1530-9290.2010.00266.x)
Livesay F, 'The Need for a New Understanding of Manufacturing and Industrial Policy in Leading Economies' [2012] 7(3) *Innovations: Technology, Governance, Globalization* 193–202
Lovelock C and Gummesson E, 'Whither Services Marketing? In Search of a New Paradigm and Fresh Perspectives' [2004] 7(1) *Journal of Service Research* (https://doi.org/10.1177/1094670504266131)
Manzini E and Vezzoli C, 'Product-Service Systems and Sustainability: Opportunities for Sustainable Solutions' [2002] UNEP <https://wedocs.unep.org/bitstream/handle/20.500.11822/8123/-Product-Service%20Systems%20and%20Sustainability_%20Opportunities%20for%20Sustainable%20Solutions-20021192.pdf?sequence=2&isAllowed=y>
Manzini E, Vezzoli C and Clark G, 'Product-Service Systems. Using an Existing Concept as a New Approach to Sustainability' [2001] 1(2) *Journal of Design Research* (https://doi.org/10.1504/JDR.2001.009811)
Marks G, Hooghe L and Blank K, 'European Integration from the 1980s: State-Centric v. Multi-Level Governance' [1996] 34(3) *JCMS: Journal of Common Market Studies* 341 (https://doi.org/10.1111/j.1468-5965.1996.tb00577.x)
Mathieu V, 'Product Services: From a Service Supporting the Product to a Service Supporting the Client' [2001] 16(1) *Journal of Business & Industrial Marketing* (https://doi.org/10.1108/08858620110364873)
Maxwell D and Van der Vorst R, 'Developing Sustainable Products and Services' [2003] 11(8) *Journal of Cleaner Production* (https://doi.org/10.1016/S0959-6526(02)00164-6)
Mayers K et al, 'Implementing Individual Producer Responsibility for Waste Electrical and Electronic Equipment through Improved Financing' [2013] 17(2) *Journal of Industrial Ecology* (https://doi.org/10.1111/j.1530-9290.2012.00528.x)
Miller M, *The Internet of Things: How Smart TVs, Smart Cars, Smart Homes, and Smart Cities Are Changing the World* (Indianapolis, IN, Que 2015)
Moeller S and Wittkowski K, 'The Burdens of Ownership: Reasons for Preferring Renting' [2010] 20(2) *Managing Service Quality: An International Journal* (https://doi.org/10.1108/09604521011027598)
Möhlmann M, 'Collaborative Consumption: Determinants of Satisfaction and the Likelihood of Using a Sharing Economy Option Again' [2015] 14(3) *Journal of Consumer Behaviour* (https://doi.org/10.1002/cb.1512)
Mont O, 'Clarifying the Concept of Product–Service System' [2002] 10(3) *Journal of Cleaner Production* (https://doi.org/10.1016/S0959-6526(01)00039-7)
Mont O, 'Editorial for the Special Issue of the Journal of Cleaner Production on Product Service Systems' [2003] 11 *Journal of Cleaner Production* (https://doi.org/10.1016/S0959-6526(02)00163-4)
Neely A, 'The Servitization of Manufacturing: An Analysis of Global Trends' (14th European Operations Management Association Conference, Ankara, January 2007)
Neely A, 'Exploring the Financial Consequences of the Servitization of Manufacturing' [2009] 1 *Operations Management Research* (https://doi.org/10.1007/s12063-009-0015-5)
Nolden C, Sorrell S and Polzin F, 'Innovative Procurement Frameworks for Energy Performance Contracting in the UK Public Sector' (2015) SPRU Working Paper Series 2015-31 <https://www.sussex.ac.uk/webteam/gateway/file.php?name=2015-31-swps-nolden-et-al.pdf&site=25>
Oliva R and Kallenberg R, 'Managing the Transition from Products to Services' [2003] 14(2) *International Journal of Service Industry Management* (https://doi.org/10.1108/09564230310474138)
Opresnik D and Taisch M, 'The Value of Big Data in Servitization' [2015] 165 *International Journal of Production Economics* (https://doi.org/10.1016/j.ijpe.2014.12.036)
Pigosso D C A et al, 'Is the Industrial Product-Service System Really Sustainable?' in [editors], *Proceedings of the 2nd CIRP IPS2 Conference 2010* (Linköping, Linköping University Electronic Press 2012), 59–65
Plepys A, Heiskanen E and Mont O, 'European Policy Approaches to Promote Servicizing' [2015] 97 *Journal of Cleaner Production* (https://doi.org/10.1016/j.jclepro.2014.04.029)

Porter M E and Ketels C H M, 'UK Competitiveness: Moving to the Next Stage' (2003) DTI Economics Paper n. 3 <https://www.hbs.edu/ris/Publication%20Files/file14771_83b42e5a-7e88-49be-9d33-2fc7585a87d9.pdf>

Reisch LA and Thøgersen J, *Handbook of Research on Sustainable Consumption* (Cheltenham Glos, Edward Elgar Publishing 2017)

Sachs N, 'Planning the Funeral at the Birth: Extended Producer Responsibility in the European Union and the United States' [2006] 30 *Harvard Environmental Law Review* 51–98

Schettkat R and Yocarini L, 'The Shift to Services Employment: A Review of the Literature' [2006] 17(2) *Structural Change and Economic Dynamics* (https://doi.org/10.1016/j.strueco.2005.04.002)

Schmidt-Bleek F, *Wieviel Umwelt Braucht Der Mensch?: MIPS – das Mass Für ökologisches Wirtschaften* (Basel, Springer-Verlag 2013)

Smith L, Maull R and Ng I C L, 'Servitization and Operations Management: A Service Dominant-Logic Approach' [2014] 34(2) *International Journal of Operations & Production Management* 242 (https://doi.org/10.1108/IJOPM-02-2011-0053)

Smith W, 'The Uber-All Economy of the Future' [2016] 20 *Independent Review* 383

Stahel W R, 'Product Life as a Variable: The Notion of Utilization' [1986] 13 *Science and Public Policy* (https://doi.org/10.1093/spp/13.4.185)

Sterling B, *Shaping Things* (The MIT Press 2005)

Tether B and Bascavusoglu-Moreau E, 'Servitization: The Extent and Motivations for Service Provision amongst UK Manufacturers' (2012) IMP Innovation, Strategy and Sustainability Working Paper <https://research.manchester.ac.uk/en/publications/servitization-the-extent-and-motivations-for-service-provision-am>

Thoben K-D, Eschenbächer J and Jagdev H, 'Extended Products: Evolving Traditional Product Concepts' (7th International Conference on Concurrent Enterprising, Bremen, June 2001)

Tietze F and Hansen E, 'To Own or to Use? How Product Service Systems Facilitate Eco-Innovation Behavior' (2013) Centre for Technology Management working paper series 4 <https://www.ifm.eng.cam.ac.uk/uploads/Working_paper/16_04_Tietze_Hansen.pdf>

Toffel M W, 'Contracting for Servicizing' (2008) Harvard Business School Technology & Operations Mgt. Unit Research Paper 08-063 <https://papers.ssrn.com/sol3/papers.cfm?abstract_id=1090237>

Tukker A, 'Eight Types of Product–Service System: Eight Ways to Sustainability? Experiences from SusProNet' [2004] 13(4) *Business Strategy and the Environment* (https://doi.org/10.1002/bse.414)

Tukker A, 'Product Services for a Resource-Efficient and Circular Economy: A Review' [2015] 97 *Journal of Cleaner Production* (https://doi.org/10.1016/j.jclepro.2013.11.049)

Tukker A and Tischner U, *New Business for Old Europe: Product-Service Development, Competitiveness and Sustainability* (Sheffield, Greenleaf 2006)

van Acker L, 'The New Ecodesign Package: An Important Step towards a Circular Economy' [2020] 69 *Ars Aequi: Juridisch Studentenblad* 793–801

Vandermerwe S and Rada J, 'Servitization of Business: Adding Value by Adding Services' [1988] 6(4) *European Management Journal* (https://doi.org/10.1016/0263-2373(88)90033-3)

Walther G et al, 'Implementation of the WEEE-Directive—Economic Effects and Improvement Potentials for Reuse and Recycling in Germany' [2010] 47 *International Journal of Advanced Manufacturing Technology* 461 (https://doi.org/10.1007/s00170-009-2243-0)

WDMA European Office, '"Industrie 4.0: Mastering the Transition", 10 Key Recommendations for a European Framework for the Successful Digital Transition in Industry' [2016] *VDMA European Office* 10 <https://www.pac.gr/bcm/uploads/industrie-4-0-mastering-the-transition-december-2015.pdf>

Weber R H and Weber R, *Internet of Things: Legal Perspectives* (Zürich, Springer Science & Business Media 2010)

White A L, Mark Stoughton and Linda Feng, 'Servicizing: The Quiet Transition to Extended Product Responsibility' [1999] *US Environmental Protection Agency* <https://www.tellus.org/pub/Servicizing%20-%20The%20Quiet%20Transition%20to%20Extended%20Product%20Responsibility.pdf>

Wikhamn B R, Ljungberg J and Styhre A, 'Enacting Hard and Soft Product Offerings in Mature Industries: Moving towards Servitisation in Volvo' [2013] 17(4) *International Journal of Innovation Management* (https://doi.org/10.1142/S136391961350014X)

Windahl C and Lakemond N, 'Integrated Solutions from a Service-Centered Perspective: Applicability and Limitations in the Capital Goods Industry' [2010] 39(8) *Industrial Marketing Management* (https://doi.org/10.1016/j.indmarman.2010.03.001)

27

PRIVATE STANDARDISATION FOR SUSTAINABILITY

Opportunities and Cases

Yayun Shen[1] *and Michael Faure*[2]

27.1 Introduction

Increasing consideration has been given to private law in achieving sustainability. The problem of externalities justifies government intervention in environmental governance, which may resort to command and control (e.g., fines and sanctions) and public spending (e.g., taxation and subsidies).[3] Nevertheless, public regulation has its limits due to the high costs and imperfect information needed to design and implement the instruments.[4] For instance, the additionality problem arises when subsidies are given to those who would buy green products anyway, regardless of the monetary incentives.[5] With incomplete information, the government may not accurately determine variables for a tax based on expected outcomes.[6] As a result, the revenue collected may be less than the loss caused by the actual harm, and the expected harm will be excessive when the untaxed variables also contribute to the harm.[7] Command and control, e.g., sanctions and fines, can punish non-compliance, but again, because of the high information and administrative cost, governments are not able to identify non-compliance in all cases.[8] Therefore, private law instruments, such as property rules, liability rules, and contracting, are taken as complements to public regulation in environmental governance.[9] Private law instruments allow bargaining towards the optimal level of care or that of activity and make use of private information (or expertise).[10] Liability rules not only deter non-compliance *ex ante* but can also compensate for environmental losses *ex post*.[11]

Nevertheless, the role of private standardisation in environmental governance is underexplored. Environmental standardisation concerns standard-setting and implementation, which are often achieved with public regulation but are also possible with private regulation. Private actors may show greater willingness to comply with private standards than with public standards imposed by the government if private standard-setting involves private stakeholders and is supported by the community of stakeholders.[12] Companies running global businesses may voluntarily comply with international environmental standards such as ISO 14001, LEED[13] and the Forest Stewardship Council (FSC).[14] Yet, private standard-setting also raises concerns, depending on how private standards are institutionalised via private regulation. For example, private standard-setting achieved via self-regulation can be ruled by private interests and would not sufficiently focus on the public interests at which environmental governance aims.[15] Environmental governance may only benefit from private standardisation if the risks of private interests taking over can in some way be remedied.

DOI: 10.4324/9781032662046-32

This chapter aims to sketch the possibilities offered by private standards, the variety of standards existing in practice (Sections 27.2.1 and 27.2.2), and the theoretical advantages of private standardisation (Section 27.2.3) and to provide examples in a variety of contexts where private standards have been used, through which we equally assess the relative successes of private standards but also the challenges that do arise (Section 27.3). In the light of the theories and examples, the study then explains the possibilities and the necessary conditions for private standards to achieve sustainability (Section 27.4). Section 27.5 concludes.

27.2 Private Standardisation for Sustainability: The Possibilities

27.2.1 Environmental Standards: An Overview

Environmental standards can be generally categorised into target standards, performance standards, and specification standards.[16] Target standards focus on the quality of the environment to be achieved, e.g., in water surfaces, and are therefore also referred to as environmental quality standards.[17] It can also be costly for firms to determine how they should perform to meet the target standards, which usually do not specify compliance at the organisational or individual level.[18] The environmental quality to be achieved can depend on the actions of many, and it is often difficult to identify causality between the actions of one generator and the effect on the quality of the environment. Therefore, private standard-setting focuses more on performance standards and specification standards than on target standards.[19] Performance standards are outcome-based, determining the quality or the care level that individual emitters need to achieve. Specification standards require or prohibit the use of a specific production method or other inputs (i.e., process-based) and mostly regulate the economic actor such as manufacturers.[20] Compared to performance standards, specification standards are generally easier and cheaper for industry parties to formulate because profession associations can utilise their expertise to create detailed technical specifications.[21] However, specification standards, due to their specificity, can unintentionally stifle green technology innovation.[22] When strict technical requirements are mandated, manufacturers might focus solely on meeting those requirements, even if there are more innovative and environmentally friendly alternatives available. This can slow down the adoption of new technologies that could be greener or more efficient but do not fit within the prescribed specifications. Also, stringent specification standards can lead to anti-competitive effects by favouring incumbent manufacturers who are already equipped to meet the standards. Newcomers or smaller companies might struggle to meet the technical requirements, creating a barrier to entry. This can reduce competition in the market and limit the potential for disruptive innovations from emerging players.[23]

This section explores the possibilities of private standardisation in environmental governance on a theoretical basis. Section 27.2.2 sketches the different types of private standard-setting and implementation (i.e., certification), whereas Section 27.2.3 identifies the theoretical advantages of private standardisation over public regulation and distinguishes private standardisation from self-regulation.

27.2.2 The Variety of Private Standards

As far as the contents are concerned, private standards mainly include standards regulating quality or outputs (product standards) and standards specifying processes (process standards).[24] Product standards set criteria for product attributes or emissions from the life cycle of a product. By definition, product standards are similar to performance standards regulating pollutants. Private standards of this type include ISO 14001 (requiring environmental performance evaluation of sites and organisations),[25] NSF/ANSI 244 (requiring public water supplies to be microbiologically safe),[26] the LEED standards for sustainable buildings,[27] the Carbon Trust PAS2060 standards for carbon neutrality,[28] and the Cradle-to-Cradle Certified standards that assess the sustainability performance of products.[29]

Process standards specify how products are made, outlining criteria for the methods or places of production. By definition, process standards are akin to specification standards, which may discourage green technology innovation and may not follow the equi-marginal principle of abatement.[30] Compared to emissions reduction, resource management focuses more on process standards, as with eco-labelling for food and wood products. For instance, seafood eco-labels, e.g., MSC, Friend of the Sea, Fairtrade, usually specify where fish and seafood come from and how they are harvested or farmed.[31] Another example could be the Forest Management Certification (FSC) standard, which lays down requirements for paper and timber production processes. The FSC certification is to provide assurance that those timber products originating from a forest are managed in a responsible and sustainable way and that the paper and timber supply in the market does not contribute to environmental destruction or loss of habitat.[32]

Compliance with private standards needs to be underpinned by certification, a process in which a qualifying certifier provides formal and written assurance reflecting that a product or process conforms to particular standards. There are generally three types of certification schemes[33]:

i Self-declaration or first-party certification, by which a firm or group sets personalised standards, evaluates, and self-declares its own compliance. In this case, the standards apply only to the standard setters (i.e., individual firms within the group) unless the standards are incorporated into industry-based standards or public regulation;
ii Internal auditing or second-party certification, where professional associations make private standards, and compliance is certified by internal audit procedure or by recruiting certifiers to audit compliance.
iii Third-party certification, where compliance with the standards is verified by an independent and accredited certification body, which is not involved in the standard-setting and does not have any conflicts of interest either.

It is noteworthy that self-regulation differs from the multi-governance model of private regulation in terms of external oversight. Industry-based standards (e.g., standards from professional associations) show the characteristics of self-regulation, with the standard-setter and the certifier coinciding with one group of industry actors. Standardisation with NGOs or third parties represents the multi-governance model of private forms of regulation, using the market power of consumers or pro-environmental communities and social groups.[34] Increasingly, the multi-governance model has been applied to standardisation across borders, i.e., transnational private regulation (TPR), which not only includes industries but also NGO-led regulators and multi-stakeholder organisations.[35] External oversight may also come from public regulation. For instance, public regulations for building energy efficiency incorporate the contents of private standards[36] or refer to the LEED certification scheme.[37]

In either case, compliance with the LEED standards is monitored by public regulation although the standards are made by a professional association (i.e., the US Green Building Council) (Table 27.1).[38]

Table 27.1 Characteristics of public and private environmental standards

	Public	*Private*		
		Self-enforced	*Industry-based*	*Multi-governance*
Standard-setting	Government-led standard-making via due process of law	Personalised standards made or approved by a group of firms via agreements	Professional associations in industries via a formal process	Standard-setting via a formal process that aligns with the codes or principles of good governance (e.g., GAL)
Stakeholder engagement	Public authorities; interest groups	The regulated firms or organisations	Professional associations; The regulated firms	(International) NGOs, scientists, industry representatives, parties of interest (e.g., consumers)
Compliance and monitoring	Mandatory compliance enforced via command and control; Voluntary compliance supported by public funding programmes	Voluntary and Self-declared compliance	Voluntary compliance certified by internal audits or independent accredited certification bodies	Voluntary compliance certified by independent accredited certification bodies; Monitoring from NGOs or public regulators

Notes: The table summarises the main characteristics of the different types of standardisation. However, the differences are not clear cut and there could be mixed models of standardisation in practice.

27.2.3 The Appeal of Private Standardisation: Lower Administrative Costs, Flexibility, and Diversity

Private standardisation may have the advantages of private regulation or co-regulation identified in the law and economics literature. One of the merits is that the administrative costs of private standardisation are lower compared to standardisation via public regulation. Administrative costs consist of the cost of standard-setting and enforcement.

On the one hand, private parties may have better information than government agencies that are often not specialised in a particular area.[39] With expertise in the field, private parties have prior knowledge, as a result of which the costs of finding out the optimal standards would be lower than under public standard-setting. For example, the Chinese central government issued a national standard for Drinking Water Safety and Sanitation to improve access to drinking water for rural residents in China. But the government failed to develop detailed rating standards due to a lack of expertise and the red tape of establishing public standards.[40] The Chinese Hydraulic Engineering Society, a professional association, therefore, developed new rating standards for water sanitation in rural areas. The private standard has been promoted by the central government and applied to 832 public authorities at the county level or lower.[41]

On the other hand, the implementation costs could be lower with private standards than with public standards as private parties may detect or predict the potential risks of non-compliance more easily, given their information advantage.[42] This could be the case with third-party certification, where the certifier can spot non-compliance without any conflict of interest or perverse incentives to greenwash. For instance, a mandatory certification scheme (PHPL) for sustainable forests was introduced in the early 2000s in Indonesia. All logging concessions were required to comply with the PHPL standard, and compliance was certified by independent auditors (LP) accredited by the government (i.e., The Ministry of Forests in Indonesia).[43] Compared to private forest certifications (e.g., FSC and the Indonesian Eco-label Institute) in Indonesia, the PHPL was under-enforced because it was paper-based and short of independence and transparency.[44] The implementation of private standards can be cost-effective when there is a greater incentive for compliance with private standards than with public standards. Private parties may be more willing to comply with private standards that are made by the regulated community sharing common interests rather than by a distant public agency.[45] Besides, firms or product manufacturers may voluntarily pursue private environmental certification to gain access to (international) markets or to enjoy access to particular consumer groups via product differentiation and labelling.[46]

Another advantage of private standardisation is flexibility in establishing standards. Compared to public regulation, private standardisation is less likely to suffer from the red tape of bureaucracy and could be more quickly adapted to risks. Private standard-setting does not have to follow the strict due process required by domestic administrative law and can lead to a lower administrative cost.[47] But private standards can still be effective within the framework of the Global Administrative Law (GAL). GAL recognises that much of global governance is characterised by 'soft law' instruments such as codes of conduct, and standards.[48] GAL helps address the challenges posed by the increasing role of international organisations in making standards that have implications on a global scale. On the one hand, GAL emphasises the importance of involving affected parties in the decision-making process.[49] Private standards can benefit from inclusive stakeholder participation, which can lead to more balanced and effective standards that address diverse concerns.[50] On the other hand, GAL promotes the establishment of review mechanisms for administrative decisions.[51] In private environmental standard-making, review mechanisms can involve independent audits, certification bodies, and third-party verification to ensure compliance with standards.[52]

Moreover, private standardisation may not significantly limit consumer choices and reduce the diversity of preferences.[53] Various private standard-setters have to compete with each other to create standards that best fit the needs of market participants and local circumstances.[54] The diversity of private standards and certification schemes not only provides consumers with diverse options but also reduces the information costs for consumers when public regulation fails.[55] For instance, there was an outbreak of food poisoning cases in 2013 in China, including diseases caused by tainted rice grown on polluted land.[56] Consequently, Chinese consumers became sceptical about public food certification in ensuring food safety, especially the outcome-based certification for animal-based products.[57] Some Chinese consumers shifted to imported organic food and eco-friendly food.[58] In the early days, organic food promotion generally involved the private sector rather than the government.[59] Thus, those private certifications meet the demand of Chinese consumers for 'green' products and, at the same time, push businesses to adopt private certifications, e.g., Hazard Analysis Critical Control Point (a food safety management system included in the international standard ISO 22000 set by international trading partners).[60] The case presented here was to highlight the potential of private standards in expanding consumer choices. While addressing environmental health risks or food safety, there are shared interests between consumer organisations and environmental groups. However, it's important to acknowledge that conflicts of interest may arise between private stakeholders.[61] For instance, consumer and environmental interests may not be aligned in standards or regulation concerning packaging. In 2014, the American Progressive Bag Alliance (APBA) did not support a proposed plastic bag ban in California, arguing that reusable bags can harbour bacteria and that plastic bags are more convenient for consumers.[62] However, environmental groups supported the ban, emphasising the need to reduce plastic pollution in oceans and waterways.[63]

27.3 Examples

We will now provide some examples of the use of private standards and specifically of the wide variety of private standards indicated above (in Section 27.2.2). We start by discussing self-enforced private standards in voluntary environmental agreements (VEAs) in Section 27.3.1; we then discuss industry-dominated standards with a low level of external oversight (Section 27.3.2) and finally we address private standardisation with third-party certification and external oversight (Section 27.3.3).

27.3.1 Self-Enforced Private Standards in Voluntary Environmental Agreements

Deforestation is one of the focuses of self-imposed standards for supply chain management. The Carbon Disclosure Program requires its members to disclose how they address forest risk commodities in their supply chain.[64] In the farming industry, the Responsible Leather Round Table (RLRT) was initiated by farmers who participate in a credit trading system and purchase sustainable farming credits.[65] Consumers also sign agreements, e.g., the Soy Moratorium[66] and the G4 Cattle Agreement,[67] to boycott products with deforestation risks. Among other voluntary environmental agreements, the zero-deforestation cattle agreements have increased in the Amazonian States, where NGOs go against food suppliers for purchasing from areas with illegal deforestation.[68] In response, meat companies signed the Terms of Adjustment of Conduct ('MPF-TAC') agreements and the zero-deforestation agreement ('G4') with Greenpeace. Greenpeace Brazil launched investigations into the beef industry in Brazil and released an international report that recognised the industry's involvement in deforestation. The Greenpeace report led to the signing of the G4 Cattle agreement by major beef processors, who agreed to comply with human rights and environmental standards concerning their sourcing from farms, e.g., preventing sales from areas facing deforestation.[69] Yet, G4 applies to properties selling directly to suppliers (e.g., slaughterhouses) rather than indirectly supplying properties such as calving and breeding ranches.[70] The implementation of private standards in the G4 agreement mainly relied on the self-enforced agreement among the regulated.

G4 confirms some of the advantages of private standardisation predicted in theory. G4 may reduce information costs among stakeholders in dealing with deforestation in the supply chain. In theory, the lower the degree of separation between private regulators and the regulated (to reduce conflicts of interest and capture), the lower information costs.[71] The slaughterhouses are often located in the deforestation area. Compared to the distant government, those slaughterhouses have bargaining power by restricting market access and hence, are able to oversee and influence farmers' and ranchers' behaviour. It is found that the G4 agreement increased productivity more in municipalities closer to G4 slaughterhouses since ranchers faced the highest potential loss if they were excluded from the supply chain following deforestation.[72] Because of the market forces, G4 provides incentives for meatpackers or ranchers to comply with the standard voluntarily, as the slaughterhouses will exclude unregistered properties from the supply chain. Using data from beef supply chains from the Brazilian Amazon state of Pará, a study shows that slaughterhouses avoid purchasing from properties in deforestation areas after committing to G4.[73] Nearly 85% of surveyed ranchers indicated that G4 drove them to register their properties in the public environmental registry.[74] The rates of deforestation during 2010–2012 were 50% lower on post-agreement supplying properties when compared to pre-agreement suppliers not engaged in the agreements.[75] Apart from environmental compliance, G4 also yields economic benefits. There has been evidence showing that G4 increased productivity by 5.6% in municipalities in the Amazon biome with G4-owned slaughterhouses and encouraged investment in inputs and capital specific to livestock.[76]

Despite the achievements, the reputation of G4 has been falsely enhanced with insufficient monitoring. G4 in Brazil led to cattle laundering as the agreement applies only to ranchers and properties selling directly to slaughterhouses.[77] In cattle laundering, ranchers who are not direct suppliers to the G4 slaughterhouses may raise their cattle on farms after illegal deforestation. Then the cattle are moved to compliant farms before selling to slaughterhouses.[78] Leakage to the unregulated supply chains may occur when cattle produced on deforested ranches are sold to slaughterhouses not involved in the monitoring system.[79] As a result, the effect on forest conservation may diminish with insufficient monitoring and the limited number of participants in G4.[80] A solution to the false information is to build up databases that track the movement of cattle between farms. Also, private standards in voluntary environmental agreements can be enforced with oversight from public regulators (i.e., enforced self-regulation).[81] However, as was already expected from a theoretical perspective, this model leads to high information and enforcement costs of external oversight from public regulation. In Brazil the high cost resulted in approximately half of the cattle and properties not being monitored at all.[82]

27.3.2 Industry-Dominated Standards with a Low Level of External Oversight

Compared to VEAs that apply to a limited range of firms, industry-based standardisation shows collective and regulatory features, where professional associations dominate standard-setting, and the enforcement relies on the self-regulatory agency. One example of industry-based standards is the LEED certification for sustainable buildings. The LEED standards require a life-cycle environmental assessment of buildings regarding energy efficiency, land use, water use, indoor air quality, and waste disposal.[83] Buildings with the LEED Silver, Gold, or Platinum certificates may, on different levels, be environmentally superior to those without certificates.

Some of the advantages of private standardisation can be observed with LEED. Based in the US, the United States Green Building Council (USGBC) plays a key role in stimulating innovation and investment in sustainable buildings in the country.[84] In the early days, the public subsidy programmes for sustainable buildings were loosely knit and barely had any significant impact on the building industry.[85] The establishment of the USGBC provided a more collective approach to sustainable buildings. The USGBC, as a self-regulator, not only created one of the most important green building certification systems ('LEED') but is also devoted to green building technology development and professional training. As the USGBC became more influential, the LEED system and green building compliance were better recognised in the industry. In the past decade, the LEED certification led to competition and comparison in the real estate market and became one of the driving forces for the adoption of sustainable building standards across countries.[86] For instance, a large number of commercial builders in China choose LEED over domestic certification because LEED is recognised by the international supply chain.[87] With the LEED certificate, building developers and landlords can sell themselves as green and get market differentiation and benefit from a rental premium.[88]

LEED standards also reduce the information or enforcement costs of public standard-setting for sustainable buildings. The US government has combined the LEED standards with policy instruments for sustainable buildings, such as tax reduction,[89] public procurement,[90] or even the liability regime. The Seattle Land Use Code has provided projects certified by LEED with an expedited permit for land use of new buildings and the demolition of residential buildings.[91] Since the LEED standard usually does not differentiate across jurisdictions, the local governments may tailor industry-made standards to local needs. For instance, the green building regulation in Illinois requires the industry rating system incorporated to be analysed and evaluated by the Capital Development Board.[92] A study which compared cities of similar size and environmental preferences, found that the LEED standard prevails twice as fast among private developers in cities with public procurement policies compared to those without such policies.[93]

The case of LEED equally shows that private standardisation may consist overwhelmingly of the industry or professionals, who can come together as influential lobbies for regulatory capturing and anti-competition.[94] On the one hand, the industry-based GB certification and the products it refers to may become dominant in the market when GB compliance is required in public regulation. For instance, in an old version of the LEED system, the USGBC required that the wood used in a LEED project should be verified by the FSC, which resulted in conflicts between the FSC and the Sustainable Forestry Initiative (SFI) programme.[95] The LEED's exclusive use of the FSC rating system has been accused of violating antitrust laws when adopted by the municipal governments.[96] Legal actions against those restrictions of competition may not always be workable as the US Supreme Court has established a state-action immunity doctrine, whereby industry entities making the GB regulations can justify their anti-competitive activities by proving that the state authorised the activities.[97]

On the other hand, self-regulators may water down GB standards for members of the professional association.[98] Private regulatory competition can result in diverse standards and certifications within an industry in response to consumer preferences, market demands, or specific industry challenges.[99] However, the existence of multiple competing standards can lead to fragmentation and inconsistencies.[100] Also, intense competition among private regulators is likely to be a 'race to the bottom', where regulatory capturing may result in standards lower than the optimal level.[101] One example could be the various GB certifications competing with each other. In order to boost their business, some professional associations may make the certification easier to obtain, which may lead to compliance with exaggerated GB standards.[102] In the meantime, the professional associations may lobby government agents to get their certification incorporated into GB regulations. In that case, GB rating systems providing easier points may succeed when 'the voice of those who benefit from lax regulation is strong'.[103] The USGBC, which was born to pursue building sustainability, tried to water down the LEED standards. The USGBC offered easy points for developers, meanwhile lobbying to make the LEED part of the state and local regulations on building sustainability.[104]

27.4 Private Standardisation with Third-Party Certification and External Oversight

Private standardisation has increasingly involved third-party certification and external monitoring, distinguishing private standardisation from industry-dominated self-regulation. External oversight can be achieved by having different parties responsible for standard-making and standard enforcement. The 'tripartite' cooperation takes shape between governments, industry, and environmental NGOs. Such tripartite cooperation can create extra-regulation compliance by the industry, in which case companies may adopt higher performance standards and not just scant efforts.[105] This section investigates the FSC and the Marine Stewardship Council (MSC) as two influential private certification schemes in the timber and fishery industries, respectively.

27.4.1 FSC

FSC was initiated by the World Wildlife Fund (WWF) and other environmental NGOs, with the engagement of timber product retailers.[106] The FSC is one of the leading members of to the International Social and Environmental Accrediting and Labelling (ISEAL) Alliance.[107] ISEAL has developed a set of Codes of Good Practice that outline the principles and criteria for credible sustainability standards systems. These codes cover areas such as standard-setting, assurance (certification and accreditation), impact assessment, monitoring, and evaluation.[108] FSC, as a member of ISEAL, aligns its practices with these codes to ensure the legitimacy and credibility of its certification system. FSC uses a qualified and independent third-party certifier to evaluate compliance with the standard for forest conservation across countries such as the US, China, Indonesia, and Bolivia.[109] Also, ISEAL provides a platform for collaboration and learning among its member organisations. FSC is able to share experiences, best practices, and lessons learned with other sustainability standards systems via the ISEAL platform. This collaborative approach helps to bridge the gaps between the Programme for the Endorsement of Forest Certification (PEFC) and the FSC and, thus, improves the overall quality and impact of sustainability standards.[110] In addition, ISEAL's Codes of Good Practice emphasise the need for a rigorous and inclusive standard-setting process. FSC's standard-setting process includes multiple rounds of consultation, expert input, and consensus-building among diverse stakeholders.[111] The FSC consists of three chambers representing various interests: social and indigenous groups, environmental organisations, and economic bodies.[112] This framework reduces conflicts of interest between those overseeing and those being overseen, as the regulatory body includes both the regulated entities and those benefiting from its decisions.[113]

At the domestic level, interaction between the FSC standard and public regulation may reduce the administrative costs for the government in dealing with deforestation. For instance, the WWF's Global Forest and Trade Network engaged in implementing a pilot launched by China's State Forestry Administration in 2006. The involvement of the Global Forest and Trade Network made the Yong'an Forest Group the first enterprise with over 1,000,000 hectares of FSC-certified forest in southern China.[114] FSC and the Indonesian Eco-label Institute (LEI) established a Joint Certification Program (JCP), which helps to gain public trust for the LEI certification in Indonesia.[115] Evidence shows the positive synergy effect between public and private regulation in an FSC-certified community forest.[116] Furthermore, the FSC may work better in reducing administrative costs when the forest certification schemes are compatible with public regulation. The FSC-BC standards closely mimic provincial policies in the US, and the FSC-PC standard resembles an average of the legal requirements in the three states it covers in the US.[117] Many states in the US have developed monitoring programmes to track compliance with Best Management Practices.[118] Incorporating FSC provides extra monitoring for compliance with riparian buffer-zone rules. Private verifiers conduct audits for the performance of forest operators against the FSC standards, and hence can provide a beneficial complement to public regulation, which suffered from insufficient funding and staff.[119] The coordination between public and private regulators helps to reduce monitoring costs.[120]

Though FSC reduces enforcement costs for public authorities, it may equally result in higher compliance costs for small-scale suppliers. Both the FSC and the LEI have established tailored schemes (i.e., the Small and Low Intensity Management Forest Standards and the Sustainable Community-Based Forest Management System in Indonesia) to the local needs in Indonesia. However, the bespoken standards face very limited participation among local communities.[121] One of the reasons could be that those local communities are incapable of developing cooperatives to fulfil the requirement for certification and remain competitive among large-scale suppliers.[122] Also, certification costs remain high for communities without donor support.[123] Though FSC could bring economic benefits and behavioural change to forest conservation,[124] the expected price premium[125] or added value is not guaranteed, with some communities already engaged in government-led forestry programmes before being FSC-certified.[126] Similarly, the national certification scheme in Bolivia is deemed to be less costly and more accessible as local communities lack the financial capacity to seek FSC certification.[127] Local communities in Bolivia have to create community enterprises or resort to collaboration with companies to comply with the standards endorsed by the FSC and operate formally.[128]

Moreover, FSC can have limited applications if the standards do not fit into the local institutional framework, which includes formal rules (e.g., the property right regime) and informal rules (e.g., culture and social norms).[129] FSC, along with the Indonesian forest certification (i.e., LEI), requires clear and secure land tenure as a requisite to grant certification.[130] Normally, companies need to obtain 'free and prior informed consent' (FPIC) from the communities, which could be difficult in Indonesia as the logging permits are often issued before the start of consultation with local communities.[131] As a result, the coverage of FSC and LEI in Indonesia is still low though both have been working there since the late 1990s.[132] In Bolivia, the forest certification schemes have not provided solutions for the public regulatory system in Indonesia that fails to create secure forest tenure and to incentivise sustainable forest management. Currently, local communities are still marginalised in the current land tenure system and are not prepared for the certification scheme.[133] FSC auditing tends to focus on documentation, monitoring, environmental issues, and compliance with standards for labour rights which can be immediately satisfied.[134] The FSC standards, however, have little respect for land tenure, indigenous people's rights, and ameliorating the relationship with local communities, which are better addressed in a domestic forest certification (i.e., *Certificación Foresta Voluntaria*) in Bolivia.[135] Because of the cultural differences between local communities and the market economy, FSC does not provide sufficient incentives for local communities in Bolivia to engage in forest management.[136] Once the external support stopped, few of them could maintain their own forest management and certification.[137]

27.4.2 MSC

The MSC is a non-profit organisation independent of its initial founding partners. The MSC adopts centralised standards for sustainable fishery management and the chain of custody, rather than having national affiliates develop their own national standards.[138] States, fishermen's associations, and industry associations of quota holders or processors can all become clients of MSC.[139] Similar to

FSC in the timber industry, MSC is also a member of ISEAL and has incorporated into the MSC certification programme the requirements laid down in ISEAL's Code of Good Practice. ISEAL supports the use of independent third-party verification and stakeholder engagement to ensure that standards are being implemented and adhered to accurately. In MSC certification programmes, fisheries gain certification from an independent assessor (i.e., Conformity Assessment Bodies), and NGOs, governments, and other bodies are involved in the assessment process.[140] Moreover, ISEAL's code emphasises that sustainability standards should consider the context of specific regions and encourages its members to focus on achieving positive impacts on social, environmental, and economic aspects. For instance, in South Africa, the MSC standards for sustainable fishing need to interact with domestic institutional frameworks, e.g., property rules for marine fishery resources, and compliance with the MSC standards had provided the South African fishing industry with a competitive advantage and price premiums.[141]

MSC may lower the administrative costs of fishery governance for public regulators. On the one hand, MSC induces voluntary compliance across borders for its recognition in the global supply chain. An illustrative case is that of the South Africa Hake Fishery, which successfully secured MSC recertification in both 2010 and 2015. This accomplishment facilitated the fishery's engagement in international trade, particularly to high-demand markets such as the UK, Germany, Switzerland, and Australia, where the demand for MSC-endorsed products is robust.[142] Similarly, the Hoki Fishery Management Company (HFMC) of New Zealand proactively pursued MSC certification in response to requirements stipulated by clients in the US and the European Union. This underscores the tangible economic value attributed to MSC certification, as it enhances market opportunities and trade relations for certified fisheries. A corollary benefit of MSC certification has been observed in the mitigation of adverse ecological externalities. For instance, the certification has been associated with a notable reduction in seabird mortalities, highlighting its efficacy in engendering ecologically responsible fishing practices. On the other hand, the MSC certification can help lower the information costs of monitoring in fishery governance that merely rely on public regulation. Compared to most fisheries in Mexico, those certified trawl rights holders face a higher possibility of inspections by the public authority.[143] The MSC certification adds an additional layer of monitoring. The MSC auditors conduct field studies rather than merely relying on existing documents from the government and their clients.[144] The field studies allow the auditor to estimate the actual impacts of the fishery on the ecosystem and the catch.[145] Also, the MSC pre-assessment involves multiple stakeholders like NGOs for external oversight. For example, the NGO and ecologists engaged in the monitoring and the MSC pre-assessment for crab fishery in South Africa.[146]

Nevertheless, the MSC may not be that cost-effective for those regulated and consumers, though it generally reduces administrative costs for public regulation. MSC certification can incur higher compliance costs for those regulated. Certified fisheries are subject to the high costs associated with the assessment procedure and the ongoing costs of annual surveillance.[147] The cost for a full assessment, necessary for the MSC certification, can vary between $200,000 and $500,000, depending on the type of fishery.[148] The high costs make the MSC less popular with small-scale fisheries in developing countries.[149] Besides, the MSC weakens fisheries' property rights due to the lengthy and costly certification procedure and standards which are more favourable to industrial fishing.[150] The products from small-scale fisheries are processed by small companies and retailers. It could increase the information costs of monitoring the supply chain if those small-scale fisheries are not engaged in the MSC certification scheme.[151] The high costs could be passed on to consumers who are willing to pay more for the eco-labelled seafood.[152] When consumer demand for sustainably sourced fish products is low, the economic value of MSC certification for fisheries is diminished. Certain certified fisheries do not command price premiums for their products, rendering the cost-effectiveness of certification questionable for these enterprises.[153] In some instances, certified fisheries continue to maintain their certified status without actively marketing their products with the MSC label, indicative of a divergence between the anticipated market benefits and realised economic gains.[154] For example, MSC has not been very successful in Japan, as most Japanese fish products are consumed domestically. Those domestic consumers are not aware of the MSC certification; hence, their willingness to pay a price premium is still low.[155] As a result, compliance with the MSC standard does not seem to bring the expected price premium to the certified fisheries.[156]

Furthermore, the MSC standard may not always be sufficient to ensure sustainable fishing. There are still concerns around MSC-certified fisheries concerning the sustainability of the stock in the deep sea, catches, and uncertainty of the impact on benthic habitat.[157] The involvement of NGOs can push compliance with higher standards than the MSC requirements. For instance, the first and second Hoki MSC assessments have faced objections from WWF New Zealand and a local NGO. The NGOs have special concerns over the influence of fishing on bycatch (sea birds and fur seals) and the ecosystem.[158] To reduce the risks mentioned, the industry has taken various measures, such as institutionalising the Ecological Risk Assessment Strategy, establishing operational procedures for seabirds and fur seals, and establishing voluntarily closed areas for bottom trawls.[159]

Last, the MSC certification may create entry barriers and monopolies. The MSC certification does not include standards for fishing rights. However, it allowed those trawl sectors to consolidate their fishing rights while keeping out new entrants and marginalising the longline sector.[160] For instance, in the reallocation of long-term fishing rights in 2006 in South Africa, 'the proportion of the total allowable catch allocated to the longlining industry was not increased', and 'no new entrants were assigned quotas, and some of the smaller existing quotas were not renewed'.[161] It is argued that the hake fishing industry may become more concentrated in the medium term.[162] In addition, the MSC certification per se has been a monopoly in the certification market, but some states and the sustainable seafood movement require sufficient alternatives to provide sustainable products.[163] Increasingly, some domestic certifications have been established in Iceland, the US, and Canada to reduce the high certification costs and to prevent a monopoly of the MSC in the market.[164]

27.5 Discussion

It is found from the above that the possibilities of private standardisation for sustainability have been realised in real-life examples. Private standardisation complements public regulation by reducing the administrative costs of standard-making and implementation. On the one hand, private actors can have more expertise in the field and capability of communication in the supply of cross-border private standards. Industry-based standards can be more effective in stimulating innovation diffusion compared to voluntary environmental agreements that have limited applications. One example is the USGBC and its LEED standard, which play a key role in stimulating innovation and investment in sustainable buildings across countries.[165] The USGBC, as a professional association, not only created one of the most important green building certifications but is also devoted to GB technology development and professional training. As the USGBC became more influential, the LEED standard for sustainable buildings was better recognised in the industry. Furthermore, the emergence of transnational private meta-regulators can help reduce information costs by reaching recognition and harmonisation in standard-setting.[166] One prominent private meta-regulator is the ISEAL establishing platforms that allow FSC to draw on best practices and standards from other sustainable forests certification schemes.

On the other hand, private standardisation may reduce the information costs of monitoring compliance for public regulators. For instance, the G4 deforestation agreement involves slaughterhouses, which are often embedded in the deforestation area. In that way, those slaughter-houses can push farmers' compliance using the power of market access and hence, oversee and influence farmers' and ranchers' behaviour.[167] In the US, FSC certifiers conduct audits for the performance of forest operators against the FSC standards and hence can provide a beneficial complement to public regulation, which suffered from insufficient funding and staff and high monitoring costs.[168] Similarly, the MSC auditors add another layer of monitoring for sustainable fishing by conducting field studies, which allow the auditor to estimate the actual impacts of the fishery on the ecosystem and bycatch.[169] Besides, private standards provide incentives for voluntary compliance if required in the global supply chain. For instance, fisheries in South Africa actively sought MSC certification to export their products to their global clients from the UK, Germany, Switzerland and Australia, the US, and the EU.[170]

The analysis equally shows the limits of private standardisation. It is argued in theory that private standard-setting provides diverse options with various standards competing with each other, which may not be guaranteed in reality. For instance, The MSC certification has been a monopoly in the certification market, resulting in insufficient alternatives to provide sustainable products.[171] In some cases, the use of private standards could amount to an agreement between industries to follow particular standards, thus restricting competition in the market.[172] Similarly, in the fishery industry, the trawl sectors misused the MSC certification to keep out new entrants and marginalise the longline sector.[173] Moreover, restrictions on competition are more likely to occur when public regulation incorporates private standards and de facto delegates rule-making power to industry parties. It is observed that the USGBC has (successfully) lobbied to make their LEED standard incorporated into the state and local regulations in the US, meanwhile offering easy reference points for developers.[174]

Another challenge facing private standardisation is capturing and greenwash. Private standardisation may be abused in the interests of the regulated community that dictates the standards.[175] Individuals in any group attempting collective action will have incentives to take a free ride on the efforts of others if the group is working to produce public goods.[176] Without benefits for individual parties, private standardisation is unlikely to occur even when large groups of people are interested in environmental protection. The very complexity of public interest decision-making in environmental governance may make standard-setting particularly susceptible to private interest influence.[177] The capturing and greenwash become even more severe if the private standard-setting is industry-dominated and there is no countervailing power against industry lobbying. The PEFC programme and the Marine Eco-Label Japan programme (MEL) in Japan are examples of those.[178] The USGBC offered easy points for developers to be LEED-certified, meanwhile lobbying to make the LEED part of the state and local regulations on building sustainability.[179] There were also concerns around MSC-certified fisheries, and the involvement of NGOs, i.e., WWF New Zealand, pushed two fisheries in New Zealand to comply with higher standards than the MSC requirements and to take measures to protect the marine ecosystem.[180] Relative to the industry-dominated standardisation, the multi-governance framework offers a potential solution for mitigating greenwashing through a strategy known as 'structural separation',[181] which refers to a separation of roles among entities responsible for various stages. This can be observed in examples such as the FSC and LEED certifications, where a clear division exists between those setting the standards and those overseeing their execution and assessment (such as third-party evaluators). The risk of manipulation is reduced by the implementation of meticulous regulations related to consultation processes, equitable distribution of authority among stakeholders and certifying bodies (or auditors).[182] It is important to note, however, that this strategy may lead to an increase in administrative expenses within private standardisation processes.

In addition, private certification schemes can have limited applications due to high compliance costs and incompatibility with local regimes. The high costs associated with the assessment procedure make the MSC less popular with small-scale fisheries in developing countries.[183] Similarly, the certification costs remain high for communities without donor support.[184] Though FSC could bring economic benefits and behavioural change towards forest conservation, the expected price premium is not guaranteed.[185] As a result, small-scale suppliers may go for domestic certification schemes deemed to be less costly and more accessible as the NGOs lack the financial capacity to seek FSC certification.[186] International private certification schemes may incorporate uniform standards that do not fit into the local institutional framework consisting of formal rules (e.g., laws and regulations) and informal rules (e.g., culture and social norms). For instance, FSC does not provide sufficient incentives for local communities in Bolivia because of the cultural differences between local communities and the market economy.[187] The MSC certification has not been very successful in Japan, as most Japanese fish products are consumed domestically. Those domestic consumers are not aware of the MSC certification, and hence, their willingness to pay a price premium is still low.[188] Also, the requirements in private certification schemes may not be compatible with domestic property regimes. The FSC standard requires clear and secure land tenure as a requisite to grant certification, which could be difficult in Indonesia as the logging permits are often issued before the start of consultation with local communities.[189]

The theory and the examples presented above may provide an educated guess on the conditions under which private standards might work to promote sustainability. The first lesson is that the use of private standardisation should be differentiated, distinguishing private regulation from pure self-regulation. Voluntary environmental agreements or industry-based standardisation can be effective in cases where operators go further than merely internalising negative externalities (regulated by public standards) and engage, for example, in green technology innovation. The generation of those positive externalities, supplementing public regulation, can be entrusted to private standardisation. Nonetheless, the problem of externalities in environmental governance makes standard-setting particularly susceptible to private interest influence.[190] Therefore, external oversight from governments and public interest groups (e.g., environmental NGOs) may reduce the problem of capturing and greenwashing. Second, voluntary compliance can be encouraged by increasing market demands for certified products and lowering the transaction costs associated with private certifications. Extra financial support may be needed for small-scale suppliers who scarcely have any incentives to start using private certification when the price premium from certified products is not guaranteed. Third, the diversity of private certification schemes may help reduce monopolies and incompatibility with local contexts. On the one hand, (uniform) private standards may have more applications if the standards fit into the local institutional framework (e.g., property rules or cultural and social norms). The examples of FSC or MSC indicate that compatibility with local regimes is particularly crucial in private standardisation for common pool resource governance. Diversity, on the other hand, may lead to competition between private certification schemes. But the law and economics theory also warns that private standard-setters may not compete with higher standards but rather with standards that please the regulated community.[191] In that case, standards in favour of the regulated community will survive the market, crowding out stringent standards for emissions reduction.[192] Last, cross-sector meta-regulators or transnational private regulators may reduce the problem of capturing or anti-competition when stakeholders (e.g., public and private regulators) within a sector are being monopolists and external oversight can hardly be implemented in the sector. Yet, the meta-regulation needs to include multi-stakeholders representatives in the organisational governance and the regulatory process.[193]

27.6 Conclusion

Standardisation includes standard-setting and implementation, which can be done via private regulation. This chapter looked into how private standards may contribute to sustainability and under which precise conditions, based on the law and economic theories and some illustrative cases concerning various private standardisation schemes.

We found that private standardisation can complement public regulation in achieving sustainability. Private standardisation reduces administrative costs and stimulates innovation diffusion in industry. Also, private standards provide incentives for voluntary compliance if required in the global supply chain. However, we equally showed the limits of private standardisation in achieving sustainability. The complexity of public interest decision-making in environmental governance may make standard-setting particularly susceptible to private interest influence, leading to the problems of greenwashing regulatory capturing, and reduced competition. In addition, private certification schemes can have limited applications due to high compliance costs and incompatibility with local regimes, especially in common-pool resource governance. The possibilities and limits of private standardisation suggest several conditions under which private standards may help to achieve sustainability. First, the employment of private standardisation should be differentiated, distinguishing private regulation from pure self-regulation. Where pure self-regulation by industry has the greatest dangers of restrictions of competition capturing and greenwashing, the same should not necessarily be the case with private regulation, especially when (private) standards are set and enforced by third parties (in other words with external oversight). Second, voluntary compliance can be encouraged by increasing market demands for certified products and lowering the transaction costs associated with private certifications. Third, the diversity of private certification schemes may help reduce monopolies and incompatibility with local requirements.

It should be acknowledged that the way in which private standards in the field of sustainability are introduced is far from homogenous. We merely focused on several examples (e.g., LEED for green buildings, FSC, and MSC) to illustrate how private regulation works. We do realise, however, that there are other examples we could not discuss within the scope of this chapter, such as the role of private standards in emissions reduction, or more generally, standards against climate change. Including more examples would also allow a comparative analysis, pointing out the successes and failures of the different regimes. The limited analysis we provided in this chapter did not allow for such an approach, which would undoubtedly be a valuable subject of future research.

Notes

1 Postdoctoral researcher at Tsinghua University (Beijing, China).

2 Professor of comparative and international environmental law at Maastricht University and professor of comparative private law and economics at Erasmus School of Law (Rotterdam), both in the Netherlands.

3 SJ Callan and JM Thomas, *Environmental Economics & Management: Theory, Policy, and Applications* (6th ed.) (South-Western Publishing Company, 2013) 53–54.

4 S Shavell, 'A Fundamental Enforcement Cost Advantage of the Negligence Rule over Regulation' (2013) 42 (2) The Journal of Legal Studies 275; S Shavell, 'Liability for Accidents' in AM Polinsky and S Shavell (eds), *Handbook of Law and Economics* (Amsterdam, Elsevier, 2007) 151.

5 WJ Baumol, & WE Oates, *The Theory of Environmental Policy* (Cambridge, Cambridge University Press, 1988) 211–230.
6 S Shavell, 'The Corrective Tax versus Liability as Solutions to the Problem of Harmful Externalities' (2011) 54 (4) The Journal of Law & Economics 250–265.
7 S Shavell, 'The Corrective Tax versus Liability as Solutions to the Problem of Harmful Externalities' (2011) 54 (4) The Journal of Law & Economics 250–265.
8 J Grand, 'The Theory of Government Failure' (1991) 21 (4) British Journal of Political Science 438; B Orbach, 'What Is Government Failure' (2013) 30 (44) Yale Journal on Regulation 45; MG Faure, 'The Complementary Roles of Liability, Regulation, and Insurance in Safety Management: Theory and Practice' (2014) 17 (6) The Journal of Risk Research 690.
9 MG Faure, & Y Shen, 'Environmental Liability as a Tool to Promote Sustainability' in B Akkermans and G van Dijck (eds), *Sustainability and Private Law* (The Hague, Eleven International Publishing 2020) 79–115.
10 S Shavell, 'A Fundamental Enforcement Cost Advantage of the Negligence Rule over Regulation' (2013) 42 (2) The Journal of Legal Studies 275.
11 L Kaplow and S Shavell, 'Property Rules versus Liability Rules: An Economic Analysis' (1996) 109 Harvard Law Review 715.
12 NJ Philipsen, 'The Role of Private Actors in Preventing Work-Related Risks: A Law and Economics Perspective' (2018) 24 (3) European Public Law 545.
13 Y Zou, 'Certifying Green Buildings in China: LEED vs. 3-Star' (2018) 10 Journal of Cleaner Production 208 https://doi.org/10.1016/j.jclepro.2018.10.204.
14 Forest Stewardship Council (FSC), 'FSC Forum with the Yong'an Forestry Group on Forests Management Standard' (1 August 2019), available at https://cn.fsc.org/cn-cn/newsroom/id/222; J Zhao et al., 'Current Status and Problems in Certification of Sustainable Forest Management in China' (2011) 48 (6) Environmental Management 1086.
15 AI Ogus, 'Self-Regulation' in B Bouckaert and G De Geest (eds), *Encyclopedia of Law and Economics* (Vol. IX) (Edward Elgar, 2000) 591–592.
16 AI Ogus, *Regulation: Legal Form and Economic Theory* (Oxford, Oxford University Press 1994) 166–171.
17 AI Ogus, *Regulation: Legal Form and Economic Theory* (Oxford, Oxford University Press 1994) 166. Target standards are particularly relevant for the government in dealing with transboundary externalities. At the EU level, for instance, there have been target standards concerning GHGs emissions, renewable energy, climate neutrality, and the use of pesticides. See, e.g., Council Regulation (EU) 2021/1119 of 30 June 2021 establishing the framework for achieving climate neutrality and amending Regulations (EC) No 401/2009 and (EU) 2018/1999 ('European Climate Law') OJ L 243/1 , Council Directive (2000/60/EC) 23 October 2000 establishing a framework for Community action in the field of water policy OJ L 327/1, Council Directive (2018/2001/EU) of 11 December 2018 on the promotion of the use of energy from renewable sources OJ L 328/82.
18 AI Ogus, *Regulation: Legal Form and Economic Theory* (Oxford, Oxford University Press, 1994) 151, 166.
19 Y Shen, & M Faure, 'Private Standards for the Public Interest? Evidence from Environmental Standardization in China (2021) 30 (3) Review of European, Comparative & International Environmental Law 405.
20 AI Ogus, *Regulation: Legal Form and Economic Theory* (Oxford, Oxford University Press, 1994) 167.
21 AI Ogus, *Regulation: Legal Form and Economic Theory* (Oxford, Oxford University Press, 1994) 167.
22 C Boussalis, Y Feldman, & HE Smith, 'Experimental Analysis of the Effect of Standards on Compliance and Performance' (2018) 12 (2) Regulation and Governance 277.
23 M Mataija, 'Standard Setting, Competition, and Trade' in *Private Regulation and the Internal Market: Sports, Legal Services, and Standard Setting in EU Economic Law* (Oxford Academic, online edn, 2016) 224–253.
24 S Washington, & L Ababouch, Private Standards and Certification in Fisheries and Aquaculture: Current Practice and Emerging Issues. FAO Fisheries and Aquaculture Technical Paper. No. 553 (Rome, FAO, 2011) 8–9 (Published in Economics 2011).
25 ISO, 'ISO 14001 and Related Standards: Environmental Management,' available at https://www.iso.org/iso-14001-environmental-management.html, accessed 25 September 2023.

26 NSF, 'NSF/ANSI 244 Technical requirements', available at https://d2evkimvhatqav.cloudfront.net/documents/dw_technical_requirements_of_nsf_ansi_244.pdf?v=1594928613 accessed 25 September 2023.
27 USGBC, 'LEED Rating System,' available at https://www.usgbc.org/leed, accessed 25 September 2023.
28 Carbon Trust, available at https://www.carbontrust.com/what-we-do/assurance-and-labelling/carbon-neutral-certification, accessed 25 September 2023.
29 The Cradle-to-Cradle Certified standards evaluate products across five categories of sustainability performance, namely material health, product circularity, clean air and climate protection, water and soil stewardship, and social Fairness. For more information. For more information, see 'Introducing the Cradle to Cradle Certified Product Standard Version 4.0', available at https://www.c2ccertified.org/get-certified/cradle-to-cradle-certified-version-4, accessed 25 September 2023.
30 SJ Callan, & JM Thomas, *Environmental Economics & Management: Theory, Policy, and Applications* (6th ed.) (South-Western Publishing Company, 2013) 98–100.
31 S Washington, & L Ababouch, Private Standards and Certification in Fisheries and Aquaculture: Current Practice and Emerging Issues. FAO Fisheries and Aquaculture Technical Paper. No. 553 (Rome, FAO, 2011) 8–9 (Published in Economics, 2011) 8–9.
32 FSC, 'The 10 FSC Principles,' available at https://fsc.org/en/fsc-standards, accessed 25 September 2023.
33 See, e.g., ISO 9001:2015 Requirements, available at http://iso9001.certification-requirements.com/what_are_first-__second-__and_.html; S Washington & L Ababouch, Private Standards and Certification in Fisheries and Aquaculture: Current Practice and Emerging Issues. FAO Fisheries and Aquaculture Technical Paper. No. 553 (Rome, FAO. 2011) 8–9 (Published in Economics, 2011) 10.
34 B Cashore, GC van Kooten, L Vertinsky, G Auld, & J Affolderbach, 'Private or Self-Regulation? A Comparative Study of Forest Certification Choices in Canada, the United States and Germany' (2005) 7 (1) Forest Policy and Economics 54.
35 F Cafaggi, 'New Foundations of Transnational Private Regulation' (2011) 38 (1) Journal of Law and Society 21.
36 SR Miller, 'Enforcement of Local Green Building Ordinances Integrating Third-Party Rating Systems' (2009) 27 (1) California Real Property Journal 57.
37 SB Schinder, 'Following Industry's LEED: Municipal Adoption of Private Green Building Standards' (2010) 62 (2) Florida Law Review 317.
38 MA Wolf, 'A Yellow Light for "Green Zoning": Some Words of Caution About Incorporating Green Building Standards into Local Land Use Law' (2011) 43 (4) Urban Lawyer 951.
39 AI Ogus, 'Self-Regulation' in B Bouckaert and G De Geest (eds), *Encyclopedia of Law and Economics* (Vol. IX) (Edward Elgar, 2000) 591.
40 China National Institute of Standardization (CNIS), *Report on Standardization by Social Groups in China 2020* (Beijing, China Quality and Standards Publishing & Media Co., Ltd., 2020) 134–137.
41 China National Institute of Standardization (CNIS), *Report on Standardization by Social Groups in China 2020* (Beijing, China Quality and Standards Publishing & Media Co., Ltd., 2020) 134–137.
42 NJ Philipsen, 'The Role of Private Actors in Preventing Work-Related Risks: A Law and Economics Perspective' (2018) 24 (3) European Public Law 544; AI Ogus, 'Self-Regulation' in B Bouckaert and G De Geest (eds), *Encyclopedia of Law and Economics* (Vol. IX) (Edward Elgar, 2000) 594–596.
43 DR Nurrochmat et al., 'Contesting National and International Forest Regimes: Case of Timber Legality Certification for Community Forests in Central Java, Indonesia' (2014) 68 (C) Forest Policy and Economics 54–64.
44 A Wells, 'Verification of Legal Compliance in Indonesia' in D. Brown et al. (eds), *Legal Timber: Verification and Governance in the Forest Sector* (London, ODI, 2008) 173–186, 181–82.
45 NJ Philipsen, 'The Role of Private Actors in Preventing Work-Related Risks: A Law and Economics Perspective' (2018) 24 (3) European Public Law 545.
46 Coldwell Banker Richard Ellis (CBRE), 'Toward Excellence: Market Performance of Green Commercial Buildings in the Greater China Region' (CBRE Research 2016), available at https://kapost-files-prod.s3.amazonaws.com/uploads/direct/1467740025-63-7415/China_Report_FINAL.pdf 3–5, accessed 25 September 2023.
47 C Stefanadis, 'Self-Regulation, Innovation, and the Financial Industry' (2003) 23 (1) Journal of Regulatory Economics 5.
48 S Cassese, 'Global Administrative Law: The State of the Art' (2015) 13 (2) International Journal of Constitutional Law 466 https://doi.org/10.1093/icon/mov022.

49 N Krisch, & B Kingsbury, 'Introduction: Global Governance and Global Administrative Law in the International Legal Order' (2006) 17 (1) European Journal of International Law 6–7.
50 W Mattli, & T Büthe, 'Global Private Governance: Lessons from a National Model of Setting Standards in Accounting' (2005) 68 (3) Law and Contemporary Problems 252–253.
51 K-H Ladeur, 'Global Administrative Law: A Transnational Perspective' in P Zumbansen (ed), *The Oxford Handbook of Transnational Law* (Oxford University Press, 2021) 157–178.
52 G Auld, & LH Gulbrandsen, 'Private Regulation in Global Environmental Governance' in R Falkner (ed), *The Handbook of Global Climate and Environment Policy* (John Wiley & Sons, 2013) 394–411.
53 NJ Philipsen, 'The Role of Private Actors in Preventing Work-Related Risks: A Law and Economics Perspective' (2018) 24 (3) European Public Law 545.
54 R van den Bergh, 'Farewell Utopia? Why the European Union Should Take the Economics of Federalism Seriously' (2016) 23 (6) Maastricht Journal of European and Comparative Law 937 https://doi.org/10.1177/1023263X160230060; SR Miller, 'Enforcement of Local Green Building Ordinances Integrating Third-Party Rating Systems' (2009) 27 (2) California Real Property Journal 57.
55 P Grajzl, & P Murrell, 'Allocating Lawmaking Powers: Self-Regulation vs. Government Regulation' (2007) 35 (3) Journal of Comparative Economics 520 https://doi.org/10.1016/j.jce.2007.01.001.
56 K Chen, X Wang, & H Song, 'Food Safety Regulatory Systems in Europe and China: A Study of How Co-Regulation Can Improve Regulatory Effectiveness' (2015) 14 (11) Journal of Integrative Agriculture 2205 https://doi.org/10.1016/S2095-3119(15)61113-3.
57 R Moruzzo et al., 'Urban Consumer Trust and Food Certifications in China' (2020) 9 (9) Foods 1153 https://doi.org/10.3390/foods9091153.
58 B Wong, 'China's Packaged Food Market: Defining Characteristics' (HKTDC Research, 27 March 2018), available at https://research.hktdc.com/en/article/NDMyMjA5MDI2, accessed 25 September 2023.
59 International Trade Center (ITC), 'Organic Food Products in China: Market Overview' (ITC 2011) 11–12.
60 L Bai et al., 'Implementation of HACCP System in China: A Survey of Food Enterprises Involved' (2007) 18 (9) Food Control 1108 https://doi.org/10.1016/j.foodcont.2006.07.006; K Chen, X Wang, & H Song, 'Food Safety Regulatory Systems in Europe and China: A Study of how Co-Regulation Can Improve Regulatory Effectiveness' (2015) 14 (11) Journal of Integrative Agriculture 2205 https://doi.org/10.1016/S2095-3119(15) 61113-3.
61 L Krämer, 'On the Interrelation between Consumer and Environmental Policies in the European Community' (1993) 16 (3–4) Journal of Consumer Policy 455–467 https://doi.org/10.1007/BF01018767.
62 RJ Kis, 'Using Legislation to Reduce One-Time Plastic Bag Usage' (2018) 38 (2) Economic Affairs 224–239; S Gibbens, 'See the Complicated Landscape of Plastic Bans in the US,' August16, 2019, *National Geographic*, available at https://www.nationalgeographic.com/environment/article/map-shows-the-complicated-landscape-of-plastic-bans, accessed 25 September 2023.
63 RJ Kis, 'Using Legislation to Reduce One-Time Plastic Bag Usage' (2018) 38 (2) Economic Affairs 224–239; S Gibbens, 'See the Complicated Landscape of Plastic Bans in the US,' August16, 2019, *National Geographic*, available at https://www.nationalgeographic.com/environment/article/map-shows-the-complicated-landscape-of-plastic-bans, accessed 25 September 2023.
64 Carbon Disclosure Project, 2013, available at http://www.cdproject.net/en-US/Programmes/Pages/forests.aspx, accessed 25 September 2023.
65 A Mammadova, M Masiero, & D Pettenella, 'Embedded Deforestation: The Case Study of the Brazilian–Italian Bovine Leather Trade' (2020) 11 (4) Forests 472.
66 Soy Moratorium, 2012, available at http://www.mvo.nl/Portals/0/publicaties/Magazine/2012/17/02%20press%20release%20ABIOVE.pdf, accessed 25 September 2023.
67 J Miller, 'JBS Recommits to Cattle Agreement in the Amazon,' 20 December 2012, available at https://www.greenpeace.org/usa/jbs-recommits-to-cattle-agreement-in-the-amazon/, accessed 25 September 2023.
68 Greenpeace International (2006), 'Eating up the Amazon,' available at URL http://www.greenpeace.org/usa/en/media-center/reports/ eating-up-the-amazon/; Greenpeace International (2009), 'Slaughtering the Amazon.' available at http://www.greenpeace.org/international/en/publications/ reports/slaughtering-the-amazon/, accessed 25 September 2023.
69 Greenpeace, 'Slaughtering the Amazon', October, 2009, available at https://www.greenpeace.org/usa/wp-content/uploads/legacy/Global/usa/planet3/PDFs/slaughtering-the-amazon-part-1.pdf, accessed 25

September 2023; A Barrionuevo, 'Giants in Cattle Industry Agree to Help Fight Deforestation', October, 6, 2009, press release on The New York Times, available at https://www.nytimes.com/2009/10/07/world/americas/07deforest.html, accessed 25 September 2023.

70 N Walker, S Patel, F Davis, S Milledge, & J Hulse, *Demand-Side Interventions to Reduce Deforestation and Forest Degradation* (International Institute for Environment and Development, Gray's Inn Road, 2013) 80–86.

71 F Cafaggi, & A Renda, 'Measuring the Effectiveness of Transnational Private Regulation,' Strijbis Foundation Report, 2014, available at https://hdl.handle.net/1814/39566, accessed 25 September 2023.

72 F Moffette, M Skidmore, & HK Gibbs, 'Environmental Policies That Shape Productivity: Evidence from Cattle Ranching in the Amazon' (2021) 109 (484) Journal of Environmental Economics and Management 102490 https://doi.org/10.1016/j.jeem.2021.102490.

73 HK Gibbs et al., 'Did Ranchers and Slaughterhouses Respond to Zero-Deforestation Agreements in the Brazilian Amazon?' (2016) 9 (1) Conservation Letters 32–42 https://doi.org/10.1111/conl.12175.

74 HK Gibbs et al., 'Did Ranchers and Slaughterhouses Respond to Zero-Deforestation Agreements in the Brazilian Amazon?' (2016) 9 (1) Conservation Letters 32–42 https://doi.org/10.1111/conl.12175.

75 HK Gibbs et al., 'Did Ranchers and Slaughterhouses Respond to Zero-Deforestation Agreements in the Brazilian Amazon?' (2016) 9 (1) Conservation Letters 39 https://doi.org/10.1111/conl.12175.

76 F Moffette, M Skidmore, & HK Gibbs, 'Environmental Policies That Shape Productivity: Evidence from Cattle Ranching in the Amazon' (2021) 109 (484) Journal of Environmental Economics and Management 102490 https://doi.org/10.1016/j.jeem.2021.102490.

77 HK Gibbs et al., 'Did Ranchers and Slaughterhouses Respond to Zero-Deforestation Agreements in the Brazilian Amazon?' (2016) 9 (1) Conservation Letters 32–42 https://doi.org/10.1111/conl.12175 39.

78 ME Skidmore et al., 'Cattle Ranchers and Deforestation in the Brazilian Amazon: Production, Location, and Policies' (2021) 68 (May) Global Environmental Change 102280 https://doi.org/10.1016/j.gloenvcha.2021.102280.

79 N Walker, S Patel, F Davis, S Milledge, & J Hulse, *Demand-Side Interventions to Reduce Deforestation and Forest Degradation* (International Institute for Environment and Development, Gray's Inn Road, 2013) 80–86.

80 HK Gibbs et al., 'Did Ranchers and Slaughterhouses Respond to Zero-Deforestation Agreements in the Brazilian Amazon?' (2016) 9 (1) Conservation Letters 32–42 https://doi.org/10.1111/conl.12175 32.

81 I Ayres, & J Braithwaite, *Responsive Regulation: Transcending the Deregulation Debate* (Oxford University Press, 1992) 103.

82 J Lee, G Gereffi, & J Beauvais, 'Global Value Chains and Agrifood Standards: Challenges and Possibilities for Smallholders in Developing Countries' (2012) 109 (31) Proceedings of the National Academy of Sciences of the United States of America 12326–12331 https://doi.org/10.1073/pnas.0913714108; I Ayres, & J Braithwaite, *Responsive Regulation: Transcending the Deregulation Debate* (Oxford University Press 1992), 120.

83 USGBC, 'LEED Rating System,' available at https://www.usgbc.org/leed, accessed 25 September 2023.

84 K Hirokawa, 'At Home with Nature: Early Reflections on Green Building Laws and the Transformation of the Built Environment' (2012) 39 (3) Environmental Law 526.

85 MacGraw-Hill Construction, 'World Green Building Trends: Business Benefits Driving New & Retrofit Market Opportunities in over 60 Countries' (Published Feb., 2013) 41, available at http://naturalleader.com/wp-content/uploads/2016/04/WorldGreenBuildingTrendsSmartMarketReport-2013-Final-Full.pdf, accessed 25 September 2023.

86 MacGraw-Hill Construction, 'World Green Building Trends: Business Benefits Driving New & Retrofit Market Opportunities in over 60 Countries' (Published Feb., 2013) 41, available at http://naturalleader.com/wp-content/uploads/2016/04/WorldGreenBuildingTrendsSmartMarketReport-2013-Final-Full.pdf, accessed 25 September 2023.

87 CBRE, 'Toward Excellence: Market Performance of Green Commercial Buildings in the Greater China Region.' CBRE research report (July 2016) 3. Full text available at https://www.cbre.com/research-and-reports/China-Major-Report---Towards-Excellence---Market-Performance-of-Green-Commercial-Buildings-in-the-Greater-China-Region 7, accessed 25 September 2023.

88 CBRE, 'Toward Excellence: Market Performance of Green Commercial Buildings in the Greater China Region.' CBRE research report (July 2016) 3. Full text available at https://www.cbre.com/research-and-reports/China-Major-Report---Towards-Excellence---Market-Performance-of-Green-Commercial-Buildings-in-the-Greater-China-Region 7, accessed 25 September 2023, 3–5.

89 For more information about the state and local tax reductions for GB in the US, see USGBC, 'Better Buildings, Better Policy: A Compilation of Green Building Policy Adoptions in the United States (2011–2014)', the USGBC resources, 19 August 2014, full text, available at http://www.usgbc.org/sites/default/files/STATE_WINS_REPORT_2014-3.pdf, last visited 1March 2017.
90 RCW 39.35D.080.
91 JH Adler, 'Free & Green: A New Approach to Environmental Protection' (2001) 34 Harvard Journal of Law & Public Policy 690–691.
92 20 ILCS 3130/1, Sec. 15 (g).
93 T Simcoe, & MW Toffel, 'Government Green Procurement Spillovers: Evidence from Municipal Building Policies in California' (2014) 68 (3) Journal of Environmental Economics and Management 413 https://doi.org/10.1016/j.jeem.2014.09.001.
94 AI Ogus, 'Self-Regulation' in B Bouckaert and G De Geest (eds), *Encyclopedia of Law and Economics* (Vol. IX) (Edward Elgar, 2000) 591–592.
95 S Del Percio, 'Revisiting Allied Tube and Noerr: The Antitrust Implications of Green Building Legislation and Case Law Considerations for Policymakers' (2009) 34 William & Mary Environmental Law and Policy Review 239–342; C Overdevest & J Zeitlin, 'Assembling an Experimentalist Regime: Transnational Governance Interactions in the Forest Sector' (2014) 8 (1) Regulation & Governance 31–34.
96 See USGBC, 'USGBC Announces New LEED Pilot ACP Designed to Help Eliminate Irresponsibly Sourced Materials—Like Illegal Wood—From the Building Material Supply Chain,' Published on 5 April 2016, available at http://www.usgbc.org/articles/usgbc-announces-new-leed-pilot-acp-designed-help-eliminate-irresponsibly-sourced-materials—, last visited 1 March 2017; USGBC, 'LEED 2009 for New Construction and Major Renovations', p. 55,available at http://www.usgbc.org/Docs/Archive/General/Docs5546.pdf, accessed 25 September 2023.
97 DA Prum, RJ Aalberts, & S Del Percio, 'In Third Parties We Trust? The Growing Antitrust Impact of Third-Party Green Building Certification Systems for State and Local Governments' (2012) 27 (1) Journal of Environmental Law and Litigation 222.
98 JA Grodsky, 'Certified Green: The Law and Future of Environmental Labelling' (1993) 147 Yale Journal on Regulation 167–169.
99 M Mataija, 'Standard Setting, Competition, and Trade' in *Private Regulation and the Internal Market: Sports, Legal Services, and Standard Setting in EU Economic Law* (Oxford Academic, online edn, 2016) 224–253.
100 C Overdevest, & J Zeitlin, 'Assembling an Experimentalist Regime: Transnational Governance Interactions in the Forest Sector' (2014) 8 (1) Regulation & Governance 34.
101 G Porter, 'Trade Competition and Pollution Standards: "Race to the Bottom" or "Stuck at the Bottom' (1999) 8 (2) The Journal of Environment & Development 133–151; K. Alter, & S Meunier, 'The Politics of International Regime Complexity' (2009) 7 (1) Perspectives on Politics 13–24.
102 See USA Today, 'In US Building Industry: Is It Too Easy to be Green?', 24 October 2012, available at https://www.usatoday.com/story/news/nation/2012/10/24/green-building-leed-certification/1650517/, accessed 25 September 2023.
103 J Stiglitz, 'Regulation and Failure' in D Moss and J Cisternino (eds), *New Perspectives on Regulation* (Cambridge, Tobin Project, 2009) 18.
104 See USA Today, 'In US Building Industry: Is it Too Easy to be Green?', published on 24 October 2012, available at https://www.usatoday.com/story/news/nation/2012/10/24/green-building-leed-certification/1650517/, accessed 25 September 2023.
105 I Ayres, & J Braithwaite, *Responsive Regulation: Transcending the Deregulation Debate* (Oxford University Press, 1992) 57 and 70.
106 FSC, 'Standard-Setting in FSC – FSC-RP-Standard Setting V1-1', 2016, pp. 3–4, available at https://ic.fsc.org/preview.standard-setting-in-fsc2016.a-6098.pdf, accessed 25 September 2023.
107 ISEAL Alliance is an organisation that brings together a diverse range of sustainability standards systems, including those related to environmental, social, and economic issues. Its members include standards-setting organisations, certifiers, NGOs, and other stakeholders involved in sustainability certification. ISEAL, 'About ISEAL', available at https://www.isealalliance.org/about-iseal/who-we-are, accessed 25 September 2023.
108 A Loconto, & E Fouilleux, 'Politics of Private Regulation: ISEAL and the Shaping of Transnational Sustainability Governance' (2014) 8 (2) Regulation & Governance 166–185.
109 FSC, 'Annual Report 2021', available at https://annual-reports.fsc.org/, accessed 25 September 2023.

110 Overdevest & Zeitlin (n96) 35; J Michal et al., 'Analysis of Socioeconomic Impacts of the FSC and PEFC Certification Systems on Business Entities and Consumers' (2019) 11 (15) Sustainability 4122 https://doi.org/10.3390/su11154122.

111 S Wolff, & J Schweinle, 'Effectiveness and Economic Viability of Forest Certification: A Systematic Review' (2022) 13 (5) Forests 798 https://doi.org/10.3390/f13050798.

112 Forest Stewardship Council (FSC), 'FSC Forum with the Yong'an Forestry Group on Forests Management Standard' (1 August 2019), available at https://cn.fsc.org/cn-cn/newsroom/id/222.

113 F Cafaggi, 'New Foundations of Transnational Private Regulation' (2011) 38 (1) Journal of Law and Society 21 36.

114 Forest Stewardship Council (FSC), 'FSC Forum with the Yong'an Forestry Group on Forests Management Standard' (1 August 2019), available at https://cn.fsc.org/cn-cn/newsroom/id/222; WWF, 'Certified Chinese Forest Reaches Million Hectares' (7 January 2009),available at https://wwf.panda.org/?156347/Certified-Chinese-forest, accessed 25 September 2023.

115 M Faure, P Mascini, & J Liu, Environmental Governance and Common Pool Resources: A Comparison of Fishery and Forestry (1st ed.) (Routledge, 2017) 40.

116 K Harada, & W Wiyono, 'Certification of a Community-Based Forest Enterprise for Improving Institutional Management and Household Income: A Case from Southeast Sulawesi, Indonesia' (2014) 13 (1) Small-Scale Forestry 47–64 https://doi.org/10.1007/s11842-013-9240-8.

117 CL McDermott, E Noah, & B Cashore, 'Differences That "Matter"? A Framework for Comparing Environmental Certification Standards and Government Policies' (2008) 10 (1) Journal of Environmental Policy and Planning 47–70.

118 P Lee, C Smyth, & S Boutin, 'Quantitative Review of Riparian Buffer Width Guidelines from Canada and the United States' (2004) 70 (2) Journal of Environmental Management 175 https://doi.org/10.1016/j.jenvman.2003.11.009.

119 M Faure, P Mascini, & J Liu, *Environmental Governance and Common Pool Resources: A Comparison of Fishery and Forestry* (1st ed.) (Routledge, 2017) 80.

120 M Faure, P Mascini, & J Liu, *Environmental Governance and Common Pool Resources: A Comparison of Fishery and Forestry* (1st ed.) (Routledge, 2017) 81.

121 M Faure, P Mascini, & J Liu, *Environmental Governance and Common Pool Resources: A Comparison of Fishery and Forestry* (1st ed.) (Routledge, 2017) 61.

122 A Hinrichs, DR Muhtaman, & N Irianto, *Forest Certification on Community Land in Indonesia* (Deutsche Gesellschaft für Technische Zusammenarbeit, 2008) 56.

123 A Hinrichs, DR Muhtaman, & N Irianto, *Forest Certification on Community Land in Indonesia* (Deutsche Gesellschaft für Technische Zusammenarbeit, 2008) 58–61.

124 K Harada, & W Wiyono, 'Certification of a Community-Based Forest Enterprise for Improving Institutional Management and Household Income: A Case from Southeast Sulawesi, Indonesia' (2014) 13 (1) Small-Scale Forestry 47–64 https://doi.org/10.1007/s11842-013-9240-8.

125 M Faure, P Mascini, & J Liu, *Environmental Governance and Common Pool Resources: A Comparison of Fishery and Forestry* (1st ed.) (Routledge, 2017) 41.

126 M Faure, P Mascini, & J Liu, *Environmental Governance and Common Pool Resources: A Comparison of Fishery and Forestry* (1st ed.) (Routledge, 2017) 46-47.

127 O Espinoza & MJ Dockry, 'Forest Certification in Bolivia: A Status Report and Analysis of Stakeholder Perspectives' (2014) 64 (3–4) Forest Products Journal 80–89.

128 PPacheco, D Barry, P Cronkleton, & AM Larson, *The Role of Informal Institutions in the Use of Forest Resources in Latin America* (Bogor: Center for International Forestry Research, CIFOR, 2008) 40–44.

129 DC North, *Transaction Costs, Institutions, and Economic Performance* (International Center for Economic Growth Publication Press, 1992).

130 M Faure, P Mascini, & J Liu, Environmental Governance and Common Pool Resources: A Comparison of Fishery and Forestry (1st ed.) (Routledge, 2017) 46.

131 T Bartley, 'Transnational Governance as the Layering of Rules: Intersections of Public and Private Standards' (2011) 12 (2) Theoretical Inquiries in Law 517–542 https://doi.org/10.2202/1565-3404.1278.

132 A Hinrichs, DR Muhtaman, & N Irianto, *Forest Certification on Community Land in Indonesia* (Deutsche Gesellschaft für Technische Zusammenarbeit, 2008) 26.

133 M Faure, P Mascini, & J Liu, *Environmental Governance and Common Pool Resources: A Comparison of Fishery and Forestry* (1st ed.) (Routledge, 2017) 46–47.

134 G Nebel, L Quevedo, J Bredahl, & F Helles, 'Development and Economic Significance of Forest Certification: The Case of FSC in Bolivia' (2005) 7 (2) Forest Policy and Economics 175–186 https://doi.org/10.1016/S1389-9341(03)00030-3.
135 G Nebel, L Quevedo, J Bredahl, & F Helles, 'Development and Economic Significance of Forest Certification: The Case of FSC in Bolivia' (2005) 7 (2) Forest Policy and Economics 175–186 https://doi.org/10.1016/S1389-9341(03)00030-3.
136 K De Pourcq, E Thomas, & P van Damme, 'Indigenous Community-Based Forestry in the Bolivian Lowlands: Some Basic Challenges for Certification' (2009) 11 (1) International Forestry Review 12–26; M Markopoulos, 'The Role of Certification in Community-Based Forest Enterprise,' in E Meidinger, C Elliott & G Oesten (eds), *Social and Political Dimensions of Forest Certification* (Berlin, Verlag 2002) 105–131.
137 K De Pourcq, E Thomas, & P van Damme, 'Indigenous Community-Based Forestry in the Bolivian Lowlands: Some Basic Challenges for Certification' (2009) 11 (1) International Forestry Review 12–26; M Markopoulos, 'The Role of Certification in Community-Based Forest Enterprise' in E Meidinger, C Elliott & G Oesten (eds), *Social and Political Dimensions of Forest Certification* (Berlin, Verlag 2002) 105–131.
138 MSC, 'MSC Standard-Setting Procedure,' 5 July 2018, available at https://www.msc.org/docs/default-source/default-document-library/msc-standard-setting-procedure.pdf?sfvrsn=dfda000b_14, accessed 25 September 2023.
139 LH Gulbrandsen, *Transnational Environmental Governance: The Emergence and Effects of the Certification of Forest and Fisheries* (Edward Elgar, 2010) 125.
140 Council MS. MSC Fisheries Certification Requirements and Guidance. Version. 2014 Oct 1;2:81–87; MSC, 'What Does the Blue Label Mean?', available at https://www.msc.org/what-we-are-doing/our-approach/what-does-the-blue-msc-label-mean?gclid=EAIaIQobChMImeWt_Ink-wIVC6-WCh1g9QgoEAAYASADEgK8aPD_BwE, accessed 25 September 2023.
141 P Lallemand, M Berg, M Hansen, C Venter, & M Purves, 'An Analysis of the Economic Benefits of MSC Certification for the South African Hake Fishery' (2014) 6–15, available at http://ir.library.oregonstate.edu/xmlui/bitstream/handle/ 1957/55162/Pur ves%20392%20IIFET.pdf?sequence=1, accessed 25 September 2023.
142 B Herzog, *New Zealand Experience with Rights-Based Fisheries Management – Unfinished Business* (Eburon, 2002) 176.
143 M Faure, P Mascini, & J Liu, *Environmental Governance and Common Pool Resources: A Comparison of Fishery and Forestry* (1st ed.) (Routledge, 2017) 130.
144 MSC, Reassessment Report for South Africa Trawl Fishery, 2007, condition 4.
145 M Faure, P Mascini, & J Liu, *Environmental Governance and Common Pool Resources: A Comparison of Fishery and Forestry* (1st ed.) (Routledge, 2017) 152.
146 B Phillips, T, Ward, & C Chaffee, *Eco-Labelling in Fisheries: What Is It All About* (New Jersey, Wiley-Blackwell, 2008) 173.
147 C Roheim, F Asche, & J Santos, 'The Elusive Price Premium for Ecolabelled Products: Evidence from Seafood in the UK Market' (2011) 62 (3) Journal of Agricultural Economics 655–668 https://doi.org/10.1111/j.1477-9552.2011.00299.x; F Asche et al., 'Pricing of Eco-Labels with Retailer Heterogeneity' (2015) 53 (C) Food Policy 82–93.
148 A Pierucci, S Columbu, & IT Kell, 'A Global Review of MSC Certification: Why Fisheries Withdraw?' (2022) 143 Marine Policy 105124 https://doi.org/10.1016/j.marpol.2022.105124; M Wakamatsu, & H Wakamatsu, 'The Certification of Small-Scale Fisheries' (2017) 77 Marine Policy 97–103 https://doi.org/10.1016/j.marpol.2016.12.016.
149 M Pérez-Ramírez, B Phillips, D Lluch-Belda, & S Lluch-Cota, 'Perspectives for Implementing Fisheries Certification in Developing Countries' (2012) 36 (2) Marine Policy 297–302 https://doi.org/10.1016/j.marpol.2011.06.013; S Ponte, 'The Marine Stewardship Council (MSC) and the Making of a Market for "Sustainable Fish"' (2012) 12 (2–3) Journal of Agrarian Change 300–315 https://doi.org/10.1111/j.1471-0366.2011.00345.x.
150 Ibid.
151 AIzawa, & M Makino, 'MSC Certification and Its Implementation for Japan's Fisheries – Its Role and Issues' (2005) 9 (2) Global Environmental Research 151–156, 154.
152 G Sogn-Grundvåg, TA Larsen, & JA Young, 'The Value of Line-Caught and Other Attributes: An Exploration of Price Premiums for Chilled Fish in UK Supermarkets' (2013) 38 Marine Policy 41–44 https://

doi.org/10.1016/j.marpol.2012.05.017; C-HJ Sun, FS Chiang, M Owens, & D Squires, 'Will American Consumers Pay More for Eco-Friendly Labelled Canned Tuna? Estimating us Consumer Demand for Canned Tuna Varieties Using Scanner Data' (2017) 79 Marine Policy 62–69 https://doi.org/10.1016/j.marpol.2017.02.006.

153 LM Bellchambers, BF Phillips, & M Pérez-Ramírez, 'From Certification to Recertification the Benefits and Challenges of the Marine Stewardship Council (msc): A Case Study Using Lobsters' (2016) 182 (C) Fisheries Research 88–97 https://doi.org/10.1016/j.fishres.2015.08.029; J Blomquist, V Bartolino, & S Waldo, 'Price Premiums for Eco-Labelled Seafood: Effects of the msc Certification Suspension in the Baltic Sea Cod Fishery' (2020) 47 (1) European Review of Agricultural Economics 50–70 https://doi.org/10.1093/erae/jby047.

154 I Van Putten et al., 'Shifting Focus: The Impacts of Sustainable Seafood Certification' (2020) 15 (5) PLoS One: Article e0233237 https://doi.org/10.1371/journal.pone.0233237.

155 D Gondor, & H Morimoto, 'Role of World Wildlife Fund (WWF) and Marine Stewardship Council (MSC) in Seafood Eco-Labelling Policy in Japan' (2011) 2 (2) Sustainability Accounting, Management and Policy Journal 214–230 https://doi.org/10.1108/20408021111185385.

156 A Izawa, & M Makino, 'MSC Certification and Its Implementation for Japan's Fisheries – Its Role and Issues' (2005) 9 (2) Global Environmental Research 151–156; H Wakamatsu, H Uchida, CA Roheim, CM Anderson, & S Managi, 'Consumer Preferences for Eco-Labelled Seafood in Japan and the Influence of Information on the Preferences' in IIFET Montpellier Proceeding 2010, pp. 1– 12, available at http://ir/library.oregonstate.edu/xmlui/handle/1957/39233, accessed 25 September 2023; D Gondor, & H Morimoto, 'Role of World Wildlife Fund (WWF) and Marine Stewardship Council (MSC) in Seafood Eco-Labelling Policy in Japan' (2011) 2 (2) Sustainability Accounting, Management and Policy Journal 214–230 https://doi.org/10.1108/20408021111185385.

157 M Faure, P Mascini, & J Liu, *Environmental Governance and Common Pool Resources: A Comparison of FisheryandForestry*(1sted.)(Routledge,2017) 127.Alsosee,availableatwww.msc.org/track-a-fishery/fisheries-in-the-program/certified/south-atlantic-indian-ocean/south-africa-hake-trawl-fishery/sustainability-notes, accessed 25 September 2023.

158 E Aalders, J Akroyd, T Ward, & K Short, 'Case Study 4: The New Zealand Hoki, A: The Fishery' in B Phillips, T Ward & C Chaffee (eds), *Eco-Labelling in Fisheries: What Is It All About?* (New Jersey, Wiley-Blackwell, 2008) 152–153; B Weeber, & C Wallace, 'The New Zealand Best Fish Guide' in T Ward & B Phillips (eds), *Seafood Ecolabelling: Principles and Practice* (New Jersey, John Wiley & Sons 2008) 367–388.

159 Deepwater Group Limited, Sustainable Management of New Zealand's Hoki Fisheries, 2011, p. 2. Full text available at www.cport.net/assets/uploads/files/Sustainable %20Management%20of%20New%20Zealand%20Hoki%20120711.pdf, accessed 25 September 2023.

160 M Faure, P Mascini, & J Liu, *Environmental Governance and Common Pool Resources: A Comparison of Fishery and Forestry* (1st ed.) (Routledge, 2017) 129.

161 S Ponte, 'Greener Than Thou: The Political Economy of Fish Ecolabeling and Its Local Manifestations in South Africa' (2008) 36 (1) World Development 168 https://doi.org/10.1016/j.worlddev.2007.02.014.

162 S Ponte, 'Greener Than Thou: The Political Economy of Fish Ecolabeling and Its Local Manifestations in South Africa' (2008) 36 (1) World Development 168 https://doi.org/10.1016/j.worlddev.2007.02.014.

163 P Foley, & E Havice, 'The Rise of Territorial Eco-Certifications: New Politics of Transnational Sustainability Governance in the Fishery Sector' (2016) 69 Geoforum 28 https://doi.org/10.1016/j.geoforum.2015.11.015.

164 P Foley, & E Havice, 'The Rise of Territorial Eco-Certifications: New Politics of Transnational Sustainability Governance in the Fishery Sector' (2016) 69 Geoforum 30 https://doi.org/10.1016/j.geoforum.2015.11.015.

165 K Hirokawa, 'At Home with Nature: Early Reflections on Green Building Laws and the Transformation of the Built Environment' (2012) 39 (3) Environmental Law 526.

166 F Cafaggi, 'A Comparative Analysis of Transnational Private Regulation: Legitimacy, Quality, Effectiveness and Enforcement. Quality, Effectiveness and Enforcement' (June 12, 2014) *EUI LAW*, 26, available at https://www.eesc.europa.eu/sites/default/files/resources/docs/a-comparative-analysis-of-transnational-private-regulation-fcafaggi_12062014.pdf, accessed 25 September 2023.

167 See Section 3.1 of this contribution.

168 See Section 4.1. of this contribution.

169 See Section 4.2 of this contribution.

170 B Hersoug, *New Zealand Experience with Rights-Based Fisheries Management – Unfinished Business* (Eburon, 2002) 176–177.
171 P Foley, & E Havice, 'The Rise of Territorial Eco-Certifications: New Politics of Transnational Sustainability Governance in the Fishery Sector' (2016) 69 Geoforum 28 https://doi.org/10.1016/j.geoforum.2015.11.015.
172 S Del Percio, 'Revisiting Allied Tube and Noerr: The Antitrust Implications of Green Building Legislation and Case Law Considerations for Policymakers' (2009) 34 William & Mary Environmental Law and Policy Review 239–342, 239; AI Ogus, 'Self-Regulation' in B Bouckaert and G De Geest (eds), *Encyclopedia of Law and Economics* (Vol. IX) (Edward Elgar, 2000) 591.
173 See Section 4.2 of this contribution.
174 Y Shen, & MG Faure, 'Smart Instrument Mixes to Promote Green Building' (2018) 49 Duke Environmental Law and Policy Forum 49–108, available at https://scholarship.law.duke.edu/delpf/vol29/iss1/2.
175 AI Ogus, 'Self-Regulation' in B Bouckaert and G De Geest (eds), *Encyclopedia of Law and Economics* (Vol. IX) (Edward Elgar, 2000) 592.
176 M Olson, *The Logic of Collective Action. Public Goods and the Theory of Groups* (Harvard University Press, 1971) 49–51.
177 AI Ogus, *Regulation: Legal Form and Economic Theory* (Oxford, Oxford University Press 1994) 205.
178 AI Ogus, 'Self-Regulation' in B Bouckaert and G De Geest (eds), *Encyclopedia of Law and Economics* (Vol. IX) (Edward Elgar, 2000) 591–592.
179 See Section 3.2 of this contribution.
180 See Section 4.2 of this contribution.
181 F Cafaggi, 'New Foundations of Transnational Private Regulation' (2011) 38 (1) Journal of Law and Society 21 46–47.
182 F Cafaggi, 'New Foundations of Transnational Private Regulation' (2011) 38 (1) Journal of Law and Society 21 50.
183 See Section 4.2 of this contribution.
184 Section 4.1 of this contribution.
185 Section 4.1 of this contribution.
186 M Faure, P Mascini, & J Liu, *Environmental Governance and Common Pool Resources: A Comparison of Fishery and Forestry* (1st ed.) (Routledge, 2017) 61.
187 Section 4.1 of this contribution.
188 See Section 4.2 of this contribution.
189 See Section 4.2 of this contribution.
190 AI Ogus, *Regulation: Legal Form and Economic Theory* (Oxford, Oxford University Press, 1994) 205.
191 GA Akerlof, 'The Market for Lemons: Quality, Uncertainty and the Market Mechanism' (1970) 84 (3) Quarterly Journal of Economics 488 https://doi.org/10.2307/1879431.
192 Y Li, & K Van Veld, 'Green, Greener, Greenest: Eco-Label Gradation and Competition' (2015) 72 (C) Journal of Environmental Economics and Management 164 https://doi.org/10.1016/j.jeem.2015.05.003.
193 F Cafaggi, 'New Foundations of Transnational Private Regulation' (2011) 38 (1) Journal of Law and Society 21, 51.

Bibliography

Aalders E, Akroyd J, Ward T, & Short K, 'Case Study 4: The New Zealand Hoki, A: The Fishery', in B Phillips, T Ward, & C Chaffee (eds.), *Eco-Labelling in Fisheries: What Is It All About?* (New Jersey, Wiley-Blackwell, 2008) 152–153

Adler J, 'Free & Green: A New Approach to Environmental Protection' (2001) 34 Harvard Journal of Law & Public Policy 690–691

Akerlof G, 'The Market for Lemons: Quality, Uncertainty and the Market Mechanism' (1970) 84 (3) Quarterly Journal of Economics 488 https://doi.org/10.2307/1879431

Alter K, & Meunier S, 'The Politics of International Regime Complexity' (2009) 7 (1) Perspectives on Politics 13–24

Asche F et al., 'Pricing of Eco-Labels with Retailer Heterogeneity' (2015) 53 (C) Food Policy, 82–93

Auld G, & Gulbrandsen L, 'Private Regulation in Global Environmental Governance' in R Falkner (ed.), *The Handbook of Global Climate and Environment Policy* (John Wiley & Sons, 2013), 394–411

Ayres I, & Braithwaite J, *Responsive Regulation: Transcending the Deregulation Debate* (Oxford University Press 1992), 103
Bai L et al., 'Implementation of HACCP System in China: A Survey of Food Enterprises Involved' (2007) 18 (9) Food Control 1108 https://doi.org/10.1016/j.foodcont.2006.07.006
Bartley T, 'Transnational Governance as the Layering of Rules: Intersections of Public and Private Standards' (2011) 12 (2) Theoretical Inquiries in Law 517–542 https://doi.org/10.2202/1565-3404.1278
Baumol W, & Oates W, *The Theory of Environmental Policy* (Cambridge, Cambridge University Press 1988) 211–230
Bellchambers L, Phillips B, & Pérez-Ramírez M, 'From Certification to Recertification the Benefits and Challenges of the Marine Stewardship Council (msc): A Case Study Using Lobsters' (2016) 182 (C) Fisheries Research 88–97 https://doi.org/10.1016/j.fishres.2015.08.029
Blomquist J, Bartolino V, & Waldo S, 'Price Premiums for Eco-Labelled Seafood: Effects of the Msc Certification Suspension in the Baltic Sea Cod Fishery' (2020) 47 (1) European Review of Agricultural Economics 50–70 https://doi.org/10.1093/erae/jby047
Boussalis C, Feldman Y, & Smith H, 'Experimental Analysis of the Effect of Standards on Compliance and Performance' (2018) 12 (2) Regulation and Governance 277
Cafaggi F, 'New Foundations of Transnational Private Regulation' (2011) 38 (1) Journal of Law and Society 21
Cafaggi F, 'A Comparative Analysis of Transnational Private Regulation: Legitimacy, Quality, Effectiveness and Enforcement. Quality, Effectiveness and Enforcement' (June 12, 2014) *EUI LAW*, 26 https://www.eesc.europa.eu/sites/default/files/resources/docs/a-comparative-analysis-of-transnational-private-regulation-fcafaggi_12062014.pdf accessed 25 September 2023.
Callan S, & Thomas J, *Environmental Economics & Management: Theory, Policy, and Applications* (6th ed.) (South-Western Publishing Company, 2013) 53–54
Cashore B, van Kooten G, Vertinsky L, Auld G, & Affolderbach J, 'Private or Self-Regulation? A Comparative Study of Forest Certification Choices in Canada, the United States and Germany' (2005) 7 (1) Forest Policy and Economics 54
Cassese S, 'Global Administrative Law: The State of the Art' (2015) 13 (2) International Journal of Constitutional Law 466 https://doi.org/10.1093/icon/mov022
Chen K, Wang X, & Song H, 'Food Safety Regulatory Systems in Europe and China: A Study of How Co-Regulation Can Improve Regulatory Effectiveness' (2015) 14 (11) Journal of Integrative Agriculture 2205 https://doi.org/10.1016/S2095-3119(15)61113-3
China National Institute of Standardization (CNIS), *Report on Standardization by Social Groups in China 2020* (Beijing, China Quality and Standards Publishing & Media Co., Ltd., 2020) 134–137
De Pourcq K, Thomas E, & van Damme P, 'Indigenous Community-Based Forestry in the Bolivian Lowlands: Some Basic Challenges for Certification' (2009) 11 (1) International Forestry Review 12–26
Del Percio S, 'Revisiting Allied Tube and Noerr: The Antitrust Implications of Green Building Legislation and Case Law Considerations for Policymakers' (2009) 34 William & Mary Environmental Law and Policy Review 239–342
Espinoza O, & Dockry M, 'Forest Certification in Bolivia: A Status Report and Analysis of Stakeholder Perspectives' (2014) 64 (3–4) Forest Products Journal 80–89
Faure M, 'The Complementary Roles of Liability, Regulation, and Insurance in Safety Management: Theory and Practice' (2014) 17 (6) The Journal of Risk Research 690
Faure M, Mascini P, & Liu J, *Environmental Governance and Common Pool Resources: A Comparison of Fishery and Forestry* (1st ed.) (Routledge, 2017) 40
Faure M, & Shen Y, 'Environmental Liability as a Tool to Promote Sustainability.' in B Akkermans & G van Dijck (eds.), *Sustainability and Private Law* (The Hague, Eleven International Publishing, 2020) 79–115
Foley P, & Havice E, 'The Rise of Territorial Eco-Certifications: New Politics of Transnational Sustainability Governance in the Fishery Sector' (2016) 69 Geoforum 28 https://doi.org/10.1016/j.geoforum.2015.11.015
Gibbs H et al., 'Did Ranchers and Slaughterhouses Respond to Zero-Deforestation Agreements in the Brazilian Amazon?' (2016) 9 (1) Conservation Letters 32–42 https://doi.org/10.1111/conl.12175
Gondor D, & Morimoto H, 'Role of World Wildlife Fund (WWF) and Marine Stewardship Council (MSC) in Seafood Eco-Labelling Policy in Japan' (2011) 2 (2) Sustainability Accounting, Management and Policy Journal 214–230 https://doi.org/10.1108/20408021111185385
Grajzl P, & Murrell P, 'Allocating Lawmaking Powers: Self-Regulation vs. Government Regulation' (2007) 35 (3) Journal of Comparative Economics 520 https://doi.org/10.1016/j.jce.2007.01.001
Grand J, 'The Theory of Government Failure' (1991) 21 (4) British Journal of Political Science 438

Grodsky J, 'Certified Green: The Law and Future of Environmental Labeling' (1993) 147 Yale Journal on Regulation 167–169

Gulbrandsen L, *Transnational Environmental Governance: The Emergence and Effects of the Certification of Forest and Fisheries* (Edward Elgar, 2010) 125

Harada K, & Wiyono W, 'Certification of a Community-Based Forest Enterprise for Improving Institutional Management and Household Income: A Case from Southeast Sulawesi, Indonesia' (2014) 13 (1) Small-Scale Forestry 47–64 https://doi.org/10.1007/s11842-013-9240-8

Hersoug B, *New Zealand Experience with Rights-Based Fisheries Management – Unfinished Business* (Eburon, 2002) 176–177

Hinrichs A, Muhtaman D, & Irianto N, *Forest Certification on Community Land in Indonesia* (Deutsche Gesellschaft für Technische Zusammenarbeit, 2008) 56

Hirokawa K, 'At Home with Nature: Early Reflections on Green Building Laws and the Transformation of the Built Environment' (2012) 39 (3) Environmental Law 526

Izawa A, & Makino M, 'MSC Certification and Its Implementation for Japan's Fisheries – Its Role and Issues' (2005) 9 (2) Global Environmental Research 151–156

Kaplow L, & Shavell S, 'Property Rules versus Liability Rules: An Economic Analysis' (1996) 109 Harvard Law Review 715

Krämer L, 'On the Interrelation between Consumer and Environmental Policies in the European Community' (1993) 16 (3–4) Journal of Consumer Policy 455–467 https://doi.org/10.1007/BF01018767

Krisch N, & Kingsbury B, 'Introduction: Global Governance and Global Administrative Law in the International Legal Order' (2006) 17 (1) European Journal of International Law 6–7

Lee J, Gereffi, G, & Beauvais J, 'Global Value Chains and Agrifood Standards: Challenges and Possibilities for Smallholders in Developing Countries' (2012) 109 (31) Proceedings of the National Academy of Sciences of the United States of America 12326–12331 https://doi.org/10.1073/pnas.0913714108

Lee P, Smyth C, & Boutin S, 'Quantitative Review of Riparian Buffer Width Guidelines from Canada and the United States' (2004) 70 (2) Journal of Environmental Management 175 https://doi.org/10.1016/j.jenvman.2003.11.009

Li Y, & Van Veld K, 'Green, Greener, Greenest: Eco-Label Gradation and Competition' (2015) 72 (C) Journal of Environmental Economics and Management 164 https://doi.org/10.1016/j.jeem.2015.05.003

Loconto A, & Fouilleux E, 'Politics of Private Regulation: ISEAL and the Shaping of Transnational Sustainability Governance' (2014) 8 (2) Regulation & Governance 166–185

Mammadova A, Masiero M, & Pettenella D, 'Embedded Deforestation: The Case Study of the Brazilian–Italian Bovine Leather Trade' (2020) 11 (4) Forests 472

Markopoulos M, 'The Role of Certification in Community-Based Forest Enterprise' in E Meidinger, C Elliott & G Oesten (eds.), *Social and Political Dimensions of Forest Certification* (Berlin, Verlag, 2002) 105–131

Mataija M, 'Standard Setting, Competition, and Trade' in *Private Regulation and the Internal Market: Sports, Legal Services, and Standard Setting in EU Economic Law* (Oxford Academic, online edn, 2016) 224–253

Mattli W, & Büthe T, 'Global Private Governance: Lessons from a National Model of Setting Standards in Accounting' (2005) 68 (3) Law and Contemporary Problems 252–253

McDermott C, Noah E, & Cashore B, 'Differences That "Matter"? A Framework for Comparing Environmental Certification Standards and Government Policies' (2008) 10 (1) Journal of Environmental Policy and Planning 47–70

Michal J et al., 'Analysis of Socioeconomic Impacts of the FSC and PEFC Certification Systems on Business Entities and Consumers' (2019) 11 (15) Sustainability 4122 https://doi.org/10.3390/su11154122

Miller S, 'Enforcement of Local Green Building Ordinances Integrating Third-Party Rating Systems' (2009) 27 (2) California Real Property Journal 57

Moffette F, Skidmore M, & Gibbs H, 'Environmental Policies That Shape Productivity: Evidence from Cattle Ranching in the Amazon' (2021) 109 (484) Journal of Environmental Economics and Management 102490 https://doi.org/10.1016/j.jeem.2021.102490

Moruzzo R et al., 'Urban Consumer Trust and Food Certifications in China' (2020) 9 (9) Foods 1153 https://doi.org/10.3390/foods9091153

Nebel G, Quevedo L, Bredahl J, & Helles F, 'Development and Economic Significance of Forest Certification: The Case of FSC in Bolivia' (2005) 7 (2) Forest Policy and Economics 175–186 https://doi.org/10.1016/S1389-9341(03)00030-3

North D, *Transaction Costs, Institutions, and Economic Performance* (International Center for Economic Growth Publication Press, 1992)

Nurrochmat D et al., 'Contesting National and International Forest Regimes: Case of Timber Legality Certification for Community Forests in Central Java, Indonesia' (2014) 68 (C) *Forest Policy and Economics* 54–64

Ogus A, *Regulation: Legal Form and Economic Theory* (Oxford, Oxford University Press 1994) 166–171.

Ogus A, 'Self-Regulation' in B Bouckaert & G De Geest (eds.), *Encyclopedia of Law and Economics* (Vol. IX) (Edward Elgar, 2000) 591–592

Olson M, *The Logic of Collective Action. Public Goods and the Theory of Groups* (Harvard University Press, 1971) 49–51.

Orbach B, 'What Is Government Failure' (2013) 30 (44) Yale Journal on Regulation 45

Overdevest C, & Zeitlin J, 'Assembling an Experimentalist Regime: Transnational Governance Interactions in the Forest Sector' (2014) 8 (1) Regulation & Governance 31–34

Pacheco P, Barry D, Cronkleton P, & Larson A, *The Role of Informal Institutions in the Use of Forest Resources in Latin America* (Bogor: Center for International Forestry Research, CIFOR, 2008) 40–44

Pérez-Ramírez M, Phillips B, Lluch-Belda D, & Lluch-Cota S, 'Perspectives for Implementing Fisheries Certification in Developing Countries' (2012) 36 (2) Marine Policy 297–302 https://doi.org/10.1016/j.marpol.2011.06.013

Philipsen N, 'The Role of Private Actors in Preventing Work-Related Risks: A Law and Economics Perspective' (2018) 24 (3) European Public Law 545

Phillips B, Ward T, & Chaffee C, *Eco-Labelling in Fisheries: What Is It All About* (New Jersey, Wiley-Blackwell, 2008) 173

Pierucci A, Columbu S, & Kell L, 'A Global Review of MSC Certification: Why Fisheries Withdraw?' (2022) 143 Marine Policy 105124 https://doi.org/10.1016/j.marpol.2022.105124

Ponte S, 'Greener Than Thou: The Political Economy of Fish Ecolabeling and Its Local Manifestations in South Africa' (2008) 36 (1) World Development 168 https://doi.org/10.1016/j.worlddev.2007.02.014

Ponte S, 'The Marine Stewardship Council (MSC) and the Making of a Market for "Sustainable Fish"' (2012) 12 (2–3) Journal of Agrarian Change 300–315 https://doi.org/10.1111/j.1471-0366.2011.00345.x

Porter G, 'Trade Competition and Pollution Standards: "Race to the Bottom" or "Stuck at the Bottom' (1999) 8 (2) The Journal of Environment & Development 133–151

Prum D, Aalberts R, & Del Percio S, 'In Third Parties We Trust? The Growing Antitrust Impact of Third-Party Green Building Certification Systems for State and Local Governments' (2012) 27 (1) Journal of Environmental Law and Litigation 222

R Kis J, Using Legislation to Reduce One-Time Plastic Bag Usage' (2018) 38 (2) Economic Affairs 224–239

Roheim C, Asche F, & Santos J, 'The Elusive Price Premium for Ecolabelled Products: Evidence from Seafood in the UK Market' (2011) 62 (3) Journal of Agricultural Economics 655–668 https://doi.org/10.1111/j.1477-9552.2011.00299.x

Schinder S, 'Following Industry's LEED: Municipal Adoption of Private Green Building Standards' (2010) 62 (2) Florida Law Review 317

Shavell S, 'Liability for Accidents,' in AM Polinsky & S Shavell (eds.), *Handbook of Law and Economics* Amsterdam (Elsevier 2007) 151

Shavell S, 'The Corrective Tax versus Liability as Solutions to the Problem of Harmful Externalities' (2011) 54 (4) The Journal of Law & Economics 250–265

Shavell S, 'A Fundamental Enforcement Cost Advantage of the Negligence Rule over Regulation' (2013) 42 (2) The Journal of Legal Studies 275

Shen Y, & Faure M, 'Smart Instrument Mixes to Promote Green Building' (2018) 49 Duke Environmental Law and Policy Forum 49–108 https://scholarship.law.duke.edu/delpf/vol29/iss1/2

Shen Y, & Faure M, 'Private Standards for the Public Interest? Evidence from Environmental Standardization in China (2021) 30 (3) Review of European, Comparative & International Environmental Law 405

Simcoe T, & Toffel M, 'Government Green Procurement Spillovers: Evidence from Municipal Building Policies in California' (2014) 68 (3) Journal of Environmental Economics and Management 413 https://doi.org/10.1016/j.jeem.2014.09.001

Skidmore M et al., 'Cattle Ranchers and Deforestation in the Brazilian Amazon: Production, Location, and Policies' (2021) 68 (May) Global Environmental Change 102280 https://doi.org/10.1016/j.gloenvcha.2021.102280

Sogn-GrundvÅg G, Larsen T, & Young J, 'The Value of Line-Caught and Other Attributes: An Exploration of Price Premiums for Chilled Fish in UK Supermarkets' (2013) 38 Marine Policy 41–44 https://doi.org/10.1016/j.marpol.2012.05.017

Stefanadis C, 'Self-Regulation, Innovation, and the Financial Industry' (2003) 23 (1) Journal of Regulatory Economics 5

Stiglitz J, 'Regulation and Failure' in D Moss & J Cisternino (eds.), *New Perspectives on Regulation* (Cambridge, Tobin Project, 2009) 18

Sun C-HJ, Chiang F, Owens M, & Squires D, 'Will American Consumers Pay More for Eco-Friendly Labeled Canned Tuna? Estimating us Consumer Demand for Canned Tuna Varieties Using Scanner Data' (2017) 79 Marine Policy 62–69 https://doi.org/10.1016/j.marpol.2017.02.006

Van den Bergh R, 'Farewell Utopia? Why the European Union Should Take the Economics of Federalism Seriously' (2016) 23 (6) Maastricht Journal of European and Comparative Law 937 https://doi.org/10.1177/1023263X160230060

Van Putten I et al., 'Shifting Focus: The Impacts of Sustainable Seafood Certification' (2020) 15 (5) PLoS One: Article e0233237 https://doi.org/10.1371/journal.pone.0233237

Wakamatsu M, & Wakamatsu H, 'The Certification of Small-Scale Fisheries' (2017) 77 Marine Policy 97–103 https://doi.org/10.1016/j.marpol.2016.12.016

Walker N, Patel S, Davis F, Milledge S, & Hulse J, *Demand-Side Interventions to Reduce Deforestation and Forest Degradation* (International Institute for Environment and Development, Gray's Inn Road, 2013) 80–86

Washington S, & Ababouch L, Private Standards and Certification in Fisheries and Aquaculture: Current Practice and Emerging Issues. FAO Fisheries and Aquaculture Technical Paper. No. 553 (Rome, FAO. 2011) 8–9 (Published in Economics, 2011)

Weeber B, & Wallace C, 'The New Zealand Best Fish Guide' in T Ward & B Phillips (eds.), *Seafood Ecolabelling: Principles and Practice* (New Jersey, John Wiley & Sons, 2008) 367–388

Wells A, 'Verification of Legal Compliance in Indonesia' in D Brown et al. (eds.), *Legal Timber: Verification and Governance in the Forest Sector* (London, ODI, 2008) 173–186, 181–182

Wolf M, 'A Yellow Light for "Green Zoning": Some Words of Caution About Incorporating Green Building Standards into Local Land Use Law' (2011) 43 (4) Urban Lawyer 951

Wolff S, & Schweinle J, 'Effectiveness and Economic Viability of Forest Certification: A Systematic Review' (2022) 13 (5) Forests 798 https://doi.org/10.3390/f13050798

Wong, B, 'China's Packaged Food Market: Defining Characteristics' (HKTDC Research, 27 March 2018) https://research.hktdc.com/en/article/NDMyMjA5MDI2 accessed 25 September 2023

Zhao J et al., 'Current Status and Problems in Certification of Sustainable Forest Management in China' (2011) 48 (6) Environmental Management 1086

Zou Y, 'Certifying Green Buildings in China: LEED vs. 3-Star' (2018) 10 Journal of Cleaner Production 208 https://doi.org/10.1016/j.jclepro.2018.10.204

PART 6

Dealing with Climate Justice and Corporate Responsibility

28
SUSTAINABILITY AND PRIVATE INTERNATIONAL LAW

Ralf Michaels and Samuel Zeh[1]

28.1 Introduction

The greatest challenges facing our planet are not confined by national borders, but the role of private international law in the quest for sustainability has long been underestimated. Neither the UN Sustainable Development Goals (SDGs) nor the European Green Deal address private international law explicitly.[2] While the transformative potential of *substantive* private law is increasingly being recognised, private *international* law has only recently been begun to be scrutinised in this regard.[3]

Sustainability across national borders presents a challenge for private law. In this chapter, we show how this challenge makes the inclusion of private international law indispensable in any analysis (Section 28.2). Through the three functions of the discipline of private international law – coordination, facilitation, and regulation – we then demonstrate more closely the functional relationship between it and sustainability (Section 28.3). These theoretical findings are then translated into practice on the examples of two case studies, one from climate change litigation and another from supply chain governance (Section 28.4). We conclude by laying out some of the many elements that still require further research (Section 28.5).

28.2 Sustainability as a Transborder Challenge for Private Law

The Brundtland report defines sustainability as 'meet[ing] the needs of the present without compromising the ability of future generations to meet their own needs'.[4] This famous definition hints at the inherent tension between sustainability and development, which raises difficult questions: is sustainability compatible with growth? Or does development inevitably create negative externalities for future generations?

Development thereby presents specific challenges for private law. It has long relied on private law and private ordering – property rights, contracts, access to courts, and so on. But in its focus on two-party relations, private law is often systemically blind to its negative externalities, and to the way in which local welfare gains often come at the cost of third parties, the environment, or future generations. The social, economic, and environmental dimensions of sustainability[5] are thus closely linked to private parties and private law: corporations generate economic growth, yet they

 DOI: 10.4324/9781032662046-34

pollute the environment and contribute to global climate change. Supply chains can enhance prosperity for producers and consumers alike, yet they also threaten the exploitation of labourers and the environment. Migrants are led away from poverty, but migration often leads to clashes between different legal cultures in the field of family law and beyond.

Private law must, and often does, respond to these challenges. But private law is national law, and it is therefore limited by national borders. By contrast, sustainability is in fact frequently a transborder challenge for private law. The examples just mentioned illustrate that. Emissions of CO_2 impact the atmosphere everywhere, with effects far from the place of emission. Supply chains are frequently international because they make use of comparative advantages. Migration spans borders and therefore crosses legal orders. Private law can only make a significant contribution towards sustainability if it has transborder effectivity, and its full operation can only be understood by accounting for its transborder dimension.

This transborder dimension is the domain of private international law, and an analysis of private law that does not take private international law into account is likely incomplete. Take for example the well-known RWE case currently pending before the German courts, in which a Peruvian farmer is demanding compensation from RWE in the form of measures to protect his house from flooding. The claimant alleges that RWE has contributed significantly to CO_2 emissions, and thereby to global climate change, which, in turn, has resulted in glaciers melting and the meltwater threatening his house.[6] One important question of private law concerns the issue of causation,[7] which different legal systems resolve in different ways. The differences are important: the German court in the first instance found that the claimant had not established causation,[8] while the district court of The Hague reached a different result in a comparable case, holding that 'every emission of CO2 and other greenhouse gases, anywhere in the world and caused in whatever manner, contributes to [imminent environmental damage in the Netherlands]'.[9] In other words, it matters what court hears a case and what law is applied, and so the private law question cannot be discussed without consideration of private international law.

The three classical domains of private international law are jurisdiction, choice of law, and the recognition and enforcement of foreign judgments. Private international law rules of jurisdiction determine whether the plaintiff in the RWE case can bring suit before a court in Germany or Peru (or either one), or indeed in still other, possibly more favourable countries. Private international law choice-of-law rules determine which law a court will apply to the claim – the *lex fori* (the law of the court's own jurisdiction) or some other law. Private international law rules of international civil procedure will determine how the parties can be served with process and how evidence can be gathered across borders.[10] And private international law rules of recognition and enforcement of foreign judgments will determine whether a favourable judgment rendered by a court in one jurisdiction can be recognised in another jurisdiction in which the defendant may have assets. Private international law, in other words, is responsible for both the limits and the effects of substantive private law across borders. Private international law is what makes national private law transnationally effective.

Importantly, although private international law organises the transnational effects of private law, its own rules are themselves traditionally national rules. There is no unified set of jurisdictional rules or a body of tort law that applies to Peru and Germany at the same time. Only a few areas of private international law are regulated in international treaties, most importantly under the leadership of the Hague Conference. Within the European Union, however, most areas of private international law are now unified, such as in the Rome I Regulation on contractual obligations[11] and the Rome II Regulation on non-contractual obligations,[12] or the Brussels I Regulation.[13]

28.3 The Functional Relationship between Sustainability and Private International Law

28.3.1 The Coordinative Function

Given the centrality of private international law, why is it regularly ignored in discussions on sustainable private law? One reason may be that private international law is often still viewed as a 'merely technical' discipline, neatly separated from substantive private law, that determines the appropriate jurisdiction to deal with a matter, without regard to the content of that jurisdiction's substantive law. For many, this is still the core of private international law, though awareness is growing of its broader significance.

From this perspective, the function of private international law is the coordination of different private law regimes.[14] Rules of jurisdiction coordinate among potential courts, determining whether a court is an available forum and the relationships between different available courts[15] while leaving regulation of the actual proceedings to the courts' own law. In the hope of minimising conflicts between legal orders, choice-of-law rules assign the resolution of a particular legal issue to a single substantive law but leave substantive decisions to that law. Rules on recognition and enforcement of foreign judgments determine whether a judgment is portable but do not affect the content of the judgment.

This coordinative function is, of course, important. It can help achieve the formal ideal of decisional harmony – to make sure that the outcome of litigation does not depend on the court before which the action is brought. This avoids forum shopping, minimising the ability of parties to seek the law most favourable to them through their choice of court. In addition, private international law can enable cooperation between legal systems. For example, private international law instruments facilitate the service of documents[16] and the taking of evidence[17] in an international context, and they enable the circulation of judicial and arbitral decisions beyond their country of origin. And private international law can ensure that enforcement does not depend on where a suit was brought. For all three purposes, regional or global unification of private international law appears ideal. Short of such unification, private international law rules ought to be formulated in a way that makes them potentially universalisable.

The coordinative function of private international law plays a role with regard to SDG 16 (Peace, Justice and Strong Institutions) and especially to its target 16.3, to '...ensure equal access to justice for all'.[18] Moreover, global unification of private international law, even outside the framework of the Hague Conference on Private International Law,[19] would benefit SDG 17 (Partnership for the Goals).[20] But SDGs 16 and 17 themselves are ancillary to other, 'substantive' SDGs – necessary for their fulfilment, but not specific to sustainability. Similarly, as a tool of coordination, private international law is relevant to sustainability, but not in specific ways.

28.3.2 The Facilitative Function

Coordination is not the only function of private international law, however.[21] A second function is its facilitation of the movement of persons and of transactions and commerce. This is, of course, a core function of private law, and as such, it is well-known by laypeople as well as lawyers. Private law facilitates private ordering through property and contracts, and it facilitates individuality by granting status. Yet this facilitation cannot be fully understood without accounting for private international law because private international law enables the spaces created by private law to reach across borders.

For example, a property right established under the law of one country – typically that of the place where the property is located (*lex rei sitae*) – will in principle be recognised by other legal systems. A contract concluded in one country becomes binding elsewhere through the force of private international law. A status validly acquired under one country's law is, in principle, afforded validity in other countries as well – protection of particular importance for migrants.[22] A corporation established validly under one country's law can potentially operate worldwide. Private international law thereby gives a transborder dimension to the facilitation accorded by substantive private law, and it allows us to cross borders without losing our rights and status.

In this regard, private international law is an indispensable part of private law. On its own, substantive private law could facilitate private ordering only to a limited degree, the limit being the borders of the jurisdiction. Private international law defines the territorial scope of this facilitation, and in this sense, it is an indispensable element of private law.

In addition to this first facilitative function, which is an expansion of substantive law, there is a second facilitative function specific to private international law: private international law facilitates cross-border movement and relations through party autonomy, that is, by allowing parties to choose the competent court and the applicable law.[23] Such choices can be made directly, through choice-of-court and choice-of-law agreements, or indirectly, through manipulating the relevant connecting factors. Parties can decide to be active in a chosen country or to designate a specific place of performance. Under certain conditions, they can thereby influence what courts have jurisdiction and what law applies. Moreover, private international law allows parties to opt out of national courts altogether and have their disputes resolved through private arbitration. In arbitration, they are even allowed to opt out of national laws altogether by subjecting their disputes to non-state law.

This facilitative function is specific to private international law because it goes beyond freedom of contract. Freedom of contract cannot overcome mandatory rules, but selecting the applicable law can; parties cannot choose what procedural rules a court will follow, but private international law allows them to choose a court that follows procedural laws they like; or it allows them to choose arbitration, an adjudicatory system within which they can indeed choose the procedural rules. Therefore, no account of the room the law leaves for private ordering is complete if it ignores party autonomy.

The facilitative function of private international law is directly relevant to sustainability.[24] Insofar as private law and private ordering enhance sustainability, private international law can add to their effect by facilitating transborder movement. A more detailed inquiry into such possible positive effects would be desirable.

More often, however, facilitation threatens sustainability. Insofar as transnational commerce accelerates the destruction of the world's ecosystems and communities, private international law, by facilitating commerce, can become complicit in it. Insofar as legal stability and predictability favour economic growth, private international law may help bring about such growth, and if that growth is unsustainable, private international law will be complicit in that as well. Party autonomy may enable corporations to evade regulations and create negative externalities.[25] The recognition of property rights across borders can limit the ability of states to regulate with an eye to sustainability. Ironically, what was long considered a major strength of private international law – that it facilitates transborder mobility of goods, capital, and people – can thus become a burden on sustainability.

28.3.3 The Regulatory Function

The domain of private ordering that substantive private law offers is not limitless, of course. Private law also has a regulatory function. Some private law regulates directly, through commands and restrictions: we must register our marriages, we are barred from selling opioids. Some private law regulates indirectly, such as in the way well-tailored liability rules can set incentives.

Just as with the facilitative function of private international law, this regulatory function comes in two variants.[26] First, private international law gives regulatory private law effectivity across borders, thereby ensuring its extraterritorial application. Thanks to private international law, a *numerus clausus* principle in the property rights regime established by the *lex rei sitae* will serve worldwide as a valid restriction on the creation of new security interests in an asset. Private international law gives transnational force to a standard of liability established at the place where a tort is committed. And an internationally mandatory rule will override even the private ordering of party autonomy. Just as with facilitation, private international law gives cross-border effectivity to regulations. Or, to put it differently, private international law is what enables States to regulate across borders, and it is necessary to prevent under-regulation due to the territorial restrictions of law.

At the same time, private international law can also prevent over-regulation because of too many internationally mandatory rules all calling for application. Private international law can help resolve conflicts between different regulatory laws[27] too, which, notably, is one of its emerging functions. But as long as private international law remains mostly confined to nonregulatory substantive law, the application of foreign regulatory law is still anomalous. Moreover, conflicts between regulatory laws are often solved not through private international law mechanisms but rather by unilateral rules designating the territorial scope of one's own law.[28]

The second regulatory function is specific to private international law. Private international law itself regulates human conduct by limiting certain effects of private ordering across borders and by incentivising or disincentivising the use of one legal system or another.[29] This regulatory function is of particular importance for sustainability. Regulatory private international law rules can protect not only weaker parties but also the environment, thereby furthering the core ideals of the SDGs and the EU Green Deal.

Private international law can also support the pursuit of sustainability by providing substantive limits to the applicable law if this law is incompatible with regulatory goals. Firstly, a court can refuse to apply a law if it finds such application to be manifestly incompatible with the public policy (*ordre public*) of the forum (Article 21 Rome I Regulation, Article 26 Rome II Regulation). This requires that the application of the law would violate the fundamental values of a legal system and that there be a sufficient domestic relationship or proximity to the forum state.[30] The public policy exception can, in principle, be used to refuse the application of a foreign private law that is manifestly incompatible with the forum's conception of human rights.[31] However, it is often difficult to derive a specific duty of care directly from human rights provisions like the UDHR: How many smoke detectors is a factory required to have to avoid a violation of the human right to physical integrity in the event of a fire?[32] Furthermore, in a civil liability case in Germany, the mere fact that the seat of the defendant is located there does not establish a sufficient connection to the forum State.[33] Therefore, claimants from the Global South usually cannot rely on the public policy exception if they suffered harm at their habitual residence. Beyond this, *ordre public* can only be invoked in exceptional cases,[34] and so its regulatory potential should not be overestimated.[35]

If the forum wants to enforce its substantive regulatory standards generally, it has to resort to overriding mandatory provisions instead (Article 9 Rome I Regulation, Article 16 Rome II Regulation).[36] These rules are applicable irrespective of the *lex causae* and its specific application. A current example is corporate civil liability for human rights violations and environmental damage as stipulated in the European Commission's proposal for a directive on corporate sustainability (see Section 28.4.2).

Both the public policy exception and overriding mandatory provisions are traditionally seen as national instruments. Articles 16 (overriding mandatory provisions) and 26 (public policy of the forum) of the Rome II Regulation expressly refer to the values or rules of the forum. This conflicts with sustainability, which is a global challenge. If every nation only enforces its own understanding of material justice and equity in cross-border cases, sustainable development can hardly be achieved. The result instead will be the same regulatory conflicts that private international law aims to resolve in the first place. Even if the approaches of various states share the same primary goals, the differences between them can lead to under-regulation or over-regulation.

This indicates that the concepts behind both public policy and overriding mandatory provisions need to be better elaborated if they are supposed to figure importantly in the quest for sustainability. A first step in that direction would be to broaden the notion of public policy. EU law already influences *ordre public* at the national level, which could eventually lead to the development of a truly European public policy.[37] On a global level, fundamental values, shared across national borders, could be seen as expressions of a transnational *ordre public*.[38] This concept is used most often in the field of international arbitration, in which the values of any one state are irrelevant. It can be invoked in cases that involve such issues as slavery, drug trafficking or corruption.[39] In light of the SDGs and the Paris climate accords, sustainability could also be considered a fundamental value that nations share. The concept of a transnational *ordre public* could also be utilised, perhaps by State courts, if the otherwise applicable law is in violation of international human rights[40] or is manifestly incompatible with the protection of our natural resources. However, a transnational *ordre public* would have many of the same limitations as its national counterparts: it is difficult to derive specific civil legal duties from such a broad concept. Further research is required to define a transnational *ordre public* more clearly and to explore its potential relevance for sustainability.

Furthermore, some connecting factors generally benefit either weaker parties or the environment. For example, Article 7 of the Rome II Regulation allows the person seeking compensation for environmental damage to choose between the application of the law of the country in which the injury occurs and the law of the country in which the event giving rise to the injury occurred.[41] Thus, the injured party can opt for the more stringent law.[42] The explicit goal behind this is to enhance environmental protection (Recital 25 of the Rome II Regulation).[43]

Pursuant to Article 6(1) of the Rome I Regulation, a consumer contract is governed by the law of the country in which the consumer has his habitual residence. The substantive rules will not necessarily be the most consumer-friendly, but at least they will be familiar to the weaker party. In addition, private international law can limit the application of connecting factors that contravene regulatory goals. Party autonomy can be a source of inequality (see SDG 10: Reduced Inequalities),[44] and restricting its scope can be necessary to protect weaker parties. For example, Article 6(2) of the Rome I Regulation does not allow a choice of law that results in consumers being deprived of the protections afforded to them by provisions that the parties, by virtue of the otherwise applicable law, have no power to derogate from. The same applies to individual employment contracts (Article 8(1) of the Rome I Regulation).

28.4 The Role of Private International Law in Practice

Two prominent examples, one from contractual and one from non-contractual relations, demonstrate the crucial role of private international law with regard to sustainability in practice.

28.4.1 Climate Change Litigation

First of all, in accordance with SDG 13, Climate Action, climate change litigation can play an important role in advancing sustainability.[45] If companies face liability for their contribution to global climate change, they will have a strong financial incentive to reduce their carbon dioxide emissions.[46] Private international law has a crucial role to play here.

In the EU, a corporation can be sued in the courts of the Member State in which it has its statutory seat, central administration, or principal place of business (Articles 4(1), 63(1) of the Brussels I Regulation). In torts cases, the claimant can alternatively opt for the courts of the place where the harmful event occurred or may occur (Article 7(2) of the Brussels I Regulation). It was under these provisions that the German courts had jurisdiction to adjudicate the claim against RWE (see above at Section 28.2), as the defendant was seated in Germany.[47] Similarly, Royal Dutch Shell, the defendant in the case before the district court of the Hague, was based in the Netherlands.[48] In the Peruvian farmer's case, the claimant could alternatively have sued RWE in Peru, where the melting of glaciers is threatening his house. But especially in litigation between defendants from the Global North and plaintiffs from the Global South, plaintiffs often prefer Northern courts,[49] and it therefore matters that these courts do in fact exercise jurisdiction. While the provisions of the Brussels I Regulation are mandatory, common-law courts are sometimes able to dismiss cases over which they could otherwise exercise jurisdiction by reference to a more appropriate court elsewhere under the doctrine of *forum non conveniens*. Although this is a sensible device in theory, it often means that plaintiffs lose access to effective courts in the Global North, with no equivalent substitute available elsewhere.

Moving to the choice of law, the law applicable to liability is determined by Article 7 of the Rome II Regulation (environmental damage), which also applies to climate change litigation.[50] The regulation allows the plaintiff to choose between the law of the place of injury and that of the place of conduct. This discrimination in favour of plaintiffs is explicitly justified under the regulatory goals of Article 191 of the TFEU (ex-Article 174 TEC) concerning environmental protection. In the RWE case, this means that the claimant has effectively been able to choose the law of Germany, where the defendant emitted greenhouse gases, as the law of the place of conduct.[51] If the claimant does not choose to have that law apply, then the law of the country in which the injury occurred is applied instead.[52] For the defendant, who may potentially be subject to the law of any country in which injuries from climate emissions are felt, this poses a significant risk.[53]

Although the regulation of emissions mainly concerns the regulatory function of private international law, its facilitative function comes into play, too. Operating licences and emissions certificates both confer a right to produce a certain amount of emissions and depending on the applicable law, they may limit liability. But the right to emit greenhouse gases would be worth little unless such licences and certificates are recognised internationally. According to Article 17 of the Rome II Regulation, the rules of safety and conduct of the place of the event allegedly giving rise to the liability are to be considered in assessing the conduct. This could include operating licences and emissions certificates.[54] The degree to which such rights are recognised internationally is important for the effectiveness of climate change litigation.[55] Arguably, only a narrow interpretation of Article 17 of the Rome II Regulation can ensure that private litigation can make a significant contribution to the fight against global climate change. If foreign-issued licences were automatic bars to liability, the environment (or the plaintiff) might not be sufficiently protected.[56] This suggests that permits issued under public laws should limit civil legal claims only if the global climate impact of the permitted activity was been sufficiently considered in the permitting process.[57] Moreover, there is a need for clarification regarding the specific effects an operating licence could have on the civil liability of the polluter.[58] Thus, operating licences should only affect climate change liability if this is stipulated by the applicable law (as determined by Article 7 of the Rome II Regulation).[59]

Once a favourable judgment has been rendered, it needs to be enforced. If the claim was not brought in the defendant's home jurisdiction (as it was in the RWE case), enforcement might require recognition and enforcement of the judgment in a jurisdiction in which the defendant has assets. Within the EU, such recognition and enforcement are virtually automatic. By contrast, a Peruvian judgment might be harder to enforce elsewhere. This can be seen from the example of the Chevron case, in which Ecuadorian plaintiffs were largely unable to have a multi-billion-Euro Ecuadorian judgment enforced internationally.[60]

These discussions should suffice to demonstrate the crucial role of private international law in climate change litigation. Private international law is important for plaintiffs' strategic decision-making: they can choose a suitable forum within the confines of the rules of jurisdiction, opt for the application of a favourable substantive law (Article 7 of the Rome II Regulation), and then apply to have the judgment enforced in another country where assets are located. And the discussion should also demonstrate its importance to regulators aiming for sustainability.

28.4.2 Corporate Social Responsibility and Supply Chain Liability

Another example that illustrates the regulatory potential of private international law is corporate social responsibility, especially with regard to supply chain governance.[61] Global supply chains are ambivalent from the perspective of sustainability; while they can create growth and ease poverty, they can also facilitate the exploitation of labourers and lead to significant environmental damage in producer countries with low regulatory standards.

Corporate social responsibility directly relates to several key aspects of sustainability: environmental protection is necessary to ensure the availability of clean water and sanitation (SDG 6)[62] and to preserve life, both on land (SDG 15)[63] and underwater (SDG 14).[64] The environmental impact of international corporations extracting resources in the Global South can be seen in cases like *Vedanta*, in which Zambian claimants alleged that a copper mine had polluted their local water supply and demanded damages from the mine owner's English parent company.[65]

Target 8.8 of SDG 8 (Decent Work and Economic Growth)[66] expressly calls for the protection of labour rights and secure working environments. Labour issues figured in the KiK case, which arose out of a factory fire in Pakistan in which hundreds of workers died.[67] Relatives of the victims and some of the survivors sought compensation from the German clothing company KiK, whose local supplier had manufactured textiles at the factory. The claimants argued that KiK had breached its duty of care by not ensuring sufficient fire safety.

Notably central to these matters is private international law. Injured employees or third parties receive compensation only if a competent court decides that the applicable law provides for a remedy. In cases of supply chain liability, claimants can often choose among different courts: generally, European corporations can be sued in an EU Member State in which the defendant corporation has its statutory seat, central administration, or principal place of business (Article 4(1), 63(1) of the Brussels I Regulation). In tort cases, the courts of the place where the harmful event occurred or may occur have jurisdiction as well (Article 7(2) of the Brussels I Regulation). Employers domiciled in a Member State can also be sued (among other places) in the courts of the place where employees habitually carry out their duties (Article 21(1)(b)(i) of the Brussels I Regulation), which of course does not help employees in Pakistan.

The choice-of-law framework largely runs parallel to the conflict of jurisdiction. Absent a choice-of-law agreement, the law applicable to tort claims is generally the law of the country in which the injury occurred (Article 4(1) Rome II Regulation). Employment contracts are generally governed by the law of the country in which an employee habitually carries out their duties in the performance of the contract (Article 8(2) Rome I Regulation).[68]

In the case of the factory fire in Pakistan, the jurisdiction of the German courts was based on Articles 4(1), 63(1) of the Brussels I Regulation, as the seat of the defendant corporation was located in Germany.[69] Alternatively, the claimants could have sued KiK in Pakistan, as the place where the harmful event occurred.[70] KiK's tort liability was determined based on Pakistani law (Article 4(1) Rome II Regulation). Such a determination under the law of a non-forum state can result in a lack of legal protection for victims, particularly when international corporations, to save costs, purposely choose countries with low safety standards. In the case before the district court of Dortmund, KiK escaped liability due to the short limitations period in Pakistani law.[71]

One solution would be to assign a duty of care to the international corporation that profits the most from the value chain, which is what France's 2017 *loi de vigilance* provides.[72] However, a substantive duty of care is only relevant in practice if it is applicable in cross-border cases; if a German corporation's liability for a fire at a Pakistani factory is determined only according to the law of the country in which the damage occurred (Article 4(1) Rome II Regulation),[73] it does not matter whether German law stipulates a duty of care.[74] This problem could be addressed via new general conflict-of-laws rules, for example by introducing a connecting factor for global supply chains or human rights violations.[75] Alternatively, the injured parties could be afforded the right to choose the law most favourable to them,[76] as is already the case with environmental damages (Article 7 Rome II Regulation, see above).

However, the European Commission's 2022 proposal for a Directive on Corporate Sustainability and Due Diligence chose a different path.[77] This proposed Directive aims to protect human rights and the environment, in accordance with the European Green Deal and the SDGs, by imposing a new duty of care on companies.[78]

Under Articles 7 and 8 of the proposed Directive, Member States must ensure that companies take appropriate measures to prevent or put a stop to actual or potential adverse human rights and environmental impacts. If companies fail to comply with these obligations, they are liable for any resulting damages (Article 22(1)).[79] Pursuant to Article 22(5) and Recital 61, this liability is supposed to be of overriding mandatory application in cases in which the law applicable to such claims is not that of a Member State. Therefore, the new rules on liability would have to be applied regardless of the *lex causae*.[80]

This would be very effective at ensuring that the Directive achieves its substantive goals in all cross-border cases.[81] But at the same time, overriding mandatory provisions cause friction in the traditional private international law framework. They can result in competing approaches and in conflicts between different laws.[82] For example, different countries could impose very different (or even incompatible) mandatory legal duties and liabilities on international corporations, causing legal uncertainty and imposing significant additional costs. On top of that, one country or bloc enforcing its own legal standards worldwide could be seen as legal imperialism.[83]

Whether the proposed liability mechanism will prove effective remains to be seen. Other jurisdictions' legislators have chosen different approaches: Germany's Supply Chain Due Diligence Act (*Lieferkettensorgfaltspflichtengesetz* – LkSG)[84] only contains public law sanctions for human rights violations and expressly does not provide for civil liability (§ 3 (3) LkSG).[85]

However, the benefits of private enforcement should not be disregarded. Overriding mandatory rules of civil liability can be a powerful tool in the struggle for sustainability. Although such rules must remain the exception, the fact is that protecting human rights and the environment is exceptionally important.

Nor is the deterrent effect of private enforcement limited to the jurisdiction in which the original judgment is delivered. Within the EU, judgments of one Member State's courts are generally recognised and enforced in other Member States (see 28.4.1 above). Whether countries outside the EU will also recognise judgments from EU Member States based on an overriding mandatory provision remains to be seen. Instead, they might find that such broad liability rules are incompatible with their respective *ordre public*. This is a potential downside of using private international law instruments to unilaterally enforce European standards. But the risk should not be overstated, either, because the public policy exception is defined very narrowly. Beyond this, the proposed EU directive could eventually serve as a model for other countries as well, which would restore decisional harmony between different countries' courts.

28.5 Conclusion

This chapter has shown that, with regard to transborder challenges such as sustainability, private law cannot be properly addressed without private international law. While private international law plays a role in coordinating different jurisdictions' substantive private laws, its main relevance for sustainability actually lies in its facilitative and regulatory functions.

Of those two functions, the regulatory function is of particular importance. Preventing the destruction of ecosystems and communities will only be possible if harmful behaviour is discouraged on a global scale. Regulatory private international law can make an important contribution to this discouragement, particularly in the context of climate change litigation and supply chain liability. Overriding mandatory provisions could help protect the environment and human rights worldwide against private ordering, and more generally the concept of transnational public policy could perhaps be expanded to include sustainability. But other instruments should also bear further exploration, particularly in the area of supply chains, where the regulatory potential of various tools of private international law has not yet been exhausted. New connecting factors that take sustainability into account more directly are another option that remains to be considered.

At the same time, the facilitative function of private international law also deserves more attention than it has traditionally received. Granted, though it is traditionally regarded as one of the main advantages of private international law, the facilitative function has become suspect; by facilitating cross-border transactions and commerce, private international law can create negative externalities, too. However, while not all regulation advances sustainability, nor can sustainability be achieved without private ordering. The potential of transnational private ordering deserves more careful consideration. One promising example of sustainable private ordering is envisaged in the current ABA project on model contract clauses to protect workers in international supply chains.[86]

Finally, discussions of the role of private international law, like discussions of sustainability more generally, must account for deep global inequalities, especially between the Global North and Global South. Private international law in its current form is mostly a Northern discipline,[87] and the doctrines and institutions discussed in this chapter are those of the Global North. From a global perspective, this is obviously insufficient.

What this implies is that to properly understand and appreciate the role of private international law in any shift towards sustainability, a broader understanding of private international law itself, including its functions, is required. Realising the inseparable relationship between private international law and substantive private law may be a necessary first step that makes all other steps possible.

Notes

1 Ralf Michaels: Director at the Max Planck Institute for Comparative and International Private Law, Hamburg, Germany; Chair in Global Law at Queen Mary University London, UK; Professor of Law at Hamburg University, Germany. Samuel Zeh: Doctoral student at the University of Hamburg and former research assistant at the Max Planck Institute for Comparative and International Private Law, Hamburg, Germany.

2 United Nations, 'A/RES/70/1 – Transforming our world: the 2030 Agenda for Sustainable Development' (25 September 2021) <https://sdgs.un.org/goals> accessed 17 July 2023; *European Commission*, 'A European Green Deal' (2019) <https://commission.europa.eu/strategy-and-policy/priorities-2019-2024/european-green-deal_en> accessed 17 July 2023.

3 Especially Ralf Michaels, Verónica Ruiz Abou-Nigm and Hans van Loon, *The Private Side of Transforming Our World – UN Sustainable Development Goals 2030 and the Role of Private International Law* (Cambridge, Intersentia 2021); see also Jaakko Salminen and Mikko Rajavuori, 'Private International Law, Global Value Chains and the Externalities of Transnational Production: Towards Alignment?' [2021] 12(2) *Transnational Legal Theory* 230 (https://doi.org/10.1080/20414005.2021.1970470); Hans van Loon, 'Private International Law in Support of Sustainable Development – Impulses from Osnabrück' in Helmut Grothe, Peter Mankowski and Frederick Rieländer (eds), *Europäisches und internationales Privatrecht – Festschrift für Christian von Bar zum 70. Geburtstag* (München, C. H. Beck 2022) 197; Thu Thuy Nguyen, 'Transnational Corporations and Environmental Pollution in Vietnam – Realising the Potential of Private International Law in Environmental Protection' [2022] 18(2) *Journal of Private International Law* 238 (https://doi.org/10.1080/17441048.2022.2096059); Sabrina Robert Cuendet and Pascale Ricard, *White Paper 18 – SDGs Beyond 2030* (Paris, ILA 2023) 53.

4 World Commission on Environment and Development, *Our Common Future* (1987) ("Brundtland Report"), Part I. Chapter 2 IV.1. Regarding this definition see, e.g., Arnaud Diemer, 'Six Key Drivers for Sustainable Development' [2019] 18 *International Journal of Environmental Sciences & Natural Resources* 128 (http://dx.doi.org/10.19080/IJESNR.2019.18.555994); Alexander Hellgardt and Victor Jouannau, 'Nachhaltigkeitsziele und Privatrecht' (2022) 222 *Archiv für die civilistische Praxis* 163, 167–170 (https://doi.org/10.1628/acp-2022-0008); Jan-Erik Schirmer, *Nachhaltiges Privatrecht* (Tübingen, Mohr Siebeck 2023) 15–17.

5 See, e.g., Arnaud Diemer, 'Six Key Drivers for Sustainable Development' [2019] 18 *International Journal of Environmental Sciences & Natural Resources* 128 (http://dx.doi.org/10.19080/IJESNR.2019.18.555994); Ben Purvis, Yong Mao and Darren Robinson, 'Three Pillars of Sustainability: In Search of Conceptual Origins' [2019] 14 *Sustainability Science* 681 (https://doi.org/10.1007/s11625-018-0627-5).

6 See OLG Hamm (30 November 2017), I-5 U 15/17, ZUR 2018, 118 f. The judgment of the court of first instance: LG Essen (15 December 2016), 2 O 285/15, NVwZ 2017, 734. For an English summary of the case see, e.g., Daniel Ennöckl, 'Climate Change Litigation in Austria and Germany: Climate Change Litigation in Germany and Austria – Recent Developments' [2020] 14(4) *Carbon & Climate Law Review (CCLR)* 306, 309–311 (https://doi.org/10.21552/cclr/2020/4/9).

7 See, e.g., Will Frank, 'Störerhaftung für Klimaschäden?' [2017] *Neue Zeitschrift für Verwaltungsrecht* 664, 665–669; David A. Grossman, 'Warming up to a Not-So-Radical Idea: Tort-Based Climate Change Litigation' [2003] 28 *Columbia Journal of Environmental Law* 1, 22–33; Erik Pöttker, *Klimahaftungsrecht – Die Haftung für die Emission von Treibhausgasen in Deutschland und den Vereinigten Staaten von Amerika* (Tübingen, Mohr Siebeck 2014) 140–252, 306–368; Jan-Erik Schirmer, *Nachhaltiges Privatrecht* (Tübingen, Mohr Siebeck 2023) 178–208.

8 LG Essen (15 December 2016), 2 O 285/15, NVwZ 2017, 734, 736. The case is still pending before the court of appeals, see OLG Hamm (30 November 2017), I-5 U 15/17, ZUR 2018, 118 f.

9 Hague District Court (26 May 2021), Milieudefensie et al v RDS, ECLI:NL:RBDHA:2021:5337, ECLI:NL:RBDHA:2021:5339 (English version) para 4.4.37.

10 The German court of appeal apparently considered the plaintiff's claim strong enough to justify collection of evidence in Peru. Tobias Lutzi, 'German Judges Travel to Peru in Climate-Change Trial' (*ConflictofLaws.net*, 31 May 2022) <https://conflictoflaws.net/2022/german-judges-travel-to-peru-in-climate-change-trial/> accessed 17 July 2023.

11 Regulation (EC) No 593/2008 of the European Parliament and of the Council of 17 June 2008 on the law applicable to contractual obligations [2008] OJ L177/6.

12 Regulation (EC) No 864/2007 of the European Parliament and of the Council of 11 July 2007 on the law applicable to non-contractual obligations [2007] OJ L199/40.

13 Regulation (EU) No 1215/2012 of the European Parliament and of the Council of 12 December 2012 on jurisdiction and the recognition and enforcement of judgments in civil and commercial matters (recast) [2012] OJ L351/1.

14 On this function, see, e.g., Kurt Siehr, 'Coordination of Legal Systems in Private International Law' in Talia Einhorn and Kurt Siehr (eds), *Intercontinental Cooperation Through Private International Law – Essays in Memory of Peter E. Nygh* (The Hague, T.M.C. Asser Press 2004) 325.

15 See Ralf Michaels, 'Two Paradigms of Jurisdiction' [2007] 27 *Michigan Law Review* 1003.

16 For example, the Regulation (EU) 2020/1784 of the European Parliament and of the Council of 25 November 2020 on the service in the Member States of judicial and extrajudicial documents in civil or commercial matters (service of documents) (recast) [2020] OJ L405/40.

17 For example, the Regulation (EU) 2020/1783 of the European Parliament and of the Council of 25 November 2020 on cooperation between the courts of the Member States in the taking of evidence in civil or commercial matters (taking of evidence) (recast) [2020] OJ L405/1.

18 Cf. Christopher A Whytock, 'Transnational Access to Justice' [2020] 38 *Berkeley Jounal of International Law* 154; Hans van Loon, 'Access to Justice and Legal Aid in Cross-border Situations' in Department of Justice – The Government of Hong Kong Special Administrative Region (org), *Hong Kong Legal week 2020 – Inaugural Rule of Law Congress – Proceedings – Towards 2030: A Decade of Action for Rule of Law* (Hong Kong, Department of Justice Hong Kong SAR 2021) 48–59; Sabrina Robert Cuendet and Pascale Ricard, *White Paper 18 – SDGs beyond 2030* (Paris, ILA 2023) 53, 55–56.

19 William Brydie-Watson, 'The Three Sisters of Private International Law: An Increasingly Co-Operative Family Rather Than Sibling Rivals' in Thomas John, Rishi Gulati and Ben Köhler (eds), *The Elgar Companion on the Hague Conference on Private International Law* (Cheltenham, Edward Elgar Publishing 2020) 23.

20 Fabricio B. Pasquot Polido, 'SDG 17: Partnership for the Goals' in Ralf Michaels, Verónica Ruiz Abou-Nigm and Hans van Loon (eds), *The Private Side of Transforming our World – UN Sustainable Development Goals 2030 and the Role of Private International Law* (Cambridge, Intersentia 2021) 541.

21 Ralf Michaels, Verónica Ruiz Abou-Nigm and Hans van Loon, 'Introduction: The Private Side of Transforming Our World – UN Sustainable Development Goals 2030 and the Role of Private International Law' in Ralf Michaels, Verónica Ruiz Abou-Nigm and Hans van Loon (eds), *The Private Side of Transforming our World – UN Sustainable Development Goals 2030 and the Role of Private International Law* (Cambridge, Intersentia 2021) 13.

22 Sabine Corneloup and Jinske Verhellen, 'SDG 16: Peace, Justice and Strong Institutions' in Ralf Michaels, Verónica Ruiz Abou-Nigm and Hans van Loon (eds), *The Private Side of Transforming our World – UN Sustainable Development Goals 2030 and the Role of Private International Law* (Cambridge, Intersentia 2021) 505.

23 See Horatia Muir Watt, '"Party Autonomy" in International Contracts: From the Makings of a Myth to the Requirements of Global Governance' (2010) 6 *European Review of Contract Law* 250 (https://doi.org/10.1515/ercl.2010.250); more generally on the topic of party autonomy see, e.g., Alex Mills, *Party Autonomy in Private International Law* (Cambridge, Cambridge University Press 2018).

24 Cf. A. Halfmeier, 'Nachhaltiges Privatrecht' (2016) 216 *Archiv für die civilistische Praxis* 717, 742–743 who argues that (substantive) private law's emphasis on 'Freiheit und Selbstbestimmung' ('freedom and self-determination') can be an important aspect of sustainability.

25 See, e.g., Horatia Muir Watt, 'Party Autonomy' in Jürgen Basedow, Gisela Rühl, Franco Ferrari and Pedro De Miguel Asensio (eds), *Encyclopedia of Private International Law* (Cheltenham, Edward Elgar Publishing 2017) 1336, 1340–1341.

26 Ralf Michaels, Verónica Ruiz Abou-Nigm and Hans van Loon, 'Introduction: The Private Side of Transforming Our World – UN Sustainable Development Goals 2030 and the Role of Private International Law' in Ralf Michaels, Verónica Ruiz Abou-Nigm and Hans van Loon (eds), *The Private Side of Transforming Our World – UN Sustainable Development Goals 2030 and the Role of Private International Law* (Cambridge, Intersentia 2021) 13.

27 See, e.g., Annelise Riles, 'Managing Regulatory Arbitrage: A Conflict of Laws Approach' [2014] 47 *Cornell Journal of International Law* 63; Ralf Michaels, 'Towards a Private International Law for Regulatory Conflicts?' [2016] 59 *Japanese Yearbook of International Law* 175 (https://doi.org/10.1628/00372515X14373846829991); Matthias Lehmann, 'Regulation, Global Governance and Private International Law: Squaring the triangle' [2020] 16(1) *Journal of Private International Law* 1 (https://doi.org/10.1080/17441048.2020.1744255).

28 See Hannah Buxbaum and Ralf Michaels, 'Reasonableness as a Limitation on the Extraterritorial Application of U.S. Law: From 403 to 405 (via 404)', in: Paul B. Stephan and Sarah A. Cleveland (eds), *The Restatement and Beyond: The Past, Present, and Future of U.S. Foreign Relations Law* (Oxford, Oxford University Press 2020) 283.

29 Jacco Bomhoff and Anne Meuwese, 'The Meta-Regulation of Transnational Private Regulation' [2011] 38(1) *Journal of Law and Society* 138 (https://doi.org/10.1111/j.1467-6478.2011.00538.x); Robert Wai, 'Transnational Liftoff and Juridical Touchdown: The Regulatory Function of Private International Law in an Era of Globalization' [2002] 40 *Columbia Journal of Transnational Law* 209; Horatia Muir Watt, 'Private International Law Beyond the Schism' [2011] 2(3) *Transnational Legal Theory* 347 (https://doi.org/10.5235/204140011800664120).

30 See, e.g., Olaf Meyer, 'A Flexible System in Flux: On the Realignment of Public Policy' in Olaf Meyer (ed), *Public Policy and Private International Law – A Comparative Guide* (Cheltenham, Edward Elgar Publishing 2022) 1, 17–23; Alex Mills, 'The Dimensions of Public Policy in Private International Law' [2008] 4(2) *Journal of Private International Law* 201, 207–218 (https://doi.org/10.1080/17536235.2008.11424339); Ioanna Thoma, 'Public Policy (*ordre public*)' in Jürgen Basedow, Giesela Rühl, Franco Ferrari and Pedro de Miguel Asensio (eds), *Encyclopedia of Private International Law* (Cheltenham, Edward Elgar Publishing 2017) 1453.

31 See, e.g., Marie von Falkenhausen, *Menschenrechtsschutz durch Deliktsrecht – Unternehmerische Pflichten in internationalen Lieferketten* (Tübingen, Mohr Siebeck 2020) 328–329; Marc-Philippe Weller, Luca Kaller and Alix Schulz, 'Haftung deutscher Unternehmen für Menschenrechtsverletzungen im Ausland' [2016] 216(3–4) *Archiv für die civilistische Praxis* 387, 394–395 (https://doi.org/10.1628/000389916X14646106714428); Chris Thomale and Leonhard Hübner, 'Zivilgerichtliche Durchsetzung völkerrechtlicher Unternehmensverantwortung' [2017] 72(8) *JuristenZeitung* 385, 392–393 (https://doi.org/10.1628/002268817X14888930582415); Christoph Wendelstein, '"Menschenrechtliche" Verhaltenspflichten im System des Internationalen Privatrechts' [2019] 83(1) *Rabels Zeitschrift für ausländisches und internationales Privatrecht* 111, 144–150 (https://doi.org/10.1628/rabelsz-2019-0004).

32 Christoph Wendelstein, '"Menschenrechtliche" Verhaltenspflichten im System des Internationalen Privatrechts' [2019] 83(1) *Rabels Zeitschrift für ausländisches und internationales Privatrecht* 111, 148–150 (https://doi.org/10.1628/rabelsz-2019-0004).

33 Christoph Wendelstein, '"Menschenrechtliche" Verhaltenspflichten im System des Internationalen Privatrechts' [2019] 83(1) *Rabels Zeitschrift für ausländisches und internationales Privatrecht* 111, 147–148 (https://doi.org/10.1628/rabelsz-2019-0004). However, some authors argue that the international jurisdiction of German courts establishes a sufficient connection to the German forum, at least in case of a grave violation of human rights, see, e.g., Marc-Philippe Weller, Luca Kaller and Alix Schulz, 'Haftung deutscher Unternehmen für Menschenrechtsverletzungen im Ausland' [2016] 216(3–4) *Archiv für die civilistische Praxis* 387, 396–397 (https://doi.org/10.1628/000389916X14646106714428).

34 Chris Thomale and Leonhard Hübner, 'Zivilgerichtliche Durchsetzung völkerrechtlicher Unternehmensverantwortung' [2017] 72(8) *JuristenZeitung* 385, 392–393 (https://doi.org/10.1628/002268817X14888930582415).

35 Cf. Marie von Falkenhausen, *Menschenrechtsschutz durch Deliktsrecht – Unternehmerische Pflichten in internationalen Lieferketten* (Tübingen, Mohr Siebeck 2020) 326–331.

36 See, e.g., Mike Wilderspin, 'Overriding Mandatory Provisions' in Jürgen Basedow, Gisela Rühl, Franco Ferrari and Pedro De Miguel Asensio (eds), *Encyclopedia of Private International Law* (Cheltenham, Edward Elgar Publishing 2017) 1330.

37 See, e.g., Wolfgang Wurmnest, 'Public Policy in European Private International Law' in Olaf Meyer (ed), *Public Policy and Private International Law – A Comparative Guide* (Cheltenham, Edward Elgar Publishing 2022) 25.

38 This concept was most notably developed by Pierre Lalive, 'Transnational (or Truly International) Public Policy in International Arbitration' in Pieter Sanders (ed), *Comparative Arbitration Practice and Public Policy in Arbitration* (Deventer, Kluwer Law International 1987) 258. See also Catherine Kessedjan, 'Transnational Public Policy' in Albert Jan van den Berg (ed), *International Council for Commercial Arbitration, International Arbitration 2006: Back to Basics?* (Alphen aan den Rijn, Kluwer Law International 2007) 857.

39 See, e.g., Andrew Barraclough and Jeff Waincymer, 'Mandatory Rules of Law in International Commercial Arbitration' (2005) 6 *Melbourne Journal of International Law* 205, 218; Jeffrey M. Waincymer,

'International Commercial Arbitration and the Application of Mandatory Rules of Law' [2009] 5(1) *Asian International Arbitration Journal*, 1, 36 (https://doi.org/10.54648/aiaj2009001).

40 See on this, e.g., Marc-Philippe Weller, Luca Kaller and Alix Schulz, 'Haftung deutscher Unternehmen für Menschenrechtsverletzungen im Ausland' [2016] 216(3–4) *Archiv für die civilistische Praxis* 387, 395–396 (https://doi.org/10.1628/000389916X14646106714428), although the authors reject this idea.

41 See, e.g., Drossos Stamboulakis and Jay Sanderson, 'SDG 15: Life on Land' in Ralf Michaels, Verónica Ruiz Abou-Nigm and Hans van Loon (eds), *The Private Side of Transforming Our World – UN Sustainable Development Goals 2030 and the Role of Private International Law* (Cambridge, Intersentia 2021) 482.

42 See, e.g., Eduardo Álvarez-Armas, 'SDG 13: Climate Action' in Ralf Michaels, Verónica Ruiz Abou-Nigm and Hans van Loon (eds), *The Private Side of Transforming Our World – UN Sustainable Development Goals 2030 and the Role of Private International Law* (Cambridge, Intersentia 2021) 423.

43 The first sentence of Recital 25: 'Regarding environmental damage, Article 174 of the Treaty, which provides that there should be a high level of protection based on the precautionary principle and the principle that preventive action should be taken, the principle of priority for corrective action at source and the principle that the polluter pays, fully justifies the use of the principle of discriminating in favour of the person sustaining the damage'.

44 Tinus Kruger, 'SDG 10: Reduced Inequalities' in Ralf Michaels, Verónica Ruiz Abou-Nigm and Hans van Loon (eds), *The Private Side of Transforming Our World – UN Sustainable Development Goals 2030 and the Role of Private International Law* (Cambridge, Intersentia 2021) 326.

45 See, e.g., Eduardo Álvarez-Armas, 'SDG 13: Climate Action' in Ralf Michaels, Verónica Ruiz Abou-Nigm and Hans van Loon (eds), *The Private Side of Transforming our World – UN Sustainable Development Goals 2030 and the Role of Private International Law* (Cambridge, Intersentia 2021) 409; Matthias Lehmann and Florian Eichel, 'Globaler Klimawandel und Internationales Privatrecht – Zuständigkeit und anzuwendendes Recht für transnationale Klagen wegen klimawandelbedingter Individualschäden' (2019) 83(1) *Rabels Zeitschrift für ausländisches und internationales Privatrecht* 77.

46 See, e.g., Eduardo Álvarez-Armas, 'SDG 13: Climate Action' in Ralf Michaels, Verónica Ruiz Abou-Nigm and Hans van Loon (eds), *The Private Side of Transforming Our World – UN Sustainable Development Goals 2030 and the Role of Private International Law* (Cambridge, Intersentia 2021) 430–432; Marc-Philippe Weller and Mai-Lan Tran, 'Klimawandelklagen im Rechtsvergleich – private enforcement als weltweiter Trend?' [2021] 3 *Zeitschrift für Europäisches Privatrecht* 573.

47 However, the court did not explicitly mention this issue in the judgment. The judgment of the court of first instance: LG Essen (15 December 2016), 2 O 285/15, NVwZ 2017, 734. The case is still pending before the court of appeal, see OLG Hamm (30 November 2017), I-5 U 15/17, ZUR 2018, 118 f. For an English summary of the case see, e.g., Daniel Ennöckl, 'Climate Change Litigation in Austria and Germany: Climate Change Litigation in Germany and Austria – Recent Developments' [2020] 14(4) *Carbon & Climate Law Review (CCLR)* 306, 309–311 (https://doi.org/10.21552/cclr/2020/4/9).

48 Hague District Court (26 May 2021), *Milieudefensie et al v RDS*, ECLI:NL:RBDHA:2021:5337, ECLI:NL:RBDHA:2021:5339 (English version).

49 See the statistics in Joana Setzer and Catherine Higham, *Global Trends in Climate Change Litigation: 2022 Snapshot* (Grantham Research Institute on Climate Change and Environment) <https://www.lse.ac.uk/granthaminstitute/wp-content/uploads/2022/08/Global-trends-in-climate-change-litigation-2022-snapshot.pdf> accessed 17 July 2023; on the need for capacity expansion for litigation in the Global South, see Joana Setzer and Lisa Benjamin, 'Climate Litigation in the Global South: Constraints and Innovations' [2020] 9(1) *Transnational Environmental Law* 77 (https://doi.org/10.1017/S2047102519000268).

50 Eduardo Álvarez-Armas, 'SDG 13: Climate Action' in Ralf Michaels, Verónica Ruiz Abou-Nigm and Hans van Loon (eds), *The Private Side of Transforming Our World – UN Sustainable Development Goals 2030 and the Role of Private International Law* (Cambridge, Intersentia 2021) 423; Eva-Maria Kieninger, 'Conflicts of Jurisdiction and the Applicable Law in Domestic Courts' proceedings' in Wolfgang Kahl and Marc-Philippe Weller (eds), *Climate Change Litigation – A Handbook* (München, C. H. Beck 2021) 119, 140; Matthias Lehmann and Florian Eichel, 'Globaler Klimawandel und Internationales Privatrecht – Zuständigkeit und anzuwendendes Recht für transnationale Klagen wegen klimawandelbedingter Individualschäden' [2019] 83(1) *Rabels Zeitschrift für ausländisches und internationales Privatrecht* 77, 93–95 (https://doi.org/10.1628/rabelsz-2019-0003).

51 Eva-Maria Kieninger, 'Conflicts of Jurisdiction and the Applicable Law in Domestic Courts' Proceedings' in Wolfgang Kahl and Marc-Philippe Weller (eds), *Climate Change Litigation – A Handbook* (München, C. H. Beck 2021) 119, 141; Matthias Lehmann and Florian Eichel, 'Globaler Klimawandel

und Internationales Privatrecht – Zuständigkeit und anzuwendendes Recht für transnationale Klagen wegen klimawandelbedingter Individualschäden' [2019] 83(1) *Rabels Zeitschrift für ausländisches und internationales Privatrecht* 77, 96 (https://doi.org/10.1628/rabelsz-2019-0003). However, other locations could potentially also 'give rise' to the damage, see, e.g., Madeleine Petersen Weiner and Marc-Philippe Weller, 'The "Event Giving Rise to the Damage" under art. 7 Rome II Regulation in CO2 Reduction Claims – A break through an empty Shell?' (*Conflict of Laws*, 2 January 2023) <https://conflictoflaws.net/2023/the-event-giving-rise-to-the-damage-under-art-7-rome-ii-regulation-in-co2-reduction-claims-a-break-through-an-empty-shell/> accessed 17 July 2023. In the RWE case, the court of first instance applied German law without a choice-of-law analysis: LG Essen (15 December 2016), 2 O 285/15, NVwZ 2017, 734.

52 Eva-Maria Kieninger, 'Conflicts of Jurisdiction and the Applicable Law in Domestic Courts' Proceedings' in Wolfgang Kahl and Marc-Philippe Weller (eds), *Climate Change Litigation – A Handbook* (München, C. H. Beck 2021) 119, 140; Matthias Lehmann and Florian Eichel, 'Globaler Klimawandel und Internationales Privatrecht – Zuständigkeit und anzuwendendes Recht für transnationale Klagen wegen klimawandelbedingter Individualschäden' [2019] 83(1) *Rabels Zeitschrift für ausländisches und internationales Privatrecht* 77, 96 (https://doi.org/10.1628/rabelsz-2019-0003).

53 Matthias Lehmann and Florian Eichel, 'Globaler Klimawandel und Internationales Privatrecht – Zuständigkeit und anzuwendendes Recht für transnationale Klagen wegen klimawandelbedingter Individualschäden' [2019] 83(1) *Rabels Zeitschrift für ausländisches und internationales Privatrecht* 77, 96 (https://doi.org/10.1628/rabelsz-2019-0003). However, for a different perspective see Eva-Maria Kieninger, 'Conflicts of jurisdiction and the applicable law in domestic courts' proceedings' in Wolfgang Kahl and Marc-Philippe Weller (eds), *Climate Change Litigation – A Handbook* (München, C. H. Beck 2021) 119, 142–144.

54 See, e.g., Eduardo Álvarez-Armas, 'SDG 13: Climate Action' in Ralf Michaels, Verónica Ruiz Abou-Nigm and Hans van Loon (eds), *The Private Side of Transforming Our World – UN Sustainable Development Goals 2030 and the Role of Private International Law* (Cambridge, Intersentia 2021) 434–436; Eva-Maria Kieninger, 'Conflicts of Jurisdiction and the Applicable Law in Domestic Courts' Proceedings' in Wolfgang Kahl and Marc-Philippe Weller (eds), *Climate Change Litigation – A Handbook* (München, C. H. Beck 2021) 119, 141–144; Matthias Lehmann and Florian Eichel, 'Globaler Klimawandel und Internationales Privatrecht – Zuständigkeit und anzuwendendes Recht für transnationale Klagen wegen klimawandelbedingter Individualschäden' [2019] 83(1) *Rabels Zeitschrift für ausländisches und internationales Privatrecht* 77, 97–105 (https://doi.org/10.1628/rabelsz-2019-0003).

55 Matthias Lehmann and Florian Eichel, 'Globaler Klimawandel und Internationales Privatrecht – Zuständigkeit und anzuwendendes Recht für transnationale Klagen wegen klimawandelbedingter Individualschäden' [2019] 83(1) *Rabels Zeitschrift für ausländisches und internationales Privatrecht* 77, 99–101 (https://doi.org/10.1628/rabelsz-2019-0003).

56 Matthias Lehmann and Florian Eichel, 'Globaler Klimawandel und Internationales Privatrecht – Zuständigkeit und anzuwendendes Recht für transnationale Klagen wegen klimawandelbedingter Individualschäden' [2019] 83(1) *Rabels Zeitschrift für ausländisches und internationales Privatrecht* 77, 99 (https://doi.org/10.1628/rabelsz-2019-0003).

57 Eva-Maria Kieninger, 'Conflicts of Jurisdiction and the Applicable Law in Domestic Courts' Proceedings' in Wolfgang Kahl and Marc-Philippe Weller (eds), *Climate Change Litigation – A Handbook* (München, C. H. Beck 2021) 119, 144.

58 Eva-Maria Kieninger, 'Conflicts of Jurisdiction and the Applicable Law in Domestic Courts' Proceedings' in Wolfgang Kahl and Marc-Philippe Weller (eds), *Climate Change Litigation – A Handbook* (München, C. H. Beck 2021) 119, 142–144; Matthias Lehmann and Florian Eichel, 'Globaler Klimawandel und Internationales Privatrecht – Zuständigkeit und anzuwendendes Recht für transnationale Klagen wegen klimawandelbedingter Individualschäden' [2019] 83(1) *Rabels Zeitschrift für ausländisches und internationales Privatrecht* 77, 101–105 (https://doi.org/10.1628/rabelsz-2019-0003).

59 Eva-Maria Kieninger, 'Conflicts of Jurisdiction and the Applicable Law in Domestic Courts' Proceedings' in Wolfgang Kahl and Marc-Philippe Weller (eds), *Climate Change Litigation – A Handbook* (München, C. H. Beck 2021) 119, 144.

60 See on this, e.g., Lorenzo Pellegrini, Murat Arsel, Martí Orta-Martínez and Carlos F. Mena, 'International Investment Agreements, Human Rights, and Environmental Justice: The Texaco/Chevron Case from the Ecuadorian Amazon' [2020] 23(2) *Journal of International Economic Law* 455 (https://doi.org/10.1093/jiel/jgaa016).

61 Jaakko Salminen and Mikko Rajavuori, 'Private International Law, Global Value Chains and the Externalities of Transnational Production: Towards Alignment?' [2021] 12(2) *Transnational Legal Theory* 230 (https://doi.org/10.1080/20414005.2021.1970470); Sabrina Robert Cuendet and Pascale Ricard, *White Paper 18 – SDGs beyond 2030* (Paris, ILA 2023) 53, 56–57, 82–85.

62 See on this Goal, e.g., Richard Frimpong Oppong, 'SDG 6: Clean Water and Sanitation' in Ralf Michaels, Verónica Ruiz Abou-Nigm and Hans van Loon (eds), *The Private Side of Transforming Our World – UN Sustainable Development Goals 2030 and the Role of Private International Law* (Cambridge, Intersentia 2021) 189.

63 See on this Goal, e.g., Drossos Stamboulakis and Jay Sanderson, 'SDG 15: Life on Land' in Ralf Michaels, Verónica Ruiz Abou-Nigm and Hans van Loon (eds), *The Private Side of Transforming Our World – UN Sustainable Development Goals 2030 and the Role of Private International Law* (Cambridge, Intersentia 2021) 463.

64 See on this goal, e.g., Tajudeen Sanni, 'SDG 14: Life below Water' in Ralf Michaels, Verónica Ruiz Abou-Nigm and Hans van Loon (eds), *The Private Side of Transforming Our World – UN Sustainable Development Goals 2030 and the Role of Private International Law* (Cambridge, Intersentia 2021) 441.

65 Vedanta Resources PLC and another v Lungowe and others [2019] UKSC 20. On this case and others that concern the pollution of watercourses see Richard Frimpong Oppong, 'SDG 6: Clean Water and Sanitation' in Ralf Michaels, Verónica Ruiz Abou-Nigm and Hans van Loon (eds), *The Private Side of Transforming Our World – UN Sustainable Development Goals 2030 and the Role of Private International Law* (Cambridge, Intersentia 2021) 189, 190, 205–215. Generally on human rights protection in international supply chains Marie von Falkenhausen, *Menschenrechtsschutz durch Deliktsrecht – Unternehmerische Pflichten in internationalen Lieferketten* (Tübingen, Mohr Siebeck 2020).

66 See on this Goal, e.g., Ulla Liukkunen, 'SDG 8: Decent Work and Economic Growth' in Ralf Michaels, Verónica Ruiz Abou-Nigm and Hans van Loon (eds), *The Private Side of Transforming Our World – UN Sustainable Development Goals 2030 and the Role of Private International Law* (Cambridge, Intersentia 2021) 245.

67 LG Dortmund (10 January 2019), 7 O 95/15, BeckRS 2019, 388 (first instance); OLG Hamm (21 May 2019), I-9 U 44/19, NJW 2019, 3527 (court of appeals). See on this, e.g., Patrick Ostendorf, '(Kollisionsrechtliche) Stolpersteine bei Haftungsansprüchen gegen deutsche Unternehmen für Menschenrechtsverletzungen ihrer ausländischen Zulieferer' [2019] 39(4) *Praxis des Internationalen Privat- und Verfahrensrechts* (IPRax) 297. For an English summary of the case see, e.g., Marie Elaine Schäfer, 'Germany's Approach to Cross-Border Corporate Social Responsibility of Enterprises: Latest Developments' (*Conflict of Laws*, 30 April 2020) <https://conflictoflaws.net/2020/germanys-approach-to-cross-border-corporate-social-responsibility-of-enterprises-latest-developments/> accessed 17 July 2023.

68 See, e.g., Ulla Liukkunen, 'SDG 8: Decent Work and Economic Growth' in Ralf Michaels, Verónica Ruiz Abou-Nigm and Hans van Loon (eds), *The Private Side of Transforming Our World – UN Sustainable Development Goals 2030 and the Role of Private International Law* (Cambridge, Intersentia 2021) 245, 252 ff.

69 LG Dortmund (10 January 2019), 7 O 95/15, BeckRS 2019, 388.

70 Some litigation did indeed take place in Pakistan. See Faisal Siddiqi, 'Paradoxes of Strategic Labour Rights Litigation: Insights from the Baldia Factory Fire Litigation' in Miriam Saage-Maaß, Peer Zumbansen, Michael Bader, Palvasha Shahab (eds), *Transnational Legal Activism in Global Value Chains – The Ali Enterprises Factory Fire and the Struggle for Justice* (Cham, Springer 2021) 59.

71 LG Dortmund (10 January 2019), 7 O 95/15, BeckRS 2019, 388 (first instance); OLG Hamm (21 May 2019), I-9 U 44/19, NJW 2019, 3527 (court of appeals).

72 Loi n° 2017-399 du 27 mars 2017 relative au devoir de vigilance des sociétés mères et des entreprises donneuses d'ordre, see, e.g., Laura Nasse, 'Devoir de vigilance – Die neue Sorgfaltspflicht zur Menschenrechtsverantwortung für Großunternehmen in Frankreich' [2019] 4 *Zeitschrift für Europäisches Privatrecht* 774; Laura Nasse, *Loi de vigilance: Das französische Lieferkettengesetz* (Tübingen, Mohr Siebeck 2022) 69 ff.

73 Cf. the case LG Dortmund (10 January 2019), 7 O 95/15, BeckRS 2019, 388 (first instance); OLG Hamm (21 May 2019), I-9 U 44/19, NJW 2019, 3527 (court of appeals). The claims were unsuccessful due to a Pakistani statute of limitations which was deemed to be compatible with German public policy.

74 Cf. Leonhard Hübner, Victor Habrich and Marc-Philippe Weller, 'Corporate Sustainability Due Diligence – Der EU-Richtlinienentwurf für eine Lieferkettenregulierung' [2022] 25(14) *Neue Zeitschrift für Gesellschaftsrecht* 644, 649.

75 See also the different options explored by Jaakko Salminen and Mikko Rajavuori, 'Private International Law, Global Value Chains and the Externalities of Transnational Production: Towards Alignment?' [2021] 12(2) *Transnational Legal Theory* 230, 242–245 (https://doi.org/10.1080/20414005.2021.1970470).

76 Jaakko Salminen and Mikko Rajavuori, 'Private International Law, Global Value Chains and the Externalities of Transnational Production: Towards Alignment?' [2021] 12(2) *Transnational Legal Theory* 230, 244 (https://doi.org/10.1080/20414005.2021.1970470).

77 Commission, 'Proposal for a Directive of the European Parliament and of the Council on Corporate Sustainability Due Diligence and amending Directive (EU) 2019/1937' COM/2022/71 final. Regarding the proposed directive see, e.g., Tobias Bomsdorf and Berthold Blatecki-Burgert, 'Lieferketten-Richtlinie und Lieferkettensorgfaltspflichtengesetz' [2022] *Zeitschrift für Rechtspolitik* 141; Leonhard Hübner, Victor Habrich and Marc-Philippe Weller, 'Corporate Sustainability Due Diligence – Der EU-Richtlinienentwurf für eine Lieferkettenregulierung' [2022] 25(14) *Neue Zeitschrift für Gesellschaftsrecht* 644; Sebastian Lutz-Bachmann and K. Vorbeck, 'Neue Sorgfaltspflichten durch die EU-Lieferkettenrichtlinie – Zur Bedeutung des Entwurfs der EU-Lieferkettenrichtlinie für die Automobilindustrie' [2022] *Recht Automobil Wirtschaft* 122; Claire Methven O'Brien and Olga Martin-Ortega, *Commission Proposal on Corporate Sustainability Due Diligence: Analysis from a Human Rights Perspective* (Brussels, Directorate-General for External Policies, 2022) (in-depth analysis requested by the European Parliament's Subcommittee on Human Rights); Michael Nietsch and Michael Wiedmann, 'Der Vorschlag zu einer europäischen Sorgfaltspflichten-Richtlinie im Unternehmensbereich (Corporate Sustainability Due Diligence Directive)' [2022] 5 *Corporate Compliance Zeitschrift* 125; Gerald Spindler, 'Der Vorschlag einer EU-Lieferketten-Richtlinie' [2022] 43(16) *ZIP – Zeitschrift für Wirtschaftsrecht* 765. See also Rui Dias, 'CSDD and PIL: Some Remarks on the Directive Proposal' (*Conflict of Laws*, 2 June 2022) <https://conflictoflaws.net/2022/csdd-and-pil-some-remarks-on-the-directive-proposal/> accessed 17 July 2023. The Commission's proposal is based on a Resolution by the European Parliament from 10 March 2021, see on this, e.g., Tinus Kruger, 'SDG 10: Reduced Inequalities' in Ralf Michaels, Verónica Ruiz Abou-Nigm and Hans van Loon (eds), *The Private Side of Transforming our World – UN Sustainable Development Goals 2030 and the Role of Private International Law* (Cambridge, Intersentia 2021) 326, 335–345.

78 The explanatory memorandum explicitly refers to the European Green Deal and the UN Sustainable Development Goals, see Proposal for a Directive of the European Parliament and of the Council on Corporate Sustainability Due Diligence and amending Directive (EU) 2019/1937, COM/2022/71 final, 1.

79 Tobias Bomsdorf and Berthold Blatecki-Burgert, 'Lieferketten-Richtlinie und Lieferkettensorgfaltspflichtengesetz' [2022] *Zeitschrift für Rechtspolitik* 141, 142–143; Leonhard Hübner, Victor Habrich and Marc-Philippe Weller, 'Corporate Sustainability Due Diligence – Der EU-Richtlinienentwurf für eine Lieferkettenregulierung' [2022] 25(14) *Neue Zeitschrift für Gesellschaftsrecht* 644, 648–649; Michael Nietsch and Michael Wiedmann, 'Der Vorschlag zu einer europäischen Sorgfaltspflichten-Richtlinie im Unternehmensbereich (Corporate Sustainability Due Diligence Directive)' [2022] 5 *Corporate Compliance Zeitschrift* 125, 132–134.

80 Tobias Bomsdorf and Berthold Blatecki-Burgert, 'Lieferketten-Richtlinie und Lieferkettensorgfaltspflichtengesetz' [2022] *Zeitschrift für Rechtspolitik* 141, 142–143; Leonhard Hübner, Victor Habrich and Marc-Philippe Weller, 'Corporate Sustainability Due Diligence – Der EU-Richtlinienentwurf für eine Lieferkettenregulierung' [2022] 25(14) *Neue Zeitschrift für Gesellschaftsrecht* 644, 649–650; Michael Nietsch and Michael Wiedmann, 'Der Vorschlag zu einer europäischen Sorgfaltspflichten-Richtlinie im Unternehmensbereich (Corporate Sustainability Due Diligence Directive)' [2022] 5 *Corporate Compliance Zeitschrift* 125, 133–134; Jaakko Salminen and Mikko Rajavuori, 'Private International Law, Global Value Chains and the Externalities of Transnational Production: Towards Alignment?' [2021] 12(2) *Transnational Legal Theory* 230, 244–245 (https://doi.org/10.1080/20414005.2021.1970470); Gerald Spindler, 'Der Vorschlag einer EU-Lieferketten-Richtlinie' [2022] 43(16) *ZIP – Zeitschrift für Wirtschaftsrecht* 765, 776. The civil liability in the French *loi de vigilance* is also an overriding mandatory provision, see, e.g., Laura Nasse, *Loi de vigilance: Das französische Lieferkettengesetz* (Tübingen, Mohr Siebeck 2022) 256–271.

81 Leonhard Hübner, Victor Habrich and Marc-Philippe Weller, 'Corporate Sustainability Due Diligence – Der EU-Richtlinienentwurf für eine Lieferkettenregulierung' [2022] 25(14) *Neue Zeitschrift für Gesellschaftsrecht* 644, 650–651.

82 Regarding the proposed directive see, e.g., Michael Nietsch and Michael Wiedmann, 'Der Vorschlag zu einer europäischen Sorgfaltspflichten-Richtlinie im Unternehmensbereich (Corporate Sustainability Due Diligence Directive)' [2022] 5 *Corporate Compliance Zeitschrift* 125, 134.

83 Cf. Jaakko Salminen and Mikko Rajavuori, 'Private International Law, Global Value Chains and the Externalities of Transnational Production: Towards Alignment?' [2021] 12(2) *Transnational Legal Theory* 230, 245–247 (https://doi.org/10.1080/20414005.2021.1970470).

84 Gesetz über die unternehmerischen Sorgfaltspflichten zur Vermeidung von Menschenrechtsverletzungen in Lieferketten vom 16. Juli 2021 (BGBl. I S. 2959). For an overview in English see <https://www.csr-in-deutschland.de/EN/Business-Human-Rights/Supply-Chain-Act/supply-chain-act.html;jsessionid=9F61D2EB0229E04899F47D97CCF2535F> accessed 17 July 2023.

85 See, e.g., Laura Nasse, 'Das neue Lieferkettensorgfaltspflichtengesetz – Pflicht zur Entkopplung der deutschen Wirtschaft vom chinesischen Markt?' [2022] *Recht Automobil Wirtschaft* 3; Walter G Paefgen, 'Haftung für die Verletzung von Pflichten nach dem neuen Lieferkettensorgfaltspflichtengesetz' [2021] 42(39) *ZIP – Zeitschrift für Wirtschaftsrecht* 2006; Giesela Rühl and Constantin Knauer, 'Zivilrechtlicher Menschenrechtsschutz? Das deutsche Lieferkettengesetz und die Hoffnung auf den europäischen Gesetzgeber' [2022] 77(3) *JuristenZeitung* 105 (https://doi.org/10.1628/jz-2022-0056); Adam Sagan and Alexander J. Schmidt, 'Das Lieferkettensorgfaltspflichtengesetz – Ein Überblick aus der Perspektive des Arbeitsrechts' [2022] 27(6) *NZA-Rechtsprechungs-Report Arbeitsrecht* 281; Eric Wagner and Marc Ruttloff, 'Das Lieferkettensorgfaltspflichtengesetz – Eine erste Einordnung' [2021] *Neue Juristische Wochenschrift* 2145.

86 David V Snyder, Susan A Maslow and Sarah Dadush, 'Balancing Buyer and Supplier Responsibilities: Model Contract Clauses to Protect Workers in International Supply Chains, Version 2.0' (American Bar Association, 28 April 2021) <https://businesslawtoday.org/2021/04/balancing-buyer-supplier-responsibilities-model-contract-clauses-protect-workers-international-supply-chains-version-2-0/> accessed 17 July 2023.

87 Maria Julia Ochoa Jiménez, 'Exploring a Minefield: Private International Law in Latin America, Its Neocolonial Character, and Its Potentialities' [2021] 8(2) *Critical Analysis of Law* 87 (https://doi.org/10.33137/cal.v8i2.37855).

Bibliography

Álvarez-Armas E, 'SDG 13: Climate Action' in Michaels R, Ruiz Abou-Nigm V and van Loon H (eds), *The Private Side of Transforming Our World – UN Sustainable Development Goals 2030 and the Role of Private International Law* (Cambridge, Intersentia 2021) 409

Barraclough A and Waincymer J, 'Mandatory Rules of Law in International Commercial Arbitration' [2005] 6 *Melbourne Journal of International Law* 205

Bomhoff J and Meuwese A, 'The Meta-Regulation of Transnational Private Regulation' [2011] 38(1) *Journal of Law and Society* 138 (https://doi.org/10.1111/j.1467-6478.2011.00538.x)

Bomsdorf T and Blatecki-Burgert B, 'Lieferketten-Richtlinie und Lieferkettensorgfaltspflichtengesetz' [2022] 49 *Zeitschrift für Rechtspolitik* 141

Brydie-Watson W, 'The Three Sisters of Private International Law: An Increasingly Co-Operative Family Rather Than Sibling Rivals' in John T, Gulati R and Köhler B (eds), *The Elgar Companion on the Hague Conference on Private International Law* (Cheltenham, Edward Elgar Publishing 2020) 23

Buxbaum H and Michaels R, 'Reasonableness as a Limitation on the Extraterritorial Application of U.S. Law: From 403 to 405 (via 404)' in Stephan PB and Cleveland SA (eds), *The Restatement and Beyond: The Past, Present, and Future of U.S. Foreign Relations Law* (Oxford, Oxford University Press 2020) 283

Corneloup S and Verhellen J, 'SDG 16: Peace, Justice and Strong Institutions' in Michaels R, Ruiz Abou-Nigm V and van Loon H (eds), *The Private Side of Transforming Our World – UN Sustainable Development Goals 2030 and the Role of Private International Law* (Cambridge, Intersentia 2021) 505

Dias R, 'CSDD and PIL: Some Remarks on the Directive Proposal' (*Conflict of Laws*, 2 June 2022) <https://conflictoflaws.net/2022/csdd-and-pil-some-remarks-on-the-directive-proposal/>

Diemer A, 'Six Key Drivers for Sustainable Development' [2019] 18 *International Journal of Environmental Sciences & Natural Resources* 128 (http://dx.doi.org/10.19080/IJESNR.2019.18.555994)

Ennöckl D, 'Climate Change Litigation in Austria and Germany: Climate Change Litigation in Germany and Austria – Recent Developments' [2020] 14(4) *Carbon & Climate Law Review (CCLR)* 306 (https://doi.org/10.21552/cclr/2020/4/9)

Frank W, 'Störerhaftung für Klimaschäden?' [2017] 36 *Neue Zeitschrift für Verwaltungsrecht* 664

Frimpong Oppong R, 'SDG 6: Clean Water and Sanitation' in Michaels R, Ruiz Abou-Nigm V and van Loon H (eds), *The Private Side of Transforming Our World – UN Sustainable Development Goals 2030 and the Role of Private International Law* (Cambridge, Intersentia 2021) 189

Grossman DA, 'Warming Up to a Not-So-Radical Idea: Tort-Based Climate Change Litigation' [2003] 28 *Columbia Journal of Environmental Law* 1

Hellgardt A and Jouannau V, 'Nachhaltigkeitsziele und Privatrecht' (2022) 222 *Archiv für die civilistische Praxis* 163, 167–170 (https://doi.org/10.1628/acp-2022-0008)

Hübner L, Habrich V and Weller MP, 'Corporate Sustainability Due Diligence – Der EU-Richtlinienentwurf für eine Lieferkettenregulierung' [2022] 25(14) *Neue Zeitschrift für Gesellschaftsrecht* 644

Kessedjan C, 'Transnational Public Policy' in van den Berg AJ (ed), *International Council for Commercial Arbitration, International Arbitration 2006: Back to Basics?* (Alphen aan den Rijn, Kluwer Law International 2007) 857

Kieninger EM, 'Conflicts of Jurisdiction and the Applicable Law in Domestic Courts' Proceedings' in Kahl W and Weller MP (eds), *Climate Change Litigation – A Handbook* (München, C. H. Beck 2021) 119

Kruger T, 'SDG 10: Reduced Inequalities' in Michaels T, Ruiz Abou-Nigm V and van Loon H (eds), *The Private Side of Transforming our World – UN Sustainable Development Goals 2030 and the Role of Private International Law* (Cambridge, Intersentia 2021) 326

Lalive P, 'Transnational (or Truly International) Public Policy in International Arbitration' in Sanders P (ed), *Comparative Arbitration Practice and Public Policy in Arbitration* (Deventer, Kluwer Law International 1987) 258

Lehmann M, 'Regulation, Global Governance and Private International Law: Squaring the Triangle' [2020] 16(1) *Journal of Private International Law* 1 (https://doi.org/10.1080/17441048.2020.1744255)

Lehmann M and Eichel F, 'Globaler Klimawandel und Internationales Privatrecht – Zuständigkeit und anzuwendendes Recht für transnationale Klagen wegen klimawandelbedingter Individualschäden' [2019] 83(1) *Rabels Zeitschrift für ausländisches und internationales Privatrecht* 77, 93–95 (https://doi.org/10.1628/rabelsz-2019-0003)

Liukkunen U, 'SDG 8: Decent Work and Economic Growth' in Michaels R, Ruiz Abou-Nigm V and van Loon H (eds), *The Private Side of Transforming our World – UN Sustainable Development Goals 2030 and the Role of Private International Law* (Cambridge, Intersentia 2021) 245

Lutz-Bachmann S and Vorbeck K, 'Neue Sorgfaltspflichten durch die EU-Lieferkettenrichtlinie – Zur Bedeutung des Entwurfs der EU-Lieferkettenrichtlinie für die Automobilindustrie' [2022] 10 *Recht Automobil Wirtschaft* 122

Lutzi T, 'German Judges Travel to Peru in Climate-Change Trial' (*Conflict of Laws*, 31 May 2022) <https://conflictoflaws.net/2022/german-judges-travel-to-peru-in-climate-change-trial/>

Methven O'Brien C and Martin-Ortega O, *Commission Proposal on Corporate Sustainability Due Diligence: Analysis from a Human Rights Perspective* (Brussels, Directorate-General for External Policies 2022)

Meyer O, 'A Flexible System in Flux: On the Realignment of Public Policy' in Meyer O (ed), *Public Policy and Private International Law – A Comparative Guide* (Cheltenham, Edward Elgar Publishing 2022) 1

Michaels R, 'Two Paradigms of Jurisdiction' [2007] 27 *Michigan Law Review* 1003

———, 'Towards a Private International Law for Regulatory Conflicts?' [2016] 59 *Japanese Yearbook of International Law* 175 (https://doi.org/10.1628/00372515X14373846829991)

———, 'Introduction: The Private Side of Transforming our World – UN Sustainable Development Goals 2030 and the Role of Private International Law' in Michaels R, Ruiz Abou-Nigm V and van Loon H (eds), *The Private Side of Transforming Our World – UN Sustainable Development Goals 2030 and the Role of Private International Law* (Cambridge, Intersentia 2021) 13

Michaels R, Ruiz Abou-Nigm V and van Loon H, *The Private Side of Transforming Our World – UN Sustainable Development Goals 2030 and the Role of Private International Law* (Cambridge, Intersentia 2021)

Mills A, 'The Dimensions of Public Policy in Private International Law' [2008] 4(2) *Journal of Private International Law* 201, 207–218 (https://doi.org/10.1080/17536235.2008.11424339)

———, *Party Autonomy in Private International Law* (Cambridge, Cambridge University Press 2018)

Muir Watt H, '"Party Autonomy" in International Contracts: From the Makings of a Myth to the Requirements of Global Governance' (2010) 6 *European Review of Contract Law* 250 (https://doi.org/10.1515/ercl.2010.250)

———, 'Private International Law Beyond the Schism' [2011] 2(3) *Transnational Legal Theory* 347 (https://doi.org/10.5235/204140011800664120)

———, 'Party Autonomy' in Basedow J, Rühl G, Ferrari F and De Miguel Asensio P (eds), *Encyclopedia of Private International Law* (Cheltenham, Edward Elgar Publishing 2017) 1336

Nasse L, 'Devoir de vigilance – Die neue Sorgfaltspflicht zur Menschenrechtsverantwortung für Großunternehmen in Frankreich' [2019] 4 *Zeitschrift für Europäisches Privatrecht* 774

———, 'Das neue Lieferkettensorgfaltspflichtengesetz – Pflicht zur Entkopplung der deutschen Wirtschaft vom chinesischen Markt?' [2022] 10 *Recht Automobil Wirtschaft* 3

———, *Loi de vigilance: Das französische Lieferkettengesetz* (Tübingen, Mohr Siebeck 2022)

Nietsch M and Wiedmann M, 'Der Vorschlag zu einer europäischen Sorgfaltspflichten-Richtlinie im Unternehmensbereich (Corporate Sustainability Due Diligence Directive)' [2022] 5 *Corporate Compliance Zeitschrift* 125

Ochoa Jiménez MJ, 'Exploring a Minefield: Private International Law in Latin America, Its Neocolonial Character, and Its Potentialities' [2021] 8(2) *Critical Analysis of Law* 87 (https://doi.org/10.33137/cal.v8i2.37855)

Ostendorf P, '(Kollisionsrechtliche) Stolpersteine bei Haftungsansprüchen gegen deutsche Unternehmen für Menschenrechtsverletzungen ihrer ausländischen Zulieferer' [2019] 39(4) *Praxis des Internationalen Privat- und Verfahrensrechts* (IPRax) 297

Paefgen WG, 'Haftung für die Verletzung von Pflichten nach dem neuen Lieferkettensorgfaltspflichtengesetz' [2021] 42(39) *ZIP – Zeitschrift für Wirtschaftsrecht* 2006

Pasquot Polido FB, 'SDG 17: Partnership for the Goals' in Michaels R, Ruiz Abou-Nigm V and van Loon H (eds), *The Private Side of Transforming Our World – UN Sustainable Development Goals 2030 and the Role of Private International Law* (Cambridge, Intersentia 2021) 541

Pellegrini L, Arsel M, Orta-Martínez M and Mena CF, 'International Investment Agreements, Human Rights, and Environmental Justice: The Texaco/Chevron Case from the Ecuadorian Amazon' [2020] 23(2) *Journal of International Economic Law* 455 (https://doi.org/10.1093/jiel/jgaa016)

Petersen Weiner M and Weller MP, 'The "Event Giving Rise to the Damage" under Art. 7 Rome II Regulation in CO2 Reduction Claims – A Break through an Empty Shell?' (*ConflictofLaws.net*, 2 January 2023) <https://conflictoflaws.net/2023/the-event-giving-rise-to-the-damage-under-art-7-rome-ii-regulation-in-co2-reduction-claims-a-break-through-an-empty-shell/>

Pöttker E, *Klimahaftungsrecht – Die Haftung für die Emission von Treibhausgasen in Deutschland und den Vereinigten Staaten von Amerika* (Tübingen, Mohr Siebeck 2014)

Purvis B, Mao Y and Robinson D, 'Three Pillars of Sustainability: In Search of Conceptual Origins' [2019] 14 *Sustainability Science* 681 (https://doi.org/10.1007/s11625-018-0627-5)

Riles A, 'Managing Regulatory Arbitrage: A Conflict of Laws Approach' [2014] 47 *Cornell Journal of International Law* 63

Robert Cuendet S and Ricard P, *White Paper 18 – SDGs Beyond 2030* (Paris, ILA 2023) 53

Rühl G and Knauer C, 'Zivilrechtlicher Menschenrechtsschutz? Das deutsche Lieferkettengesetz und die Hoffnung auf den europäischen Gesetzgeber' [2022] 77(3) *JuristenZeitung* 105 (https://doi.org/10.1628/jz-2022-0056)

Sagan A and Schmidt AJ, 'Das Lieferkettensorgfaltspflichtengesetz – Ein Überblick aus der Perspektive des Arbeitsrechts' [2022] 27(6) *NZA-Rechtsprechungs-Report Arbeitsrecht* 281

Salminen J and Rajavuori M, 'Private International Law, Global Value Chains and the Externalities of Transnational Production: Towards Alignment?' [2021] 12(2) *Transnational Legal Theory* 230 (https://doi.org/10.1080/20414005.2021.1970470)

Sanni T, 'SDG 14: Life below Water' in Michaels R, Ruiz Abou-Nigm V and van Loon H (eds), *The Private Side of Transforming Our World – UN Sustainable Development Goals 2030 and the Role of Private International Law* (Cambridge, Intersentia 2021) 441

Schäfer ME, 'Germany's Approach to Cross-Border Corporate Social Responsibility of Enterprises: Latest Developments' (*Conflict of Laws*, 30 April 2020) <https://conflictoflaws.net/2020/germanys-approach-to-cross-border-corporate-social-responsibility-of-enterprises-latest-developments/>

Schirmer JE, *Nachhaltiges Privatrecht* (Tübingen, Mohr Siebeck 2023)

Setzer J and Benjamin L, 'Climate Litigation in the Global South: Constraints and Innovations' [2020] 9(1) *Transnational Environmental Law* 77 (https://doi.org/10.1017/S2047102519000268)

Setzer J and Higham C, *Global Trends in Climate Change Litigation: 2022 Snapshot* (Grantham Research Institute on Climate Change and Environment) <https://www.lse.ac.uk/granthaminstitute/wp-content/uploads/2022/08/Global-trends-in-climate-change-litigation-2022-snapshot.pdf>

Siddiqi F, 'Paradoxes of Strategic Labour Rights Litigation: Insights from the Baldia Factory Fire Litigation' in Saage-Maaß M, Zumbansen P, Bader M and Shahab P (eds), *Transnational Legal Activism in Global Value Chains – The Ali Enterprises Factory Fire and the Struggle for Justice* (Cham, Springer 2021) 59

Siehr K, 'Coordination of Legal Systems in Private International Law' in Einhorn T and Siehr K (eds), *Intercontinental Cooperation Through Private International Law – Essays in Memory of Peter E. Nygh* (The Hague, T.M.C. Asser Press 2004)

Snyder DV, Maslow SA and Dadush S, 'Balancing Buyer and Supplier Responsibilities: Model Contract Clauses to Protect Workers in International Supply Chains, Version 2.0' (American Bar Association, 28 April 2021) <https://businesslawtoday.org/2021/04/balancing-buyer-supplier-responsibilities-model-contract-clauses-protect-workers-international-supply-chains-version-2-0/>

Spindler G, 'Der Vorschlag einer EU-Lieferketten-Richtlinie' [2022] 43(16) *ZIP – Zeitschrift für Wirtschaftsrecht*, 765

Stamboulakis D and Sanderson J, 'SDG 15: Life on Land' in Michaels R, Ruiz Abou-Nigm V and van Loon H (eds), *The Private Side of Transforming Our World – UN Sustainable Development Goals 2030 and the Role of Private International Law* (Cambridge, Intersentia 2021) 463

Thoma I, 'Public Policy (*Ordre Public*)' in Basedow J, Rühl G, Ferrari F and de Miguel Asensio P (eds), *Encyclopedia of Private International Law* (Cheltenham, Edward Elgar Publishing 2017) 1453

Thomale C and Hübner L, 'Zivilgerichtliche Durchsetzung völkerrechtlicher Unternehmensverantwortung' [2017] 72(8) *JuristenZeitung* 385 (https://doi.org/10.1628/002268817X14888930582415)

Thuy Nguyen T, 'Transnational Corporations and Environmental Pollution in Vietnam – Realising the Potential of Private International Law in Environmental Protection' [2022] 18(2) *Journal of Private International Law* 238 (https://doi.org/10.1080/17441048.2022.2096059)

van Loon H, 'Access to Justice and Legal Aid in Cross-Border Situations' in Department of Justice – The Government of Hong Kong Special Administrative Region (org), *Hong Kong Legal week 2020 – Inaugural Rule of Law Congress – Proceedings – Towards 2030: A Decade of Action for Rule of Law* (Hong Kong, Department of Justice Hong Kong SAR 2021)

van Loon H, 'Private International Law in Support of Sustainable Development – Impulses from Osnabrück' in Grothe H, Mankowski P and Rieländer F (eds), *Europäisches und internationales Privatrecht – Festschrift für Christian von Bar zum 70. Geburtstag* (München, C. H. Beck 2022) 197

von Falkenhausen M, *Menschenrechtsschutz durch Deliktsrecht – Unternehmerische Pflichten in internationalen Lieferketten* (Tübingen, Mohr Siebeck 2020)

Wagner E and Ruttloff M, 'Das Lieferkettensorgfaltspflichtengesetz – Eine erste Einordnung' [2021] 74 *Neue Juristische Wochenschrift* 2145

Wai R, 'Transnational Liftoff and Juridical Touchdown: The Regulatory Function of Private International Law in an Era of Globalization' [2002] 40 *Columbia Journal of Transnational Law* 209

Waincymer JM, 'International Commercial Arbitration and the Application of Mandatory Rules of Law' [2009] 5(1) *Asian International Arbitration Journal* 1, 36 (https://doi.org/10.54648/aiaj2009001)

Weller MP, Kaller L and Schulz A, 'Haftung deutscher Unternehmen für Menschenrechtsverletzungen im Ausland' [2016] 216(3–4) *Archiv für die civilistische Praxis* 387 (https://doi.org/10.1628/000389916X14646106714428)

Weller MP and Tran ML, 'Klimawandelklagen im Rechtsvergleich – Private Enforcement als Weltweiter Trend?' [2021] 3 *Zeitschrift für Europäisches Privatrecht* 573

Wendelstein C, '"Menschenrechtliche" Verhaltenspflichten im System des Internationalen Privatrechts' [2019] 83(1) *Rabels Zeitschrift für ausländisches und internationales Privatrecht* 111, 144–150 (https://doi.org/10.1628/rabelsz-2019-0004)

Whytock CA, 'Transnational Access to Justice' [2020] 38 *Berkeley Journal of International Law* 154

Wilderspin M, 'Overriding Mandatory Provisions' in Basedow J, Rühl G, Ferrari F and De Miguel Asensio P (eds), *Encyclopedia of Private International Law* (Cheltenham, Edward Elgar Publishing 2017) 1330

Wurmnest W, 'Public Policy in European Private International Law' in Meyer O (ed), *Public Policy and Private International Law – A Comparative Guide* (Cheltenham, Edward Elgar Publishing 2022) 25

29
INDIVIDUAL REDRESS AND ECOLOGICAL CLAIMS – NEW WINE INTO OLD WINESKINS?

Sandra Passinhas[1]

29.1 Introduction

The consumer has always been considered an important actor in the politics of the transition to a more sustainable development,[2] namely a more sustainable consumption.[3]

The EU Action Plan for the Circular Economy,[4] in 2015, expressly qualified economic actors, such as businesses and consumers, as key actors in driving this process. The circular economy promised to boost the EU's competitiveness by protecting businesses against scarcity of resources and volatile prices, helping to create new business opportunities and more efficient ways of producing and consuming. The New Circular Economy Action Plan For a cleaner and more competitive Europe[5] reiterated, in 2020, to provide a future-oriented agenda for achieving a cleaner and more competitive Europe in co-creation with economic actors, consumers, citizens, and civil society organisations, building on circular economy actions implemented since 2015.[6]

Later, the Consumers Agenda,[7] drafted in the aftermath of the COVID-19 Pandemic, assumed the need to take account of consumer protection requirements in the formulation and implementation of other policies and activities (Article 12 of the TFUE), to complement other EU initiatives, such as the European Green Deal[8] and the Circular Economy Action Plan, and to support relevant international frameworks, such as the United Nations' 2030 Agenda for Sustainable Development.[9] Empowering consumers and providing them with cost-saving opportunities is a key building block of the sustainable product policy framework. Therefore, the concern with consumer protection develops twofold.[10]

Firstly, consumer protection must ensure that trustworthy and relevant information on products at the point of sale, including on their lifespan and on the availability of repair services, spare parts, and repair manuals, is provided. This should also include consumer protection against misleading ecological claims, namely, greenwashing (information that is not true or presented in a confusing or misleading way to give the inaccurate impression that a product or enterprise is more environmentally sound) and minimum requirements for labelling schemes (e.g., the EU Ecolabel, the official European Union voluntary label for environmental excellence)[11] and for information tools (e.g., the updated labels that provide information on products and appliances covered by the ecodesign directive).[12] The information must be adequate both in content and in context. In order to make sustainable consumption choices, consumers need better and more reliable information

DOI: 10.4324/9781032662046-35

on the ecological characteristics of goods and services, while avoiding information overload. In that sense, digital transformation also offers new opportunities to provide more targeted and understandable information. Digital information is qualified to empower consumers to check the reliability of the information, make comparisons between products and be informed about environmental impacts.

Secondly, for consumers, a sustainable consumption policy might intend to provide high-quality, functional, and safe products, which are efficient and affordable, last longer and are designed for reuse, repair, and high-quality recycling. Access to repair, availability of spare parts and, in the case of ICT and electronics, of upgrading services, should be promoted. Consumer remedies, such as giving preference to repair over replacement, extending the minimum liability period for new or second-hand goods, and restarting the liability period after repair, should be strengthened.[13] Early obsolescence (i.e., premature failures of goods) must be fought against. These initiatives are meant to promote better value retention, prioritise safer and more durable products, and keep materials in the economic cycle (refuse, reduce, repair, reuse, and recycle) for as long as possible.

This chapter will concentrate on the first dimension described, namely consumers' decision-making processes as manifested in the choice of goods and services, in a normative framework that favours information disclosure and expect this to be used by consumers in a reasonable manner. It will start by pointing out that consumers are required to adopt more sustainable individual consumption practices. This requires better information and stronger protection against unfair misleading commercial practices concerning the ecological characteristics of the products. Then, it will describe the related Proposal for a Directive as regards empowering consumers for the green transition and underline the important role of Directive 29/2005/CE,[14] since it outlaws misleading unfair commercial practices, and offers a general and far-reaching rule for considering a misleading omission, where 'material information to take an informed transactional decision is omitted'. The impact of such a provision is better understood in the framework of Article 11a, introduced by Directive (UE) 2019/2061,[15] providing individual redress for consumers, which will be analysed in point 4. The focus will impinge mainly upon the fact that consumers can now rely on protection in a dispute regarding the rights and obligations arising from the contract itself. Finally, a thorough reassessment of the remedies introduced by the Directive, price reduction and termination, will be offered and their inadequacy will be demonstrated. Price reduction and termination are remedies for defective goods or non-performance and aim at correcting the imbalance between parties' rights and obligations. The shortcomings of these choices are, in my opinion, evident. My argument will rest upon the observation that unfair commercial practices impair the consent of consumers to enter a contract, and therefore it must be considered invalid. Accordingly, individual redress of consumers whose consent was impaired should be anchored in the (in)validity of the contract, and rescission is the remedy adequate to protect consumers.

29.2 From Consumer Choice to Consumer Engagement

The green and ethical consumer – the purchaser of goods and services that reward environmentally and socially responsible innovation and provide resources to efficient products and services[16] – is seen as the driving force of market transformation. Although it would be very interesting, a comprehensive review of theories of consumer motivation is beyond the scope of this chapter. Nonetheless, as a primary assumption it must be recalled that it is axiomatic in neo-classical economics that individuals are rational utility-maximisers, that is to say, they calculate and follow the course of economic action which brings them the most utility (benefit, pleasure or satisfaction) that they can afford. However, it seems uncontroversial that consumers' choices are imbued with social and economic implications.[17] Consumers do not act as isolated individuals, but rather as people in society; they do not simply respond to their innate wants and desires, but also to influence peers' and fellow citizens' unconscious habitual routines and social norms.[18]

Consumers are encouraged to move towards more sustainable consumption patterns.[19] Individual citizens are encouraged to weigh up concerns of sustainability alongside (perhaps more compelling) concerns of convenience, availability, self-expression, taste, or affordability (e.g., it should be considered that sustainable textiles or reusable and recyclable packaging might be economically more expensive). Although it is considered that socio-cultural processes and shared rules, values and networks, play a strong role in determining environmental commitment, it may be of concern that the value-action gap, between what people claim to care about and what they act on, reaches to sabotage that commitment.[20] Although research in the areas of sustainable and ethical consumption is often framed by the problem of motivating consumers to adjust their behaviour towards more responsible patterns,[21] this is not the case for European consumers. On the contrary, consumers across Europe are showing a growing interest in contributing personally to achieving climate neutrality, preserving natural resources and biodiversity, and reducing water, air, and soil pollution.[22]

Within the consumer sovereignty narrative, information on the ecological characteristics of products, and information on how to use these products in a more efficient way, is of utmost importance. Respondents to the public consultation for Consumers Agenda pointed to a lack of information and to concerns regarding the reliability of environmental claims and information on products as important obstacles to greater uptake of sustainable consumption choices.[23] Therefore, European efforts to promote sustainable consumption should rely on initiatives to correct market failures and ensure that individuals have greater information to enact their consumer sovereignty, i.e., empowering citizens to pursue their own goals in a way that is favourable to collective outcomes in markets. The recent Proposal for a Directive of the European Parliament and of the Council as regards empowering consumers for the green transition through better protection against unfair practices and better information[24] aims to contribute to a circular, clean, and green EU economy, by enabling consumers to make informed purchasing decisions and therefore contribute to more sustainable consumption. It also targets unfair commercial practices that mislead consumers away from sustainable consumption choices. These purposes could be achieved by amending the Unfair Commercial Practices Directive 2005/29/EC and the Consumer Rights Directive (EU) 2011/83.[25]

Further analysis of the Proposal will be presented below. At this point, attention will be devoted to the European regime of unfair commercial practices.

29.3 Information Paradigm and Misleading Commercial Practices

29.3.1 The Directive 2005/29/EC on Unfair Commercial Practices

Article 5, n. 1 of Directive 2005/29/EC contains a general provision outlawing business-to-consumer unfair commercial practices, according to a threefold structure. It will be briefly described, according to the standard of proof that is required to consider a practice unfair.

Firstly, Annex I was added to the Directive, according to which 31 individually listed misleading and aggressive commercial practices were *per se* unfair, that is to say, they did not require an individual assessment.

Secondly, commercial practices which are misleading in terms of Article 6 (misleading actions) and Article 7 (misleading omissions) or aggressive in terms of Article 8 (aggressive commercial practices) and Article 9 (harassment, coercion, and undue influence), were also deemed unfair. We will now focus on misleading practices. According to Article 6(1), a commercial practice shall be regarded as misleading if it contains false information and is therefore untruthful or in any way deceives or is likely to deceive the average consumer, in relation to one or more of the elements listed therein, and in either case causes or is likely to cause consumers to take a transactional decision that they would not have taken otherwise. The fact that the information is factually correct does not prevent the qualification of commercial practice as misleading. Furthermore, paragraph 2 states that a commercial practice shall also be regarded as misleading if, in its factual context, taking account of all its features and circumstances, it causes or is likely to cause the average consumer to take a transactional decision that he would not have taken otherwise, and it involves: any marketing of a product, including comparative advertising, which creates confusion with any products, trademarks, trade names or other distinguishing marks of a competitor; non-compliance by the trader with commitments contained in codes of conduct by which the trader is bound.

Article 7, governing misleading omissions,[26] rules, in paragraph 1, that a commercial practice shall be regarded as misleading if, taking account of all its features and circumstances and the limitations of the communication medium, it omits material information that the average consumer needs, to take an informed transactional decision. It also qualifies, in paragraph 2, as a misleading omission the practice of a trader that hides or provides in an unclear, unintelligible, ambiguous or untimely manner such material information, or fails to identify the commercial intent of the commercial practice if not already apparent from the context, and where this causes or is likely to cause the average consumer to take a transactional decision that he would not have taken otherwise.[27] According to paragraph 5, information requirements established by Community law in relation to commercial communication including advertising or marketing, a non-exhaustive list of which is contained in Annex II, shall be regarded as material.

Thirdly, the Unfair Commercial Practices Directive introduced, in Article 5, a general clause to recognise unfair commercial practices. Pursuant to Article 5(2), a commercial practice shall also be deemed unfair if it is contrary to the requirements of professional diligence, and if it materially distorts, or is likely to materially distort, the economic behaviour towards the product of the average consumer whom it reaches or to whom it is addressed, or of the average member of the group, when a commercial practice is directed to a particular group of consumers. The requirements set out in Article 5(2) must be satisfied cumulatively for the practice to be considered unfair.[28]

29.3.2 The Proposal for a Directive Empowering Consumers for the Green Transition

The Proposal for a Directive empowering consumers for the green transition, aimed at encouraging consumers to make more environmentally sustainable purchasing decisions, endorses two main problems: consumers usually lack reliable information at the point of sale to make environmentally sustainable consumption choices[29] and face misleading commercial practices related to the sustainability of products.[30]

Article 1 of the Proposal amends Directive 2005/29/EC in the following way. It added to Article 2 several definitions, such as: '(o) 'environmental claim',[31] including (p) 'explicit environmental claim'[32] and (q) 'generic environmental claim'[33]; (r) 'sustainability label'[34]; (s) 'certification scheme'[35]; (t) 'sustainability information tool'[36]; (u) 'recognised excellent environmental performance'[37]; (v) durability[38]; (w) 'software update'[39]; (x) consumable[40]; and (y) 'functionality'.[41]

Article 6 of the Directive, on misleading actions, is amended so that the assessment of the misleading character of a commercial practice's environmental or social impact shall be included in the characteristics of the considered product. In paragraph 2 of Article 6, it is added that a commercial practice shall also be regarded as misleading if, in its factual context, taking account of all its features and circumstances, it causes or is likely to cause the average consumer to take a transactional decision that he would not have taken otherwise, and it involves: '(d) making an environmental claim related to future environmental performance without clear, objective and verifiable commitments and targets and without an independent monitoring system; (e) advertising benefits for consumers that are considered as a common practice in the relevant market.'

In Article 7, on misleading information, paragraph 7 is added:

> Where a trader provides a service which compares products, including through a sustainability information tool, information about the method of comparison, the products which are the object of comparison and the suppliers of those products, as well as the measures in place to keep that information up to date, shall be regarded as material.

Annex I to Directive 2005/29/EC is substantially strengthened. Ten commercial practices which are in all circumstances unfair are inserted: '2a. Displaying a sustainability label which is not based on a certification scheme or not established by public authorities';

> 4a. Making a generic environmental claim for which the trader is not able to demonstrate recognised excellent environmental performance relevant to the claim; 4b. Making an environmental claim about the entire product when it actually concerns only a certain aspect of the product.

'10a. Presenting requirements imposed by law on all products within the relevant product category on the Union market as a distinctive feature of the trader's offer';

> 23d. Omitting to inform the consumer that a software update will negatively impact the use of goods with digital elements or certain features of those goods even if the software update improves the functioning of other features; 23e. Omitting to inform the consumer about the existence of a feature of a good introduced to limit its durability; 23f. Claiming that a good has a certain durability in terms of usage time or intensity when it does not; 23g. Presenting goods as allowing repair when they do not or omitting to inform the consumer that goods do not allow repair in accordance with legal requirements; 23h. Inducing the consumer into replacing the consumables of a good earlier than for technical reasons is necessary; 23i. Omitting to inform that a good is designed to limit its functionality when using consumables, spare parts or accessories that are not provided by the original producer.

In which respects, the following definitions are inserted into Article 2 of the Directive 2011/83/EU: (3a) 'energy-using good'[42]; (14a) 'commercial guarantee of durability'[43]; (14b) durability[44]; (14c) 'producer'[45]; (14d) 'reparability score'[46]; (14e) 'software update'.[47]

Information requirements for contracts other than distance or off-premises contracts are also reinforced. Before the consumer is bound by a contract other than a distance or an off-premises contract, or any corresponding offer, the trader shall provide the consumer with the following information in a clear and comprehensible manner, if that information is not already apparent from the context: (ea) for all goods, where the producer makes it available, information that the goods benefit from a commercial guarantee of durability and its duration in units of time, where that guarantee covers the entire good and has a duration of more than two years; (eb) for energy-using goods, where the producer does not make available the information referred to in point (ea), information that the producer has not provided information on the existence of a commercial guarantee of durability of more than two years. This information shall be at least as prominent as any other information about the existence and the conditions of after-sales services and commercial guarantees provided; (ec) for goods with digital elements, where the producer makes such information available, the minimum period in units of time during which the producer provides software updates unless the contract provides for a continuous supply of the digital content or digital service over a period of time. Where information about the existence of a commercial guarantee of durability is provided in accordance with point (ea), the information on the updates shall be provided if those updates are supplied for a longer period than the commercial guarantee of durability; (ed) for digital content and digital services, where their provider is different from the trader and makes such information available, the minimum period in units of time during which the provider provides software updates unless the contract provides for a continuous supply of the digital content or digital service over a period of time. The following information is also required: (i) where applicable, the reparability score for the goods; (ij) when point (i) is not applicable, information made available by the producer about the availability of spare parts, including the procedure of ordering them, and about the availability of a user and repair manual."

In respect of distance and off-premises contracts (Article 6, paragraph 1), before the consumer is bound by a distance or off-premises contract, or any corresponding offer, the trader shall provide the consumer with the following information in a clear and comprehensible manner:

> (ma) for all types of goods, where the producer makes it available, information that the goods benefit from a commercial guarantee of durability and its duration in units of time, where that guarantee covers the entire good and has a duration of more than two years; (mb) for energy-using goods, where the producer does not make available information referred to in point (ma), information that the producer has not provided information on the existence of a commercial guarantee of durability of more than two years. This information shall be at least as prominent as any other information about the existence and the conditions of after-sales services and commercial guarantees provided in accordance with point (m); (mc) for goods with digital elements, where the producer makes such information available, the minimum period in units of time during which the producer provides software updates unless the contract provides for a continuous supply of the digital content or digital service over a period of time. Where information about the existence of a commercial guarantee of durability is provided in accordance with point (ma), the information on the updates shall be provided if those updates are supplied for a longer period than the commercial guarantee of durability; (md) for digital content and digital services, where their provider is different from the trader and makes such information available, the minimum period in units of time during which the provider provides software updates unless the contract provides for a continuous supply of the digital content or digital service over a period of time.

Similarly, the following information:

> (u) where applicable, the reparability score for the goods; (v) when point (u) is not applicable, information made available by the producer about the availability of spare parts, including the procedure of ordering them, and about the availability of a user and repair manual,

is also mandatory. Accordingly, Article 8(2), the first subparagraph is replaced by the following:

> If a distance contract to be concluded by electronic means places the consumer under an obligation to pay, the trader shall make the consumer aware in a clear and prominent manner, and directly before the consumer places his order, of the information provided for in Article 6(1), points (a), (e), (ma), (mb), (o) and (p).

Even though the Proposal on empowering consumers for the green transition amends both Directives 2005/29/EC and 2011/83/EU, it should be emphasised that the Unfair Commercial Practices Directive is of utmost importance against non-compliance with information duties or misleading commercial practices concerning the ecological characteristics of the products. In the first place, it outlaws unfair commercial practices. But even more importantly, it also offers a general remedy for substantial misleading omissions.[48] According to Article 7, paragraph 5, information requirements established by European law in relation to commercial communication including advertising or marketing, a non-exhaustive list contained in Annex II, shall be regarded as material. Therefore, Article 7 might be applicable as a rule whenever non-compliance with information duties occurs. This far-reaching provision, positively protecting consumers' interests, constitutes an effective tool for ensuring consumers' ability to make an informed decision and make a transactional decision according to their preferences.

29.4 Individual Redress

Individual redress was not a subject under the Directive on Unfair Commercial Practices.[49] Private law remedies for individual cases, in particular damages claims put forward by individual consumers that have suffered as a result of an unfair business practice, were not mentioned at all in the Directive.[50] The enforcement provision in Article 11 dealt with court actions and administrative measures to force a business to cease an unfair commercial practice or to prohibit such practices in the first place (requiring that an injunction procedure should be available), and the penalties referred to in Article 13 were of a criminal or administrative law nature. The focus of the Directive was on what one could call the collective protection of consumers against unfair commercial practices.[51] It depended on whether and to what extent the Member States were willing to grant individual consumers the right to sue traders to combat unfair commercial practices.[52]

For example, Portugal[53] has combined the two sorts of protection against unfair commercial practices. Consumer protection bodies (whether public or private) were authorised to intervene in the interests of preventing the use of unfair commercial practices against consumers,[54] as collective preventive or remedial measure. Furthermore, consumers could also rely on protection in a dispute regarding the rights and obligations arising from the contract itself, given that any contract failing with the requirement of fairness was voidable. In that sense, rescission is the remedy which enables the contract to be retrospectively treated as though it had never come into existence. If a voidable contract is not rescinded it remains effective but continues to be eligible to be rescinded, except where one of the bars to rescission applies. Rescission is usually conducted by communicating the request to the other party to the contract and without requiring a court order to effect rescission. Once the claimant has announced that he has rescinded the contract, it will be treated as invalid *ab initio*. If the contract has not yet been fully performed or completed (*executório*), there will be no other consequences. If any part of the contract has been executed by the claimant, he will then be able to bring an action to recover what has been transferred to the other party, and vice-versa. Alternatively, Article 14 of Decree-Law 57/2008 provided that the consumer could require equitable modification of the contract.[55]

Furthermore, Portuguese law, as claimed by Collins,[56] also regarded an unfair commercial practice as a type of statutory tort or wrong, which should always give rise to a compensatory remedy on proof of damage, according to the general rules of tortious and contractual liability. In this context,[57] the question of whether the law governing any possible claim for damages because of these same 'practices' by the other party to a contract may be considered contractual but may also be considered tortious or extracontractual, is not relevant.[58]

The objection raised by doctrine, concerning the ability and willingness of a consumer to bring an action,[59] that consumers tend to step back from taking legal action as the damage is minor and the assertion of their rights is not worth its prosecution, does not hold ground in Portugal. Since 2019 a strong scheme of alternative dispute resolution is available, and arbitrage is mandatory on the unilateral decision of the consumer, upon 5.000,00 Euros.[60]

The European regulation changed with Directive (EU) 2019/2161, on better enforcement and modernisation of Union consumer protection rules. This Directive assumed that[61] there were insufficient individual remedies for consumers harmed by breaches of national legislation transposing Directive 2005/29/EC, and that Member States should ensure that remedies were available for consumers harmed by unfair commercial practices in order to eliminate all the effects of those unfair practices. Consequently, Article 3 of Directive Omnibus inserted Article 11a (Redress) to Directive 29/2005:

> 1. Consumers harmed by unfair commercial practices shall have access to proportionate and effective remedies, including compensation for damage suffered by the consumer and, where relevant, a price reduction or the termination of the contract[62] Member States may determine the conditions for the application and effects of those remedies. Member States may consider, where appropriate, the gravity and nature of the unfair commercial practice, the damage suffered by the consumer and other relevant circumstances. 2. Those remedies shall be without prejudice to the application of other remedies available to consumers under Union or national law.

Price reduction and termination of the contract, the remedies elected by the Directive, are remedies against non-performance of the contract. Price reduction refers to an abatement of the contract price to remedy insufficient performance by one party that does not conform to the contract. The aggrieved party retains the defective performance offered by the party in breach, while the latter receives only a reduced payment to account for the defect in its performance. Price reduction is usually ordered by the Court, following a claim by the non-breaching creditor. Termination refers to the right of an innocent party to treat himself as discharged from any future obligation to perform because of the breach of her contractual partner. However, any refund to the consumer may be reduced on account of the consumer's use of the goods, although this is subject to qualifications. The right to terminate is, usually, prospective; where it is the performance of a contractual obligation which is defective, this should only affect the contract from the time when the innocent party accepts the breach.[63]

Both remedies are, of course, adequate for Directives (UE) 770/2019 and 771/2019, where the professional is obliged to deliver a good, supply content or provide service in conformity with the contract. But can we make the same assessment for unfair commercial practices?

29.5 Critical Analysis and Concluding Remarks

The Directive 29/2005/CE seeks to tackle unfair practices that distort the economic behaviour of consumers so that they take transactional decisions they would not otherwise have taken. Material distortion of the economic behaviour of the consumer[64] is defined in Article 2(e) as using a 'commercial practice to appreciably impair the consumer's ability to make an informed decision, thereby causing the consumer to take a transactional decision that he would not have taken otherwise'. According to Article 2(k) a 'transactional decision' means any decision taken by a consumer concerning whether, how and on what terms to purchase, make payment in whole or in part for, retain or dispose of a product or to exercise a contractual right in relation to the product, whether the consumer decides to act or to refrain from acting.[65] These definitions evidence the aim of the Directive of bolstering autonomy and emphasising individual responsibility of consumers who are confident and proactive in gathering and processing information before making transactional decisions.[66]

When transactional decisions of consumers are distorted by a flow of incorrect and misleading information and can therefore be assumed to be taken against real consumer preferences,[67] the consequences should be the invalidity of the contract. Civil law countries traditionally accept three defects of consent (vices du contentment, *Willensmiinget wilsgebreken*): mistake, fraud, and threat. The starting point of English law is different. It does not begin with the (defective) intention of the party in error, but from the party who caused the error. This party may have made false statements or – exceptionally – remained silent while it ought to have spoken. English law, therefore, requires a so-called misrepresentation in order for the contract to be nullified by the innocent party.[68] Misrepresentation can be defined as a spoken or written untrue statement of fact which induces a party to conclude a contract (silence can only constitute misrepresentation in the rare cases in which there is a duty to speak).[69] Other relevant aggravating factors include duress or undue influence of the claimant to enter into the contract.[70]

In common, a contract can be rescinded if there was some relevant flaw at the time the contract was concluded, which can be considered as a matter of law to have impaired the consent of one of the parties to the contract so that their consent to enter into the contract can be treated as vitiated.[71] Where the objectionable conduct preceded the formation of the contract, the law should seek to restore the innocent party to the position he was in before he relied on the misrepresentation; in other words, the right to rescind is retrospective. Once a contract is rescinded,[72] it is nullified and everything which has been done under the contract can be undone by operation of law so that the parties are restored to their original pre-contractual position; there is a giving and a taking back on both sides. The rescission does not depend on the party who seeks the remedy also being required to plead a claim for damages as well.[73]

This difference between rescission and termination in effect follows from the location of the objectionable conduct. Since the party has factually consented to enter the transaction but her consent can be considered to have been impaired by virtue of some defect operating at the time the contract was formed, so that the contract may not have been properly made.[74] Individual redress of consumers whose consent was impaired by unfair commercial practices should be anchored in the validity of the contract and not in performance. Instead of price reduction and termination, the adequate remedy would be rescission. In fact, remedies for performance are suitable when referring to a significant imbalance in the parties' rights and obligations arising from the contract. They are not adequate, however, when a contract concluded by a consumer in reliance upon an untruth is at stake. In this situation, a consumer should have the right to free himself from any disadvantageous contractual consequences, e.g., to have returned all payments made based on the contract. Therefore, consumers should have the option to get the contract rescinded and once rescinded, be treated as though it was void.

In a context where consumers are invited to more sustainable individual consumption and provided with better and more reliable information on ecological characteristics of goods and services, and stronger protection against unfair misleading commercial practices concerning the characteristics of the products, this claim is utterly important.

Unfair commercial practices impair the consent of consumers to a contract and arguably prevent the consumer from making more ecological or sustainable choices. The focus of the remedies should, therefore, be on the consent of the consumer. As already mentioned, consumers are asked to move towards more sustainable consumption patterns and encouraged to weigh up concerns of sustainability alongside concerns of convenience, availability, or affordability. It should also be recalled that European consumers are showing a growing interest in contributing personally to achieving climate neutrality, preserving natural resources and biodiversity, and reducing water, air, and soil pollution.

Consequently, when the consumer's consent is restrained, distorted, or impaired, the contract should be treated as *ab initio* invalid. Rescission should be the available remedy to retrospectively extinguish or annihilate the contract, as though it had never come into existence.

Price reduction and termination of the contract, the remedies elected by the Directive on unfair commercial practices, and that are preserved by the Proposal Directive empowering consumers for the green transition, are remedies against non-performance of the contract. Price reduction refers to an abatement of the contract price to remedy a performance by one party that does not conform to the contract. The aggrieved party retains the defective performance offered by the party in breach, while the latter receives only a reduced payment to account for the defect in its performance. It restores the equilibrium of the contract, but it might not be enough. It should be borne in mind that the consumer, arguably, does not want to pay less for a not sustainable product or service; on the contrary, it should be assumed that the consumer would not want to acquire that same product or service.

A similar reasoning applies to termination. On the one hand, requirements for termination might not be available, mainly when there is compliance with what has been agreed upon between parties. On the other hand, when termination is an actual remedy, the right to terminate is, usually, prospective, and any refund to the consumer may be reduced on account of the consumer's use of the goods. In both cases, consumers are still bound by a contract that they would have not to conclude if their consent was not impaired.

Price reduction and termination are remedies aiming at correcting the imbalance between parties' rights and obligations. Consumers whose consent was impaired by a misleading commercial practice should not be forced to remain bound by a contract; on the contrary, individual redress of consumers whose consent was impaired should be anchored in the (in)validity of the contract, and, thus, in the retrospective extinction the contract. Rescission is, therefore, an adequate remedy to fully protect consumers. Such a solution would also pave the way for a more contracted behaviour of professionals in their commercial practices and, consequently, would contribute to a more sustainability-friendly market.

The foreseen Directive empowering consumers for the green transition should take this into consideration. Otherwise, it would be just putting new wine into old wineskins.

Notes

1 Civil Law Professor at the University of Coimbra, Portugal.

2 Sustainable development was defined by the World Commission on Environment and Development (1987), widely known as the Brundtland Commission and Development, by its chairperson, as 'Development that meets the needs of the present without compromising the ability of future generations to meet their own needs'.

3 The term 'sustainable consumption' entered the international policy arena in Agenda 21, the action plan for sustainable development adopted by 179 heads of state at the 1992 Rio Earth Summit.

4 Commission, 'Closing the Loop – An EU action plan for the Circular Economy' [Communication from the Commission to the European Parliament, the Council, the European and Social Committee and the Committee of the Regions] COM (2015) 0614 final.

5 Commission, 'A new Circular Economy Action Plan for a cleaner and more competitive Europe' [Communication from the Commission to the European Parliament, the Council, the European and Social Committee and the Committee of the Regions] COM (2020) 98 final, 11 March 2020.

6 The search for new consumption patterns goes together with the definition of new production patterns.

7 Commission, 'New Consumer Agenda Strengthening consumer resilience for sustainable recovery' [Communications from the Commission to the European Parliament and the Council] COM (2020) 696 final, 13 November 2020.

8 Commission, 'The European Green Deal' [Communication from the Commission to the European Parliament, the European Council, the Council, the European Economic and Social Committee and the Committee of the Regions] COM (2019) 640 final.

9 For a general overview of the UN 2030 Agenda for Sustainable Development, Ana Cândido Muniz Cipriano, 'Consumer Law and the United Nations' in Alberto do Amaral Júnior, Lucila Almeida, and Luciane Klein Vieira (eds), *Sustainable Consumption* (Springer 2020) 154–165.

10 Gordon R. Foxall, *Understanding Consumer Choice* (Palgrave Macmillan 2005) 15, consumer behaviour includes all the activities of buyers, ex-buyers, and potential buyers from pre-purchase deliberation to post-purchase evaluation, and from continued consumption to discontinuance.

11 Council Regulation (EC) No 66/2010 of 25 November 2009 on the EU Ecolabel [2010] OJ L 27/1, as amended.

12 Council Directive 2009/125/EC of 21 October 2009 establishing a framework for the setting of ecodesign requirements for energy-related products (recast) [2009] OJ L 285/ 10, as amended.

13 As pointed out by Thierry Bourgoignie, 'Sustainable Consumption and Obsolescence of Consumer Products' in Alberto do Amaral Júnior, Lucila Almeida, and Luciane Klein Vieira (eds), *Sustainable Consumption* (Springer 2020) 29, at the individual level, the concern is to protect the purchasing power of the consumer against repeated, successive, and often useless purchases of products that are likely to be still fit for use even after they are thrown away. At the collective level, the stakes are to reduce waste and to prevent or reduce the negative impacts of overconsumption on the environment.

14 Council Directive 2005/29/EC of 11 May 2005 concerning unfair business-to-consumer commercial practices in the internal market and amending Council Directive 84/450/EEC, Directives 97/7/EC, 98/27/EC, and 2002/65/EC of the European Parliament and of the Council and Regulation (EC) No 2006/2004 of the European Parliament and of the Council [2005] OJ L 149/22.

15 Council Directive (EU) 2019/2161 of 27 November 2019 amending Council Directive 93/13/EEC and Directives 98/6/EC, 2005/29/EC, and 2011/83/EU of the European Parliament and of the Council as regards the better enforcement and modernisation of Union consumer protection rules [2019] OJ L 328/7.

16 Doris Fuchs, Marlyne Sahakian, Tobias Gumbert, Antonietta Di Giulio, Michael Maniates, Sylvia Lorek, and Antonia Graf, *Consumption Corridors* (Routledge 2021) 61.

17 On the political role of the green consumer, see Henrik Paul Bang, 'Critical Theory in a Swing: Political Consumerism between Politics and Policy' in Mark Bevir and Frank Trentmann (eds), *Governance, Consumers and Citizen* (Palgrave Macmillan 2007) 225; Felix Ekardt, *Sustainability – Transformation, Governance, Ethics, Law* (Springer 2020) 182; Doris Fuchs et al, *Consumption Corridors* (Routledge 2021) 62–83.

18 Gill Seyfang and David Elliott, *The New Economics of Sustainable Consumption* (Palgrave MacMillan 2009) 13; Grégoire Wallenborn, 'How to Attribute Power to Consumers? When Epistemology and Politics Converge' in Edwin Zaccaï (ed), *Sustainable Consumption, Ecology and Fair Trade* (Routledge 2007) 68.

19 Marteen A Hajer, *The Politics of Environmental Discourse: Ecological Modernization and the Policy Process* (Clarendon Press 1995) 264ff; Doris Fuchs, Sylvia Lorek, Antonietta Di Giulio, and Rico Defila, 'Sources of Power for Sustainable Consumption: Where to Look' in Cindy Isenhour, Mari Martiskainen, and Lucie Middlemiss (eds), *Power and Politics in Sustainable Consumption Research and Practice* (Routledge-SCORAI 2019) 63.

20 Gill Seyfang and David Elliott, *The New Economics of Sustainable Consumption* (Palgrave MacMillan 2009) 11.

21 Alice Malpass, Clive Barnett, Nick Clarke, and Paul Cloke, 'Problematizing Choice: Responsible Consumers and Sceptical Citizens' in Mark Bevir and Frank Trentmann (eds), *Governance, Consumers and Citizens* (Palgrave Macmillan 2007) 244.

22 'Attitudes of European citizens towards the environment' (*Special Eurobarometer 501*, March 2020) <https://ec.europa.eu/commfrontoffice/publicopinion/index.cfm/survey/getSurveydetail/nstruments/special/surveyky/2257> accessed 3 July 2023.

23 'Factual summary report – public consultation for New Consumer Agenda' (*European Commission*, 13 November 2020) <https://ec.europa.eu/info/files/factual-summary-report-public-consultation-new-consumer-agenda_en> accessed 3 July 2023.

24 Commission, 'Proposal for a Directive of the European Parliament and of the Council amending Directives 2005/29/EC and 2011/83/EU as regards empowering consumers for the green transition through better protection against unfair practices and better information' COM (2022) 143 final.

25 Council Directive 2011/83/EU of 25 October 2011 on consumer rights, amending Council Directive 93/13/EEC and Directive 1999/44/EC of the European Parliament and of the Council and repealing Council Directive 85/577/EEC and Directive 97/7/EC of the European Parliament and of the Council [2011] OJ L 304/64.

26 As pointed out by Thomas Wilhelmsson, 'Misleading Practices' in Geraint Howells, Hans Micklitz, and Thomas Wilhelmsson (eds), *European Fair Trading Law – The Unfair Commercial Practices Directive* (Ashgate 2006) 125, there is no clear line, however, to be drawn between misleading actions and misleading omissions.

27 See paragraphs 3 and 4, for specifications depending on the medium used to communicate and in the case of an invitation to purchase.

28 Article 5(2) and Article 2(e) are not entirely compatible, as underlined by Hans-W. Micklitz, 'The General Clause on Unfair Practices' in Geraint Howells, Hans Micklitz, and Thomas Wilhelmsson (eds), *European Fair Trading Law – The Unfair Commercial Practices Directive* (Ashgate 2006) 102.

29 The European solution is drawn according to the information paradigm. It is, however, commonly accepted that information by itself does not always lead to utility-maximising decisions because (a) information may not reach the consumers; (b) consumers may not want to look for information; (c) consumers may have difficulties in understanding and using information; (d) consumers may not have alternative solutions; (e) and consumers' culture makes information less efficient. Substantive, content-oriented rules, and prohibitions might also be strongly considered.

30 Evelyne Terryn, 'Lutter contre l'écoblanchiment est necessaire mais ne suffit pas pour atteindre une consummation responsible' (2022) 47(1) *Revue Juridique de l'Environnement* 73–80; Terryn Evelyne and Vanessa Mak, 'Circular Economy and Consumer Protection: The Consumer as a Citizen and the Limits of Empowerment through Consumer Law' (2020) 43 *Journal of Consumer Policy* 227–248.

31 'Environmental claim' means any message or representation, which is not mandatory under Union law or national law, including text, pictorial, graphic or symbolic representation, in any form, including labels, brand names, company names or product names, in the context of commercial communication, which states or implies that a product or trader has a positive or no impact on the environment or is less damaging to the environment than other products or traders, respectively, or has improved their impact over time.

32 This means an environmental claim that is in textual form or contained in a sustainability label.

33 This means any explicit environmental claim, not contained in a sustainability label, where the specification of the claim is not provided in clear and prominent terms on the same medium.

34 This means any voluntary trust mark, quality mark or equivalent, either public or private, that aims to set apart and promote a product, a process, or a business with reference to its environmental or social aspects or both. This does not cover any mandatory label required in accordance with Union or national law.

35 This means a third-party verification scheme that is open under transparent, fair, and non-discriminatory terms to all traders willing and able to comply with the scheme's requirements, which certifies that a product complies with certain requirements, and for which the monitoring of compliance is objective, based on international, Union, or national standards and procedures and carried out by a party independent from both the scheme owner and the trader.

36 This means software, including a website, part of a website or an application, operated by or on behalf of a trader, which provides information to consumers about environmental or social aspects of products, or which compares products on those aspects.

37 This means environmental performance compliant with Regulation (EC) 66/2010, with national or regional EN ISO 14024 type I ecolabelling schemes officially recognised in accordance with Article 11 of Regulation (EC) 66/2010, or top environmental performance in accordance with other applicable Union law.

38 This means the ability of the goods to maintain their required functions and performance through normal use, as defined in Council Directive (EU) 2019/771 of 20 May 2019 on certain aspects concerning contracts for the sale of goods, amending Regulation (EU) 2017/2394 and Directive 2009/22/EC, and repealing Directive 1999/44/EC [2019] (OJ L 136/28), Article 2, point (13).

39 This means a free update, including a security update, that is necessary to keep goods with digital elements, digital content and digital services in conformity in accordance with Council Directive (EU) 2019/770 of 20 May 2019 on certain aspects concerning contracts for the supply of digital content and digital services [2019] OJ L 136/1 and Council Directive (EU) 2019/771 of 20 May 2019 on certain aspects concerning contracts for the sale of goods, amending Regulation (EU) 2017/2394 and Directive 2009/22/EC, and repealing Directive 1999/44/EC [2019] OJ L 136/28.

40 This means any component of a good that is used up recurrently and needs to be replaced for the good to function as intended.

41 This means the ability of the goods to perform their functions having regard to their purpose, as defined in point (9) of Article 2 of Directive (EU) 2019/771.

42 This means any good that depends on energy input (electricity, fossil fuels and renewable energy sources) to work as intended.

43 This means a producer's commercial guarantee of durability referred to in Article 17 of Directive (EU) 2019/771, under which the producer is directly liable to the consumer during the entire period of that guarantee for repair or replacement of the goods.

44 As defined in Article 2, point (13), of Directive (EU) 2019/771, the ability of the goods to maintain their required functions and performance through normal use.

45 As defined in Article 2, point (4), of Directive (EU) 2019/771, a manufacturer of goods, an importer of goods into the Union or any person purporting to be a producer by placing its name, trademark, or other distinctive sign on the goods.

46 This means a score expressing the capacity of a good to be repaired, based on a method established in accordance with Union Law.

47 This means a free update, including a security update, that is necessary to keep goods with digital elements, digital content, and digital services in conformity in accordance with Directives (EU) 2019/770 and (EU) 2019/771.

48 As underlined by Giuseppe B. Abbamonte, 'The Unfair Commercial Practices Directive and Its General Prohibition' in Stephen Weatherill and Ulf Bernitz (eds), *The Regulation of Unfair Commercial Practices under EC Directive 2005/29 – New Rules and New Techniques* (Hart Publishing 2007) 18, the Directive may be described as a gap-filler, since it will fill the consumer protection lacunae in existing sector-specific legislation.

49 Mateja Durovic, *European Law on Unfair Commercial Practices and Contract Law* (Hart Publishing 2016) 159ff.

50 Recital 9. Jules Stuyck, Evelyne Terryn, and Tom Van Dyck, 'Confidence through Fairness? The New Directive on Unfair Business-to-Consumer Commercial Practices in the Internal Market' (2006) 43 *Common Market Law Review* 136.

51 Thomas Wilhelmsson, 'Scope of the Directive' in Geraint Howells, Hans Micklitz, and Thomas Wilhelmsson (eds), *European Fair Trading Law – The Unfair Commercial Practices Directive* (Ashgate 2006) 51.

52 See Thomas Wilhelmsson, 'Scope of the Directive' in Geraint Howells, Hans Micklitz, and Thomas Wilhelmsson (eds), *European Fair Trading Law – The Unfair Commercial Practices Directive* (Ashgate 2006) 52, mentioning that the thorough preparatory works on the Unfair Commercial Practices Directive have caused the European Commission to raise the question of whether the Community should introduce a compensation claim to the benefit of individual consumers. This idea did not pass through the legislative process, which led to the conclusion that the Unfair Commercial Practices Directive strove for collective protection – for collective legal redress – of only the consumer interest. Also, Geraint Howells, 'Introduction', in Geraint Howells, Hans Micklitz, and Thomas Wilhelmsson (eds), *European Fair Trading Law – The Unfair Commercial Practices Directive* (Ashgate 2006) 21; Hans-W. Micklitz, 'Legal Redress' in *European Fair Trading Law – The Unfair Commercial Practices Directive* (Ashgate 2006) 220.

53 Decree-Law 57/2008, of March 26.

54 As underlined by Dörte Poelzig, 'Private or Public Enforcement of the UCP Directive? Sanctions and Remedies to Prevent Unfair Commercial Practices' in Willem van Boom, Amandine Garde, and Orkun Akseli (eds), *Introduction to 'The European Unfair Commercial Practices Directive'* (Ashgate 2014) 257: 'by combining public and private enforcement, deficits of one system may be compensated by the other system'.

55 This solution, however, was criticised by the doctrine. All the arguments that will ground criticism on the actual legal remedies are also convincing here. See Luís Menezes Leitão, 'As práticas Comerciais Desleais nas Relações de Consumo' in Alda Pellegrini Grinover et al (eds), *Liber Amicorum Mário Frota: a causa dos direitos dos consumidores* (Coimbra, Livraria Almedina 2012) 369–386 <https://portal.oa.pt/upl/%7B29c2d451-b398-486b-9842-a5cad254e869%7D.pdf> accessed 22 December 2022.

56 Hugh Collins, 'The Unfair Commercial Practices Directive' (2005) 4 *ERCL* 426.

57 Simon Whittaker, 'The Relationship of the Unfair Commercial Practices Directive to European and National Contract Laws' in Stephen Weatherill and Ulf Bernitz (eds), *The Regulation of Unfair Commercial Practices under EC Directive 2005/29 – New Rules and New Techniques* (Hart Publishing 2007) 147.

58 Under Portuguese law, protection against unfair commercial practices is a matter of Consumer contract law, and not only the regulation of business behaviour in the interests of consumers. According to the distinction of Simon Whittaker, 'The Relationship of the Unfair Commercial Practices Directive to European and National Contract Laws' in Stephen Weatherill and Ulf Bernitz (eds), *The Regulation of Unfair Commercial Practices under EC Directive 2005/29 – New Rules and New Techniques* (Hart Publishing 2007) 147.

59 Hans-W. Micklitz, 'Legal Redress' *European Fair Trading Law – The Unfair Commercial Practices Directive* (Ashgate, 2006) 219; Simon Whittaker, 'The Relationship of the Unfair Commercial Practices Directive to European and National Contract Laws' in Stephen Weatherill and Ulf Bernitz (eds), *The Regulation of Unfair Commercial Practices under EC Directive 2005/29 – New Rules and New Techniques* (Hart Publishing 2007) 141.

60 For a general overview, António Pinto Monteiro, Maria Raquel Guimarães and Sandra Passinhas, 'Enforcement and Effectiveness of Consumer Law in Portugal: Filling the Gap between the Law on the Books and the Law in Action', Hans-W. Micklitz and Geneviève Saumier (eds), *Enforcement and Effectiveness of Consumer Law* (Springer 2018) 457–477.

61 Recitals 3 and 16 of Directive (EU) 2019/2161.

62 In Spanish 'una reducción del precio o la resolución del contrato'; in Italian: 'la riduzione del prezzo o la risoluzione del contratto'; in French 'une réduction du prix ou la fin du contrat'; in German 'Preisminderung oder Beendigung des Vertrags'.

63 Roger Halson, *Contract Law* (2nd edn, Pearson 2013) 45–46.

64 Hans-W. Micklitz, 'The General Clause on Unfair Practices' in Geraint Howells, Hans Micklitz, and Thomas Wilhelmsson (eds), *European Fair Trading Law – The Unfair Commercial Practices Directive* (Ashgate 2006) 106. In order to define whether the outward freedom of action of the consumer is distorted, the modalities and the circumstances which relate to the conclusion of the contract, the contents and the conditions of the contract as well as after-sales services need to be considered. There is a link between the concept of autonomy, which unifies economic behaviour, the informed and transactional decision, and the term 'commercial practices' which accompanies the commercial transaction from the cradle to the grave, including advertising, pre-contractual negotiations, and aftersales service.

65 The transactional decision in Article 2(k) is not to be treated in the same way as the decision to conclude a contract even if this might often be the case in practice. The Directive is also concerned with commercial practices which concern the exercise by a consumer of rights which he or she enjoys under the contract. So, a consumer's exercise of his or her contractual right may be affected by misleading information or aggressive behaviour by the business to constitute an unfair commercial practice within the meaning of the 2005 Directive. See Christian Twigg-Flesner, Deborah Parry, Geraint Howells, and Annette Nordhausen, *An Analysis of the Application and Scope of the Unfair Commercial Practices Directive* (Report for the Department of Trade and Industry, 18 May 2005), 2.39, 14 <http://www.dti.gov.uk/ccp/consultpdf/final_report180505.pdf> accessed 3 July 2023. The following considerations do not apply here.

66 Bert Keirsbilck, *The New European Law of Unfair Commercial Practices and Competition Law* (Hart Publishing 2011) 271ff. As underlined by Micklitz, id 103, the threefold terminology, as introduced by the Community legislator – economic behaviour (Article 5(2)), informed decision (Article 2(e)), transactional decision (Article 2(k)) – should not be overestimated in their significance and function.

67 Dörte Poelzig, 'Private or Public Enforcement of the UCP Directive? Sanctions and Remedies to Prevent Unfair Commercial Practices' in Willem van Boom, Amandine Garde, and Orkun Akseli (eds), *Introduction to 'The European Unfair Commercial Practices Directive'* (Ashgate 2014) 5.

68 Jan M. Smits, *Contract Law – A Comparative Introduction* (Elgar 2014) 160ff. Simon Whittaker, 'The Relationship of the Unfair Commercial Practices Directive to European and National Contract Laws' in Stephen Weatherill and Ulf Bernitz (eds), *The Regulation of Unfair Commercial Practices under EC Directive 2005/29 – New Rules and New Techniques* (Hart Publishing 2007) 147: Article 3(2) refers to 'contract law and, in particular, to the rules on the validity, formation or effect of a contract'. The particular emphasis of this provision on 'validity' and 'formation' of contracts within contract law is explicable by the substantive content of the requirements of the 2005 Directive, for the two broad limbs of unfair legislative practices which it sets out concern the provision or failure to provide information and the use of improper pressure, practices the traditional contractual expressions of which (such as misrepresentation, undue influence and duress in English law or *dol* and *violence* in French law) go to the validity of the contract.

69 Jan M. Smits, *Contract Law – A Comparative Introduction* (Elgar 2014) 174.

70 Common Law presents other potential tools. For a general overview, Solène Rowan, *Remedies for Breach of Contract – A Comparative Analysis of the Protection of Performance* (Oxford University Press 2012) 70ff.

71 See James Devenney, 'The Consumer Rights Act 2015 and Related Reforms: An Epic Disappointment?' in Roger Halson and David Campbell (eds), *Research Handbook on Remedies in Private Law* (Elgar Publishing 2019) 269, about the overlap with the general law of misrepresentation.

72 Neil Andrews, 'Breach and Performance' in Andrew Tettenborn, Malcolm Clarke, Neil Andrews, and Graham Virgo (eds), *Contractual Duties Performance: Breach, Termination and Remedies* (2nd edn, Sweet and Maxwell 2017) 78: In view of the modern analytical distinction between rescission for misrepresentation (or some other preliminary ground of vitiation) and termination for breach, the term 'rescission' should be confined to the process of setting aside retrospectively a contract which is vitiated by reason of misrepresentation, or other grounds of initial invalidity. Most commentators recognise the need for terminological precision and have abandoned the old ambiguous language of 'rescission for breach'.

73 Graham Virgo, 'Rescission' in Andrew Tettenborn, Malcolm Clarke, Neil Andrews, and Graham Virgo (eds), *Contractual Duties Performance: Breach, Termination and Remedies* (2nd edn, Sweet and Maxwell 2017) 3.

74 As Micklitz pointed out, in 'The General Clause on Unfair Practices' in Geraint Howells, Hans Micklitz, and Thomas Wilhelmsson (eds), *European Fair Trading Law – The Unfair Commercial Practices Directive* (Ashgate 2006) 110, although it seems that Article 2(e) establishes a causal link, requiring the commercial practice to be such that it causes the consumer to take a transactional decision that he would not have taken otherwise, the relationship between distortion and decision-making is to be considered normative. To this extent 'causality' is contained already in the decision whether the autonomy of the consumer is distorted or is likely to be distorted by a commercial practice.

Bibliography

Abbamonte G, 'The Unfair Commercial Practices Directive and Its General Prohibition' in Stephen Weatherill and Ulf Bernitz (eds), *The Regulation of Unfair Commercial Practices under EC Directive 2005/29 – New Rules and New Techniques* (Hart Publishing 2007) 18

Andrews N, 'Breach and Performance' in Andrew Tettenborn, Malcolm Clarke, Neil Andrews, and Graham Virgo (eds), *Contractual Duties Performance: Breach, Termination and Remedies* (2nd edn, Sweet and Maxwell 2017) 78

Bang H, 'Critical Theory in a Swing: Political Consumerism between Politics and Policy' in Mark Bevir and Frank Trentmann (eds), *Governance, Consumers and Citizen* (Palgrave Macmillan 2007) 225

Bourgoignie T, 'Sustainable Consumption and Obsolescence of Consumer Products' in Alberto do Amaral Júnior, Lucila Almeida, and Luciane Klein Vieira (eds), *Sustainable Consumption* (Springer 2020) 29

Cipriano A, 'Consumer Law and the United Nations' in Alberto do Amaral Júnior, Lucila Almeida, and Luciane Klein Vieira (eds), *Sustainable Consumption* (Springer 2020) 154–165

Collins H, 'The Unfair Commercial Practices Directive' (2005) 4 *ERCL* 426

Devenney J, 'The Consumer Rights Act 2015 and Related Reforms: An Epic Disappointment?' in Roger Halson and David Campbell (eds), *Research Handbook on Remedies in Private Law* (Elgar Publishing 2019) 269

Durovic M, *European Law on Unfair Commercial Practices and Contract Law* (Hart Publishing 2016) 159ff

Ekardt F, *Sustainability – Transformation, Governance, Ethics, Law* (Springer 2020) 182

Evelyne T and Mak V, 'Circular Economy and Consumer Protection: The Consumer as a Citizen and the Limits of Empowerment through Consumer Law' (2020) 43 *Journal of Consumer Policy*, 227–248.

Foxall G, *Understanding Consumer Choice* (Palgrave Macmillan 2005) 15

Fuchs D, Lorek S, Di Giulio A, and Defila R, 'Sources of Power for Sustainable Consumption: Where to Look' in Cindy Isenhour, Mari Martiskainen, and Lucie Middlemiss (eds), *Power and Politics in Sustainable Consumption Research and Practice* (Routledge-SCORAI 2019) 63

Fuchs D, Sahakian M, Gumbert T, Di Giulio A, Maniates M, Lorek S, and Graf A, *Consumption Corridors* (Routledge 2021) 61

Hajer M, *The Politics of Environmental Discourse: Ecological Modernization and the Policy Process* (Clarendon Press 1995) 264 ff

Halson R, *Contract Law* (2nd edn, Pearson 2013) 45–46

Howells G, 'Introduction' in Geraint Howells, Hans Micklitz, and Thomas Wilhelmsson (eds), *European Fair Trading Law – The Unfair Commercial Practices Directive* (Ashgate, 2006) 21

Keirsbilck B, *The New European Law of Unfair Commercial Practices and Competition Law* (Hart Publishing 2011) 271ff

Leitão L, 'As práticas Comerciais Desleais nas Relações de Consumo' in Alda Pellegrini Grinover et al (eds), *Liber Amicorum Mário Frota: a causa dos direitos dos consumidores* (Coimbra, Livraria Almedina 2012) 369–386 <https://portal.oa.pt/upl/%7B29c2d451-b398-486b-9842-a5cad254e869%7D.pdf> accessed 22 December 2022.

Malpass A, Barnett C, Clarke N, and Cloke P, 'Problematizing Choice: Responsible Consumers and Sceptical Citizens' in Mark Bevir and Frank Trentmann (eds), *Governance, Consumers and Citizens* (Palgrave Macmillan 2007) 244

Micklitz H, 'The General Clause on Unfair Practices' in Geraint Howells, Hans Micklitz, and Thomas Wilhelmsson (eds), *European Fair Trading Law – The Unfair Commercial Practices Directive* (Ashgate 2006) 102

Micklitz H, 'Legal Redress' in Geraint Howells, Hans Micklitz, and Thomas Wilhelmsson (eds), *European Fair Trading Law – The Unfair Commercial Practices Directive* (Ashgate, 2006) 220

Monteiro A, Guimarães M, and Passinhas S, 'Enforcement and Effectiveness of Consumer Law in Portugal: Filling the Gap between the Law on the Books and the Law in Action' Hans-W. Micklitz and Geneviève Saumier (eds), *Enforcement and Effectiveness of Consumer Law* (Springer 2018) 457–477

Poelzig D, 'Private or Public Enforcement of the UCP Directive? Sanctions and Remedies to Prevent Unfair Commercial Practices' in Willem van Boom, Amandine Garde, and Orkun Akseli (eds), *Introduction to 'The European Unfair Commercial Practices Directive'* (Ashgate 2014) 257

Rowan S, *Remedies for Breach of Contract – A Comparativer Analysis of the Protection of Performance* (Oxford University Press 2012) 70ff

Seyfang G and Elliott D, *The New Economics of Sustainable Consumption* (Palgrave MacMillan 2009) 13

Smits J, *Contract Law – A Comparative Introduction* (Elgar 2014) 160ff

Stuyck J, Terryn E, and Van Dyck T, 'Confidence through Fairness? The New Directive on Unfair Business-to-Consumer Commercial Practices in the Internal Market' (2006) 43 *Common Market Law Review* 136

Terryn E, 'Lutter contre l'écoblanchiment est necessaire mais ne suffit pas pour atteindre une consummation responsible' (2022) 47(1) *Revue Juridique de l'Environnement* 73–80

Twigg-Flesner C, Parry D, Howells G, and Nordhausen A, *An Analysis of the Application and Scope of the Unfair Commercial Practices Directive* (Report for the Department of Trade and Industry, 18 May 2005), 2.39, 14 <http://www.dti.gov.uk/ccp/consultpdf/final_report180505.pdf> accessed 3 July 2023

Virgo G, 'Rescission' in Andrew Tettenborn, Malcolm Clarke, Neil Andrews, and Graham Virgo (eds), *Contractual Duties Performance: Breach, Termination and Remedies* (2nd edn, Sweet and Maxwell 2017) 3

Wallenborn G, 'How to Attribute Power to Consumers? When Epistemology and Politics Converge' in Edwin Zaccaï (ed), *Sustainable Consumption, Ecology and Fair Trade* (Routledge 2007) 68

Whittaker S, 'The Relationship of the Unfair Commercial Practices Directive to European and National Contract Laws' in Stephen Weatherill and Ulf Bernitz (eds), *The Regulation of Unfair Commercial Practices under EC Directive 2005/29 – New Rules and New Techniques* (Hart Publishing 2007) 147

Wilhelmsson T, 'Misleading Practices' in Geraint Howells, Hans Micklitz, and Thomas Wilhelmsson (eds), *European Fair Trading Law – The Unfair Commercial Practices Directive* (Ashgate 2006) 125

Wilhelmsson T, 'Scope of the Directive' in Geraint Howells, Hans Micklitz, and Thomas Wilhelmsson (eds), *European Fair Trading Law – The Unfair Commercial Practices Directive* (Ashgate 2006) 5

30
THE EU GREEN DEAL AND THE SUSTAINABILITY OF CORPORATE ACTIVITIES

Hugo-Maria Schally

30.1 Introduction

The European Green Deal (EGD)[1] introduced by President of the European Commission Ursula van der Leyen in 2019 is a comprehensive economic and environmental strategy proposed by the European Commission, designed to make the European Union carbon neutral by 2050. It covers a wide range of sectors including energy, agriculture, circular economy, transport, and biodiversity. The Green Deal contributes to achieving several of the Sustainable Development Goals (SDGs)[2] such as on Climate Action (SDG 13), Affordable and Clean Energy (SDG 7), Life Below Water and Life on Land (SDGs 14 and 15), but most particularly Responsible Consumption and Production (SDG 12) which aims for sustainable consumption and production patterns. The achievement of this goal is linked to an element of the EGD that is key for the EU contribution i.e., the successive EU Circular Economy Action Plans and the initiatives thereunder which promote the sustainable use of resources and the minimisation of the generation of waste.

These initiatives target business and industry and companies as economic actors together with their customers and consumers as they play a critical role in achieving these goals. Companies are the primary sources of emissions and waste, but also have the resources and innovation capabilities to develop sustainable solutions. This contribution attempts to show that the transition towards a more sustainable European economy – one of the main objectives of the EGD – will be greatly facilitated if private law tools (with transparency and contractual disclosure among the main elements) are combined with public law enforcement (through compliance and reporting duties placed upon companies and enforced by state agencies implementing EU law). The description of the relevant elements of the EGD as they have emerged since 2019 (or in some cases even before) will highlight the relevance of the interplay and synergies between public and private law in the pursuit of the overarching goals of relevant EU policies, especially in the fields of industrial and environmental policy. The contribution will look at the elements of the Green Deal that will have an impact on companies' operations and discuss the question of how does the European Union Green Deal influence the sustainability strategies and practices of corporations within the EU. It will also look at how the EU Green Deal has shaped corporate legal obligations under public and private law, and how companies, but also other stakeholders may use and adapt to these new legal requirements.

 DOI: 10.4324/9781032662046-36

This contribution will focus on an analysis of the EU Green Deal and related EU legislation. This entails describing legislation and policy itself and to identify the specific legal obligations placed on corporations. In addition, the contribution will also attempt to provide some indications on possible developments in the link to private law instruments and challenges that could be based on the transparency requirements that are put on market participants, without however going into a wider conceptual discussion on the potential of certain approaches.

30.2 The European Green Deal: Political Context

The Green Deal aims at strengthening corporate sustainability in several ways.

The EGD was announced and has been implemented as a European growth strategy, aiming at transforming the European economy and societies. Since 2019, the EGD has led to an unprecedented number of EU initiatives both policy and legislative.[3] These initiatives, many of which have already been transposed into legislation applicable in the EU, comprise proposals for legislation, involving the implementation of stricter norms and standards, pushing corporations to reduce their environmental footprints and adopt more sustainable practices. It foresees financial incentives, mobilising at least one trillion euros in sustainable investments until 2025, helping companies that invest in sustainable technologies and practices, supporting innovation by recognising the importance of research and innovation in achieving sustainability goals and supporting businesses developing new technologies and solutions. Such legislation also covers important elements of transparency and reporting, emphasising the importance of transparency in corporate sustainability. It promotes approaches that encourage and/or require businesses to report on their environmental impact and sustainability efforts, which, in turn, can help attract investors and customers who value sustainability.

The Communication[4] introducing the EGD highlighted, in particular, that

> Sustainability should be further embedded into the corporate governance framework, as many companies still focus too much on short-term financial performance compared to their long-term development and sustainability aspects. At the same time, companies and financial institutions will need to increase their disclosure of climate and environmental data so that investors are fully informed about the sustainability of their investments.

It further stated that it was important to facilitate sustainable investment

> (..) by making it easier for companies to identify sustainable investments and ensuring that they are credible. This could be done via clear labels for retail investment products and by developing an EU green bond standard that facilitates sustainable investment in the most convenient way.

30.2.1 Sustainability and Transparency

By aligning their strategies with the EGD, corporations can contribute significantly to the achievement of the SDGs. They can reduce their environmental footprints, develop sustainable technologies, and contribute to a more sustainable economy. Corporate reporting plays a crucial role in promoting environmental sustainability. It involves the disclosure of information about a company's environmental policies, practices, and impacts, allowing stakeholders to assess its environmental performance and commitment to sustainability. There are several ways in which corporate transparency and reporting can contribute to environmental sustainability:

a *Transparency*: Corporate reporting provides transparency about a company's environmental impact. It helps stakeholders understand the environmental risks and opportunities associated with a company's operations, products, and services. This transparency can encourage companies to improve their environmental performance.
b *Accountability*: Regular reporting holds corporations accountable for their environmental impact. It allows stakeholders to track a company's progress towards its environmental goals and compare its performance with industry standards and peers. This can motivate companies to reduce their environmental footprint and invest in sustainable practices.
c *Decision-Making*: Corporate reporting provides valuable information for decision-making. It can help investors, consumers, and policymakers make informed decisions that support environmental sustainability. For instance, investors can use this information to invest in companies with strong environmental performance, consumers can choose to support businesses that prioritise sustainability, and policymakers can develop regulations based on the information provided.
d *Risk Management*: Corporate reporting can help companies manage environmental risks. It can identify areas of potential risk, such as regulatory compliance, resource scarcity, or climate change impacts. By addressing these risks, companies can improve their resilience and long-term sustainability.
e *Stakeholder Engagement*: Corporate reporting can facilitate stakeholder engagement. It provides a platform for dialogue between a company and its stakeholders, including investors, employees, customers, and communities. This engagement can lead to a better understanding of stakeholder expectations, improved reputation, and stronger relationships.
f *Benchmarking and Improvement*: Reporting allows for benchmarking and continuous improvement. By tracking and reporting key environmental metrics over time, companies can identify areas for improvement, set goals, and measure progress. This data-driven approach can lead to significant environmental improvements.

There are various frameworks for corporate sustainability reporting at the global level, such as the Global Reporting Initiative (GRI),[5] the Sustainability Accounting Standards Board (SASB),[6] and the Task Force on Climate-related Financial Disclosures (TCFD).[7] These provide guidelines for reporting on a range of environmental topics, including greenhouse gas emissions, energy use, water management, waste and pollution, and biodiversity. By adhering to these standards, corporations can contribute significantly to environmental sustainability. All these frameworks, however, are based on voluntary action by companies, which limit the extent to which such corporate behaviour will become mainstream. Since adherence to such standards generates costs it invites 'free riders', which see economic opportunities in non-compliant behaviour.

30.2.2 Product-Related Initiatives

The numerous legislative and policy initiatives targeting the production, consumption and disposal of specific materials and products and covering a number of sectors ranging from industry to agriculture create a policy basis for the transformation of the economy and the transition to a circular economy. One of the main initiatives in this context is the continuation and expansion of work towards the transition to a circular economy in Europe and at the global level. As part of the 2015 package, the European Commission had already committed to present a comprehensive implementation report five years after the adoption of the 2015 Action Plan. This report was presented as a working document of the European Commission in the Spring of 2020.[8] It took a first stock of developments since 2015 and provided the basis for further reflection on measures for the transition to a circular economy. It was also the basis for the new Circular Economy Action Plan presented by the European Commission on the same day.[9]

Since March 2020, the European Commission has already put forward several proposals to help achieve these goals. These are proposals aimed at: ensuring that products placed on the EU market are designed to last longer and are easier to repair, retrofit, recycle and reuse; improving the sorting, reuse, and recycling of textiles, making it easier for consumers to choose sustainable textiles, and applying the eco-design principle also to clothing; phasing out single-use products where possible and replacing them with durable or reusable products; incentivising the product-as-a-service model; limiting and ultimately eliminating the targeted addition of microplastics to products and materials and avoiding the unplanned separation of microplastics at all relevant stages of the life cycle of products.

In a dense sequence, the European Commission proposed a new regulation on sustainable batteries[10] to lower the limit values for highly harmful chemicals in waste,[11] new rules on shipments of waste,[12] a new Eco-design Regulation,[13] a Construction Products Regulation,[14] a Directive to strengthen consumer rights for the green transition,[15] an EU strategy for sustainable and circular textiles the revision of the Industrial Emissions and Landfill Directive the establishment of an Industrial Emissions Portal.[16] Once enacted, these initiatives will have a major impact on how products will be designed, produced and what will be consumed in the EU.

30.3 Transparency-Related Initiatives

The EU has also recognised that it will be important to provide all economic actors with information about the environmental and social performance and impact of economic activities. It also recognises that it is important that companies monitor and evaluate the impact of their operations on society and the environment. That is an important driver for encouraging companies to adopt sustainable practices as it makes them accountable to stakeholders. Such transparency also promotes responsible decision-making by investors and consumers. Moreover, the availability of such information supports and facilitates stakeholder engagement and collaboration. In addition, transparency, and reporting on the environmental impacts of business activities will strengthen consumer rights by providing information to make informed purchasing decisions based on environmental considerations and raising awareness about the environmental impact of consumer choices and behaviour. It can also promote demand for sustainable products and services, thus facilitating a shift towards more sustainable business practices.

Reporting and transparency requirements therefore serve as a key mechanism for driving sustainability and ensuring that companies are held accountable for their environmental impact. The policy development over the last years has placed a strong emphasis on reporting and transparency obligations as a means of promoting sustainability and driving action towards a more sustainable future.

The first Circular Economy Action Plan of 2015[17] had been the starting point for a development in Europe that moves away from a linear economic model that is highly resource intensive and provided the basis for the wider agenda presented under the EGD. It was already a recognition of the fact that in a world where meeting the needs of the population is associated with a huge increase in pressure on resources such as land, water, food, feed, raw materials, and energy, it was no longer possible to continue to rely on an economic approach based on the principle of 'taking, producing, using, throwing away', which is not only inappropriate from an environmental sustainability perspective but also from an economic point of view.

Setting rules for communicating to the public, informing the buyers and consumers about the characteristics of products and how to report on the environmental impacts of economic activities plays an important role in the implementation of the EGD. This pertains to all sectors of the economy and society and action and should help investors, civil society organisations, consumers, policymakers, and other stakeholders to evaluate the environmental performance of companies and encourage these companies to develop a responsible approach to business.

That was however only one further step in the development of a comprehensive framework. The framework was complemented by a new Directive on Corporate Sustainability Reporting,[18] which modernises and strengthens the rules concerning the social and environmental information that companies must report. A broader set of large companies, as well as listed SMEs – approximately 50,000 companies in total – will now be required to report on sustainability. The new rules will ensure that investors and other stakeholders have access to the information they need to assess investment risks arising from climate change and other sustainability issues. They will also create a culture of transparency about the impact of companies on people and the environment. Finally, reporting costs will be reduced for companies over the medium to long term by harmonising the information to be provided. Companies will have to apply the new rules for the first time in the 2024 financial year, for reports published in 2025. Companies subject to the Corporate Sustainability Reporting Directive (CSRD) will have to report according to European Sustainability Reporting Standards (ESRS). The draft standards are developed by the EFRAG, previously known as the European Financial Reporting Advisory Group, an independent body bringing together various stakeholders. The standards will be tailored to EU policies while building on and contributing to international standardisation initiatives. The Commission adopted the first standard, based on the draft standards published by EFRAG in November 2022[19] by delegated act on 31st July 2023.[20] The CSRD also makes it mandatory for companies to have an audit of the sustainability information that they report. In addition, it provides for the digitalisation of sustainability information. The rules introduced by the Non-Financial Reporting Directive[21] remain in force until companies have to apply the new rules of the CSRD.

This has now also been recognised in the international field. The 15th Conference of the Parties of the Convention on Biological Diversity adopted the Kunming-Montreal Biodiversity Framework (GBF),[22] which contains a clear obligation for all Parties to the CBD regarding transparency and reporting on environmental impacts of economic activities. Targets 15 and 16 of the GBF stipulate the obligation to ensure that companies report on the environmental impacts of their activities and that consumers are encouraged and enabled to make sustainable consumption choices.

There are several EU policy and legislative initiatives that have been introduced over the last years that aim at fully implementing these political intentions. This contribution will try to identify a few of them and discuss their contribution to the transformation of the European policy landscape and the overall rules governing the activities of companies.

A very specific example of adding transparency and accountability requirements for a whole sector is the EU Deforestation Regulation,[23] which was adopted by the EU in April 2023, and which will be applicable in 2025. It includes increased reporting requirements through mandatory due diligence requirements. Companies are required to report on their supply chains and the origins of the commodities they use, particularly those linked to deforestation such as soy, palm oil, beef, and cocoa and companies will need to disclose more information about their supply chains. Companies must ensure that their supply chains are not contributing to deforestation, which necessitates regular audits, third-party certifications, and greater collaboration with suppliers to verify sourcing practices. Companies must identify, prevent, and mitigate deforestation risks in their supply chains and need to develop and implement action plans to address these risks and report on their progress. The regulation also leads to more comprehensive ESG (Environmental, Social, and Governance) reporting, as deforestation is a significant environmental issue. It will also impact companies' ESG ratings, which are increasingly used by investors to evaluate companies. The increased transparency will also lead to greater scrutiny from stakeholders, including consumers, investors, and NGOs, which will drive companies to improve their sustainability practices and engage more actively with stakeholders on deforestation issues. As non-compliance with the regulation can result in legal penalties, damaged reputations, and potential loss of investment, there is a strong incentive for companies to enhance their transparency and sustainability practices.

30.4 Consumer Protection as a Tool

The European Union has been continuously working on consumer protection and rights. One significant strategy document related to consumer rights and protection around that time is the 'New Deal for Consumer' adopted by the European Commission in April 2018.[24] The 'New Deal for Consumers' aimed to modernise consumer law, ensure its enforcement, and provide a system for redress. It consisted of essentially two proposals: the Directive on better enforcement and modernisation of EU consumer protection rules, which aimed at improving conditions for consumers – especially in terms of transparency – when shopping online (including on digital marketplaces), dual quality of products, and penalties for infringements of consumer law; and the Directive on representative actions for the protection of the collective interests of consumers, which aimed at establishing a collective redress mechanism and would allow qualified entities to bring representative actions on behalf of consumers against infringements of consumer law.

While these proposals were initially part of the response to the Dieselgate scandal and other unfair commercial practices, and aimed to increase consumer trust and ensure a high level of consumer protection by 2020, these proposals had evolved into the Consumer Agenda for 2020–2025,[25] which provides the framework for EU consumer policy focusing on five key areas: green transition, digital transition, effective enforcement of consumer rights, specific needs of certain consumer groups, and international cooperation.

These proposals on improved transparency aim at empowering consumers to make informed and environment-friendly choices when buying products were a first step in that direction. Consumers now have the right to be better informed about how long the product is made to last and if it can be repaired. They have more rights against misleading practices related to greenwashing or to premature obsolescence of products. Sellers must provide information on whether the producer of a consumer good offers a commercial guarantee of durability of two plus years, and whether software updates are provided, for goods with digital elements, digital content, and digital services. Information about a product's reparability score, where applicable, or other repair information given by the producer (availability of spare parts, repair manual, etc.) must be made available.

The new rules, therefore, aim at ensuring that traders do not mislead consumers about the environmental and social impacts, durability, and reparability of products. They ban making vague claims where environmental excellence cannot be demonstrated ('eco', 'green', etc.), environmental claims about the entire product when they only concern parts of the product and claiming that a good has a certain durability when it does not.

Together with the enhanced system for market surveillance, it will also be possible for consumer representatives to engage in collective legal action as initially proposed by the European Commission in 2018.[26] This was agreed among the co-legislators in 2020 and Member States had until December 2022 to amend their domestic law to meet the minimum requirements[27] under the EU Directive 2020/1828, which was adopted in late 2020. As the website of an international law office proclaimed 'The representative actions directive: get set for a new wave of European class actions'.[28]

The Commission made the next step through the publication of a third package of proposals under the Circular Economy Package on the 22nd of March 2023. In addition to proposing new consumer rights for easier and cheaper options to repair goods – fully in line with the goals set out in the EGD[29] – that package included a proposal for a so-called 'Green Claims Directive'.[30]

The proposal once enacted would require companies to respect minimum norms when they choose to make a green claim about their products or services. Such claims will need to be independently verified and proven with scientific evidence. This will result in more clarity for consumers, better quality information enabling consumers to reliably choose environment-friendly products, and less unfair competition for businesses that make a genuine effort to produce environmentally sustainable products.

The proposal would also bring more clarity to the confusing world of green labels. Consumers are bombarded with information about up to 230 different ecolabels on the EU market. To control the proliferation of green labels, new public labelling schemes will not be allowed unless developed at the EU level. Any new private schemes will need to show higher environmental ambition than existing ones and get a pre-approval to be allowed. There are detailed rules about environmental labels in general: they must also be reliable, transparent, independently verified, and regularly reviewed.

30.5 Influencing Public and Private Investment

In addition to enhancing reporting and transparency, it is also vital to direct investments towards sustainable projects and activities. This is why the European Union's Action Plan on Sustainable Finance although adopted in 2018 that is before the announcement of the EGD is a key part of the EU's efforts to connect finance with sustainability. The plan is intended to reorient capital flows towards sustainable investment, manage financial risks stemming from climate change, environmental degradation, and social issues, and foster transparency and long-termism in financial and economic activity. The action plan includes a set of proposed measures, including: establishing a clear and detailed EU classification system (or taxonomy) for sustainable activities; creating standards and labels for green financial products; fostering investment in sustainable projects; incorporating sustainability into financial advice; and developing sustainability benchmarks.

Once fully implemented, these measures will have significant implications for both corporate and investor behaviour.

Regarding corporate behaviour, the action plan promotes transparency, requiring corporations to disclose more detailed information about their environmental, social, and governance (ESG) practices. As the EU taxonomy defines what is considered a 'sustainable' activity, companies have an incentive to adjust their strategies and operations to align with these criteria, in order to attract investments. With the increased focus on sustainability-related risks, corporations are encouraged to better manage these risks and potentially invest more in sustainable projects and processes.

Regarding investor behaviour, the EU taxonomy and the green labels make it easier for investors to identify sustainable investments. This can drive capital towards more sustainable companies and projects. The focus on sustainability-related disclosures helps investors better assess risks related to environmental and social issues, which can influence their investment decisions. The action plan encourages a longer-term perspective among investors, with the recognition that sustainable investments often yield returns over the long term.

By promoting sustainable finance, the EU Action Plan has the potential to drive significant change in corporate and investor behaviour, helping to transition towards a more sustainable economy.

One of the significant legislative initiatives relates to an initiative for the creation of a common classification system for sustainable economic activities, or an 'EU taxonomy'. To achieve this, a common language, and a clear definition of what is 'sustainable' is needed. This was the basis for the proposal to create a common classification system for sustainable economic activities, or an 'EU taxonomy' mainly directed at the sector of investment and finance. The EU taxonomy as enshrined in the Taxonomy Regulation[31] is a classification system, establishing a list of environmentally sustainable economic activities aiming at helping the EU to scale up sustainable investment and implement the EGD. The EU taxonomy provides companies, investors, and policymakers with appropriate definitions for which economic activities can be considered environmentally sustainable. In this way, it creates security for investors, protects private investors from greenwashing, helps companies to become more climate-friendly, mitigates market fragmentation and helps shift investments where they are most needed.

30.6 Conclusions

The initiatives described above aim at establishing a coherent policy framework to make sustainable goods, services, and business models the norm and to transform consumption patterns in a more sustainable direction. This will over time also have significant impacts on the rights of consumers and the contractual relationships between sellers of products and services and their customers as it will further exacerbate the vital role that reporting and transparency play in private, contractual, and consumer protection legislation. In the context of private, contractual, and consumer protection legislation, reporting and transparency are not just about compliance with laws and regulations. They also contribute to trust, accountability, and better decision-making, benefiting businesses, consumers, and society as a whole. The new rules that will apply in the EU – as described above will affect several relationships, such as:

a *Consumer Trust and Confidence*: Transparency in business practices helps build consumer trust. When a business openly shares information about its products, services, and operations, consumers can make informed decisions. For example, disclosing product ingredients or production methods can assure consumers about product safety, quality, and ethics.

b *Contractual Obligations*: In contractual relationships, transparency is critical. Parties to a contract need clear, accurate, and timely information to ensure that they understand their rights, responsibilities, and the terms of the contract. This reduces the risk of disputes and non-compliance, but in the cases described also gives additional rights to buyers and consumers.

c *Consumer Protection*: Transparency is a cornerstone of consumer protection legislation. Businesses are now more often than not required to disclose important information about their products or services, such as pricing, product specifications, terms of service, and refund policies. This enables consumers to compare offerings, avoid misleading practices, and exercise their rights.

d *Risk Management and Compliance*: For businesses, transparency and reporting can help ensure compliance with laws and regulations. By maintaining clear records and openly reporting information, businesses can demonstrate their compliance to regulators, reducing the risk of penalties and avoiding situations of non-compliance.

e *Responsible Business Practices*: Reporting will also drive more responsible business practices. When businesses are required to report on their social and environmental impacts, it will provide an incentive to adopt more sustainable and ethical practices.

f *Market Efficiency*: Transparency improves market efficiency. When businesses disclose accurate and timely information, it allows for a fair and efficient functioning of the market, ensuring that prices accurately reflect the true value of products or services.

From the elements described above one can infer that there will be an increasing impact on private law relationships between individuals and companies. As companies are increasingly legally required to include sustainability information in their annual reports, non-compliance could result in penalties or other legal consequences. If a company misrepresents its sustainability efforts in its reporting (greenwashing), it may be liable for damages to shareholders, customers, or other stakeholders who relied on these misrepresentations. Companies often share trade secrets or proprietary information in their sustainability reports. This information may need to be protected by copyright or other intellectual property laws. In contracts between businesses (B2B), especially in supply chains, there will increasingly be clauses requiring disclosure of sustainability performance. Failure to meet these obligations may result in a breach of contract. Before entering a contract, businesses may perform due diligence, which may include reviewing the sustainability reports of potential partners. Sustainability reporting, therefore, can influence the formulation of contracts. Contracts between businesses and consumers (B2C) will also be influenced by sustainability reporting. For instance, a company's environmental commitments may form part of the 'ethical terms' in a contract.

The impacts are however not limited to rules and regulations about transparency and reporting, but also linked to the numerous new rules about products and materials as introduced under the EGD. Design and product quality requirements are integral components of both private and contract laws.

Two major components of private law, tort law and intellectual property law, directly relate to design and product quality.

Regarding Tort Law, in the context of product liability, manufacturers can be held legally responsible for defective products that harm consumers. If a product's design is faulty or the product fails to meet regulated quality standards, causing harm, the manufacturer could face a lawsuit.

As for Intellectual Property Law, design elements can be protected under intellectual property rights, such as design rights or patents. Companies often register their designs to prevent other entities from copying them. If another company infringes these rights, the holder can take legal action to stop the infringement and seek damages. In contracts between businesses, such as manufacturing or supply agreements, the design and quality of products to be delivered are typically specified. If a party fails to meet these specifications, it could be a breach of contract, potentially leading to termination of the contract, damages, or other remedies.

When a business sells a product to a consumer, there's an implied contract that the product will be of satisfactory quality and fit for its intended purpose. In both private and contract laws, clear communication about design and product quality is crucial. It can help prevent disputes and ensure all parties understand their rights and obligations. It can also foster trust, helping build long-term business relationships and enhance consumer confidence.

These are just a few ways public law rules on both product and material quality and their sustainability as well as on corporate transparency and sustainability reporting will increasingly interact with private, company, intellectual property, tort, and contract law.

Notes

1 EU Commission, 'The European Green Deal' (Communication) COM (2019) 640 final <https://eur-lex.europa.eu/legal-content/EN/TXT/?qid=1576150542719&uri=COM%3A2019%3A640%3AFIN> accessed 29.01.2023.
2 UNGA Resolution, 'Transforming our world: the 2030 Agenda for Sustainable Development' A/RES/70/1, adopted on 25.09.2015 <https://sdgs.un.org/goals> last accessed 20.08.2023.
3 Commission, 'A European Green Deal Striving to be the first climate-neutral continent' <https://commission.europa.eu/strategy-and-policy/priorities-2019-2024/european-green-deal_en> last accessed 29.11.2022.
4 EU Commission, 'The European Green Deal' (Communication) COM (2019) 640 final <https://eur-lex.europa.eu/legal-content/EN/TXT/?qid=1576150542719&uri=COM%3A2019%3A640%3AFIN> accessed 29.01.2023.
5 Global Reporting Initiative <https://www.globalreporting.org/> last accessed 29.11.2022.
6 Sustainability Accounting Standards Board <https://www.sasb.org/> last accessed 29.11.2022.
7 Task Force on Climate-related Financial Disclosures <https://www.fsb-tcfd.org/> last accessed 29.11.2022.
8 Commission, 'Leading the way to a global circular economy: State of play and outlook' (Staff Working Document) SWD (2020) 100 final <https://data.consilium.europa.eu/doc/document/ST-6767-2020-INIT/en/pdf> last accessed 24.09.2023.
9 Commission, 'A new Circular Economy Action Plan – For a cleaner and more competitive Europe' (Communication) COM (2020) 98 final <https://eur-lex.europa.eu/legal-content/EN/TXT/?uri=COM%3A2020%3A98%3AFIN> last accessed 24.09.2023.
10 Commission, 'Proposal for a Regulation of the European Parliament and of the Council concerning batteries and waste batteries, repealing Directive 2006/66/EC, and amending Regulation (EU) 2019/1020' COM (2020) 798 final <https://eur-lex.europa.eu/legal-content/EN/TXT/?uri=celex%3A52020PC0798> last accessed 24.09.2023.
11 Commission, 'Proposal for a Regulation of the European Parliament and of the Council amending Annexes IV and V to Regulation (EU) 2019/1021 of the European Parliament and of the Council on persistent organic pollutants' COM (2021) 656 final <https://eur-lex.europa.eu/legal-content/EN/TXT/?uri=celex%3A52021PC0656> last accessed 24.09.2023.
12 Commission, 'Proposal for a Regulation of the European Parliament and of the Council on shipments of waste and amending Regulations (EU) No 1257/2013 and (EU) 2020/1056' COM (2021) 709 final <https://eur-lex.europa.eu/legal-content/EN/TXT/?uri=CELEX%3A52021PC0709> last accessed 24.09.2023.
13 Commission, 'Proposal for a Regulation of the European Parliament and of the Council establishing a framework for setting eco-design requirements for sustainable products and repealing Directive 2009/125/EC' COM (2022) 142 final <https://eur-lex.europa.eu/legal-content/EN/TXT/?uri=CELEX%3A52022PC0142> last accessed 24.09.2023.
14 Commission, 'Proposal for a Proposal for a Regulation of the European Parliament and of the Council laying down harmonised conditions for the marketing of construction products, amending Regulation (EU) 2019/1020 and repealing Regulation (EU) No 305/2011' COM (2022) 144 final <https://eur-lex.europa.eu/legal-content/EN/TXT/?uri=celex%3A52022PC0144> last accessed 24.09.2023.
15 Commission, 'Proposal for a Proposal for a Regulation of the European Parliament and of the Council amending Directives 2005/29/EC and 2011/83/EU as regards empowering consumers for the green transition through better protection against unfair practices and better information' COM (2022) 143

final <https://eur-lex.europa.eu/legal-content/EN/TXT/?uri=CELEX%3A52022PC0143> last accessed 24.09.2023.

16 Commission, 'Proposal for a Proposal for a Regulation of the European Parliament and of the Council on the reporting of environmental data from industrial installations and the establishment of an Industrial Emissions Portal' COM (2022) 157 final <https://eur-lex.europa.eu/legal-content/EN/TXT/?uri=celex%3A52022PC0157> last accessed 24.09.2023.

17 Commission, 'Closing the loop – An EU Action Plan for the Circular Economy' (Communication) COM (2015) 614 final.
Commission, 'Proposal for a Directive amending Directive 2008/98/EC on waste' COM (2015) 595 final <https://eur-lex.europa.eu/legal-content/EN/TXT/?uri=CELEX%3A52015PC0595> last accessed 25.09.2023.
Commission, 'Proposal for a Directive amending Directive 94/62/EC on packaging and packaging waste' COM (2015) 596 final <https://eur-lex.europa.eu/legal-content/EN/TXT/?uri=celex%3A52015PC0596> last accessed 25.09.2023.
Commission, 'Proposal for a Directive amending Directive 1999/31/EC on the landfill of waste' COM (2015) 594 final <https://eur-lex.europa.eu/legal-content/EN/TXT/?uri=CELEX:52015PC0594> last accessed 25.09.2023.
Commission, 'Proposal for a Directive amending Directives 2000/53/EC on end-of-life vehicles, 2006/66/EC on batteries and accumulators and waste batteries and batteries, and 2012/19/EU on waste electrical and electronic equipment' COM (2015) 593 final <https://eur-lex.europa.eu/legal-content/EN/TXT/?uri=CELEX%3A52015PC0593> last accessed 25.09.2023.

18 Council Directive (EU) 2022/2464 of 14 December 2022 amending Regulation (EU) No 537/2014, Directive 2004/109/EC, Directive 2006/43/ EC, and Directive 2013/34/EU, as regards corporate sustainability reporting (2022) OJ L 322.

19 European Reporting Lab, 'Proposals for a Relevant and Dynamic Eu Sustainability Reporting Standard-Setting. Final Report' (2021) <https://finance.ec.europa.eu/system/files/2021-03/210308-report-efrag-sustainability-reporting-standard-setting_en.pdf> last accessed 29.11.2022.

20 COMMISSION DELEGATED REGULATION (EU) …/… of 31.7.2023 supplementing Directive 2013/34/EU of the European Parliament and of the Council as regards sustainability reporting standards (Text with EEA relevance).

21 Council Directive 2014/95/EU of 22 October 2014 amending Directive 2013/34/EU as regards disclosure of non-financial and diversity information by certain large undertakings and groups (2014) OJ L 330/1 <https://eur-lex.europa.eu/legal-content/EN/TXT/?uri=celex%3A32014L0095> last accessed 25.09.2023.

22 UNEP, Convention on Biological Diversity, 'Kunming-Montreal Global Biodiversity Framework' CBD/COP/DEC/15/4, 19 December 2022 <https://www.cbd.int/doc/decisions/cop-15/cop-15-dec-04-en.pdf> last accessed 23.09.2023.

23 Council Regulation (EU) 2023/1115 on the making available on the Union market as well as export from the Union of certain commodities and products associated with deforestation and forest degradation and repealing Regulation (EU) No 995/2010 (2023) OJ L 150/206 <https://eur-lex.europa.eu/legal-content/EN/TXT/?uri=CELEX%3A32023R1115> last accessed 25.09.2023.

24 Commission, 'A New Deal for Consumers' (Communication) COM (2018) 183 final <https://eur-lex.europa.eu/legal-content/EN/TXT/?uri=CELEX%3A52018DC0183> last accessed 24.09.2023.

25 Commission, 'New Consumer Agenda. Strengthening consumer resilience for sustainable recovery' COM (2020) 696 final <https://eur-lex.europa.eu/legal-content/EN/TXT/?uri=CELEX%3A52020DC0696> last accessed 24.09.2023.

26 Commission, 'Proposal for a Directive of the European Parliament and of the Council on representative actions for the protection of the collective interests of consumers, and repealing Directive 2009/22/EC' (2018) 184 final <https://eur-lex.europa.eu/legal-content/EN/TXT/?uri=CELEX%3A52018PC0184> last accessed 25.09.2023.

27 Council Directive (EU) 2020/1828 of 25 November 2020 on representative actions for the protection of the collective interests of consumers and repealing Directive 2009/22/EC (2020) OJ L 409/1.

28 Herbert Smith Freehills, 'The Representative Actions Directive: get set for a new wave of European class actions' [2022] Legal Briefings <https://www.herbertsmithfreehills.com/insight/the-representative-actions-directive-get-set-for-a-new-wave-of-european-class-actions> last accessed 25.09.2023.

29 Commission, 'Proposal for a Directive on common rules promoting the repair of goods and amending Regulation (EU) 2017/2394, Directives (EU) 2019/771 and (EU) 2020/1828' COM (2023) 155 final <https://eur-lex.europa.eu/legal-content/EN/TXT/?uri=celex%3A52023PC0155> last accessed 25.09.2023.
30 Commission, 'Proposal for a Directive on substantiation and communication of explicit environmental claims' COM (2023) 166 final <https://eur-lex.europa.eu/legal-content/EN/TXT/?uri=COM%3A2023%3A0166%3AFIN> last accessed 25.09.2023.
31 Council Regulation (EU) 2020/852 of 18 June 2020 on the establishment of a framework to facilitate sustainable investment and amending Regulation (EU) 2019/2088 (2020)] OJ L 198/13.

31

CORPORATE PURPOSE AS A CONDUIT FOR SUSTAINABILITY IN CORPORATE GOVERNANCE

Iain MacNeil and Irene-marié Esser[1]

31.1 Introduction

Corporate purpose has emerged as a key focal point in the debate on how corporate governance might contribute to sustainability.[2] In earlier research,[3] we examined the link between sustainability and corporate governance and proposed an entity model of ESG (Environment, Social and Governance considerations) which encompassed the role of corporate purpose in the transition to sustainability.[4] It focuses on the internalisation of externalities and suggests various mechanisms that can be adopted by companies to enhance stakeholder engagement and participation such as board committees, procedural due diligence requirements and enhancement of non-financial reporting. While corporate purpose was referenced in that model, it was not the main focus of attention and so we now take the opportunity to elaborate on its potential role, with a particular focus on UK law and practice while referencing developments in the EU and at the global level.

In this chapter, our focus is therefore to determine the nature and parameters of corporate purpose as a legal principle and how it might promote sustainability in corporate governance. We argued before that the key issues have been whether corporate purpose encompasses objectives other than profit maximisation and whether, and if so, how, interests other than those of shareholders can legitimately be considered in board decision-making. We argued that the default position is that it falls to the board of directors to determine corporate purpose in accordance with the standard process of majority decision-making. However, the emergence of a de facto norm of shareholder value maximisation constrained any inclination on the part of boards to act in that way. Some commentators, therefore, suggest that law reform or at least shareholder consent would be necessary to alter corporate purpose. Others see the framing of directors' duties in UK company law as broad enough to enable adjustment to purpose through board decision-making.

In the second section, we distinguish three different contexts in which the concept of corporate purpose has played a foundational role in the framing of the law. The first relates to entity selection; the second to shareholder and creditor protection in the transactional context; and the third is focused more on the contemporary debate around purpose and its role in stakeholder protection. In the third section, we examine the discourse and proposals of academics and policymakers on corporate purpose. In the final section, we map corporate purpose onto the entity model of ESG that we proposed in earlier research.

 DOI: 10.4324/9781032662046-37

31.2 The Concept of Corporate Purpose

Corporate purpose as a concept has served different purposes over time. We distinguish three different contexts in which the concept has played a foundational role in the framing of the law. The first relates to entity selection, where corporate purpose represents a division between commercially focused entities and those with a more social orientation. The second relates to shareholder and creditor protection in the transactional context and emerged from concerns that directors might use their powers to expose shareholders and creditors to risks that went beyond the business for which the company was established. The third, focused more on the contemporary debate around purpose, is the role of corporate purpose in stakeholder protection. While the third instance dominates the contemporary discourse, the other two instances remain relevant in terms of understanding purpose as a legal technique and how it might contribute to stakeholder protection. In particular, the first two instances of the use of corporate purpose may help to clarify the role of private ordering in promoting sustainability in corporate governance as they differentiate entities by reference to social purpose and implement a form of stakeholder protection (respectively).

Some of the most influential proposals[5] address corporate purpose directly on the basis that it is a variable that could be directly adjusted. An alternative view is that it is an outcome that depends on other variables for its calibration.[6] On that view, constituent elements such as capital, profit and governance are key drivers of purpose and more amenable to adjustment through recognised legal techniques than the more amorphous concept of purpose. We explore that aspect through the role of corporate purpose in entity selection, where adjustment to the constituent elements already plays a key role. That analysis also identifies the outer limits of workable adjustment to mainstream commercial companies and the point at which the selection of an alternative entity is the more appropriate option.

A second concern is that the amorphous nature of purpose may result in attempts to adjust it directly ultimately proving futile, as techniques are likely to develop to circumvent the objective. That experience is evident to some extent in the history of the objects clause in UK company law, which was an early form of a mandatory purpose provision until the Companies Act 2006. The same concern has been expressed in connection with proposals for a modern version of an objects clause focused on stakeholder interests. A related issue is the capacity of soft law instruments such as the UK Corporate Governance Code to drive changes in purpose that conflict with statutory corporate law when the latter is premised on shareholder primacy.

31.2.1 Purpose and Entity Selection

Corporate purpose is well-established as a legal technique for differentiating entities with a commercial focus from those with some element of social focus. Entity selection enables the founding investors to select a legal entity that is aligned with their vision for how the business will operate and how it will balance the interests of shareholders and other stakeholders.[7] The availability of different models has expanded over time, with the result that founders are less reliant on adjusting the standard corporate model to achieve their objectives. We refer to three well-established alternative models to illustrate the options that are available.[8]

In the case of UK charities, charitable purpose is the key driver of the legal framework. The alternative role of capital, profit and governance in charities demonstrates the centrality of these concepts to fundamental change in organisational purpose. The change results from the non-profit status of charities and the limitations on the use of charitable funds (equivalent to capital) for charitable purposes. In other words, the charitable purpose is given effect by eliminating the pursuit of profit in the conventional sense and applying an 'asset lock' to ensure that funds cannot normally be withdrawn other than for the purpose of transfer to another charity.

Charitable status is limited to entities that have a charitable purpose and that means the activities must satisfy the test of public benefit.[9] Moreover, the effect of charitable purpose is that while the concept of profit remains relevant for the purposes of accounting, its conventional meaning in the context of commercial companies is altered as the profits and assets are available only for charitable purposes. This choice of entity stands at the opposite end of the spectrum from standard commercial companies, or at least those companies where purpose is understood to refer to the creation of shareholder value. Community interest companies, introduced in the UK in 2005, combine a minor element of commercial focus with a major focus on community interest as the purpose of the entity. In this instance the 'asset lock' is less severe as some of the profits may be distributed to members.

In the US, the Delaware Public Benefit Corporation is a popular choice for entities that want to integrate public benefit into purpose. It requires the entity to pursue a specific social purpose (with public benefit) stated in its articles of association and to produce a statement addressing the social impact at least biennially; and directors must balance the economic interests of shareholders with the interests of constituents materially affected by the business and with the specific social purpose.[10] While this approach responds to the concern that fiduciary duty in the standard corporate model prioritises shareholder interests, stakeholders are not granted enforceable legal rights and thus the model resembles the Enlightened Shareholder Value (ESV) model[11] in the UK in that respect.

The B Corp framework can be described as private and transnational in the sense that it represents a (global) certification approach to the integration of stakeholder-focused purpose into the entity.[12] It also represents a hybrid that is amenable to public listing and investment by mainstream institutional investors. For a UK-incorporated company, the key requirement for B Corp certification is to amend the articles of association to commit the company to: creating a material positive impact on society and the environment through its business and operations; and considering 'stakeholder interests' – including shareholders, employees, suppliers, society, and the environment.[13]

These provisions may serve to distinguish B Corps from mainstream commercial companies but the extent to which they do that and the consistency of the indicators across B Corps has been questioned.[14]

These examples draw attention to the central role of capital, profit and governance structure with regard to purpose. As the role of profit changes, governance must also change as shareholder-focused governance is largely premised on control over the generation and application of profit by directors operating under duties that prioritise shareholder interest. While logic might suggest that the availability of alternative models (above) would set the appropriate scope for such adjustment, there may be instances in which other considerations (such as taxation and access to external finance) may constrain the options.

Looking beyond entity selection, the standard corporate model in the UK offers the potential to adjust purpose but not standard partnerships or limited liability partnerships.[15] In the US context, adjustments to the standard model have been characterised as creating a segment of 'profit-with-purpose businesses' that espouse a mission to advance the common good, authorising directors to consider the adopted social purpose and report on their social impact using standard metrics. We consider in Section 31.3 various proposals that have been made in this context and in Section 31.4 we consider adjustments that align with the 'entity model' of ESG governance that we proposed in earlier research.[16]

31.2.2 Shareholder and Creditor Protection

In contrast with the more recent focus of corporate purpose discourse on stakeholder interests and potential regulatory intervention, the historic focus on corporate purpose in the UK was on shareholder and creditor protection. The objective was to mitigate the risk that directors would operate the business in a manner that deviated from the venture that the shareholders and creditors had funded and thereby expose them to financial risk. Two legal principles in particular gave effect to this policy: the objects clause, which related to the scope and validity of company actions; and the proper purposes rule, which is part of the fiduciary duty of company directors to act within their powers. Both principles are linked to the modern debate on corporate purpose as they control the manner in which a company operates and whose interests are to be protected in that context.[17] They remain relevant as examples of how mandatory and default rules respectively might operate as a conduit for sustainability. Both rules are framed as mandatory but their operation can be adjusted through the power of shareholders to control the articles of association and thereby define the objects and the purpose of the company.

31.2.2.1 Objects Clause

In the past, British company law required a company to state its objects in its memorandum of association and the courts limited the legal capacity of a company to its objects. In that sense, the objects were treated as the purpose of the company. Any action outside the scope of the objects was void and third parties contracting with a company were thereby faced with considerable risk. However, the Companies Act 2006 now provides that a company's objects are unrestricted and that there is no need to have an objects clause[18]; it also sets aside any issue of validity of acts as a result of anything in the constitution (and so objects clauses are not effective in restricting corporate capacity).[19] As noted by Davies in this context, both the policy and legal technique associated with the objects clause were misplaced: diversification into new activities might well prove profitable, and companies could avoid the restrictive effects of the law by drafting wide and subjectively worded objects clauses.[20] Thus, the history of the objects clause as a technique to define and delimit the purpose of a company serves as a cautionary tale about the risks and limitations of relying on a mandatory purpose clause for other purposes, such as expanding stakeholder protection. We return to this issue in Section 31.3.

31.2.2.2 The Proper Purposes Rule

The proper purposes rule is part of the fiduciary duty of loyalty and requires directors to act in accordance with the company constitution and exercise powers for the purposes for which they are conferred.[21] The salience of corporate purpose for sustainability was highlighted in a recent case claiming that the corporate trustee of the UK's largest pension scheme was in breach of the proper purposes rule by failing to develop a Paris Agreement-aligned plan for sustainable investing.[22] But even so, the proper purposes rule does not clarify the purposes for which powers are generally conferred. Case-law has clarified the matter with regard to some specific contexts,[23] but the broader issue of purpose is only addressed indirectly in the Companies Act 2006 by s.172, which implements the so-called ESV Model.[24] That leaves open the option for companies to specify purpose in their articles and for s.172 to be interpreted by reference to such a provision, but there is little evidence of companies adopting this approach. The implication seems to be that purpose did not require precise specification because its meaning was already clear in practice and reinforced by the explicit reference in s.172 to shareholders as the primary beneficiaries of the success of the company and the only stakeholder capable of enforcing s.172. Nor has the more recent focus on sustainability changed that pattern. Thus, the proper purposes rule does not advance the current debate around purpose in any meaningful way, albeit it could provide effective protection to directors if the purpose were specified in the articles or regulatory obligations with a similar effect (e.g., corporate due diligence) were imposed. In that sense the proper purposes rule has an unrealised potential to act as a conduit for sustainability in corporate governance.[25] We elaborate below on some specific proposals that might realise that potential.

31.2.3 Stakeholder Protection

A high-level overview of the stakeholder protection objectives underpinning the contemporary debate on purpose can be drawn from some of the leading definitions of purpose that have been put forward. In 2019, Larry Fink, CEO of the Global Asset Manager Blackrock, proposed that

> Corporate purpose is the higher purpose of a company that goes beyond the sole profit orientation. The purpose is to define and deliver a long-term value-creating promise, either in the company's local environment or in the global market environment, that is directly related to the company's value creation.[26]

That concept of purpose as something that goes beyond the goal of profit was evident later in the definition proposed by the British Academy Future of the Corporation programme which characterised corporate purpose as 'producing profitable solutions for problems of people and planet, not profiting from producing problems for either'.[27]

These definitions have been influential in driving the debate as well as the policy responses and proposals that are considered in Section 31.3. So far as the UK is concerned, they can be linked to the ESV model of stakeholder protection that was introduced by the CA 2006. The general consensus is that this is a weak form of stakeholder protection, with no legal rights for stakeholders and potentially less discretion for the board than under the earlier statutory provisions.[28] The nature and application of the ESV model was clarified recently in the case of *BTI 2014 LLC v Sequana SA & Ors*[29] on the scope and content of directors' duties with regard to creditors when a company is nearing insolvency. In that context, the Supreme Court of Appeal considered the duty of directors to act, in good faith, for the benefit of the company and what that means. In other words, who are the ultimate beneficiaries of the duty to act in good faith? While the case focused primarily on the point in time when directors should consider the interests of creditors, which may be different from those of the shareholders, it also provides confirmation and clarification of shareholder primacy.

Referencing the common law, Lord Reed stated that the director's fiduciary duty to act in good faith in the interests of the company means the interests of its members as a whole.[30] With respect to the codification of the duty in s.172 of the Companies Act 2006 the Court held that since this duty is focused on promoting the success of the company 'for the benefit of its members as a whole', the shareholders are the intended beneficiaries of that duty albeit that it is owed to the company. The Court did refer to the secondary importance of other considerations that directors need to take into account under s.172 but held that the primary duty remains focused on promoting the success of the company for the benefit of its members.[31] To that extent, 'the common law approach of shareholder primacy is carried forward into the 2006 Act'.[32] It seems clear that the Court's approach confirmed shareholder primacy as the position under the common law but also under the codified statutory provisions in s.172.

Also relevant in this context is the shift we noted in earlier research from an ethics-driven model of CSR to a financial model of ESG in which portfolio selection and engagement with companies (often driven by stewardship codes) has effectively shifted the focus away from the tension between shareholders and stakeholders implicit in s.172. In that sense, the precise framing of s.172 may matter less than the overall balance of the law, social norms and market practice between shareholder and stakeholder interests.

31.3 Discourse and Proposals

In this section, we review and evaluate some of the key contributions to the debate on corporate purpose with a view to understanding the problems with the current role of purpose in corporate law and the policy solutions that are proposed to resolve them. We also address the linkage between purpose and sustainability, which has become a key driver of initiatives to reform purpose and has some potential as an overarching principle linking corporate governance to sustainable objectives. We note in this context that the role of purpose in this sense has not been addressed directly by corporate governance codes, albeit the UK Corporate Governance Code has recently moved in that direction. For example, the OECD Principles of Corporate Governance (Preface)[33] refer to the purpose of corporate governance but not the company itself, while the ICGN Global Governance Principles[34] make no mention of purpose. Thus, while the concept of purpose already has some meaning and application in different legal contexts (as set out in Section 31.2), the contemporary debate linked to sustainability represents a re-purposing of the concept to address both old (shareholder primacy) and new (sustainability) concerns, and in particular, the concern that the former might constrain approaches to solving the latter.

31.3.1 Diagnosis of the Problems

Corporate purpose is intrinsically linked to the debate relating to sustainable companies and better stakeholder protection but there are potential challenges in this context, not least how to define corporate purpose. We addressed the challenge in defining corporate purpose in the previous section, arguing that one should not attempt to define purpose in itself, but instead adjust various elements linked to purpose (its 'foundational framework' in a sense). Another problematic issue, where there is ample disagreement, is how corporate purpose fits into the current company law structure, and especially whether it can be aligned with s.172 of the Companies Act 2006.

Edmans, in his book 'Grow the Pie', argues that it is clear that companies should serve wider society, but what is not clear is how they should do that. He defines corporate purpose as 'why a company exists – who it serves, its reason for being, and the role that it plays in the world'.[35] In other words, how a company can genuinely make changes, focusing on more than just a 'public relation exercise'. He argues that one needs an approach where companies can create both profit for investors and value for society.[36] However, without profits, shareholders would not finance companies and companies would not finance investors. Any reform to businesses that ignores the role of profit in society is unlikely to be implemented. So, a solution is needed that will work for both business and society. He proposes a 'grow the pie' mentality focusing on creating value for society, which can, but will not necessarily, lead to profit maximisation over the long term.

Davies refers to corporate purpose in the context of directors' duties and identifies the often problematic framing of purpose within a shareholder-orientated company law framework.[37] In that context, commentators are not always in agreement when it comes to (1) the potential for setting a corporate purpose encompassing broader, sustainable goals within the current legal framework; and (2) the interaction with the codified directors' duty in s.172 of the Companies Act 2006. Under a shareholder primacy approach, it is therefore unclear to what extent directors could set a corporate purpose, through strategy and policies, which has the potential to deviate from shareholder value. And with shareholder primacy being confirmed it would be interesting to see how far directors are willing to go to agree on a sustainable corporate purpose that will add value to society, but not necessarily to the shareholders as a collective group.

The Financial Reporting Council (FRC), in their April 2022 report on corporate purpose and ESG, states that purpose deals with 'why the company exists, what the company does, in what market the company operates, what the company is seeking to achieve and how the company will achieve it'.[38] The FRC claims that while companies need to take into account the interests of a broad range of stakeholders, it is only through sustained growth and healthy profits that can they have sufficient resources for redistribution and 'doing good'. The purpose should not become an aim on its own but should be underpinned by ethics – 'it should serve as a "moral compass" through which innovation and profitability are pursued'.[39] Detail on what this should entail and how it should link to the current company law framework is however missing.

31.3.2 Proposals

In recent years, calls for companies to have a 'corporate purpose' and for businesses to contribute to sustainability have grown significantly. Various proposals are made to achieve this. As mentioned earlier Edmans proposes a 'grow the pie' mentality in which the pie represents social value, not profits – profits are only one slice of the pie. Under the pie-growing mentality, a company's primary objective is social value rather than profits. He provides various examples where this approach ended up more profitable than if profits were the end goal. That's because it enables many investments to be made by the company, which ended up delivering substantial long-term payoffs. But since these payoffs could not have been forecasted from the outset, the projects would never have been approved under a traditional shareholder value framework.

Mayer argues that profit must be derivative from solving a problem, it should not create or produce a problem: he proposes rethinking corporate purpose by identifying or determining purpose and then implementing it, and this is something that the board should oversee. Barby et al. suggest that this process can be supported by changes to the role of capital and profit by comparison with the conventional model, focusing on the need to value and report on human, social and natural capital.[40] They propose a three-stage approach via 'Motives, Metrics and Money', arguing that both enterprise cost and social valuation approaches should be available for this purpose. The former would be modified to incorporate negative externalities as costs and positive externalities as investments. The latter would aim to measure impacts and value them for inclusion in the reported accounts by reference to metrics related to non-material and non-financial assets and liabilities.[41]

Mayer and Barby et al.'s approach represents a fundamental recalibration of key elements of the accounting and reporting framework for companies and would likely face considerable difficulties in its adoption as the conventional framework is deeply embedded in corporate culture and practice. It has already started to evolve to encompass sustainability considerations: for example, the EU has extended non-financial reporting (NFR) to encompass reporting on sustainable economic activities as defined by the EU Taxonomy. In our view, it is, nevertheless, consistent with a fundamental change in corporate purpose, as it adjusts the conceptual foundations of capital and profit to align with its vision for corporate purpose. That is a key step that has been omitted from many other proposals for fundamental change in corporate purpose.

Davies argues that s.172 Companies of the Act 2006 should not be seen as an obstacle when directors want to set sustainable policies (only policies with no benefit to shareholders over the long term are really ruled out).[42] It gives a wide discretion to the board over the setting of policies, and although it refers to the members as ultimate beneficiaries, it does not restrict the setting of a corporate purpose that benefits stakeholders other than just shareholders. Rather, it puts corporate purpose at the heart of directors' duties. Davies, therefore, does not propose to reform s.172, but instead evaluates mandatory and voluntary purpose provisions (as proposed by Mayer[43]), particularly whether the adoption of a stated purpose should be mandatory for (at least large) companies and whether the chosen purpose should be required to contain a social or communal element. Davies argues that such a proposal is not desirable as a mandatory, broad-purpose requirement, as it would likely be either largely ineffective or largely unnecessary. To be effective, substantial changes would be needed either in corporate law (which would downgrade the governance rights of shareholders significantly) or in the investment goals of shareholders.

The Better Business Act (BBA) initiative proposes to change business behaviour through reform of directors' duties. It proposes law reform to ensure that businesses are legally required to benefit workers, customers, communities, and the environment while delivering profit. Under this initiative, the BBA would change the default position of shareholder primacy for all companies so that directors would be empowered to advance the interests of their shareholders alongside the interests of the wider society and the environment. In situations where a director has to choose between the company's intention to create positive social or environmental impacts and the interests of shareholders, the directors would no longer be compelled to default to prioritising shareholders. It is however stated that this Act would not create new rights for stakeholders. The position would remain the same in that directors owe their duties only to the company, and only the company or a shareholder acting on behalf of it can take action against a director for a breach of their duty. It, therefore, remains unclear how this will improve the position of stakeholders and how it is different from the current position.

It is clear therefore that commentators hold diverse views on setting a corporate purpose encompassing broader sustainability goals within the current legal framework. Two differences are particularly significant. The first relates to whether corporate purpose itself should be adjusted or whether key levers such as the framing and reporting of capital and profit should be adjusted. The second issue, which follows on from the first, is whether and if so to what extent explicit adjustment of corporate purpose to promote sustainability will require a more fundamental recalibration of corporate law. These issues remain unresolved in the UK context (and elsewhere) and have not been addressed by policy responses to date.

31.3.3 Policy Responses

In the UK, there have been various policy responses linked to corporate purpose, in particular through the UK Corporate Governance Code and work undertaken by the FRC, as well as a Green Paper on corporate governance in 2016 followed by a government response in 2017. The Code does not define corporate purpose, nor does the FRC provide any detailed discussion of purpose in its prior consultation documents, reports, or speeches. The 2018 revised Guidance on Board Effectiveness does, however, provide that 'a company's purpose is the reason for which it exists'.[44] This is a rather cryptic and circular definition. It also suggests that it is a matter for the company, i.e. shareholders, to decide on purpose, but then it states, later, that corporate purpose is the responsibility of the board.[45]

Principle B of the UK Corporate Governance Code (2020) provides that the board should establish the company's purpose, alongside values and strategy, and satisfy itself that they are aligned with its culture. Kershaw and Schuster refer to the position adopted by the Government in its response to the Green Paper,[46] confirming that a change to a company's purpose is fundamental and should involve a change to the company's articles (i.e., a special resolution requiring a 75% majority vote). However, they also go on to recognise that delegation of 'all the powers of the company' as per the standard articles would too allow the board to set the corporate purpose. As they recognise, there are some doubts over this interpretation and whether or not the delegation to the board of directors can stretch that far, even allowing for the fact that the UK Corporate Governance Code recognises that the board should set the corporate purpose of the company. This seems, at first sight, to be in contrast to the position of the Companies Act 2006, as confirmed in the Green Paper response of the Government. Perhaps the difference lies in the degree of change: if the articles are to be changed, shareholder approval is indeed required and it is doubtful if delegation to the board on that would be sufficient – whereas the Code's reference to setting of purpose by the board may envisage decisions implicating the operational side of the business, in respect of which the role of the board is not disputed.

31.4 Purpose in an Entity Model of ESG

In earlier research we proposed an entity model of ESG which encompassed the role of corporate purpose in the transition to sustainability.[47] By referencing the older model of CSR it focuses more on the role of the board and operational decision-making by comparison with the financial model of ESG that is more focused on portfolio selection and stewardship. The entity approach allows for the contextual calibration of purpose by reference to the operational context, whereas the financial model of ESG is generally less able to do that – for instance, investment and stewardship policies are more generic and do not respond precisely to context. The entity model is more attuned to the policy of flexibility in corporate governance, which was promoted by the UK and followed in models around the world,[48] and is conducive to the voluntary framing of purpose provisions in companies.

In this section we follow the schema we adopted earlier to outline the entity model to elaborate on how we see corporate purpose fitting in to that model. Our ESG entity model provides an alternative, process-driven solution, rather than a reform of the law. It focuses on the internalisation of externalities,[49] and suggests various mechanisms that can be adopted by companies to enhance stakeholder engagement and participation such as board committees, procedural due diligence requirements and enhancement of non-financial reporting. While corporate purpose was referenced in that model, it was not the main focus of attention and hence, we now take the opportunity to elaborate on its potential role.

31.4.1 Focus

In Section 31.2.3, we referenced some prominent definitions of corporate purpose so as to clarify the link with stakeholder protection. In the context of the entity model, we would propose the following definition: 'Corporate purpose is the articulation of how a company's business model links to its environmental and social responsibilities'. This would link well with our claim that the entity model aligns better with the old CSR model rather than the more recent ESG model. Our proposed definition is narrower than that of the British Academy Project[50] but more manageable as a first step. It does not require reform of directors' duties or even adjustment of the articles, but as outlined below, would be facilitated by reforms in the governance process and due diligence. While those elements align it with the approach of the International

Sustainability Standard Board (ISSB), our approach differs in the sense that it aims for a more prominent role for stakeholder interests in corporate decision-making and reporting (double rather than single materiality).[51]

We also see a more prominent role for entity selection in resolving the tensions implicit in the standard corporate model. This issue has attracted relatively little attention in the academic and policy debates on corporate purpose but has the potential to satisfy investor and consumer demand for changes to corporate purpose in a manner that provides a structural solution to the conflict of shareholder and stakeholder interests.

31.4.2 Channel

In our earlier specification of the entity model of ESG, our discussion on this element focused on the choice between a financial model of ESG driven by investors and an entity model in which board decision-making at the entity level plays a more prominent role and integrates stakeholders into decision-making. In the more specific context of corporate purpose, two key issues have emerged: who should decide on corporate purpose, the shareholders or the board; and how should corporate purpose be operationalised? The two issues are linked in the sense that corporate law allocates powers between the shareholders and the board and in the UK the 'delegation' model of board powers means that shareholders ultimately control the division (as well as having strong dismissal powers over directors).[52] Thus, from the legal perspective, the technique (the 'how') should be considered first and then the appropriate organ for decision and action (the 'who'). But since we considered 'Channel' before 'Implementation' in the specification of the entity model of ESG, we follow the same presentation here for the sake of consistency.

On the basis that we favour implementation techniques for corporate purpose that generally fall within the (typically broad) powers delegated to the board in the UK,[53] we propose that the board should be responsible for articulating corporate purpose. That is the approach adopted by the UK Corporate Governance Code, and while it may conflict with the shareholder primacy basis of corporate law if fundamental change to purpose is pursued,[54] less dramatic and evolutionary change is possible within the ESV model of s.172 of the Companies Act 2006. Davies for example views s.172 as providing wide discretion to directors to set the company's policies.[55] Thus, even though s.172 embeds the ESV approach, it should not be interpreted as a restriction on policy setting. Directors can give greater weight to the interests of stakeholders as the duty is subjective and there is no limit on the time horizon, nor is there any constraint to consider issues such as the potential for externalities to be priced by the market in due course and increase the company's costs in the absence of mitigating action. In that sense, purpose is already sufficiently embedded in the duties of directors and there is some alignment of shareholder and stakeholder interests.

The inclusion of risk management within the ambit of duties falling within s.172 (duty of care) provides further support for the role of the board in this context. Risk management does not require adjustment to purpose so long as the risk is financially material, and for some externalities that will be the case. This aligns with Davies' approach that some voluntary ESG measures are already incentivised and would not require mandatory regulation.[56] It also locates responsibility with the board, alongside other forms of risk management. Inevitably, of course, there remain externalities (e.g. biodiversity) that are not priced by the market and where stakeholder and shareholder interests will continue to diverge even as capital markets take greater cognisance of sustainability risks. Private ordering (including board decision-making) would likely not respond to resolve these issues and so in that context a stronger case can be made for mandatory regulation (either of the relevant domain, such as environment or biodiversity, or of corporate governance and finance).

31.4.3 Implementation

In our entity model of ESG we proposed three techniques that would help change corporate purpose even if no changes were made to directors' duties. They included a separate stakeholder committee as a sub-committee of the board; a (mandatory) duty of due diligence, framed in terms of process and without enforcement rights for stakeholders; and voluntary purpose provisions in the articles (and as a variant 'say on purpose' provisions providing a vote to shareholders analogous to 'say on pay').

We argued in that context that a more prominent role for board decision-making opens the potential for more effective stakeholder integration into the process by comparison with the financialised model of ESG. As part of the mechanisms linked to our ESG entity model we looked at the potential role that a stakeholder board committee might play to fulfil an oversight function or even be required to sign off on the due diligence. We sketched the contours of our suggested model, and it included some key legal reforms, in particular a duty of due diligence, which has already been implemented in some countries, with the EU likely to follow soon.[57] The model also included a role for soft law and voluntary action in recognition of the important steps that have already been taken in that direction by investors. Our proposal, framed in the context of the UK system, avoided the need for changes to directors' duties, but could nevertheless drive significant change through more effective integration of stakeholders in corporate decision-making.

We specifically focused on the UK voluntary (or soft law) mechanism provided by Provision 5 in the UK Corporate Governance Code on workforce engagement mechanisms. It takes the stakeholder agenda a step forward by requiring listed companies to adopt one of three options (i.e., a designated non-executive director, a formal employee advisory council, or a director from the workforce) for workforce engagement.[58] We view this development as a key stage in the evolution of the UK Corporate Governance Code in terms of integrating stakeholder interests into board decision-making. We also, by way of comparison, considered the social and ethics committee (SEC) in the South African context, that is compulsory for certain companies.[59] Provision 5 is a step in the right direction, and it is interesting to note that in this context, the majority of companies opted for a designated non-executive director, an advisory panel, or a combination of the two. No company opted for a director from the workforce.[60] Compliance with Provision 5 is therefore high, as far as companies opting for one of the options or explaining what they have in place. However, our research did not find much evidence of how workforce interests are integrated into decision-making or strategy, nor is there much evidence that such considerations changed outcomes relative to what they would have otherwise been.

In the context of the approach we propose, in which variables other than purpose (i.e., capital, profit and governance) are adjusted, we are of the view that mandatory, or even voluntary, purpose statements or amendments to the articles are perhaps not the best way forward. It is likely, as argued by Davies, that such mandatory purpose provisions could be circumvented (e.g., by broad drafting techniques) and would therefore not represent a reliable technique for changing the balance in board decision-making as between commercial factors (prioritising shareholder interests) and other factors (such as environmental and social). A related issue is that while purpose drives the contextual application of directors' duties in the Companies Act 2006 (i.e., duties must be applied relative to purpose in the context of each company) it is not necessarily the case that this creates a need to frame purpose explicitly and instrumentally. That has not occurred in the form envisaged by the current debate for good reason, as it attempts to capture an elusive concept that is better addressed through the levers of purpose that are already prominent in the legal infrastructure of companies i.e., capital, profit, and governance.

In 2022, the European Commission adopted a proposal for a directive on Corporate Sustainability Due Diligence (CSDD) introducing an environmental and human rights due diligence duty for financial and non-financial companies.[61] The CSDD Directive will apply to a company's own operations, its subsidiaries and their supply chains, including direct and indirect business relationships. The Directive introduces duties for directors to set up and oversee the implementation of due diligence processes and integrate due diligence into corporate strategy. As part of the duty to act in the best interest of the company, directors will now have to take into account the human rights, climate change and environmental impacts of business activities. The CSDD operates in tandem with the Corporate Sustainability Reporting Directive (CSRD). The latter is a reporting mechanism relating to the sustainable activities of companies (including social and human rights aspects). Both proposals reference the UN Guiding Principles (UNGPs) on Business and Human Rights and the OECD Guidelines for Multinational Enterprises (OECD Guidelines) as foundational international due diligence frameworks.[62]

31.4.4 Metrics

We noted in earlier research that sustainability metrics largely evolved in tandem with the financialised model of ESG, whereas the older CSR model was focused more on ethical responsibility and much less driven by metrics. Nevertheless, we proposed that the World Economic Forum's initiative on stakeholder capitalism could provide a set of metrics that has attracted widespread support from the auditing and actuarial professions.[63] Such metrics would support the entity model of ESG by providing a clearer framework for directors' decision-making as well as the due diligence standard, discussed above.

However, the use of such metrics fits better with the conventional model of reporting aligned to shareholder primacy than to an adjusted model of corporate purpose that focuses more on stakeholder protection and integration of stakeholders into board decision-making. As discussed above, adjustments to purpose are likely to involve changes to the role of capital, profit, and governance by comparison with the conventional model. We address the last aspect by reference to our proposal for a stakeholder committee and a due diligence obligation linked to board decision-making. The first two levers of purpose (capital and profit) are more problematic and less amenable to explicit adjustment in a model that envisages evolutionary change. As mentioned earlier, they are the subject of proposed changes to the measurement of capital and profit proposed by Mayer and Barby et al., focusing on the need to value and report on human, social and natural capital, possibly by building on existing non-financial reporting frameworks, although some work is also required to move beyond traditional cost-based accounting and valuation. According to prevailing standards, accounts currently do not provide all the information that is relevant to promoting responsible, purposeful business practices because, against that benchmark, profit is overstated where companies cause detriments to other parties and not fully recognised where they confer benefits.

31.4.5 Reporting

Through the process of reporting companies disclose the various risks that they face, especially with regard to the social dimension and the environment. In this context, the role of materiality as a threshold for disclosure plays a key role. Single materiality requires disclosure by reference to the impact on the company (and by implication shareholders), whereas double materiality requires disclosure by reference to the external impact of corporate activities, and in particular environmental and social impacts.

The EU's non-financial reporting directive (NFRD)[64] led the way in the global development of reporting based on the double-materiality perspective. Its provisions were encompassed in our earlier study on the impact of non-financial reporting in the UK.[65] We concluded that disclosure plays a crucial role, but on its own it is not enough to improve stakeholder participation. For stakeholders to be fully engaged they need adequate information, in the form of disclosure, but they also need mechanisms to facilitate participation from their side. The prioritisation of shareholder value and the lack of focus on stakeholder interests could be the main reason for the lack of engagement with stakeholders. Stakeholder integration into decision-making is therefore needed to ensure that reporting drives action.

On 5 January 2023 the Corporate Sustainability Reporting Directive (CSRD) entered into force and will extend the scope of non-financial reporting in the EU to a wider range of companies.[66] It will, inter alia, require external assurance (audit) of sustainability information and end the option of 'comply or explain' that previously applied under the NFRD.

At the global level, the Taskforce on Climate-related Financial Disclosures (TCFD) provides a framework for corporate disclosure on environmental issues.[67] UK premium-listed companies are required since 1 January 2021 to state, in their annual report, whether their disclosures are consistent with TCFD recommendations, or to explain why they are not. The UK Government is also making TCFD-aligned disclosure mandatory for over 1,300 of the largest UK-registered companies and financial institutions, making it the first G20 country to do so.[68]

The TCFD focus is mainly on single materiality as are the evolving standards that are being developed by the International Sustainability Standards Board, which emerged as a key global standard-setter following commitments made at COP26. There are other global frameworks (notably the Global Reporting Initiative[69] and emerging EU standards) that follow a double materiality approach, but single materiality now seems to be in the ascendancy outside the EU, and so it is still uncertain how this form of regulatory competition will be resolved. In the context of corporate purpose, single materiality in reporting tends to support shareholder primacy (at least in the absence of government intervention on issues such as carbon pricing), whereas double materiality is more aligned with stakeholder integration. In that sense, voluntary adoption of a double materiality reporting approach could provide support to evolutionary change in purpose by expanding the scope of accountability associated with board decision-making.

31.5 Conclusion

Corporate purpose as a legal principle serves different purposes in different contexts. We noted that purpose has been used to: differentiate commercial entities from those with social purposes; protect shareholders and creditors from transactions in which directors attempted to diversify the scope of the business defined by the constitution; specify the objectives for which directors could validly exercise powers delegated to them; and specify whose interests are relevant for board decision-making. While the last aspect dominates current discourse, we note the potential for the first to take a more prominent role in the current debate in the light of expansion in the options for entity selection. Meanwhile the second example provides some relevant experience as to how corporate purpose can be used to protect specific stakeholders and control the activities of companies.

In recent times, attention has focused in particular on the role of purpose in the context of sustainability and stakeholder protection. But even in that context its meaning and potential for changing corporate behaviour remains contested. This is reflected in proposals to change corporate purpose extending across a wide spectrum from incremental to transformative: some proposals involve legislative change while others claim that meaningful change is possible within the current legal framework or soft law codes; some focus on the board of directors as the agents of change while others view the board as trapped in a paradigm of shareholder primacy that necessitates mandatory regulatory intervention; and some view shareholders as potential agents of change through stewardship and the sustainability preferences of underlying investors.

What does seem clear is that the role of corporate purpose in driving change is contingent on parallel developments related to sustainability in adjacent domains. In the EU and UK context, developments linked to sustainability such as non-financial reporting, financial market and product disclosures, as well as due diligence are particularly relevant. To the extent that they incentivise or require corporate strategy and behaviour to become more sustainable, they potentially limit the burden that falls onto corporate purpose as a driver of change. In that sense, the movement towards sustainability can be seen as shared across the corporate and financial or investment channels that we highlighted in previous research. It also encompasses the global evolution of stewardship by institutional investors, which we have previously characterised as an accelerator of the rise of the salience of ESG as a driver of corporate governance. Moreover, as per the evolution of EU policy, it seems likely that mandatory and market-led actions need to move in tandem, at least if global rather than just national or regional change is envisaged.

The emergence of corporate purpose draws attention to the need to adjust board decision-making and corporate strategy to adapt to sustainability at the operational level, rather than relying on the more indirect effects that may be transmitted through the financial investor channel. In that sense it aligns very clearly with our earlier entity model of ESG, which drew attention to the limitations of the financial channel and the benefits of focusing more clearly on entities. In this chapter, we elaborated on the role of corporate purpose in that model, focusing on the case for adjusting the levers of purpose rather than purpose itself and arguing that incremental change is more realistic than transformational change. Against that background, it will be interesting to observe how corporate purpose evolves in the years ahead and to what extent law can be a driver of change.

Notes

1 This chapter reflects the law as at 31st December 2022.

2 See generally Elizabeth Pollman and Robert B Thompson, *Research Handbook on Corporate Purpose and Personhood* (Edward Elgar, Cheltenham, 2021).

3 Iain MacNeil and Irene-marie Esser, 'From a Financial to an Entity Model of ESG' (2022) 23(1) EBOLR 9.

4 Ibid.

5 See for example Colin Mayer, *Prosperity: Better Business Makes the Greater Good* (OUP, Oxford 2018).

6 This approach can also be linked with the idea of the need for a supportive 'ecology' in corporate law proposed by Kershaw and Schuster in their discussion of the UK Corporate Governance Code provision on corporate purpose: D Kershaw and E Schuster, 'The Purposive Transformation of Corporate Law' (2021) 69 *American Journal of Comparative Law* 478, 489.

7 See Colin Mayer, *Prosperity: Better Business Makes the Greater Good* (OUP, Oxford 2018) 225: 'Beyond requiring companies to prioritize their purposes, corporate law should enable companies to adopt diverse forms of ownership and governance that empower different parties to the firm and give them the means to enforce their rights'.

8 For a concise introduction to 'not for profit' companies in the UK (noting that it is not a term of art in UK company law) see Paul Davies, Sarah Worthington and Chris Hare, *Gower, Principles of Modern Company Law* (11th ed, Sweet & Maxwell, London 2021) 8–12. See also *Social Enterprise Law Tracker* <https://socentlawtracker.org/#/map> accessed 15 February 2023 for data on US 'not for profit' entities and their distribution across states.

9 See generally the Charities Act 2011 (c.25 England & Wales) and the Charities and Trustee Investment (Scotland) Act 2005 (Acts of the Scottish Parliament 10). The activities of charities must be exclusively

charitable: mixed purposes are not possible albeit that trading activities may be subsumed in the activities of the charity so long as they support charitable purpose and public benefit objectives.

10 See further Orrick et al., *Balancing Purpose and Profit* (2014) <https://www.trust.org/contentAsset/raw-data/2435372d-3e6d-42d7-a726-a1fe21cba400/file> accessed 15 February 23 referencing the Delaware legislation as well as the Model Benefit Corporation Legislation advocated by the non-profit organisation B Lab. Unlike the majority of states, Delaware lacks a constituency statute authorising directors to consider stakeholder interests.

11 See further Companies Act 2006 s.172.

12 Certification is provided by B Lab, a non-profit network created in 2006 with the mission to inspire and enable people to use business as a force for good. B Lab UK is a charity launched in 2015.

13 See further *B Lab United Kingdom* <https://bcorporation.uk/b-corp-certification/how-to-certify-as-a-b-corp/legal-requirement/> accessed 15 February 23.

14 See A Tucker and D Brakman Reiser, 'Equity Finance for Social Enterprises', chapter 7 in Iris Chiu and Iain MacNeil (eds), *Research Handbook on Global Capital Markets Law* (Edward Elgar, London 2023).

15 In both instances, it is a statutory requirement that the entity is established 'with a view to profit' (Partnership Act 1890 s.1(1) and Limited Liability Partnerships Act 2000 s.2(1)). But the combination of profit with other purposes is not ruled out.

16 Iain MacNeil and Irene-marie Esser, 'From a Financial to an Entity Model of ESG' (2022) 23(1) *EBOLR* 9.

17 Paul Davies, 'Shareholder Voice and Corporate Purpose: The Purposeless of Mandatory Corporate Purpose Statements' (2022) European Corporate Governance Institute –Law Working Paper No. 666/2022, available at <https://ssrn.com/abstract=4285770> accessed 15 February 2023.

18 S.31 of the Companies Act 2006.

19 S.39 of the Companies Act 2006.

20 Paul Davies, Sarah Worthington and Eva Micheler, *Gower and Davies' Principles of Modern Company Law* (8th ed., Sweet & Maxwell, London 2008) para 7–2.

21 Companies Act 2006 s.171.

22 See *McGaughey v USSL* [2022] EWHC 1233 (Ch) and on appeal [2023] EWCA Civ 873. In both instances, the claim was dismissed, albeit mainly because it did not meet the standards required for a derivative action. The Court of Appeal left open the possibility of a direct action based on breach of trust.

23 See e.g., the cases discussed in Paul Davies, Sarah Worthington and Chris Hare, *Gower, Principles of Modern Company Law* (11th ed, Sweet & Maxwell, London 2021), paras 10-018 to 10-021.

24 It remains the case, however, that observance of s.172 does not excuse a breach of s.171: see *Fairford Water Ski Club Ltd. v Cohoon* [2020] EWHC 290 (Comm) at [51]. On the contrary, observance of s.171 logically precedes the analysis of compliance with s.172.

25 For a more definitive view of the limited transformative potential of corporate purpose see Brian Cheffins, 'The Past, Present and Future of Corporate Purpose' (2023) Delaware Journal of Corporate Law, University of Cambridge Faculty of Law Research Paper No. 15/2023, European Corporate Governance Institute – Law Working Paper No. 713/2023, available at SSRN: <https://ssrn.com/abstract=4420800> or http://doi.org/10.2139/ssrn.4420800 accessed 15 February 23.

26 See L Fink, 'Profit & Purpose' (2019) <https://corpgov.law.harvard.edu/2019/01/23/purpose-profit/> accessed 15 February 2023.

27 See *Principles for Purposeful Business* <https://www.thebritishacademy.ac.uk/publications/future-of-the-corporation-principles-for-purposeful-business> accessed 15 February 2023.

28 See more generally on ESV, Andrew Keay, *The Enlightened Shareholder Value Principle and Corporate Governance* (Routledge, Oxfordshire 2013).

29 BTI 2014 LLC v Sequana SA & Ors, [2022] UKSC 25

30 Paras 11, 17 and 21.

31 Para 67.

32 Para 65.

33 See *G20/OECD Principles of Corporate Governance* <https://www.oecd-ilibrary.org/docserver/9789264236882-en.pdf?expires=1674814975&id=id&accname=guest&checksum=ED22BD719C23BC04B8AB7FA82281DCFF> accessed 15 February 2023 at 1: 'The purpose of corporate governance is to help build an environment of trust, transparency and accountability necessary for fostering long-term investment, financial stability and business integrity, thereby supporting stronger growth and more inclusive societies'.

34 See *ICGN Global Governance Principles* <https://www.fsa.go.jp/en/refer/councils/corporategovernance/reference/icgn.pdf> accessed 15 February 2023.
35 Alex Edmans, *Grow the Pie – How Great Companies Deliver Both Purpose and Profit* (Cambridge University Press, 2021) 3.
36 See this Tom Gosling, Interview with Alex Edmans, Professor of Finance at London Business School, ECGI Conversations (ECGI, 17 January 2023) on the trade-off between profits and purpose: <https://www.youtube.com/watch?v=hc3HntlnHT8&ab_channel=ECGI> accessed 15 February 2023.
37 Paul Davies, 'Shareholder Voice and Corporate Purpose: The Purposeless of Mandatory Corporate Purpose Statements' (2022) European Corporate Governance Institute – Law Working Paper No. 666/2022 <https://ssrn.com/abstract=4285770> accessed 15 February 2023.
38 See *In Focus: Corporate Purpose and ESG* <https://www.frc.org.uk/getattachment/Directors/The-Culture-Project/FRC-In-Focus-Corporate-Purpose_April-2022.pdf> accessed 15 February 2023. This Report followed on from a previous report on culture, which argued that it is important to consider culture with ESG and corporate purpose, as they are interlinked.
39 See *In Focus: Corporate Purpose and ESG* <https://www.frc.org.uk/getattachment/Directors/The-Culture-Project/FRC-In-Focus-Corporate-Purpose_April-2022.pdf> 4, accessed 15 February 2023.
40 C Barby et al., 'Measuring Purpose, An Integrated Framework', available at: <https://ssrn.com/abstract=3778161> accessed 15 February 2023.
41 C. Barby et al., 'Measuring Purpose, An Integrated Framework', available at: <https://ssrn.com/abstract=3778161> 6, accessed 15 February 2023, '[…] cost-based accounting and valuation answer different questions. Cost-based accounts establish the resources that companies have to expend in correcting detriments and/or delivering positive outcomes outside as well as within their legal boundaries, whereas valuations capture the net benefits or detriments of the firm's activities and impacts on its stakeholders as well as its shareholders'.
42 Paul Davies, 'Shareholder Voice and Corporate Purpose: The Purposeless of Mandatory Corporate Purpose Statements' (2022) European Corporate Governance Institute – Law Working Paper No. 666/2022, 35, available at <https://ssrn.com/abstract=4285770> accessed 15 February 2023.
43 Colin Mayer, *Prosperity: Better Business Makes the Greater Good* (Oxford University Press, 2018) 201.
44 Financial Reporting Council, *Guidance on Board Effectiveness* (2018) para 12, 3.
45 Financial Reporting Council, *Guidance on Board Effectiveness* (2018) para 14, 3.
46 D Kershaw and E Schuster, 'The Purposive Transformation of Corporate Law' (2021) 69 *American Journal of Comparative Law* 478, 489. See <https://www.gov.uk/government/consultations/corporate-governance-reform> accessed 15 February 2023.
47 Iain MacNeil and Irene-marie Esser, 'From a Financial to an Entity Model of ESG' (2022) 23(1) *EBOLR* 9, section 4.
48 Iain MacNeil and Irene-marie Esser, 'The Emergence of 'Comply or Explain' as a Global Model for Corporate Governance Codes' (2022) 33(1) *European Business Law Review* 1–56.
49 For a rare example of explicit reference to need for investors and companies to internalise externalities, see *The Natural Capital Declaration* <https://www.unepfi.org/fileadmin/documents/ncd_booklet.pdf> accessed 15 February 2023. The linkage of such action to fiduciary duty was not addressed and the GFANZ/NZAMI frameworks that emerged from COP26 contain caveats with respect to government policy and fiduciary duty. See further T Gosling and I MacNeil, 'Can Investors Save the Planet? – NZAMI and Fiduciary Duty' (2023) 18(2) *Capital Markets Law Journal* 172–193. Available at SSRN: <https://ssrn.com/abstract=4277960> accessed 15 February 2023.
50 'Producing profitable solutions for problems of people and planet, not profiting from producing problems for either'. See *Principles for Purposeful Business* <https://www.thebritishacademy.ac.uk/publications/future-of-the-corporation-principles-for-purposeful-business/> accessed 15 February 2023.
51 See *ISSB describes the concept of sustainability and its articulation with financial value creation* <https://www.ifrs.org/news-and-events/news/2022/12/issb-describes-the-concept-of-sustainability/> accessed 15 February 2023 (Report of 14 Dec 2022).
52 See generally Paul Davies, Sarah Worthington and Chris Hare, *Gower, Principles of Modern Company Law* (11th ed, Sweet & Maxwell, London 2021) paras 9–002 to 9–005.
53 See the Companies (Model Articles) Regulations 2008, SI 2008/3229.
54 D Kershaw and E Schuster, 'The Purposive Transformation of Corporate Law' (2021) 69 *American Journal of Comparative Law* 478, 489.

55 Paul Davies, 'Shareholder Voice and Corporate Purpose: The Purposeless of Mandatory Corporate Purpose Statements' European Corporate Governance Institute – Law Working Paper No. 666/2022, 35, available at <https://ssrn.com/abstract=4285770> accessed 15 February 2023.

56 Paul Davies, 'Shareholder Voice and Corporate Purpose: The Purposeless of Mandatory Corporate Purpose Statements' European Corporate Governance Institute – Law Working Paper No. 666/2022, 35, available at <https://ssrn.com/abstract=4285770> accessed 15 February 2023.

57 Currently, due diligence rules are in place only in a few EU countries, such as France, while in Germany, companies with more than 3,000 employees will need to carry out mandatory due diligence from January 2023 under the Supply Chain Due Diligence Act (*Lieferkettensorgfaltspflichtengesetz* – 'LkSG').

58 Katarzyna Chalaczkiewicz-Ladna, Irene-marie Esser and Iain MacNeil, 'Workforce Engagement and the UK Corporate Governance Code' in A Martínez-Echevarría and García de Dueñas (eds) *Corporate Governance, Sustainability and Reputation = Gobierno Corporativo, Sostenibilidad Y Reputación,* Thomson Reuters Aranzadi. 2022; and Part 2, Katarzyna Chalaczkiewicz-Ladna, Irene-marie Esser and Iain MacNeil, 'The Workforce Engagement Mechanisms in the UK: A Way Towards More Sustainable Companies? (Part 2)' (2022) <https://papers.ssrn.com/sol3/papers.cfm?abstract_id=4296944> accessed 15 February 2023.

59 For a detailed discussion of this committee, see, Irene-marie Esser and Piet Delport, 'The Protection of Stakeholders: The South African Social and Ethics Committee and the United Kingdom's Enlightened Shareholder Value Approach: Part 1' (2017) 50(1) De Jure 97–110; and for Part 2: 50(2) 221–240; Gavin Hinks, 'Companies 'need new mechanism' to integrate stakeholder interests' (2020) *Board Agenda* <https://boardagenda.com/2020/09/04/companies-need-new-mechanism-to-integrate-stakeholder-interests/> accessed 15 February 2023.

60 Gavin Hinks, 'Top Companies Dismiss Concept of Workers on Boards' (2023) *Board Agenda* <https://boardagenda.com/2023/01/06/top-companies-dismiss-concept-of-workers-on-boards/> accessed 15 February 2023 and Gavin Hinks, 'FTSE 100 rejects idea of employee director to boost worker engagement' 2021 *Board Agenda* <https://boardagenda.com/2021/04/29/ftse-100-rejects-idea-of-employee-director-to-boost-worker-engagement/> accessed 15 February 2023.

61 Commission, Proposal for a Directive of the European Parliament and of the Council on Corporate Sustainability Due Diligence and amending Directive (EU) 2019/1937, COM/2022/71 final.

62 See Ashley Smith-Roberts, 'Everything You Need to Know about the EU CSDD & EU CSRD' (*Levin Sources*, 3 May 2022) <https://www.levinsources.com/knowledge-centre/insights/eu-csdd-eu-csrd-mining> accessed 15 February 2023.

63 In 2020 the *World Economic Forum white paper* <https://www.weforum.org/reports/measuring-stakeholder-capitalism-towards-common-metrics-and-consistent-reporting-of-sustainable-value-creation> accessed 15 February 2023 set out 21 core metrics and 34 expanded metrics, covering environmental, social and governance (ESG) issues ranging from emissions to social factors such as pay and gender ratios and governance targets.

64 Council Directive 2014/95/EU of 22 October 2014 amending Directive 2013/34/EU as regards disclosure of non-financial and diversity information by certain large undertakings and groups [2014] OJ L 330/1.

65 Irene-marie Esser, Katarzyna Chalaczkiewicz-Ladna and Iain MacNeil, 'Engaging Stakeholders in Corporate Decision-making through Strategic Reporting: An Empirical Study of FTSE 100 Companies' (2018) 29(5) European Business Law Review, Irene-marie Esser, Katarzyna Chalaczkiewicz-Ladna and Iain MacNeil, 'Engaging Stakeholders in Corporate Decision-Making through Strategic Reporting: An Empirical Study of FTSE 100 Companies (Part 2)' (2020) 31(2) *European Business Law Review* 209–242.

66 See Council Directive (EU) 2022/2464 of 14 December 2022 amending Regulation (EU) No 537/2014, Directive 2004/109/EC, Directive 2006/43/EC and Directive 2013/34/EU [2022] OJ L322/15, as regards corporate sustainability reporting. See also <https://finance.ec.europa.eu/capital-markets-union-and-financial-markets/company-reporting-and-auditing/company-reporting/corporate-sustainability-reporting_en> accessed 15 February 2023.

67 *Task Force on Climate-Related Financial Disclosures* <https://www.fsb-tcfd.org/> accessed 15 February 2023. TCFD was established in 2015 by the Financial Stability Board (FSB), an international body formed in response to the global financial crisis.

68 The Companies (Strategic Report) (Climate-related Financial Disclosure) Regulations 2022 (SI 2022/31) came into force on April 6, 2022. The Regulations amend the Companies Act 2006 and largely implement the TCFD recommendations.

69 *Global Reporting Initiative* <https://www.globalreporting.org/> accessed 15 February 2023.

References

Barby C et al., 'Measuring Purpose, an Integrated Framework', available at: <https://ssrn.com/abstract=3778161> accessed 15 February 2023

Chalaczkiewicz-Ladna K, Esser I and MacNeil I, 'Workforce Engagement and the UK Corporate Governance Code', in A. Martínez-Echevarría y García de Dueñas (eds) *Corporate Governance, Sustainability and Reputation = Gobierno Corporativo, Sostenibilidad Y Reputación*, Thomson Reuters Aranzadi, 2022, 159–184

Chalaczkiewicz-Ladna K, Esser I and MacNeil I, 'The Workforce Engagement Mechanisms in the UK: A Way Towards More Sustainable Companies? (Part 2)' (2022) available at <https://papers.ssrn.com/sol3/papers.cfm?abstract_id=4296944> accessed 15 February 2023

Cheffins B, 'The Past, Present and Future of Corporate Purpose (April 17, 2023). Delaware Journal of Corporate Law, 2023, University of Cambridge Faculty of Law Research Paper No. 15/2023', European Corporate Governance Institute – Law Working Paper No. 713/2023, available at SSRN: <https://ssrn.com/abstract=4420800> or http://dx.doi.org/10.2139/ssrn.4420800

Davies P, 'Shareholder Voice and Corporate Purpose: The Purposeless of Mandatory Corporate Purpose Statements', *European Corporate Governance Institute – Law Working Paper No. 666/2022*, available at <https://ssrn.com/abstract=4285770> accessed 15 February 2023

Davies P, Worthington S and Hare C, *Gower, Principles of Modern Company Law*, 11th ed., Sweet & Maxwell, London 2021, 8–12

Davies P, Worthington S and Micheler E, *Gower and Davies' Principles of Modern Company Law*, 8th ed., Sweet & Maxwell, London 2008, para 7–2

Edmans A, *Grow the Pie – How Great Companies Deliver Both Purpose and Profit*, Cambridge University Press, 2021, 3

Esser I and Delport P, 'The Protection of Stakeholders: The South African Social and Ethics Committee and the United Kingdom's Enlightened Shareholder Value Approach: Part 1', *De Jure* (2017), 50(1), 97–110; and for Part 2: 50(2), 221–240

Esser I, MacNeil I and Chalaczkiewicz-Ladna K, 'Engaging Stakeholders in Corporate Decision-Making through Strategic Reporting: An Empirical Study of FTSE 100 Companies' (2018) 29(5) *European Business Law Review* 729–772

Fink L, 'Profit & Purpose' (2019) at accessed 15 February 2023 <https://corpgov.law.harvard.edu/2019/01/23/purpose-profit/>

Gosling T and MacNeil I, 'Can Investors Save the Planet? – NZAMI and Fiduciary Duty' (2023) 18(2) *Capital Markets Law Journal* 172–193

Hinks G, 'Companies 'Need New Mechanism' to Integrate Stakeholder Interests', *Board Agenda*, 2020 available at <https://boardagenda.com/2020/09/04/companies-need-new-mechanism-to-integrate-stakeholder-interests/> accessed 15 February 2023

Hinks G, 'FTSE 100 Rejects Idea of Employee Director to Boost Worker Engagement', *Board Agenda*, 2021, available at <https://boardagenda.com/2021/04/29/ftse-100-rejects-idea-of-employee-director-to-boost-worker-engagement/> accessed 15 February 2023

Hinks G, 'Top Companies Dismiss Concept of Workers on Boards', *Board Agenda*, 2023 available at <https://boardagenda.com/2023/01/06/top-companies-dismiss-concept-of-workers-on-boards/> accessed 15 February 2023

Keay A, *The Enlightened Shareholder Value Principle and Corporate Governance*, Routledge, Oxfordshire, 2013

Kershaw D and Schuster E, 'The Purposive Transformation of Corporate Law' (2021) 69 *American Journal of Comparative Law* 478, 489

MacNeil I and Esser I, 'From a Financial to an Entity Model of ESG' (2022) 23(1) *European Business Organization Law Review* 9

MacNeil I and Esser I, 'The Emergence of 'Comply or Explain' as a Global Model for Corporate Governance Codes' (2022) 33(1) *European Business Law Review* 1–56

Mayer C, *Prosperity: Better Business Makes the Greater Good*, OUP, Oxford, 2018

Orrick et al., *Balancing Purpose and Profit* (2014), available at <https://www.trust.org/contentAsset/raw-data/2435372d-3e6d-42d7-a726-a1fe21cba400/file> accessed 15 February 2023

Pollman E and Thompson R, *Research Handbook on Corporate Purpose and Personhood*, Edward Elgar, Cheltenham, 2021

Tucker A and Brakman Reiser D, 'Equity Finance for Social Enterprises', chapter 7 in I. Chiu and I. MacNeil (eds) *Research Handbook on Global Capital Markets Law*, Edward Elgar, London, 2023, 107–123

32

ONLINE PLATFORMS AND SUSTAINABILITY

How to Engage Digital Intermediaries in Sustainability Goals?

Carlos Gómez Ligüerre and Lela Mélon[1]

32.1 Scope and Purpose

In the age of rapid digital transformation, online platforms have emerged as transformative forces, reshaping the way we interact, conduct business, and address societal challenges. As the global community confronts the urgent need to tackle climate change and promote sustainability, it becomes paramount to examine the role and impact of online platforms in fostering a sustainable future. This academic exploration aims to shed light on the link between platform economy, sustainability, and climate change, with a specific focus on analysing this relationship in the context of the Digital Services Act and Digital Markets Act.

The Sustainable Development Goal 12 promoted by United Nations is about ensuring sustainable consumption and production patterns. Target 12.6 aims at 'Encourage companies, especially large and transnational companies, to adopt sustainable practices and to integrate sustainability information into their reporting cycle', and target 12.7 seeks to 'Promote public procurement practices that are sustainable, in accordance with national policies and priorities'. In the current state of development of what has been called the digital economy, it is unthinkable that these objectives can be achieved without the active participation of online platforms.

Climate change, driven by unsustainable resource consumption and greenhouse gas emissions, poses significant threats to our planet's ecological balance. Understanding the potential influence of online platforms in addressing climate change and sustainability is crucial, given their pervasive impact on production and consumption patterns, energy usage, waste generation, and transportation. Therefore, it is essential to critically examine how the Digital Services Act and Digital Markets Act, proposed legislative frameworks in the European Union, intersect with sustainability and climate change considerations.

The Digital Services Act and Digital Markets Act are comprehensive regulatory initiatives designed to govern digital platforms' operations, market power, and responsibilities. As we embark on this exploration, we will analyse how these acts (could) address sustainability and climate change concerns within the realm of online platforms. Through this analysis, we aim to identify the opportunities, challenges, and potential strategies embedded within the Digital Services Act and Digital Markets Act that can enhance the role of online platforms in combating climate change and promoting sustainability. This exploratory exercise should serve for further

DOI: 10.4324/9781032662046-38

academic discussion and calls for incorporating sustainability-related concerns in the regulation of the platform economy.[2]

While recognizing the potential benefits, we also acknowledge that tensions and complexities arise when considering the intersection of online platforms, sustainability, and climate change. Ethical considerations, privacy concerns, digital divide issues, and the impact of platform economies on equity and governance require careful examination. Our analysis will briefly address these challenges, aiming to provide insights into mitigating risks and promoting responsible practices within the digital platform landscape.

32.2 So How Could Online Platforms Ideally Contribute to Sustainability Goals? Best Practices

Digital platforms have the potential to contribute significantly to sustainability efforts in various ways. Platforms can serve as hubs for collaboration, innovation, and knowledge sharing among diverse stakeholders, including businesses, researchers, and policymakers.

The landscape of online platforms is as vast as the services and products they offer. Their diversity undoubtedly makes it difficult to regulate them. In all cases, however, there is one common element: online platforms facilitate contracting by reducing transaction costs. They are, therefore, agents that promote efficiency (in terms of time and cost reduction) in the contractual process. This same advantage can also promote more sustainable procurement of goods and services. Online platforms can contribute to the reduction of transaction costs that lead to more sustainable consumption and production.[3]

Recently, the European Commission has published the Proposal for a Directive establishing common rules to promote the repair of goods (COM/2023/155/final). Although this is also only a proposal, it seems timely to give notice of the initiative, as it adds a new function to online platforms that may have direct sustainability implications.

The Proposal was announced in the frame of the Action Plan for the Circular Economy that the European Commission (COM/2020/98/final) published in March 2020. The Plan is one of the actions promoted by the European Commission to meet the internal market objectives set out in the European Green Deal. In this context, the Proposal contributes to the environmental objectives of the European Union. It is, as recognized in the report accompanying the Proposal, an additional instrument to the rest of the European regulatory projects to achieve the necessary climate neutrality of the Union. The text has, however, an additional significance to that of other initiatives promoted by the European Commission in the field of sustainability or environmental protection. Its provisions highlight the relevance of contract law for the achievement of this objective and the importance of contractual mechanisms to achieve a sustainable economy.

The Proposal imposes on the Member States the obligation to 'ensure' that 'at the consumer's request', the producer (or the person corresponding to him) repairs the product. The quoted expressions are taken from Article 5 of the Proposal. This obligation of the Member States is specified in the Proposal in the provision of a centralized information system on repairers, through an online platform on 'repairs and goods subject to reconditioning', in Article 7 of the Proposal. Such information should facilitate the conclusion of a 'contract for the provision of repair services' with such marked features in Article 7 of the Proposal that it may end up being a specific category of service contract involving online platforms devoted to promoting repair.

The Proposal takes advantage of the possibilities of contracting through online platforms to promote the repair of goods. It is an example of the efficiency of digital platforms and their potential to streamline procedures necessary for sustainability.

In addition, by fostering a collaborative ecosystem, platforms can accelerate the development and adoption of sustainable technologies, business models, and policies. Some key aspects in which online platforms can make a positive impact could be the following:

Resource Efficiency: online platforms can enable resource optimization by facilitating the sharing economy and collaborative consumption. Through peer-to-peer sharing, platforms can promote the efficient use of underutilized resources, such as cars, accommodation, and tools. This reduces the need for excessive production and consumption, leading to a decrease in resource extraction, waste generation, and environmental degradation.

Sustainable Supply Chains: online platforms can furthermore play a vital role in promoting sustainable supply chains. By providing transparency and traceability throughout the supply chain, platforms enable consumers to make informed choices about the sustainability credentials of products and services. This encourages businesses to adopt environmentally friendly practices, including responsible sourcing, fair labour conditions, and reduced carbon emissions.

Renewable Energy Integration: many online platforms operate massive data centres and infrastructure that require substantial energy consumption. To mitigate their environmental footprint, platforms can transition to renewable energy sources, such as solar or wind power, to power their operations. This shift can significantly reduce greenhouse gas emissions associated with electricity consumption and contribute to the development of a low-carbon economy.

Environmental Awareness and Education: digital platforms have the potential to raise awareness about environmental issues and educate users on sustainable practices. Through targeted advertising, content sharing, and community engagement, platforms can promote sustainable behaviours, provide information on eco-friendly products and services, and foster a culture of environmental responsibility.

Access to Sustainable Products and Services: online platforms can provide convenient access to a wide range of products and services, including sustainable alternatives. They can highlight and promote environmentally friendly options, such as organic products, eco-certified goods, and energy-efficient appliances. This helps consumers make sustainable choices easily and encourages businesses to offer more sustainable options.

It is important to note that while online platforms have the potential to contribute to sustainability, there are also challenges and potential negative impacts associated with their operations. These include issues of data privacy, the digital divide, e-waste management, and the overall carbon footprint of digital technologies. To maximize the positive impact, it is crucial for platforms to adopt responsible practices, collaborate with stakeholders, and embrace sustainability as a core principle in their operations and strategies.

There already exist practices in the European and global markets that are harvesting their positive role in the quest for sustainable practices and fostering circular and sustainable solutions. By way of example, while causing some social sustainability challenges, Airbnb as a leading platform for short-term accommodation rentals has taken steps to promote sustainability within the tourism industry. It has implemented the 'Green Airbnb' initiative, encouraging hosts to adopt eco-friendly practices such as using renewable energy, providing recycling facilities, and promoting sustainable transportation options to guests. It also supports community-driven sustainability projects and provides resources for hosts to reduce their environmental impact.[4] Another example is the OLIO platform, a food-sharing platform that connects individuals and businesses with surplus food to those in need. The platform aims to reduce food waste by encouraging sharing and redistributing excess food items that would otherwise be discarded. OLIO's community-driven approach promotes sustainability by fostering local connections and minimizing food waste's environmental impact.[5] Very similar to OLIO is the Too Good To Go platform that connects users with restaurants, cafes, and grocery stores to rescue surplus food that would otherwise go to waste. It helps reduce food waste and promotes sustainable consumption by offering discounted meals and products that are close to their expiration date. BlaBlaCar is another example of a sustainability-related online platform as a ride-sharing platform that enables individuals to share car journeys, reducing the number of vehicles on the road and promoting more efficient transportation. By maximizing the use of existing vehicles, BlaBlaCar helps reduce carbon emissions and congestion. The platform also encourages travellers to offset their carbon footprint by making voluntary contributions towards environmental projects.[6]

These examples highlight the diverse ways in which online platforms can contribute to sustainability, whether through promoting responsible consumption, reducing waste, supporting renewable energy, or facilitating more efficient resource utilization. While these platforms are to some large extent exceptions, as in many cases their mission and vision are intrinsically informed by sustainability concerns, the question remains for the rest of digital platforms if, in the absence of their sustainable mission and vision, the existing EU instruments could insert certain responsibility for sustainable transition, even if only indirectly through the incorporation of pertinent EU sustainability-related legislation?

To date, the central issue in relation to online platforms has been their contractual and non-contractual liability, namely, that of the liability of platforms in relation to consumers and users of goods and services, and the victims of such goods.[7] Also, such liability relates to the obligations of the platform with respect to the persons offering the services that are contracted through the platform.[8]

It is essential for online platforms to recognize that their influence extends beyond the digital realm and into the physical world, with profound implications for the environment and society at large. By embracing sustainable practices, they can contribute to the achievement of the United Nations Sustainable Development Goals and support the objectives outlined in initiatives such as the European Green Deal. Through collaboration with stakeholders across sectors, including governments, businesses, and civil society organizations, online platforms can drive collective action towards a more sustainable future. Recognition of this influence by online platforms should lead them to adopt 'good practices' or 'codes of conduct'. Despite their importance, the legal qualification of these practices is, however, difficult, especially when they affect third parties outside the contracting of the good or service.[9]

In any case, the relevance of sustainability objectives and the preponderance of online platforms in the provision of goods and services make a comprehensive legal approach to the phenomenon advisable.[10] A specific regulation of online platforms that imposes sustainability duties on them seems necessary. The urgency and relevance of the issue do not allow us to entrust it exclusively to self-regulation or to the indirect requirement of duties.

32.3 Reality: From 'Safe Harbours' to a General Liability Clause

Online platforms have become a dominant force in the procurement of goods and services, significantly impacting sustainability considerations. They permeate all economic sectors and engage various stakeholders. Online platforms cater to both businesses (P2B) and consumers (P2C) and even facilitate direct contracting between consumers (C2C). Recognizing their significance, references to online platforms are pervasive in legislative initiatives pertaining to the Digital Single Market, encompassing areas such as digital goods and services, copyright protection, collaborative economy, and sustainability. As sustainability concerns continue to gain prominence, it becomes imperative to examine the role of online platforms in advancing sustainable practices across these sectors.[11]

The concept of intermediation in commercial transactions has a long-standing history, with implications for sustainability considerations. Over time, supply chains have relied on intermediaries to bridge the gap between manufacturers and end users. The magnitude of intermediaries involved varies depending on factors such as product characteristics, industry dynamics, and transportation requirements. Understanding and addressing the sustainability challenges arising from these intermediary networks is essential for fostering responsible and eco-conscious practices throughout the supply chain.

Traditionally, European law has paid limited attention to online platforms, perceiving them as akin to conventional intermediaries. This view, while acknowledging similarities, has resulted in a restrained focus on developing targeted regulatory frameworks for online platforms. However, the emergence of sustainability challenges necessitates reevaluating this approach. The unique characteristics of online platforms, combined with their widespread influence, require a more robust regulatory framework that addresses sustainability concerns and fosters responsible platform practices.

The most remarkable manifestation of this vision has been the general exemption from liability for information society services, including online platforms, insofar as their activity is limited to the mere transmission ('mere conduit') of information.[12] However, in light of sustainability concerns, this exemption raises challenges. The conventional perspective, rooted in the notion of 'safe harbours', has traditionally followed the principle of 'non-liability for intermediaries'.[13] While this approach has facilitated the growth of online platforms, it has also inadvertently overlooked many, but especially the environmental and social implications of their operations. Addressing sustainability concerns within the online platform ecosystem requires reevaluating the liability framework and encouraging platforms to take greater responsibility for the environmental and social impact of the information they transmit. Balancing the benefits of innovation and the need for sustainable practices is essential in shaping a platform economy that aligns with the broader goals of sustainability.

However, online platforms introduce distinct characteristics that necessitate a shift in the conventional paradigm. These unique attributes stem from the successful integration of emerging technologies with intermediary services. These distinguishing factors elucidate why online platforms are now recognized as significant players in the realm of e-commerce.[14] Consequently, online platforms are gradually departing from a regime that grants them exemption from potential liability concerning their role as intermediaries or their assumption of functions akin to those of a seller or an importer.[15]

The significance of online platforms within the European market for goods and services has not escaped the attention of European law.[16] The European Union's strategic interest in spearheading the regulation of digital markets has been apparent for some time. Online platforms bridge the gap between digital markets and more conventional sectors. Consequently, regulations governing their operations have implications not only for the digital realm but also for tangible goods and services. Recognizing the interplay between online platforms and sustainability is pivotal in ensuring that regulations promote sustainable practices across these domains.

The Digital Markets Act[17] and the Digital Services Act[18] represent the two most consequential legislative endeavours in this domain. Both acts place significant emphasis on regulating online platforms. The former primarily concentrates on market regulation, aligning more closely with competition law than with the regulation of contracts and legal relationships involving platforms, users, and end-users. The latter establishes rules for overseeing the activities of online platforms, imposing proactive responsibilities on platforms, with potential administrative sanctions and civil liability for non-compliance. These legislative developments unequivocally reflect the evolving perception of online platforms. They are gradually shedding the label of mere intermediaries and assuming the role of gatekeepers responsible for controlling access to content and services. Integrating sustainability considerations within these regulatory frameworks is crucial for ensuring that online platforms contribute to sustainable development.

The Digital Services Act, while primarily focused on regulatory aspects of online platforms, also acknowledges the importance of sustainability concerns in its definition and categorization of online platforms.[19] According to Article 3, the letter (i) of the Act, an online platform is defined as a hosting service that stores and disseminates information to the public at the recipient's request, except in cases where this activity is minor and purely ancillary to another service or a minor functionality of the principal service, and cannot be used without that other service due to objective and technical reasons, with no intention to evade the applicability of the Regulation.

This definition demonstrates an awareness of the potential environmental and social impacts associated with online platforms. By recognizing the integrated nature of certain services and functionalities, the Act aims to prevent platform operators from circumventing sustainability obligations by treating sustainability-related features as minor or ancillary. This approach ensures that sustainability considerations are not overlooked or downplayed in the design and operation of online platforms.

Furthermore, the Act's emphasis on the integration of sustainability-related features into the core services of online platforms underscores the need for platforms to actively contribute to the promotion of sustainable practices. This can include features such as facilitating the trade of sustainable products, providing information on the environmental impact of products and services, or offering options for sustainable transportation or accommodation services. By integrating sustainability into their core functionalities, online platforms can play a vital role in fostering a more sustainable digital economy.

The definition and classification provided by the Digital Services Act represent just one perspective among several possible approaches. However, it is important to acknowledge a notable criticism regarding its alignment with the intuitive understanding of average users of online platforms. These users primarily utilize these platforms to access goods or services from third parties or directly engage with platform-offered services. The legal categories employed in the Act often rely on technical criteria that may not align with the users' perception of the contracted service. Nonetheless, the introduction of legal categories and a defined framework through the Digital Services Act[20] implies a significant technical advancement compared to the previous state. This development highlights the recognition of online platforms as more than mere intermediaries, but rather as significant contributors to the provision of goods and services. This shift had already been initiated by Regulation 2019/1150 of 20 June 2019, which aimed to promote fairness and transparency for business users of online intermediation services. In the context of sustainability, such legislative measures indicate a growing acknowledgement of the role online platforms play in facilitating the supply of sustainable products and services.

Articles 4–6 of the Digital Services Act play a crucial role in defining and regulating the liability framework for online platforms, particularly in their role of transmitting information in a communication network ('mere conduit'), 'caching', or 'hosting' activities. However, it is important to note that these articles primarily outline the conditions under which online platforms can be exempt from liability. As highlighted in Recital 17 of the Digital Services Act, these provisions aim to strike a balance between fostering innovation and ensuring accountability:

> The rules on liability of providers of intermediary services set out in this Regulation should only establish when the provider of intermediary services concerned cannot be held liable in relation to illegal content provided by the recipients of the service. Those rules should not

> be understood to provide a positive basis for establishing when a provider can be held liable, which is for the applicable rules of Union or national law to determine. Furthermore, the exemptions from liability established in this Regulation should apply in respect of any type of liability as regards any type of illegal content, irrespective of the precise subject matter or nature of those laws.

When considering the context of sustainable products and services, these liability provisions become even more significant. Online platforms that facilitate the dissemination of information relating to sustainable products, such as eco-friendly alternatives or ethically sourced goods, can benefit from the liability exemptions provided by the Digital Services Act. This encourages the promotion and availability of sustainable options to a wider audience, contributing to the overall sustainability goals.

Moreover, the liability framework outlined in the Digital Services Act serves as a guide for online platforms to navigate their responsibilities when hosting or providing access to content related to sustainable practices. By clarifying the boundaries of liability, it provides a level of legal certainty that can foster innovation and investment in sustainable product and service offerings on these platforms.

However, it is crucial to strike a balance between liability exemptions and the need for responsible oversight. While the Digital Services Act acknowledges the importance of fostering innovation and providing legal protections for online platforms, it also recognizes the necessity of addressing any potential negative impacts that may arise from the dissemination of misleading or harmful information regarding sustainability.

Therefore, in the realm of sustainable products and services, the liability provisions outlined in the Digital Services Act serves as a foundation for ensuring accountability while encouraging the growth and accessibility of sustainable alternatives. By promoting transparency and responsible practices among online platforms, these provisions contribute to the advancement of sustainability objectives and facilitate the transition towards a more sustainable economy.

Moreover, in all cases, it is not excluded the possibility for a judicial or administrative authority, in accordance with a Member State's legal system, to require the service provider to terminate or prevent an infringement.

The Digital Services Act provides a framework that allows online platforms to potentially avoid liability for damages and legal infringements related to data hosting, caching, or a mere transmission of information services, which could have implications for sustainable products and services. However, it should be noted that beyond the scope of liability exclusion, online platforms may still be held accountable under applicable national regulations. This is reinforced by the general principle of compensation for damages outlined in Article 54 of the Digital Services Act, which grants recipients of the service the right to pursue compensation from intermediary service providers in accordance with Union and national laws for any harm or loss incurred as a result of the providers' failure to fulfil their obligations under the regulation.

32.4 Specific Duties for Very Large Online Platforms and Very Large Search Engines

The Digital Services Act introduces a comprehensive framework that addresses the unique challenges posed by very large online platforms (VLOPs) and search engines (VLSEs). These platforms and search engines, as defined in Article 33.1 of the Digital Services Act, are characterized by having a substantial user base, with average monthly active recipients of the service in the

Union equal to or exceeding 45 million. This threshold serves as a criterion for identifying VLOPs and VLSEs, ensuring that the regulatory measures apply specifically to these significant players in the digital landscape.

With the increasing prominence of sustainable products and services, the specific duties assigned to VLOPs and VLSEs under the Digital Services Act can have a profound impact on fostering sustainability within the online ecosystem. Recognizing the influential role these platforms and search engines play in shaping consumer choices and market dynamics, the legislation should set forth targeted obligations to promote responsible practices in relation to sustainability, which it indirectly does through the requirement of the platforms needing to comply with pertinent EU law. In light of the ever-growing sustainability-related obligations for corporate actors, sustainability-related obligations (especially the ones requiring heightened transparency) will be increasingly growing also in the digital market.

Among the key provisions, Article 33.4 of the Digital Services Act outlines a dedicated procedure for identifying and monitoring VLOPs and VLSEs. This procedure involves assessing the average monthly active recipients of the service, ensuring that only the largest platforms and search engines fall within the scope of the specific duties. By singling out these entities, the legislation aims to strike a balance between promoting innovation and competition while ensuring effective oversight of the most influential players. The size of these players gives them significant market power, which could (or better said should) be used to facilitate and support sustainable products and services.

Firstly, VLOPs and VLSEs can contribute to sustainability by implementing measures that promote transparency and accountability regarding sustainable products and services. They can provide clear and accurate information to users about the environmental impact of products, thereby enabling informed decision-making. By facilitating access to eco-friendly alternatives and promoting sustainable consumption patterns, these platforms and search engines can encourage users to make choices that align with their sustainability values.

Furthermore, VLOPs and VLSEs can support sustainable products and services by fostering innovation within their platforms. They can create dedicated spaces or features that highlight sustainable businesses and products, making it easier for users to discover and engage with them. By actively promoting and supporting sustainable entrepreneurs and businesses, these platforms can help drive the adoption of sustainable practices and products on a larger scale.

In addition to promoting sustainability directly, VLOPs and VLSEs can contribute by ensuring effective oversight and regulation. The Digital Services Act's procedure for identifying and monitoring these large platforms and search engines is designed to strike a balance between fostering innovation and competition, while also providing appropriate oversight of the most influential players. This oversight can help prevent the misuse or abuse of market dominance, ensuring fair competition and creating a level playing field for sustainable businesses to thrive.

It is important to note that the specific actions and strategies employed by VLOPs and VLSEs to facilitate and support sustainable products and services may vary. They may involve partnerships with sustainable organizations, the development of dedicated sustainability programmes, or the integration of sustainability criteria into their algorithms and ranking systems. By leveraging their reach and influence, these platforms and search engines have the potential to significantly impact the adoption and growth of sustainable practices and products in the digital marketplace.

Taking a look at one of the central duties placed upon VLOPs and VLSEs as the obligation to implement robust mechanisms for promoting transparency and accountability regarding products and services, the importance of their role in terms of sustainability transition is undeniable, given their sheer size and the number of consumers using their services. This duty includes providing clear and accurate information to users about products, which should be used to facilitate access to eco-friendly alternatives and promoting sustainable consumption patterns. By doing so, these platforms and search engines can empower users to make informed choices that align with their sustainability values.

Article 34 of the Digital Services Act imposes an additional duty of care on these large players to assess the systemic risks that their operation or design may cause in the territory of the European Union. While the Act identifies various systemic risks, it is important to highlight the need to incorporate sustainability concerns into this framework. Sustainable products and services are critical in addressing global challenges such as climate change, resource depletion, and environmental degradation. By promoting sustainable practices, online platforms and search engines can contribute to the achievement of the Sustainable Development Goals outlined in the European Green Deal.

To fully embrace their pivotal role as catalysts for sustainability, very large online platforms and search engines should go beyond mere compliance with regulatory obligations and actively prioritize the integration of sustainable practices and principles throughout their operations. The Digital Services Act, with its provisions aimed at addressing systemic risks and ensuring accountability, provides a significant opportunity for these platforms to assume leadership in driving positive environmental and social change.

In addition to assessing and mitigating the dissemination of illegal content related to environmental harm, online platforms should proactively foster environmental stewardship by implementing measures that promote sustainability. This could include partnering with sustainable brands and businesses, facilitating the discovery and promotion of eco-friendly products and services, and implementing innovative features that encourage users to adopt sustainable consumption patterns. By incorporating sustainability as a core value, online platforms can leverage their vast user base and influence to drive meaningful impact and create a more sustainable digital ecosystem.

Furthermore, as key actors in the digital landscape, these platforms have the capacity to shape user behaviour and raise awareness about pressing sustainability issues. By providing clear and accurate information about the environmental footprint of products and services, online platforms can empower users to make informed choices aligned with their sustainability values. They can also leverage their vast data resources and analytics capabilities to generate insights and recommendations that promote sustainable living and consumption.

The systemic risks identified in the Digital Services Act, such as their potential impact on fundamental rights, public health, gender-based violence, and consumer protection, should be viewed through the lens of sustainability. Online platforms should proactively consider the broader implications of these risks on environmental sustainability and social well-being. By developing robust mechanisms to address these risks and mitigate their negative impact, platforms can foster a digital environment that promotes sustainability, inclusivity, and responsible use of technology.

It can therefore be claimed that the Digital Services Act presents a critical juncture for very large online platforms and search engines to embrace their responsibility as drivers of sustainability. By actively integrating sustainable practices, promoting eco-friendly alternatives, and leveraging their influence to foster sustainable consumption patterns, these platforms can spearhead positive change and contribute to a more sustainable and inclusive digital ecosystem. Embracing sustainability as a guiding principle is not only a moral imperative but also a strategic opportunity for platforms to enhance their reputation, build user trust, and create long-term value for their stakeholders. And the stakeholder pressure for sustainable products and services, and the transparency to that effect, is growing by the day.

32.5 Online Platforms as Indirect Gatekeepers to Sustainability

The absence of a general sustainability obligation in the Digital Services Act for online platforms can be attributed to the diverse range of activities and sectors in which these platforms operate. The multifaceted nature of their operations may have made it challenging to establish a comprehensive sustainability requirement that applies uniformly across all platforms. However, this diversity does not diminish the importance of addressing sustainability concerns within the digital ecosystem.

While the identification of systemic risks in the Digital Services Act may have been influenced by the variety of platform activities, this disparity did not prevent the inclusion of specific obligations for large platforms. The drafters of the Act may have exercised caution in imposing additional obligations on an industry that is difficult to regulate and largely comprised companies located outside the European Union. Striking a balance between fostering innovation and ensuring effective regulation is a delicate task, and over-regulation could potentially stifle the growth and development of the online platform sector.

Nevertheless, it is crucial to examine the extent to which the new legal framework for online platforms allows for their involvement in promoting sustainability objectives. While sustainability is not explicitly mandated, it indirectly enters the realm of online platform activities through the obligations on traceability outlined in Article 30 of the Digital Services Act. This article imposes the responsibility on platforms that enable consumers to engage in distance contracts with traders to ensure that these traders exclusively offer products or services that comply with the relevant provisions of Union law.

Although Article 30 primarily pertains to product safety regulations, its broad language does not exclude other legal obligations tied to sustainability in specific sectors. Thus, there is potential for sustainability obligations in various fields, such as environmental protection, labour rights, and social responsibility, to intersect with the activities of online platforms. By upholding their duty to verify compliance with applicable Union law, platforms can indirectly contribute to promoting sustainability objectives by supporting the provision of goods and services that meet sustainability criteria.

Furthermore, the evolving nature of sustainability regulations and the increasing global focus on environmental and social challenges necessitate ongoing scrutiny of the legal framework governing online platforms. As the sustainability landscape continues to evolve, it is essential for policymakers to consider the potential role of online platforms in advancing sustainability goals. Striking the right balance between regulatory oversight and fostering innovation is crucial to ensure that platforms are not only compliant with existing sustainability obligations but also encouraged to actively contribute to sustainability efforts through their operations and services.

All in all, while the Digital Services Act does not explicitly impose a general sustainability obligation on online platforms, there are avenues within the legislation where sustainability considerations can come into play. Platforms have a responsibility to verify compliance with Union law and, in doing so, can indirectly contribute to the promotion of sustainability objectives. As the digital ecosystem evolves, policymakers should continuously evaluate the legal framework to ensure that it remains responsive to emerging sustainability challenges and encourages platforms to play an active role in fostering a more sustainable future. By engaging online platforms as partners in sustainability, we can leverage their influence and reach to drive positive change and address pressing environmental and social issues.

The control mechanism outlined in Article 30 of the Digital Services Act, which requires online platforms to ensure the compliance of products and services offered by merchants, has some limitations. Firstly, it excludes small- and medium-sized enterprises (SMEs), as defined by

Recommendation 2003/361/EC, from carrying out this control. This exemption leaves a portion of online platform activities outside the scope of such compliance checks. Secondly, the Digital Services Act missed the opportunity to impose specific sustainability duties directly on online platforms themselves, in terms of both their operation and design. Instead, the Act relies on them as indirect controllers, responsible for verifying compliance with legal requirements by third-party merchants.

However, there are further developments and measures being proposed within the EU to strengthen the fight against unsustainable corporate practices and enhance consumer protection. One significant instrument is the Proposal for a Directive aimed at empowering consumers for the green transition through better protection against unfair practices and improved information.[21] This proposal seeks to amend the Unfair Commercial Practices Directive 2005/29/EC and the Consumer Rights Directive 2011/83/EU. This Proposal places obligations on producers to provide accurate and precise information regarding the sustainability aspects of their products. This includes details on the existence and length of a producer's commercial guarantee of durability for all types of goods, information on product reparability, a ban on displaying sustainability labels not based on certification schemes or established by public authorities, and a prohibition on using generic environmental claims in marketing towards consumers without proper substantiation.

Moreover, the Proposal directly imposes obligations on traders themselves. Traders are prohibited from misleading consumers about the environmental and social impacts, durability, and reparability of products. They are also required to provide clear commitments when making environmental claims related to future environmental performance. Failure to meet these obligations can have legal consequences, such as the requirement to provide detailed data on the method of comparing products in terms of sustainability, the products and suppliers covered, and measures to keep the information up to date. Additionally, traders are prohibited from making environmental claims about the entire product when it only applies to a specific aspect and from presenting legal requirements applicable to all products in a category as a distinctive feature of their offer.

The Proposal exemplifies how Article 30 of the Digital Services Act paves the way for sustainability to enter directly into the regulatory framework. Given the increasing focus on sustainable corporate conduct and practices in the EU, it is likely that trader obligations will continue to expand and strengthen in line with broader corporate responsibility for their sustainable or unsustainable practices.

This supports the claim that although the sustainability of products may not be explicitly outlined as a guiding principle of the Digital Services Act, it is expected to become an increasingly important obligation for traders to consider when offering products on online platforms. As regulatory efforts in sustainable corporate practices continue to evolve, sustainability will find its place among the obligations and requirements imposed on traders, shaping the landscape of online platform activities and promoting a more sustainable digital ecosystem.

The regulatory landscape surrounding sustainability and consumer protection is constantly evolving within the EU. The Proposal for a Directive aimed at empowering consumers for the green transition is just one example of the ongoing efforts to strengthen and broaden corporate responsibility for sustainable practices. As such, online platforms should remain attentive to these developments and proactively adapt their operations to align with emerging sustainability obligations and expectations.

In conclusion, while the Digital Services Act may not explicitly mandate sustainability obligations for online platforms, the broader context and evolving regulatory environment emphasize the importance of incorporating sustainability considerations into their activities. By proactively embracing sustainability as a guiding principle, online platforms can contribute to a greener and more inclusive digital ecosystem, empower consumers to make sustainable choices and support the achievement of the EU's sustainable development objectives.

32.6 The Desirability of Going beyond Indirect Control

The solution proposed by the Digital Services Act involves a return to the distinction between first and secondary liability in relation to online platforms. Primary liability refers to obligations directly imposed on the online platform. Secondary liability generally refers to others' obligations whose infringement may be imposed on the online platform if the primary obligor fails to perform. Such a secondary liability implies making liable 'the-person-in-between' for infringements done by third people. In the traditional approach, online platforms, as well as other providers of online services, have been excluded from such a liability.

The distinction should be considered overcome. It belongs to the idea of online platforms as mere intermediaries. In fact, it has come under fire in many of the examples referred to in previous pages, where the online platform becomes a guarantor of legality on which duties of control and liability for non-compliance can be imposed.

The Commission Notice on the market surveillance of products sold online (2017/C 250/01), regarding the application of Directive 2005/29, of 11 May 2005, on Unfair Commercial Practices, stated that

> as regards third party economic operators acting on a platform, the platform itself should take appropriate measures to enable third-party traders to comply with EU consumer and marketing law in conjunction with EU product legislation and/or product safety law requirements.

The statement approaches online platforms' liabilities to those to be required to gatekeepers. It is not legally binding but clearly shows the overcoming of the traditional distinction between primary and secondary liabilities when tackling OP's liabilities.

In the same vein, and clearly showing the erosion of the 'intermediation principle' regarding OP's liabilities, Recital 23 of Directive 2019/771, of 20 May 2019, on certain aspects concerning contracts for the sale of goods, sets forth that:

> Platform providers could be considered to be sellers under this Directive if they act for purposes relating to their own business and as the direct contractual partner of the consumer for the sale of goods. Member States should remain free to extend the application of this Directive to platform providers that do not fulfil the requirements for being considered a seller under this Directive.

The indirect control of legality by online platforms is abandoned in two relevant proposals on the subject. Both are different in their nature. One corresponds with an official proposal for a new European Regulation drafted by the European Commission. The other is a private initiative, though very influential in the European arena since it comes from the European Law Institute (ELI). In spite of their different qualification from a legal perspective, both evidence the existence of a potential trend to enhance the prominence of online platforms. Therefore, to increase the responsibilities (and corresponding liabilities) that online platforms should assume in providing goods and services.

The first is the one contained in the Proposal for a Regulation on General Product Safety (COM/2021/346/final).[22] Its Article 20 on 'Specific obligations of online marketplaces related to product safety' extends to online platforms the duties of control and communication regarding the safety of the products they offer.

The second is the 'Model Rules on Online Platforms' drafted by the European Law Institute in 2019.[23] Articles 19 and 20 of the Model Rules suggest extending consumer protection rules, even in favour of professional users of online platforms, when the platforms are not transparent enough about their involvement in the provision of goods or services and when they exert a 'predominant influence' over the users of the platform for providing goods and services.[24]

It seems that there is nothing to prevent the same pattern from being applied to online platforms for the promotion of sustainability and the fulfilment of the Sustainable Development Goals.

32.7 To Conclude

In this chapter, we have examined the role of online platforms in driving sustainability and the implications of relevant regulations and proposals. Our analysis has revealed the opportunities and challenges faced by platforms in assuming their responsibility towards sustainability and the transition to a more environmentally conscious future.

Throughout our exploration, we have emphasized the crucial influence of very large online platforms and search engines in shaping both environmental outcomes and social well-being. We have underscored the need for platforms to assess and mitigate potential negative effects, including the dissemination of illegal content related to environmental harm and the impact on fundamental rights, public health, gender-based violence, and consumer protection.

Moreover, we have explored the provisions of the Digital Services Act and their potential for platforms to actively contribute to sustainability. By integrating features that promote eco-friendly alternatives, providing accurate information on the environmental impact of products and services, and encouraging responsible consumption patterns, platforms can foster a sustainable digital ecosystem.

While we have observed the absence of explicit mention of sustainability in the systemic risks outlined by the Act, we argue that online platforms now hold a crucial role in driving sustainable development objectives. This recognition is evident in initiatives such as the European Green Deal, which calls for the alignment of platform activities with sustainability principles.

Additionally, we have examined the Proposal for a Directive aimed at empowering consumers for the green transition through improved protection against unfair practices and enhanced information. This proposal highlights the importance of accurate and precise sustainability information, the prohibition of misleading claims, and the promotion of transparency in environmental and social impacts. It signifies a significant step towards broadening the obligations of traders and platforms, aligning them with the growing emphasis on sustainable corporate conduct.

In light of our findings, we offer several recommendations for further research and policymaking. Policymakers should strengthen regulatory frameworks, foster collaboration and dialogue among platforms and stakeholders, encourage technological innovations that support sustainability, promote user empowerment through accurate information and sustainable choices, and address jurisdictional challenges through international cooperation.

In conclusion, our work demonstrates that online platforms possess transformative potential in driving sustainability and facilitating positive change. Despite the challenges, platforms can become key contributors to a greener and more inclusive digital landscape through proactive measures, transparent operations, and responsible practices. By embracing sustainability as a guiding principle, platforms can play a vital role in shaping a more sustainable future.

Notes

1 The Authors are affiliated with Universitat Pompeu Fabra (Barcelona, Spain). Authors participate as research members in the project PID2021-126354OB-I00/MICIN/AEI/10.13039/501100011033/FEDER, UE, on 'Contractual and Non-Contractual Liability of Online Platforms'. The authors would like to thank the editors for their review of the text and the peer reviewers for their suggestions and comments.

2 This paper aims to contribute to the debate on the relevance of private law in meeting the Sustainable Development Goals. Among the most recent contributions in the area Sabrina Lanni, *Greening the Civil Codes: Comparative Private La wand Environmental Protection* (New York, Routledge 2023).

3 See, Alain Strowel and Wouter Vergote, 'Digital Platforms: To Regulate or Not to Regulate? Message to Regulators: Get the Economics Right First, the Focus on the Right Regulation' in Bran Devolder (ed), *The Platform Economy. Unravelling the Legal Status of Online Intermediaries* (Cambridge, Intersentia 2019) 3–30.

4 *See* Airbnb, 'Airbnb Corporate Sustainability Update Corporate Emissions Footprint and Progress Against Goals' [2022] <https://s26.q4cdn.com/656283129/files/doc_downloads/governance_doc_updated/2022/Airbnb-Corporate-Sustainability-Update_September-2022.pdf> accessed 29 May 2023.

5 OLIO Platform, available at <https://olioapp.com/en/> last accessed on 29 May 2023.

6 BlaBla Car Platform, available at <https://www.blablacar.com/> last accessed on 29 May 2023.

7 See, Bram Devolder, 'Contractual Liability of the Platform' in Bran Devolver (ed), *The Platform Economy. Unravelling the Legal Status of Online Intermediaries* (Cambridge, Intersentia 2019) 31–88.

8 Such was the debate in the ECJ Judgments Case C-434/15 *Elite Taxi* [2017] ECR not yet published (ECLI:EU:C:2017:981) and Case C-320/16 *Uber France* [2018] ECR not yet published (ECLI:EU:C:2018:221).

9 On the subject, among others, Mateusz Grochowski, 'Default Rules Beyond a State: Special Purpose Lawmakers in the Platform Economy' in Stefan Grundmann and Mateusz Grochowski (eds), *European Contract La wand the Creation of Norms* (Cambridge, Intersentia 2021) 227–254, and Roger Brownsword, Rob van Gestel and Hans Micklitz, *Contract and Regulation: A Handbook on New Methods of Law Making in Private Law* (Cheltenham, Edward Elgar 2017).

10 Similar to the approach made by Christoph Busch et al, 'Discussion Draft of a Directive on Online Intermediaries Platforms' [2016] 5(4) *EuCML* 164.

11 See, Célia Mak, 'Regulating Online Platforms: The Case of Airbnb' in Stefan Grundmann (ed), *European Contract Law in the Digital Age* (Cambridge, Intersentia 2018) 87–102; Alexandre de Streel, Miriam Buiten and Martin Peitz, 'Liability of Online Hosting Platforms. Should Exceptionalism End?' [2018] *Centre on Regulation in Europe Report* <https://cerre.eu/publications/liability-online-hosting-platforms-should-exceptionalism-end/> accessed 30 April 2023; Esther Arroyo Amayuelas, 'Article 14. Hosting' and 'Article 15. No General Obligation to Monitor' in Dirk Schulze and Dirk Staudenmayer (eds), *EU Digital Law (Article-by-Article Commentary)* (Baden-Baden, C.H. Beck Hart Nomos 2020) 475–494; Gerald Spindler, 'The Liability System of Art. 17 DSMD and National Implementation. Contravening Prohibition of General Monitoring Duties?' [2019] 10(3) *Journal of Intellectual Property, Information Technology and E-Commerce Law* 344; Christiane Wendehorst, 'Platform Intermediary Services and Duties under the E-Commerce Directive and the Consumer Rights Directive' [2016] 5(1) *EuCML* 30–33.

12 See, Célia Mak, 'Regulating Online Platforms: The Case of Airbnb' in Stefan Grundmann (ed), *European Contract Law in the Digital Age* (Cambridge, Intersentia 2018) 87–102; Alexandre de Streel, Miriam Buiten and Martin Peitz, 'Liability of Online Hosting Platforms. Should Exceptionalism End?' [2018] *Centre on Regulation in Europe Report* <https://cerre.eu/publications/liability-online-hosting-platforms-should-exceptionalism-end/> accessed 30 April 2023; Esther Arroyo Amayuelas, 'Article 14. Hosting' and 'Article 15. No General Obligation to Monitor' in Dirk Schulze and Dirk Staudenmayer (eds), *EU Digital Law (Article-by-Article Commentary)* (Baden-Baden, C.H. Beck Hart Nomos 2020) 475–494; Gerald Spindler, 'The Liability System of Art. 17 DSMD and National Implementation. Contravening Prohibition of General Monitoring Duties?' [2019] 10(3) *Journal of Intellectual Property, Information Technology and E-Commerce Law* 344; Christiane Wendehorst, 'Platform Intermediary Services and Duties under the E-Commerce Directive and the Consumer Rights Directive' [2016] 5(1) *EuCML* 30–33.

13 See Directive 2000/31/EC of the European Parliament and of the Council of 8 June 2000, on certain legal aspects of information society services, in particular electronic commerce, in the Internal Market (Directive on electronic commerce) [2000] OJ L178/1, art. 12, paragraphs 1 and 2.

14 The evolution is well explained and assessed in the document by Philippe Madiega, 'Reform of the EU Liability for Online Intermediaries' [2020] *European Parliamentary Research Services* <https://www.europarl.europa.eu/RegData/etudes/IDAN/2020/649404/EPRS_IDA(2020)649404_EN.pdf> accessed 29 May 2023.

15 Like it has been included in many sections of the Proposal for a Directive of the European Parliament and of the Council on Liability for Defective Products, COM/2022/495 final. In the Recital 28 of the Proposal it can be read that: 'When online platforms perform the role of manufacturer, importer or distributor in respect of a defective product, they should be liable on the same terms as such economic operators'. As the Commission, 'A European Agenda for the Collaborative Economy' COM(2016) 356 final sets out 'there may be cases in which collaborative platforms can be considered as offering other services in addition to the information society services that they offer to intermediate between the providers of underlying services and their users. In particular, in certain circumstances, a platform may also be a provider of the underlying service (e.g., transport or short-term rental service). In such a case, collaborative platforms could be subject to the relevant sector-specific regulation, including business authorisation and licensing requirements generally applied to service providers' (p. 6).

16 It needs to be noticed that Regulation 2019/1150, of 20 June 2019, on promoting fairness and transparency for business users of online intermediation services [2019] OJ L186/57, introduces a range of obligations for online intermediaries. The Regulation primarily focuses on duties related to the disclosure of terms and conditions (art. 3), the conditions for restricting or suspending services provided to specific providers through the intermediation system (art. 4), the calculation and publication of provider rankings (art. 5), the access to data (art. 9), and the requirement to offer an internal complaint system (art. 11). These obligations specifically apply to online intermediaries. In case of non-compliance, the remedies available are those typically provided to address violations of mandatory rules within the legal systems of each Member State (art. 15), with specific recognition given to consumer associations and public bodies to initiate legal proceedings for the protection of collective interests (art. 14).

17 Regulation (EU) 2022/1925 of the European Parliament and of the Council of 14 September 2022 on contestable and fair markets in the digital sector and amending Directives (EU) 2019/1937 and (EU) 2020/1828 (Digital Markets Act) [2022] OJ L265/1.

18 Regulation (EU) 2022/2065 of the European Parliament and of the Council of 19 October 2022 on a Single Market For Digital Services and amending Directive 2000/31/EC (Digital Services Act) [2022] OJ L277/1.

19 On the subject of definition and taxonomy problems, relevant scholarly works include Joana Campos Carvalho, 'Online Platforms: Concept, Role in the Conclusion of Contracts and Current Legal Framework in Europe' in Esther Arroyo Amayuelas and Sergio Cámara Lapuente (eds), *El Derecho Privado en el nuevo paradigma digital* (Madrid, Marcial Pons 2020) 239–254; Esther Arroyo Amayuelas, 'El Derecho de las plataformas en la Unión Europea' in Esther Arroyo Amayuelas et al (eds), *Servicios en plataformas. Estrategias regulatorias* (Madrid-Barcelona, Marcial Pons 2021) 21–67 and María Teresa Álvarez Moreno, *La contratación electrónica mediante plataformas en línea* (Madrid, Reus 2021).

20 It seems appropriate to recall that, sometime before the enactment of the Digital Services Act, the important and very instructive report Andrea Bertolini, 'Liability of Online Platforms, European Parliamentary Research Services' [2021] *European Parliamentary Research Service* <https://www.europarl.europa.eu/RegData/etudes/STUD/2021/656318/EPRS_STU(2021)656318_EN.pdf> accessed 29 May 2023, after proposing various criteria for the classification of online platforms, it was summarized the state of the art in this way: 'Science-engineers commonly use this term to denote a set of technologies or interfaces available to a broad base of users who build processes, applications, technologies and business models with it and on it. On the contrary, social scientists, policymakers, as well as the general public, normally employ it to indicate triangular digital infrastructures allowing interactions between different subjects and actors, covering – by extension – (i) the infrastructure itself (ii) the entity (legal subject) who runs it, and (iii) the economic model adopted by the latter. Under this approach – which primarily focuses on the economic structure realised OPs are described as software-based facility[ies] providing two – or multi-sided markets where providers and users of content, goods and services can meet', and understood as 'a broad label for numerous types of [digital] multi-sided business models' (p. 9).

21 Proposal for a Directive of the European Parliament and of the Council amending Directives 2005/29/EC and 2011/83/EU as regards empowering consumers for the green transition through better protection against unfair practices and better information, COM/2022/143 final.

22 Proposal for a Regulation of the European Parliament and of the Council on general product safety, amending Regulation (EU) No 1025/2012 of the European Parliament and of the Council, and repealing

Council Directive 87/357/EEC and Directive 2001/95/EC of the European Parliament and of the Council, COM/2021/346 final.

23 European Law Institute, 'Model Rules on Online Platforms' [2019] *European Law Institute* <https://www.europeanlawinstitute.eu/fileadmin/user_upload/p_eli/Publications/ELI_Model_Rules_on_Online_Platforms.pdf> accessed 29 May 2023.

24 See, Florian Busch, 'European Model Rules for Online Intermediary Platforms' in Uwe Blaurock, Martin Schmidt-Kessel, and Katharina Erler (eds), *Platformen: Geschäftsmodell und Vertrag* (Baden-Baden, Nomos 2018) 37–57.

Bibliography

Álvarez Moreno MT, *La contratación electrónica mediante plataformas en línea* (Madrid, Reus 2021)

Arroyo Amayuelas E, 'Article 14. Hosting' and 'Article 15. No General Obligation to Monitor' in Schulze D and Staudenmayer D (eds), *EU Digital Law (Article-by-Article Commentary)* (Baden-Baden, C.H. Beck Hart Nomos 2020) 475–494

Arroyo Amayuelas E, 'El Derecho de las plataformas en la Unión Europea' in Arroyo Amayuelas E et al (eds), *Servicios en plataformas. Estrategias regulatorias* (Madrid-Barcelona, Marcial Pons 2021) 21–67

Bertolini A, 'Liability of Online Platforms, European Parliamentary Research Services' [2021] *European Parliamentary Research Service* <https://www.europarl.europa.eu/RegData/etudes/STUD/2021/656318/EPRS_STU(2021)656318_EN.pdf>

Brownsword R, van Gestel R and Micklitz H, *Contract and Regulation: A Handbook on New Methods of Law Making in Private Law* (Cheltenham, Edward Elgar 2017)

Busch C et al, 'Discussion Draft of a Directive on Online intermediaries Platforms' [2016] 5(4) *EuCML* 164–169

Busch F, 'European Model Rules for Online Intermediary Platforms' in Blaurock U, Schmidt-Kessel M and Erler K (eds), *Plattformen: Geschäftsmodell und Vertrag* (Baden-Baden, Nomos 2018) 37–57

Campos Carvalho J, 'Online Platforms: Concept, Role in the Conclusion of Contracts and Current Legal Framework in Europe' in Arroyo Amayuelas E and Cámara Lapuente S (eds), *El Derecho Privado en el nuevo paradigma digital* (Madrid, Marcial Pons 2020) 239–254

de Streel A, Buiten M and Peitz M, 'Liability of Online Hosting Platforms. Should Exceptionalism End?' [2018] *Centre on Regulation in Europe Report* <https://cerre.eu/publications/liability-online-hosting-platforms-should-exceptionalism-end/>

Devolder B, 'Contractual Liability of the Platform' in Devolver B (ed), *The Platform Economy. Unravelling the Legal Status of Online Intermediaries* (Cambridge, Intersentia 2019) 31–88

European Law Institute, 'Model Rules on Online Platforms' [2019] European Law Institute <https://www.europeanlawinstitute.eu/fileadmin/user_upload/p_eli/Publications/ELI_Model_Rules_on_Online_Platforms.pdf>

Grochowski M, 'Default Rules Beyond a State: Special Purpose Lawmakers in the Platform Economy' in Grundmann S and Grochowski M (eds), *European Contract La wand the Creation of Norms* (Cambridge, Intersentia 2021) 227–254

Lanni S, *Greening the Civil Codes: Comparative Private La wand Environmental Protection* (New York, Routledge 2023)

Madiega P, 'Reform of the EU Liability for Online Intermediaries' [2020] *European Parliamentary Research Services* <https://www.europarl.europa.eu/RegData/etudes/IDAN/2020/649404/EPRS_IDA(2020)649404_EN.pdf> accessed 29 May 2023

Mak C, 'Regulating Online Platforms: The Case of Airbnb' in Stefan Grundmann (ed), *European Contract Law in the Digital Age* (Cambridge, Intersentia 2018) 87–102

Spindler G, 'The Liability System of Art. 17 DSMD and National Implementation. Contravening Prohibition of General Monitoring Duties?' [2019] 10(3) *Journal of Intellectual Property, Information Technology and E-Commerce Law* 344–374

Strowel A and Vergote W, 'Digital Platforms: To Regulate or Not to Regulate? Message to Regulators: Get the Economics Right First, the Focus on the Right Regulation' in Bran Devolder (ed), *The Platform Economy. Unravelling the Legal Status of Online Intermediaries* (Cambridge, Intersentia 2019) 3–30

Wendehorst C, 'Platform Intermediary Services and Duties under the E-Commerce Directive and the Consumer Rights Directive' [2016] 5(1) *EuCML* 226–232

33
THE SUSTAINABILITY OBLIGATION IN GLOBAL VALUE CHAIN CONTRACTS

Jakub Błażej Zwierzchowski[1] *and Ewa Rott-Pietrzyk*[2]

33.1 Introduction

The inclusion of sustainability contract clauses (SCCs) in commercial contracts is a relatively new practice that has intensified since the events in 2013 in Rana Plaza Bangladesh.[3] SCCs are voluntarily adopted in more than half of all written contracts concluded within the Global Value Chain (GVC).[4] At the same time, the increasing prevalence of SCCs has led the concept of 'sustainable development' to be considered in scholarly debate as an implied term.[5]

SCCs are strictly connected with due diligence and to Corporate Social Responsibility (CSR) instruments,[6] but it should also be seen (from the perspective of both positive and negative externalities) as a private instrument supporting the SDGs under the United Nations Resolution Transforming our World: the 2030 Agenda for Sustainable Development (2030 UN Agenda).[7] Due diligence and CSR lie in the private, self-regulatory domain, but an intensification of legislative activities in this area can be observed. From a European perspective, voluntary due diligence and the concept of 'sustainable development' enter the regulatory orbit and the realm of commercial contracts with the future adoption of the EU Proposal for a Directive on Corporate Sustainability Due Diligence (EU Proposal).[8]

However, when analysing the fundamental framework for 'sustainable development' as set out in the 2030 UN Agenda, it is clear that the SDGs cover a wide variety of issues. These are: social (such as poverty eradication and decent work wages, gender equality and inclusion), environmental (e.g., climate catastrophe, protection of biodiversity and environmental pollution), economic (such as the use of materials and responsible consumption) and political (for example, reconciling the development of the Global South and the Global North). The diverse nature of 'sustainable development', taken together with the principles of contract law[9] and the sectoral particularities of GVCs, can thus cause a challenge to the structure and scope of SCCs. In this context, 'sustainability' can be taken either as a 'multi' or a 'mono' obligation. The idea of a 'multi' obligation emerges from the material scope of the SCC[10] and, depending on its scope, is linked to one or other SDG. Our considerations, however, will be carried from the perspective of 'sustainability' as a 'mono' obligation, containing certain general characteristics, categories of behaviours, rights and duties relating to GVC contracts. This means that this text aims to reveal a common core of the 'sustainability obligations' in international commercial contracts, across the SCC's material scope, the

DOI: 10.4324/9781032662046-39

sector of the GVC (e.g., agriculture, technology, clothing or luxury industry) and type of contract (e.g., sale, delivery, production or manufacturing).

Therefore, the main research question that arises here is: how can we shape a 'sustainability obligation' in line with commercial practice and regulatory proposals?

Our claim is that imposing a default 'sustainability obligation' can be done both through business practice and legislative activity: the former can be done by conceptualising sustainable development as an implied term[11]; the latter, by adopting obligatory human and environmental due diligence regulations in contract law.

In order to walk the reader through our reasoning, we will first analyse the characteristics and associated challenges of the SCCs. Then, we will reconstruct the regulatory model of contractual due diligence. This will allow for an evaluation of how the regulatory activity has implications on the SCCs characteristics and how the core of the default 'sustainability obligation' in the GVC contracts looks in respect of practice and the regulatory model under the EU Proposal.

The SCCs will be analysed in the context of the GVC contractual relationship. However, as there is scarce empirical information available on the contractual structure and governance of the GVC, there is also little empirical research available on the SCCs. Thus, the analysis will take the form of secondary research based on qualitative studies and doctrinal research, and based upon the standard contract terms of companies on the FTSE 100 London Stock Exchange. The SCCs will be detached from their subject matter, which will allow for a determination of the theoretical implications of the SCCs regardless of their connection with SDGs. The reconstruction of the models will be conducted based on selected provisions of the EU Proposal that may be of particular importance for GVC contractual relationships.[12] In this paper, we do not aim to analyse the feasibility of contracts as a due diligence mechanism or to assess the potential of the EU Proposal for private law in general. However, we wish to show how the EU Proposal reflects the commercial practice and to answer the challenges it brings.

33.2 Global Value Chain Contracts and Sustainability: Characteristics of Sustainability Contract Clauses

GVCs are responsible for two thirds of global exports and more than half of global imports.[13] As Klaas Hendrik Eller stated, '(…) GVCs need to be understood as an organizational arrangement and simultaneously as a stage in the evolution of a global political economy.'[14] Therefore, GVCs are also responsible for both positive and negative externalities across the globe. Positive externalities include economic upgrading, exchange of know-how, employment growth and the rise of environmental standards. The negative externalities, mainly on host-state communities, are the increase of inequality, deterioration of labour standards and environmental damage. Negative externalities have recently started to progressively come under public scrutiny, leading to a debate about who is responsible and how to effectively hold them liable.[15]

Due to the complex structure and governance of GVCs, enforcement has proven to be a challenge. Multinational enterprises (MNEs) have emerged as important players and the 'leaders' of GVCs. In simplified terms, MNEs establish business relationships with leading suppliers in their sector, then leading suppliers rely on local systems through sub-suppliers and further exploit market flexibility for economic profit.[16] However, this hierarchical chain structure not always exists: GVCs are becoming increasingly complex and their actors are overlapping, as, for example, the same suppliers can have relations with multiple different buyers. This means that GVCs resemble more a 'network' than an actual chain,[17] as the GVC is more multifaceted than it may seem at first. Spread across the globe and responsible for various stages of production, the GVC ties the various entities to each other through the conclusion of contracts. In addition to serving economic purposes, contracts have also become an area of strategic corporate engagement and activism,[18] which can have an effect on human rights and the environment.[19] On the one hand, the political economy of the contracting practice and contract law are becoming preeminent not only within the GVCs, but also on a broader, global level encompassing social, environmental, cultural and economic interests. On the other hand, contracts also allow the strategic and instrumental use of concepts of private law in order to limit risk, responsibility and liability in relation to various externalities.[20] Some authors even conclude that, in this way, contract laws (covering the GVCs) contribute to systemic hermeneutical injustice, which has a bearing on the fundaments of contract law.[21] It seems that GVC shape the need for a novel direction of research on contracts, because the legal frameworks of GVCs are fragmented and composed of many disciplines (private law – mainly contracts – public law, economics etc.). GVCs are seen as the social institution of the value chain itself.[22]

It is worth underlying that, even though there is scarce empirical evidence on the contractual structure and governance of GVCs, there is information that suggests that more than half of the written contracts between GVC buyers and suppliers now include SCCs.[23] The European Commission places this figure at about 69%,[24] Ecovadis and Affectio Mutandis claim 73%[25] and Katerina Mitkidis et al. have given a figure of 72%.[26]

A few words on the characteristics of SCCs.

It has been said that SCCs should be included in contracts for various reasons: in order to prevent operational and financial risks, reputational damage, consumer boycotts,[27] or, in general, as a response to societal demand for a 'sustainability' agenda.[28] K. Mitkidis defines SCCs as

> (...) contractual provisions covering social and environmental obligations that are not directly connected to the subject matter of a specific contract and which pursue long-term business objectives and public interests, by frequently aiming to extend their applicability to third parties and employing relational monitoring and enforcement tools.[29]

The concept of sustainability can refer to both products and production processes and so do SCCs. Nonetheless, SCCs do not generally concern issues related to the physical qualities of goods[30] and SCCs are more associated with the performance of contracts. Therefore, the aim of SCCs is to shape parties' obligations (buyers and suppliers) so that the performance will meet certain requirements. The SCC's involve, in particular, the imposition of a duty on the supplier to protect social and ecological interests affected by the chain's activities.[31] Here, it must be emphasised that SCCs arise under the contract concluded between parties and are completely dependent on the parties' will. The freedom to shape the content of the contract empowers the parties to establish, change, and abolish the norms of conduct determining the legal relations between them.[32] Nevertheless, in the context of the buyer-supplier relation under a GVC, SCCs can easily be imposed on parties with weaker negotiating power and little resources. This creates a situation in which SCCs impose substantive rights and duties exceeding those provided under national law, only buyers can enforce them but ultimately don't due to a lack of financial incentives.[33] Consequently, internal tensions are created in the chain. On the one hand, SCCs can lead to the accomplishment of SDGs that go beyond the commercial interests of the parties while, on the other, SCCs can be unfair and discriminatory, thus creating a need to protect the supplier as the weaker party in the chain.[34] SCCs are stated as being 'general', so they do not determine the subject matter of the contract or set out the specific rights and duties.[35] What this means is that parties simply state that the performance of a contract should be conducted according to the 'best practices', 'highest standards' or in line with the below-indicated documents.[36] SCCs are constructed in a way that they refer to other documents, both hard law and soft law, as well as transnational standardisation instruments, e.g. ISO,[37,38] most of which consist of transnational private law.[39] SCCs tend to refer to company policies, namely corporate codes (private self-regulation).[40] These corporate codes are an example of how private actors manage to establish a framework of considerable binding effects.[41] Codes are either annexed to contracts or concluded in separate contractual assurances.[42,43] However, it remains theoretically a problem as how this form of exercise of private autonomy could be seen as a legally binding and legitimate source of law.[44] Therefore, the fact that SCCs are 'general' and do not specify rights and duties causes challenges to both the interpretation and enforcement of SCCs.

Although SCCs are not connected with the primary economic purpose of the business relationship, they are most often connected with a remedy in case of non-performance. This construction, in principle, increases the supplier's awareness that SCCs are binding, enforceable and are an essential part of the contract. According to some studies, almost half of SCCs used involve the remedy of termination of contract, though buyers rarely use it for economic reasons.[45] Additionally, basing the remedy on the termination of contract does not solve negative externalities. Therefore, the postulated remedy for SCCs should be gradable and should include, among other things, an action plan to correct the wrongdoings, the suspension of the contract, contractual penalties and then, finally, termination.[46]

Due to the connection between SCCs and the remedies, the assessment of whether the SCCs' have been fulfilled needs to be made. Such an assessment can be conducted by the buyer and the supplier or – more popularly – by third-party auditing. This structure is assumed to transform 'sustainability' from an 'obligation of means',[47] as classified in public international law, into an 'obligation of result.' Classifying SCCs as an obligation of result implies they involve more than just a duty of care. However, the verification of this performance inevitably involves additional costs for suppliers.[48] The transmission of verification costs is also visible in the self-regulatory documents,[49] which illustrate that the phenomenon of non-participation in costs by buyers is a popular practice open to abuse.[50]

An analysis of the literature reveals that the most challenging consequences of SCCs appear not to be for the relationship between the parties but for third-party beneficiaries[51]: the transfer of obligations onto third parties operates a departure from the theory of privity of contract.[52] Indeed, in light of the principles of private law the recognition of third-party beneficiaries poses a challenge. As stated by Vibe Ulfbeck and Ole Hansen: '(…) the fact that developing sustainability adjusted contract law to some extent implies looking at the contractual relationship as a trilateral relationship between the contracting parties and external, societal interests.'[53]

Therefore, except for pragmatic corporate reasons, SCCs are not supposed to protect parties to the contract, but workers, local communities and the general population.[54] Moreover, SCCs are also focused on extending obligations related to the UN 2030 Agenda SDGs to commercial third parties[55]: contracts between buyers and suppliers include SCCs which create a duty to transfer (cascade) sustainability-oriented duties from primary suppliers to their suppliers or contractors (sub-suppliers, sub-contractors). The literature emphasises that the majority of SCCs are limited to second tier suppliers, i.e. sub-suppliers,[56] but their effectiveness is most often limited to the first tier.[57] In practice, SCCs seem not only to set a duty to transfer obligations onto the second tier, but unspecified further tiers, which raises questions about the depth of the chain covered by the SCCs.[58] Despite the interesting theoretical implications regarding third-party issues, it should be noted that SCCs tend not to be enforced in the *inter partes* relationship,[59] so can be deemed not helpful for harmed third parties,[60] and standard contract terms generally exclude third-party beneficiaries.[61] Therefore, any discussion regarding third-party issues invariably leads to questions regarding the reorientation of principles and primary aims of private law, which traditionally is to balance opposing interests of the parties, leaving out third-party-related interests.

The considerations set out above give grounds to specify the main characteristics of SCCs. They: are connected with the performance of contract; are 'general'; are linked to a remedy; establish a verification duty and are related to third parties.

33.3 Global Value Chains' Contracts and Due Diligence: Regulatory Model Under the EU Proposal

As demonstrated, GVCs can potentially be powerful vehicles for affecting social change at an international level.[62] Contracts containing SCCs are able to address GVC-driven externalities and thus be understood as the '(...) vessels for transporting (...) norms across borders' which '(...) directly regulate the conduct of private actors transacting internationally.'[63] However, so far, CSR and due diligence instruments have been somewhat a response to public expectations and not fully answering GVC externalities.[64] CSR and due diligence are essentially corporate voluntary and self-regulation activities, and national laws are not fully addressing overseas corporate misconduct. By contrast, public international law also has shortcomings because it applies primarily to governmental and not corporate conduct. Therefore, national laws have begun to address cross-border corporate activities at both public and private levels. The actions taken so far are often fragmented and sector-specific,[65] and the approach of private law varies. In some countries, such as France, due diligence is covered by the civil liability regime,[66] while in others, such as Germany, civil liability is excluded.[67] Nevertheless, there is an increase in regulatory activity affecting the most popular private law due diligence instruments, also concerning contractual relationships. This activity can be observed in the EU, which, in order to harmonise the EU Member States' approach to corporate due diligence, has drawn up the EU Proposal.[68] This proposal goes beyond a mere sustainability reporting and aims to establish companies' full duty of care to stakeholders in the GVC regarding human rights and environmental impacts, and influence the GVC contractual relationship. As stated above, SCCs are a popular private regulatory instrument, and the EU Proposal contains provisions that might directly influence the chain's contractual governance and the process of adopting SCCs.

Regarding the regulatory model of contractual due diligence under the EU Proposal, it sets out the obligation for the Member States to ensure that companies take appropriate measures to prevent or mitigate potential adverse impacts on human rights and environmental, and bring to an end adverse impacts.[69] One of the proposed solutions in the proposal is to oblige the companies to seek the conclusion of contractual assurances from their direct business partners.[70] Such assurances are meant to ensure that during the performance of the business relationship business partners will comply with the company's code of conduct or the company's mitigation plan.[71] This way, the EU Proposal gives the grounds to refer to the relevant self-regulatory framework, and strengthens the position of corporate codes in B2B GVC relations.

However, in the event where prevention, mitigation or minimisation and elimination of impacts is not possible, companies should refrain from entering new relations or from extending ones with the partners in connection to or in the value chain of which the impact has arisen. Moreover, the EU Proposal establishes a gradable remedy system, where the company will either temporarily suspend commercial relations while pursuing efforts to minimise or eliminate the externalities,[72] or will terminate the business relationship with respect to the activities concerned, if the adverse impact is considered severe.[73, 74]

According to the EU Proposal, besides seeking contractual assurances from their direct business partners, companies should also seek, when necessary, corresponding contractual assurances from their other (indirect) partners. This establishes a premise for transferring obligations arising under codes of conduct and contract 'cascading'.[75] 'Cascading' exists when companies may seek to conclude a contract with an indirect partner to achieve compliance with the company's code of conduct.[76] Finally, compliance with contractual assurances should be accompanied with the appropriate verification measures.[77]

The model affects third parties in a limited way; its solutions focus on the B2B relation and third-party issues regard only indirect partners – sub-suppliers, excluding workers, for example, who are the most interested third parties to be beneficiaries. The model does not establish an obligation that codes of conduct should be in any way communicated to the supplier's workers or create a framework so the entities affected by the impacts of GVCs could enforce their rights arising from the codes of conduct.

However, it should be noted that the EU Proposal does, at least in theory, strengthen the position of weaker entities in the GVC, namely SMEs. SMEs are supported in the situation where imposed compliance with self-regulation could jeopardise their viability.[78] Contractual balance is also promoted by the EU Proposal by requiring a fair, reasonable and non-discriminatory scope of duties when dealing with SMEs, and by shifting the burden of verification costs to companies, where verification is carried out in relation to SME.[79] In terms of language, the provisions of the EU Proposal supporting weaker entities are similar to those established in unfair contract terms regulations, which could be an impulse towards adopting a general, multi-sectoral regulation[80] concerning unfair contract terms as part of GVC.

These considerations allow for the reconstruction of a regulatory model of contractual due diligence under the EU Proposal oriented towards the performance of business relationships, referencing self-regulation, including a two-step gradable remedy system and an obligation to verify imposed duties, related to business third parties and supporting SMEs.

Most importantly, the EU Proposal regulatory model does not aim at creating new states of affairs or new forms of behaviour,[81] but rather aims at impose a duty to use contracts as a route for respecting human rights and ensuring environmental due diligence.[82]

The EU Proposal model is not based on a contract between buyer and supplier,[83] as is the case with procurement or supply agreements. The model the EU Proposal is based on involves contractual assurances and corporate self-regulation (codes of conduct).[84] One the one hand, the EU Proposal model provides both flexibility and stability, because within the 'established business relationship' multiple contracts may be concluded between one company and one supplier, and the model does not require the creation of separate 'sustainability' frameworks for one business relationship. Therefore, the entire business relationship between the buyer and supplier will be subject to one regime under the company's code of conduct. On the other hand – apart from the discussions on the different modes of implementation of company codes of conduct to make them theoretically effective –, it opens the problem of the very content and shape of corporate self-regulation. At the moment, codes of conduct are freely regulated by private actors. The objective scope, the degree of detail in the code, as well as the shape of the provisions, e.g., bearing the burden of the duties it imposes, are not standardised.[85]

There is, however, an obvious loophole affecting the effective functioning of the model in the EU Proposal. In theory, both Articles 7 and 8 of the EU Proposal provide for civil liability, which could entail a broader category of harm. Nonetheless, the reservation under Article 22.2 of the EU Proposal excludes liability for damages arising from concluded contracts and contractual assurances. Such a solution certainly limits the potential consequences that could result from the

EU Proposals' duty to use contracts as a route for maintaining human rights and environmental due diligence.[86] The fact that the EU Proposal strengthens the contractual due diligence mechanism should be viewed positively, but it raises concerns about whether opting for the contracts with the reservations outlined in Article 22 of the EU Proposal might diminish the significance of contractual due diligence, rendering it another item from the checklist that needs to be filled by the corporate legal compliance department. SCCs have so far not recognised the joint and several liability of GVC entities, nor have they fully supported aggrieved parties or provided for proper compensation for aggrieved parties. The functioning of the model considering Article 22 will not change the *status quo*.

Finally, it should be noted that Article 12 of the EU Proposal '(...) sets out the obligation for the Commission to adopt guidance about non-binding model contract clauses (...).' At the moment of preparing this paper, the model clauses have not been presented.[87] However, it should be expected that they will affect the practice of concluding SCCs, thus supporting their harmonisation process.[88] Nevertheless, given the significance of self-regulation under the EU Proposal, it seems surprising that the Commission is to prepare a model of contract clauses and not a model of provisions for codes of conduct. Therefore, on the one hand, the EU Proposal should lead to the harmonisation of due diligence activities, while on the other, it bases them on codes of conduct, which may merely deepen the discrepancy in the due diligence.

33.4 Conclusion

In this text, based on empirical studies, doctrinal research, and corporate sources, we analysed SCCs to describe their challenges and present characteristics. In addition, we selected provisions under the EU Proposal that are likely to have influence on contractual relations in GVCs. This allowed us to reconstruct its regulatory model of contractual due diligence. At their core, the proposed regulatory changes are related to the key characteristics of SCCs, which means that we were able to juxtapose SCC characteristics with the model (Table 33.1).

Therefore, answering the main research question in light of commercial practice and proposed regulatory changes, it appears the 'sustainability obligation' at its main core is an obligation strongly associated with third parties, which is based on self-regulation, set towards the verifiable performance of a commercial relationship, under the threat of various, gradable sanctions.

Both SCCs and solutions arising from the EU Proposal model establish self-regulation (codes of conduct) as a basis and point of reference for performing business relationships and interpreting contracts in the GVC. Additionally, the model links duties arising from contractual assurances to verify compliance in a way that is comparable to what SCCs do. As do SCCs, the model combines

Table 33.1 Juxtaposition of SCCs characteristics with the model of contractual due diligence under the EU Proposal

Characteristics of sustainability contract clauses:	*Model of contractual due diligence under the EU Proposal:*
Connection with the performance of the contract	Oriented towards the performance of business relationships
'Generality'	Referencing to self-regulation
Linked to remedy	Two-step gradable remedy system
Verification duty	Obligation to verify imposed duties
Related to third parties	Related to business third parties
–	Supporting SMEs

the performance of the obligations with the remedies. The EU proposal model applies to two remedial solutions, namely termination and suspension. Finally, the model's solutions – like those of SCCs – aim to take obligations arising under codes of conduct and to 'cascade' those obligations onto third parties (sub-suppliers, sub-contractors).

The most important observation from our analysis is that legislative activity clearly follows the steps and emulates business practice. The EU Proposal model correctly identifies the SCCs practice and follows business practice and the emerging trade customs to a significant extent. Certainly, in the perspective of enacting the EU Proposal, regulatory activity will strengthen existing practices and increase the percentage of the SCCs included in GVC contracts. From a political viewpoint, this should be looked on positively, as it gives grounds for establishing an obligatory, contractual route for 'sustainable development' in the commercial relationship.

From a theoretical viewpoint, there are certain questions that must be futurely addressed, namely: To what extent should regulations affect the very principles of contract law, e.g., limit the principle of autonomy or the principle of privity of contract in a GVC? Which self-regulatory elements – in terms of the UN 2030 Agenda SDGs – should be directly regulated, i.e., by specifying the minimum material scope of codes of conduct? Should the shape of the provisions under codes of conduct be regulated, e.g., so they are not to be shaped in an aspirational way but as an obligation of result? Should abusive clauses for other GVC sectors, besides the agriculture chain, also be regulated, and should they be extended to self-regulation?

For true change to come about, the last question may be of particular importance, as overreliance on contractual instruments imposing 'sustainability' in the GVC might not be as feasible or as effective as hoped. In practice, the buyers avoid – or, up to a certain extent, minimise – their responsibility and liability for duties arising from SCCs. A similar result is also likely be achieved through the mandatory contractual assurances in connection with Article 22 of the EU Proposal. Therefore, knowing the core of 'sustainability obligation' and taking the concept of 'sustainability' to heart, maybe we should look to the future and focus on regulating the purchasing practices of leading buyers, and not their liability for yet another signed document.

Notes

1 PhD candidate University of Silesia, LLM Comparative Law China University of Political Science and Law.

2 Professor of Law, Institute of Legal Sciences, Faculty of Law and Administration of the University of Silesia in Katowice, arbitrator.

3 Javed Siddiqui, Kenneth McPhail, and Sharmin Shabnam Rahman, 'Private governance responsibilisation in global supply chains: the case of Rana Plaza' (2020) 33(7) *Accounting, Auditing & Accountability Journal*, pp. 1569–1594.

4 Naturally, there is no comprehensive study that would indicate specific percentages of the occurrence of SCCs in GVC contracts, but most studies show they occur in more than half of all written contracts (see Chapter 2 of this paper).

5 Cristina Poncibò, 'The contractualisation of environmental sustainability' (2016) 12(4) *European Review of Contract Law*, pp. 335–355; Vibe Ulfbeck and Ole Hansen, 'Sustainability clauses in an unsustainable contract law?' (2020) 16(1) *European Review of Contract Law*, pp. 186–205.

6 Anna Beckers, 'Using contracts to further sustainability? A contract law perspective on sustainable public procurement' in Beate Sjåfjell and Anja Wiesbrock (eds), *Sustainable Public Procurement under EU Law. New Perspectives on the State as Stakeholder*, Cambridge University Press, Cambridge 2015, pp. 206–229.

7 UNGA Resolution, 'Transforming our world: the 2030 Agenda for Sustainable Development' A/RES/70/1, adopted on 25.09.2015. The agenda distinguishes 17 Goals and 169 specific Targets <https://sdgs.un.org/goals> last accessed: 20.08.2023.

8 EU Commission, 'Proposal for a Directive of the European Parliament and of the Council on Corporate Sustainability Due Diligence and amending Directive (EU) 2019/1937' COM(2022) 71 final, Brussels, 23.2.2022 COM(2022) 71 final <https://eur-lex.europa.eu/legal-content/EN/TXT/?uri=CELEX%3A52022PC0071> last accessed: 19.08.2023.

9 Namely the principles of autonomy, freedom of contract and privity of contract.

10 The most typical SCCs concern: child labour, working hours, freedom of association, collective bargaining, the use of the precautionary principle, reporting GHG emissions, control of soil and water contamination, energy savings, appropriate waste disposal, and logistics. See, Robert McCorquodale, Lise Smit, Stuart Neely and Robin Brooks, 'Human rights due diligence in law and practice: good practices and challenges for business enterprises' (2017) 2(2) *Business and Human Rights Journal*, pp. 195–224; Fabrizzio Cafaggi, 'Regulation through contracts: supply-chain contracting and sustainability standards' (2016) 12(3) *European Review of Contract Law*, pp. 218–258.

11 Our understanding of 'implied terms' is the same as in Ole Lando and Hugh Beale, *Principles of European Contract Law* (Kluwer Law International, Hague 2000), Article 6:102; and the Draft Common Frame of Reference Article II.-9:101(2) in Christian von Bar, Eric Clive and Hans Schulte-Nolke (eds), *Principles, Definitions and Model Rules of European Private Law. Draft Common Frame of Reference (DCFR)* (Outline Edition, Sellier European Law Publishers, Munich 2009); compare Article 5.1.2. UNIDROIT Principles of International Commercial Contracts (2016), Principle No. IV.6.1 of Transnational Commercial Law <https://www.trans-lex.org/928900/highlight_express_or_implied/express-and-implied-obligations> last accessed: 31.08.2023; see also Frederick Wilmot-Smith, 'Express and Implied Terms' (2023) 43(1) *Oxford Journal of Legal Studies*, pp. 54–75.

12 We base our reconstructed model on the EU Proposal, but acknowledge the EU Parliament's proposed amendments and will refer to them in the footnotes; see EU Parliament, 'Amendments on the proposal for a directive of the European Parliament and of the Council on Corporate Sustainability Due Diligence and amending Directive (EU) 2019/1937' P9_TA(2023)0209 1.06.2023 <https://www.europarl.europa.eu/doceo/document/TA-9-2023-0209_EN.html> last accessed: 30.08.2023.

13 With the reference to the 2013 United Nations study, UNCTAD, 'Global Value Chains and Development. Investment and value-added trade in the global economy. A preliminary analysis' UNCTAD/DIAE/2013/1, <https://unctad.org/system/files/official-document/diae2013d1_en.pdf> last accessed: 19.08.2023.

14 Klaas Hendrik Eller, 'Is 'global value chain' a legal concept? Situating contract law in discourses around global production' (2020) 16(1) *European Review of Contract Law*, pp. 3–24.

15 On several of these issues, Sarah Joseph, 'Taming the Leviathans: multinational enterprises and human rights' (1999) 46(2) *Netherlands International Law Review*, pp. 171–203; Vibe Ulfbeck, Alexandra Andhov and Katerina Mitkidis (eds), *Law and Responsible Supply Chain Management: Contract and Tort Interplay and Overlap*, Routledge 2019; Jaakko Salminen and Mikko Rajavuori, 'Private international law, global value chains and the externalities of transnational production: towards alignment?' (2021) 12(2) *Transnational Legal Theory*, pp. 230–248.

16 This approach is popular for GVCs producing consumer goods. See, Anne-Titia Bové and Steven Swartz, *Starting at the Source: Sustainability in Supply Chains*, McKinsey&Company 2016 <https://www.mckinsey.com/capabilities/sustainability/our-insights/starting-at-the-source-sustainability-in-supply-chains> last accessed: 19.08.2023.

17 Veronica H. Villena and Dennis A. Gioia, 'A more sustainable supply chain' March–April 2020 *Harvard Business Review*, p. 86.

18 Klaas Hendrik Eller and Jaakko Salminen, 'Reimagining contract in a world of global value chains' (2020) 16(1) *European Review of Contract Law*, pp. 1–2.

19 See EU-funded Research and Innovation Project (Horizon 2020) 'Contracts 2.0' <https://www.project-contracts20.eu/> last accessed: 25.08.2023, with special reference to the findings on contracts associated with agricultural GVCs; see <https://www.project-contracts20.eu/best-practice-examples/> last accessed: 25.08.2023.

20 Paul Verbruggen, 'New liabilities in global value chains: an introduction' (2022) 13(4) *European Journal of Risk Regulation*, pp. 541–547.

21 Lyn K.L. Tjon Soi Len, 'Hermeneutical injustice, contract law, and global value chains' (2020) 16(1) *European Review of Contract Law*, pp. 139–159.

22 Klaas Hendrik Eller and Jaakko Salminen, 'Reimagining contract in a world of global value chains' (2020) 16(1) *European Review of Contract Law*, pp. 1–2.

23 This information comes from the business literature, which omits certain details on legal issues such as contractual terms. Moreover, companies do not disclose such information due to trade secret policies. See similar observations in Jaakko Salminen, 'Towards a genealogy and typology of governance through contract beyond privity' (2020) 16(1) *European Review of Contract Law*, pp. 25–43.

24 European Commission, Directorate-General for Justice and Consumers, F. Torres-Cortes, C. Salinier, H. Deringer, et al., 'Study on due diligence requirements through the supply chain: final report' Publications Office 2020, p. 63 <https://data.europa.eu/doi/10.2838/688> last accessed: 1.09.2023.

25 Ecovadis and Affectio Mutandi, 'Sustainability Clauses in Commercial Contracts: The Key to Corporate Responsibility. Study of CSR Contractual Practices Among Buyers and Suppliers' (2018) p. 13 <https://resources.ecovadis.com/whitepapers/sustainability-clauses-commercial-contracts-key-corporate-responsibility> last accessed: 1.09.2023.

26 Kateřina Mitkidis, Sonja Perkovic and Panagiotis Mitkidis, 'Tendencies in contractual governance to promote human and labour rights in transnational supply chains' (2019) 23(4) *Competition & Change*, with a special reference to the Table 2, p. 13.

27 David V Snyder, 'Contracts as an instrument of international management and governance' in Susan A. Maslow and David V. Snyder (eds), *Contracts for Responsible and Sustainable Supply Chains: Model Contract Clauses, Legal Analysis, and Practical Perspectives*, 2023 <https://papers.ssrn.com/sol3/papers.cfm?abstract_id=4530887> last accessed: 29.08.2023.

28 Vibe Ulfbeck and Ole Hansen, 'Sustainability clauses in an unsustainable contract law?' (2020) 16(1) *European Review of Contract Law*, pp. 186–205.

29 Katerina Mitkidis, *Sustainability Clauses in International Business Contracts*, Eleven Publishing, The Netherlands 2015, p. 75.

30 Issues at the intersection of 'sustainable development' and products generally concern the consumer-seller rather than the buyer-supplier relationship. This applies to claims based on consumer and sales laws – issues such as planned obsolescence, products' ecological inefficiency, their design, or remedies for their lack of conformity (repair, replacement), or competition and advertisement laws (e.g., greenwashing and labelling schemes). See, Fryderyk Zoll, Johnaton Watson, Katarzyna Południak-Gierz et al., 'Academic proposal for amending the directive (EU) 2019 771 of the European Parliament and of the council of 20 May 2019 on certain aspects concerning contracts for the sale of goods' (2022) 3 *Transformacje Prawa Prywatnego*, pp. 109–159; Katarzyna Południak-Gierz, 'Eco-reasonableness. Possibilities of incorporating green principles into general private law clauses' in Mads Andenas, Maren Heidemann (eds) *Quo vadis Commercial Contract? Reflections on Sustainability, Ethics and Technology in the Emerging Law and Practice of Global Commerce* (Springer International Publishing 2023, pp. 107–127). The socio-environmental non-sustainability of the production process could also be considered a good's lack of conformity and provide warranty rights or other claims. It could thus, theoretically, create a trilateral relation of a consumer-buyer-supplier. In modern realities, creating the possibility of enforcing such claims may put pressure on the main buyers (sellers of the product) to have certain standards on their value chains.

31 Marta Andrecka and Katerina Mitkidis, 'Sustainability requirements in EU public and private procurement-a right or an obligation' (2017) 1 *Nordic Journal of Commercial Law*, pp. 55–89.

32 Marek Safjan, 'Źródła prawa prywatnego' in Marek Safjan (ed) *System Prawa Prywatnego. Prawo cywilne – część ogólna* (Book 1, C.H. Beck, Instytut Nauk Prawnych PAN, Warszawa 2012, pp. 182–184); Radosław Strugała, 'Art. 3531' in Piotr Machnikowski (ed) *Zobowiązania Tom I Przepisy ogólne i powiązane przepisy Księgi I KC, Komentarz* (CH Beck, Warszawa 2022, pp. 1513–1524).

33 Charity Ryerson, Dean Pinkert and Avery Kelly, 'Seeking justice: the state of transnational corporate accountability' (2022) *Yale Law Journal Forum*, pp. 787–813 <https://www.yalelawjournal.org/pdf/F7.CALFinalDraftWEB_puwuyish.pdf> accessed 15.08.2023.

34 Vibe Ulfbeck and Ole Hansen, 'Sustainability clauses in an unsustainable contract law?' (2020) 16(1) *European Review of Contract Law*, pp. 186–205.

35 Ecovadis and Affectio Mutandi, 'Sustainability Clauses in Commercial Contracts: The Key to Corporate Responsibility. Study of CSR Contractual Practices Among Buyers and Suppliers' (2018) p. 14 <https://resources.ecovadis.com/whitepapers/sustainability-clauses-commercial-contracts-key-corporate-responsibility> last accessed: 1.09.2023.

36 Rentokil Supplier General Terms and Conditions Point 18.1 'The Seller shall establish and maintain appropriate business standards, procedures and controls to ensure compliance with the Company's Supplier Code of Conduct (…) and also with all environmental regulations, labour laws and best practice in the Seller's industry.'; and Point 18.3 'In performing its obligations under this agreement the Seller shall

(a) comply with all applicable anti-slavery and human trafficking laws, statutes, regulations and codes from time to time in force.' <https://www.rentokil.ie/termsandconditions/supplier-terms-conditions/> last accessed: 20.08.2023. Glaxosmithkline Terms and Conditions of Purchase Point 8.1 [Standard of Services] 'Supplier warrants and represents to GSK that any Services performed by Supplier or duly appointed sub-contractor: 8.1.1. shall be performed in a good and workmanlike fashion and with all due speed, care, skill, and diligence; and 8.1.2. shall be carried out in accordance with the Agreement, with current industry standard codes of practice, and the highest standards prevailing in Supplier's industry.' <https://www.gsk.com/media/2807/gsk-uk-terms-and-conditions-goods-and-services.pdf> last accessed: 19.08.2023.

37 E.g., ISO 26000 guidance standard on social responsibility <https://www.iso.org/iso-26000-social-responsibility.html> accessed 15.08.2023.

38 Katerina Mitkidis, *Sustainability Clauses in International Business Contracts*, Eleven Publishing, The Netherlands 2015, pp. 65–83, Anna Beckers, 'Regulating corporate regulators through contract law? The case of corporate social responsibility codes of conduct' (2016) 12 *EUI Working Papers – Max Weber Programme*, pp. 1–27 <https://hdl.handle.net/1814/41485> last accessed: 25.08.2023.

39 This claim is further supported by the study Robert Eadie and Shane Rafferty, 'Do corporate social responsibility clauses work? A contractor perspective' (2014) 7(1) *International Journal of Procurement Management*, pp. 19–34 with reference to Table 1. On the weakness of the human rights obligation for business entities under international and domestic laws, Markus Krajewski, 'A nightmare or a noble dream? Establishing investor obligations through treaty-making and treaty-application' (2020) 5(1) *Business and Human Rights Journal,* pp. 106–129.

40 Smiths Group PLC Standard Purchasing Terms and Conditions for Goods and Services Point 12.1 'You will conduct your business ethically and lawfully and in accordance with our Supplier Code of Business Ethics (…) or an equivalent code of ethics.' <https://www.smiths.com/-/media/files/purchasing-terms-and-conditions/smith-group-plc--july-2018--poland-procurement-terms--dual-language.pdf> last accessed 20.08.2023.

41 Martin Herberg, 'Global legal pluralism and interlegality: environmental self-regulation in multinational enterprises as global law-making' in Olaf Dilling, Martin Herberg and Gerd Winter (eds) *Responsible Business: Self-Governance and Law in Transnational Economic Transactions*, Hart Publishing, Oxford and Portland, OR 2008, pp. 17–40.

42 See the example of contractual assurance, NationalGrid Supplier Code of Conduct Acknowledgement <https://www.nationalgridus.com/media/procurement/supplier_ethics_acknowledgement.pdf> last accessed 15.08.2023.

43 On the general issue of legal enforcement of sustainability-oriented duties under corporate codes; Anna Beckers, *Enforcing Corporate Social Responsibility Codes: On Global Self-Regulation and National Private Law*, Hart Publishing, Oxford 2015.

44 Stefan Grundmann and Mateusz Grochowski, 'The creation of norms: an evolutionary view on European contract law' in Stefan Grundmann and Mateusz Grochowski (eds), *European Contract Law and the Creation of Norms*, Intersentia 2021, pp. 5–6; Florian Möslein, 'Ambiguities of self-regulation: some illustrative examples of "good" companies' certification' in Stefan Grundmann and Mateusz Grochowski (eds), *European Contract Law and the Creation of Norms*, Intersentia 2021, pp. 175–198. See also the older literature, the findings of which remain valid in terms of the re-conceptualisation of *lex mercatoria* and establishment of transnational law; pp. 177–192; Clive Schmitthoff, 'International business law. A new law merchant' (1961) 2 *Current Law and Social Problems*, p. 129; Philip Jessup, *Transnational Law*, Yale University Press, New Haven, CT 1956.

45 Ecovadis and Affectio Mutandi, 'Sustainability clauses in commercial contracts: the key to corporate responsibility. Study of CSR contractual practices among buyers and suppliers' (2018) p. 48 <https://resources.ecovadis.com/whitepapers/sustainability-clauses-commercial-contracts-key-corporate-responsibility> last accessed: 1.09.2023.

46 Jaap Baaij and Alex Geert Castermans, 'The potential of contractual assurances to advance supply chain due diligence' (2023) 28 *EUI RSC Working Paper Global Governance Programme*, p. 19 <https://hdl.handle.net/1814/75509> last accessed 15.08.2023; Fabrizio Cafaggi, 'Regulation through contracts: supply-chain contracting and sustainability standards' (2016) 12(3) *European Review of Contract Law*, pp. 224, 246.

47 Virginie Barral, 'Sustainable development in international law: nature and operation of an evolutive legal norm' (2012) 23(2) *European Journal of International Law*, pp. 377–400.

48 Stefano Ponte, 'The hidden costs of environmental upgrading in global value chains' (2022) 29(3) *Review of International Political* Economy, pp. 818–843. AstraZeneca General terms and conditions of purchase [Germany] Point 12 [Auditing] 'Following a reasoned request from AstraZeneca, the contractor must allow [for a verification compliant with] appropriate ethical standards in accordance with the requirements of this Agreement. If AstraZeneca requires that the audit be carried out by a specified third party, the contractor agrees to carry out this audit and to pay the fees incurred for the third party's audit.'; <https://www.astrazeneca.de/content/dam/az-de/pdfs/General-terms-and-conditions-of-purchase-GCP.pdf> last accessed: 20.08.2023.

49 Rolls Royce Global Supplier Code of Conduct Point 5.2 [Global Supplier Code of Conduct adherence] '(...) On our request, suppliers shall complete evidence-based assessments relating to their performance on environmental, social and governance matters and will share the results of such assessments with us.' <https://www.rolls-royce.com/~/media/Files/R/Rolls-Royce/documents/sustainability/global-supplier-code-of-conduct-en-2023.pdf > last accessed: 20.08.2023.

50 According to Ecovadis and Affectio Mutandis, only in 25% of SCCs the cost of the verification is shared between buyers and suppliers, with the rest charging the suppliers. Ecovadis and Affectio Mutandi, 'Sustainability Clauses in Commercial Contracts: The Key to Corporate Responsibility. Study of CSR Contractual Practices Among Buyers and Suppliers' (2018) p. 15 <https://resources.ecovadis.com/whitepapers/sustainability-clauses-commercial-contracts-key-corporate-responsibility> last accessed: 1.09.2023.

51 Katerina Mitkidis, 'Enforcement of sustainability contractual clauses in supply chains by third parties' in Vibe Ulfbeck, Alexandra Andhov, Katerina Mitkidis (eds), *Law and Responsible Supply Chain Management Contract and Tort Interplay and Overlap*, Routledge 2019, pp. 65–88. James Thuo Gathii, 'Incorporating the third party beneficiary principle in natural resource contracts' (2014) 43 *Georgia Journal of International & Comparative Law*, pp. 94–137.

52 Cristina Poncibò, 'The contractualisation of environmental sustainability' (2016) 12(4) *European Review of Contract Law*, p. 354. Edoardo Marcenaro and Pietro Galizzi, 'The Sustainable Contract' (2021) 14(1) *International In-House Counsel Journal*, Springer, p. 4.

53 Vibe Ulfbeck and Ole Hansen, 'Sustainability clauses in an unsustainable Contract Law?' (2020) 16(1) *European Review of Contract Law*, p. 205.

54 Sarah Dadush, 'Prosocial contracts: making relational contracts more relational' (2022) 82 *Law & Contemporary Problems*, pp. 156–157.

55 NEXT Terms and Conditions of Purchase Point 8 [Code of Practice] '8.1 All Product supplied to NEXT will be manufactured in accordance with the NEXT Code of Practice Manual. The Seller is responsible for ensuring any subcontractors engaged in any part of the manufacturing, finishing or packaging processes are compliant with the NEXT Code of Practice as set out in the Code of Practice Manual, as updated from time to time.' <https://xcdn.next.co.uk/PDFS/Next_Terms_and_Conditions.pdf> last accessed: 18.08.2023.

56 Ecovadis and Affectio Mutandi, 'Sustainability clauses in commercial contracts: the key to corporate responsibility. Study of CSR contractual practices among buyers and suppliers' (2018) p. 16 <https://resources.ecovadis.com/whitepapers/sustainability-clauses-commercial-contracts-key-corporate-responsibility> last accessed: 1.09.2023.

57 Miriam M. Wilhelm et al., 'Sustainability in multi-tier supply chains: understanding the double agency role of the first-tier supplier' (2016) 41 *Journal of operations management*, pp. 42–60.

58 Rio Tinto Global Purchase Order Conditions for Goods and Services Point 26.6 'The Supplier must ensure that it and its subcontractors (and to the extent practicable, its other suppliers and business partners) will comply with all applicable laws, statutes and regulations in force pertaining to modern slavery (which is deemed to include forced labour, human trafficking and child labour) and take appropriate steps to meet international standards around modern slavery where these set a higher standard than domestic law.' <https://www.riotinto.com/-/media/Content/Documents/Suppliers/Terms/RT-Purchase-order-global-goods-and-services-EN.pdf> last accessed: 19.08.2023.

59 In the case of private and not judicial enforcement as stated above.

60 Observations from the KiK lawsuit. For an in-depth analysis see Jan Erik Schirmer, *Nachhaltiges Privatrecht*, Jus Privatum 270, Mohr Siebeck, 2023, pp. 274–368.

61 Experian Supplier Standard Terms and Conditions Point 12.3 'Persons or entities who are not a party to the Agreement (other than Experian's Affiliates or Experian Indemnified Parties, as applicable) shall not have any rights under the Agreement and the Parties hereby agree that nothing in the Agreement shall be construed as creating a right that is enforceable by any person or entity that is not a Party to

the Agreement or a permitted assignee of such Party.' <https://www.experian.com/assets/corporate/supplier-terms-and-conditions.pdf> last accessed: 20.08.2023.

62 Fabrizio Cafaggi and Paola Iamiceli, 'Regulating contracting in global value chains. Institutional alternatives and their implications for transnational cContract law' (2020) 16(1) *European Review of Contract Law*, pp. 44–73.

63 Sarah Dadush, 'Prosocial contracts: making relational contracts more relational' (2022) 82 *Law & Contemporary Problems*, p. 158.

64 The Tobacco industry can serve as an example here, demonstrating that voluntary CSR does not advance human rights and environmental protection and mainly serves as a tool for making profit. Daniele van Kalmthout et al., 'Mandatory environmental and human rights due diligence' (2021) 19 *Tobacco Induced Diseases*.

65 E.g., EU Parliament and Council Regulation (EU) No 995/2010 of 20 October 2010 laying down the obligations of operators who place timber and timber products on the market [2010] OJ L295/23.

66 LOI no 2017-399 du 27 mars 2017 relative au devoir de vigilance des sociétés mères et des entreprises donneuses d'ordre (1) NOR: ECFX1509096L, Journal Officiel de la Republique Française Texte 1 sur 99, 28.03.2017 <https://www.legifrance.gouv.fr/jorf/id/JORFTEXT000034290626/> last accessed: 20.08.2023.

67 Gesetz über die unternehmerischen Sorgfaltspflichten in Lieferketten, Bundesgesetzblatt Teil I, 2021, Nr. 46 vom 22.07.2021 <https://dip.bundestag.de/vorgang/.../275852> last accessed: 20.08.2023.

68 EU Commission, 'Proposal for a Directive of the European Parliament and of the Council on Corporate Sustainability Due Diligence and amending Directive (EU) 2019/1937' COM(2022) 71 final, Brussels 23.2.2022 <https://eur-lex.europa.eu/legal-content/EN/TXT/?uri=CELEX%3A52022PC0071> last accessed: 19.08.2023.

69 Art. 7.1 EU Proposal, Art. 8.1 EU Proposal.

70 Doubts can be raised by the use of term 'business relationship' in the French Loi de vigilance. Nevertheless, the use of this term extends due diligence to other indirect business partners beyond the contractual relationship (unspecified n-tier suppliers). Through the use of this term, the EU Proposal can be also applied to the entities with whom buyers do not have written contract. This is a practice in some GVCs and in one third of the business relationships. See INWORK International Labour Office, 'Issue Brief No.10. Purchasing practices and working conditions in global supply chains: Global Survey results' Geneva 2017, p. 3 <https://www.ilo.org/travail/info/fs/WCMS_556336/lang--en/index.htm> last accessed: 25.08.2023.

71 Art. 7.2(b), Art. 8.3(c) EU Proposal.

72 Art. 7.5(a), 8.6(a) EU Proposal.

73 Art. 7.5(b), 8.6(b) EU Proposal.

74 However, Article 7.6 and 8.7 EU Proposal unfortunately establishes that regarding financial services they are not obliged to terminate the loan, loan or other financial services agreement.

75 Art. 7.2(b), 8.3(c) EU Proposal.

76 Art. 7.3, Art. 8.4 EU Proposal.

77 Art. 7.4, 8.5 EU Proposal.

78 Art. 7.2(d), Art. 8.3(e) EU Proposal.

79 Art. 7.4, Art. 8.5 EU Proposal.

80 Unlike the current EU Parliament and Council Directive (EU) 2019/633 of the of 17 April 2019 on unfair trading practices in business-to-business relationships in the agricultural and food supply chain [2019] OJ L 111 <http://data.europa.eu/eli/dir/2019/633/oj> last accessed: 19.08.2023.

81 Jaap Baaij and Alex Geert Castermans, 'The potential of contractual assurances to advance supply chain due diligence' (2023) 28 *EUI RSC Working Paper Global Governance Programme*, pp. 10–11 <https://hdl.handle.net/1814/75509> last accessed 15.08.2023.

82 Nonetheless, this duty is imposed only on the 'largest' companies, i.e. with more than 500 employees and a net worldwide turnover of more than EUR 150 million in the preceding financial year, as well as companies from the three sectors (textiles, agriculture and extraction of minerals) with more than 250 employees and net worldwide turnover of more than EUR 40 million; *vide* Art. 2.1(a) and (b) EU Proposal. Additionally, EU Proposal applies both to EU and non-EU companies.

83 The EU Parliaments proposed an amendment which considers 'contractual provisions' and not 'contractual assurances' as the main instrument, which is more in line with the current business practice. See EU Parliament, 'Amendments on the proposal for a directive of the European Parliament and of the Council on Corporate Sustainability Due Diligence and amending Directive (EU) 2019/1937' P9_TA(2023)0209

1.06.2023, pp. 96–116 <https://www.europarl.europa.eu/doceo/document/TA-9-2023-0209_EN.html> last accessed: 30.08.2023.

84 However, the EU Parliaments amendments downscales the importance of self-regulation. See EU Parliament, 'Amendments on the proposal for a directive of the European Parliament and of the Council on Corporate Sustainability Due Diligence and amending Directive (EU) 2019/1937' P9_TA(2023)0209 1.06.2023, pp. 96–116 <https://www.europarl.europa.eu/doceo/document/TA-9-2023-0209_EN.html> last accessed: 30.08.2023.

85 As shown in Chapter 2, shifting the costs onto weaker entities within GVC is a popular practice and obvious challenge from the perspective of SCCs and the principles of private law. With this measure, that applies primarily to SMEs, the EU Proposal model tries to achieve a contractual balance and to support weaker entities.

86 For various perspectives on the assessment of the EU Proposal in light of private law, see European Company Law Experts Group, 'Legal certainty and the directive on Corporate Sustainability Due Diligence', The ECGI Blog 2.08.2022 <https://ecgi.global/blog/legal-certainty-and-directive-corporate-sustainability-due-diligence> last accessed: 15.08.2023, Claire Methven O'Brien and Olga Martin-Ortega, 'Sustainable corporate governance: submission to Consultation on European Commission's proposal for a Directive on corporate sustainability due diligence COM (2022) 71 final. (2022)' 23.05.2022 <https://www.researchgate.net/profile/Claire-Obrien-12/publication/360845316_Sustainable_corporate_governance_Submission_to_Consultation_on_European_Commission's_proposal_for_a_Directive_on_corporate_sustainability_due_diligence_COM202271_final> last accessed: 19.08.2023, Charity Ryerson, Dean Pinkert and Avery Kelly, 'Seeking justice: the state of transnational corporate accountability' (2022) *Yale Law Journal Forum*, pp. 787–813 <https://www.yalelawjournal.org/pdf/F7.CALFinalDraftWEB_puwuyish.pdf> accessed 15.08.2023.

87 Some entities already produced model SCCs. See American Bar Association <https://www.americanbar.org/groups/human_rights/business-human-rights-initiative/contractual-clauses-project/> last accessed: 15.08.2023; The Chancery Lane Project < https://chancerylaneproject.org/> last accessed: 15.08.2023. On the evaluation of model SCCs from the perspective of Sustainable Development Goal 12 UN 2030 Agenda, see Nevena Jevremović, 'Enforcing sustainability clauses in global supply chains with a focus on proactive and relational contract rheory: case of SDG 12' (2022) 25(2) *ZEuS Zeitschrift für Europarechtliche Studien*, pp. 393–418.

88 Model clauses '(…) may be helpful in terms of clarity and harmonisation, their effectiveness and efficacy remain limited because of their non-binding nature for recipient companies and third parties, since compliance with the guidelines does not constitute a safe harbour'. Alain Pietrancosta, 'Codification in company law of general CSR requirements: pioneering recent French reforms and EU perspectives' (2022) *European Corporate Governance Institute-Law Working Paper* no. 639, p. 36 <https://ssrn.com/abstract=4083398> last accessed: 25.08.2023.

Bibliography

Andrecka M and Mitkidis K, 'Sustainability requirements in EU public and private procurement—a right or an obligation' (2017) 1 *Nordic Journal of Commercial Law*, pp. 55–89.

Baaij J and Castermans AG, 'The potential of contractual assurances to advance supply chain due diligence' (2023) 28 *EUI RSC Working Paper Global Governance Programme*, p. 19 <https://hdl.handle.net/1814/75509> last accessed 15.08.2023.

Barral V, 'Sustainable development in international law: nature and operation of an evolutive legal norm' (2012) 23(2) *European Journal of International Law*, pp. 377–400.

Beckers A, *Enforcing Corporate Social Responsibility Codes: On Global Self-Regulation and National Private Law*, Oxford Hart Publishing 2015.

Beckers A, 'Using contracts to further sustainability? A contract law perspective on sustainable public procurement' in Beate Sjåfjell and Anja Wiesbrock (eds), *Sustainable Public Procurement under EU Law. New Perspectives on the State as Stakeholder*, Cambridge University Press, Cambridge 2015, pp. 206–229.

Beckers A, 'Regulating corporate regulators through contract law? The case of corporate social responsibility codes of conduct' (2016) 12 *EUI Working Papers – Max Weber Programme*, pp. 1–27 <https://hdl.handle.net/1814/41485> last accessed: 25.08.2023.

Bové A-T and Swartz S, *Starting at the Source: Sustainability in Supply Chains*, McKinsey&Company 2016 <https://www.mckinsey.com/capabilities/sustainability/our-insights/starting-at-the-source-sustainability-in-supply-chains> last accessed: 19.08.2023.

Cafaggi F, 'Regulation through contracts: supply-chain contracting and sustainability standards' (2016) 12(3) *European Review of Contract Law*, pp. 224, 246.
Cafaggi F and Iamiceli P, 'Regulating contracting in global value chains. Institutional alternatives and their implications for transnational contract law' (2020) 16(1) *European Review of Contract Law*, pp. 44–73.
Dadush S, 'Prosocial contracts: making relational contracts more relational' (2022) 82 *Law & Contemporary Problems*, pp. 156–157.
Eadie R and Rafferty S, 'Do corporate social responsibility clauses work? A contractor perspective' (2014) 7(1) *International Journal of Procurement Management*, pp. 19–34.
Eller KH, 'Is 'global value chain' a legal concept? Situating contract law in discourses around global production' (2020) 16(1) *European Review of Contract Law*, pp. 3–24.
Eller KH, Jaakko Salminen, 'Reimagining contract in a world of global value chains' (2020) 16(1) *European Review of Contract Law*, pp. 1–2.
Gathii JT, 'Incorporating the third party beneficiary principle in natural resource contracts' (2014) 43 *Georgia Journal of International & Comparative Law*, pp. 94–137.
Grundmann S and Grochowski M, 'The creation of norms: an evolutionary view on European contract law' in Stefan Grundmann and Mateusz Grochowski (eds), *European Contract Law and the Creation of Norms*, Intersentia 2021, pp. 5–6.
Herberg M, 'Global legal pluralism and interlegality: environmental self-regulation in multinational enterprises as global law-making' in Olaf Dilling, Martin Herberg and Gerd Winter (eds), *Responsible Business: Self-Governance and Law in Transnational Economic Transactions*, Hart Publishing, Oxford and Portland, OR 2008, pp. 17–40.
Jessup P, *Transnational Law*, Yale University Press, New Haven, CT 1956.
Jevremović N, 'Enforcing sustainability clauses in global supply chains with a focus on proactive and relational contract theory: case of SDG 12' (2022) 25(2) *ZEuS Zeitschrift für Europarechtliche Studien*, pp. 393–418.
Joseph S, 'Taming the leviathans: multinational enterprises and human rights' (1999) 46(2) *Netherlands International Law Review*, pp. 171–203.
Krajewski M, 'A nightmare or a noble dream? Establishing investor obligations through treaty-making and treaty-application' (2020) 5(1) *Business and Human Rights Journal*, pp. 106–129.
Lando O and Beale H, *Principles of European Contract Law*, Kluwer Law International, Hague 2000.
Lyn KL and Len TS, 'Hermeneutical injustice, contract law, and global value chains' (2020) 16(1) *European Review of Contract Law*, pp. 139–159.
Marcenaro E and Galizzi P, 'The sustainable contract' (2021) 14(1) *International In-House Counsel Journal*, Springer, p. 4.
McCorquodale R, Smit L, Neely S and Brooks R, 'Human rights due diligence in law and practice: good practices and challenges for business enterprises' (2017) 2(2) *Business and Human Rights Journal*, pp. 195–224.
Mitkidis K, *Sustainability Clauses in International Business Contracts*, Eleven Publishing, The Netherlands 2015, pp. 65–83.
Mitkidis K, 'Enforcement of sustainability contractual clauses in supply chains by third parties' in Vibe Ulfbeck, Alexandra Andhov and Katerina Mitkidis (eds), *Law and Responsible Supply Chain Management Contract and Tort Interplay and Overlap*, Routledge 2019, pp. 65–88.
Mitkidis K, Perkovic S and Mitkidis P, 'Tendencies in contractual governance to promote human and labour rights in transnational supply chains' (2019) 23(4) *Competition & Change*, pp. 394–419.
Möslein F, 'Ambiguities of self-regulation: some illustrative examples of "good" companies' certification' in Stefan Grundmann and Mateusz Grochowski (eds), *European Contract Law and the Creation of Norms*, Intersentia 2021, pp. 175–198.
O'Brien CM, and Martin-Ortega O, 'Sustainable corporate governance: submission to Consultation on European Commission's proposal for a Directive on corporate sustainability due diligence COM (2022) 71 final. (2022)' 23.05.2022 <https://www.researchgate.net/profile/Claire-Obrien-12/publication/360845316_Sustainable_corporate_governance_Submission_to_Consultation_on_European_Commission's_proposal_for_a_Directive_on_corporate_sustainability_due_diligence_COM202271_final> last accessed: 19.08.2023.
Pietrancosta A, 'Codification in company law of general CSR requirements: pioneering recent French reforms and EU perspectives' (2022) European Corporate Governance Institute-Law Working Paper no. 639, p. 36 <https://ssrn.com/abstract=4083398> last accessed: 25.08.2023.

Południak-Gierz K, 'Eco-reasonableness. Possibilities of incorporating green principles into general private law clauses' in Mads Andenas and Maren Heidemann (eds), *Quo vadis Commercial Contract? Reflections on Sustainability, Ethics and Technology in the Emerging Law and Practice of Global Commerce* (Springer International Publishing 2023), pp. 107–127.

Poncibò C, 'The contractualisation of environmental sustainability' (2016) 12(4) *European Review of Contract Law*, pp. 335–355.

Ponte S, 'The hidden costs of environmental upgrading in global value chains' (2022) 29(3) *Review of International Political Economy*, pp. 818–843.

Ryerson C, Pinkert D and Kelly A, 'Seeking justice: the state of transnational corporate accountability' (2022) *Yale Law Journal Forum*, pp. 787–813 <https://www.yalelawjournal.org/pdf/F7.CALFinalDraftWEB_puwuyish.pdf> accessed 15.08.2023.

Safjan M, 'Źródła prawa prywatnego' in Marek Safjan (ed), *System Prawa Prywatnego. Prawo cywilne – część ogólna* Book 1, C.H. Beck, Instytut Nauk Prawnych PAN, Warszawa 2012, pp. 182–184.

Salminen J, 'Towards a genealogy and typology of governance through contract beyond privity' (2020) 16(1) *European Review of Contract Law*, pp. 25–43.

Salminen J and Rajavuori M, 'Private international law, global value chains and the externalities of transnational production: towards alignment?' (2021) 12(2) *Transnational Legal Theory*, pp. 230–248.

Schmitthoff C, 'International business law. A new law merchant' (1961) 2 *Current Law and Social Problems*, p. 129.

Snyder DV, 'Contracts as an instrument of international management and governance' in Susan A Maslow and David V Snyder (eds), *Contracts for Responsible and Sustainable Supply Chains: Model Contract Clauses, Legal Analysis, and Practical Perspectives* 2023 <https://papers.ssrn.com/sol3/papers.cfm?abstract_id=4530887> last accessed: 29.08.2023.

Strugała R, 'Art. 3531' in Piotr Machnikowski (ed), *Zobowiązania Tom I Przepisy ogólne i powiązane przepisy Księgi I KC, Komentarz*, CH Beck, Warszawa 2022, pp. 1513–1524.

Ulfbeck V, Andhov A and Mitkidis K (eds), *Law and Responsible Supply Chain Management: Contract and Tort Interplay and Overlap*, Routledge, 2019.

Ulfbeck V and Hansen O, 'Sustainability clauses in an unsustainable contract law?' (2020) 16(1) *European Review of Contract Law*, pp. 186–205.

van Kalmthout D et al., 'Mandatory environmental and human rights due diligence' (2021) 19 *Tobacco Induced Diseases*, pp. 1–10.

Verbruggen P, 'New liabilities in global value chains: an introduction' (2022) 13(4) *European Journal of Risk Regulation*, pp. 541–547.

Villena VH and Gioia DA, 'A more sustainable supply chain' March–April (2020) *Harvard Business Review*, p. 86. https://hbr.org/2020/03/a-more-sustainable-supply-chain

von Bar C, Clive E and Schulte-Nolke H (eds), *Principles, Definitions and Model Rules of European Private Law. Draft Common Frame of Reference (DCFR)* (Outline Edition, Sellier European Law Publishers, Munich 2009).

Wilhelm MM et al., 'Sustainability in multi-tier supply chains: understanding the double agency role of the first-tier supplier' (2016) 41 *Journal of Operations Management*, pp. 42–60.

Wilmot-Smith F, 'Express and implied terms' (2023) 43(1) *Oxford Journal of Legal Studies*, pp. 54–75.

Zoll F, Watson J, Południak-Gierz K et al., 'Academic proposal for amending the Directive (EU) 2019 771 of the European Parliament and of the Council of 20 May 2019 on certain aspects concerning contracts for the sale of goods' (2022) 3 *Transformacje Prawa Prywatnego*, pp. 109–159.

34

CONTRACT LAW AND SUSTAINABILITY OF GLOBAL VALUE CHAINS

Assessing the Proposal for an EU Corporate Sustainability Due Diligence Directive from a Contract Law Perspective

Rebecca Ravalli

34.1 Introduction

Private law has increasingly become a key tool for the regulation of sustainability. In the lack of a commonly agreed definition of sustainability, the notion has been translated into a variety of different standards – of private or public origin – and definitions, which serve the purpose of orienting the activities of both private and public actors towards the accomplishment of the UN Sustainable Development Goals (SDGs).[1]

Private law, and contract law more specifically, have assumed a role in guiding this action, especially with regard to the promotion of sustainability in processes of production through the regulation of global value chains (GVCs) for the prevention, mitigation and remediation of adverse impacts on people and the environment.[2]

In this regard, the EU is currently working on the adoption of a Directive on Corporate Sustainability Due Diligence in order to harmonise national rules imposing mandatory due diligence for companies.[3]

This chapter aims at providing a preliminary assessment of the proposed Directive in light of the role that contract law plays in regulating sustainability in GVCs. First, it provides a description of the main functions played by contracts in global production processes; then it provides a description of the main elements of the proposed Directive and finally concludes by assessing its provisions from a contract law and sustainability perspective.

34.2 Contract as a Tool for Regulating Production Processes

In the last 30 years, modes of production have been witnessing a profound transformation. Most of final goods are produced through GVCs, in which production is divided into several fragments, which can be carried out in different countries. This was not the case until the end of the 20th century when production was still mostly carried out in single firms. If trade can be thought of as

DOI: 10.4324/9781032662046-40

always having been somehow characterised by a 'global' extension, the level of capillarisation and fragmentation is a characteristic of more recent modes of organisation of production.

The fragmentation of the operational business level has been accompanied by an increasing centrality of private law instruments in supporting the operational infrastructure. Fragmentation has indeed gone hand in hand with a process of outsourcing production to external suppliers,[4] which has ultimately resulted in a contractualisation of processes of production, characterised by the separation between the ownership of the product's brand and the ownership of the production process.[5]

34.2.1 The Move to Contractualisation

The move to contractualisation has played two key roles in the expansion of global production:

Contractualisation, together with the development and distribution of information and communication technologies (ICTs),[6] allowed companies to offshore fragments of production processes in less-industrialised economies characterised by lower costs of production. At the same time, the cost reduction of processes of production was then paired with the possibility to contribute to global economic development to the extent that trade-in value was considered a way of distributing economic gains along the actors involved in the chain.[7] In this manner, contracts made it possible to connect private economic actors with a less costly supply of labour. Allegedly to the benefit of both sides.

Off-shoring production also led to the territorial extension of GVCs, implying the need to deal with multiple national jurisdictions. From this perspective, the adoption of contracts as 'operational tool' of global production processes became crucial to solve the issues deriving from different national legal orders.[8] Contracts became the regulatory regime, offering a regulatory layer to private actors operating cross-border activities.[9] This extra-regulatory layer has provided economic actors with the capacity to pursue different economic and normative purposes: On the one hand to manage their business and on the other to enforce rights which may not be evenly protected in the different jurisdictions covered by the production, such as intellectual property (IP) rights or sustainability-related matters.[10]

34.2.2 Sustainability and Contract Law

The form of private regulation granted by contracts in cross-border activities has been put to use by private actors in the last few years, especially with regard to sustainability-related matters. In practice, obligations pertaining to sustainability are typically included in broader Master Supplier Agreements (MSAs) through which the buyers mostly define the conditions for suppliers to become their business partners.[11] Sustainability requirements are then second-order performance establishing the way in which the main performance must be carried out, thereby creating a multi-level structure of correlated performances within the contract.

The obligations related to sustainability requirements may be derived from a company's internal policy (such as, for instance, a code of conduct) or from the company's subscription to a voluntary sustainability standard (VSS). Such standards provide for principles and requirements that the company must comply with in order to maintain a label or certification. Compliance is typically through a process of verification and accreditation which relies on regular auditing. In principle, a company subscribes to certain standards and contract offers the means through which the VSS is implemented along the entire value chain, both for the actors in the production and distribution tiers, through the so-called 'contractual cascading'. As a matter of fact, by being *manufacturers without factories*, 'contractual cascading' may sometimes be the only mechanism for companies to achieve compliance with such standards.

However, contracts as tools for self-regulation are not necessarily neutral when it comes to sustainability-related matters, insofar the obligation voluntarily assumed by the leader of the chain may be transferred, through contracts, onto suppliers who are then liable to ensure compliance with such obligation. As a result, the company that adopts voluntary sustainability standards is not the actor that must comply with them *de facto*: the obligation to comply is transferred to their contractors which might not be operating under the economic and social conditions necessary to comply with sustainability standards.[12] This practice, known as contractual cascading, may be instrumentalised by companies so as to avoid bearing the responsibility derived from the adoption of voluntary self-regulation against externalities of production as well as the consequences of the lack of compliance with such standards.

Related to the latter point, there is another dimension under which the regulation of sustainability-related matters via contract is not neutral: when standards are given the shape of contractual obligations, leaders of the chain are free to terminate the contract in case the sub-contractor does not fulfil the obligation. At the same time, because including sustainability standards in agreements with sub-contractors is sufficient, leaders of GVCs can claim their products to be sustainable and to meet sustainability requirements, whereby getting by with compliance to the voluntary standards they have adopted for themselves.

34.3 The Proposal for a Corporate Sustainability Due Diligence Directive (CSDDD)

In February 2022, the European Commission adopted the text of the proposal for a Corporate Sustainability Due Diligence Directive (hereinafter the Directive).[13] The proposed Directive will impose on (certain) companies the obligation to adopt due diligence policies in order to prevent and mitigate the potential and actual adverse impacts of their activities on human rights and the environment. While the title of the proposed Directive hints at the measures it adopts in terms of corporate governance, the text of the law is far-reaching as it provides for rules on the entire value chain of the addressed companies, on civil liability and on contractual assurance between business partners.

This section illustrates some of the core elements of the proposed Directive to provide a broad picture of its content. It will then provide an assessment of its provisions in light of the contractual nature of processes of production as illustrated above.

34.3.1 Scope

As the text of the proposal currently stands, the subjective scope of the proposed Directive is defined by two core criteria: the number of employees and the net turnover.[14] Regarding the former, the companies subject to the Directive must have had at least 500 employees on average in the last two financial years and a net turnover of at least 150 million euros in the last financial year. The norm is complemented by specific – and lower – thresholds for companies which operate in high-risk sectors, such as the textile and food industries and for activities related to the extraction of mineral resources.[15]

The global character of a business value chain requires considering also actors that are not based on the EU territory. The proposed Directive addresses this issue by including in its subjective scope of application also non-EU companies. These companies are subjected to the same obligations as the EU ones as long as they generate a net turnover of more than 150 million euros in the Union in the financial year preceding the last financial year. Lower thresholds are again applicable to companies operating in the high-risk sectors mentioned above.

34.3.2 Objective

The core objective of the proposed Directive is that of improving corporate governance practices in order to integrate risk management and mitigation processes of human rights and environmental adverse impacts stemming from the business' value chain into corporate strategies.

In its General Approach adopted on 1st December 2022, the Council expressed its stance in favour of the risk-based approach adopted by the Commission in the proposed text. The risk-based approach entails that the obligations included in the proposed Directive are not to be read as obligations of results but merely as obligations of means. Hence, companies that will be covered by the scope of the proposed Directive will not be required to guarantee that the adverse impacts will never occur. The approach adopted by the proposed Directive is hence based on procedural guarantees ('appropriate measures' in the text of the proposal), which aim at ensuring the prevention of adverse impacts and their mitigation when it is not possible to bring to an end those that occur despite the procedural safeguards adopted in the corporate policies.[16]

34.3.3 Core Obligations/Appropriate Measures

The core obligations set out in the proposed Directive concern the adoption of 'appropriate measures' for the prevention and mitigation of potential and actual adverse impacts along the business value chain. In the text agreed by the Council, the term value chain has been replaced by 'chain of activities' so as to avoid misinterpretation of the actual objective scope of the obligations adopted in the Directive. The activities covered by this revised terminology cover all those occurring in the supply side of the chain, while the Council has opted to exclude consumers' disposal of products from those included in the downstream side.[17] Radically different is the position adopted by the European Parliament in its position of 1st June 2023, where the notion of value chain encompasses both upstream and downstream activities, including those related to waste management of the product by individual consumers.[18]

One of the core measures that companies may adopt in order to fulfil their obligations is the request for contractual assurance from their direct business partners. The contractual assurance aims at requiring business partners to comply with the company's code of conduct and may extend to requests to such a business partner to include the same assurance from its direct business partners, so as to generate a contractual cascading, which should ensure that the company's code of conduct is abided by all actors included in the company's value chain.

When contractual assurances are requested from business partners, the company is required to also set up appropriate measures to verify compliance, which may include suitable industry initiatives or third-party verification.[19]

34.3.4 Administrative Sanctions

The proposed Directive instructs Member States to designate supervisory authorities for the supervision of companies' compliance with the obligations imposed by the proposed Directive. The supervisory authority designated by the Member State will have competence over companies having a registered office in that Member State. For non-EU companies, the competent supervisory authority is the one of the Member State where the non-EU company has a branch or generates most of its net turnover in cases it has branches in more than one Member State or no branches on the EU territory. Competent supervisory authorities will be conferred with powers to investigate companies' compliance and to charge non-compliance with remedial action, when possible, and administrative penalties. The investigations may be taken by the supervisory authority's own motion as well on the basis of a 'substantiated concern' which may be submitted by any natural or legal person.[20]

34.3.5 Private Law Remedies

Accompanying the administrative sanctions is the rule governing civil liability for infringement of the obligations included in the proposed Directive. The rule is included in Article 22 of the proposal as it stands and requires Member States to make sure that companies are held liable for damages caused to a natural or legal person as a result of intentional or negligent failure to comply with the due diligence requirements for the prevention and mitigation of those human rights and environmental adverse impacts protected by the international instruments listed in Annex I of the proposed Directive.[21]

The *locus standi* is granted upon the existence of two elements: (a) the unattended right, prohibition, or obligation is listed in Annex I and is aimed to protect the natural or legal person; (b) the company's failure to comply results in damage to the protected natural and legal person.

The article goes on to state that the victims must be entitled to receive full compensation – without resulting in over-compensation – but it does not elaborate further on the burden of proof, which is one of the main obstacles in obtaining compensation for damages.

From a GVC perspective, one important element to consider is that the Article excludes the company's liability when the damage is caused exclusively by one of its business partners. Victims may be left without compensation if the business partner is located outside the EU territory.[22] This is a particularly delicate dimension of the nature of the adverse impacts in a company's value chain as through the contractualisation of processes of production companies may externalise costs, including liability costs. By not providing any guidelines on how to assess that damage is caused *only* by the company's business partner, the proposed Directive, as it currently stands, leaves a vacuum on the coordination between the rule on civil liability and the contractual nexus linking the company's value chain, which may weaken the effectiveness of the requirement for full compensation of damaged natural and legal persons.

Furthermore, the rule on civil liability has been amended by the Council, which has removed the last paragraph of the text proposed by the Commission according to which 'the liability provided for in provisions of national law transposing this Article [Article 22] is of overriding mandatory application in cases where the law applicable to claims to that effect is not the law of a Member State'.[23] If confirmed in the trialogue, the removal of the overriding mandatory application of Article 22 will prevent the application of Article 16 of the Rome II Regulation, according to which overriding mandatory rules must be applied instead of the law of the country of the subsidiary or business partner, which would be the applicable law under the general rule of Article 4(1) of the Rome II Regulation.[24]

34.4 Assessing the CSDDD in the Light of the Relationship between Contract and Sustainability

It is now time to provide some considerations on the current text of the proposed Directive with respect to the relationship between contract and human rights and environmental adverse impacts.

A short reference on the absence of coordination between the contractual nature of the linkages between a company's contractual value chain and the rule on civil liability has already been made. The Directive, however, does not provide private law remedies only in relation to civil liability for victims of an adverse impact. It also mentions contractual remedies with respect to the contractual relationship between a company and its business partners.

These remedies complement the rule on the contractual assurance that a company may seek from its business partners in order to extend the scope of the due diligence obligations further along the chain. Should a company seek contractual assurance from a business partner, the proposed Directive requires that it also manages the verification of compliance with the obligations included in that contractual assurance from its direct business partners. This section analyses how the Directive addresses the topic of contractual/private regulation of a company's value chain and to what extent these rules concur with the effective prevention and mitigation of human rights and environmental adverse impacts.

34.4.1 Verification of Compliance

When a company aims at fulfilling its due diligence obligations by seeking contractual assurance from its direct business partners, Member States should also require companies to adopt appropriate measures to verify compliance. The content of the contractual assurance may reflect the company's code of conduct and the prevention and correction action plans when necessary.

In order to extend the scope of its due diligence obligations to indirect business partners, the company may require the direct partners to replicate the same contractual assurance from their own suppliers, thereby triggering contractual cascading. The proposed Directive also foresees the possibility for the company to seek the same contractual assurance from an indirect business when limiting such a request to the direct ones should not be sufficient to effectively prevent or mitigate the human rights and environmental adverse impacts.

The compliance with such contractual obligations should be verified by means of industry initiatives or third-party verification. Article 2 of the proposed Directive defines 'industry initiative' as 'a combination of voluntary due diligence procedures in the chains of activities, tools and mechanisms, including independent third-party verifications, developed and overseen by governments, industry associations or groupings of interested organisations' and 'independent third-party verification' as the

> means verification of the compliance by a company, or parts of its value chain, with human rights and environmental requirements resulting from the provisions of this Directive by an expert which is independent of the company, free from any conflicts of interests, has experience and competence in environmental and human rights matters and is accountable for the quality and reliability of the verification.[25]

In this way, the proposed Directive seems to legitimise the adoption of voluntary sustainability standards schemes and code of conduct as appropriate measures to ensure human rights and environmental due diligence along a company's value chain, in line with the market trend that sees an increasing popularisation and adoption of such voluntary schemes.[26] However, the reliability and effectiveness of third-party verification or industry initiative in ensuring compliance with human rights and environmental due diligence obligations is debatable.[27]

34.4.2 Termination and Suspension of the Contract

Finally, the Directive addresses the remedies for contractual liability by providing guidance on the conditions which may determine the suspension or termination of the contractual relationship. Companies may have recourse to such remedies in case their business partner is not compliant with the obligation included in the contractual assurance. Termination and suspension of the contract, however, should be used as a last resort, due to the disruptive effects which these remedies may have on business partners, as well as on the supply of goods. Termination and suspension may in fact have socially adverse impacts, which may be even more severe than those that a failure to comply may cause, as the COVID-19 pandemic has demonstrated in the last few years with many workers left without a living wage or leading to insolvency of smaller suppliers.[28]

Termination of contractual relationships has typically been one of the weak spots for effectively regulating sustainability in GVCs mainly because the impacts of ending or suspending the contract may be harsher on people than maintaining the relationship. This is why, in many voluntary companies' 'terms of engagement' – through which the company establishes the terms to be observed by their contractors – termination of the contractual relationship is often associated with the most severe violations (for instance, child or forced labour). For less severe violations, companies may keep the contractual relation alive and support the supplier with the adoption of measures to improve or remedy the failure to comply with the terms of engagement. The proposed Directive seems to favour this approach to the extent it mentions the company's influence on suppliers as an alternative – and preferred – solution to repair the failure of complying with the contractual due diligence assurances.[29]

As it has been observed earlier, however, these forms of self-regulation provide an extra layer of regulation which does not necessarily affect the company's power to terminate the contract at any time, to the extent that a company typically is in a stronger market/contractual position vis-à-vis its contractors. From a 'relational perspective',[30] this unbalance may lead to abuse of market powers to the detriment of smaller suppliers, especially if located in less-industrialised economies, an issue that has been addressed and regulated for certain sectors such as the food sector.[31]

In this regard, it must be noticed that the proposed Directive also mentions the obligation for companies that seek contractual assurance from small and medium enterprises (SMEs) to use fair, reasonable and non-discriminatory terms, despite the fact that the matter of unfair terms in B2B relationships is not harmonised at the EU level.[32] In order to bypass this obstacle, Article 12 of the proposed Directive requires the Commission to adopt model contract clauses that companies may voluntarily adopt in order to seek contractual assurance.[33] This latter provision resembles the initiative undertaken by the American Bar Association with the later version of Model Contract Clauses (MCC).

Both the European CSDD Directive and the American MCC, however, do not offer a solution to the practice of contractual cascading whereby the obligation to prevent and mitigate adverse impacts in processes of production rests with suppliers.[34]

In this sense, both initiatives fail to address the implications that the formal dimension of contract law has on its *de facto* application in GVCs, i.e., the contracting practices emerging in GVCs.

Some authors have expressed this concern and stressed the need to enhance the collaborative nature of contracts to achieve positive outcomes and the need to a fairer allocation of liability between buyers and suppliers.[35] These solutions do require reflection on the economic and political factors that shape contractualisation in GVCs and that have historically contributed to the increasing emergence of horizontal structures of production organisation. These factors can only enter the picture if regulators are willing to see the institutional role played by contracts in regulating the economic order of GVCs: this means not only regulating them contract by contract but rather to have a holistic look at the aims which are pursued by means of contracts and at the capacity of each actor to achieve them.

A broader perspective would have the benefit of seeing sustainability due diligence in different terms from the formal ones that are currently under discussion. For instance, strict liability, consumer remedies and capacity building are only a few of the regulatory strategies that may enter the broader picture and effectively complement the tools provided by due diligence in order to achieve sustainable systems of production.

34.5 Conclusive Remarks

The proposal for a Corporate Sustainability Due Diligence Directive officially aims at the harmonisation of national rules on mandatory due diligence. The legislative instrument, however, goes further than this in their regulatory objectives by providing guidance on the use of contract as a tool to foster sustainability in companies' value chain.

In this respect, the proposed Directive acknowledges and at the same time legitimises mechanisms that most companies already adopt on a voluntary basis for ensuring the prevention, mitigation, and remediation of adverse impacts on sustainability factors, such as the adoption of voluntary sustainability standards and their implementation along the chain via contractual cascading.

The recognition of contract as an instrument for sustainability also implies the recognition that the law of contract is not neutral insofar as it entails an allocation between the parties of liability for fulfilling the contractual obligations. In this regard, the proposed Directive regulates contract law directly by requiring that remedies such as contract termination are used only as a last resort.

While the regulation of contractual cascading and contract termination in the proposed Directive promise to be effective measures of addressing one of the soft spots in achieving sustainable value chains, the proposed Directive does not address the issue of voluntary standards as directly. Will the adoption of standards be enough to fulfil the due diligence requirements imposed by the Directive? Or will the standards adapt to the requirements set by the proposed Directive, thereby contributing to shaping a concept of sustainability in production processes?

In this respect, how the relation between public and private regulation develops remains to be seen and the upcoming months of negotiations among the bodies in the trialogue will be crucial to assure the effectiveness of due diligence requirements stemming from the Directive. What is absolutely clear is that contract law and private regulation will continue to play a crucial role in achieving sustainability in global processes of production.

Notes

1 United Nations, Transforming Our World: The 2030 Agenda for Sustainable Development (2015) A/RES/70/1.

2 A. Marx, 'Global Governance and the Certification Revolution: Types, Trends and Challenges' (2010) Leuven Centre for Global Governance Studies Working Paper No. 53 < http://dx.doi.org/10.2139/ssrn.1764563> accessed 25 September 2023.

3 This paper is based on a version of the Directive prior to the amendments that have been made in the first half of 2024, after the European Parliament and the Council failed to adopt the Directive in the first round of negotiation.

4 J. Salminen, 'From National Product Liability to Transnational Production Liability: Conceptualizing the Relationship of Law and Global Supply Chains' (DPhil thesis, University of Turku 2017), p. 105.

5 The distinction has gone thus far so as to lead to the emergence of the so-called 'manufacturers without factories', an expression coined by Gereffi, which effectively depicts the process of fragmentation/

contractualisation of production processes. See, Gerrefi, 'International Trade and Industrial Upgrading in the Apparel Commodity Chain' (1999) 48:1 *Journal of International Economics*, pp. 37–70.

6 R. Baldwin, *The Great Convergence: Information Technology and the New Globalization*, The Belknap Press of Harvard University Press, Cambridge 2016.

7 World Trade Organisation (WTO), Global Value Chains in a Changing World (2013), <https://www.wto.org/english/res_e/booksp_e/aid4tradeglobalvalue13_e.pdf>; Global Value Chains Development Report: Beyond Production (2021) <https://www.wto.org/english/res_e/booksp_e/00_gvc_dev_report_2021_e.pdf> accessed 25 September 2022.

8 D.V. Snyder, 'Contracting for Process' (2022) 85 *Law and Contemporary Problems*, pp. 255–276.

9 F. Cafaggi, 'The Regulatory Function of Transnational Commercial Contracts: A New Architecture' (2013) 36:6 *Fordham International Law Journal*, pp. 1556–1618; H. Collins, *Regulating Contracts*, Oxford University Press Oxford 2002; A. Beckers, 'The Invisible Networks of Global Production: Re-Imagining the Global Value Chain in Legal Research' (2020) 16:1 *European Review of Contract Law*, pp. 95–117; P. Zumbansen, 'The Law of Society: Governance Through Contract' (2007) 14 *Indiana Journal of Global Legal Studies*, p. 191; F. Snyder, 'Governing Economic Globalisation: Global Legal Pluralism and European Law' (1999) 5:4 *European Law Journal*, pp. 334–374; G. Teubner, *Networks as Connected Contracts: Edited with an Introduction by H. Collins, International Studies in the Theory of Private Law* (7), Hart Enterprises, Oxford 2011; O. Hansen, C.S. Petersen and V.G. Ulfbeck, 'Private Governance and the Potential of Private Law' (2020) 28:2 *European Review of Private Law*, pp. 333–374; V.G. Ulfbeck and O. Hansen, 'Sustainability Clauses in an unsustainable Contract Law?' (2020) 16:1 *European Review of Contract Law*, pp. 186–205; V.G. Ulfbeck, O. Hansen and A. Andhov, 'Contractual Enforcement of CSR Clauses and the Protection of Weak Parties in the Supply Chain' in V. Ulfbeck, A. Andhov and K. Mitikidis (eds), *Law and Responsible Supply Chain Management: Contract and Tort Interplay and Overlap*, Routledge, New York 2019, pp. 46–64; P. Verbruggen, 'Introduction: Regulating Private Regulators: Understanding the Role of Private Law' (2019) 2 *European Review of Private Law*, pp. 175–196.

10 L. Bernstein, 'Beyond Relational Contracts: Social Capital and Network Governance in Procurement Contracts', (2015) 7 *Journal of Legal Analysis*, p. 561; E. Van der Zee, 'Regulatory Structure of Standards Underlying Sustainability Labels. Laying the Foundations for Effective Intervention Strategies to Increase Consumer Confidence in Sustainability Labels' in H. Bremmers and K. Purnhagen, *Regulating and Managing Food Safety in the EU: A Legal-Economic Perspective*, Springer, Cham 2018, pp. 39–56.

11 L. Bernstein, 'Beyond Relational Contracts: Social Capital and Network Governance in Procurement Contracts' (2015) 7 *Journal of Legal Analysis*, p. 561.

12 S. Dadush, 'Contracting for Human Rights: Looking to Version 2.0 of the ABA Model Contract Clauses' (2019) 68:5 *American University Law Review*, pp. 1519–1554.

13 Commission, 'Proposal for a Directive (EU) on Corporate Sustainability Due Diligence and amending Directive (EU) 2019/1937' COM (2022) 71 final. On 1st December 2022, the Council made some amendments to the text proposed by the European Commission. The compromised text was adopted by the Council as its General Approach, which provides the President with the authorisation to enter negotiations with the European Parliament and the European Commission, see Council General Approach 2022/0051(COD). In the text adopted by the Council, the term 'value chain' has been replaced with ' chain of activities ', which however was not adopted by the European Parliament in its position adopted on 1st of June 2023.

14 Art. 2, Proposal for a Directive (EU) on Corporate Sustainability Due Diligence and amending Directive (EU) 2019/1937 COM (2022) 71 final.

15 The identification of the sectors for which specific rules are applies is based on those sectors for which the OECD has provided sector-specific guidelines on due diligence.

16 European Law Institute (ELI), Business and Human Rights: Access to Justice and Effective Remedies [2022] <https://www.europeanlawinstitute.eu/fileadmin/user_upload/p_eli/Publications/ELI_Report_on_Business_and_Human_Rights.pdf> p. 25.

17 Commission, 'Proposal for a Directive (EU) on Corporate Sustainability Due Diligence and amending Directive (EU) 2019/1937' COM (2022) 71 final – General Approach, 2022/0051(COD), p. 38.

18 European Parliament, Report on the proposal for a directive of the European Parliament and of the Council on Corporate Sustainability Due Diligence and amending Directive (EU) 2019/1937 [2022], available at <https://www.europarl.europa.eu/doceo/document/A-9-2023-0184_EN.html#_section1>.

19 Art. 7 para. 5 and Art. 8 para 5, Proposal for a Directive (EU) on Corporate Sustainability Due Diligence and amending Directive (EU) 2019/1937, COM (2022) 71 final.

20 Art. 19, Proposal for a Directive (EU) on Corporate Sustainability Due Diligence and amending Directive (EU) 2019/1937, COM (2022) 71 final.
21 The element of fault was absent in the text proposed by the European Commission, it has been added by the Council in its General Approach, Proposal for a Directive (EU) on Corporate Sustainability Due Diligence and amending Directive (EU) 2019/1937, COM (2022) 71 final.
22 European Law Institute (ELI), Business and Human Rights: Access to Justice and Effective Remedies (2022) < https://www.europeanlawinstitute.eu/fileadmin/user_upload/peli/Publications/ELI_Report_on_Business_and_Human_Rights.pdf> p. 46.
23 Commission, 'Proposal for a Directive (EU) on Corporate Sustainability Due Diligence and amending Directive (EU) 2019/1937', COM (2022) 71 final.
24 Art. 16, Council 'Directive (EC) No 864/2007 of 11 July 2007 on the law applicable to non-contractual obligations (Rome II)' (2007) OJ L 199/40; European Law Institute (ELI), Business and Human Rights: Access to Justice and Effective Remedies (2022) < https://www.europeanlawinstitute.eu/fileadmin/user_upload/p_eli/Publications/ELI_Report_on_Business_and_Human_Rights.pdf> p. 50.
25 Art. 2, Proposal for a Directive (EU) on Corporate Sustainability Due Diligence and amending Directive (EU) 2019/1937, COM (2022) 71 final.
26 Only the Ecolabel Index is currently counting the existence of 456 ecolabels in 199 countries and 25 sectors. See Ecolabel Index Website <https://www.ecolabelindex.com/> accessed 18 September 2023.
27 M. Tampe, 'Leveraging the Vertical: The Contested Dynamics of Sustainability Standards and Labour in Global Production Networks: Leveraging the Vertical' (2018) 56 *British Journal of Industrial Relations*, p. 43. One case known to the general public is the collapse of the Rana Plaza factory in Bangladesh where the recommendations by the auditors to evacuate the building were ignored by the factory owner, resulting into the death and injury of thousands of workers. See, Clean Clothes Campaign, Rana Plaza <https://cleanclothes.org/campaigns/past/rana-plaza> accessed 18 September 2023.
28 The Business and Human Rights Resource Centre has been monitoring particularly the effects of Covid-19 on workers in the food and beverage supply chains and those in the electronics supply chains. See, for example, Business & Human Rights Resource Centre, 'COVID-19: Food and beverage supply chain workers' <https://www.business-humanrights.org/en/big-issues/covid-19-coronavirus-outbreak/covid-19-food-and-beverage-supply-chain-workers/> accessed 18 September 2023; Business & Human Rights Resource Centre, 'Impacts of COVID-19 on supply chain worker in the electronics sector' (8 June 2020) <https://www.business-humanrights.org/en/latest-news/impacts-of-covid-19-on-supply-chain-workers-in-the-electronics-sector/> accessed 18 September 2023.
29 Art. 7 para 1 le t. (c) and Art. 8 para 1 le t. (c), Proposal for a Directive (EU) on Corporate Sustainability Due Diligence and amending Directive (EU) 2019/1937 COM (2022) 71 final .
30 P. Dicken, P.F. Kelly, K. Olds and H. Wai-Chung Yeung, 'Chains and Networks, Territories and Scales: Towards a Relational Framework for Analysing the Global Economy' (2001) 1 *Global Networks: A Journal of Transnational Affairs*, p. 89.
31 In the food sector the EU has adopted a dedicated act to regulate trading practices between companies and suppliers: Commission 'Directive (EU) 2019/633 on unfair trading practices in business-to-business relationships in the agricultural and food supply chain' (2019) OJ L 111/59.
32 Some jurisdictions, like Germany and Belgium, have specific norms for regulating unfair terms also in B2B contractual relationships, while some others, like Italy, lack this form of protection for small and medium enterprises.
33 The initiative will mirror (or compete) the one launched by the American Bar Association, see D.V. Snyder, S.A. Maslow and S. Dadush, *Balancing Buyer and Supplier Responsibilities Model Contract Clauses to Protect Workers in International Supply Chains, Version 2.0* (2021) 77 *Business Lawyer (ABA)* __ *(Winter 2021–2022)*, American University, WCL Research Paper No. 2021-15, Rutgers Law School Research Paper No. Forthcoming, <https://cadmus.eui.eu/bitstream/handle/1814/73799/MCC%20 2.0%20as %20pub.pdf?sequence=2&isAllowed=y>.
34 S. Dadush, 'Contracting for Human Rights: Looking to Version 2.0 of the ABA Model Contract Clauses' (2019) 68:5 *American University Law Review*, pp. 1519–1554.
35 S. Dadush, 'Contracting for Human Rights: Looking to Version 2.0 of the ABA Model Contract Clauses' (2019) 68:5 *American University Law Review*, pp. 1519–1554; J. Saloranta and A. Hurmerinta-Haanpää, 'Proactive Contract Theory in the Context of Corporate Sustainability Due Diligence' (2023) *Journal of Strategic Contracting and Negotiation*, pp. 1–16.

Bibliography

Baldwin R, *The Great Convergence: Information Technology and the New Globalization* (Cambridge, The Belknap Press of Harvard University Press 2016).

Beckers A, 'The Invisible Networks of Global Production: Re-Imagining the Global Value Chain in Legal Research' (2020) 16:1 *European Review of Contract Law* 95.

Bernstein L, 'Beyond Relational Contracts: Social Capital and Network Governance in Procurement Contracts' (2015) 7 *Journal of Legal Analysis* 561.

Cafaggi F, 'The Regulatory Function of Transnational Commercial Contracts: A New Architecture' (2013) 36:6 *Fordham International Law Journal* 1556.

Collins H, *Regulating Contracts* (Oxford, Oxford University Press 2002).

Dadush S, 'Contracting for Human Rights: Looking to Version 2.0 of the ABA Model Contract Clauses' (2019) 68:5 *American University Law Review* 1519.

Dicken PK, Olds KPF and Wai-Chung Yeung H, 'Chains and Networks, Territories and Scales: Towards a Relational Framework for Analysing the Global Economy' (2001) 1 *Global Networks: A Journal of Transnational Affairs* 89.

European Law Institute (ELI), Business and Human Rights: Access to Justice and Effective Remedies (2022) <https://www.europeanlawinstitute.eu/fileadmin/user_upload/p_eli/Publications/ELI_Report_on_Business_and_Human_Rights.pdf>.

Gerrefi G, 'International Trade and Industrial Upgrading in the Apparel Commodity Chain' (1999) 48:1 *Journal of International Economics* 37.

Hansen O, Petersen CS and Ulfbeck VG, 'Private Governance and the Potential of Private Law' (2020) 28:2 *European Review of Private Law* 333.

Marx A, 'Global Governance and the Certification Revolution: Types, Trends and Challenges' (2010) Leuven Centre for Global Governance Studies Working Paper No. 53, <http://dx.doi.org/10.2139/ssrn.1764563>

Salminen J, 'From National Product Liability to Transnational Production Liability: Conceptualizing the Relationship of Law and Global Supply Chains' (DPhil thesis, University of Turku 2017) 105.

Saloranta J and Hurmerinta-Haanpää A, 'Proactive Contract Theory in the Context of Corporate Sustainability Due Diligence' (2023) 6:3–4 *Journal of Strategic Contracting and Negotiation* 1.

Snyder DV, 'Contracting for Process' (2022) 85 *Law and Contemporary Problems* 255.

Snyder F, 'Governing Economic Globalisation: Global Legal Pluralism and European Law' (1999) 5:4 *European Law Journal* 334.

Snyder DV, Maslow SA and Dadush S, 'Balancing Buyer and Supplier Responsibilities Model Contract Clauses to Protect Workers in International Supply Chains, Version 2.0' (2021) 77 *Business Lawyer (ABA) __ (Winter 2021–2022)*, American University, WCL Research Paper No. 2021-15, Rutgers Law School Research Paper No. Forthcoming, <https://cadmus.eui.eu/bitstream/handle/1814/73799/MCC%20 2.0%20as %20pub.pdf?sequence=2&isAllowed=y>

Tampe M, 'Leveraging the Vertical: The Contested Dynamics of Sustainability Standards and Labour in Global Production Networks: Leveraging the Vertical' (2018) 56 *British Journal of Industrial Relations* 43.

Teubner G, *Networks as Connected Contracts*: *Edited with an Introduction* (Oxford, Hart 2011).

Ulfbeck VG and Hansen O, 'Sustainability Clauses in an Unsustainable Contract Law?' (2020) 16:1 *European Review of Contract Law* 186.

Ulfbeck VG, Hansen O and Andhov A, 'Contractual Enforcement of CSR Clauses and the Protection of Weak Parties in the Supply Chain' in Ulfbeck VG, Hansen O and Mitikidis K (eds), *Law and Responsible Supply Chain Management: Contract and Tort Interplay and Overlap* (New York, Routledge 2019) 46.

Van Der Zee E, 'Regulatory Structure of Standards Underlying Sustainability Labels. Laying the Foundations for Effective Intervention Strategies to Increase Consumer Confidence in Sustainability Labels' in Bremmers H and Purnhagen K (eds), *Regulating and Managing Food Safety in the EU: A Legal-Economic Perspective* (Cham, Springer 2018) 39.

Verbruggen P, 'Introduction: Regulating Private Regulators: Understanding the Role of Private Law' (2019) 2 *European Review of Private Law* 175.

Zumbansen P, 'The Law of Society: Governance Through Contract' (2007) 14 *Indiana Journal of Global Legal Studies* 191.

35

ENVIRONMENTAL LITIGATION

Is It a Matter of Public or Private Law? An Exploration of Cross-Fertilisation of Environmental Law Concepts in the Context of French Case Law

Marc Clément[1]

35.1 Introduction

When asked whether environmental litigation is a matter of public or private law, the obvious answer is that it is clearly both. The question would not need a lengthy discussion if the division did not raise difficult litigation strategy questions. To which judge should one turn and what to ask for?

The analysis we propose is based on the French situation and, more specifically, on the relationships, in this national context, between the judicial courts – civil and criminal courts – and the administrative courts as revealed by the case law. This division is specific to France, even though many countries have separate judicial orders or specialisations that deal specifically with public law.[2] Nevertheless, the questions this study addresses have a broader scope since they involve analysing the place of private and public laws in environmental issues, which is relevant in all national contexts.

In this respect, the climate disputes highlight the multiplicity of remedies available under either private or public law. In France, the administrative law litigation's first case was brought before the Council of State by the municipality of Grande Synthe and its mayor – the so-called 'Grande Synthe' case, and a parallel case was brought before the administrative court of Paris by the four non-governmental organisations which identified the case as the 'L'Affaire du Siècle' case. In both cases, the complainants want France's action against climate change to be more vigorous.[3] At the same time, legal action is being taken against Total (involved in oil production) because of the company's inadequate commitments to the Paris Agreements, and against Casino (that deals with retail) because of deforestation in Brazil, and Colombia because of its beef purchases.[4]

All these actions have the same objective: to obtain much stronger commitments in the fight against climate change. What they have in common is that they are targeting entities – the State or large companies – that have significant means of action. The State can require economic actors to reduce their carbon footprint, but companies can also choose to modify their development strategy.

This dualism of action is not specific to climate change. In fact, in a legal system in which the State and the European Union play a major role in their regulatory action, environmental litigation thus has two dimensions which manifest themselves: in the authorisation regimes granted by

DOI: 10.4324/9781032662046-41

public authorities (1) associated with the activities of companies and individuals which lead to liability claims (2).

35.2 Challenging Authorisations Granted by Administrations as a Major Avenue for Environmental Litigation

In the 17th century, trees were needed for shipbuilding. In 1669, Louis XIV's minister, Colbert, thus introduced rules that imposed a mode of forest exploitation on all owners.[5] The example of forest exploitation shows that long-term management – it takes several decades for a tree to produce large pieces – conflicts with the immediate interests of those who prefer to have short-term revenues. The Colbert Forest Regulation involved the enactment of rules that limited the free disposal of forests by owners. The owner had to inform authorities of any tree cutting; if they wished to derogate from the strict obligations of forest preservation to clear plots of land, they had to ask for a permit.[6]

In addition to this example, in France authorisations were given to operations because of their effects on the neighbourhood: in 1810, Napoleon I issued a decree aimed at regulating emerging industrial activities[7] and the decree required that establishments 'spreading an unhealthy or inconvenient odour' obtain a permit. It was then a question of arbitrating between the uses that owners could make of their property. The State was thus the guarantor of this arbitration representing the general interest.

In both these cases, an authorisation system was developed which was the precursor of modern environmental authorisation law: the aim was to ensure that citizens or companies respected environmental constraints. The example of the Imperial Decree of 1810 shows that it was primarily public health issues that mobilised the State's action. A similar situation can be found today in developments relating to air quality: in order to comply with the provisions of Directive 2008/50/EC on air quality, Member States must put in place measures that create substantial constraints for both companies and citizens. This directive is based on air quality standards whose justification is based on public health concerns, in particular by reference to the standards laid down by the World Health Organisation.

35.2.1 Environmental Authorisation as a Screen for Private Law Relationships

The intervention of the State shifts litigation about neighbourhood disputes – e.g., air emissions of polluting substances – from the sphere of private concern to the sphere of public law: the question is then whether the State has taken the right measures and whether it is capable of enforcing them.

The need for authorisation by competent authorities is present in European environmental law in many areas. This is the case for industrial installations with Article 4 of Directive 2010/75/EU,[8] which makes the issuance of a permit the first of the conditions imposed by Directive 2008/50/EC on air quality. The Directive, therefore, brings the installations concerned within the scope of public law. Similarly, the protection of groundwater requires administrative authorisations prior to abstraction or groundwater recharge.[9] It should be noted, however, that not all activities are covered by this authorisation regime. Directive 2010/75/EU covers specific installations either because they fall under its Annex I or categories such as combustion installations.

Since an authorisation is delivered by a public authority, the litigation is based primarily on challenging the legality of this authorisation, which is facilitated since the Aarhus Convention,[10] which aims to allow a broad access to justice and imposes an effective control by the judge. Indeed, Article 9 paragraph 2 of this Convention gives wide access to the courts against decisions taken by a public authority: without going as far as an *actio popularis*, associations whose purpose is the protection of the environment are recognised as having a possibility of bringing a case before the courts beyond the rights of access to courts they do not necessarily have access to in another context.

Thus, the switch from a private law system to a public law system makes it much easier to initiate legal proceedings. In France, the admissibility of a complaint is decided on the basis of an interest in acting which is not limited to the violation of subjective rights. The procedure before the administrative judge is particularly simple. The situation is much more complex when it comes to initiating proceedings before the civil or criminal courts. The French civil procedure requires the parties to comply with much stricter rules than administrative procedure, such as the need for the parties themselves to master the communication of procedural documents. The extensive use of these rules by parties wishing to postpone the outcome of the proceedings is one of the causes of the often-long delays. As far as criminal proceedings are concerned, the burden of proof is, in itself, a difficult threshold to meet for the victims of environmental crimes.

35.2.2 What Remains of Private Law Relationships in a Permit Regime or on Its Margins

In order to act as a screen between private parties, an environmental permit must be sought. This is the case, for example, of exemptions that must be obtained in the event of damage to protected species.

Under French law, the violation of protected species is punishable by law. The strict protection regime for species is often the result of European legislation – Directive 2009/147/EC and Directive 92/43/EEC – and the damage may be direct to individuals of a species as a result of deliberate actions (such as hunting or poaching) or unintentional, as a result of the development of activities such as the operation of wind turbines, which will have an effect on bats or birds.

The French Environmental Code – using the same criteria as Directive 92/43/EEC – only allows protected species to be affected for limited reasons and this requires a derogation (Article 16 of the Directive). It should be noted that we are in exactly the same regime as we were under the 1669 Royal Ordinance on Forests: protection is strict and exceptions require an administrative measure.

But in the absence of a requested derogation, the operator of an installation is deprived of the screening role that the derogation can play. The Court of Cassation took up this reasoning in a recent ruling in the case of wind turbines.[11] A nature protection association challenged the operator of a wind farm in the Hérault Département, noting the destruction of kestrels (*falco naumanni*) between 2012 and 2016. The association had asked the *Préfet*[12] to take measures and in 2014 the *Préfet* of the Hérault Département intervened with orders to impose measures on the operator to reduce the impact on the kestrel. However, the damage continued. The Court of Cassation held that these 2014 measures did not allow the entire dispute to be brought under public law, since the operator did not have any derogation allowing it to harm the species. The Court of Cassation developed its reasoning by relying on the strict protection regime resulting from the European directives and on the fact that the measures taken by the administration – the initial operating permit, which was not accompanied by a derogation, and the additional measures taken in 2014 – did not implicitly constitute a position on the need for a derogation. The Court of Cassation thus noted that the orders issued by the *Préfet* of Hérault were not based on the provisions of the Environmental Code relating to the protection of species and therefore deduced that the *Préfet* had not intervened in this area.

Thus, the Court of Cassation intends to preserve the possibility of raising disputes between private individuals, including for activities that are covered by authorisations: it reads authorisations as being limited to their purpose, and as soon as the authorisation is silent on the protection of a species, the judicial judge remains competent for this part of a potential breach of environmental law. This position is remarkable since, in the French case, the Law of 16–24 August 1790 and the Decree of 16 Fructidor An III lay down the principle that the judicial judge is prohibited from substituting his assessment for that of the administration. The French Revolution, which wanted to limit the powers of the judiciary, wished to avoid its interference in administrative action. It can therefore be seen that, even in the French case, private law relationships remain even if authorisation is granted.

The advantage for environmental protection associations of going to judicial courts is that they can base their action on a criminal offence – in this case, the destruction of species – in order to lead to liability for compensation. In addition, environmental criminal law remains a powerful incentive to ensure nature protection, particularly since the Court of Cassation in its ruling of 30 November 2022 ruled that the offence of undermining the conservation of a protected species is constituted as soon as a single specimen is affected.

Another dimension of penal litigation as regard industrial installations relies on disregarding the requirements imposed by the administration for an installation that does not require formal authorisation. French law establishes a special category of installations that are not subject to authorisation, but which must nevertheless be declared to the authorities. Without going into the details of this system, it should be noted that even in the absence of an authorisation – within the meaning of Directive 2010/75/EU – it is likely that an activity will be subject to administrative requirements. This is either because these requirements are part of a general regulation or because the administrative authority, following the declaration, imposes requirements on the operation even though in French law it does not constitute an 'authorisation'. In this case, it is possible to initiate a penal summary procedure which allows any useful safeguard measures to be taken.[13]

Criminal Law will also come into play when the requirements of a permit are not complied with, for instance, in the event that the exemption from species protection obtained requires compensation or restoration measures. This was the case in the judgement[14] in which the operator of a gas transmission pipeline obtained authorisation to clear 40 hectares of forest for the construction of a gas pipeline on condition that the land be restored by reforestation. As these areas are resting or breeding grounds for protected species, the damage to these areas is equivalent to damaging the species concerned. The Court of Cassation thus considered that the offence was constituted by the fact that the operator's reforestation did not take place more than two years after the expiry of the time limit given.

The civil or criminal judge will, therefore, intervene as soon as the protective effects of the administrative authorisation can no longer be applied. It should be emphasised that French criminal law covers a very large number of situations involving damage to nature.

35.2.3 Litigation Concerning the Control of Prescriptions: The Complementarity Roles of Private and Public Laws

When authorisation is granted, the administrative authority is responsible for monitoring compliance with the conditions set out in the authorisation. Directive 2010/75/EU requires environmental inspections to be carried out and they are an essential part of this regime.[15] The administration, therefore, does not only intervene to issue standards but also has obligations to monitor and control.

The administration should also amend the requirements of a permit if necessary. Directive 2010/75/EU provides in its Article 21 that a periodic review of permit conditions must be carried out, in addition to the need to review permit conditions when new 'best available techniques' are published. Finally, it should be added that specific obligations cover the end of the activity and the closure of an industrial site (Article 22 of the Directive). The administration is thus involved in the entire activity cycle of the industrial site.[16]

But what can be done when the administration fails? Under French Administrative Law, it is possible to ask the administration to act, which is a first recourse. A resident close to an installation or an environmental protection NGO can ask the administration to act based on its control and policing powers. In the absence of a response from the administration or in the event of an explicit refusal, it is, then, possible to refer the matter to the administrative judge. The mechanism of the implicit refusal decision, which automatically arises after two months, allows the matter to be referred to the judge. This is a procedural legal artefact that prevents the administration from remaining silent and thus preventing litigation before the judge since the French administrative judge can only be seized based on an administrative decision.

But this possibility of acting before the administrative judge does not deprive the neighbours of an installation or an association of bringing a case before the judicial judge in a classic dispute over neighbourhood disturbances. The delicate question then arises for the judicial judge as to whether the authorisation that has been obtained is a barrier to his intervention. Logically, the Court of Cassation[17] holds that if the operator complies with the requirements of the authorisation, then it is before the administrative judge that the appeal must be presented. Conversely, by not respecting the requirements of the authorisation obtained, an operator may be held civilly and criminally liable since he has deliberately placed himself outside the administrative regime.

First, it should be remembered that the administration itself, through the action of the environmental inspectors, has the possibility of referring to the criminal courts in the case of infringements. The threat of repressive action is one of the elements making it possible to ensure the effectiveness of controls. In addition, the administration can impose administrative sanctions, which can range from the closure of an installation to the imposition of precautionary measures. But, as already mentioned, the administration is not always diligent in this matter.

A recent official report for the French Ministry of Justice[18] analysed in detail the environmental litigation presented before the criminal courts. It led to the creation in France of regional environmental courts so that this litigation could be entrusted to specialised judges.[19] The report notes that environmental litigation remains a marginal issue for the judicial judge and that its technical nature requires the specialisation of the judges responsible for handling it. The report also specifies that creating a court of judges who considers the interest of the environment is the same approach as the already existing specialised judges taking into account the interests of minors. It should also be emphasised that a large proportion of this environmental criminal litigation does not concern major media cases but rather everyday crime ranging from waste dumping to the dumping of pollutants in rivers or even poaching. This makes it difficult for the largely clogged up French criminal justice system to deal effectively with these cases, which seem minor compared to other criminal cases. Thus, in 75% of cases, the criminal sanction results in alternative measures to prosecution and only 24% of cases result in a criminal hearing. The report stresses that this very limited recourse to criminal sanctions is very probably the result of insufficient consideration of environmental issues. However, the report also stresses that administrative sanctions are no more effective, particularly in the case of urban planning law. There is, indeed, an interest in better combining administrative and criminal sanctions to ensure the effectiveness of environmental law.

Here again, associations play an important role and Article L. 142-2 of the Environment Code recognises the possibility for environmental associations to act as private parties for facts such as pollution resulting from an industrial installation. The Environmental Code even allows very a broad access to the criminal courts for certain associations, covering all offences related to the environment, urban planning, pollution, nuclear safety or even commercial practices and misleading advertising. Case law based on this text has a liberal approach of access to justice, the theoretical motivation of which is based on the principles laid down by the Aarhus Convention.

If it is essential for associations to be able to intervene in this area either by asking the administrative authority to take measures or by referring the matter to the civil or criminal courts, it is because they play an indispensable role as watchdogs. It should be remembered that nature protection associations have expertise that is often lacking in the administrations, particularly in terms of knowledge of biodiversity. For example, the inventory of ornithological areas to be protected, which led to the designation of Special Protection Areas under the Directive 92/43/EEC, was carried out by the BirdLife association on a European scale.

35.3 Litigation on Liability: Repairing Nature by Mobilising Private and Public Laws

The complex articulation between public law and civil law is also illustrated in the litigation of liability and compensation of damages.

Directive 2004/35/EC on environmental liability has been slowly implemented in the EU Member States and in France specifically.[20] It must first be emphasised that the European legislator has ensured that this regime is very far from a general regime, but on the contrary, it was strongly framed[21]: the directive covers certain environmental damage and concerns damage resulting from certain activities mentioned in Annex III without restriction and for activities not listed in Annex III only if the operator is at fault or negligent. The White Paper on Environmental Liability[22] already excluded an application of the regime to diffuse pollution and thus excluded climate change from the scope of the text.

It should be stressed that the directive explicitly excludes the liability of the operator who benefits from a permit in the event of damage to species. Furthermore, the compensation for damage relies on a system which gives priority to compensation in kind. The idea is to correct the damages caused to nature and not to financially compensate for such damages.

35.3.1 Repairing Environmental Damage through Administrative Action

The question of repairing environmental damage is one that goes to the heart of the specificities of each national law and directly affects the modalities of these mechanisms that have been developed over the centuries. The diversity between the Member States is very great in this respect. This is the main reason for the difficulties in developing international and European law in this area, as shown by the lack of effect of the 1993 Lugano Convention,[23] which has not been ratified by any State.

The system of compensation for ecological damage as defined by Directive 2004/35/EU is a mixed system combining public and private laws. An operator who notices environmental damage is obliged to inform the competent administrative authority (Article 6 of the Directive), and he must put in place measures to correct the damage under the supervision of such an authority (Article 7 of the Directive). The involvement of the administrative authority in the event of environmental damage turns it into an actor in the remediation and monitoring of the measures, who will therefore necessarily be challenged in case of insufficiency.

By taking up the terms of Article 9 of the Aarhus Convention regarding access to justice, Article 12 of Directive 2004/35/EU thus makes it possible to refer a matter to the competent authorities in order to request that they take action or complete their action. This referral is supplemented by the possibility of referring the matter to the judge (Article 13 of Directive 2004/35/EU).

Directive 2004/35/EU thus explicitly invites a shift to an administrative law regime regarding the repair of environmental damage.

35.3.2 Ecological Damage: A Concept Adopted by the Judicial Judge That Is Spreading to the Whole of French Law

The ship Erika was chartered by Total International Limited, registered in Bermuda, to transport 30,000 tonnes of heavy fuel oil from Dunkirk (France) to Livorno (Italy). It sank on 12 December 1999 off the tip of Penmarc'h (Brittany, France) while the vessel was in the French Exclusive Economic Zone. The 20,000 tonnes of oil spilt over 400 km of French coastline. The environmental impact was considerable, with several thousand birds killed and the marine flora and fauna being polluted. The affected coastal municipalities and nature protection associations sought redress before the courts. Their main aim was to ensure that Total, which acts through several subsidiaries, is held responsible and ensures the restoration of the environment, either directly or through its financial participation.

The first difficulty with the case lies in the fact that it involves several players, illustrating the economic reality of international trade and, consequently, the extreme difficulties faced by nature defenders when it comes to challenging these arrangements. Although Total International Limited is the charterer of the vessel, Total SA sold the 30,000 tonnes of fuel oil to Total International Limited, which is then sold to Enel. Furthermore, the vessel belongs to the Maltese company Tevere Shipping Company Limited, was operated by the Bahamas-registered company Selmont International and had been inspected by the Italian classification society Rina.

The second difficulty is to determine how the International Law of the Sea – in this case, the Convention on Civil Liability (CLC)[24] and the Fund Convention[25] – relates to European and national law. However, without going into the details of the regimes provided for by these conventions, the limitation of liability that they entail led the claimants in the Erika case to explore complementary avenues to expand liability beyond international law. The question was whether liability could be incurred under the ordinary law of civil liability or by applying the polluter-pays principle under European waste legislation in order to better cover environmental damage and obtain greater compensation.

Several municipalities in Brittany affected by the oil spill thus referred the matter to the courts both on the basis of criminal liability and on the basis of Total's civil liability. On the criminal side, the Court of Cassation[26] confirmed the reasoning of the Paris Court of Appeal[27] by basing Total SA's liability on the additional commitments to monitor ships – vetting – that the company had put in place and that it had not respected in terms of the frequency of ship inspections. As a result, Total had committed a *faute*. This approach makes it possible to avoid the limits set by international conventions, particularly regarding the level of compensation. By placing itself outside the scope of the international law of the sea, the victims can benefit from the full compensation of damages provided for by French law. As a result, Total SA was fined 300,000 euros for the criminal penalty and, more importantly, nearly 200 million euros for compensation for damages. On the civil side, because the accidentally spilt hydrocarbons are waste, Total SA was convicted under the legislation on waste. The polluter-pays principle applies and the chain of liability provided for in the Waste Framework Directive means that the owner of the oil can be held liable.

In ruling on the dispute, the Court of Cassation recognised the existence of ecological damage. The aim is to compensate for damage to nature without seeking to link this damage to other types of damage, particularly property damage. The ecological damage recognised by the Court of Cassation also goes beyond that potentially compensated by Directive 2004/35/EC since it covers all environmental damage.

This recognition by the Court of Cassation of ecological damage led to its introduction into the French Civil Code in 2013.[28] These provisions were completed in 2016,[29] when a regime of reparation in kind was established as a priority and only in the form of financial compensation if the damage – or part of the damage – was not repairable. A recent application of these provisions led, for example, to the Poitiers Court of Appeal ordering a company to pay a municipality 34,870.81 euros corresponding to the costs incurred with clean-up, in addition to 10,000 euros in compensation for the damage suffered: the clean-up costs correspond to compensation in kind for the damage and the sum of 10,000 euros compensates for the remainder of the ecological damage which cannot be directly repaired.[30]

In the chronology leading to the appearance and implementation of the concept of ecological damage, we can see a fertilisation movement between the positions taken by the judicial judges followed by the legislator, who will then impose the implementation of ecological damage on the administrative jurisdiction.

35.3.3 Ecological Damage: An Emerging Concept for French Administrative Law

The introduction of ecological harm into the Civil Code ha hads, in turn, an effect on administrative jurisdiction. Until recently, the administrative judge has been reluctant to compensate for environmental damage.[31] This concept aims to deal with damage to ecosystems.[32] It is intended to deal with environmental damage that cannot be repaired in any other way: property rights can indeed make it possible to deal with some of the environmental damage, but this reparation only imperfectly takes account of this damage because the economic value of ecosystems is only a very limited approach to the question. The French administrative judge had not had the opportunity to apply the new provisions resulting from the 2016 law.

He was invited to do so by the 'Affaire du siècle' case brought by environmental associations before the Paris administrative court, which aims to have ecological damage recognised as a result of the State's inaction.[33] This is the first case in which this damage has been recognised by the French administrative judge and it leads to the condemnation of the State for climate inaction. The associations – unlike in the 'Grande-Synthe' case – based their action on the compensation for the ecological damage they believe they have suffered as a result of this climate inaction. In reality, while they aim to strengthen climate measures, this climate action injunction is presented as the logical consequence of the recognition of ecological damage: if the damage is recognised, then measures must be taken to stop it. As regard past damages due to climate inaction, the associations asked for compensation of the symbolic amount of one euro. The legal approach was particularly risky, as it was clear that this was a significant departure from compensation for clearly identified environmental damage. The Paris Administrative Court nevertheless upheld the damage. The solution reached by the Paris Administrative Court needs to be confirmed or clarified in light of subsequent litigation, but it does provide some food for thought. The opportunity for clarification may arise with a new court case related to compensation for ecological damage in the context of pesticide authorisations.[34]

In the French approach to this damage as set out in Article 1248 of the Civil Code, certain associations which are granted special accreditation by administrative authorities can claim compensation for ecological damage. It is therefore not necessary for them to establish that climate change affects them directly and specifically. They act on behalf of society as a whole. According to the Paris Administrative Court, ecological damage results from the general consequences of climate change. It refers to the work of the IPCC to hold that the increase in the earth's temperature is a reality and that projections lead to significant increases in the earth's average temperature. The link with the effects on ecosystems is not developed, but it can be assumed that it was not necessary to dwell on the issue since these phenomena are well documented in the scientific literature. It should be noted that the French administrative judge does not make the effort in this case to precisely define the effects of climate change on national ecosystems.

It is not enough that the existence of climate-related ecological damage be established for the State to be faulty and liable. Indeed, in order to be able to pass the stage leading to the compensation of the damage, it is necessary that the damage be well connected to the one whose compensation it is claimed to be necessary. There must be a recognised deficiency of the State to combat climate change.

It is at this stage that this liability litigation meets the climate inaction litigation developed in the Grande-Synthe judgement.[35] The administrative court establishes the faulty failure of the State by using the same reasoning as the Council of State. The State has obligations under its international, European, and national commitments that it has not fulfilled, as evidenced by the insufficient reduction in French emissions. It should be noted that the court does not consider the fact that the objectives of increasing the share of renewable energies in gross final energy consumption have not been met, nor the objectives relating to the improvement of energy efficiency, as elements demonstrating the State's failure. The judge considers that in sectoral policies – transport, energy, buildings – the failure to meet a target does not directly contribute to the worsening of climate effects, since this can only be observed globally. In other words, it is possible that if renewable energy targets are not met, other targets will compensate for this poor performance. Similarly, the link between French emissions and the effect on average global temperature is considered too indirect to engage the responsibility of the French State alone. At the very least, this level of responsibility should be modulated by the actual share of French emissions in the world's emissions.

It is clear that placing oneself on the ground of responsibility is a delicate matter. Indeed, while climate change is the result of collective global responsibility, identifying the share that each actor takes in the resulting ecological damage is a scientifically impossible task. All the more in court, where the burden of proof lies with the person seeking compensation. While the Grande-Synthe case did not need to go through this complex stage, the Affaire du Siècle case takes this risk.

The State is therefore only liable for the damage resulting from the failure to comply with the carbon budget it had set for the period 2015–2018. Relying on the notion of ecological damage only allows us to deal with a past situation: this is the logic of liability litigation, which can only be based on the actual damage caused and thus necessarily look at past situations. Ecological damage is assimilated to the consequences of France's carbon budget overruns during the period 2015–2018. It is only at a later stage that the judge will be able to address requests to take measures to stop the aggravation of the damage.

Once the ecological damage has been established, the first objective is to impose measures to restore ecosystems. Reparation must be thus made primarily in kind[36]; in other words, financial compensation is only granted if it is not possible to repair the damage.

The financial compensation requested by the associations consisted in asking for one euro each. This sum raises questions: if real ecological damage has been caused by the State, how can a purely symbolic sum be sufficient? Doesn't the mere fact of asking for a derisory sum reflect the associations' discomfort with their own demands? This is what the court held, indicating that the sum was necessarily unrelated to an assessment of the damage. In order for this claim to have any meaning, the associations had to take the trouble to present an economic argument linking the sums claimed to the damage suffered, even if only approximately. The lesson seems to have been assimilated by the associations, which in the new pesticide litigation are only making a request for compensation in kind.

Since it is necessary to give priority to compensation in kind, the court will first seek compensation for exceeding the carbon budget over the period 2015–2018. It requires the State to compensate for 15 Mt CO_2eq emitted in excess by the measures it considers appropriate before 31 December 2022. The court notes that a reduction in the number of CO_2 emissions after a set date is not fully equivalent to the planned reduction since these excess emissions have produced effects. But the court also acknowledges that it has no means of quantifying this effect, it does not propose any remedial measures on this specific point.

The scope of the "ecological damage" litigation is much less satisfactory for the associations. If it is a victory, it is confined to exceeding the 2015–2018 carbon budget. Moreover, the question of the trajectory of French emissions is not called into question. The State has not appealed the judgement.

35.4 Conclusion

As can be seen, litigation strategies in environmental matters are complex and difficult to control. This is particularly the case in the field of climate change litigation since international commitments are difficult to challenge in national courts. A good command of the articulation between the tools offered by private law and public law is one of the elements for the success and efficiency of environmental law's implementation. Criminal law, combined with the right to compensation before the civil courts, plays a crucial role in the implementation of environmental law because it creates a deterrent effect and a strong incentive to respect the law.

In this respect, a parallel can be drawn with Tax Law, which combines both substantial administrative sanctions and control measures, which in France are addressed by the administrative judge, and criminal sanctions. Criminal proceedings are not frequently initiated by the administration, but they help to develop tax compliance.

At a time when the challenges we are facing – the collapse of biodiversity and climate change – are immense, the issue is not so much to develop new standards or new concepts of environmental law but to ensure that the texts adopted are fully implemented. The European Commission identified this requirement in a 2016 Communication[37] and more recently in 2019 in the Green Deal.[38] This question of implementation is a difficult one: it is more rewarding to develop new legislation than to ensure, patiently and persistently, that Environmental Law is fully respected. It must be realised that the very nature of environmental legislation is to affect all areas of society, and this often involves the conjunction of multiple sources of pollution where the contribution of each is marginal but results in a catastrophic overall effect. In this respect, it is essential that all the available tools, whether they come under Administrative Law, Civil Law, or Criminal Law, be mobilised. The linkage between these fields remains difficult because the specialisation and technicality of modern law leads to the compartmentalisation of reasoning. There is also a challenge here for the training of future jurists, but also for spaces for exchanges between private and public law jurists.

Notes

1 President at the Administrative Court of Lyon, Member of the French Environmental Authority and Vice-president of the Aarhus Convention Compliance Committee.

2 In Europe (for instance, Germany and Italy), see the website of the association ACA-Europe which is integrated by the Court of Justice of the European Union, as well as the Councils of State and the Supreme administrative jurisdictions of each of the Members of the European Union, *ACA-Europe – Association of the Councils of State and Supreme Administrative Jurisdictions of the European Union* <https://www.aca-europe.eu/index.php/en/> accessed 24 May 2023.

3 French Council of State 19 November 2020 and 1 July 2021 Grande Synthe nr. 427301, Administrative Court of Paris, 3 February 2021 and 14 October 2021, Association Oxfam France, Association Notre Affaire à Tous, Fondation pour la Nature et l'Homme, Association Greenpeace France, nr. 1904967, 1904968, 1904972, 1904976. In both cases, a preliminary judgement was issued, which corresponds to the first indicated date.

4 See the website of the French NGO *Notre Affaire à Tous for details on actions* <https://notreaffaire-atous.org/actions/les-territoires-qui-se-defendent-et-si-nous-mettions-enfin-les-entreprises-face-a-leurs-responsabilites/> accessed 24 May 2023.

5 Ordinance of Louis XIV of 1669 'sur le fait des Eaux et Forêts' available on Gallica.fr <https://gallica.bnf.fr/ark:/12148/bpt6k9601493m> accessed 24 May 2023.

6 The Ordinance creates an obligation of information of the administration: 'Titre XXVI Article III Ne pourront ceux qui possèdent Bois de haute Futaye, assis à dix lieues de la Mer, & deux des rivières navigables, les vendre ou faire exploiter, qu'ils n'en ayent, six mois aupravant, donné avis au Controlleur Général des Finances, & au Grand-Maître, à peine de trois mille livres d'amende, & de confiscation des bois coupés ou vendus'. As it sets strict rules under penalty of fines and confiscation of property, it requires authorisation to depart from them.

7 Imperial Decree of 15 October 1810 'relatif aux Manufactures et Ateliers qui répandent une odeur insalubre ou incommode', see text on <https://aida.ineris.fr/reglementation/decret-imperial-15101810-relatif-manufactures-ateliers-repandent-odeur-insalubre> accessed 24 May 2023. Art. 1 imposes the operator of facilities to hold a permit.

8 Council Directive 2010/75/EU of 24 November 2010 on industrial emissions (integrated pollution prevention and control) [2010] OJ L 334/17, Art. 4 'Obligation to hold a permit: 1. Member States shall take the necessary measures to ensure that no installation or combustion plant, waste incineration plant or waste co-incineration plant is operated without a permit. (…)'.

9 Council Directive 2000/60/EC of 23 October 2000 establishing a framework for Community action in the field of water policy [2000] OJ L 327/1, art. 11 'Programme of measures (…) 3. "Basic measures" are the minimum requirements to be complied with and shall consist of: (…) (e) controls over the abstraction of fresh surface water and groundwater, and impoundment of fresh surface water, including a register or registers of water abstractions and a requirement of prior authorisation for abstraction and impoundment. These controls shall be periodically reviewed and, where necessary, updated. Member States can exempt from these controls, abstractions or impoundments which have no significant impact on water status'.

10 Aarhus Convention – Convention on Access to Information, Public Participation in Decision-making and Access to Justice in Environmental Matters, opened for signature 25 June 1998, 2161 UNTS 447.

11 French Court of Cassation 3rd Civil Chamber, 30 November 2022, nr 21-16.404, (B), FS.

12 The 'Préfet' is the State representative in the French State local administration.

13 Code de l'Environnement Article L. 216-3.

14 French Court of Cassation, Criminal Chamber, 18 October 2022, nr. 21-86.965 F-B.

15 Council Directive 2010/75/EU of the European Parliament and of the Council of 24 November 2010 on industrial emissions (integrated pollution prevention and control) [2010] OJ L 334/17, art. 23.

16 See in particular the activities of the network Impel 'European Union Network for the Implementation and Enforcement of Environmental Law', which develops cooperation between environmental inspectorates of Member States and is officially supported by the European Commission (*The European Union*

Network for the Implementation and Enforcement of Environmental Law (IMPEL) <https://www.impel.eu/en> accessed 24 May 2023.

17 French Court of Cassation, Criminal Chamber, 30 January 2007, nr. 06-82.638.

18 Report 'Une justice pour l'environnement – Mission d'évaluation des relations entre justice et environnement', October 2019 <http://www.justice.gouv.fr/art_pix/rapport_justice_pour_environnement.pdf> accessed 24 May 2023.

19 French Law of 24 December 2020 'relative au Parquet européen, à la justice environnementale et à la justice pénale spécialisée'.

20 Council Directive 2004/35/CE of 21 April 2004 on environmental liability with regard to the prevention and remedying of environmental damage [2004] OJ L 143/56. The Directive gave until 30 April 2007 for its transposition and the first evaluation by the European Commission indicates that this transposition was very slow and only achieved on 1 July 2010 (European Commission, Communication COM (2010) 582 final). The 2016 Evaluation Report COM (2016) 204 (European Commission, Communication COM (2016) 204 final) indicates that the number of annual cases opened on the basis of the Directive's provisions ranges from 1 to 95 in the EU Member States. This shows a very diverse impact of the Directive throughout the EU Member States.

21 See for instance the analysis by the European Parliament of Directive of 11 October 2017 A8-0297/2017 and also the European Parliament Resolution of 26 October 2017 on the implementation of the Directive (2016/2251(INI)).

22 Commission, European Commission, Communication COM (2000) 66 final, 9 February 2000.

23 Council of Europe, Convention on Civil Liability for Damage resulting from Activities Dangerous to the Environment (Lugano Convention) (adopted 21 June 1993). It requires the ratification by three countries to enter into force.

24 International Convention on Civil Liability for Oil Pollution Damage (adopted 29 November 1969), 973 UNTS 3.

25 International Convention on the Establishment of an International Fund for Compensation for Oil Pollution Damage (adopted 27 November 1992), 1956 UNTS 211, known as the 'Fund Convention'.

26 French Court of Cassation, Criminal Chamber, 25 September 2012, nr. 10-82.938.

27 Paris Court of Appeal, Chamber 11, 30 March 2010, nr. 08/02278.

28 See also the French report, 17 September 2013 to the Minister of Justice 'Pour la réparation du préjudice écologique'.

29 Ordinance n° 2016-131 of February 10, 2016 for the reform of contract law, in force since October 1, 2016.

30 Poitier Court of Appeal 11 October 2022 RG nr. 21/00065.

31 For example, French Council of State, 12 July 1969, Ville de Saint-Quentin, n° 74546 which refuses to compensate a fishermen's association for damage to the natural aquatic environment or French Council of State 13 November 2009 Association agrée de pêche et de protection du milieu aquatique la sauvegarde de La Moine n° 310038: the compensation for the damage is refused on the grounds that the fishermen's association is not obliged to repair the damage caused by the pollution, which is equivalent to saying that only economic damage would be compensable.

32 Art. 1247 of the French Civil Code: 'Ecological damage consisting of non-negligible damage to the elements or functions of ecosystems or to the collective benefits derived by man from the environment may be compensated under the conditions laid down in this title.'

33 French Council of State 19 November 2020 and 1 July 2021 Grande Synthe nr. 427301, Administrative Court of Paris, 3 February 2021 et 14 October 2021, Association Oxfam France, Association Notre Affaire à Tous, Fondation pour la Nature et l'Homme, Association Greenpeace France, n° 1904967, 1904968, 1904972, 1904976, in both cases a preliminary judgement was issued, which corresponds the first date which is indicated.

34 See *Justice pour le Vivant* <https://justicepourlevivant.org/> accessed 24 May 2023, case submitted 17 February 2022 to the Administrative Court of Paris.

35 French Council of State 19 November 2020 and 1 July 2021 Grande Synthe n° 427301, Administrative Court of Paris, 3 February 2021 et 14 October 2021, Association Oxfam France, Association Notre Affaire à Tous, Fondation pour la Nature et l'Homme, Association Greenpeace France, nrs. 1904967, 1904968, 1904972, 1904976, in both cases a preliminary judgement was issued, which corresponds the first date which is indicated.

36 Art. 1249 of French Code Civil: 'La réparation du préjudice écologique s'effectue par priorité en nature. En cas d'impossibilité de droit ou de fait ou d'insuffisance des mesures de réparation, le juge condamne le responsable à verser des dommages et intérêts, affectés à la réparation de l'environnement, au demandeur ou, si celui-ci ne peut prendre les mesures utiles à cette fin, à l'État. L'évaluation du préjudice tient compte, le cas échéant, des mesures de réparation déjà intervenues, en particulier dans le cadre de la mise en œuvre du titre VI du livre Ier du code de l'environnement'.

37 European Commission, Communication C/2016/8600, 'Better results through better application', OJ 19 January 2017 C 18/10.

38 European Commission, Communication COM (2019) 640 final, "The European Green Deal" (11 December 2019) <https://eur-lex.europa.eu/legal-content/EN/TXT/?uri=COM%3A2019%3A640%3AFIN> accessed 24 May 2023: 'In addition to launching new initiatives, the Commission will work with the Member States to step up the EU's efforts to ensure that current legislation and policies relevant to the Green Deal are enforced and effectively implemented.'.

EPILOGUE

A Marble World and Humans Shaking Hands

Upon concluding our journey through the challenges that sustainability brought to private law, we now summarise each contribution to this book, organised into six parts. Each segment sheds light on the intersection between private law principles, on the one hand, and the environment's needs, on the other.

In Part 1 (*Greening Private Law and in Particular the Law of Obligations*), the contributors elucidated the role of private law institutions in advancing sustainability. Barbara Pozzo initiated this volume by revealing the ongoing hybridisation of classic private law instruments to meet environmental sustainability demands. Dario Hug explained how an eco-literate interpretation could serve as an anchor point for integration into contracts. Rodosveta Vassileva explored the potential of Bulgarian Private Law aligning with the Sustainable Development Goals (SDGs), while Rosa M. Garcia-Teruel emphasised civil law's role in promoting energy efficiency. Wrapping up the first part, Candida Leone addressed some of the challenges in systematically integrating sustainability into private law education.

Part 2 (*Fostering Environmental Protection through Consumer Law*) features critical assessments of the EU Proposal for a Directive on Common Rules Promoting the Repair of Goods by Fryderyk Zoll, Katarzyna Południak-Gierz, Wojciech Bańczyk and Maciej Bujalski, and those regarding the Sale of Goods Directive by Maria Miguel Oliveira da Silva and Jorge Morais Carvalho. Lucila de Almeida and Fabrizio Esposito eloquently discussed imposing obligations on consumers emphasising a narrative of 'responsibility' over one of 'empowerment'. Mireia Artigot Golobardes highlighted the potential of consumer law as an instrument for sustainability that is not yet reflected in its rules, nor in market environments, where people are encouraged to buy unsustainably and where prices do not internalise environmental costs. Petra Weingerl then presented Slovenia's approach, stressing that the framework is still rather reliant on the information paradigm and does not yet sufficiently encourage sustainable production and consumption.

Part 3 (*Coordinating Public and Private Sustainability Law in Europe*) includes the contributions by Guido Alpa and Roberta Calvano which focused on the positivisation of sustainability into the Italian legal system and, more specifically, the Italian Constitution. Verica Trstenjak built her contribution on the role of the judicial dialogue and how the Court has been shaping, albeit indirectly, the realisation of sustainability objectives across various sectors and areas. Kristina Siig reflected on contract-based remedies in the ship recycling industry and Maria

DOI: 10.4324/9781032662046-42

Regina Redinha and Maria Raquel Guimarães concluded with an examination of the urgent need for a subjective right to a stable climate in Portugal.

Part 4 (*Conciliating Property Rights and Environmental Protection*) explores adjustments needed to promote a sustainable property law. Björn Hoops argued that exclusive ownership provided the legal foundation for extractive capitalism and must be adjusted to promote a sustainable economy and society. Francesco Zecchin reviewed critically this new paradigm of ownership, pointing out that it is no longer an individual right, but an economic one. Miha Juhart illustrated this challenge by the example of the duties of forest owners in Slovenia. Tatjana Josipović, on the one hand, and Jasper Verstappen and Tim van Zuijlen, on the other, provided on the novel legal challenges that sustainability management of immovables gives rise to. Josipović suggested addressing these by a better balance between the individual ownership rights of flat owners and the public interests aimed at the green transition, while Verstappen and van Zuijlen advocated a well-defined and limited implementation of legal tech. Finally, Willemijn van Doorn-Hoekveld and Marleen van Rijswick discussed social inequalities arising from disaster risk reduction and Flood Risk Management.

Part 5 (*Green Economy Transition: Sustainable Consumption and Production*) starts by a contribution by Marta Santos Silva and Tomàs Gabriel García-Micó who propose a 'pre-cooling off period' to counter the detrimental environmental impact of abusive consumer withdrawal, termed as 'green sludge'. Victor Mehnert and Hans-W. Micklitz advocated a revision of the Sale of Goods Directive in order to promote sustainable consumption, and also the integration of ecodesign standards into private contracts. Carl Dalhammar and Eléonore Maitre-Ekern discussed how to persuade consumers to purchase more sustainable products and thereby pursue the circular economy. Denis Philippe approached the topical issue of planned obsolescence from the French and European perspectives and Janja Hojnik highlighted the potential role of the service-based business model in fostering a circular and sustainable economy through the investment in durable products. Shen Yayun and Michael Faure suggested private standardisation as a complement to public regulation in environmental governance.

The final and largest part of the book (*Dealing with Climate Justice and Corporate Responsibility*) includes eight chapters. Ralf Michaels and Samuel Zeh guided us through a tour through sustainability and private international law. The chapter addressed general structural questions and presented two specific examples of private international law's role vis-à-vis sustainability, namely transnational climate change litigation and supply chain governance. In her contribution, Sandra Passinhas argued that consumers can only make sustainable choices if information on the ecological characteristics of the goods is smartly provided and reliable while also stressing the utmost importance of consent. Hugo-Maria Schally wrote about the EU Green Deal and the Sustainability of Corporate Activities. Admittedly based exclusively on policy documents, his contribution addressed the specific legal obligations of corporations in the transition towards a more sustainable European economy. The chapter co-authored by Iain MacNeil and Irene-marié Esser gave us a perspective on corporate purpose as a legal principle, how it might promote sustainability in corporate governance and how contingent it is on parallel developments related to sustainability in adjacent domains. Co-authors Carlos Gómez Ligüerre and Lela Mélon stressed the importance of online platforms for sustainability. They argued that the influence and reach of such platforms can be leveraged in order to drive positive change and address pressing environmental and social issues. The two subsequent contributions discussed contract law and sustainability of global value chains. Jakub Błażej Zwierzchowski and Ewa Rott-Pietrzyk determined the core of the sustainability obligation as a contractual route for imposing human rights and environmental duties in the GVC and Rebecca Ravalli discussed the advantages and disadvantages of relying on contracts to operate on global production networks. Both contributions proposed regulatory strategies to strengthen the effectiveness of EU regulation in these matters. The last chapter by Marc Clément is on the judicial practice regarding environmental litigation, having discussed whether it is a matter of public or private law and explored a cross-fertilisation of environmental law concepts in the context of French law.

Throughout the book, the planet is portrayed as a marble world, where legal principles and societal imperatives coalesce and individuals become stewards of present and future generations. We hope we have beckoned you to question how we can navigate this newfound terrain, where the handshake of responsibility binds us to the enduring promise of a sustainable future. The Co-Editors of the *Routledge Handbook on Private Law and Sustainability*,

Marta Santos Silva
Andrea Nicolussi
Christiane Wendehorst
Pablo Salvador Coderch
Marc Clément
Fryderyk Zoll

INDEX

Note: **Bold** page numbers refer to tables; *italic* page numbers refer to figures.

www.ingramcontent.com/pod-product-compliance
Ingram Content Group UK Ltd.
Pitfield, Milton Keynes, MK11 3LW, UK
UKHW050758180926
475701UK00008B/1116
* 9 7 8 1 0 3 2 6 6 2 0 2 2 *

Procida Mirabelli di Lauro A, *Immissioni e "rapporto proprietario"* (Napoli, ESI1984) 398
Rabin R, 'Environmental Liability and the Tort System' (1987) 24 Houston L.R., 27
Ronellenfitsch M, Wolf R, 'Ausbau des Individualschutzes gegen Umwelt belastungen als Aufgabe des bürgerlichen und des öffentlichen Rechts?' (Neue Juristische Wochenschrift 1986, 1960–1965.
Rose C, 'Rethinking Environmental Controls: Management Strategies for Common Resources' (1991) 1 Duke Law Journal, 1–38
Rossi S, 'Luci e ombre dei codici etici d'impresa' (2008) Rivista di Diritto Societario, 23
Roth G, *Materiellrechtliche und prozessuale Aspekte eines privatrechtlichen Umweltschutzes* (Neue Juristische Wochenschrift 1972) 921
Sale K, *The Green Revolution. The American Environmental Movement 1962–1992* (New York, Hill and Wang 1993) 24
Salvi C, 'Immissioni, ecologia, norme costituzionali' (1973) I (2) Giur. it., 798
Salvi C, *Le immissioni industriali: rapporti di vicinato e tutela dell'ambiente* (Milano, Giuffrè 1979)
Sands P, 'The "Greening" of International Law: Emerging Principles and Rules' (1994) 1 (2) Indiana Journal of Global Studies, 293
Schwartz G, 'Reality in the Economic Analysis of Tort Law: Does Tort Law Really Deter?' (1994) 42 UCLA L. REV. 377, 381–387
Segerson K, Miceli T, 'Voluntary Environmental Agreements: Good or Bad News for Environmental Protection?' (1998) 36 (2) Journal of Environmental Economics and Management, 109–130
Strand P, 'The Inapplicability of Traditional Tort Analysis to Environmental Risk: The Example of Toxic Waste Pollution Victim Compensation' (1983) 35 Stanford Law Review, 575
Tarlock A, 'The Influence of International Environmental Law on U.S. Pollution Control Law' (1997) 21 Vermont L. Rev., 759
Teubner G, 'Self-Constitutionalizing TNCs? On the Linkage of "Private" and "Public" Corporate Codes of Conduct' (2011) 18 (2) Indiana Journal of Global Legal Studies, 617
Torres J, *La responsabilidad civil objetiva por daños al medio ambiente y su regulación en México*, Secretaría General Secretaría de Servicios Parlamentarios Dirección General del Centro de Documentación, Información y Análisis
Trimarchi P, *Rischio e responsabilità oggettiva* (Milano, Giuffrè, 1961)
Van Gool E, Michel A, 'The New Consumer Sales Directive 2019/771 and Sustainable Consumption: A Critical Analysis' (2021) 10 (4) Journal of European Consumer and Market Law, 136–147
Visintini G, 'Il divieto di immissioni ed il diritto alla salute nella giurisprudenza odierna e nei rapporti con le recenti leggi ecologiche' (1980) II Riv. dir. civ., 249
Westermann H, *Die Funktion des Nachbarrechts. Zugleich eine Untersuchung der Bedeutung eines Immissionschutzgesetzes für das Privatrecht, Festschrift für Karl Larenz zum 70. Geburstag* (München, Beck 1973) 1003
Wilhelmsson T, 'Consumer Law and the Environment: From Consumer to Citizen' (1998) 21 (1) Journal of Consumer Policy, 45
Wittman D, 'Prior Regulation versus Post Liability: The Choice between Input and Output Monitoring' (1977) 6 Journal of Legal Studies, 193
Xianzhong S, 'Rezeption der westlichen Zivilrechtswissenschaft im modernen China: Analysen aus der rechtsvergleichenden Perspektive' (2010) 18 (5) European Review of Private Law, 901–913
Zatz M, Harbour S, 'The United States Environmental Protection Agency's 33/50 Program: The Anatomy of a Successful Voluntary Pollution Reduction Program' (1999) 7 (1) Journal of Cleaner Production, 17–26
Zhai T, Chang Y-C, 'The Contribution of China's Civil Law to Sustainable Development: Progress and Prospects' (2019) 11 (1) Sustainability 294 <https://doi.org/10.3390/su11101029> accessed 25 June 2023